To a new friend . . .

Shannon King

DESKTOP ENCYCLOPEDIA OF
TELECOMMUNICATIONS

McGRAW-HILL TELECOMMUNICATIONS

Desktop
Encyclopedia
of
Telecommunications
Second Edition

Nathan J. Muller

McGraw-Hill
New York   San Francisco   Washington, D.C.
Auckland   Bogotá   Caracas   Lisbon   London
Madrid   Mexico City   Milan   Montreal   New Delhi
San Juan   Singapore   Sydney   Tokyo   Toronto

McGraw-Hill

A Division of The McGraw-Hill Companies

2 3 4 5 6 7 8 9 0   DOC/DOC   0 9 8 7 6 5 4 3 2 1 0

ISBN 0-07-135893-5

The sponsoring editor for this book was Steve Chapman, the editing supervisor was Sally Glover, and the production supervisor was Pamela Pelton. It was set in Century Schoolbook by D&G Limited, LLC.

Printed and bound by R.R. Donnelley & Sons Co.

McGraw-Hill books are available at special quantity discounts to use as premiums and sales promotions, or for use in corporate training programs. For more information, please write to Director of Special Sales, McGraw-Hill, Two Penn Plaza, New York, NY 10121-2298.. Or contact your local bookstore.

Throughout this book, trademarked names are used. Rather than put a trademark symbol after every occurrence of a trademarked name, we use names in an editorial fashion only, and to the benefit of the trademark owner, with no intention of infringement of the trademark. Where such designations appear in this book, they have been printed with initial caps.

This book is printed on recycled, acid-free paper containing a minimum of 50 percent recycled de-inked fiber.

CONTENTS

Contents

Contents

Contents

Contents

Contents

Contents

Contents

Contents

Contents

Contents

Contents

Contents

Contents

Contents

Contents

PREFACE

Telecommunications encompasses many technologies and is one of the most rapidly advancing fields in the world today. Spending on telecommunications equipment and services is estimated to have reached $983 billion at year-end 1999 in Canada, Mexico, Western and Eastern Europe, Latin America, and the Asia-Pacific region combined. Spending on transport services, equipment, and support services will soar to $1.8 trillion in 2003 at a 16.7 percent compound annual growth rate. U.S. manufacturers are expected to garner $45 billion, or 12.7 percent, of the estimated $345 billion that will be spent on telecommunications equipment in the year 2003.[1] With continued deregulation and trade liberalization, telecommunications could mean global income gains of some $1 trillion over the next decade or so.

The rapid pace of innovation in the telecommunications industry is revolutionizing the world economy in other ways. New technology is eliminating barriers to marketplace entry, opening new possibilities in international trade, and transforming the sale and transfer of products and services by eliminating the middle-man, enabling companies to interface directly with their customers wherever they happen to be around the globe.

Other benefits flow from the resulting competition in telecom markets. People and nations can communicate more easily and understand one another better, unencumbered by physical distance. Consumers enjoy more choice, better quality, and lower prices for products and services. Modernization and investment are encouraged worldwide, creating more jobs and fueling the cycle of innovation.

Changes in technology have helped bring about a new era in the telecommunications marketplace in which voice-data convergence has become a priority. This convergence is blurring traditional industry boundaries and enabling companies to do business in ways that were never thought possible—and at much less cost. Convergence enables companies to take advantage of economies of scale, combine services into bundled packages, and develop ways to enter new markets and to create new ventures with fewer risks than has ever been possible.

At the same time, however, there is an unprecedented level of merger and acquisition activity in the telecommunications industry that is fast

---

[1]According to a report titled "2000 Multimedia Telecommunications Review and Forecast," which was jointly issued by the *Telecommunications Industry Association* (TIA) and the *Multimedia Telecommunications Association* (MMTA)

absorbing the entertainment and Internet industries. The result of these mega-mergers is larger, more entrenched incumbents with deeper pockets. They not only have eliminated one or more competitors in the transaction, but their size can deter future competitors from entering their markets. The challenge for regulators is how to guard competition where it already exists while also promoting competition and innovation in emerging markets.

This book provides nontechnical professionals with the essential knowledge required to succeed in the dynamic, fast growing telecommunications industry. This comprehensive volume offers readers a painless way to fill any knowledge gaps they might have while providing new insights about the operational aspects of today's increasingly complex networks. Of course, this book also makes an excellent reference for those who are outside the industry and who want to better understand how telecommunications technologies are advancing and changing our everyday lives.

The topics that are described in the following pages span local and wide-area networking environments and include coverage of both voice, data, and video. In addition to explanations of technologies, equipment and services, and network applications, there are also discussions of key regulations and standards, industry trends, and the various organizations that have contributed to the evolutionary growth of the telecommunications industry. Because technology by itself is of little benefit unless properly implemented, this book also contains management concepts that help put many of the other topics into perspective.

The information that is contained in this book, especially as it relates to specific vendors and products, is believed to be accurate at the time it was written and is, of course, subject to change with continued advancements in technology and shifts in market forces. Mention of specific products and services is for illustration purposes only and does not constitute an endorsement of any kind by either the author or the publisher.

—Nathan J. Muller

Access Charges

Access charges are the fees that incumbent and competitive *Local Exchange Carriers* (LECs) charge the *Interexchange Carriers* (IXCs) for connections to their local exchanges at both the originating and terminating ends of the call. These costs are eventually passed through to consumers and show up on their long-distance bills in the form of a higher per-minute rate. The IXCs have believed for years that they have been overcharged for access, and they have raised numerous complaints with the *Federal Communications Commission* (FCC).

Reform Efforts

In 1997, the FCC endeavored to reform the system of access charges. First, the FCC created a framework so that the rates that are charged for the components of access are more reflective of costs. Second, the FCC moved residual costs that were traditionally recovered on a per-minute basis into a more efficient, flat-rate charge system. Finally, multi-line business and multi-line residential customers picked up a greater share of costs through increased subscriber line charges and flat-rate charges.

The FCC has always treated the *Competitive Local Exchange Carriers* (CLECs) as non-dominant in the provision of terminating access service, because they did not appear to possess market power. The FCC had reserved the right to revisit the issue of regulating CLEC terminating access rates, however, if there were sufficient indications that they were imposing unreasonable terminating access charges.

Revisiting the Issue

This issue was reviewed in August 1999. The FCC noted that with originating access, the calling party has the choice of service provider, the decision to place a call, and the ultimate obligation to pay for the call. The calling party is also the customer of the IXC that purchases the originating access service. As long as IXCs can influence the choice of the access provider, a LEC's capability to charge excessive originating access rates is limited. IXCs will just shift their traffic from that carrier to a competing access provider.

The FCC noted that with terminating access, the choice of service provider for terminating access is made by the called party. The decision to place the call and payment for the call, however, lies with the calling party. The calling party, or its long-distance service provider, has little or no capability to influence the called party's choice of service provider. Furthermore, IXCs are required by statute to charge averaged rates, so not only does the calling party not choose the terminating LEC, but the IXCs are required to spread the cost of terminating access rates among all end users.

Because the paying party does not choose the carrier that terminates its interstate calls, CLECs might have an incentive to charge excessive rates for terminating access. Accordingly, the FCC tentatively concluded that terminating access might remain a bottleneck that is controlled by whichever LEC provides terminating access to a particular customer—even if competitors have entered the market. The commission also recognized, however, that excessive terminating access charges might encourage IXCs to enter the access market in order to avoid paying these charges.

Market Remedies

The FCC decided not to adopt any regulations governing CLEC terminating access charges and did not address the issue of CLECs originating access charges. Based on the available record, the commission decided to continue to treat non-incumbent LECs as non-dominant in the provision of terminating access service. Although an IXC must use the CLEC that is serving an end user in order to terminate a call, the commission found that the record did not indicate that CLECs previously had charged excessive terminating access rates or that CLECs distinguished between originating and terminating access in their service offerings. The commission concluded that it did not appear that CLECs had structured their service offerings in ways that were designed to exercise any market power over terminating access.

The commission further observed that as CLECs attempt to expand their market presence, the rates of incumbent LECs or other potential competitors should constrain the CLECs' terminating access rates. In addition, the commission found that overcharges for terminating access could encourage access customers to take competitive steps in order to avoid paying unreasonable terminating access charges.

The commission explained that although high terminating access charges might not create a disincentive for the call recipient to retain its local carrier, because the call recipient does not pay the long-distance charge, the call recipient might nevertheless respond to incentives that are offered by an IXC that has an economic interest in encouraging the end user to switch to another local carrier. Thus, the commission concluded that the possibility of competitive responses by IXCs would constrain non-incumbent LEC pricing.

Although the commission declined to adopt any regulations governing the provision of terminating access that is provided by CLECs (because CLECs did not appear to possess market power), the commission noted that it could address the reasonableness of CLECs terminating access rates in individual instances through the exercise of its authority to investigate and adjudicate complaints. Moreover, the commission stated that it would be sensitive to indications that the terminating access rates of CLECs were unreasonable.

The FCC acknowledges that CLEC access rates might, in fact, be higher due to the CLECs' high start-up costs for building new networks, their small geographical service areas, and the limited number of subscribers over which CLECs can distribute costs. Requiring IXCs to bear these costs, however, might impose unfair burdens on IXC customers who pay rates that reflect these CLEC costs—although the IXC customers might not subscribe to the CLEC. IXCs currently spread their access costs among all of their end users. The FCC is currently soliciting input in regards to this problem.

Summary

The FCC prefers to rely upon a marketplace solution in order to constrain CLEC access rates. In the event that it concludes that legal or other impediments preclude the adoption of a market-based solution, however, the FCC also seeks comments regarding a regulatory backstop in order to constrain CLEC access rates. The FCC would rather not intervene in the marketplace, particularly with respect to competitive new entrants, unless intervention is necessary to fulfill its statutory obligation to ensure just and reasonable rates. If market forces are not operating to constrain CLEC access charges, the FCC intends to institute the least-intrusive means possible in order to correct any market failures.

See Also

> *Federal Communications Commission* (FCC)
>
> Presubscribed Interexchange Carrier Charge
>
> Price Caps
>
> Regulatory Process
>
> Telecommunications Act of 1996
>
> Universal Service

Advanced Intelligent Network (AIN)

The *Advanced Intelligent Network* (AIN) provides carriers with the means to create and uniformly support telecommunications services and features via a common architectural platform. New services are created and supported through processors, software, and databases that are distributed throughout the public network. These intelligent nodes are linked via a separate high-speed messaging network called *Signaling System 7* (SS7) in order to support a variety of services and advanced call-handling features across multiple vendor domains.

By accessing these intelligent nodes, users are able to design and control their own services and customize features without telephone company involvement. Of course, this statement assumes that the carriers offer such access as a customer service. To date, such services have been few and far between, causing many in the industry to believe that the full benefits of AIN have yet to be realized.

Advantages for Carriers

The AIN is a natural extension of the flexibility that is provided by the voice-oriented *Virtual Private Networks* (VPNs) of AT&T, MCI WorldCom, and Sprint—which have been in operation since the early 1980s.

The AIN enables carriers to offer new services to subscribers and at the same time reduce their capital investment and operating costs. Carriers also have the flexibility to design and implement new services without having to rely on traditional switching vendors in order to support these new services. Most importantly, with growing competition in local exchange markets, the AIN will provide the tools that are required by carriers in order to remain viable in a competitive environment.

Traditionally, carriers purchased software and updates from the switch vendor and then loaded them into each switching system that provided the service. If the carrier had several different types of switches on its network, the process of introducing a new service was more cumbersome because the software upgrades had to be coordinated among the various switch manufacturers in order to ensure service continuity. This process delayed the availability of new services to some locations.

Because the services and features in an AIN are defined in software programs that are distributed among fewer locations (intelligent nodes), telephone companies can develop and enhance the network software and eliminate switch manufacturer involvement altogether. Once the carrier develops new services, it can immediately offer these services to customers via intelligent nodes that are distributed throughout the network. By accessing these nodes, customers can instantly obtain a uniform set of services for maximum efficiency and economy—even across multiple locations.

By enabling users to design, add, or change their own networks and services from a management terminal, carriers are relieved of much of the administrative burden that is associated with customer service. In turn, this functionality reduces the carrier's personnel requirements—and, consequently, reduces the cost of network operation.

Carriers also can control costs by aggregating customer demand in order to ensure full use of the network. This process would entail the centralization of infrequently used or highly specialized capabilities and the distribution of frequently used and highly shared capabilities over a wider area. The result is a more efficient utilization of network resources and lower operating costs. The distributed architecture of the AIN makes all of these results possible.

Advantages for Users

Rather than investing heavily in premises-based equipment in order to obtain a high level of performance and functionality via private networks, customers can tap into the service logic of intelligent nodes that are embedded in the public network. The service logic can be programmed to implement advanced calling features, such as interactive voice response or speech recognition and bandwidth on demand, in order to support multimedia applications. In some cases, customers are also able to design their own sophisticated hybrid networks without carrier involvement and to manage them from an on-premises terminal as though they were private networks.

AIN enables users to assemble the required resources in the form of functional components, in accordance with their design specifications. Users also can test the integrity of network models by simulation prior to implementation. Service provisioning is virtually instantaneous. Because telephone companies can activate new services quickly and easily at all network locations, users will not have to wait months or years for new services to become available to all of their locations. Users can even receive additional bandwidth within minutes of their request. As AIN technology and fiber-optic networks continue to evolve, large corporate users will be able to activate *Gigabit-per-Second* (Gbps) channels on demand in order to send data thousands of times faster than today's *Megabit-per-Second* (Mbps) channels.

These benefits portend substantial cost savings in up-front hardware investments and ongoing network operating costs. Ultimately, users will no longer have to settle for off-the-shelf networking solutions from carriers and hardware vendors—nor will they have to rely on high-priced private networks for the desired levels of performance and functionality.

Building Blocks of AIN

The intelligent network architecture (refer to Figure A-1) is composed of the following discrete elements that interact in order to support the creation and delivery of services:

- *Service Switching Points* (SSPs)   Distributed switching nodes that process calls by interacting with SCPs
- *Service Control Points* (SCPs)   Centralized nodes that operate based on service-control logic

Figure A-1
Intelligent network elements, tied together with SS7, provide the infrastructure that makes it possible to deliver a variety of new telecommunications services that have numerous advantages to both carriers and users.

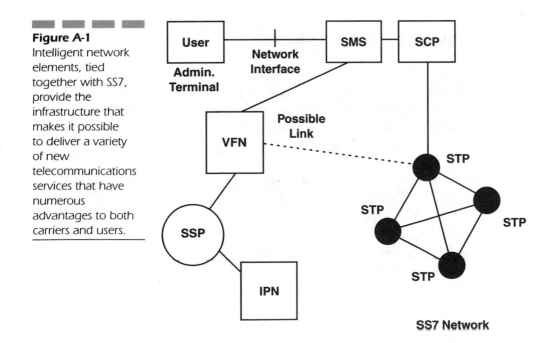

- *Signal Transfer Points* (STPs)   Tandem packet switches that route SS7 messages among SCPs and SSPs
- *Service Management System* (SMS)   A centralized operations system that is used for creating and introducing services
- *Intelligent Peripheral Nodes* (IPNs)   Intelligent nodes that provide a range of capabilities that are used in conjunction with many types of services
- *Vendor Feature Nodes* (VFNs)   Network nodes that provide a variety of value-added services that can compete with those that are offered by the local telephone company
- SS7   A separate, high-speed, packet-switched network that provides call-control signaling between intelligent network nodes.

AIN Applications

The following discussion explores some of the new business services that AIN makes possible.

Area-Wide Networking

Designed for organizations and businesses that have multiple locations, area-wide networking enables subscribers to economically link their locations into a single communications network. Although the service is a fundamentally different solution because it is defined by software and simply lays over the existing network, it can be integrated with existing equipment and systems such as Centrex, *Private Branch Exchanges* (PBXs), or key systems. Area-wide networking not only simplifies dialing between locations, but it also simplifies billing. Employees who work from home, for example, can be tied into the corporate network without extra dialing—enabling them to use office resources and to make and receive business calls as though they were using the in-house office phone system. Work-related calls are automatically billed to the company, saving the time and money that is associated with expense recording. The service offers several dialing plans, access-control options, and security features that can be customized. Future add-ons are easily accommodated without the need for expensive equipment or associated maintenance costs.

Intelligent Caller Data

Intelligent caller data is designed to appeal to businesses that depend on incoming phone calls in order to sell products or services. Intelligent caller data goes far beyond caller ID to give companies a report that has aggregate information about incoming calls. The report summarizes the quantity of calls by date, hour, area code and prefix, ZIP+4 code, and demographic code. This information can be used by businesses to establish a profile of their customers, to test the effectiveness of various forms of advertising, and to target their sales efforts for maximum effect.

Disaster Routing

An organization's need to respond to its incoming calls and to keep communications flowing must continue, even if the organization is unable to maintain operations due to a flood, fire, storm, or power outage (or because a cable to the premises is severed). With disaster routing, companies have the control that they need in order to ensure that critical calls will get through, regardless of the disaster that is confronting them. Incoming calls to multiple telephone numbers are redirected to specified locations—even if the company has a PBX system with direct inward dialing. Calls can be routed to another department, branch location emergency hot site, long-distance site, or to any other specified location—enabling the company to continue business operations.

Intelligent Call Redirection

Intelligent call redirection enables customers to have incoming calls to multiple telephone numbers forwarded, based on the time of day, day of the week, percentage of calls, or based on other variables. For example, a business can use this service to reroute calls immediately in the event of a natural disaster, such as a fire or a storm. This service can also be used to give priority service to key customers by forwarding their calls to designated salespeople. In addition, it helps businesses manage resources during peak calling hours or sends calls to alternate locations after regular business hours. The service is designed to appeal to businesses that depend on incoming phone calls, such as brokerage firms, banks, government agencies, mortgage companies, insurance claims centers, catalog companies, answering services, and hospitals.

Intelligent One-Number Calling

Intelligent one-number calling is a service for businesses that have several sites. This service enables multiple locations to be linked by a single telephone number, ensuring that incoming calls are picked up at the right location. This service also enables a business to advertise one phone number, thereby cutting overhead costs. The basic service works by identifying the ZIP codes of incoming calls and automatically routing the calls to the location that is specified for trade areas that are established by the business. Several options are available, offering a business even greater control over its calls: time-of-day/day-of-week routing, specific date routing, and allocation routing. When used in combination with intelligent caller data, businesses can also receive monthly reports that enable them to analyze incoming call volumes based on variables such as time-of-day, ZIP code, and demographic code. This information enables them to market specifically to certain customers and to target advertising dollars more efficiently.

Intelligent Call Screening

Intelligent call screening involves checking calls before completion, in order to prohibit access to corporate systems and restrict access to computers, PBXs, and facsimiles. Calls from phone numbers that have been authorized or calls from callers who have been given an access code can gain this access. All other calls are denied. Authorization is based on the calling party's number. Authorized callers can be changed, expanded, or updated via a password-protected *Personal Computer* (PC) interface at any time, or on an emergency basis, via touch-tone input into an *Interactive Voice Response* (IVR) system. Reports that detail all calling activity are available. These reports track which authorized users connected and when, and they also track unauthorized usage for rejected calls. Reports are distributed either weekly or monthly and are available on either diskette or paper. If immediate review of either the authorized telephone number list or the access code list is required, the administrator can call the IVR.

Automatic Callback

Automatic callback enables users to automatically place a call to the last number that was dialed without having to redial the full number—regardless of whether the call was answered, unanswered, or busy. This service is used to contact parties that a caller has been unable to reach or to continue an interrupted conversation. When the feature is activated, the number that the user dialed last is rung again. If the line is idle, the call goes through. If the line is busy, the caller hears a special announcement, and the switch continues to monitor the number. When the line becomes idle again, the caller hears a special ring on his or her phone or hears a confirmation tone. When the caller picks up the phone, the telephone that he or she was dialing rings.

Summary

The AIN provides the means for carriers to create, uniformly introduce, and support new services. This intelligence is derived from sophisticated software and processors that are embedded within the public network. With the means to access this intelligence, users are able to engineer their own services, customize features, and exercise more control over their communications without telephone company involvement—no matter how many different carriers are involved. Although such services have been slow in coming, as the intelligent network expands, both large and small companies will eventually be able to view the entire network as a customizable extension of everything they do in the office.

See Also

  Signaling System 7 (SS7)
  Virtual Private Networks (VPNs)

Advanced Peer-to-Peer Networking (APPN)

APPN is IBM's enhanced *Systems Network Architecture* (SNA) technology for linking devices without requiring a mainframe. Specifically, APPN is IBM's proprietary SNA routing scheme for client-server computing in multi-protocol environments. As such, APPN is part of IBM's LU 6.2 architecture, which is also known as *Advanced Program-to-Program Communications* (APPC). APPC facilitates communications between programs that are running on different platforms.

SNA Routing

Generally, APPN is used when SNA traffic must be prioritized by class of service in order to get the information to its destination with minimal delay —or when SNA traffic must be routed peer to peer without going through a mainframe.

Included in the APPN architecture are *Automatic Network Routing* (ANR) and *Rapid Transport Protocol* (RTP) features. These features route data around network failures and provide performance advantages, closing the gap with *Transfer Control Protocol/Internet Protocol* (TCP/IP). ANR provides end-to-end routing over APPN networks, eliminating the intermediate routing functions of early APPN implementations—while RTP provides flow control and error recovery. A more advanced feature called *Adaptive Rate Based* (ARB) is available with IBM's *High-Performance Routing* (HPR) for congestion prevention.

HPR can be added in order to streamline SNA traffic so that routers can move the data around link failures or outages. HPR is used when traffic must be sent through the distributed network without disruptions. HPR provides link-utilization features that are important when moving SNA traffic over the *Wide-Area Network* (WAN) and provides congestion control for optimizing bandwidth. HPR's performance gain comes from its end-to-end flow controls, which are an improvement over APPN's hop-by-hop flow controls.

The ARB feature that is available with HPR uses three inputs to determine the sending rate for data. As data is sent into the network, the rate at which the data is sent is monitored. At the destination node, that rate is also monitored and reported back to the originating node. The third input is the permitted sending rate. Together, these inputs determine the optimal throughput rate, which minimizes the potential for packet discards (and alleviates congestion). By enabling peer-to-peer communications among all network devices, APPN helps SNA users connect to LANs and more effectively create and use client-server applications. APPN supports multiple protocols (including TCP/IP) and enables applications to be independent of the transport protocols that deliver them.

APPN's other benefits include enabling information routing without a host, tracking network topology, and simplifying network configuration and changes. For users who are still supporting 3270 applications, APPN can address dependent *Logical-Unit* (LU) protocols as well as the newer LU 6.2 sessions, which both protect existing investments in applications that rely on older LU protocols.

Summary

Other SNA routing techniques are available. *Data-Link Switching* (DLSw), for example, is used in environments that consist of a large installed base of mainframes and TCP/IP backbones. DLSw assumes the characteristics of APPN and HPR routing and combines them with TCP/IP and other *Local-Area Network* (LAN) protocols. DLSw encapsulates TCP/IP and supports *Synchronous Data-Link Control* (SDLC) and *High-Level Data-Link Control* (HDLC) applications (refer to Figure A-2). DLSw also prevents session timeouts and protects SNA traffic from becoming susceptible to link failures during periods of heavy congestion.

SNA traffic can also be routed over frame relay. Like DLSw, SNA and APPN protocols are encapsulated; in this case, within frame-relay frames. Frame relay provides SNA traffic with guaranteed bandwidth through *Permanent Virtual Circuits* (PVC), and compared to DLSw, it uses little overhead in the process.

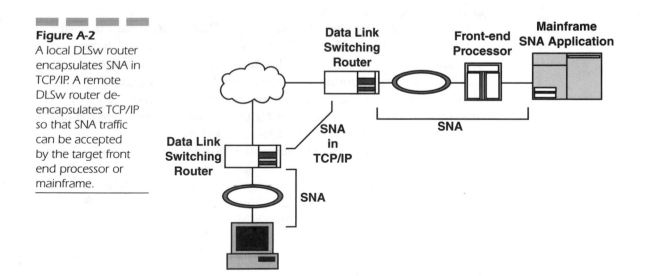

Figure A-2
A local DLSw router encapsulates SNA in TCP/IP. A remote DLSw router de-encapsulates TCP/IP so that SNA traffic can be accepted by the target front end processor or mainframe.

See Also

Advanced Program-to-Program Communications (APPC)
Frame Relay

Advanced Program-to-Program Communications (APPC)

Unlike the traditional master-slave relationship that is implied in most micro-to-mainframe link products, peer protocols support direct communication between programs. One of these peer protocols is the *Advanced Program-to-Program Communication* (APPC) facility of IBM's SNA. APPC is a set of functions that govern the way in which various categories of devices communicate as peers over the network, providing complete interoperability among the devices. In support of this objective, LU 6.2 emerged as a single, product-independent LU type, which marked a departure from previous LU types.

A logical unit is a software-defined access point through which users interact over the SNA network, rather than a physical access point (such as a port). Logical units enable communication between users without each user having to know detailed information about the other's device type and characteristics. LU 6.2 software is used to implement a set of functions that are collectively known as APPC.

The SNA network uses *Physical Units* (PUs) to indicate categories of devices and the resources that they present to the network. The resources that are associated with a particular device category include the communications links. A combination of hardware and software in the device implements the physical unit, of which there are four types. Type 1 (PU 1) devices are dumb terminals, whereas Type 2 (PU 2) devices are user-programmable and have processing capabilities. Type 4 (PU 4) refers to the host node, and Type 5 (PU 5) refers to a communications controller. No Type 3 physical unit (PU 3) exists.

The PU that implements the most comprehensive set of functions is PU Type 2, Version 1 (PU 2.1). PU 2.1 supports connections to other PU 2.1 nodes—as well as conventional hierarchical connections to the mainframe —efficiently and economically. PU 2.1 also supports simultaneous multiple links and parallel sessions over a given link. PU 2.1 is used in conjunction with LU 6.2 in implementing APPC.

The SNA network also includes *System Service Control Points* (SSCPs), which provide the services that are required to manage the network as well as establish and control the interconnections that enable users to communicate with each other. The SSCP provides a broader functionality than an LU, which represents a single user (or a physical unit), which in turn represents a device and its associated resources. The relationship of all three SNA components is illustrated in Figure A-3.

Summary

Among other things, APPC overcomes the inefficiencies that result when a PC is forced to emulate a 3270 terminal in order to access data on the mainframe. In terminal emulation mode, the PC and mainframe must devote processing resources to servicing screen-by-screen data transfers. The PC can be appropriately equipped with an emulation board that uses its own processor to handle the increased load. The mainframe can become bogged down, however, when it is forced to handle requests from many PCs in emulation mode. A popular solution to this problem is to use departmental or workgroup systems in order to service local PC users, thus offloading the

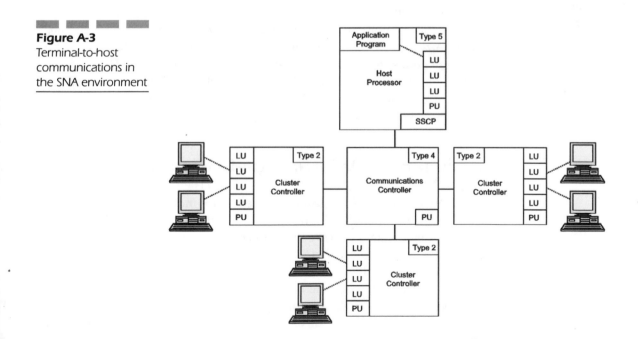

Figure A-3
Terminal-to-host
communications in
the SNA environment

mainframe. Adding APPC capabilities in the distributed environment makes computing and networking even more efficient and economical.

See Also

  Advanced Peer-to-Peer Networks (APPN)

Air-Ground Radiotelephone Service

With the air-ground radiotelephone service, a carrier uses wireless technology to provide telephone service to subscribers during an aircraft flight. Two versions of this service exist: one for general aviation, and one for commercial aviation.

General Aviation Air-Ground Stations

Air-ground radiotelephone service has been available to general aviation for more than 30 years. The service is implemented through general aviation air-ground stations, which comprise a network of independently licensed stations. These stations employ a standardized duplex analog technology called *Air-Ground Radiotelephone Automated Service* (AGRAS) that provides telephone service to subscribers who are flying over the United States or Canada in private aircraft, from small single-engine craft to corporate jets. Because there are only 12 channels available for this service, this service is not available to passengers who are on commercial airline flights.

Commercial Aviation Air-Ground Systems

Commercial aviation air-ground systems are nationwide systems that employ various analog or digital wireless technologies to provide telephone service to passengers who are flying in commercial aircraft over the United States, Canada, and Mexico. Passengers use credit cards or prearranged accounts to make telephone calls from bulkhead-mounted telephones (or, in larger jets, from seatback-mounted telephones). This service was available from one company on an experimental basis during the 1980s and began regular competitive operations in the early 1990s. Three operating systems currently exist—one of which is GTE AirFone.

When an AirFone call is placed over North America, information is sent from the phone handset to a receiver in the plane's belly and then down to one of the 135 strategically placed ground radio base stations. From there, the information is sent to one of three main ground switching stations and then over to the public telephone network to the receiving party's location. When an AirFone call is placed over water, information is sent first to an orbiting satellite. From there, the call transmission path is similar to the North American system, except that calls are sent to a satellite Earth station instead of a radio base station. Calls can be placed to any domestic or international location.

To receive calls aboard aircraft, the passenger must have an activation number. In the case of AirFone, an activation number can be obtained by dialing 0 toll-free on board or 1-800-AIRFONE from the ground. For each flight segment, the activation number will be the same; however, the passenger must activate the phone for each flight segment and must include his or her seat number. The person who is placing the call from the ground dials 1-800-AIRFONE and follows the voice prompts to enter the passenger's activation number. If the call is accepted, the passenger is prompted to slide a calling card or credit card through the phone's card reader in order to pay for the call.

The following steps are involved when receiving a call:

- The phone will ring on the plane, and the screen will indicate a call for the seat location.
- The passenger enters a *Personal Identification Number* (PIN) to ensure that no one else can answer the call.
- The phone number of the calling party is displayed on the screen.
- If the call is accepted, the passenger is prompted to slide a calling card or credit card to pay for the call.
- Once the call has been accepted, the passenger is automatically connected to the party on the ground.
- If the passenger chooses not to accept the call, he or she follows the screen prompts—and no billing will occur.

Air-to-ground calls are extremely expensive. The cost to place domestic calls using GTE's AirFone service, for example, is $2.99 to connect and $3.28 per minute or partial minute, plus applicable tax. By comparison, AT&T's Inflight Calling costs $2.99 to connect plus $2.99 per minute. These rates apply to all data/facsimile and voice calls. Even calls to 800 and 888 numbers—which are normally toll-free on the ground—are charged at the same rate as regular AirFone and Inflight calls. No billing ever occurs for the ground party. The charges for international calls are higher. For example, AT&T charges $5 to connect and $5 per minute.

Summary

The growing popularity of cellular phones has no impact on the demand for air-ground radiotelephone service, because the use of cell phones during flight is prohibited. The radio frequencies that are used by cellular phones might interfere with communications and other equipment on the aircraft. The frequency that is assigned to commercial services (such as AirFone) is unique, so the service does not interfere with the plane's navigation, communications, or data-processing equipment. Instead, airborne phones use two separate, narrow bandwidths between 849 and 851 MHz for air-to-ground communications and between 894 and 896 MHz for receiving calls.

See Also

Cellular Communications

Alliance for Telecommunications Industry Solutions (ATIS)

Alliance for Telecommunications Industry Solutions (ATIS) is a North American standards body that leads the development of telecommunications standards, operating procedures, and guidelines through its sponsored committees and forums. ATIS member companies are North American and Caribbean providers of telecommunications services and include telecommunications service providers, competitive local carriers, cellular carriers, interexchange companies, local exchange companies, manufacturers, software developers, resellers, enhanced service providers, and providers of operations support.

The telecommunications industry routinely turns to ATIS and its sponsored committees for solutions to numerous new challenges—the most recent of which include Y2K interoperability testing, global wireless communications, number portability, improved data transmission, toll fraud, and the convergence of new technologies.

ATIS was established at the divestiture of the Bell System in 1984. As industry competition grew and new technologies developed, the role of ATIS expanded. Today, ATIS is one of the world's leading standards development bodies for telecommunications, dedicating itself to the following goals:

- Actively promoting the timely resolution of national and international issues involving telecommunications standards and the development of operational guidelines

- Acting as an information resource to its members, to forum participants, to federal and state agencies, and to other interested parties

- Promoting industry process and harmony with minimal regulatory or legislative intervention

Through their close affiliation with ATIS, member companies demonstrate a commitment to an industry-regulated approach towards standards development that encourages competition and the swift rollout of new telecommunications products and services to consumers.

Headquartered in Washington, D.C., ATIS has professional staff members that provide industry forum secretariat services, committee management and administration, legal counsel, public relations and media outreach, technical assistance, Internet services and Web site maintenance, and printing and publishing facilities (among other services). ATIS staff offers both the technical expertise and group facilitation skills to effectively administer the activities of the committees and forums.

Organization

More than 3,000 industry representatives participate each year in ATIS committees and forums, following strict guidelines that ensure an industry-wide consensus regarding the development of standards and procedures. ATIS is organized into the following committees and forums:

- *Carrier Liaison Committee* (CLC)   Provides executive oversight for the *Network Interconnection Interoperability Forum* (NIIF), the *Ordering and Billing Forum* (OBF), the *Industry Numbering Committee* (INC), and the *Toll Fraud Prevention Committee* (TFPC).

  - *Network Interconnection Interoperability Forum* (NIIF)   The NIIF provides an open forum to encourage the discussion and resolution (on a voluntary basis) of industry-wide issues that are associated with telecommunications network interconnection and interoperability. This forum involves network architecture, management, and testing and operations and facilitates the exchange of information.

  - *Ordering and Billing Forum* (OBF)   The OBF provides a venue for customers and providers in the telecommunications industry to identify, discuss, and resolve national issues that affect ordering, billing, provisioning, and the exchange of information about access services and other connectivity and related matters.

- *Industry Numbering Committee* (INC) The INC provides an open forum to address and resolve industry-wide issues that are associated with the planning, administration, allocation, assignment, and use of resources and related dialing considerations for public telecommunications within the *North American Numbering Plan* (NANP) area.

- *Toll Fraud Prevention Committee* (TFPC) The TFPC is dedicated to the identification and prevention of toll fraud vulnerabilities in the national PSN and resolves issues involving fraud that are pertinent to the telecommunications industry.

- Committee T1—Telecommunications Accredited by the *American National Standards Institute* (ANSI), Committee TI develops technical standards for telecommunications network interconnection, interoperability, and performance. More than 1,500 scientists and engineers bring their technical expertise to Committee T1's six technical subcommittees.

- Committee O5—Wood Poles Develops standards and specifications for industry use in areas dealing with wood poles, cross-arms, and other wood products

- *Internetwork Interoperability Test Coordination* (IITC) Committee Provides strategic planning, industry funding, and management mechanisms for test coordination, as well as forward-looking test-coordination programs in recognition of the expanding issues of increased network interconnection. The IITC Committee oversees the *Network Testing Committee* (NTC).

- *Network Reliability Steering Committee* (NRSC) Performs analyses of network outages and provides recommendations for corrective actions. NRSC issues quarterly and annual reports to the industry and to the FCC in liaison with the FCC's *Network Reliability Council* (NRC).

- *Protection Engineers Group* (PEG) Provides guidance in efforts to improve the safety and reliability of telecommunications networks, in addition to recommending standards for electrical protection of communications facilities (including broadband service architectures and cellular systems)

- *SONET Interoperability Forum* (SIF) Resolves interoperability issues in order to promote the wide deployment of SONET, including validation of specifications for conformance testing and establishing requirements and applications for remote login specifications.

- *Telecommunications Industry Forum* (TCIF)   Promotes electronic commerce (e-commerce), electronic data interchange, improvements in bar coding of telecommunications products for inventory control, and electronic bonding

- *Data-Aware Transport Activity* (DATA)   Addresses multi-vendor interoperability, focusing on interface compatibility, service, internetworking, and end-to-end management

Summary

Standards Committee T1, a key U.S. standards development organization for the telecommunications sector that is sponsored by ATIS, has been approved by the *International Telecommunication Union* (ITU) to provide draft and final standards for use as reference documents to *ITU Telecommunication Sector* (ITU-T) study groups. Additionally, ITU-T study group leaders will be able to consider Committee T1's documents for use when preparing their own draft recommendations. Standards Committee T1 was granted approval under the ITU's Series A recommendations A.5 and A.6, which enable standards development organizations to provide ITU-T working groups with access to their work for use as part of ITU-T draft recommendations. The Series A recommendations are recognized within the standards development community as a way to effectively streamline and accelerate the coordination of international standards development. Standards Committee T1 is one of only five standards development organizations to formally receive the Series A qualification.

See Also

> *American National Standards Institute* (ANSI)
> *International Telecommunication Union* (ITU)

Amateur Radio Service

Amateur radio service is defined by the FCC as "a radio communication service for the purpose of self-training, intercommunication and technical investigations carried out by amateurs; that is, duly authorized persons interested in radio technique solely with a personal aim and without pecuniary interest."[1]

Amateur radio stations are licensed by the FCC and can engage in domestic and international communications, both two way and one way. Applications for new licenses or for a change in operator class are filed through a *Volunteer Examiner-Coordinator* (VEC). Operators can use their station equipment as soon as they see that the information regarding their amateur operator/primary station license grant appears in the amateur service database. New operators do not need to have the license document in their possession in order to commence the operation of an amateur radio station.

Because amateur stations must share the airwaves, each station licensee and each control operator must cooperate in order to select transmitting channels and to make the most effective use of the amateur service frequencies. A specific transmitting channel is not assigned for the exclusive use of any amateur station.

Types of Communications

With regard to two-way communications, amateur stations are authorized to exchange messages with other stations in the amateur service—except those in any country whose administration has given notice that it objects to such communications.[2] In addition, transmissions to a different country must be made in plain language. Communication is limited to messages of a technical nature (that relate to tests) and remarks of a personal nature (for which, by reason of their unimportance, use of public telecommunications services is not justified).

Amateur radio stations can also engage in one-way communications. For example, they are authorized to transmit auxiliary, beacon, and distress signals. Specifically, an amateur station can transmit the following types of one-way communications:

---

[1]Two exceptions to this rule exist. A person can accept compensation when in a teaching position and when the amateur station is used as part of classroom instruction at an educational institution. The other exception is when the control operator of a club station is transmitting telegraphy practice or information bulletins.

[2]As of year-end 1999, no administration in another country had given notice that it objected to communications between the amateur radio stations.

■ Brief transmissions that are necessary to make adjustments to the station

■ Brief transmissions that are necessary for establishing two-way communications with other stations

■ Transmissions that are necessary to provide emergency communications

■ Transmissions that are necessary for learning or improving proficiency in the use of international Morse Code

■ Transmissions that are necessary to disseminate an information bulletin that is of interest to other amateur radio operators

■ Telemetry

Prohibited Communications

Although the FCC does not provide a list of communications that are suitable or unsuitable for the amateur radio service, there are several types of amateur-operator communications that are specifically prohibited, including the following:

■ Transmissions that are performed for compensation

■ Transmissions that are performed for the commercial benefit of the station control operator

■ Transmissions that are performed for the commercial benefit of the station control operator's employer

■ Transmissions that are intended to facilitate a criminal act

■ Transmissions that include codes or ciphers that are intended to obscure the meaning of the message

■ Transmissions that include obscene or indecent words or language

■ Transmissions that contain false or deceptive messages, signals, or identification

■ Transmissions on a regular basis that could reasonably be furnished alternatively through other radio services

Broadcasting information that is intended for the general public is also prohibited. Amateur stations cannot engage in any form of broadcasting or in any activity that is related to program production or newsgathering for broadcasting purposes. The one exception is when communications are directly related to the immediate safety of human life or to the protection of property. This information can be provided by amateur stations to broadcasters for dissemination to the public when no other means of communication is reasonably available before or at the time of the event.

Amateur stations are not afforded privacy protection. In other words, the content of the communications by amateur stations might be intercepted by other parties and divulged, published, or utilized for another purpose.

Summary

In August 1999, the FCC's *Wireless Telecommunications Bureau* (WTB) began the use of the *Universal Licensing System* (ULS) for all application and licensing activity in the amateur radio services. The ULS is an interactive licensing database that was developed by the WTB in order to consolidate and replace 11 existing licensing systems that were used to process applications and grant licenses in wireless services. ULS provides numerous benefits, including fast and easy electronic filing, improved data accuracy through automated checking of applications, and enhanced electronic access to licensing information.

See Also

Citizens Band Radio Service

Telegraphy

Alternative Exchange Points

Alternative exchange points, which are sometimes referred to as bandwidth hotels, provide service providers with floor space for their network equipment within a secure building. The building is typically equipped with appropriate *Heating, Ventilation, and Air Conditioning* (HVAC) as well as electrical connections and a diesel-powered generator to guard against commercial power failures. These bandwidth hotels offer service providers a low-cost alternative to colocation agreements with incumbent telephone companies and peering arrangements with major *Network Access Points* (NAPs) on the Internet backbone. Often, the alternative exchange provider will have business partnerships and strategic alliances in place to offer its customers a complete outsourcing solution for their equipment and bandwidth procurement, engineering, network interconnection, maintenance, and security needs.

Provider Profile

One bandwidth hotel provider is Telehouse, which offers dedicated or shared colocation space and uninterruptible power with battery and generator backup, as well as continuous site monitoring by technical staff and site security. In addition, Telehouse offers a portfolio of services that meet unique customer requirements, including circuits on a resale basis to international locations.

The company's *New York International Internet Exchange* (NYIIX), an international peering service for *Internet Service Providers* (ISPs), is a fully supported gateway and peering facility that serves the international and metropolitan regional ISP market. Once connected to the high-speed NYIIX switch, customers have immediate connectivity to any and all customers that also peer at the facility. Ethernet switch technology is used for peering services, and the peering arrangements are negotiated privately between each party. Carrier selection is also left to the discretion of each customer. Telehouse is carrier neutral and supports any carrier that wants to better serve its clients.

Telehouse offers a complete range of services—the core of which is physically secure, power-protected data-center space. The company encourages the colocation of any equipment that makes a customer's network easier to manage. Telehouse provides colocation service for routers and servers. The backup power generators and batteries, combined with Telehouse technical support, enhance the overall up time of customer services.

Customers can even assign their own staff at the Telehouse facility and maintain all responsibility for the management and operation of their equipment. Telehouse also offers a flexible service arrangement in which customers rely on its staff for specific activities, such as night-shift operations, maintenance, and backups. The company can provide any gradient of support services up to (and including) complete outsourcing of technical operation and site management.

For small to mid-size housing requirements, Telehouse offers the use of prebuilt data-center space. The space can be as small as one equipment cabinet for use as a network node or as large as 1,000 square feet for a mid-range computer installation. The data site is operated by Telehouse, with access controlled by Telehouse staff. For added security, the customer's equipment can be enclosed in a cage. Power consumption, air conditioning, and fire protection are all included in the fixed monthly charge.

For data centers of 1,000 square feet or more in size, Telehouse offers dedicated sites where the customer controls the access. Telehouse provides pre-built sites with raised floors, fire protection, air conditioning, and continuous site monitoring. Customers have their own room with a locked door and have complete control over who has access and under what circumstances. A portion of the space can be allocated to the customer's own on-site staff.

All space that is offered by Telehouse comes with a staff of building engineers who are charged with monitoring and maintaining all elements of the infrastructure, including generators, batteries, and fire systems. The entire environment is continuously monitored for such things as temperature and humidity. These and other variables have a predefined acceptable range. If anything goes out of range, alerts are sent to the engineer on duty, who investigates the situation and takes the appropriate corrective action.

Issues to Consider

When evaluating alternative exchanges, you should choose an exchange that is completely neutral and that has an open-door policy with respect to carriers that can offer and provision bandwidth at a low cost. Neutrality enables organizations to choose from the largest number of product and service providers. This flexibility provides the most cost savings, eases training and staffing burdens, and enables the organization to take advantage of faster, more economical circuit provisioning and interconnection—all of which results in a faster response to event-driven and sustained customer bandwidth demands.

An alternative exchange provider that offers a research and development environment might be of particular value to organizations that depend on serving their customers with leading-edge technologies. For example, an organization can work with the alternative exchange provider and its other customers in order to test and ultimately select the best technologies and solutions for the next-generation requirements of Internet businesses. Among the specific R&D projects that might prove beneficial include monitoring and caching technologies, multicast networks and systems, new protocols, and various switching products. Products and technologies are evaluated and tested with simulated production-level traffic, resulting in a reduced deployment risk and a faster time to market.

Summary

Despite new colocation rules that were issued by the FCC in 1999 and streamline and speed up the processes of incumbent telephone companies opening their central offices to colocation arrangements, many competitors want the freedom, flexibility, and neutrality that alternative exchange arrangements offer. Furthermore, participation in alternative colocation arrangements can also reduce the expense of provisioning voice and data services to customers and can result in the faster rollout of services and adoption of new technologies.

See Also

Colocation Arrangements

Interconnection Agreements

American National Standards Institute (ANSI)

Founded in 1918, the *American National Standards Institute* (ANSI) is a federation of standards-developing organizations in the United States. ANSI represents the interests of about 1,400 corporate, organization, government agency, institutional, and international members through its offices in New York City and Bethesda, Maryland. The organization's primary goal is the enhancement of global competitiveness of U.S. businesses and the American quality of life by promoting and facilitating voluntary consensus standards and conformity assessment systems.

ANSI does not develop standards itself; rather, it facilitates the development of standards by establishing a consensus among qualified groups. ANSI ensures that its guiding principles—consensus, due process, and openness—are followed by the more than 175 distinct entities that are currently accredited by ANSI. These accredited organizations are committed to supporting the development of standards that address technological innovation, marketplace globalization, and regulatory reform issues.

ANSI is the sole U.S. representative and dues-paying member of the two major non-treaty international standards organizations: the *International*

Organization for Standardization (ISO), and via the *United States National Committee* (USNC), the *International Electrotechnical Commission* (IEC).

ANSI is a founding member of the ISO and plays an active role in its governance. ANSI is one of five permanent members to the governing ISO Council and one of four permanent members of ISO's Technical Management Board. U.S. participation through the USNC is equally strong in the IEC. The USNC is one of 12 members on the IEC's governing Committee of Action.

Through ANSI, the United States has immediate access to the ISO and IEC standards-development processes. ANSI participates in almost the entire technical program of both the ISO (78 percent of all ISO technical committees) and the IEC (91 percent of all IEC technical committees) and administers many key committees and subgroups (16 percent in the ISO; 17 percent in the IEC). As part of its responsibilities as the United States member body to the ISO and the IEC, ANSI accredits *United States Technical Advisory Groups* (U.S. TAGs) or USNC *Technical Advisors* (TAs). The U.S. TAG's (or TA's) primary purpose is to develop and transmit, via ANSI, United States positions on activities and ballots of the international technical committee. In many cases, United States standards are taken forward, through ANSI or its USNC, to the ISO or IEC—where they are adopted in whole or in part as international standards.

Summary

ANSI's program for accrediting third-party product certification has experienced significant growth in recent years, and the program continues to work toward the worldwide acceptance of product certifications that are performed in the United States and toward promoting reciprocal agreements between U.S. accreditors and certifiers.

See Also

   Alliance for Telecommunications Industry Solutions (ATIS)

   Institute of Electrical and Electronics Engineers (IEEE)

   International Electrotechnical Commission (IEC)

   International Organization for Standardization (ISO)

   International Telecommunication Union (ITU)

Analog Line Impairment Testing

Despite the trend toward digital lines, many analog lines are still in place —particularly in the local loop. Analog lines are used for low-speed data via modems, as well as for voice via telephones. Whether analog circuits are switched or dedicated or two-wire or four-wire, they must meet some basic performance guidelines. By taking impairment measurements, a technician can determine whether carriers are complying with their stated levels of performance. If not, the carrier can work to solve the problem.

The ability to test for impairments is especially important when an organization is using conditioned leased lines. A conditioned line is a line that has been selected for its desirable characteristics—signal-to-noise ratio, intermodulation distortion, phase jitter, attenuation distortion, and envelope delay distortion-or that has been treated with equalizers in order to improve the user's ability to transmit data at higher speeds than would normally be possible over ordinary analog private lines. Because conditioning is provided by the carrier at an extra cost, periodically testing these facilities enables users to verify that they are indeed receiving the level of performance for which they are paying. The test equipment that is used for this purpose is often referred to as the *Transmission Impairment Measurement Set* (TIMS).

A TIMS can make a few basic measurements by passively bridging into a circuit. These measurements require sending reference tones down the line and receiving them again. By analyzing the difference between what is sent and what is received, the TIMS calculates the level of impairment. TIMS can measure a variety of *Voice Frequency* (VF) impairments, including the following:

- Overall signal quality
- VF transmit level
- VF receive level
- Data carrier detect loss
- Dropouts
- Signal-to-noise ratio
- Gain hits
- Phase hits
- Impulse hits

- Frequency offset
- Phase jitter
- Non-linear distortion

The measurements for these selected parameters can be compared against the performance thresholds that are set by the network manager. If these thresholds are exceeded, data traffic might have to be rerouted to another facility until the primary line can be brought back into specification or until the rate of transmission can be downspeeded to avoid the corrupting effects of the line impairment.

See Also

Line Conditioning

Protocol analyzers

Applications Service Providers (ASP)

Applications Service Providers (ASP) host business-class applications in their data centers and make them available to customers on a subscription basis over the network. Among the applications that are commonly outsourced in this way are e-mail, Web hosting of an e-business, scheduling, workgroup collaboration, and *Enterprise Resource Planning* (ERP). Customers are charged a fixed monthly fee for the applications.

The idea of outsourcing applications is especially appealing to small to mid-size businesses that are struggling under the weight of *Information Technology* (IT) infrastructure costs. Such firms are looking for a way to leverage public-network resources and centralized applications development and support expertise. Today, a company can only grow as fast as its IT infrastructure allows. IT managers spend as much as 25 percent of their time recruiting talent when they should be using that time to think about strategic initiatives.

To address this situation, many types of companies are setting themselves up as ASPs in this relatively new market, including long-distance carriers, telephone companies, computer firms, ISPs, software vendors, integrators, and business management consultants. Intel Corporation, for example, is building data centers around the world in order to be ready for hosting e-business sites for millions of businesses that will embrace the Internet within five years.

As an ASP, Corio enables businesses to obtain best-of-breed applications at the lowest possible cost. Corio is responsible for maintaining and managing the applications and ensuring their availability to its customers from its data centers. For a fixed monthly fee for the suite of integrated business applications and services, businesses can achieve a 70 percent reduction in the average of *Total Cost of Ownership* (TCO) in the first year (versus traditional models) and a 30 percent to 50 percent TCO reduction over a five-year period.

Corio has set up caged servers in Exodus Communications and Concentric Network data centers. This arrangement enables Corio to better focus on its expertise: deploying, maintaining, and monitoring applications. With regard to application availability, Corio offers a service-level guarantee of 99.9 percent.

New companies have also started up specifically to address the needs of ASPs. Exodus Communications, for example, is a high-end collocation and *Internet Protocol* (IP) network provider that also offers management services in addition to carrier-grade data centers in order to house its customers' equipment. As of year-end 1999, the company had 21 data centers. Although Exodus itself does not offer applications hosting services, it provides data center facilities and management services to ASPs such as Corio.

Market Growth

Application outsourcing has been around for nearly 30 years under the concept of the service bureau. In the service bureau arrangement, business users rented applications running the gamut from rudimentary data processing to high-end proprietary payroll. Companies such as *Electronic Data Systems* (EDS) and IBM hosted the applications at centralized sites for a monthly fee and typically provided access via low-speed, private-line connections.

In an early 1990s incarnation of the service bureau model, AT&T rolled out its hosted Lotus Notes and Novell NetWare services—complete with continuous monitoring and management. Users typically accessed the applications over a frame relay or dedicated private line. AT&T's Notes hosting effort failed and was discontinued in early 1996. The carrier lacked the expertise that was needed to provide application-focused services and did not offer broad enough access to these applications.

Where AT&T failed, smaller companies such as Interliant succeeded. Since 1994, Interliant has been providing Lotus Notes service and has expanded its business with a suite of applications ranging from sales automation to collaboration tools. The lesson is that large telecommunications companies are focused on building public networks for efficiency, which is different than offering business users software to make their businesses run more smoothly.

To help sell the benefits of applications outsourcing, 25 companies have formed the Applications Service Provider Industry Consortium. The consortium includes a wide range of companies, including AT&T and UUNET, an MCI Worldcom company, on the service-provider side. Compaq, IBM, and Sun Microsystems are representative of the systems and software vendors. Interconnect companies such as Cisco Systems are also members. The consortium's goals include education, common definitions, research, standards, and best practices.

Several trends have come together to rekindle the market for applications outsourcing. The rise of the Internet as an essential business tool, the increasing complexity of software programs, and the shortage of IT expertise have created a ready environment for carriers and other companies to enter the applications-hosting business.

Summary

As more corporations realize they have to do business on the Web, they are turning to ASPs in order to supply the applications and management expertise in an outsourcing arrangement. Although the market consists largely of Internet-centric companies, a growing number of companies are outsourcing traditional enterprise applications so that they can devote more resources to core business issues. The economics of outsourcing are fairly compelling, and new companies are being created to deal with customers' emerging outsourcing requirements. A true ASP, however, supports a range of enterprise applications. When companies outsource, they want someone to manage all of their corporate applications, not just one. This situation gives ASPs a competitive advantage over firms that specialize in providing only Oracle, SAP, or PeopleSoft applications, for example.

Area Codes

In 1947, AT&T and Bell Laboratories designed the *North American Numbering Plan* (NANP), which included area codes. A telephone number consists of 10 digits. The first three digits are the area code; the second three digits are the central office or exchange; and the last four digits are the individual telephone line numbers. Ten thousand possible combinations of these digits exist within each exchange. When a customer makes a telephone call, the network uses the area code and exchange to determine where to send the call. The area code tells the network the geographic area in which the called party lives, and the exchange indicates the particular switch within that area code to which the call should be routed. This system for routing calls requires numbers to be given out in blocks of 10,000, because each exchange contains 10,000 telephone numbers.

In each area code, 7,920,000 numbers are available. In recent years, a shortage of telephone numbers has caused an increase in the number of area codes in order to expand the supply of telephone numbers. Not only has the subscriber base grown since 1947, but more subscribers also want extra lines to handle facsimiles and Internet access. Add to that the surging demand for pagers and cellular phones, both of which require telephone numbers, and you can easily see how the quantity of available telephone numbers is being depleted.

The Telecommunications Act of 1996 marked the beginning of competition for local telephone customers. Competing local telephone companies, wireless telephone companies, and paging companies all need inventories of numbers before they offer services to customers. Because of the system that was set up by the telecommunications industry for a monopoly environment, telephone numbers are currently given out in blocks of 10,000. In each area code, there are 792 blocks of 10,000 (792 x 10,000 = 7,920,000 available numbers). Some numbers are not available, such as seven-digit numbers that start with 0, 1, or 911. Therefore, each area code has somewhat fewer than eight million usable numbers. Even if a small percentage of the available telephone numbers in an area code is in use, an area code can run out of numbers if all 792 number blocks are spoken for.

If all of the available area codes become used up, the national dialing pattern would need to be expanded by one or more digits. The switches that are used by telephone companies to route calls are designed to handle seven-digit phone numbers and three-digit area codes. Altering the existing design of switches would be extremely expensive and would significantly increase the cost of telephone service. To prevent such disruptions, the FCC has proposed a number of ways to preserve the life of the current 10-digit dialing pattern for as long as possible.

Assignment

A total of 680 usable area codes are available for assignment. Of that 680, 215 were in service in the United States as of June 1999. More than 70 of these regions might need new area codes within a year or two. In addition, more than 40 area codes are in service in the other countries that participate in the NANP, including Canada and a number of Caribbean nations. By comparison, there were 119 area codes in service in the United States at the end of 1991.

Federal and state regulators share the responsibility for administering area codes. The NANP administrator is responsible for day-to-day administration, assignment, and management of area codes in the United States. Congress gave the FCC jurisdiction over telephone number administration in 1996. The FCC delegated to the states the authority to decide when, and in what form, to introduce new area codes. Area codes can be added by either a geographic split or by an overlay. The FCC receives advice regarding number administration issues from the *North American Numbering Council* (NANC), an advisory body that is made up of industry participants, consumer advocates, and state regulators.

Solutions

Most area codes are added by way of a geographic split. The geographic area that is covered by an existing area code is split into two or three different areas. One of the sections retains the existing area code, while others receive new area codes. The benefit of a geographic split is that an area code remains defined as a geographic area. Customers know something about the location of the people whom they are calling. The down side of a geographic split is that many customers must cope with the inconvenience of changing their area code.

An overlay is an alternative way of adding an area code. As the name suggests, the new area code overlays the pre-existing area code, most often serving the identical geographic area. The benefit of an overlay is that customers retain their existing area codes, because only new lines get the new area code. An overlay requires all customers, including those who have telephone numbers in the pre-existing area code, to dial area codes for local calls.

Area code overlays can result in two different homes in the same geographic area with the same seven-digit local number, but with two different area codes. To route calls to the right destinations, customers must dial 10 digits. The FCC has required 10-digit dialing with area code overlays in order to level the playing field, so that new telephone companies can offer their services without suffering a competitive disadvantage. Without 10-digit dialing, established telephone companies might have an advantage over new telephone companies. Customers could find it less attractive to choose a new telephone company if doing so would mean always dialing 10 digits, when choosing an established telephone company would enable them to dial only seven digits. In addition, 10-digit dialing permits a fuller use of all of the numbers within an area code—thereby extending the life of the area code. At this writing, the state of Maryland and certain cities in California, Colorado, Florida, Georgia, Pennsylvania, and Texas have 10-digit dialing. Other states will follow as the need for more telephone numbers becomes an issue.

The FCC is examining several ideas for improving number utilization. These ideas include requiring telephone companies to prove that they need new numbers as well as other more-technical solutions, such as giving telephone numbers to companies in smaller blocks of 1,000 telephone numbers. This situation would result in less waste, because telephone companies that need small quantities of numbers will be allocated 1,000 numbers, not 10,000—significantly reducing the amount of unused numbers.

Telephone number portability has been introduced in many U.S. cities and enables consumers to change local telephone companies without having to change their telephone numbers. You can use the same databases that make number portability possible in order to apportion telephone numbers to local telephone companies in smaller blocks, such as blocks of 1,000. This process is known as number pooling. The FCC is currently seeking public input concerning whether to use this process to alleviate area code exhaust.

Summary

AT&T designed the area code system in the 1940s to make it possible to route long-distance calls automatically. Although the area code system was designed when the telecommunications industry was a monopoly, that system is still in use in today's increasingly competitive telecommunications marketplace. In recent years, the demand for telephone numbers has resulted in a shortage, requiring such remedies as geographic splits and overlays to expand the number of area codes (and, consequently, the quantity of available telephone numbers). These solutions are only temporary, however. Issuing telephone numbers in blocks of 1,000 rather than 10,000 might provide a longer-term solution.

See Also

Dialing Parity

Local Number Portability

ARCnet

The *Attached Resource Computer Network* (ARCnet), introduced by Datapoint Corporation in 1977, was the first LAN technology. Although Ethernet and token-ring architectures are often supported by a wiring hub, the wiring hub is an integral part of ARCnet's design and made it more reliable, flexible, and economical than Ethernet and the token-ring structure. Over the years, ARCnet has been overshadowed by higher-speed LAN technologies. Today, ARCnet is embedded in many companies' products, but these products are not advertised as ARCnet. Unlike other network technologies, there is no upward migration path for ARCnet to higher speeds.

Operation

ARCnet operates at 2.5 Mbps and use a token-passing protocol. When an ARCnet node receives the token, it is permitted to send packets of data to other stations. While the token-ring protocol passes its token around a physical cable ring, ARCnet passes its token from node to node in the order of each node's address.

Nodes could be located up to 2,000 feet from an ARCnet hub through the use of RG-62 coaxial cabling. The total end-to-end length of the network could be 20,000 feet—nearly four miles. With twisted-pair wiring, each node could be located up to 400 feet from an ARCnet hub. As many as 254 connections can be supported on a single ARCnet LAN via interconnected, active hubs.

ARCnet was originally designed to be configured as a distributed star, which entailed each node being directly connected to a hub (refer to Figure A-4) and several hubs being connected to each other. This design suited organizations that terminated cables in centrally located wiring closets. Later, an Ethernet-like bus topology was introduced for ARCnet, which enabled nodes to be interconnected via a single run of cable. The two topologies could even be combined for maximum configuration flexibility. For example, instead of connecting a single ARCnet node to a port on an ARCnet hub, a bus cable with a maximum of eight nodes attached could be connected to the hub.

ARCnet makes use of two types of hubs: passsive and active. Passive hubs are small, four-port devices (non-powered) that support workstations at distances of up to 100 feet using coaxial cabling. Active hubs are eight-port units that support workstations at distances of up to 200 feet using

Figure A-4
ARCnet configuration incorporating both passive and active hubs

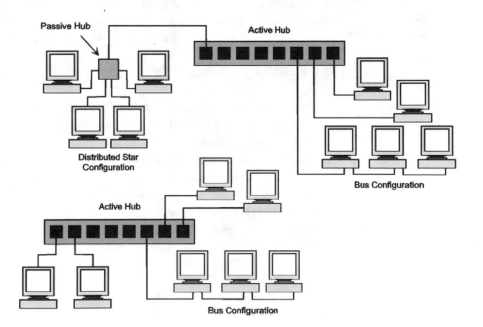

coaxial cabling and up to 400 feet with twisted-pair wiring. By attaching passive hubs to each of an active hub's eight ports, a single active hub can support 24 workstations.

The primary advantages of this distributed star arrangement include cost savings on cable installation and hub ports. A central, active hub that uses twisted-pair wiring offers the best protection against network failure, because each station has its own dedicated connection to the active hub. Furthermore, the central-hub approach makes all of the wiring accessible at one point, which simplifies troubleshooting, fault isolation, and network expansion.

From the beginning, ARCnet used internal transceivers in its hubs. Under Datapoint's concept of conjoint networks, there was no need for bridges and routers to interconnect multiple LANs. Selected workstations and/or file servers could be configured to participate directly in up to six LANs at the same time. Access to each LAN or group of LANs was effectively controlled through hardware.

Although ARCnet is rarely considered today by companies that are seeking LAN solutions, the technology is still employed for niche applications such as data acquisition, plant monitoring and control, closed-circuit cameras, commercial heating and air conditioning, waterway controls, automotive navigation systems, building automation, and motor drive communications. The vendors of these applications use ARCnet as the network for linking various product components.

Summary

A proposal for an improved version of ARCnet was circulated in 1992. So-called ARCnet Plus offered 20 Mbps data transmission and claimed to be downward compatible with ARCnet, enabling a mix of ARCnet and ARCnet Plus equipment to share the same network. At network startup, each ARCnet node would broadcast its capabilities, letting the others know that it can handle higher-speed transmissions. This situation permitted two ARCnet Plus nodes to communicate at the maximum rate. If an ARCnet Plus node encountered a slower ARCnet node, the network would step down to 2.5 Mbps. ARCnet Plus used packet lengths of up to 4,096 bytes, versus 508 bytes under ARCnet. ARCnet Plus also supported more nodes (2,047, compared to ARCnet's 254). Although some vendors offer non-standard ARCnet products that run at up to 5 Mbps, the proposal for the 20-Mbps version of ARCnet was never implemented in commercial products.

See Also

Ethernet

StarLAN

Token ring

Asset Management

Asset management entails the proper accounting of various hardware and software assets within the organization. Without a thorough understanding of an organization's assets, you cannot accurately plan departmental budgets or allocate costs. The failure to account for and manage assets such as desktop hardware and software, in-house cabling, and network lines has other ramifications, as well. This failure can lead to cost overruns on projects, can leave the door open to employee theft (asset shrinkage), and can lead to misuse or abuse of the network. The lack of controls can expose the company to financial penalties for copyright infringement, such as when employees copy software.

Types of Assets

Organizations must track several types of assets. These assets fall into the general categories of hardware, software, network, and cable assets.

Hardware Assets   Hardware inventory starts with identifying the major kinds of systems that are in use in the distributed computing environment —from the servers all the way down to the desktop, laptop, and hand-held computers, as well as their various components, including the *Central Processing Unit* (CPU), memory, boards, and disk drives. The asset-management utilities that come with servers generally scan connected computers for this kind of information.

Most asset management products provide the following basic hardware configuration information:

- The CPU model and vendor
- The type of memory (extended or expanded) and the amount of memory (in kilobytes or megabytes)

- The amount and percentage of hard disk space that is used and available, and the volume number and directories
- Ports that are available and that are in use

Most hardware identification is based on the premise that if a driver is loaded, then the associated hardware must be present. Many of these drivers go unused and are not removed, however, resulting in inaccurate inventory. This situation is remedied by industry standards, such as the *Desktop Management Interface* (DMI) and *Plug and Play* (PnP).

Hardware inventories can be updated automatically on a scheduled basis: daily, weekly, or monthly. Typically included as part of the hardware inventory is the physical location of the unit, the owner (workgroup or department), and the name of the user. Other information might include vendor contact information and the unit's maintenance history. All of this information is manually entered and updated.

In addition to providing inventory and maintenance management, some products provide procurement management, as well. They maintain a catalog of authorized products from preferred suppliers, as well as list and discount prices. They track all purchase requests, purchase orders, and deliveries. With some products, even the receiving of new equipment can be automated—with the system collecting information from scans of asset tags and bar codes. Warranty information can also be added.

Still other asset-management packages accommodate additional information for financial reporting, such as the following:

- *Cost*   Purchase price of the unit and add-in components
- *Payment schedule*   Principal and interest
- *Depreciation*   One-time expense or multi-year schedule
- *Taxes*   Local, state, federal (as applicable)
- *Lease*   Terms and conditions
- *Chargeback*   Cost that is charged against the budgets of departments, workgroups, users, or projects

This kind of information is manually entered and updated in the asset management database. Depending on the product, this information can be shared with spreadsheets and with other financial applications and can be used for budget monitoring, expense planning, and tax preparation.

Software Assets   Another technology asset that must be tracked is client software. Not only can software tracking (also called applications metering) reduce support costs, it can also protect the company from litigation resulting from claims of copyright infringement, such as when users copy and distribute software on the network in violation of the vendor's license agreement.

Asset-management products that support software tracking automatically discover what software is being used on each system on the network by scanning local hard drives and file servers for all installed software. They perform this action by looking for the names of all executable files and arranging them in alphabetical order. They determine how many copies of the executable files are installed and provide the product name and the publisher. Files that cannot be identified absolutely are listed as found but are flagged as unidentified. Once the file is eventually identified, the administrator can fill in the missing information. The accumulated asset information can be used to build a software distribution list. The administrator can then automatically install future upgrades on each workstation that appears on the distribution list.

The administrator can also monitor the usage status of all software on the network in order to enforce license compliance. If several copies of an application are not being used, they can be made available to other users. If all copies of an application are in use, a queue is started—and the application is made available on a first-come, first-served basis.

Network Assets   The kinds of network assets that must be monitored, controlled, and accounted for in an inventory include repeaters, bridges, routers, gateways, hubs, and switches (i.e., any device that is used to implement the network). These types of equipment usually can be centrally managed and controlled anyway.

The entire chassis of a hub, for example, can be viewed via the network-management system, showing the types of cards that are inserted into each slot. With a zoom feature, any card can be isolated—and a representation of the ports, *Light Emitting Diodes* (LEDs), and configuration switches can be displayed at the management console. A list of devices that are attached to any given port can also be displayed and/or printed. Some of the other views that are available from a hub's management system include the following:

- *Configuration view*   Organizes the configuration values for a device and its model, including model name, IP address, security string and device name, type, location, and firmware version
- *Resource view*   A special-purpose view for endpoint devices that shows where the endpoint device accesses its application resources, such as primary and secondary print servers, e-mail servers, and file servers

- *Cablewalk view*   Illustrates the connections that exist along a segment of cable [e.g., Ethernet, token ring, and *Fiber Distributed Data Interface* (FDDI)] and the devices that are connected to each segment

- *Diagnostic view*   Organizes diagnostic and troubleshooting information for a device, including errors, collisions, events, and alarms

- *Performance view*   Displays performance statistics, including load, hard and soft errors, and frame traffic

- *Port performance views*   For each individual port or board, the port performance views summarize port-specific or board-specific performance statistics.

- *Application view*   Organizes device application information, including device IP and *Internet Control Message Protocol* (ICMP) statistics

- *Assigns view*   Enables a specific technician to be assigned to devices that are owned by network users

The hub's management system might also provide a method for documenting the equipment and cable plant inventory through a third-party cable-asset management system. This application and other third-party applications are typically integrated with *Application Programming Interfaces* (APIs).

Cable Assets   Among the assets that also must be managed is the cabling that connects all of the devices on the network, including coaxial cable (thick and thin), twisted-pair (shielded and unshielded), and optical fiber (single-mode and multi-mode). A number of specialized applications are available to keep track of the wiring that is associated with connectors, patch panels, cross-connects, and wiring hubs. They use a graphical library of system components in order to display a network. Clicking any system component brings up the entire data path and all of its connection points. These cable-management products offer color maps and floor plans that are used to illustrate the cabling infrastructure in one or more buildings. A zoom feature can isolate backbone cables within a building, on a floor, or within an office.

Some cable-asset management products can generate work orders for moving equipment or rewiring. Managers can create both logical and physical views of their facilities and even view a complete data path simply by clicking a connection. Some cable-asset management products automatically validate the cabling architecture by checking the continuity of the data paths and the type of network for every wire. In addition, a complete picture of the connections can be generated and printed. With this information, network administrators and technicians know where new equipment should go, what needs to be disconnected, and what should be reconnected. Some cable-asset management products can even calculate network load statistics in order to facilitate proactive management and troubleshooting.

Similar to other types of asset-management applications, cable-management applications can be run as stand-alone systems or can be integrated with help-desk products, hub management systems, and major enterprise management platforms (such as IBM's NetView, Hewlett-Packard's Open-View, and SunSoft's Solstice SunNet Manager).

Summary

Several approaches to asset management are available. Organizations can buy one or more software packages, use an integrated approach that is available with some help-desk or network-management systems, or out-source the asset-management task to a systems integrator or computer vendor. In addition to containing the cost of technology acquisitions and reining in hidden costs, such asset management can improve help-desk operations, enhance network management, assist with technology migrations, minimize asset shrinkage, and provide essential information for planning a corporate reengineering strategy.

See Also

Electronic Software Distribution

Network Design

Network-Management Systems

Asynchronous Transfer Mode (ATM)

Asynchronous Transfer Mode (ATM) is a protocol-independent, cell-switching technology that offers *Quality of Service* (QoS) guarantees for the support of data, voice, and video traffic at multi-megabit-per-second speeds. ATM is also highly scalable, making it equally suited for interconnecting legacy systems and LANs and for building WANs over today's high-performance fiber-optic infrastructures. ATM-based networks can be accessed through a variety of standard interfaces, including frame relay.

Applications

Many applications are particularly well suited for ATM networks, including the following:

- *LAN internetworking*   ATM can be used to interconnect LANs over the WAN. Special protocols make the connection-oriented ATM network appear as a connectionless Ethernet or as a token-ring LAN segment.

- *Videoconferencing or broadcasting*   ATM can be provisioned for interactive video conferencing between two or more locations or to support point-to-multi-point video broadcasts.

- *Telemedicine*   With ATM, large amounts of bandwidth can be provisioned to support the rapid exchange of high-resolution diagnostic images and multimedia patient records while permitting interactive consultations among medical specialists at different locations.

- *Private-line connectivity*   An ATM virtual circuit can be used to provide a more economical way to provision a T1 leased line on the WAN. ATM protocols can even emulate N x 64 Kbps DS0 transport.

- *PBX voice trunking*   An ATM virtual circuit can be used to interconnect PBXs and to maintain full PBX feature support, call routing, and switching. Voice trunking combines multiple calls onto a single virtual circuit for further bandwidth optimization, reduced delay, and lower cost. PBX voice trunking requires an integrated access device at the customer premises, between the PBX and ATM switch, that performs the protocol conversions necessary to extend feature signaling across the ATM network.

ATM also offers a consolidation solution for any company that maintains separate networks for voice, video, and data. The reason for separate networks is to provide appropriate bandwidth and to preserve quality standards for the different applications. ATM provides a unified platform for multi-service networking that meets the bandwidth and QoS needs of all applications. Although the startup cost for ATM is high, the economics of network consolidation mean that customers do not have to wait long to realize the returns on their investments.

QoS Categories

ATM efficiently serves a broad range of applications by enabling an appropriate QoS to be specified for each application. Various categories have been developed to help characterize network traffic, each of which has its own QoS requirements. These categories and QoS requirements are summarized in Table A-1.

Of these QoS categories, GFR is the newest—having been approved by the ATM Forum in mid-1999. GFR is designed for applications that might require a minimum rate guarantee and that can benefit from accessing additional bandwidth dynamically in bursts, without requiring adherence to a flow-control protocol. Users can send any rate up to the peak cell rate, but the network commitment can only be sent to the minimum cell rate. Possible applications include aggregating TCP/IP traffic between two routers.

Operation

QoS enables ATM to admit a CBR voice connection while protecting a VBR connection for a transaction-processing application, and it enables an ABR or UBR data transfer to proceed over the same network. Each virtual circuit will have its own QoS contract, which is established at the time of connection setup at the *User-to-Network Interface* (UNI). The network will not permit any new QoS contracts to be established if they will adversely affect the network's capability to meet existing contracts. In such cases, the application will not have the capability to get on the network until the network is fully capable of meeting the new contract.

When the QoS is negotiated with the network, there are performance guarantees that accompany the negotiation: maximum cell rate, available cell rate, cell transfer delay, and cell loss ratio. The network reserves the resources that are needed to meet the performance guarantees, and the user is required to honor the contract by not exceeding the negotiated parameters. Several methods are available to enforce the contract. Among them is traffic policing and traffic shaping.

Traffic policing is a management function that is performed by switches or routers on the ATM network. To police traffic, the switches or routers use a buffering technique that is referred to as a leaky bucket. This technique entails traffic flowing (leaking) out of the buffer (bucket) at a constant rate (the negotiated rate), regardless of how fast the traffic flows into the buffer. If the traffic flows into the buffer too fast, the cells will be allowed onto the network only if enough capacity is available. If there is not enough capacity, the cells are discarded and must be retransmitted by the sending device.

Traffic shaping is a management function that is performed at the UNI of the ATM network. Traffic shaping ensures that traffic matches the contract that is negotiated between the user and network during connection

Table A-1

Quality of Service Requirements					
Category	Application	Bandwidth Guarantee	Delay Variation Guarantee	Throughput Guarantee	Congestion Feedback
Constant Bit Rate (CBR)	Provides a fixed virtual circuit for applications that require a steady supply of bandwidth, such as voice, video, and multimedia traffic	Yes	Yes	Yes	No
Variable Bit Rate (VBR)	Provides enough bandwidth for bursty traffic, such as transaction processing and LAN interconnection (as long as rates do not exceed a specified average)	Yes	Yes	Yes	No
Unspecified Bit Rate (UBR)	Makes use of any available bandwidth for routine communications between computers but does not guarantee when or if data will arrive at its destination	No	No	No	No
Available Bit Rate (ABR)	Makes use of available bandwidth and minimizes data loss through congestion notification. Applications include e-mail and file transfers.	Yes	No	Yes	Yes
Guaranteed Frame Rate (GFR)	Provides a minimum rate guarantee to virtual circuits at the frame/packet level and enables the fair usage of any extra network bandwidth	Yes	Yes	Yes	Yes

setup. Traffic shaping also helps guard against cell loss in the network. If too many cells are sent at once, cell discards can result, which will disrupt time-sensitive applications. Because traffic shaping regulates the data transfer rate by evenly spacing the cells, discards are prevented.

Cell Structure

Voice, video, and data traffic is usually comprised of bytes, packets, or frames. These larger payloads are chopped up into smaller fixed-length cells by the customer's router or the carrier's ATM switch. ATM cells are fixed at 53 octets[3] and consist of a five-octet header and a 48-octet payload (refer to Figure A-5).

The cell header contains the information that is needed to route the information field through the ATM network. The header supports five functions:

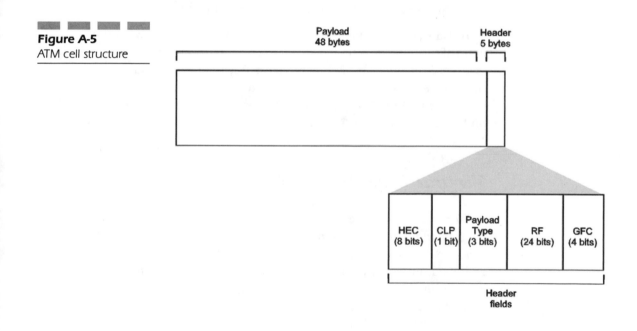

Figure A-5
ATM cell structure

---

[3]This odd cell size was the result of a compromise among international standards bodies. The United States wanted the cell's data payload size to be 64 bytes, and Europe wanted a data payload of 32 bytes. The compromise was simply to average the two, which equals 48 bytes. The cell's header required five bytes, providing an overall cell size of 53 bytes.

- *Generic Flow Control* (GFC)   This four-bit field only has local significance: it enables customer-premises equipment at the UNI to regulate the flow of traffic for different grades of service.

- *Routing Field* (RF)   This 24-bit field contains a *Virtual Path Identifier / Virtual Channel Identifier* (VPI/VCI) combination (i.e., address) that is used to route the cell through the network.

- *Payload Type* (PT)   This three-bit field is used to indicate whether the cell contains user information or connection-management information. This field also provides for network congestion notification.

- *Cell Loss Priority* (CLP)   This one-bit field, when set to 1, indicates that the cell might be discarded in the event of congestion.

- *Header Error Check* (HEC)   This eight-bit field is used by the physical layer to detect and correct bit errors in the cell header. The header carries its own error check to validate the VPIs and VCIs and to prevent the misdelivery of cells to the wrong UNI at the remote end. Cells that are received with header errors are discarded. Higher-layer protocols are responsible for initiating lost cell recovery procedures.

Initially, a concern existed about the high overhead of cell relay, with its ratio of five header octets to 48 data octets. With innovations in *Wave Division Multiplexing* (WDM) to increase fiber optic's already-high capacity, however, ATM's overhead is no longer a serious issue. Instead, the focus is on ATM's unique capability to provide a QoS in support of all applications on the network.

Virtual Circuits (VCs)

ATM virtual circuits can be bidirectional or unidirectional, meaning that each *Virtual Circuit* (VC) can be configured for one-way or two-way operation. The virtual circuits can be configured as point-to-point (i.e., PVC), switched, or multi-point. They can also be symmetric or asymmetric in nature. In other words, each bidirectional VC can be configured for symmetric operation (the same speed in both directions) or asymmetric operation (different speeds in each direction).

A VC has two components: a virtual path and a virtual channel. In this simplified view of an ATM network (refer to Figure A-6), the customer has two locations that are connected by a virtual path, which contains a bundle of virtual channels. Each of the three VCs is assigned to a particular end system, such as a PBX, server, or router. The individual connections

Figure A-6
A simplified view of
virtual circuits
through an ATM
network

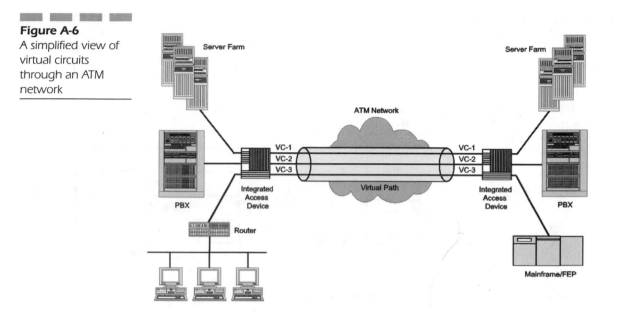

between the end devices at each location are identified by ATM addresses, which consist of a VCI/VPI.

In this example, an *Integrated Access Device* (IAD) is used to consolidate the VCs and to deliver them to the ATM switch via a dedicated line. VC-1 provides LAN users with access to a mainframe. VC-2 provides trunking between the PBXs. VC-3 provides LAN users with access to a remote server.

In a large network, there might be hundreds of virtual paths. ATM standards permit up to 65,000 VCs to share the same virtual path. This scheme simplifies network management and network recovery. When a virtual path must be reconfigured to bypass a failed port on an ATM switch, for example, all of its associated virtual connections go with it—eliminating the need to reconfigure each VC individually.

ATM Layers

Like other technologies, ATM uses a layered protocol model. ATM exists at Layer 2 of the OSI model and has only four layers (refer to Figure A-7), which typically operate above SONET at Layer 1.

The *ATM Adaptation Layer* (AAL) provides the necessary services to support the higher-layer protocols. AAL can exist in the end stations or in

ATM protocol model
in relation to the
Open Systems
Interconnection (OSI)
reference model and
Synchronous Optical
Network (SONET)
physical layer
protocols

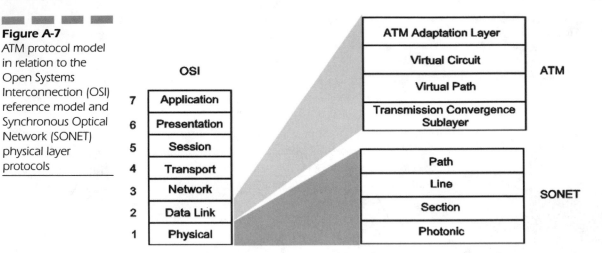

Table A-2	AAL	Application Examples	Notes
	1	Isochronous constant bit-rate services, such as audio and video streaming	Supports connection-oriented services that require constant bit rates and that have specific timing and delay requirements
	2	Isochronous variable bit-rate services, such as interactive videoconferencing	Supports connection-oriented services that do not require constant bit rates (i.e., variable bit-rate applications)
	3/4	Near real-time variable bit-rate data, such as SMDS (connectionless) or X.25 (connection-oriented).	Originally intended as two layers: one for connection-oriented services, and the other for connectionless services. Consolidation into one layer supports variable bit-rate services over both types of connections.
	5	Near real-time variable bit-rate data, such as datagrams and signaling messages	Supports connection-oriented variable bit-rate data services; a leaner AAL compared with AAL3/4 at the expense of error recovery and built-in retransmission; uses less bandwidth overhead, has simpler processing requirements, and reduces implementation complexity

the network switches. Among other tasks, this layer is responsible for segmenting the information into 53-byte cells and reassembling it back into its native format *Segmentation And Reassembly* (SAR) at the receiving end. Table A-2 describes the adaptation layers:

Summary

A solid base of standards now exists to enable equipment vendors, service providers, and end users to implement a wide range of applications via ATM. The standards will continue to evolve as new applications emerge. The rapid growth of the Internet is one area in which ATM can have a significant impact. With the Internet being forced to handle a growing number of multimedia applications—telephony, videoconferencing, faxes, and collaborative computing, to name a few—congestion and delays are becoming more frequent and prolonged. ATM backbones will play a key role in alleviating these conditions, enabling next-generation networks to be used to their full potential.

See Also

Synchronous Optical Network (SONET)

Wavelength Division Multiplexing (WDM)

Attenuation

Attenuation is the decrease or reduction in the power of a signal as the distance of a circuit increases. In other words, attenuation simply means signal loss. Attenuation is of particular concern in provisioning T1 service in the local loop. Commonly, you run 26 *American Wire Gauge* (AWG) cabling from the CO to 10,000 feet and then change to thicker 24 AWG for longer runs, in order to sustain high performance. Just as common are bridged taps, which are spurs off a circuit that run at distances of up to 5,000 feet. Wire gauge changes and bridged taps introduce attenuation, or a weakening of the signal. The average United States local loop has 22 splices, resulting in a considerable amount of attenuation. When carriers provision T1 lines, they remove bridged taps and avoid mixed-gauge wiring whenever possible.

Attenuation affects all media, even the performance of fiber-optic links. Because the glass fibers are not always perfectly clear, the amount of light coming out of the distant end of a cable is always somewhat less than the amount of light that goes into the fiber. Fiber loss might be due to several causes, including the following:

- Absorption by impurities
- Scattering by impurities or by the defects at the core-cladding interface, and Rayleigh scattering by the molecules of the medium (i.e., silica)

■ Fiber bends and micro-bends

■ Scattering and reflection at splices

All of these phenomena contribute to the degradation of the fiber transmission. Excessive attenuation results in a received signal that is too weak to be useful. This quality is usually linear, increasing in direct proportion to the length of a fiber-optic cable run.

Summary

Attenuation refers to a signal loss and is measured in decibels (dB). Attenuation is the ratio between the input power and the output power of a signal. The minimum attenuation of a standard telecommunications fiber occurs around the 1550 nanometer (nm) wavelength; it is of the order of 0.2 dB/km.

See Also

Decibel

Automatic Call Distributors (ACDs)

Automatic Call Distributors (ACDs) provide an efficient method of handling heavy volumes of incoming calls in customer service and telemarketing environments. These systems greet callers with recorded messages, provide a menu of dial options, and route calls to appropriate individuals or to the voice mail system. ACDs can also provide music while callers are on hold and make periodic announcements regarding queue status or holding time. ACDs can even be linked via private lines to form distributed call-processing systems that can route incoming calls to corporate locations in other time zones. ACDs also can handle incoming calls from the Internet.

By rapidly directing calls to agents who are available and who have the information or technical expertise that the caller requires, companies can make efficient use of their communications systems and human resources to provide high levels of customer satisfaction. Many systems also provide a range of management capabilities, including reporting functions that incorporate such statistics as the number of incoming calls handled, the number of calls abandoned, and the system's peak calling capacities. These reports can be used to determine factors such as agent productivity and the need for more equipment, lines, or staff.

ACD Operation

A primary goal of the ACD is to maintain the productivity of agents through the efficient distribution of incoming calls. Supervisor, or master, positions are points within the service center that are staffed by managerial personnel who monitor individual calls, agents, and overall system activity.

ACDs typically answer a telephone call on the first ring or after a fixed number of rings, then examine preprogrammed processing tables for routing instructions while callers who are on hold listen to recorded announcements and/or music. ACDs can also answer calls dynamically by sensing the incoming call and searching through routing schemes before answering the call. After the call is answered, other systems such as a voice response unit might gather additional information and compare that information with customer databases before passing the call to an agent position. This basic structure is common to all ACDs. Systems differ primarily by the method of call allocation, the types of system management reports, and the various control features.

Applications

Generally, ACDs are used by companies that have call centers with high incoming call volumes and that have five or more agents whose responsibilities are almost entirely restricted to handling incoming calls. Typically, ACDs are used in the following call-center environments:

- Customer service
- Help desk
- Order entry
- Credit authorization
- Reservations
- Insurance claims
- Catalog sales

A relatively new call-center environment is the Web page on corporate IP-based intranets or the Internet. Companies can set up a *connect-me* button on their Web page so that when a user clicks the link from a multimedia-equipped PC, a voice connection is established over the IP network. When the call hits the other side of an IP-to-*Public Switched Telephone Network* (PSTN) gateway, the call is transferred to the ACD, where it is routed to the next-available agent. When used in this manner, the ACD can be used to support e-commerce applications by assisting customers during the purchase decision.

IP-based ACDs can usually serve up to 50 agents in a call-center environment. These devices enable call-center agents to handle and manage Internet/intranet calls with traditional tools, including hold, retrieve, transfer, and conference. One vendor, PakNetX, offers an IP-based ACD that integrates audio, video, and data as one contact—offering a new level of personalized customer service over the Web (refer to Figure A-8).

Figure A-8
PakNetX's ACD
topology

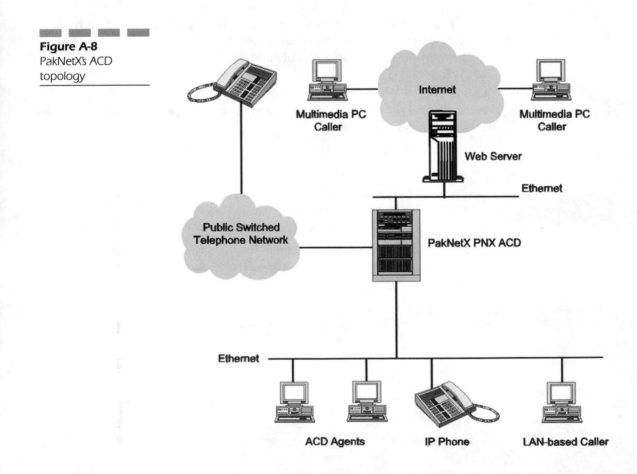

Callers typically reach the customer-service agent by clicking a *connect-me* button on the company's Web site. When the customer initiates the call, the Windows NT-based PNX ACD is actually called. H.323 call setup occurs between the customer and the ACD. When an agent is available, an outbound call is placed to the agent so that the caller and agent can conference together. All of the H.323/T.120 packets go through the ACD. Upon call arrival, the agent interface (as shown in Figure A-9) shows call-context information, which can include the customer's name, account number, currently viewed URL, and subscriber service level.

As each call is connected through the ACD, packets pass through the switch where each packet address is remapped for routing to the appropriate agent. This process enables agents to be hidden from direct public access, simplifying LAN topology management issues and eliminating potential security problems.

The PNX ACD software enables the agent to provide helpful content to the customer and to engage in shared Web browsing with the customer, as well as to communicate via text chat, e-mail, file transfer, and data collaboration. Even while a caller is waiting to be connected to a live agent, streaming audio and video can be downloaded to the company's Web page that provides information that is of interest to the customer until the call can be answered.

System Components

ACD systems consist of incoming lines, agent positions, supervisor positions, and the switch itself. Often, toll-free 800 lines are connected to ACD systems, but any type of line can also be connected [including *Integrated Services Digital Network Primary Rate Interface* (ISDN PRI) lines]. ACD lines can be routed through a PBX that provides general carrier-interface support

Figure A-9
The agent interface of PakNetX's PNX ACD

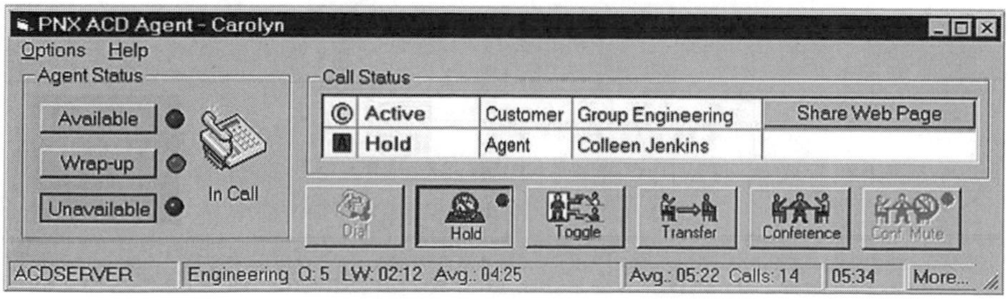

for the business. This arrangement is most common in systems that use digital T1 trunks to carry both ACD-related calls and other general call traffic.

Types of ACDs

Various types of equipment and services are available in order to provide ACD features and functions:

- *Stand-alone systems*   This system is mainly used in environments where the service center is separate from the rest of the business and where ACD functions do not require integration with the corporate telephone system.

- *Integrated systems*   ACD functionality is added to a PBX key telephone system with ACD software, providing call allocation and service supervision within the telephone system itself.

- *PC-based systems*   Software that is added to a multimedia-equipped PC that includes functions such as voice mail, interactive voice response, intelligent queue announcements, and computer integration, along with the traditional ACD distribution and routing functions. The software collects call statistics and generates management reports.

- *Automatic call sequencers*   Independent devices that perform the same type of call-to-agent station allocation as an ACD, but without complex load and time calculations. The systems rely on the PBX for routing calls, because they have no switching matrix of their own.

- *Centrex-based systems*   The telephone company provides ACD functions and features as part of its Centrex service.

- *Central office-based systems*   The telephone company provides ACD functions and features as a service that is separate from Centrex.

- *Third-party service*   Third-party firms provide ACD service to other companies and handle their own call overflow, as well. The operation is completely transparent to the caller.

Summary

A number of developments have combined to make the use of ACDs possible in almost any business that has a need to handle a large number of incoming calls. The developments include advances in call-processing technology, improvements in the public switched telephone network, developments in *Computer Telephony Integration* (CTI), the growing popularity of the Internet, and advancements in PC-related technologies—especially in the area of multimedia.

See Also

> Call Centers
>
> Centrex
>
> *Private Branch Exchange* (PBX)

Automatic Number Identification

Automatic Number Identification (ANI) is the identification of the caller when the call is inbound on dedicated network services such as T1 or ISDN PRI. A separate but related service is called *Dialed Number Identification Service* (DNIS), which enables a company to have several inbound numbers associated with a dedicated network service (such as T1 or ISDN PRI). Based on the number that is dialed by the inbound calling party via the dedicated inbound service, the call will be routed to a specific hunt group, extension, or application.

Implementation Issues

The original purpose of ANI was to enable carriers to automatically identify and bill customers for the calls that they made. ANI can also support other services, however. As the IXCs recognized the value of ANI to businesses for identifying customers and matching their telephone numbers to database records for instant retrieval, they began offering ANI as a separate service. This situation raised the privacy issue among consumer groups, who voiced their objections to the FCC.

After considering the privacy issue, the FCC decided not to prohibit carriers from delivering ANI information with their toll-free 800 service customers, even when the calling party requests privacy. For consumers, the major concern about ANI services was actually the reuse of the calling party information by 800 service customers. Accordingly, the FCC ruled that customer information that is gained from ANI services cannot be reused or sold to other companies without the consent of the calling party.

The only exception is that the information could be used to offer products or services to customers where the products or services are directly related to products or services that have been previously provided. Carriers that provide ANI services are required to include these restrictions in contracts that offer the service.

In its decision to allow ANI to be used for non-billing applications, the FCC observed that even small efficiencies regarding individual transactions become significant in an economy that averages more than one billion interstate calling minutes a day. These savings could lower the service costs for suppliers, which could lead to lower prices for consumers.

Summary

ANI enables service providers and consumers to conduct transactions over the telephone more efficiently. For example, computer services could recognize the calling party's number and either permit or deny access to a bank account or credit card database. Stockbrokers, travel agents, parts and equipment providers, and booksellers could route a call to a preprogrammed location that is closer to the calling party, in order to expedite deliveries or services. Retailers could verify credit and billing information instantaneously. Customized services that depend on the caller's individualized preferences could also be developed. Of course, ANI is useful in supporting emergency 911 systems as well, particularly when combined with *Automatic Location Identification* (ALI)—a feature that uses a computer database to associate a physical location with a telephone number.

See Also

Caller Identification

Custom Local Area Signaling Services

Signaling System 7 (SS7)

Telephone Fraud

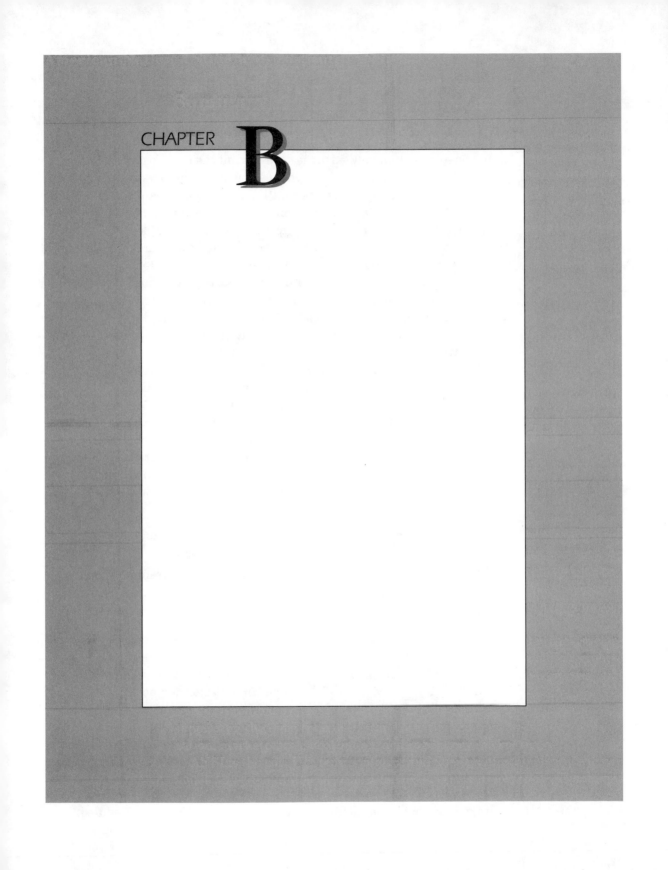

CHAPTER B

Bandwidth Management Systems

The increasing demands of multimedia and priority traffic on corporate intranets has ultimately led to the development of a new class of performance-optimization tools. These tools, which are usually referred to as bandwidth managers or packet shapers, permit the allocation of bandwidth in an IP network according to various application-performance criteria. In effect, these tools provide a relatively inexpensive method of implementing policy-based routing over IP networks without large-scale changes to network infrastructure or the installation of expensive special-purpose routers. These tools, combined with load-balanced servers and caching strategies, reduce congestion on networks, ensure the delivery of priority messages, and support real-time traffic (including packetized voice data).

IP bandwidth management solutions give network managers the tools to prioritize business-critical traffic over less-important traffic, such as e-mail, Web browsing, *File Transfer Protocol* (FTP) file transfers, and external push applications. The capability to prioritize traffic improves the performance of vital applications, increases the number of transactions per second, optimizes bandwidth usage, and improves user satisfaction.

Bandwidth managers employ a variety of traffic-control techniques—including TCP rate control, queuing, and policy definition—in order to ensure that essential traffic reaches its destination during periods of network congestion. The need for these capabilities depends on several factors, such as the following:

- The variety of traffic types that are run over an intranet or TCP/IP WAN service
- Whether certain traffic is of more value than other traffic
- Whether any traffic has delay characteristics that require it to be prioritized

If an organization is running traffic that is all of equal value, these capabilities will not be needed. If the organization finds itself running more multimedia applications on its intranet, however, such as streaming audio or video, IP telephony, and collaborative computing, then adding such capabilities will improve network performance more economically than upgrading the network with more bandwidth.

Bandwidth managers are positioned at the edge of corporate networks, because they decide which traffic receives preferential treatment before reaching the WAN router. The networks are usually deployed between the WAN and the LAN, but they can sit on a LAN that is connected to a WAN-linked router. Policies define how various types of traffic are handled.

Operation

Bandwidth managers identify and manipulate traffic classes by looking at the *Type of Service* (TOS) bit in the IP header, the IP address, the TCP or *User Datagram Protocol* (UDP) port number, the *Domain Name Service* (DNS), the application, or the *Uniform Resource Locator* (URL). Traffic shapers, by using the TOS bit, can identify traffic without adding to the IP header, which eliminates the need to change routers and switches. When prioritizing traffic, they use a variety of different strategies—including queuing, changing the size of the TCP/IP window (TCP rate control), or a combination of both. Some can also handle protocols other than IP [such as Internetwork Packet Exchange (IPX)], and some offer automatic traffic identification.

Queuing techniques can be used separately or with TCP rate control. Queuing types include priority, weighted, and class-based. Priority queuing sets queues for high and low-priority traffic and empties high-priority queues first. Cisco's *Weighted Fair Queuing* (WFQ), for example, assigns traffic to priority queues and also assigns an appropriate amount of bandwidth. Its *Class-Based Queuing* (CBQ) guarantees a transmission rate to a queue, and other queues can borrow from unused bandwidth in order to support their traffic.

While queuing technology helps prioritize traffic, it also has some drawbacks. For example, it only works on outbound traffic, and packets that are delayed in queues either get dropped and/or timeout, requiring retransmissions that cause a significant reduction in network efficiency. TCP rate control takes a more efficient and precise approach to bandwidth allocation: it applies TCP rate-based flow-control policies to both individual traffic flows and to classes of flows. This function results in predictable service-level control for true IP QoS. Because this rate control does not rely on a queue and thus provides bidirectional bandwidth management for both inbound and outbound traffic, TCP rate control increases network efficiency by avoiding retransmissions and packet loss.

Bandwidth management systems have emerged as a quick fix for intranet congestion problems, providing a way to introduce policy-based routing at a relative low cost and with minimal impact upon the existing infrastructure. The future of these devices and software systems, however, is toward direct incorporation of these capabilities in routers and switches and bandwidth management integration with directory services. Such changes in infrastructure will be long in coming, even with new standards such as the *Differentiated Service* (DiffServ) standard that was developed by the *Internet Engineering Task Force* (IETF). The development of separate bandwidth management systems, however, demonstrates the need for policy-based routing in order to improve the performance of corporate intranets.

Types of Solutions   Bandwidth management products come in the form of hardware-only and software-only solutions, or a combination of hardware and software. Hardware-based bandwidth managers tend to offer the best performance, because they rely on *Application-Specific Integrated Circuits* (ASICs) and dedicated memory in order to implement bandwidth management. They can be difficult to upgrade, however. Software systems provide considerable flexibility and are easy to upgrade but also have a performance penalty. Combined hardware/software systems have a performance and feature set that is somewhere between the two. Bandwidth management software and devices are also starting to be integrated with other network-management components, such as firewalls and integrated-access devices.

Packeteer's PacketShaper is one of the oldest bandwidth management solutions. Specifically, PacketShaper is a policy-based, application-adaptive bandwidth management hardware solution that controls the use of WAN bandwidth and delivers end-to-end quality of service on a per-application-flow basis (refer to Figure B-1). This hardware/software solution supports all

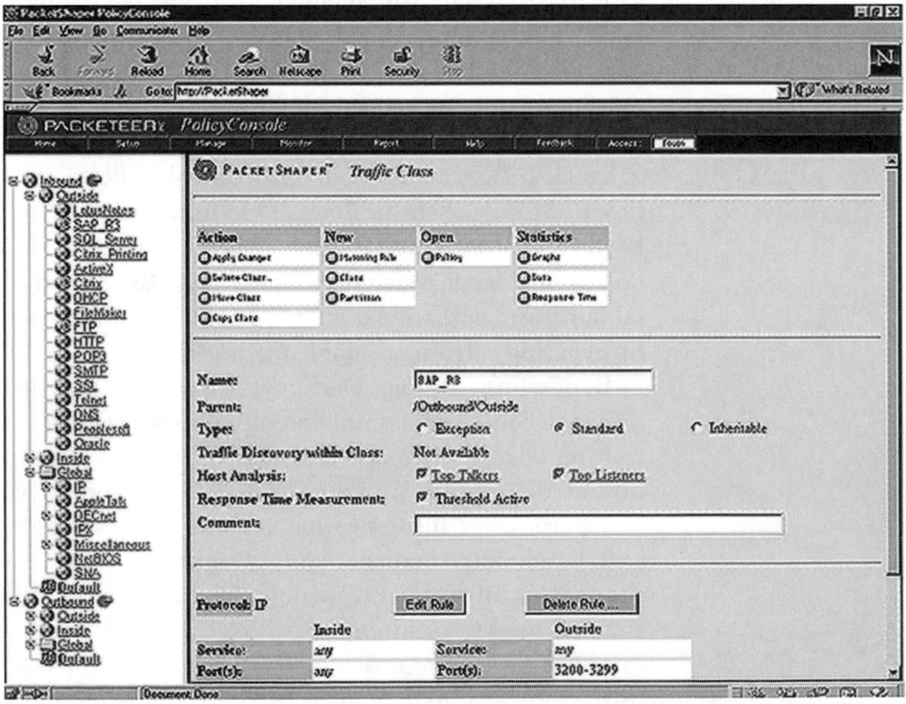

Figure B-1
Packeteer's PacketShaper PolicyConsole software enables a network manager to configure, control, and monitor the PacketShaper remotely via a standard Web browser. The traffic tree on the left side of the Manage screen shows both inbound and outbound traffic being automatically discovered by PacketShaper.

four elements that are essential for ensuring the performance of business-critical applications: automatic traffic discovery and classification; in-depth performance analysis; rate-based traffic control; and application service-level management. The solution employs a TCP rate control mechanism that enables network administrators to go beyond simple traffic priorities in order to set kilobit-per-second partitions for each classified traffic flow. This solution can also show a graphical representation of the top 10 types of traffic that are running through a WAN.

Orchestream's Enterprise software solution uses policy control as a means of implementing QoS. This solution enables network managers to set rules in order to control the users and applications that can access the different classes on the network. Policies are based on rules that simply relate the capability of the subjects (users and/or applications) to access network resources (data, systems, and other resources). The power of the policy-based approach is that it is possible to manage the access controls for thousands of individuals—such as a finance group—as easily as for a single user. The rules can be varied dynamically, depending on the business's needs.

Orchestream offers this functionality on any of today's installed base of routers and automates the configuration of all relevant network devices so that traffic is seamlessly managed as it passes through each router in the network. Whether the organization needs to manage network QoS, security, or both, it can do so from one point of control in the network.

Allot Communications' AC family of hardware products is designed to be installed at the LAN/WAN boundary of an enterprise or *Internet Service Provider* (ISP) network, in order to enforce administrator-defined policies through advanced QoS traffic-shaping algorithms. This solution enables organizations to define tiered user services, to support time-sensitive multimedia, and to give priority treatment to vital applications.

Load Balancing

Another way to improve the handling of IP traffic on corporate intranets is to make use of load-balancing systems. While bandwidth management tools enable the allocation of portions of available bandwidth to different users, load balancers operate on the server side, routing traffic to the best server that is available to handle a job. In a load-balanced network, incoming traffic is distributed among replicated servers, thus permitting server clusters to share the processing load, to provide failover capability, and to speed up the response time for users.

With Allot Communications' NetEnforcer, for example, advanced load-balancing policies can be defined that reflect the capabilities of individual servers on the network. NetEnforcer is a software module for the company's AC series of bandwidth managers. The module enables policies to be defined that redirect traffic based on the capabilities of the servers. For example, all video can be redirected to the video server; all Web traffic can be redirected to the Web server; and all employees in marketing can be redirected to the marketing server. NetEnforcer continuously adjusts both the flow and prioritization of applications through the network and the distribution of those applications to servers.

Traffic can be balanced between available servers by using algorithms such as the following:

- *Round-robin*   Each server will be treated with equal priority.
- *Weighted round-robin*   Each server can be given an individual weight or priority that is based on its capability to deliver specific applications.
- *Maintenance rerouting*   Traffic can be rerouted to another server when the originally targeted server becomes unavailable.

The selection of the right load-balancing approach results in efficient, optimized traffic throughout the network.

Network Caching

Network caching offers an effective and economical way to relieve some of the massive bandwidth demand. This function, coupled with the company's own intranets, maintains an active cache of the most-often visited Web sites so that when these pages are requested again, the download occurs from the locally maintained cache server (instead of the request being routed to the actual server). The result is a faster download speed.

Caches can reside at various points in the network. For enterprises, caches can be deployed on servers throughout campus networks and in remote and branch offices. Within enterprise networks, caches are starting to become as ubiquitous as IP routers. Just about every large company now depends on Web caches in order to keep their intranets running smoothly.

Two types of cache techniques are available: passive and active. With the passive technique, the cache waits until a user requests the object again before sending a refresh request to the server. If the object has not changed, the cached object is served to the requesting user. If the object has changed, the cache retrieves the new object and serves it to the requesting user. This approach, however, forces the end user to wait for the refresh request, which can take as long as the object retrieval itself. This approach also consumes bandwidth for unnecessary refresh requests.

With active caching, the cache performs the refresh request before the next user request (if the object is likely to be requested again and is likely to have changed on the server). This automatic and selective approach keeps the cache up to date so that the next request can be served immediately. Network traffic does not increase, because an object in the cache is refreshed only if the object is likely to be requested again—and only if there is a statistically high probability that the object has changed on the source server.

For example, the Web page of a major broadcast network might contain a logo object that never changes, while the "Breaking News" object changes often. If this page is popular among corporate users, the "Breaking News" object will be refreshed prior to the next user's request. When only refreshing the content that is likely to change, users are served with the most updated information without putting unnecessary traffic on the network.

By contrast, previous generations of cache technology do not accommodate the individual nature of cached objects. They rely on global settings, which treat all objects equally (thereby severely limiting the hit ratio). Because the passive cache requires frequent, redundant refresh traffic, it induces significant response-time delays.

Active caches can achieve hit ratios of up to 75 percent, meaning that a greater percentage of user requests can be served by the cache. If the requested data is in the cache and is up to date, the cache can serve the data to the user immediately upon request. If not, the user must wait while the cache retrieves the requested data from the network. Passive caches, on the other hand, typically achieve hit rates of only 30 percent, thereby forcing users to go to the network two-and-a-half times more often in order to obtain the information that they need.

Some kinds of objects on a Web page cannot be cached and are individually marked by their Web server as such. One object of this type is a database-driven object, such as a real-time stock quote. While this particular object is not cacheable, the rest of the objects in the page usually are cacheable. For example, a Web page that delivers stock quotes might contain 30 other objects, and only one of those objects—the stock ticker—might not be cacheable. If all of the remaining objects can be cached, a significant performance benefit will result.

Summary

Organizations that are seeking bandwidth management solutions should give preference to those solutions that can be easily integrated into their current management platform or that can be added to existing firewalls and routers. Depending on the traffic volume between the corporate intranet and the public Internet, a bandwidth management solution that is integrated into a firewall might be justified. Ideally, the selected solution will integrate seamlessly into the existing network without requiring new protocols, standards, topologies, or hardware changes. Along with bandwidth management, users should give consideration to implementing load balancing and network caching solutions. These solutions can economically provide dramatic improvements in response time and make more scarce bandwidth available in order to keep vital applications running smoothly.

See Also

Firewalls

Network Management Systems (NMS)

Routers

Bell Labs

Bell Labs is a unit of Lucent Technologies. More than any other single R&D institution, Bell Labs has helped weave the technological fabric of modern society. Bell Labs has received more than 25,000 patents since 1925 and today averages more than three patents every business day. Currently, the mission of Bell Labs is to create, obtain, and rapidly deliver innovative technologies to Lucent's customers.

Bell Labs has 30,000 employees in 20 countries—4,000 of whom hold doctorate degrees. About 80 percent of Bell Labs employees are integrated into Lucent's business units. The other 20 percent, called the Central Labs, support all of Lucent's businesses. Central Labs consists of three major technical divisions:

- Research, which focuses on fundamental research in physical sciences and engineering, computing, mathematical sciences, and communication sciences
- Advanced Technologies, which focuses on software, advanced communications, and design automation

■ Technology Officer Division, which has intellectual property and standards work and has a technology integration role with the chief technical officers of each of Lucent's businesses

History

Bell Labs was created from the merger of the research and development operations of AT&T and Western Electric in 1925. Bell Labs' guiding principle is that innovations result from constant interaction between basic and applied research, between research and development, and between technologists and people who are responsible for marketing and business management. This philosophy has produced a variety of innovations over the past 75 years.

In the 1920s and 1930s, Bell Labs researchers demonstrated long-distance television transmission and the electrical digital computer and led the development of sound motion pictures and the artificial larynx. Two fundamental information-age inventions, the transistor and information theory, were products of Bell Labs research in the 1940s. The solar cell, which is the concept behind the laser, and the communications satellite were among Bell Labs' many significant contributions during the 1950s and early 1960s.

Bell Labs produced the UNIX operating system, the C and C++ programming languages, the Plan 9 distributed operating system (which advances client/server applications over public and private networks), and Inferno networking software (which supports interactive applications over any communications network). Major advances at the R&D unit in fault-tolerant network software have enabled Lucent Technologies to achieve high levels of system reliability regarding network equipment.

Bell Labs research has contributed to numerous major innovations in networking technologies. Bell Labs invented stored-program control switching, automated network management, and the intelligent network. Bell Labs also invented cellular mobile communications and has developed leading systems for digital cellular, *Personal Communications Services* (PCS), mobile computing, and wireless LANs. Bell Labs' efforts in broad-band networking have led from the earliest research in *Asynchronous Transfer Mode* (ATM) switching to the offering of the first large ATM switch to handle the multimedia transmission of voice, data, and video.

Bell Labs is a leader in algorithms for high-quality speech and audio at low bit rates, high-definition television, and data, image, and video compression in multimedia communications. These innovations have contributed to speech-processing applications, including text-to-speech synthesis, speech recognition, and automatic translation of speech from one language to another. Bell Labs is also a leader in the development of *Digital Signal Processors* (DSPs), which are the microchips that implement key algorithms for PCS, digital cellular, digital audio, and television. The microelectronics industry began with Bell Labs' invention of the transistor and critical processes, such as zone refining and molecular beam epitaxy.

Bell Labs' advances in photonics, or light-wave technology, extend from the first semiconductor lasers that could operate at room temperature to advanced lasers that are used in today's broad-band multimedia transmission systems. They extend as well from early fiber-optic research to today's optical amplifiers, high-capacity fiber, and super high-speed transmission systems (including terabit light-wave transmission).

Summary

Since 1996, Lucent Technologies has been creating new ventures based on Bell Labs technologies. These ventures help Lucent get the most from the substantial technical capabilities of Bell Labs and enable Lucent to be more market-driven. The new venture companies are highly focused on specific emerging markets. With an entrepreneurial spirit, they have low overhead costs, fewer management levels, and their own board of directors; therefore, they can operate faster and respond more quickly to changing market environments. Lucent creates about one new venture every quarter.

See Also

Bell Communications Research, Inc. (Bellcore)

Bell Communications Research, Inc. (Bellcore)

As one of the outcomes of the breakup of AT&T in 1984, *Bell Communications Research, Inc.* (Bellcore) was established to provide engineering, administrative, and other services to the newly created seven regional holding companies: Ameritech, Bell Atlantic, Bell South, NYNEX, Pacific Telesis, SBC Communications, and US WEST.

Eighty percent of the United States' public telecommunications network depends on software that is invented, developed, implemented, and/or maintained by Bellcore. Its employees are recognized leaders in the creation/development of technologies such as *Asymmetrical Digital Subscriber Line* (ADSL), *Advanced Intelligent Network* (AIN), *Asynchronous Transfer Mode* (ATM), *Integrated Services Digital Network* (ISDN), Frame Relay, PCS, *Switched Multimegabit Data Service* (SMDS), *Synchronous Optical Network* (SONET), and video-on-demand.

Over the years, Bellcore has accumulated more than 600 domestic and foreign patents for technical innovations. Bellcore pioneered many of the telecommunications services that are commonplace today, such as caller ID, call waiting, and toll-free service. Bellcore also developed the network systems that handle every single 800 and 888 call that is placed in the United States each day. Consulting services include systems integration, local number portability, unbundling and interconnection, network integrity and reliability, fraud management, and pricing and costing analyses.

As a result of changing developments in the telecommunications industry and the owners' diverging strategies and business plans, Bellcore was purchased by *Science Applications International Corporation* (SAIC) in 1997 and was renamed Telcordia Technologies in 1999. Today, Telcordia is the largest provider of *Operations Support Systems* (OSSs), network software and consulting, and engineering services to the telecommunications industry. Headquartered in Morristown, New Jersey, the company employs more than 6,000 professionals and has revenues of more than $1.2 billion. Telcordia maintains offices throughout the United States, Europe, Central and South America, and Pacific Asia.

Telcordia customers include the United States *Regional Bell Operating Companies* (RBOCs) and other major carriers; new entrants into the telecommunications marketplace, including *Competitive Local Exchange Carriers* (CLECs), cable providers, and wireless communications providers; governments; and international telecommunications carriers. The company's expertise ranges from creating bulletproof telecommunications operating systems to creating gateways that connect disparate networks, as well as providing innovative wireless data and communications business-analysis solutions.

Summary

Telcordia is the nation's largest research consortium, performing fundamental research in computer science, phototonics, materials science, network architecture, and services. With the combined assets of SAIC, the two companies serve markets where their combined skills can be leveraged, including large software projects, advanced network designs, secure networks, Internet technologies, wireless communications, and other advances in telecommunications software systems and technology.

See Also

Bell Labs

Branch Office Routers

As companies become more decentralized, there is an increasing need to tie remote branch offices into the corporate backbone network. With the wider availability of circuit-switched digital services (including ISDN) and dedicated services such as *Digital Subscriber Line* (DSL), network planners now have an array of choices that are available for meeting the diverse data communications and LAN interconnectivity needs of branch offices.

Connection Alternatives

In many cases, dialup connections over ordinary phone lines are the most economical means of accessing the corporate network from remote locations. Because current modem speed is limited to 56Kbps, this method of access is best suited for occasional use, such as retrieving e-mail or downloading small files. Although it is possible for multiple users to share the same dialup connection, application performance slows to a crawl if two or more users access the network at the same time.

Although private lines such as T1 can be used for the connection, many remote branch offices might not have enough traffic to justify the cost of installation, maintenance, and high monthly charges. For many companies, more economical digital services (such as switched 56Kbps, DSL, frame relay, or ISDN) are the answer, especially when mission-critical applications are involved. How economical these services are depends on the amount of traffic that flows through each remote node. Instead of modems, small, inexpensive routers are used at the branch office location (refer to Figure B-2). These devices are connected to a larger backbone router that is equipped with appropriate WAN interfaces, providing legacy integration and bandwidth-management tools.

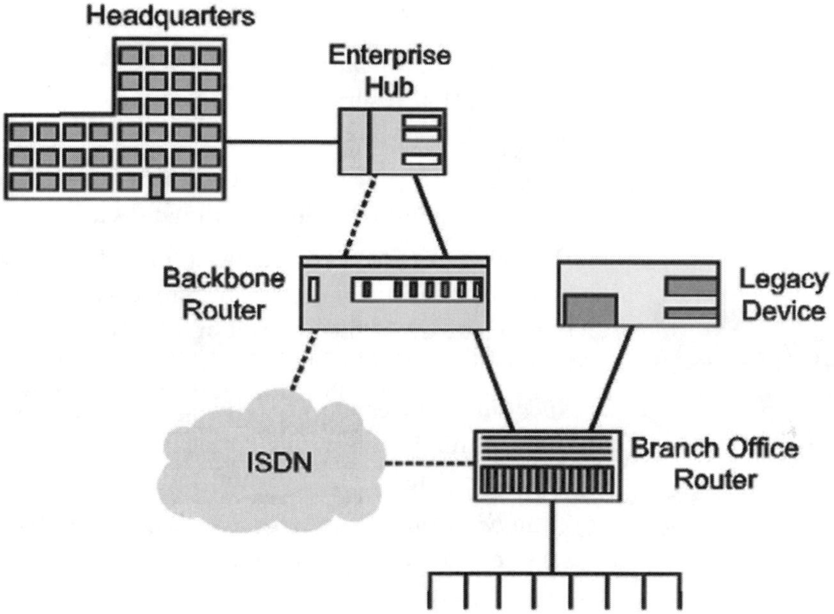

Figure B-2

Figure B-2
The use of branch-office routers saves money without sacrificing connectivity by delivering primary service via leased lines, with backup or additional bandwidth obtained through switched services such as ISDN. Connecting a legacy device (such as a cluster controller or X.25 switch) to the branch router's second serial port provides another data link to headquarters.

Headquarters

Enterprise Hub

Backbone Router

Legacy Device

ISDN

Branch Office Router

Features

Similar to backbone routers, branch-office routers can transmit data to other sites by using transmission methods such as ISDN, DSL, Frame Relay, and X.25. They can route TCP/IP, IPX, and AppleTalk traffic, and some vendors' products support SMDS and ATM via their respective *Data Exchange Interfaces* (DXIs). They can also bridge non-routable protocols transparently by using methods such as source route and translation bridging. An increasing number of branch-office routers support IBM's *Data Link Switching* (DLSw) architecture for routing *Systems Network Architecture* (SNA) and NetBIOS traffic over TCP/IP.

The difference between branch-office routers and backbone routers is mainly the hardware; specifically, the number of ports and types of interfaces that are available. The larger routers are also highly scaleable, whereas the configuration of branch routers is usually fixed. Scaleability can be achieved by purchasing stackable routers for branch offices. The main advantage of the stackable approach is that it enables network planners to start small with their router installations and to add capacity in affordable increments when growth warrants.

Depending on the vendor, one or more of the following features might be available in a branch-office router:

- *Bandwidth-on-demand*   Optimizes the use of WAN availability by ensuring that a service is only being used when required and is not being paid for when there is no data to be sent

- *Bandwidth augmentation*   Optimizes performance by combining additional channels when extra bandwidth is required

- *Data compression*   Minimizes information that is sent across the WAN by implementing two-to-one or four-to-one data compression, enabling two to four times the amount of data to be sent over the available bandwidth

- *Spoofing*   The capability to keep chatty protocols such as IP and IPX from running up service charges by filtering out overhead frames at one end of the link and emulating them at the other

- *Time-based tariff management*   The capability to specify which WAN service should be used at a particular time of day, because WAN services have different telephone charges at different times

- *Transparent backup*   Activates a secondary link if the primary link fails. When the primary link comes back online, traffic is automatically transferred back without any loss of data.

- *Connection prioritization*   The capability to assign priority to applications that have an urgent need to access data remotely. For example, SNA traffic might be given a higher priority than IPX traffic, because too much delay might cause the SNA host session to time out. Other time-sensitive protocols include *Digital Equipment Corporation* (DEC), *Local Area Transport* (LAT) and IP Telnet.

Some vendors offer an ISDN budgeting system that enables network managers to set time-of-day and day-of-week operation over ISDN services. When the number of usage hours has been exceeded, further ISDN calls are denied. Other cost-containment features include reducing SNA polling across the WAN and filtering NetBIOS broadcasts across those links, in order to minimize WAN costs.

A relatively new feature of branch-office routers is their support of *Voice-Over IP* (VoIP). The routers of Cisco Systems, for example, now accommodate IP voice cards—eliminating the need for separate PC-based gateways to transport telephone calls more economically over IP-based packet networks rather than the *Public-Switched Telephone Network* (PSTN). VoIP-equipped routers transmit data, voice, and video across a single ISDN,

Frame Relay, ATM, or *Multi-Link Point-to-Point Protocol* (MLPPP) network. Managers can add IP voice capabilities when they upgrade their routers.

Management

Because branch routers are often used at remote locations where technical expertise is not available, vendors offer Windows-based tools to make configuring the devices relatively simple. Using these tools, users can configure multiple routers simultaneously from the same computer screen, as well as create so-called snap-ins—predefined configuration building blocks—in order to reduce configuration time and to reduce the chances of making errors. With these tools, interrelated configuration parameters can be viewed, checked, and changed. Alternatively, the LAN manager at the central site can dial into the branch-office routers and configure them, then use a Web interface to monitor their performance. In some cases, the administrative activities of branch routers are offloaded to a central site router, leaving remote routers to make only basic traffic-forwarding decisions.

Some router vendors offer a platform-independent, SNMP-based application that is designed expressly for simplifying router node management. They feature an intuitive, point-and-click graphical interface for simplifying network setup and expansion, real-time operations and monitoring, and real-time event and fault monitoring for efficient problem identification and isolation. Traffic priorities—to improve application response time and optimize WAN efficiency—can also be configured centrally through the management application. The same application can control all routers, from the smallest branch-office routers to the largest backbone routers—enabling a single system for monitoring and controlling the backbone internetwork and the remote office links.

Summary

Connecting remote offices to the backbone network is becoming a critical business requirement. An effective internetworking strategy can facilitate data communication and leverage centralized information resources in order to improve productivity at remote locations. Branch-office routers that are connected to backbone routers and that are equipped with appropriate WAN interfaces offer companies more flexibility than modems for interconnecting geographically dispersed offices, workgroups, and telecommuters.

See Also

>*Digital Subscriber Line* (DSL) Technologies
>*Integrated Services Digital Network* (ISDN)
>Modems
>Routers

Bridges

Bridges are used to extend or interconnect LAN segments. At one level, they are used to create an extended network that greatly expands the number of devices and services that are available to each user. At another level, bridges can be used for segmenting LANs into smaller subnets in order to improve performance, to control access, and to facilitate fault isolation and testing without impacting the overall user population.

The bridge accomplishes this task by monitoring all traffic on the subnets that it links. The bridge reads both the source and destination addresses of all of the packets that are sent through it, and if the bridge encounters a source address that is not already contained in its address table, it assumes that a new device has been added to the local network. The bridge then adds the new address to its table.

When examining all packets for their source and destination addresses, bridges build a table that consists of all local addresses. The table is updated as new packets are encountered and as addresses that have not been used for a specified period of time are deleted. This self-learning capability permits bridges to keep up with changes on the network without requiring their tables to be manually updated.

The bridge isolates traffic by examining the destination address of each packet. If the destination address matches any of the source addresses in its table, the packet cannot pass over the bridge because the traffic is local. If the destination address does not match any of the source addresses in the table, the packet is discarded onto the adjacent network. This filtering process is repeated at each bridge on the internetwork until the packet eventually reaches its destination. Not only does this process prevent unnecessary traffic from leaking onto the internetwork, but it also acts as a simple security mechanism that can screen unauthorized packets from accessing various corporate resources.

Bridges can also be used to interconnect LANs that use different media, such as twisted-pair cable, coaxial and fiber optic cabling, and various types of wireless links. In office environments that use wireless communications technologies such as spread-spectrum and infrared, bridges can function as an access point to wired LANs. On the WAN, bridges even switch traffic to a secondary port if the primary port fails. For example, a full-time wireless bridging system can establish a modem connection on the public network if the primary wireline or wireless link is lost (due to environmental interference).

In reference to the OSI model, a bridge connects LANs at the *Media Access Control* (MAC) sublayer of the *Data Link Layer* (DLL). The bridge routes by means of the *Logical Link Control* (LLC), which is the upper sublayer of the DDL (refer to Figure B-3).

Because the bridge connects LANs at a relatively low level, throughput often exceeds 30,000 *packets per second* (pps). Multi-protocol routers and gateways, which can also be used for LAN interconnection, operate at higher levels of the OSI model. When performing more protocol conversions, routers and gateways are usually slower than bridges.

See Also

Gateways

Open Systems Interconnection (OSI)

Figure B-3
Bridge functionality, in reference to the OSI model

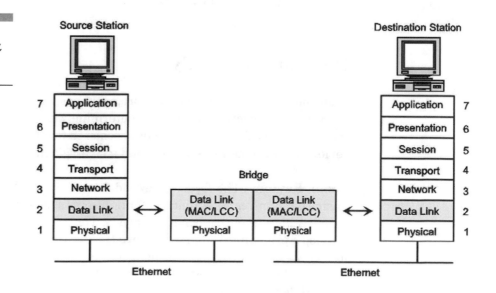

Repeaters

Routers

Business Process Re-Engineering

Business process re-engineering is a methodology that attempts to restructure business operations through the application of appropriate technologies in order to solve specific problems, as opposed to simply looking at technology from the traditional price-performance perspective. Usually, a business process re-engineering plan is developed before the organization makes any commitments to vendors or carriers for new equipment or services. Often, the plan is developed with the aid of a systems or network integrator or a third-party consulting firm.

When organizations begin planning for new systems or networks (or technology migrations), they first assess and reorganize business processes based on the idea of filtering out unnecessary procedures, eliminating redundancy, and streamlining workflow in order to enhance productivity, trim overhead costs, and improve customer response. Failure to engage in business process re-engineering before implementing an enterprise-wide technology could result in the organization wasting enormous amounts of money on new equipment, lines, and services. These items alone do not improve organizational performance.

Preliminary Analysis

A business process re-engineering project starts with a preliminary analysis of the organization, in order to determine where to apply changes for maximum impact. Several components exist in this type of analysis: a focus review, a requirements definition, and a strategic-impact analysis.

Focus Review   The focus review addresses an organization's strategic business objectives, identifying potential targets for improvement and providing a high-level cost/benefit analysis. The objective of the review is to develop a preliminary plan to improve the existing work environment. The focus review also helps management zero in on which workgroup or department would gain the most benefit from the application of new systems or network technology, based on parameters such as volume of transactions, frequency of database access, distribution of work, and contribution to the core business.

Requirements Definition   Once the decision has been made to improve a specific business process, a definition of requirements is performed. This definition builds on the preliminary input from the focus review to further define, analyze, and document the specific needs of the target workgroup or department. This definition lays the groundwork for all subsequent design and implementation activities. This step identifies the proposed project's major inputs, outputs, and volumes; defines the current workflow processes, identifying those that should be automated and those that should not; and defines all hardware, software, and network services that will be required for implementation. Often, it will be necessary to interview staff at various levels in the organization in order to obtain a clear picture of how work is performed, in addition to analyzing transaction volume and workflow patterns.

With the current manual processing methods identified and understood, appropriate steps can be recommended to eliminate inefficient or outmoded practices in order to streamline operations. This analysis ensures that the workgroup or department's total requirements have been explored and understood prior to the company's commitment to purchase new systems or network services. In addition, the requirements definition describes the types of third-party services that are required, such as consultants or integration services, which will be brought in under the direction of a project-management team. The requirements definition also provides an initial installation schedule and cost estimates for implementing the solution.

The completion of these services results in a detailed design specification, which describes the actual solution and how it will be implemented. The information that is included in this document includes the workflow analysis, system or network configuration, end-user training, acceptance test procedure, and complete project schedules and timetables.

Strategic Impact Analysis   The strategic impact analysis takes these processes a step farther, evaluating the potential effect of the proposed solution on the entire enterprise. The resulting report recommends an appropriate architecture, a method of enterprise integration, and a specific implementation approach.

Pre-Installation Planning

Once management has signed off on the proposal, the systems or network integrator uses the information gathered for pre-installation planning. This process starts with the creation of a detailed design document that defines all hardware and software up front, so that the actual installation will be performed with minimal business interruption in the shortest possible time frame. Eventually, this document will also be used to facilitate the implementation of the acceptance test plan, which enables the customer to verify that all requirements are being addressed prior to the acceptance of the new system or network.

Installation and Implementation

The system or network integrator should provide a site analysis so that equipment and software can be installed immediately upon arrival, with minimal impact on the workforce and on daily business operations. At this phase of the project's life cycle, the integrator installs the hardware and software components and brings the system or network to an operational state. The integrator then initiates a verification process in order to confirm that these components are running properly. All aspects of the new system or network are documented. Upon acceptance by the customer, the system or network is put into service in the production environment.

Post-Installation Activities

The goal of the integrator's post-installation activities is to provide for the ongoing support of the installed system or network for a predetermined time period, which ensures that unanticipated problems are resolved quickly. The mechanisms for performing this task can include any number of support services, such as remote monitoring, on-site hardware repair service, overnight shipping of replacement components and modular subsystems, and access to technical staff via a toll-free number, a dialup BBS, or a self-service Web site.

Management Services

The system or network integrator should provide management services that span the entire project's life cycle, including training. Typically, an integrator offers specialized courses that are conducted at regional training centers or on location at customer sites. Several training paths should be available, including courses for system administrators, network managers, help-desk personnel, and technicians. The training should range from basic principles of operation to administrative and technical functions.

Once the new system or network is up and running and has been accepted by the customer, the integrator's project-management team might no longer be required. Technical assistance retainer plans, however, are usually available and include the on-site services of a project manager on an ongoing or periodic basis.

Summary

Business process re-engineering seeks to analyze the current functions of a workgroup, department, or enterprise for the purpose of planning and implementing appropriate technology that will improve the efficiency of operations, the productivity of staff, and the organization's response to customers. Business process re-engineering precedes any capital investment in technology or any commitment to vendors or carriers. Often, business process re-engineering is an ongoing activity. The overriding goal of business process re-engineering is to make the organization more agile with respect to addressing the needs of a dynamic, global marketplace and eventually to secure a competitive advantage.

See Also

Downsizing

Network Integration

Workflow Automation

CHAPTER **C**

Cable Telephony

Telephone service can be implemented over two-way cable networks in two ways: via circuit switching and packet switching. In circuit switching, digitized voice signals are delivered in the traditional time-division mode of circuit switching over 6MHz radio frequency channels that are provisioned over hybrid fiber/coax networks. Anywhere from a handful to a maximum of 500 to 2,000 households share access to the service stream in any one coaxially connected service area. The trouble with circuit-switched voice-over cable is that the platforms are proprietary.

Another way to implement telephone service over cable is via packet switching. Similar to the circuit-switched version of cable telephony, IP uses shared 6 MHz channels. While IP over cable offers the advantage of standardization, the migration path to IP is so long that it often justifies the incremental cost of starting with a proprietary system. Vendors of proprietary circuit-switched solutions (i.e., Tellabs, ADC Telecommunications, and others) plan to support migration by producing gateways that connect the local cablephone host digital terminal at the head end with IP backbones. This setup would enable long-haul traffic to be transferred via IP while still retaining the circuit-switched mode for local connections within and outside the cable network.

Voice-Enabled Cable Modems

Although IP over cable has been possible for a long time, the new wrinkle is the capability to support data and voice through the same cable modem. The modem, in turn, is connected to the same coaxial cable that delivers TV service. Previously, placing voice and data on the cable network required a modem and a separate voice gateway, which is more complex and expensive.

The voice-enabled cable modem, also known as an integrated *Multimedia Terminal Adapter* (MTA), has a telephone port and an integral four-port Ethernet hub. Data speeds on the cable can be 10 Mbps or higher, depending on the cable system and how many active users are on the same shared cable subnet. The integrated MTA uses proprietary traffic-prioritization techniques in order to ensure that there is enough bandwidth to prevent voice packets from becoming delayed.

The MTA sits on the customer premises (refer to Figure C-1), where it converts phone and PC traffic into IP-formatted packets for transport over the cable network. The MTA sends the packets to a router or to a packet

Figure C-1
The route of a
telephone call over a
cable network

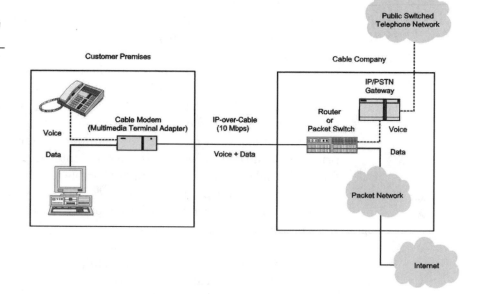

switch on the cable network, where voice and data are sorted and separated. Data is routed to the cable operator's packet network, while voice traffic is routed to an IP/PSTN gateway. The gateway decompresses the packets and returns them to analog form, so that voice can be received at a phone on the public network or on a corporate PBX.

With many corporations devising ways of implementing voice-over IP internally, the cable option enables them to extend this function to telecommuters. Cable modems already support data at the Ethernet speed of 10 Mbps, which gives telecommuters a viable way to retrieve information from the corporate LAN via their Internet access connection. Although ISDN's *Basic Rate Interface* (BRI) is more flexible in terms of bandwidth allocation and call-handling features, it only provides 128 Kbps for user data, which is not enough capacity for frequent LAN access. Furthermore, ISDN BRI is far more expensive than cable.

Local Competition

Cable operators are aggressively rolling out discount telephone services and competing with the *Incumbent Local Exchange Carriers* (ILECs) and *Competitive Local Exchange Carriers* (CLECs). MediaOne, for example, offers unlimited local calling over its cable infrastructure for a fixed monthly price that is 15 to 35 percent lower than the competition's price. The company's cable telephone service includes 15 popular calling features, such as caller ID, call waiting, last call return, call forwarding, continuous redial, distinctive ring, speed calling, three-way calling, anonymous call rejection, and selective call blocking. Customers also have four different voice mail options and can order multiple cable-based phone lines. Subscribers can use their telephone, watch television, and be on the Internet all at the same time over the same broad-band network.

Another major cable operator, Cox Communications, offers competitively priced local and domestic long-distance and international telephone service for residential customers. The company's cable telephone service is priced at a flat rate of 10 cents per minute in the continental United States—all day, every day. The extensive bandwidth capacity of Cox's facilities-based network enables the company to provide customers with numerous advantages over traditional telephone service, including second lines at a significant cost savings. Cox is rolling out voice mail, caller ID, call waiting, call forwarding, and other ancillary products.

On a larger scale, AT&T *Broadband & Internet Services* (BIS), formerly known as *Tele-Communications, Inc.* (TCI), is well positioned to develop nationwide local cable telephone access services and continues to consolidate its cable properties in order to provide this functionality in a cost-effective manner. AT&T BIS has full access to the technology that is needed to provide cable subscribers with all varieties of local and long-distance telephone service. Proposed joint ventures with Time Warner, Cox Communications, and Comcast—plus its acquisition of MediaOne—give AT&T BIS cable telephone operations access to about 70 percent of all residential cable customers and all major business markets across the United States.

AT&T sees its BIS operations as a means of putting AT&T back in the local telephone business through cable telephony. To fulfill this vision, AT&T BIS continues to acquire and consolidate cable properties. By purchasing other systems, the company continues with its strategy of creating strong geographic clusters. Continually strengthening these clusters gives AT&T BIS the scale that it needs in order to offer advanced video, high-speed data, and local telephone services, combined with high-quality customer service—all at a competitive price.

Key Components

The key components that enable cable operators to quickly deploy, manage, and bill for IP-based voice and data services became available in 1999.

CableConnect solutions from Lucent Technologies, for example, streamline the deployment process by lowering the startup costs and making the process modular. Using this approach, cable operators can start with targeted deployments and build them as the demand grows. The CableConnect solutions include IP Telephony in a Box, which is a secure cabinet that can be preconfigured and dropped off at the cable operator's network site for rapid service deployment.

IP Telephony in a Box targets cable operators who are expanding or upgrading their networks but have limited space for new equipment in existing head ends. This application uses a standard, concrete Controlled Environment Cabinet that Lucent can preconfigure with the exact equipment that a cable operator needs for the services that they plan to deploy in that area, including IP switches and routers, cable modem termination systems, and *Synchronous Optical Network* (SONET) equipment. The cabinet and components are highly scaleable and can grow from one cable telephone line to more than 10,000 lines.

Lucent also offers the CableConnect *Operations Support Solution* (OSS), which is an end-to-end operations software solution that supports customer care, billing, and network management. The CableConnect OSS supports the needs of cable operators as they migrate their service offerings from video only to voice, video, and data. Lucent's software provides customer care and billing for integrated services, flow-through provisioning, and advanced fault management and trouble isolation, in addition to complying with PacketCable requirements in order to ensure interoperability in multivendor environments. PacketCable is a project that is being conducted by *Cable Television Laboratories, Inc.* (CableLabs) and its member companies, and the project is aimed at identifying, qualifying, and supporting Internet-based voice and video products over cable systems.

Summary

Converging services that combine entertainment programming, broadband Internet access, and telephone services present a challenge that could cost the voice-oriented ILECs and CLECs a substantial market share in the not-too-distant future. Early market success on the part of the two leaders that are deploying cable-telephone systems (MediaOne Group and Cox Communications) is pressuring other cable operators to enter the market. For consumers, this convergence can mean lower monthly costs, simplified billing, and in some cases, higher satisfaction with the services that they purchase.

Cable Television Networks

Cable television began around 1949 in Lansford, Pennsylvania. A local television shop owner noticed a decrease in television sales and wanted to find out why. After talking to town residents, he discovered that low sales were due largely to the poor reception in the area. The closest station was in Philadelphia, about 65 miles away, and there was a mountain that overlooked Lansford that blocked the reception.

After helping some residents in the outlying areas set up antennas to help with their reception, the shop owner came up with an idea to help with the town's reception problem. He built an antenna on top of the mountain, and with an amplifier, he boosted the signal back to full strength. Next he ran coaxial cable down the mountain and into the town, charging people a fee to connect to the cable. The first *Community Antenna Television* (CATV) system was born, and it consisted of three channels and a few hundred subscribers.

Around the same time, other towns claimed to have put the first CATV system in operation. Regardless of which town was actually first, the intent was the same: to provide television service to remote areas where over-the-air signal reception was difficult or impossible. Almost immediately, CATV began moving into metropolitan areas such as New York City, where reception was difficult because the tall buildings caused multiple signal interference or blocked the signal altogether.

By 1952, there were 70 CATV systems in operation nationwide, with 14,000 subscribers. By the 1960s, however, growth in the CATV market had all but stopped. Cable service had been installed in most of the major market areas. The existing technology also limited cable's growth. Most cable systems only had enough capacity for 12 channels until the mid-1970s.

Major growth in the cable market began to take off after 1975. The availability of satellite receivers enabled cable operators to take specific signals and insert them into their channel lineup. This functionality led to cable-only programming. Cable system operators began adding programs such as movie channels (i.e., Home Box Office, or HBO), sports channels (i.e., ESPN), shopping channels (i.e., the Home Shopping Network, or HSN), and superstations (i.e., Turner Broadcasting System, or TBS). The technology also enabled cable companies to give subscribers pay-per-view programming. With this service, a subscriber pays a one-time fee in order to view a special event, such as a concert, a sporting event, or a first-run movie.

Program Delivery

The actual video signals that are delivered to the cable system can be generated from three basic sources:

- *Satellite or microwave receivers*　Program sources include national networks such as the *Cable News Network* (CNN), HBO, and ESPN and local sources such as commercial and public television. Usually, these program sources run 24 hours a day but might be interrupted by inserting locally originated programming or commercials.

- *Videotape*　Videotape is used to deliver prerecorded material, such as commercials, infomercials, public-service programs, and movies. The use of videotape is undesirable, however, due to the labor that is involved in making the tapes, moving them to the broadcast site, and playing them. Instead, multimedia servers are increasingly being used, because they automate program delivery.

- *Multimedia servers*　Servers store and play multimedia programming that includes graphics, animation, sound, text, and digital MPEG video. These computers might accept real-time data from weather services, Internet information sources, computer databases, and satellite data networks for automated delivery on a scheduled or on-demand basis.

Today, CATV is the primary method of program distribution in the United States, with approximately 70 million subscribers accessing programming from a cable TV network. Nationwide, there are 11,000 networks and more than one million miles of cable planted. These networks pass 95 percent of all households, making information, entertainment, and education available to almost everyone who chooses to subscribe.

Subscribers pay a monthly fee for a set of basic services and can select optional packages of premium services, including Internet access and telephone service, for an additional monthly fee. In addition, subscribers can choose pay-per-view programs by calling the cable operator to request a specific program from a menu of choices that changes daily. Usually, there is a nominal extra charge for each additional television set that is set up to receive cable programming. All services are itemized on the monthly bill from the cable TV operator.

The CATV market generates about $25 billion in revenue per year from subscribers. The funds are generally split two ways: financing the operating costs of existing networks and constructing new systems, and providing payment for programming such as HBO, *Music Television* (MTV), and the Disney Channel.

Operating Environment

Cable companies operate in an industry that is undergoing rapid change due to consolidation and technological innovations. Complicating matters is the fact that many cable companies are exploring technologies with which they do not have much previous experience, such as business-class telephony and broad-band data. Businesses today expect the most from their vendors. Businesses want superior service and products that are delivered at an excellent value. If cable companies expect to succeed in these areas, they must acquire the expertise and support infrastructure that is necessary to ensure that the needs of businesses are addressed in a timely manner.

The industry is moving from the phase of consolidation to the phase of swapping. As the list of big cable properties that are for sale continues to shrink, property swapping is likely to become more common as companies seek to build regional holdings. The rush to swap is being driven by the need to assemble clusters of cable properties. Clusters enable cable companies to more economically provision new broad-band technologies and to reap a faster return on capital investments. By swapping assets, cable companies can concentrate on one particular region of the country, much like local phone companies do.

Comcast Cable, the fourth largest CATV company in the United States (with 8.2 million subscribers), has aggressively pursued clustering for greater operational efficiency and cost savings. Through trades and acquisitions, Comcast has managed to concentrate 85 percent of its subscriber base in six regional clusters. The company is now in a position to accelerate the launch and delivery of new technologies and products to its customers, including telephone service.

AT&T is pursuing cable acquisitions and technology enhancements as a means of becoming the largest purveyor of cable TV and future multimedia entertainment. In 1999, AT&T bought TCI and MediaOne, the two largest cable companies in the United States. With these acquisitions, AT&T envisions cable as the primary means for consumers and businesses to obtain high-speed access to the Internet and to emerging data services and voice communications. AT&T also sees cable as a means of getting itself back into the local telephone business through various cable telephony offerings. AT&T expects to reach about 70 percent of the nation's residents through its own cable television networks, as well as through joint ventures with other cable companies (including Comcast Cable). To give itself leverage in negotiating telephony pacts with other cable operators, AT&T will use fixed-wireless technology to deliver voice and data services over the radio spectrum, rather than traditional coaxial cable or fiber optics and *Hybrid Fiber/Coax* (HFC).

Right now, cable TV makes up 99 percent of cable operators' revenues—and cable TV is a mature business. The rest of the industry is looking to sell new digital services and value-added applications in order to justify today's high stock market valuations. This sales strategy will also enable these businesses to compete better against the incumbent Bell operating companies and AT&T as convergence of the entertainment, Internet, and telecommunications markets accelerates.

Summary

To survive in the new competitive climate that was ushered in by the Telecommunications Act of 1996, cable companies are investing billions of dollars in upgrading their networks for full-duplex operation. Deregulation enables the telecommunications companies to deliver local data services and even TV programming, in addition to enabling the cable companies to deliver local voice telephony (which, like broad-band data, requires two-way transmission and switching capabilities). Other services that CATV operators can offer are Internet access and video-on-demand. Among the technology choices for upgrading CATV networks for these advanced services are HFC and *Fiber-in-the-Loop* (FITL) systems.

See Also

    Cable Telephony

    Hybrid Fiber/Coax (HFC)

    Video-on-Demand

Call Centers

Call centers are specialized environments that are equipped, staffed, and managed to handle a large volume of incoming calls. A call center typically has an *Automatic Call Distributor* (ACD) in order to connect calls to an order taker, a customer-service representative, and a help-desk operator or some other type of agent. Calls that cannot be answered immediately are put in a queue until the next agent becomes available. While on hold, callers might listen to music or advertising and receive periodic barge-in messages that inform them of their queue status. They might also hear a menu of dialing choices so that their call can be routed in the most appropriate way.

When the call is answered, the agent addresses the caller's immediate needs and takes down relevant information about the caller. This information is then entered into a computer database. The information can be called from the database the next time the user calls and can be delivered to the agent's computer as a screen popup. In addition, this information can be used for a variety of other purposes, including the preparation of shipping labels for ordered merchandise, follow-up sales calls, direct-mail advertising, and consumer surveys.

The elements of a typical call center include the following (refer to Figure C-2):

Figure C-2
Typical call center elements

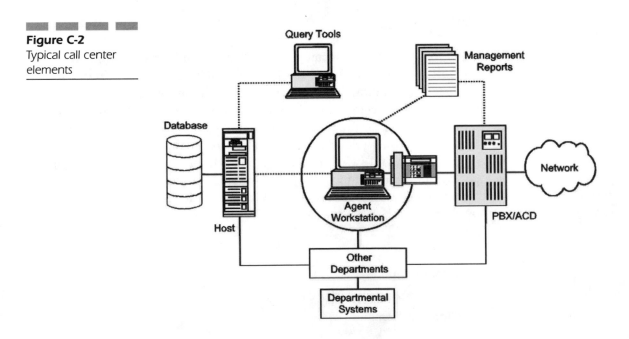

- *Telephone lines and services*   Call centers usually use digital lines such as 56K, Fractional T1 or T1, and ISDN. Services can consist of 800 toll-free numbers to take orders or 900 numbers to provide a service (the cost of which is charged to the caller's phone bill).

- *Switching system*   Most often, a standalone ACD or integrated ACD/PBX system is used to support call center operations. ACD functions can be provided by the telephone company's central office or as part of the carrier's Centrex services, however.

- *Telephone instruments*   Multi-button telephones that are equipped with a headset permit the keyboard entry of customer information by the agent.

- *Workstations*   Agents enter customer information at their workstation, usually by filling out standard forms that appear on the display.

- *Host computer and database*   Customer records are stored on a central database. The host computer can sort the records in any number of ways and generate appropriate reports.

- *Management information tools*   Managers can query the database and retrieve information that can reveal such items as new sales opportunities, levels of customer satisfaction, and call center performance.

Internet Connection

One of the newest developments in e-commerce is the integration of the Internet with traditional call centers. The Internet-enabled call center enables companies to personalize relationships with Web site visitors by providing access to a customer-service agent during a critical moment (i.e., when the visitor has a question, and the answer to that question will more than likely influence the visitor's decision to buy). With the capability to intervene in the online purchase decision and to influence the outcome, a company can realize several benefits:

- *Competitive differentiation*   Providing a convenient, value-added service that improves customers' Web experiences via live-agent assistance

- *Sales generation*   Removing obstacles in the buying process, with immediate interaction between customers and knowledgeable call center agents

■ *Increased customer satisfaction* Delivering technical support and quickly responding to customer needs with personalized, one-on-one service

One company that understands these concepts is Micron PC, whose computer Web page (www.micronpc.com) enables consumers to buy computer products online. Customers can buy standard desktop computers, notebooks, and servers or configure their own system with desired features and options. Selected items are added to a virtual shopping cart, and the site keeps a tally of the purchases. Visitors can make changes until the configuration meets a budget target. Configuration conflicts are even pointed out, giving the customer an opportunity to resolve the problem from a list of possible choices.

As the customer makes changes, the new purchase price is displayed along with the monthly lease cost, in case the customer wants to consider this finance option. When the customer is ready to buy, the shopping cart adds shipping and handling charges and the applicable state sales tax. The customer completes the transaction by entering contact and payment information and hitting the Submit button to send the purchase order to the company. Micron's secure server software encrypts credit card and personal information, ensuring that Internet transactions are private and protected. Customers can check the status of a recent purchase by entering their order number into an online search field.

Micron makes it easy for customers to request online assistance. At any time, customers can ask questions by selecting a preferred method of online communication (refer to Figure C-3). A question can be asked by entering the question into an online form, then a Micron sales representative responds via e-mail, phone, or fax. Customers can also initiate an interactive text-chat session with a sales representative or place a telephone call over the Internet to talk to a representative.

If the customer wants to initiate a voice call but is not familiar with the procedure, a help window is available that provides information about system requirements and offers step-by-step instructions for placing the call (refer to Figure C-4).

Regardless of the form of communication that is selected, all calls go to Micron's call center for handling by the next available agent or by the agent who can most effectively respond to a customer's request. While having a real-time conversation with the Web site visitor, the call center agent can push Web pages to the customer's computer with appropriate text and images that help answer complex questions or illustrate examples.

Figure C-3

Micron PC offers customers a choice of communication methods: an online form, text chat, or voice call.

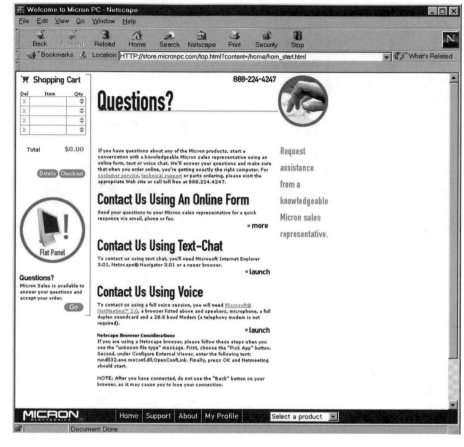

In implementing an Internet call center, offering an online configuration and pricing tool, providing secure payment with encryption, and providing an order status checker, Micron PC has not only provided online shoppers with a new level of convenience, but it has also removed key potential barriers to online sales. These barriers include customer uncertainty due to lack of decision-making information and doubt about the safety of e-commerce.

Summary

A call center can consist of only two or three agents or as many as several thousand agents. The agents might operate from a single location, or they

Figure C-4
A help window
provides customers
with assistance in
placing a voice call
over the Internet.

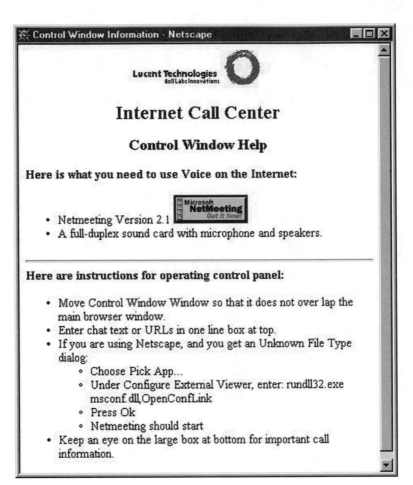

might be distributed around the world. Sometimes, calls will be automatically transferred across time zones so that the organization can provide customers with 24-hour service. The size and distribution of the call center will determine what kind of ACD system and lines that the call center will need.

See Also

> *Automatic Call Distributor* (ACD)
>
> *Automatic Number Identification* (ANI)
>
> *Computer-Telephony Integration* (CTI)
>
> *Electronic Commerce* (E-Commerce)
>
> Help Desks
>
> *Private Branch Exchange* (PBX)

Call-Detail Reporting

Call accounting systems capture detailed information about telephone calls and store them on a PC or on another collection device. The collected data is then processed into a variety of cost and usage reports. Such systems are offered in the form of stand-alone products, or they might be one of several modules within a suite of telemanagement applications that include one or more of the following functions:

- Call center management
- Moving, adding, and changing administration
- Cable management
- Inventory management
- Invoice management
- Lease management
- Trouble reporting and tracking
- Work-order management
- Traffic analysis
- Network design and optimization

Benefits

Call accounting systems represent a powerful management tool that can be used to monitor complete telephone system usage by every employee, including phone, fax, modem, and dialup Internet connection time. Such systems provide organizations with many benefits, including the following:

- *Reducing toll fraud*   Online, real-time fraud detection applications can notify telecommunications managers as the fraud is occurring, so that the action can be stopped before it gets out of hand.
- *Reducing unauthorized use*   If employees know that telephone usage is monitored, including incoming calls and local calls, the number of personal calls decreases.
- *Charge allocation*   Call charges can be billed back to the actual extensions, projects, or departments that incurred the cost.
- *Cost allocation*   Administrative charges can be allocated to departments to help pay for the telephone system and its management.

- *Tenant billing* If the organization leases office space to other companies or individuals, they can be billed directly for their phone usage.

- *Increasing productivity* Because telephone usage can be monitored, unnecessary calls are decreased—thereby increasing productivity. If the job description requires phone contact, being able to monitor the number of calls is an indicator of productivity.

Call-Detail Records

Call accounting systems produce reports from *Station Message-Detail Recording* (SMDR) devices, which capture call-detail records that are generated by PBX, Centrex, hybrid, or key telephone systems—or from tip-and-ring line scanners. Most call accounting systems compute the costs of each incoming or outgoing call, whether the call is local or long distance. Call-detail records usually consist of the following information:

- Date of the call
- Duration of the call
- Extension number
- Number dialed
- Trunk group used
- Account number (optional)

This information can be printed in detail or summarized by categories such as individual station, department, or project. Other summary categories might also include the most frequently dialed numbers, longest-duration calls, or highest-cost calls.

Standard call accounting reports can be used to locate toll fraud by identifying short, frequent calls, long-duration calls, unusual calling patterns, and unusual activity on 800 and 900 numbers. In addition, the reports can be used to identify calls that are made after hours, on weekends, or during holidays. Some vendors have introduced special toll fraud-detection packages. These programs alert managers to changes in calling patterns or breached overflow thresholds. Several types of alarms that indicate suspicious activity can be generated and sent to a printer, a local PC, a remote PC, or a pager.

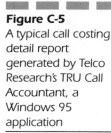

Figure C-5

A typical call costing detail report generated by Telco Research's TRU Call Accountant, a Windows 95 application

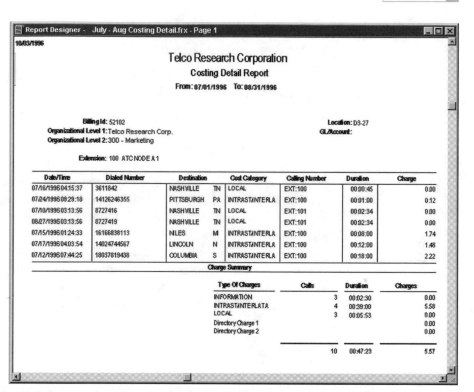

Call Costing

A key feature of call accounting systems is call costing. For each call record, the cost per call can be figured in a variety of ways in order to suit various organizational needs (refer to Figure C-5), including the following:

- Actual route cost
- Comparison route cost
- User-defined least-cost route
- What-if recosting for comparison
- Equipment charge assignment
- User-defined pricing by route group
- Tariff table pricing
- Usage-sensitive pricing
- Flat rate with percentage surcharge
- Minimum charge

- Evening and night discounts
- Operator-assisted charge

The capability to price calls according to various parameters gives the organization flexibility in allocating costs and meeting budgetary targets.

Cost Allocation

Another key feature of call accounting systems is cost allocation, which is used to distribute call costs to the appropriate internal departments, projects or clients, workgroups or subsidiaries, or external customers or individuals. With this feature, costs can be applied to calls that also include the cost for equipment, trunks, lines, maintenance, or other administrative charges. Some systems can also depreciate equipment according to organizational depreciation schedules. Most cost-allocation applications also provide an interface to the organization's general ledger.

Polling

Call records are collected from one or more locations via a polling process, which is usually done by a PC. Multiple sites can feed their data simultaneously into a single recording PC via a dialup or dedicated line. During the polling session, *Cyclic Redundancy Checking* (CRC) is used for error correction. Compression can be added in order to minimize collection time and to save on long-distance call charges. User-defined filters can be applied to exclude certain types of call records from being captured. Once the relevant data is collected, it can be processed into the appropriate report formats.

At each location, there is typically a solid-state recording unit that attaches to the PBX. These devices are available with memory capacities ranging from 256KB to 8MB. They include a battery backup that preserves data integrity for 30 to 60 days in case of a local power failure. The also have a graphical interface that is used for management. Through this interface, the system settings can be set, including the system and polling parameters and the alarm and callback schedules.

Several polling methods are available. Polling can be triggered when the SMDR device reaches 80 percent full, in order to ensure that buffers are not overwritten before the call data is retrieved. Polling sessions can occur on a scheduled basis for each location, and polling can also be initiated manually. The telecommunications manager determines the number of automatic retries (if the line is busy) and restarts (if transmission is aborted) during a polling session.

Some polling applications can collect call-detail records over TCP/IP networks, including the Internet, at a much lower cost than traditional long-distance services (refer to Figure C-6). The TRU Network Poller that is offered by Telco Research, for example, works in conjunction with a serial server and acts as a protocol converter that can translate RS-232 serial ASCII data into packets. Commands can be automatically sent from a PC over the network through a serial server to the solid-state recorder's buffer box to begin polling the call-detail records. The records move via the solid-state recorder's serial port into the serial server, where the data is packetized and is sent across the TCP/IP network to the PC, where the call accounting system resides (refer to Figure C-7). There, the call-detail records are processed into various reports.

Outsourcing An alternative to in-house call-detail record collection and processing is the third-party service bureau. The service bureau collects call records from a PC or from another type of recording device at the customer's premises via a dialup connection or a dedicated line—depending on the call record volume—on a daily, weekly, or monthly basis. The call-detail records are processed into reports for the customer. The client can choose from among a set of standard reports or can choose to have the data processed into custom reports.

Some third-party processing firms offer carrier billing verification services. Audits verify that carrier invoices accurately reflect the charges for

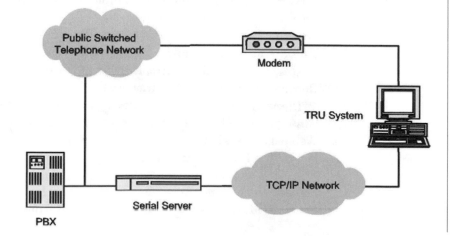

Figure C-6
A typical configuration using Telco Research's TRU Network Poller. A backup modem can be used for polling over the PSTN if the TCP/IP network is unavailable.

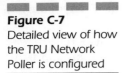

Figure C-7
Detailed view of how
the TRU Network
Poller is configured

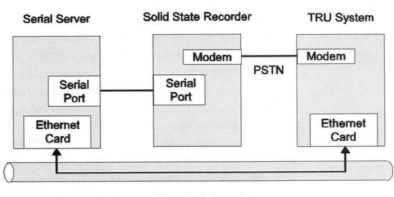

telephone services (voice and data) and for the equipment that is actually contracted for and used, ensuring that appropriate rates, taxes, and surcharges have been properly applied. Refund requests are prepared and submitted in order to ensure that customers receive the appropriate reimbursement for incorrect prior bills, as well as ensuring that future bills are accurate. Refunds of historical overcharges have produced ongoing savings to customers, generally ranging from five percent to as much as 15 percent of basic monthly service charges.

Web-based Reporting An alternative to call-detail reporting and outsourced bill processing is Web-based reporting, which is available from the carrier. MCI WorldCom, for example, enables its business customers to access account information from its Web site by using any standard Web browser that supports *Secure Sockets Layer* (SSL) encryption. All of the tools are available to customers free of charge. The tools include Electronic Billing and Reporting, Broad-Band Reporting, Price Reporting, Traffic Reporting, and Traffic Monitoring.

Electronic Billing and Reporting enables users to easily view invoices, run reports, and pay their voice and data bills monthly. After accessing the Web page, users can transfer funds from their banks to a pre-approved MCI WorldCom financial institution in order to pay their bills. Users can also print bills and mail the invoices with payment.

The Broad-Band Reporting feature enables ATM, Frame Relay, and *Switched Multi-Megabit Data Service* (SMDS) users monitor their data connections and generate monthly or daily reports on usage. The tool is also useful in determining when a link might be reaching its capacity.

The Price Reporting tool enables users to receive rated call-detail records on a daily or monthly basis, so that they can determine usage and costs. The information enables administrators to more efficiently manage voice services.

The Traffic Reporting and Traffic Monitoring tools are specifically designed for MCI WorldCom's toll-free users. The tools enable users to gather information such as peak calling times and overall network utilization for departmental charge-backs.

Summary

Call-detail reporting has been available to PBX and Centrex users for the past 15 years via third-party software packages and service bureaus. The reports help organizations account for their expenditures for voice and data communications. A new trend is the consolidation of enterprise-wide data and voice call-detail records into a single database that can be accessed by a Web browser. This functionality enables telecommunications managers or IT managers at any location to conveniently gather telephone and Internet call record information from devices such as PBXs, Centrex systems, routers, firewalls, remote access servers, and network access servers. Such tools are offered by carriers as well as by application providers and service bureaus.

See Also

Asset Management

Telecommunications Management Systems

Tariffs

Truth-in-Billing

Caller Identification

Caller identification, or caller ID, is an optional service that is offered by telephone companies for an additional fee. Caller ID enables the user to know who is calling before the user answers the phone. The name and number of the calling party is displayed on the LCD screen of a telephone or on a separate device that is connected to the phone.

While caller ID has been available in local calling areas for many years, only since 1995 has it been available for out-of-state calls. That year, the FCC issued rules governing national caller ID services. These rules give callers the choice of delivering or blocking their telephone number for any interstate call that they make.

The FCC's national caller ID rules protect the privacy of the called and calling party by requiring telephone carriers to make free, simple, and consistent per-call blocking and unblocking arrangements available. Each time a user picks up the phone to make a call, he or she can make a choice as to whether or not to block their number from being displayed to the called party. In addition, if per-line blocking is available, the caller can select that option. Per-line blocking is a service that automatically blocks the caller's telephone number from being delivered on all calls that are made from that line.

With per-call blocking, callers can block the delivery of their phone number on a call-by-call basis by dialing *67 before dialing the number. Many states enable consumers to select per-line blocking. Under this option, the telephone number will be blocked for every call, without having to dial extra digits. If the caller subscribes to the per-line blocking service but wants to allow the number to be transmitted to the called party, he or she must dial *82 before the telephone number each time a call is made.

Some services also transmit the name of the calling party. The FCC's caller ID rules require that when a caller wants his or her number to be concealed, a carrier cannot reveal the name of that subscriber. Calls to emergency lines are exempted from the FCC's caller ID rules. State rules and policies govern the obligation of carriers to honor caller privacy requests to emergency numbers.

Requesting privacy on calls to 800 and 888 numbers might or might not prevent the display of one's telephone number. When a toll-free number is dialed, the called party pays for the call. Typically, the called party can identify the telephone number of incoming calls by using a technology called *Automatic Number Identification* (ANI). When used with *Computer-Telephony Integrated* (CTI) applications, the phone number of the incoming call can be matched against a database record, enabling a customer-service representative, for example, to have information about the caller displayed on a computer terminal. This information enables the representative to have all relevant information about the customer immediately available so that the call can be handled in the most expeditious manner possible.

Summary

Nationwide caller ID offers many benefits for consumers and for the economy as a whole. Nationwide caller ID brings consumers rapid and efficient service, encourages the introduction of new technologies and new services to the public, and enables service providers and consumers to conduct transactions more efficiently. On the flip side, caller ID raises privacy concerns in that the information that is gathered by telemarketing companies via ANI can be sold to other telemarketing firms.

See Also

Automatic Number Identification (ANI)

Computer-Telephony Integration (CTI)

Custom Local Area Signaling Services (CLASS)

Calling Cards

Mobile professionals often need telecommunications services while they are away from the office, and most of them find calling cards helpful for making calls or for leaving voice messages. Most long-distance carriers offer calling cards at no extra charge and apply calls that are made with these cards to the corporate discount plan.

The caller simply dials the carrier's 800 number, followed by the telephone number and the card number. The 800 prefix ensures that a caller reaches the carrier's network from any telephone and that all eligible calls are captured and are aggregated for discount purposes. Charges are automatically posted to the account; an operator does not need to verify the call.

Calls that are placed with these cards cost less than third-party billed or collect calls and are connected without operator assistance. If a rotary telephone is used, however, the operator places the call without an additional charge for operator assistance.

Convenience Features

Depending on the carrier, extra features are available to calling card users. For example, when using a touch-tone telephone, a caller can place sequential calls without re-entering the card number. A speed-dial feature enables a cardholder to preprogram the calling card with frequently called numbers. When the called party does not answer or when the number is busy, the caller simply dials a few digits to leave a recorded message. Some cards have a magnetic strip for use at phones that are equipped with a strip reader, so that callers can place calls without manually entering their card numbers.

Callers can use their calling cards to place local toll, long-distance, international, AirFone, Inflight, and RailFone calls. Callers can place cellular calls when they are out of their immediate regional serving areas if their cellular companies use the same long-distance carrier. The card can also be used to set up teleconferences. A cardholder can even dial a call directly and charge the call to the calling card. Also, when the company migrates to another calling plan or when the company changes telephone numbers, the same calling cards can still be used.

Calling cards can also simplify record keeping and travel-expense reporting, because all card calls, domestic and international, are itemized and appear on a caller's monthly telephone bill. Billing codes can be used to identify certain calls that are placed with the calling card for chargeback purposes. The carrier offers a toll-free number where customer-service representatives can be reached to handle most card-related issues, including reporting lost cards, requesting additional cards, or reporting a problem with a card. The cards themselves can be customized with the corporate name and logo or a special design. This customization involves a setup fee and per-card charge, and depending on the design, a minimum number of cards might be required.

Security

Calling cards can be protected against fraudulent use by designating the numbers, area codes, or countries that can be called with a calling card and by specifying a dollar amount that a cardholder can charge. The card can even be limited to calling the corporate headquarters or the branch office.

An optional *Personal Identification Number* (PIN) that is embedded into the card's magnetic strip discourages fraud if the card is lost, stolen, or examined by outside parties. This strip contains pertinent information, such as the PIN and account number, which is read by most card-reader telephones. Cardholders can usually select their own PIN and choose whether or not to have the PIN printed on the front of the card.

In addition, an enhanced fraud-protection process enables the carrier to identify potential fraudulent use on a real-time basis and to quickly notify cardholders so that they can take immediate corrective action. For example, the AT&T *Fraud Analysis and Surveillance Center* (FASC) continuously monitors card calls for unusual calling activity and attempts to contact the cardholder for authorization. If the calling activity continues, prior to cardholder notification, the FASC deactivates the card.

Summary

Calling cards offer mobile professionals the means to access telecommunications services while they are away from the office. In addition to being less expensive than collect calls, calls that are placed with these cards can apply to the corporate discount plan. Numerous convenience features make today's calling cards easy to use in a variety of circumstances, and security is enhanced through the use of PINs and the carrier's own fraud-protection systems.

See Also

Prepaid Phone Cards

Calling Party Pays

Currently, when users receive calls on their wireless phones, the air-time charges are billed to them although they did not make the call. Many wireless phone users would like to change this situation so that the person who is calling them pays for the air time. Many wireless phone companies would like to give their subscribers a choice with incoming calls—whether to pay all incoming air-time charges or to charge incoming air time to the caller. This concept is known as *Calling Party Pays* (CPP).

Regulatory Issues

This issue seems simple, but some wireless companies have been reluctant to offer consumers a CPP option for several reasons. First, there is no standard for notifying callers that they will pay for the air time. Second, there is no framework in place for CPP that would set industry guidelines, establish consumer rights, and protect wireless phone companies. In addition, today's billing systems are not set up to identify the caller and interface with other carriers' systems for consolidated billing. Finally, some state regulations can be interpreted as prohibiting CPP.

In 1998, the *Cellular Telecommunications Industry Association* (CTIA) petitioned the FCC to perform the following actions:

- Declare sole authority to regulate CPP
- Establish a standard warning to callers to a CPP line that they will be billed for air time, and enabling them to hang up before the call goes through
- Foreclose all state bans and regulations prohibiting CPP

In 1999, the FCC adopted a declaratory ruling that agreed that CPP is subject to federal regulation, which also preempts state bans and regulations prohibiting CPP. The CTIA neither supports nor opposes CPP; rather, they believe that wireless phone providers should be able to offer this service option, and they should be able to allow the marketplace to determine its value. Individual wireless companies would determine the specifics of CPP plans and would devise the options that their subscribers want. For example, called parties might be able to choose whether they want to pay for some, all, or none of the calls that they receive.

Implementation Issues

In Europe and Asia, the caller generally pays, and cell phones are almost extensions of home and office phones. Most other countries still have telephone monopolies or a limited number of wireless service providers. In the United States, the intense competition in the wireless business is part of the reason why many service providers want CPP—and part of the reason why CPP will not be easy to offer.

For example, AT&T Wireless tested the idea in Minneapolis in the summer of 1999 but got so few takers that it quietly dropped plans for a national rollout. At the start of the test, AT&T Wireless charged callers 39 cents a minute, then cut the cost to 25 cents a minute in order to try to attract more customers. Even this cost cut was not enough. Other wireless companies have tried CPP over the years and have also discontinued the plans.

A few companies do have limited caller-pays options, such as billing long-distance fees to people who call from outside the receiver's local area. But billing accuracy and the inability to consistently identify the caller make CPP problematic. Also, without consolidated billing, customers will not use

CPP if it means getting a blizzard of separate bills. Service providers cannot afford that kind of billing. Just to process a bill requires $1.25 to $3, which is not profitable for a majority of calls. In other countries, *Local Exchange Carriers* (LECs) work with cellular companies on bill consolidation. In the United States, cellular calls are billed separately from regular telephone service.

Summary

The drive toward CPP in the United States comes as carriers look at the success of wireless phones in Europe, where the system is commonplace. In Europe, a person who is calling the owner of a wireless phone pays for the call, just as a caller routinely pays for a phone call over traditional telephone wires. But in the United States, cellular telephones were developed from the beginning under a system in which owners paid per minute to use their phones, whether receiving or making calls. Mobile phone companies, however, believe that this situation has kept people from using their phones —or from even keeping their phones on—for fear of running up high bills. Many carriers are waiting for the FCC to iron out the many problems that a national CPP system would create. The FCC will not mandate one billing system over the other, but its action could make it easier for the new system to go into effect on a wide scale.

See Also

Cellular Voice Communications

Federal Communications Commission (FCC)

Carterfone Decision

The Carterfone was a device invented in 1959 by Thomas F. Carter that permitted users of mobile radio systems to interconnect their land-line telephones with the radio system, in order to permit mobile and fixed users to communicate with each other. This setup seemed like a good idea at the time, but AT&T objected and advised its customers that if they used a Carterfone in conjunction with an AT&T service, they would be subject to penalties pursuant to AT&T's FCC tariff number 132, which stated the following:

"No equipment, apparatus, circuit or device not furnished by the telephone company shall be attached to or connected with the facilities furnished by the telephone company, whether physically, by induction or otherwise . . . ".

Carter and his company, the Carter Electronics Corporation, filed a private antitrust suit against AT&T, and the District Court referred the matter to the FCC. In 1968, the Commission concluded that AT&T's tariff was unreasonable and discriminatory and ordered the restrictive tariff provisions stricken from the record. The Commission was troubled by the tariff provision that would have enabled end users to install AT&T-manufactured equipment with exactly the same functionality offered by the Carterfone, but not the Carterfone itself.

The Commission determined that a customer who wanted to improve the functionality of the telephone network by interconnecting a piece of equipment that was not manufactured by the phone company should be permitted to do so, as long as that equipment did not harm the network.

The principle of consumer usage of non-telephone company-manufactured equipment with the PSTN, as outlined by the Commission in the Carterfone case, would later be codified as Part 68 of the FCC's rules. Part 68 was first adopted in 1975 as part of its *Wide Area Telecommunications Service* (WATS) rulemaking, in response to the telephone company's slowness in modifying tariffs in order to permit consumers to attach their own equipment to the public network. Part 68, which addresses the connection of terminal equipment to the public telephone network, permits consumers to connect equipment from any source to the public network if such equipment fits within the technical parameters that are outlined in Part 68.

By means of the Commission's equipment registration and certification procedures, competitive manufacturers of equipment could build and deploy a wide variety of voice and data equipment for use with the public network—without seeking prior permission from either the Commission (or more importantly, from the monopoly telephone companies). Since Part 68, the FCC has consistently applied the principle that any device that is privately beneficial without being publicly detrimental can be attached to the network.

Summary

The Carterfone right-to-attach principle established (for the first time) the consumer's right to connect devices to the PSTN that do not adversely affect the network. Through Carterfone and Part 68, the Commission opened the door to manufacturers of devices that interconnected with the telephone network and offered access to value-added services and capabilities. In the process, the FCC cleared the way for the rapid deployment of the modem. In fact, without Part 68, users of the PSTN would not have been able to connect their computers to the network, which would have stalled the development and expansion of the Internet.

See Also

Federal Communications Commission (FCC)

Cellular Data Communications

One of the oldest methods of achieving wireless access to the Internet is through wireless IP. Also known as *Cellular Digital Packet Data* (CDPD), wireless IP provides a way of passing data packets over analog cellular voice networks at speeds of up to 19.2Kbps. Although CDPD employs digital modulation and signal-processing techniques, the underlying service is still analog. The medium that is used to transport data consists of the idle radio channels that are typically used for *Advanced Mobile Phone System* (AMPS) cellular service.

Channel hopping involves automatically searching for idle channel times between cellular voice calls. Packets of data select available cellular channels and go out in short bursts, without interfering with voice communications. Alternatively, cellular carriers can also dedicate voice channels in order for CDPD traffic to meet high-traffic demands. This situation is common in dense, urban environments where cellular traffic is heaviest.

In accordance with the *Internet Protocol* (IP), the data is packaged into discrete packets of information for transmission over the CDPD network, which consists of routers and digital radios that are installed in current cell sites. In addition to addressing information, each IP packet includes information that enables the data to be reassembled in the proper order at the receiving end and to be corrected if necessary. Once the user logs on to the network the connection stays in place to send or receive data. The transmissions are encrypted over the air link for security purposes.

Although CDPD piggybacks on top of the cellular voice infrastructure, it does not suffer from the three-KHz limit on voice transmissions. Instead, it uses the entire 30KHz *Radio Frequency* (RF) channel during idle times between voice calls. Using the entire channel contributes to CDPD's faster data transmission rate. Forward error-correction ensures a high level of wireless communications accuracy. With encryption and authentication procedures built into the specification, CDPD offers more robust security than any other native wireless data transmission method. As with wireline networks, CDPD users can also customize their own end-to-end security.

To take advantage of wireless IP networks, the user should have an integrated mobile device that operates as a fully functional cellular phone and as an Internet appliance. For example, the AT&T PocketNet phone contains both a circuit-switched cellular modem and a CDPD modem, in order to provide users with fast and convenient access to two-way wireless messaging services and Internet information. GTE provides a similar service through its Wireless Data Services. Both companies have negotiated inter-carrier agreements that enable their customers to enjoy seamless CDPD service in virtually all markets across the country. AT&T's wireless IP service, for example, is available in 3,000 cities in the United States.

Among the applications for CDPD are accessing the Internet for e-mail and retrieving certain Web-based content. AT&T PocketNet phone users, for example, have access to two-way messaging, airline flight information, and financial information. Companies can also use CDPD to monitor alarms remotely, to send and/or receive faxes, to verify credit cards, and to dispatch vehicles. Although CDPD services might prove to be too expensive for heavy database access, the use of intelligent agents can cut costs by minimizing connection time. Intelligent agents gather requested information and only report back the results the next time the user logs on to the network.

Unlike voice cellular charges, which are based on call duration, CDPD fees are based on the amount of data that is transferred. In addition to monthly service plans, which vary from $8 for low usage to $49 for high usage, it costs five cents or higher per kilobyte to use the CDPD network for usage that is higher than the number of kilobytes that are included in the plan. Monthly service plans for unlimited usage are also available, and these plans range from $55 for unlimited local usage to $65 for unlimited nationwide usage. Service activation fees range from $35 to $50.

Summary

Wireless IP is an appealing method of transporting data over cellular voice networks because it is flexible, fast, widely available, compatible with a vast installed base of computers, and has security features that are not offered with other wireless data services.

See Also

Cellular Voice Communications

Internet

IP Telephony

Net Phones

Cellular Telephones

In recent years, cellular telephones have emerged as a must-have item among mobile professionals and consumers alike and have been growing in popularity every year since they were first introduced in 1983. Their widespread use for both voice and data communications is largely due to the significant progress that has been made in their portability, the availability of network services, and the declining cost for equipment and services.

System Components

Several types of cellular telephones are available. Mobile units are mounted in a vehicle. Transportable units can be easily moved from one vehicle to another. Pocket phones, weighing less than four ounces, can be conveniently carried in a jacket pocket or purse. Some cellular telephones can even be worn. Regardless of how they are packaged, cellular telephones consist of the same basic elements.

Handset/Keypad The handset and keypad provide the interface between the user and the system. The user only needs to be concerned with these components of the system. Any basic or enhanced system features are accessible via the keypad, and once a connection is established, these components provide similar handset functionality to that of any conventional telephone. Until a connection is established, however, the operation of the handset differs greatly from that of a conventional telephone.

Rather than initiating a call by first obtaining a dial tone from the network switching system, the user enters the dialed number into the unit and presses the SEND function. This process conserves the resources of the cellular system, because only a limited number of talk paths are available at any given time. The CLEAR key enables the user to correct any misdialed digits.

Once the network has processed the call request, the user will hear conventional call-progress signals (such as a busy signal or ringing). From this point on, the handset operates in a customary manner. To disconnect a call, the END function key is pressed on the keypad. The handset contains a small illuminated display that shows dialed digits and provides a navigational aid to other features. The keypad enables the storage of numbers for future use and provides access to other enhanced features, which vary according to the manufacturer.

Logic/Control The logic/control functions of the phone include the *Numeric Assignment Module* (NAM), which is used for programmable assignment of the unit's telephone number by the service provider, and the electronic serial number of the unit, which is a fixed number that is unique to each telephone. When a customer signs up for service, the carrier makes a record of both numbers. When the unit is in service, the cellular network interrogates the phone for both of these numbers in order to validate that the calling/called cellular telephone is that of an authentic subscriber.

The logic/control component of the phone also serves to interact with the cellular network protocols. Among other things, these protocols determine what control channel the unit should monitor for paging signals and what voice channels the unit should utilize for a specific connection. The logic/control component is also used to monitor the control signals of cell sites so that the phone and network can coordinate transitions to adjacent cells as conditions warrant.

Transmitter/Receiver The transmitter/receiver component of the cell phone is under the command of the logic/control unit. Powerful three-watt telephones are typically of the vehicle-mounted or transportable type, and their transmitters are understandably larger and heavier than those that are contained within lighter-weight hand-held cellular units. These powerful transmitters require significantly more input wattage than hand-held units that only transmit at power levels of a fraction of a watt, and they utilize the main battery within a vehicle or a relatively heavy rechargeable battery to do so. Special circuitry within the phone enables the transmitter and receiver to utilize a single antenna for full-duplex communication.

Antenna The antenna for a cellular telephone can consist of a flexible rubber antenna mounted on a hand-held phone, an extendible antenna on a pocket phone, or the familiar curly stub that is often seen attached to the rear window of many automobiles. Antennas and the cables that are used to connect them to radio transmitters must have electrical performance characteristics that are matched to the transmitting circuitry, frequency, and power levels. Use of antennas and cables that are not optimized for use by these phones can result in poor performance. Improper cable, damaged cable, or faulty connections can render the cell phone inoperative.

Power Sources   Cell phones are powered by a rechargeable battery. *Nickel Cadmium* (NiCd) batteries are the oldest and cheapest power source available for cellular phones. Newer *Nickel-Metal Hydride* (NiMH) batteries provide extended talk time as compared to the lower-cost, conventional nickel-cadmium units. They provide the same voltage as NiCd batteries but offer at least 30 percent more talk time than NiCd batteries and take approximately 20 percent longer to charge. Lithium ion batteries offer increased power capacity and are lighter in weight than similarly sized NiCd and NiMH batteries. These batteries are optimized for the particular model of cellular phone, which helps ensure maximum charging capability and long life.

Newer cellular phones might operate with optional high-energy AA alkaline batteries that can provide up to three hours of talk time or 30 hours of standby time. These batteries take advantage of the new Lithium/Iron Disulfide technology, which results in a 34 percent lighter weight than standard AA 1.5-volt batteries (15 grams/battery versus 23 grams/battery) and a 10-year storage life—double that of the standard AA alkaline batteries.

Vehicle-mounted cell phones can be optionally powered via the vehicle's 12-volt DC battery by using a battery eliminator that plugs into the dashboard's cigarette lighter. This component saves useful battery life by drawing power from the vehicle's battery and comes in handy when the phone's battery has run down. A battery eliminator will not recharge the phone's battery, however. Recharging the battery can only be performed with a special charger. Lead-acid batteries are used to power transportable cellular phones when the user wishes to operate the phone away from the vehicle. The phone and battery are usually carried in a vinyl pouch.

Options and Features

Cellular telephones offer many features and options, including the following:

- *Voice activation*   Sometimes called hands-free operation, this feature enables the user to establish and answer calls by issuing verbal commands. This safety feature enables the driver to control the unit without becoming visually distracted.

- *Memory functions*   Enables the storage of frequently called numbers to simplify dialing. Units might offer as few as 10 memory locations or in excess of 100, depending on the model and manufacturer.

- *Multiple numeric assignment module*   Enables a single phone to be used with multiple carriers. The phone can then be used to access the best carrier for a specific location in areas where two local service providers might have different coverage gaps. This feature is also useful in cutting the cost of roaming in other service areas where surcharges might apply.

- *Visual status display*   Conveys information about the numbers dialed, the state of the battery charge, call duration, roaming indication, and signal strength. Cellular phones differ widely in the number of characters and lines of alphanumeric information that they can display. The use of icons enhances ease-of-use by visually identifying the phone's features.

- *Programmable ring tones*   Some cellular phones enable the user to select the phone's ring tone. Multiple ring tones can be selected, with each ring tone being assigned to a different caller. A variety of ring tones can be downloaded from the Web.

- *Silent call alert*   Features include visual or vibrating notification in lieu of an audible ring tone, which can be particularly useful in locations where the sound of a ringing phone would constitute an annoyance.

- *Security features*   Includes password access via the keypad to prevent the unauthorized use of the cell phone, as well as features to help prevent access to the phone's telephone number in the event of theft.

- *Voice messaging*   Enables the phone to act as an answering machine. A limited amount of recording time (about four minutes) is available on some cell phones. Carriers also offer voice-messaging services that are not dependent on the phone's memory capacity, however. While the phone is in standby mode, callers can leave messages on the integral answering device. While the phone is off, callers can leave messages on the carrier's voice-mail system. Users are not billed for air-time charges when retrieving their messages.

- *Call restriction*   Enables the user to allow use of the phone by others to call selected numbers, local numbers, or emergency numbers without permitting them to dial the world at large and rack up air-time charges.

- *Call timers*   Call timers provide the user with information about the length of the current call and a running total of air time for all calls. This feature makes it easier for users to keep track of call charges.

- *Data transfer kit*   For cell phones that are equipped with a serial interface, there is software for the desktop PC that enables users to enter directory information via the keyboard, rather than the cell phone keypad. The information is transferred via the kit's serial cable. Through the software and cable connection, information can be synchronized between the PC and cell phone, ensuring that both devices have the most recent copy of the same information.

With the increased use of cellular telephones for personal use, the choice of color and styling is playing a greater role in the phone selection process. Cellular phones come in such diverse colors as sunstreak (yellow), dark spruce, eggplant, raspberry, regatta blue, temptation teal, and cranberry. These colors are intended to appeal to the growing population of teenage users.

Summary

Cellular phones are becoming more intelligent, as evidenced by the availability of units that are part cellular phone, part palmtop computer. These devices not only support data communications, but they also support voice messaging, e-mail, fax, and Internet access. Third-party software provides the operating system and such applications as calendaring, card files, and to-do lists. With more cellular phones supporting data communications, cellular phones are becoming available that provide connectivity to PC desktops and databases via infrared or serial RS-232 connections. Information can even be synchronized between cell phones and desktop computers, in order to ensure that the user is always accessing the most up-to-date information.

See Also

Cellular Voice Communications

Internet-Enabled Mobile Phones

Personal Communications Services (PCS)

Cellular Voice Communications

Cellular telephony provides communications service to automobiles and to hand-held portable phones and interconnects with the public telephone network by using radio transmissions that are based on a system of cells and antennas. The cellular concept was developed by AT&T's Bell Laboratories in 1947, but it was not until 1974 that the FCC set aside radio spectrums between 800MHz and 900MHz for cellular radio systems. The first cellular demonstration system was installed in Chicago in 1978, and three years later, the FCC formally authorized 666 channels for cellular radio signals and established *Cellular Geographic Servicing Areas* (CGSAs) to cover the nation's major metropolitan centers.

At the same time, the FCC created a regulatory scheme for cellular service that specified that two competing cellular companies would be licensed in each market. For each city, the commission ruled that one license would be reserved for the local telephone company (a wireline company), and the other license would be granted to another qualified applicant. When the number of applicants became prohibitively large, the Commission amended its licensing rule and specified the use of lotteries to select applicants for all but the top 30 markets. Cellular service, whether analog or digital, is now available virtually everywhere in the United States.

Applications

Cellular telephones were originally targeted at mobile professionals, enabling them to optimize their schedules by turning non-productive driving and out-of-the-office time into productive and often profitable work time. Today, cellular service is also targeted at consumers, giving them the convenience of anytime-anywhere calling, plus the security of instant access to service in times of emergency.

Cellular solutions not only facilitate routine telephone communications, but they also increase revenue potential for people who are in professions that have high-return opportunities as a direct result of being able to respond promptly to important calls. Developing countries that do not have an advanced communications infrastructure are increasingly turning to cellular technology so that they can take part in the global economy without having to go through the resource-intensive step of installing copper wire or fiber-optic cable.

Technology

Cellular networks rely on relatively short-range transmitter/receiver (transceiver) base stations that serve small sections (or cells) of a larger service area. Mobile telephone users communicate by acquiring a frequency or time slot in the cell in which they are located. A master switching center

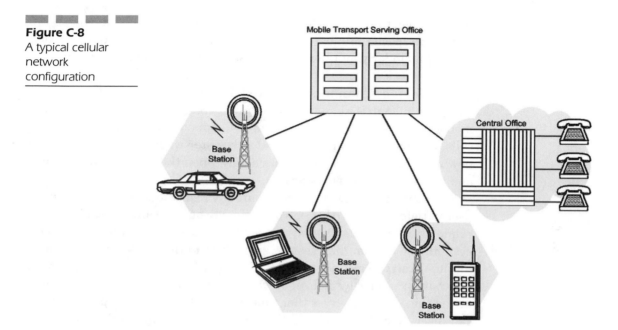

Figure C-8
A typical cellular network configuration

called the *Mobile Transport Serving Office* (MTSO) links calls between users in different cells and acts as a gateway to the PSTN. Figure C-8 illustrates the link from the MTSO to the base stations in each cell. The MTSO also has links to local telephone central offices so that cellular users can communicate with users of conventional phones.

Cell Sites   Cell boundaries are neither uniform nor constant. The usage density in the area, as well as the landscape, the presence of major sources of interference (e.g., power lines and buildings), and the location of competing carrier cells all contribute to the definition of cell size. Cellular boundaries change continuously, with no limit to the number of frequencies that are available for the transmission of cellular calls in an area. As the density of cellular usage increases, individual cells are split in order to expand capacity. By dividing a service area into small cells that have limited-range transceivers, each cellular system can reuse the same frequencies many times. Technologies such as *Code Division Multiple Access* (CDMA) and *Expanded Time Division Multiple Access* (E-TDMA) promise more capacity gains in the future.

Master Switching Center   In a typical cellular network, the master switching center operates similarly to a telephone central office and provides links to other offices. The switching center supports trunk lines to the base stations that establish the cells in the service area. Each base station supports a specific number of simultaneous calls—from three to 15, depending on the underlying technology (CDMA, TDMA, or some derivative).

Transmission Channels   Most cellular systems provide two types of channels: a control channel and a traffic channel. The base station and mobile station use the control channel to support incoming and outgoing calls, to monitor signal quality, and to register when a user moves into a new zone. The traffic channel is used only when the station is off-the-hook and is actually involved in a call.

The control and traffic channels are divided into time slots. When the user initiates access to the control channel to place a call, the mobile station randomly selects a sub-slot in a general-use time slot in order to reach the system. The system then assigns a time slot to the traffic channel. For an incoming call to a mobile station, the base station initiates conversations on the control channel by addressing the mobile station in a time slot—which, at the same time, reserves that time slot for the station's reply. If a user's call attempt collides with another user's call attempt, both instruments automatically reselect a sub-slot and try again. After repeated collisions, if no time slots are available within a predetermined time, the system rejects service requests for incoming and outgoing calls.

When a mobile telephone user places a call, the cell in which the user is traveling allocates a slot for the call. The call slot provides the user with access through the base station to the master switching center, essentially providing an extension on which the call can be placed. The master switching center, through an element of the user-to-base station connection, continuously monitors the quality of the call signal and transfers the call to another base station when the signal quality reaches an unacceptable level due to the distance traveled by the user, obstructions, and/or interference. If the user travels outside the system altogether, the master switching center terminates the call as soon as the signal quality deteriorates to an unacceptable level.

Roaming   Roaming occurs when a user moves out of the home area and into the serving area of another cellular carrier. In most cases, however, the cellular carrier belongs to an extension service such as MobiLink, which provides a service handoff between cellular carriers. Instead of calls being dropped because the user strayed beyond a service boundary, the call is handed to the next cellular carrier. The caller does not need to preregister with another service provider when traveling outside the home area. With MobiLink, for example, the subscriber merely dials a standard code to handle incoming and outgoing calls when traveling outside the home area: *18 activates roaming, and *19 deactivates roaming.

Cellular Telephones   The cellular telephone is the most visible part of the cellular system. Cellular telephones incorporate a combination of multi-access digital communications technology and traditional telephone technology and are designed to appear to the user as familiar residential or business telephone equipment. Manufacturers use miniaturization and digital signal-processing technology to make cellular phones feature-rich, yet compact and economical.

Cellular instruments consist of a transceiver, an analog/digital converter, and a supervisory/control system that manages calls and coordinates service with both the base station and the master switching center. Cellular telephones can be powered from a variety of sources, including vehicle batteries, AC adapters, and rechargeable battery sets.

Traditional cellular instrument types include hand-held, transportable, and car telephones. Advances in cellular technology are creating additional types of telephones, however, including modular and pocket phones. The trend in cellular instruments is toward multi-purpose, transportable telephones. Dual-mode cellular phones can be used with built-in wireless PBXs as well as with outside cellular service. The handset registers itself with a built-in base station and takes its commands from the wireless PBX. For out-of-building calling, the handset registers with the nearest cell site transceiver. Aside from convenience, an added benefit of the dual-mode phone is that calls made off the corporate premises can be aggregated with business calls that are made at home or on the road (for the purpose of achieving a discounted rate on all calls).

Network Optimization

Network optimization is a high priority for wireless carriers. A single cell site, including electronic equipment and a tower, can cost as much as $600,000 to build. With skyrocketing growth for wireless voice and data access in recent years, wireless service providers want to get the most efficient use from their current networks and target upgrades appropriately in order to meet customer demands. The use of network-optimization tools translates into lower wireless service costs and better coverage.

Such tools measure cell-site footprints, service areas within those footprints, and frequency assignments—all with the purpose of identifying the disruptive interference that cell sites receive from adjoining sites. By taking steps to limit interference between cells, wireless providers can maximize the bandwidth that is devoted to moving traffic. In addition to monitoring the cell sites, these network-optimization tools monitor the strength of RF signals that emanate from cell sites, as well as how calls are handed off among cell sites.

One network-optimization tool comes from ScoreBoard and relies on Microsoft Excel spreadsheets and MapInfo Corporation's MapInfo mapping application to provide carriers with analytical and graphical reports concerning how cell sites are performing and where interference is the highest. Performance data is collected by scanners that are installed in vehicles that drive through selected portions of a carrier's network. The collected data is then transmitted to ScoreBoard's data center for processing. The processed data is merged with cell-site performance data in MapInfo's location application, which is then accessed by users via a Web browser. With this information, engineers can pinpoint the portions of the wireless network that need to be upgraded in order to improve performance.

Summary

No other technology has taken the world by storm quite like cellular technology. Cellular is no longer a necessary service only for mobile professionals, but it has become necessary for the average consumer, as well. A trend exists toward replacing traditional phone service with cellular service. At year-end 1999, two percent of all U.S. wireless customers were already using their cell phones as their only phone. Customers also have shifted about 12 percent of their regular calls to wireless. Cellular systems have expanded beyond providing voice communication to supporting more sophisticated applications, such as Internet access for e-mail and accessing Web content. New Internet-enabled cellular phones feature larger displays that help make the phone an all-purpose communications appliance.

See Also

Cellular Data Communications (CDC)

Cellular Telephones

Code Division Multiple Access (CDMA)

Internet-Enabled Mobile Phones

Personal Communications Service (PCS)

Time-Division Multiple Access (TDMA)

Central Office Switches

Today's central office switches are available from several manufacturers worldwide and differ in line capacities, services and features, and the network environments that are supported. They not only switch ordinary telephone calls but also support digital voice, text, image, and data communications via ISDN. The environments that are supported can include a stand-alone office, a distributed network, collocation with an analog office, or remote configurations that enable services and features to be extended to isolated areas via remote switching units.

The switches themselves are modular, thereby making various software upgrades, card additions, and adjunct system connections possible. The switches can be equipped to provide access to Centrex/business group features and can function as a database gateway, connecting telephony functions with online databases, or as interactive services and packet networks. They can also be equipped to support LAN interconnection over the WAN. In recent years, support for a variety of broad-band technologies has been added, including Frame Relay, ATM, and SONET.

A switch can process as many as 1.5 million *Busy-Hour Call Attempts* (BHCAs). Each digital line unit can support hundreds of individual subscriber lines. A mobile exchange subsystem can be added to the central office switch to accommodate as many as 40,000 radio subscribers per exchange. The switch's service-management subsystem accommodates operator and administrative terminals. The switch also hosts data-polling systems for traffic analyses and automated billing.

Today's digital central office switches are versatile, particularly due to their support of AIN services. The switch's ISDN and AIN capabilities complement each other in that ISDN supports service access, while AIN supports service control and execution. Both facilities are based on *Signaling System 7* (SS7), and both enable users to quickly obtain and utilize new services.

Features

The central office switch offers subscribers a wealth of basic and advanced features. Basic features are those that are provided to subscribers whose facilities are not connected to ISDN lines; advanced features are those that are provided to ISDN users.

Basic features include abbreviated dialing, alarm calls, call rerouting-busy, call waiting, call rerouting-no answer, call charge indicator, toll-free calling, conference calling, direct dialing to extension, emergency call areas, hotlines, call trace, incoming call block, individual call record, outgoing traffic limitations, override block, subscriber with special services, subscriber priority, and three-way calling.

Advanced features are provided to subscribers whose systems are equipped with ISDN basic-rate (2B+D) or primary-rate (23B+D) interfaces. Advanced features include automatic callback, call forwarding, call hold, call pickup, call rerouting when busy, call waiting, charge handling, data transmission, dedicated connection, display information, incoming call block, multi-line hunt group, subscriber-programmed features, user groups, and closed user groups.

Hardware Components

The central office switch accommodates several types of hardware modules, which can be called by different names by different manufacturers:

- *Digital Line Unit* (DLU)   Interfaces subscriber traffic (including analog or ISDN signals) to the switch. These units provide analog-to-digital conversion and initial signal processing, and they concentrate traffic over T1 links to the *Line Trunk Groups* (LTGs). DLUs can accommodate any mix of single-party, dual-party, ground-start, coin-operated, direct inward dial, and ISDN lines.

- *Remote Control Unit* (RCU)   Houses multiple DLUs in a remote deployment configuration. Whatever service or feature is available from the central office is usually available to users who are connected at the RCU. The RCU can provide emergency local switching for users who are connected to the same RCU without carrying traffic to the LTGs and to the *Switching Network* (SN), however.

- *Integrated Packet Handler* (IPH)   Performs the ISDN packet-handler functions for ISDN packet subscribers, which enables ISDN-equipped customers to originate and receive packet transmissions at their desktop terminals.

- *Line Trunk Group* (LTG)   Connects subscriber and trunk lines with the *Common Channel Signaling Network Controller* (CCSNC). LTGs also function as expansion elements to the switch.

- *Switching Network* (SN)   Interfaces signals from the DLU, controls input and output switching, and provides switching control functions.

- *Coordination Processor* (CP)   Controls and coordinates the system through various operation, administration, and maintenance functions.
- *Common Channel Signaling Network Controller* (CCSNC)   Handles the transfer of SS7 messages in the distributed environment, specifically between the service switching point, the network signal transfer points, and the service control point. The switch's CCSNC application module enables the switch to separate call handling from network management.

Software Components

The operating system and applications software controls switching, administration, and maintenance of the switch and its required databases. The operating system provides organization programs for system managers (e.g., time administration, memory administration, input and output elements, and safeguarding programs). Application software can be configured to provide local, transit, and long-distance exchanges—as well as radio relay. Through software, various ISDN, intelligent network, and mobile radio system features can be incorporated into the switch on a modular basis. The switch functions that are typically provided by software include the following:

- *Local Number Portability*   Provides the means for subscribers to keep the same telephone number if they decide to switch local or long-distance service providers
- *Advanced Centrex Services*   Includes improved remote access for telecommuters to Centrex groups at the central office switch
- *Centrex Attendant Console*   Supplies additional call-handling and business-group capabilities to support *Key Telephone System* (KTS) environments
- *Enhanced KTS Business Group Applications*   Provides advanced ISDN features to analog instruments, including caller ID and a visual message waiting indicator
- *Centrex Command and Control Workstation*   Supplies a *Graphical User Interface* (GUI) for Centrex users, in order to facilitate access to and control over their facilities. Capabilities include collecting traffic, maintenance, message-detail recording data, trunk testing, and control

of authorization codes, outgoing facility routing, time-of-day redirection, alternate facility restriction, and automatic flexible routing.

Classless Switches

The primary intelligence in the PSTN is contained in Class 4 and Class 5 switches. Class 4 switches provide long-distance switching and network features, while Class 5 switches provide the local switching and telephony features. In the near future, classless central office switches will become commercially available to *Competitive Local Exchange Carriers* (CLECs). Because many of these carriers provide customers with local and long-distance services, they could do so more efficiently and economically with a single platform that could handle both types of traffic, rather than purchasing and provisioning separate Class 4 and Class 5 switches. The benefits of a classless platform can be extended even further as more voice calls are handled by IP-based packet-data networks. With IP telephony, there is no need for differently equipped switches in the network.

Salix Technologies, for example, offers the ETX5000 switch, which is distinguished by its so-called class-independence. The switch interfaces between the PSTN and PDN and supports IP, TDM, or ATM. These features and capabilities both ease and speed up a service provider's capability to deploy next-generation, enhanced telephony services. These same capabilities also facilitate the deployment of VPNs, all from a single switch. The ETX5000 scales up to 100,000 ports, thereby handling up to 50,000 calls simultaneously.

With its class-independence, the type of service that is offered is no longer tied to the location of a switch in the network, thereby making it easier to provide new services. In other words, the ETX5000 can be used for either Class 4 or Class 5 central office installations. Feature-management software enables any combination of service features to be directed to any port, eliminating the rigidity of the traditional Class 4/5 hierarchy of network switching.

For example, for IP-based carriers, the ETX5000 can function as a Class 4 switch, which means that it will perform the toll/tandem switching functions that are needed to interconnect IP networks with local-exchange telephone clients, acting as a distributed Class 4 switch. At the same time, however, the ETX5000 can support the incremental addition of new advanced-service features that are not traditionally associated with Class 4 switches, such as call redirection on busy and voice mail—features that are normally associated with Class 5 switches. This architecture permits any telephone switch feature to be distributed to any switch interface on demand.

Summary

Today's modular central office switches enable carriers to build different types of switching centers by using various hardware combinations. The switch can function as an end office, as an access tandem, or as a remote unit that is capable of serving rural communities. The switch's multiple processors and modular components simplify system modification. Upgrades or fixes are confined to a subsystem, rather than to the central switching processor; thereby reducing the likelihood that problems will occur in other parts of the network. The modular design also enables software to be updated and new processors and subsystems to be added with relative ease, creating a ready migration path for future technologies and services.

See Also

> *Advanced Intelligent Network* (AIN)
>
> Centrex
>
> *Integrated Services Digital Network* (ISDN)
>
> *Signaling System 7* (SS7)

Centrex

Centrex, which is short for central office exchange, is a service that handles business calls at the telephone company's switch, rather than through a customer-owned, premises-based *Private Branch Exchange* (PBX). Centrex provides a full complement of station features, remote switching, and network interfaces that provide an economical alternative to owning a PBX. Centrex offers remote options for businesses that have multiple locations, providing features that appear to users and to the outside world as if the remote sites and the host switch are one system.

Centrex users have access to direct inward dialing features as well as to station identification on outgoing calls. Each station has a unique line appearance in the central office, in a manner that is similar to residential telecommunications subscriber connections. A Centrex call to an outside line exits the switch in the same manner as a toll call exits a local exchange. Users dial a four-digit or five-digit number without a prefix in order to call internal extensions, and they dial a prefix (usually 9) to access outside numbers.

The telephone company operates, administers, and maintains all Centrex switching equipment for the customers. The company also supplies the necessary operating power for the switching equipment, including backup power, to ensure uninterrupted service during commercial power failures.

Centrex is also offered through resellers that buy Centrex lines in bulk from the LEC. Using its own or commercially purchased software, the reseller packages an offering of Centrex and perhaps other basic and enhanced telecommunications services in order to meet the needs of a particular business. The customer receives a single bill for all of the local, long-distance, 800, 900, and calling-card services at a fee that is less than the customer would otherwise pay.

Centrex Features

Centrex service offerings typically include *Direct Inward Dialing* (DID), *Direct Outward Dialing* (DOD), and *Automatic Identification of Outward Dialed* calls (AIOD). Advanced digital Centrex service provides all of the basic and enhanced features of the latest PBXs in the areas of voice communications, data communications, networking, and ISDN access. Commonly available features include voice mail, e-mail, message-center support, and modem pooling.

For large networks, the Centrex switch can act as a tandem switch that links a company's PBXs through an electronic tandem network. Centrex is also compatible with most private-switched network applications, including the *Federal Telecommunications System* (FTS) and the *Defense Switched Network* (DSN).

Many users subscribe to Centrex service primarily because of its networking capabilities, particularly for setting up a virtual city-wide network without major cost or management concerns. With city-wide Centrex, a business can set up a network of business locations that have a uniform dialing plan, a single published telephone number, centralized attendant service, and full-feature transparency for only an incremental cost per month over what a single Centrex site would cost.

Customer Premises Equipment

Centrex customer premises equipment is available for lease or purchase from a number of vendors, including the local telephone company. Centrex CPE combines the advanced features of a PBX with the convenience and flexibility of Centrex. Popular Centrex CPE products include multi-line phones, PC-based attendant workstations, attendant consoles, call reporting and management packages, voice-processing equipment, line monitors, and tip-and-ring scanners.

Centrex Telephones   Phones that are made specifically for Centrex service enable users to access a wealth of Centrex features with the touch of a button—without having to memorize codes. Many even provide single-button transfer capabilities to remote sites. When users have the proper equipment, the service is used more efficiently and the features are used more frequently—resulting in a better value for the company's Centrex investment dollars.

PC-Based Attendant Workstations   Screen-based attendant consoles combine Centrex access and improved calling options with Windows databases. This setup enables an attendant to work in a Windows word-processing application, for example, and to hotkey over to answer an incoming call. Some PC attendant workstations come equipped with database directories that can support multiple telephone, fax, and paging numbers for each entry and that enable single-keystroke dialing. Those that support ANI and directory-name lookup can perform screen pops of caller information, in order to give a truly personalized call-handling service. Centrex workstations that interface with LANs also enable attendants to do more than just handle calls. Hotkeying enables them to answer and transfer calls and still complete computer-based tasks over the local network.

Centrex Answering Consoles   Centrex answering consoles can take the form of multi-button telephones, conventional-looking attendant consoles, or PC-based platform systems. Centrex attendant consoles enable call handlers to perform single-button call transfers. They also have line status displays that enable the attendant know when a line is in use or when it is idle. Many consoles also enable the attendant to reprogram extensions, to access features, and to make other system rearrangements without the help of the telephone company.

Message Desk   As an option, the telephone company can provide a data line to the customer, which is called a *Station Message-Desk Interface* (SMDI). Centrex CPE interfaces with the SMDI link to give full voice mail and Centrex integration, in addition to enabling a call to an unanswered station to be rerouted directly to that person's voice mailbox without the caller having to re-enter the extension number.

Call Accounting System   Many options are available for businesses that use Centrex to obtain SMDR data in order to increase system efficiency. The telephone company can provide call-detail information to its customers, or alternatively, customers can use CPE line scanners and PC-based tele-management products in order to obtain the same functionality at a lower price. Some systems record SMDR right from the switch and store the information until it can be transferred to a PC for processing.

Administration Systems   In the past, Centrex customers had to go through the telephone company to change an extension on their Centrex service. Now, Centrex administration systems enable users to reconfigure their own Centrex service. PC-based consoles enable users to turn enhanced CO services (e.g., CO voice mail) on and off; to activate or deactivate lines; to add or reassign trunks; and to assign extensions for least-cost routing of long-distance calls.

IP Centrex

Centrex can be enhanced in order to support telephone calls over IP packet networks. The advantage of integrating Centrex with IP packet-based networks is enabling the telephone company to reduce facility costs by using the inherent efficiencies of IP networks. This functionality also enables the telephone company to offer value-added services to its Centrex customers, such as *Virtual Private Networks* (VPNs), telecommuter access, and virtual call centers. Telephone companies can also use the IP network to extend the market reach of their Centrex offerings by serving corporate locations out of their local serving area.

The integration of Centrex with IP is accomplished via a gateway that gives the existing Class 5 central office switch the capability to offer the full spectrum of voice, data, and Centrex features over an IP network (refer to Figure C-9). The gateway connects to the Class 5 switch via the Telcordia Technologies (formerly Bellcore) standard GR-303 interface. The gateway uses the existing stable and secure Class 5 infrastructure for billing, *Oper-*

Figure C-9
An IP gateway
connects to the Class
5 central office switch
via the industry-
standard GR-303
interface to
economically extend
the reach of Centrex
features.

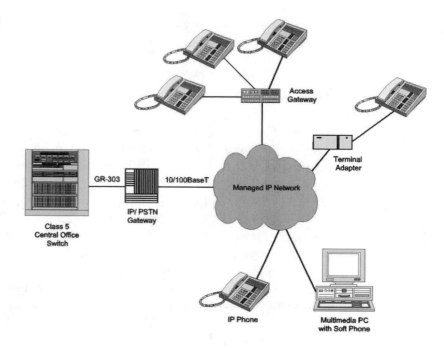

ations, Administration, Maintenance and Planning (OAM&P), signaling
(SS7), and trunking. The telephone company can increase revenues with
new IP applications that can be deployed immediately and can minimize
expenditures by providing these new services through existing equipment.

At the customer premises, a variety of equipment can be used to get voice
and data traffic through the IP network to the gateway and then to the Centrex switch. A multi-port access gateway is a concentration device that connects multiple legacy keysets to an IP network. Standard phone wiring is
needed only between desktop keysets and the gateway. Terminal adapters
connect a single legacy phone to an IP network. They perform coding/decoding
(codec) and packetization by utilizing *Digital Signal Processors* (DSPs) for
high performance. Some terminal adapters come bundled with a PC user
interface, and others have a PSTN connection as a backup.

Soft phones are telephony applications that run on a PC. Soft phones typically use the PC's main processor to perform codec and packetization, in addition to any other work that is being done on the PC. This situation sometimes
causes high latencies. To combat this problem, a few soft phones can be coupled with an add-in card (i.e., terminal adapter) in order to boost performance.

IP phones replace existing phones and connect directly to the IP network, usually via an Ethernet jack. The codec and packetization functions

are performed in the keyset. Wireless IP phones also exist, and they perform the codec function in the handset.

Summary

Centrex offers high-quality, dependable, feature-rich telephone service that supports a variety of applications. For many users, Centrex offers distinct advantages over on-premises PBX or key/hybrid systems. Centrex users save money over the short term, because there is no outlay of cash for an on-premises system, and Centrex installation charges are usually low. If the service is leased on a month-to-month basis, there is little commitment and no penalty for discontinuing the service. A company can pick up and move without worrying about reinstalling the system, which might not be right for the new location. Centrex systems can be expanded almost indefinitely by adding communication paths, memory, intercom lines, subscriber lines, and CO lines as needed. Now, with IP connectivity, Centrex can be used for VPNs, virtual call centers, and economical telecommuter access. If there is a Centrex problem, repair is immediate and inexpensive. No need exists for a company to invest in spare parts inventory, test equipment, or technical staff; rather, the telephone company is responsible for those elements. Also, as the CO switching equipment is updated, the Centrex services are also updated.

See Also

>   *Automatic Call Distributors* (ACD)
>   Call Centers
>   Central Office Switches
>   Private Branch Exchange (PBX)
>   *Voice-Over IP* (VoIP)

Channel Banks

A channel bank is a low-end multiplexer that consolidates up to 24 voice and data channels of 64Kbps each onto a higher-speed digital facility—typically, a T1 line that provides 1.544Mbps. For voice traffic, the channel bank provides analog-to-digital conversion via *Pulse-Code Modulation* (PCM).

A channel bank might be used on the front-end of an analog PBX, for example, in order to transport a bundle of voice conversations via a T1 line to another analog PBX across town. There, another channel bank receives the higher-speed digital signal and converts it back to the individual analog channels for acceptance by the PBX. Most digital PBXs now support T1, however, which eliminates the need for a channel bank and enables the PBX to connect to the span with only a *Channel Service Unit* (CSU). The CSU is required at the front end of a digital circuit in order to equalize the received signal, to filter both the transmitted and received waveforms, and to interact with the carrier's test facilities.

Channel banks are ideally suited for handling voice because their basic rate and bit pattern are matched to the management of voice signals, both for multiplexing at the DS1 rate and in preparation for additional downstream multiplexing at higher rates. In terms of applications, a channel bank is good for T1 access to long-distance carrier networks such as AT&T's Megacom, Sprint's Ultra 800/WATS, and MCI WorldCom's Prism services. Alternatively, the channel bank can be used as an economical, point-to-point system for tie lines or for *Off-Premises Extensions* (OPX) for PBXs or key systems.

Typical Configuration

A low-end channel bank typically supports either single or dual T1 connections in one chassis. The single chassis provisions up to 24 channels, while the dual T1 chassis provisions up to 48 channels. For applications in which users need growth from one to two T1 connections, the dual (or split) chassis can be equipped for 24 channels initially—leaving room for expansion to 48 channels when needed.

The channel bank supports standard voice interfaces such as two-wire loop and ground start, two-wire office connections, two-wire private-line automatic ring down, and four-wire E&M. Data interfaces include dial modem and fax connections, as well as four-wire analog leased-line modem connections. A V.35 interface enables the connection of high-speed data equipment, such as a router.

Most channel banks today are modular in design, featuring plug-in power supplies, common logic, and channel cards for fast and easy board swapping. Front panel alarm and status indicators, in conjunction with diagnostic loop-backs, enable fast and accurate troubleshooting and fault isolation. The channel bank can be optioned with a redundant, load-sharing power system in order to ensure maximum up time. Additional options include an integral CSU.

Channel Banks versus Multiplexers

Whereas channel banks were originally designed to accept analog inputs, T1 multiplexers were designed to accept digital inputs. Although T1 multiplexers can be equipped to accept analog inputs via optional plug-in cards, they are versatile devices that are especially adept at handling data streams. For example, multiplexers can compress data to increase the number of channels that are available for voice, to prioritize traffic to avoid congestion on the network, and to down-speed data in order to avoid corruption at higher rates when links experience service degradation.

Although channel-bank manufacturers have equipped their products with more data features and intelligence, resulting in entirely new devices called intelligent channel banks, T1 multiplexers still offer more functionality—especially in the area of network management. Network management information can even be inserted into each digital channel for end-to-end supervision and control. When data channels experience high error rates, the multiplexer can reroute them to other links and leave unaffected voice channels on the primary link. T1 multiplexers also are more flexible in managing the available bandwidth, implementing software-based reconfigurations for implementation by time-of-day or by events such as link failure and bit error-rate thresholds. Also, when disaster threatens to bring down the entire network, the multiplexer can implement pre-planned disaster-recovery scenarios under a cstored program control, calling into service any combination of available, private leased lines and public network services, including ISDN.

Summary

Channel banks offer an economical way to connect analog communications equipment to digital T1 service. They are ideal for static network environments in which low cost is the primary concern. Today's channel banks also offer data ports, such as V.35, for direct digital connectivity. Password-protected remote management via a modem connection enables system performance to be monitored and tests to be performed from any centralized PC or workstation, thus reducing down time.

Channel Service Units (CSUs)

Because digital transmission links are capable of transporting signals between *Data Terminal Equipment* (DTE) that is closer to their original form, there is no need for complex modulation/demodulation (modem) techniques, as is the case with sending data over analog connections. Instead, CSUs are used at the front end of the digital circuit (refer to Figure C-10) in order to equalize the received signal, to filter the transmitted and received waveforms, and to interact with both the user's and carrier's test facilities via a supervisory terminal or network-management system. The FCC's Part 68 registration rules require every T-carrier circuit to be terminated by a CSU. These devices can be used to set up a T1 line with a PBX, channel bank, T1 multiplexer, or any other DSX1-compliant DTE.

Key Functions

Line Build Out (LBO) is a functional requirement of all Part 68-registered T1 CSUs. An LBO is an electronic simulation of a length of wire line that adjusts the signal power so that it falls within a certain decibel range at both ends of the circuit. This simulation is conducted by looping a test signal over the receiver pair and measuring for signal loss. This procedure also

Figure C-10

CSUs terminate each end of the T-carrier facility.

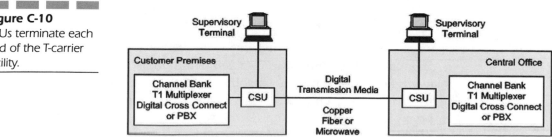

helps reduce the potential for one T1 transmitter to cross-talk into the receiver of other services within the same cable binder. Once line loss is determined, the telephone company can tell the customer what setting to use on the local CSU.

All T1 CSUs provide a repeater in order to reconstitute signals, which are susceptible to attenuation and distortion by the T1 span line and/or the customer's in-house cabling. This function is also part of the FCC Part 68 equipment registration requirements for CSUs.

In addition to equalizing the transmitted signal through the LBO and regenerating the received line signal, the CSU ensures that the user's DTE does not send signals that could possibly disrupt the carrier's network. For example, long strings of zeros do not provide timing pulses in order for the span-line repeaters to maintain synchronization. The CSU monitors the data stream from the attached DTE so that the ones density rule (e.g., the customer's data must have at least a 12.5 percent pulse density) is not violated. This rule ensures that there are sufficient pulse transitions for the span-line repeaters to maintain timing synchronization. The CSU will inject pulses if excessive zeros are being transmitted by the attached DTE. The CSU also provides isolation between the DTE and the network, in order to protect equipment and telephone company technicians from potentially harmful line voltages and lightning surges, which might propagate beyond the DTE.

The CSU provides the functionality to troubleshoot circuit and transmission problems. For proactive network management, these functionalities include LEDs that indicate both the status of the network and equipment connections and whether or not any alarm thresholds or error conditions have been detected. These indications enable technicians at either end of the circuit to isolate problems in minutes instead of hours. The CSU also contains a buffer that stores collected performance information that can be accessed by both the carrier and user for diagnostic purposes.

Summary

Because the CSU interfaces user equipment to carrier facilities, it provides a window to the network—enabling both the carrier and the customer to perform testing up to the same point. CSU access to the network has prompted vendors to equip the network with increasingly sophisticated diagnostic and network-management features. These features go a long way toward enhancing user control and ensuring the integrity of T-carrier facilities.

See Also

> *Data Service Units* (DSUs)

Citizens Band Radio Service

Citizens Band (CB) radio service is a two-way voice communication service for use in personal and business activities. The service is assigned the frequency range of 26.965MHz to 27.405MHz, and the effective communication distance is one to five miles. FCC license documents are no longer needed or issued. CB Rule 3 provides users with all of the authority that they need to operate a CB unit in places where the FCC regulates radio communications, as long as an unmodified FCC type-accepted CB unit is used. An FCC type-accepted unit has an identifying label placed on it by the manufacturer, and no age or citizenship requirement exists.

CB users might use an on-the-air pseudonym or handle of their own choosing and might operate their CB units within the territorial limits of the 50 United States, the District of Columbia, and the Caribbean and Pacific insular areas. Users can also operate their CB units on or over any other area of the world, except within the territorial limits of areas in which radio communications are regulated by another agency of the United States or within the territorial limits of any foreign government. In addition, users can use their CB units in Canada but are subject to the rules of the Canadian Department of Communications.

The power output of the CB unit might not be raised, because raising the level of radio noise would be unfair to the other users who are sharing the channel. Users also must not attach a linear amplifier or any other type of power amplifier to their CB unit or modify the unit internally. Doing so cancels its type acceptance, and the user forfeits his or her authorization to use the equipment.

No height restrictions exist for antennas that are mounted on vehicles or for hand-held units. For structures, the highest point of the antenna must not be more than 20 feet above the highest point of the building or tree on which it is mounted, or 60 feet above the ground. Lower height limits apply if the antenna structure is located within two miles of an airport.

Any of the 40 CB channels can be used on a take-turns basis. No CB channel is assigned to any specific individual or organization. Because the CB channels are shared by all users, cooperation is essential; communications should be short, with conversations never more than five minutes continuously. Users should wait at least one minute before starting another communication. Channel 9 should only be used for emergency communications or for traveler assistance.

Ten-Codes

Ten-codes are abbreviations of common questions and answers that are used on all types of radio communication. Professional CB'ers use these codes to send their message quickly and easily. Additionally, ten-codes can be readily understood by users when poor reception or language barriers must be overcome. Although the FCC authorizes CB operators to use ten-codes, it does not regulate their meaning. The most commonly used ten-codes are listed as follows:

10-1 = Receiving poorly

10-2 = Receiving well

10-3 = Stop transmitting

10-4 = Message received

10-5 = Relay message to . . .

10-6 = Busy, please stand by

10-7 = Out of service, leaving the air

10-8 = In service, subject to call

10-9 = Repeat message

10-10 = Transmission complete, standing by

10-11 = Talking too rapidly

10-12 = Visitors present

10-13 = Advise weather/road conditions

10-16 = Make pick up at . . .

10-17 = Urgent business

10-18 = Anything for us?

10-19 = Nothing for you, return to base

10-20 = My location is . . .

10-21 = Call by telephone

10-22 = Report in person to . . .

10-23 = Stand by

10-24 = Completed last assignment

10-25 = Can you contact . . .

10-26 = Disregard last information

10-27 = I am moving to channel . . .

10-28 = Identify your station

10-29 = Time is up for contact

10-30 = Does not conform to FCC rules

10-32 = I will give you a radio check

10-33 = Emergency traffic

10-34 = Trouble at this station

10-35 = Confidential information

10-36 = Correct time is . . .

10-37 = Wrecker needed at . . .

10-38 = Ambulance needed at . . .

10-39 = Your message delivered

10-41 = Please turn to channel

10-42 = Traffic accident at . . .

10-43 = Traffic tie up at . . .

10-44 = I have a message for you

10-45 = All units within range please report

10-50 = Break channel

10-60 = What is next message number?

10-62 = Unable to copy, use phone

10-63 = Net directed to

10-64 = Net clear

10-65 = Awaiting next message/assignment

10-67 = All units comply

10-70 = Fire at . . .

10-71 = Proceed with transmission in sequence

10-77 = Negative contact

10-81 = Reserve hotel room for . . .

10-82 = Reserve room for . . .

10-84 = My telephone number is . . .

10-85 = My address is . . .

10-91 = Talk closer to the microphone

10-93 = Check my frequency on this channel

10-94 = Please give me a long count (1–10)

10-99 = Mission completed, all units secure

10-200 = Police needed at . . .

Summary

At first, users were required to obtain a CB radio license and call letters from the FCC before they could go on the air. The FCC, however, was so inundated with requests for CB radio licenses that they finally abandoned formal licensing and enabled operators to buy CB radio equipment and to go on the air without any license or call letters. Although no license is required to operate a CB radio, the FCC's rules for CB radio operation are still in effect. These rules cover CB radio equipment, the ban on linear amplifiers, and the types of communications that are permitted on the air. A copy of the operating rules must be furnished by the manufacturer with each CB set that is sold.

See Also

Family Radio Service

General Mobile Radio Service

Low-Power Radio Service

Client/Server Networks

For much of the 1990s, the client/server architecture has dominated corporate efforts to downsize, restructure, and otherwise re-engineer for survival in an increasingly global economy. Frustrated with the restrictive access policies of traditional MIS managers and the slow pace of centralized, mainframe-centric applications development, client/server grew from the need to bring computing power and decision-making down to the user, so that businesses could respond faster to customer needs, competitive pressures, and market dynamics.

Architectural Model

The client/server architecture is not new. A more familiar manifestation of the architecture is the decades-old corporate telephone system, with the PBX acting as a server and the telephones acting as the clients. All of the telephones derive their features and user-access privileges from the PBX, which also processes incoming and outgoing calls. What is relatively new is the application of this model to the LAN environment, which is data-oriented. Here, an application program is broken into two parts: client and server. These parts exchange information over the network (refer to Figure C-11).

Client   The client portion of the program, or the front end, is run by individual users at their desktops and performs such tasks as querying a database, producing a printed report, or entering a new record. These functions are performed by a database specification and access language that is better known as *Structured Query Language* (SQL). SQL operates in conjunction with existing applications, and the front-end part of the program executes on the user's workstation—drawing upon its *Random Access Memory* (RAM) and *Central Processing Unit* (CPU).

Server   The server portion of the program, or the back end, resides on a computer that is configured to support multiple clients—thereby offering shared access to numerous application programs as well as to printers, file storage, database management, communications, and other resources. The

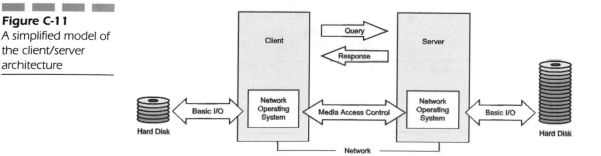

Figure C-11
A simplified model of the client/server architecture

server not only handles simultaneous requests from multiple clients, but it also performs administrative tasks such as transaction management, security, logging, database creation and updating, concurrency management, and maintaining the data dictionary. The data dictionary standardizes terminology so that database records can be maintained across a broad base of users.

Network The network consists of the transmission facility, which is usually a LAN. Among the commonly used media for LANs is coaxial cable (thick and thin), twisted-pair wiring (shielded and unshielded), and fiber optics (single-mode and multi-mode). In some cases, wireless facilities such as infrared and spread-spectrum are used to link clients and servers.

A medium-access control protocol is used to regulate access to the transmission facility. Ethernet and Token Ring are the two most popular medium-access control protocols. When linking client/server computing environments over the WAN, other communications facilities come into play, such as private T1 links (which provide a transmission rate of up to 1.544Mbps). The IP remains the most common transport protocol that is used over the WAN, while Frame Relay and other fast-packet technologies are also growing in popularity.

Client/Server Telephony One of the newest applications that leverage the client/server architecture is IP telephony. Vendors offer scaleable, IP-based PBXs that run on the Windows NT platform, which can transport packetized voice over an Ethernet LAN, over a managed IP network, or over the PSTN via an IP/PSTN gateway. Full-featured digital phones link directly to the Ethernet LAN via a 10BaseT interface, without requiring direct connection to a desktop computer. Phone features can be configured by using a Web browser (refer to Figure C-12).

Digital phones have access to the calling features that are offered through the IP PBX management software that is running on the LAN server. The call-management software supports each IP phone with functions such as call hold, call transfer, call forward, call park, and caller ID. Even advanced PBX functions such as multiple lines per phone or multiple phones per line are determined by software and can be reconfigured from any location by using a Web browser.

Summary

To date, the promises behind client/server are somewhat mixed. With client/server networks, administration and management complexity

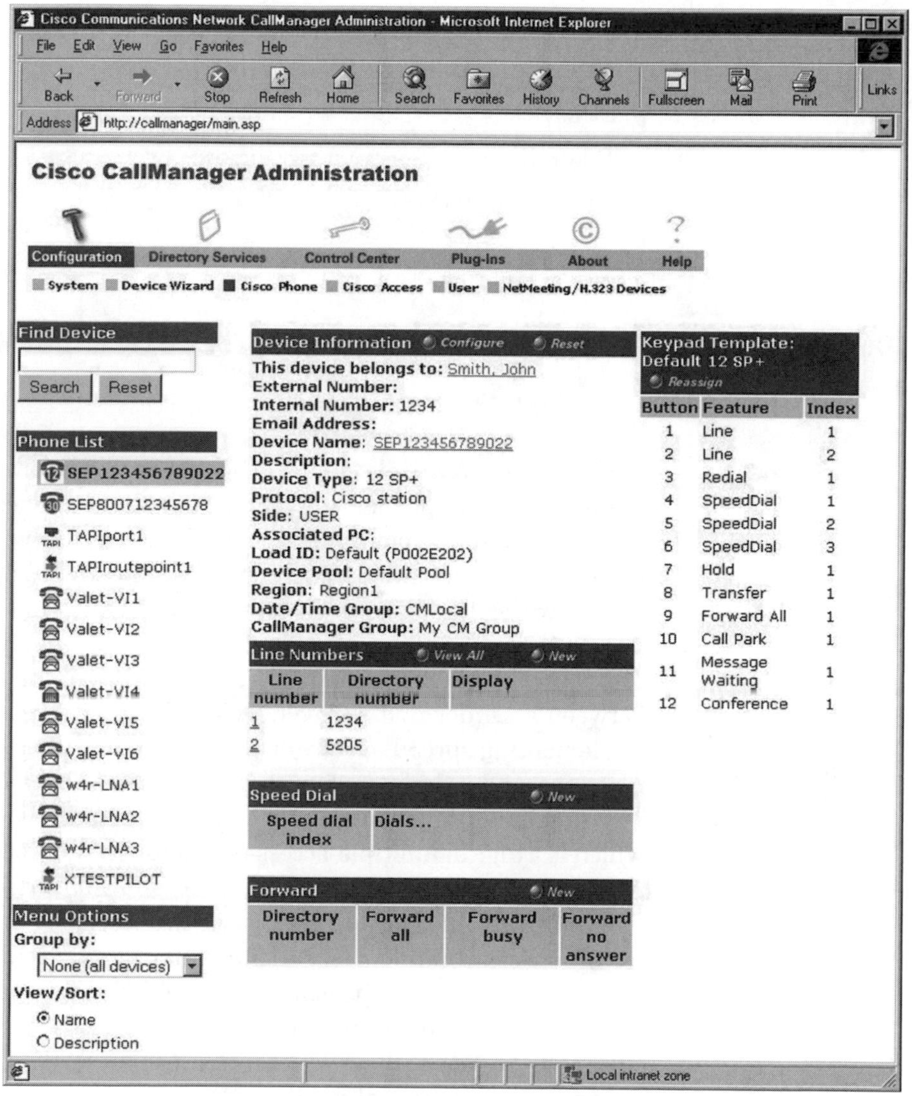

Figure C-12

A view of the administrative interface of Cisco Systems' CallManager, which enables phone features that are delivered through a server-based IP PBX to be set up from within a Web browser

increases while costs become more difficult to track. According to some industry estimates, the total cost of owning a client-server system is about three to six times greater than for a centralized mainframe system, while the software tools for managing and administering client-server cost two-and-a-half times more than comparable mainframe tools. Regardless of these factors, many organizations have successfully implemented client/server networks and have achieved significant efficiency and productivity gains.

See Also

Advanced Peer-to-Peer Networking (APPN)

Ethernet

Token Ring

Voice-Over IP (VoIP)

Code-Division Multiple Access (CDMA)

Code-Division Multiple Access (CDMA) is a spread-spectrum technology that is used for implementing cellular telephone service. Spread-spectrum is a family of digital communication techniques that were originally used in military communications and control applications. Spread-spectrum uses carrier waves that consume a much wider bandwidth than is required for simple point-to-point communication at the same data rate, which results in the carrier wave looking more like random noise than real communication between a sender and receiver. Originally, there were two motivations for implementing spread-spectrum: to resist enemy efforts to jam vital communications, and to hide the fact that communication was even taking place.

For cellular telephony, spread-spectrum technology underlies CDMA, which is a digital multiple-access technique that is specified by the *Telecommunications Industry Association* (TIA) as IS-95. Commercial applications of CDMA became possible because of two key developments: the availability of low-cost, high-density digital integrated circuits, which reduce the size, weight, and cost of the mobile phones; and the realization that optimal multiple-access communication depends on the capability of all mobile phones to regulate their transmitter power to the lowest level that will achieve adequate signal quality.

CDMA changes the nature of the mobile phone from a predominately analog device to a predominately digital device. CDMA receivers do not eliminate analog processing entirely, but they separate communication channels by means of a pseudo-random modulation that is applied and removed in the digital domain (not on the basis of frequency). This functionality enables multiple users to occupy the same frequency band, and this frequency reuse results in high spectral efficiency.

TDMA systems commonly start with a slice of spectrum, which is referred to as a carrier. Each carrier is then divided into time slots. Only one subscriber at a time is assigned to each time slot or channel. No other conversations can access this channel until the subscriber's call is finished—or until that original call is handed off to a different channel by the system. For example, TDMA systems, which are designed to coexist with AMPS systems, divide 30KHz of spectrum into three channels. By comparison, GSM systems create eight time-division channels in 200KHz-wide carriers.

CDMA systems divide the radio spectrum into carriers that are 1,250KHz (1.25MHz) wide. Unique digital codes, rather than separate RF frequencies or channels, are used to differentiate subscribers. The codes—called pseudo-random code sequences—are shared by both the mobile station (i.e., cellular phone) and the base station. All users share the same range of radio spectrum.

One of the unique aspects of CDMA is that while there are ultimate limits to the number of phone calls that can be handled by a carrier, this number is not a fixed number. Rather, the capacity of the system depends on how coverage, quality, and capacity are balanced in order to arrive at the desired level of system performance. Because these parameters are tightly intertwined, operators cannot have the best of all worlds: three times wider coverage, 40 times capacity, and high-quality sound. For example, the 13Kbps vocoder provides better sound quality but reduces system capacity, as compared to an 8Kbps vocoder. Higher capacity might be achieved through some degree of degradation in coverage and/or quality.

System Features

CDMA has been adapted for use in cellular communications with the addition of several system features that enhance efficiency and lower costs:

Mobile Station Sign-On Upon power-on, the mobile station already knows the assigned frequency for CDMA service in the local area and will tune to that frequency and search for pilot signals. Multiple pilot signals typically will be found—each with a different time offset. This time offset distinguishes one base station from another. The mobile station will pick the strongest pilot and establish a frequency reference and a time reference from that signal. Once the mobile station becomes synchronized with the base station's system time, the mobile station can then register. Registration is the process by which the mobile station tells the system that it is available for calls and notifies the system of its location.

Call Processing The user makes a call by entering the digits on the mobile station keypad and hitting the send button. If multiple mobile stations attempt a link on the access channel at precisely the same moment, a collision occurs. If the base station does not acknowledge the access attempt, the mobile station will wait for a random length of time and try again. Upon making contact, the base station assigns a traffic channel and basic information is exchanged (including the mobile station's serial number). At that point, the conversation mode is started.

As a mobile station moves from one cell to the next, another cell's pilot signal will be detected that is strong enough for it to use. The mobile station will then request a soft handoff, during which it is actually receiving both signals via different correlative elements in the receiver circuitry. Eventually, the signal from the first cell will diminish, and the mobile station will request from the second cell that the soft handoff be terminated. A base station does not hand off the call to another base station until it detects an acceptable signal strength.

This soft handoff technique is touted by CDMA advocates as a significant improvement over the handoff procedure that is used in analog FM cellular systems, in which the communication link with the old cell site is momentarily disconnected before the link to the new site is established. For a short time, the mobile station is not connected to either cell site, during which the subscriber hears background noise or nothing at all. Sometimes the mobile stations Ping-Pong between two cell sites as the links are handed back and forth between the approaching and retreating cell sites. Other times, the calls are simply dropped. Because a mobile station in the CDMA system has more than one modulator, it can communicate with multiple cells simultaneously in order to implement the soft handoff.

At the end of a call that is placed over the CDMA system, the channel will be freed and can be reused. When the mobile station is turned off, it will generate a power-down registration signal that tells the system that it is no longer available for incoming calls.

Voice Detection and Encoding   With voice activity detection, the transmitter is activated only when the user is speaking. This feature reduces interference levels—and, consequently, reduces the amount of bandwidth that is consumed—when the user is not speaking. Through interference averaging, the capacity of the system is increased, which enables systems to be designed for the average rather than the worst interference case. The IS-95 CDMA standard, however, requires that no interfering signal is received that is significantly stronger than the desired signal, because the signal would then jam the weaker signal. This situation has been called the near-far problem and means that high cell capacity does not necessarily translate into high overall system capacity.

The speech coder that is used in CDMA operates at a variable rate. When the subscriber is talking, the speech coder operates at the full rate; when the subscriber is not talking, the speech coder operates at only one-eighth of the full rate. Two intermediate rates are also defined to capture the transitions and to eliminate the effect of sudden rate changes. Because the variable rate operation of the speech coder reduces the average bit rate of the conversations, system capacity is increased.

Privacy Increased privacy is inherent in CDMA technology. CDMA phone calls will be secure from the casual eavesdropper, because unlike a conversation that is carried over an analog system, a simple radio receiver will not have the capability to pick out individual digital conversations from the overall RF radiation in a frequency band.

A CDMA call starts with a standard rate of 9.6Kbps. This rate is then spread to a transmitted rate of about 1.25Mbps. Spreading means that digital codes are applied to the data bits that are associated with users in a cell. These data bits are transmitted along with the signals of all of the other users in that cell. When the signal is received, the codes are removed from the desired signal, separating the users and returning the call to the original rate of 9.6Kbps.

Because of the wide bandwidth of a spread-spectrum signal, you will find it difficult to identify individual conversations for eavesdropping. Because a wide-band spread-spectrum signal is hard to detect, it appears as nothing more than a slight rise in the noise floor or interference level. With analog technologies, the power of the signal is concentrated in a narrower band, which makes it easier to detect with a radio receiver that is tuned to that set of frequencies.

The use of wide-band spread-spectrum signals also offers more protection against cloning—an illegal practice whereby a mobile phone's electronic serial number is taken over the air and is programmed into another phone. All calls that are made from a cloned phone are free, because they are billed to the original subscriber.

Power Control CDMA systems rely on strict control of power at the mobile station in order to overcome the so-called near-far problem. If the signal from a near mobile station were to be received at the cell site receiver with too much power, it would be overloaded by that particular mobile station. This situation would overwhelm the signals from the other mobile stations that were located farther away. The goal of CDMA is to have the signals of all mobile stations arrive at the base station with exactly the same power level. The closer the mobile station to the cell site receiver, the lower the power that is necessary for transmission; the farther away the mobile station, the greater the power that is necessary for transmission.

Two forms of adaptive power control are employed in CDMA systems: open-loop and closed-loop. Open-loop power control is based on the similarity of loss in the forward and reverse paths. The received power at the mobile station is used as a reference. If it is low, the mobile station is assumed to be far from the base station and transmits with high power. If it is high, the mobile station is assumed to be near the base station and transmits with low power. The sum of the two power levels is a constant.

Closed-loop power control is used to force the power from the mobile station to deviate from the open-loop setting. This goal is achieved by an active feedback system from the base station to the mobile station. Power control bits are sent every 1.25 milliseconds (ms) in order to direct the mobile station to increase or decrease its transmitted power by 1 decibel (dB). Lack of power control over this accuracy greatly reduces the capacity of CDMA systems.

With these adaptive power-control techniques, the mobile station transmits only enough power to maintain a link, which results in an average power requirement that is much lower than that for analog systems (which do not usually employ such techniques). CDMA's lower power requirement translates into smaller, lightweight, longer-life batteries—approximately five hours of talk time and more than two days of standby time—and makes possible smaller, lower-cost hand-held computers and hybrid computer-communications devices. CDMA phones can easily weigh in at less than eight ounces.

Spatial Diversity   Among the various forms of diversity is spatial diversity, which is employed in CDMA as well as in other multiple-access techniques (including FDMA and TDMA). Spatial diversity helps to maintain the signal during the call handoff process, when a user moves from one cell to the next. This process entails antennas in two different cell sites maintaining links with one mobile station. The mobile station has multiple correlative receiver elements that are assigned to each incoming signal and can add these elements.

CDMA uses at least four of these correlators: three that can be assigned to the link, and one that searches for alternate paths. The cell sites send the received data, along with a quality index, to the *Mobile Telephone Switching Office* (MTSO), where a choice is made regarding the better of the two signals. Not all of these features are unique to CDMA. Some can be exploited by TDMA-based systems as well, such as spatial diversity and power control. These features already exist in all TDMA standards today, while soft handoff is implemented in the European DECT cordless telecommunications standard (which is based on TDMA).

Summary

Conflicting performance claims still exist for TDMA and CDMA. Because both TDMA and CDMA have become TIA standards—IS-54 and IS-95, respectively—vendors are now aiming their full marketing efforts toward the cellular carriers. Proponents of each technology have the research to back up their claims of superior performance. Of the two, CDMA suffered a credibility problem early on because its advocates made grandiose performance claims for CDMA that could not be verified in the real-world operating environment. In some circles, this credibility problem lingers today. Of note, however, is that both technologies have been successful in the marketplace—each having been selected by many cellular carriers around the world. Both are capable of supporting emerging PCS networks and providing services such as wireless Internet access, short message service, voice mail, facsimile, paging, and video. Although TDMA-based *Global System for Mobile* (GSM) telecommunications is the dominant standard in the global wireless market, the use of CDMA is growing rapidly. GSM's head-start in the market gives it a much larger presence and practically guarantees that GSM will continue to lead the digital cellular market for the next five years.

See Also

Digital Enhanced Cordless Telecommunications

Spread-Spectrum Radio

Time-Division Multiple Access (TDMA)

Colocation Arrangements

Colocation entails the sharing of floor space within a central office, serving wire center, or other telephone company location for the purpose of placing equipment that simplifies access to the local loop by competitors. In being able to place their equipment in these locations, competitors can reduce the cost of operations and compete on a more equal footing with the *Incumbent Local Exchange Carrier* (ILEC).

Colocation arrangements are mandated by the Telecommunications Act of 1996, which is intended to create a more competitive telecommunications industry. Accordingly, the FCC issues rules concerning colocation in order to give CLECs easier access to the local loop. Colocation arrangements reduce the costs and delays that CLECs often encounter when trying to enter local markets that are dominated by the former *Bell Operating Companies* (BOCs) or ILECs. The resulting competition is supposed to give consumers access to innovative new services, provide more choice in service providers, and lower the cost of service delivery.

Colocation can also include the interconnection of facilities between the equipment of multiple service providers. CLECs, for example, can obtain floor space for their equipment within dominant carriers' central offices and have that equipment interconnected to that carrier's equipment for the purpose of offering local loop customers a variety of voice and data services. Competitors are also allowed to connect with each other's equipment without having to pay the exorbitant interconnection fees that were once charged by the dominant carrier.

Types of Arrangements

The FCC requires incumbents to make shared-cage and cageless colocation available to CLECs that request them. Shared cage is a colocation arrangement in which tightly controlled central office space can be shared by two or more competitive carriers at less cost to each. The incumbent LEC cannot increase the cost of site preparation or nonrecurring charges above the cost for provisioning such a cage of similar dimensions and material to a single colocating party. Cageless colocation enables a competitor to have its own space within the central office, but without security mechanisms such as card-entry systems and video surveillance.

When colocation space is exhausted at a particular location, the FCC requires the incumbent to permit colocation in adjacent, controlled environmental vaults or similar structures to an extent that is technically feasible. The FCC rules also enable competitors to tour the incumbent's entire central office in cases where the incumbent has denied the competitor colocation space. Incumbents must provide a list of all offices in which there is no more space. They must also remove obsolete, unused equipment in order to facilitate the creation of additional colocation space within a central office.

Incumbents are further required to enable competing carriers to establish cross-connects to the colocated equipment of other competing carriers. The competitors might even construct their own cross-connect facilities between colocated equipment without having to purchase the equipment or cross-connect capabilities from the incumbent. Often, this process is as simple as running a copper or fiber transmission facility from one colocation rack to an adjacent rack. Even when competitive equipment is colocated in the same room as the incumbent's equipment, the FCC requires the incumbent to permit the new entrant to construct its own cross-connect facilities.

These facilities-based colocation arrangements have the potential to reduce market entry costs and to speed the rollout of competitive services such as *Digital Subscriber Line* (DSL), which provides enough bandwidth to carry high-speed data and multiple voice channels over previously unused phone lines.

A competitive carrier that wants to offer DSL services to businesses or consumers, for example, can install *DSL-Access Multiplexers* (DSLAM) in the local central office. The function of the DSLAM is to terminate all of the remote DSL modems and routers at customer locations into a single location at the central office, so that the data traffic can be sent out to the Internet or to another remote destination, such as a *Virtual Private Network* (VPN). The colocation arrangement enables the competitor to leverage the local loop to offer an advanced data service that is in high demand.

Summary

Thanks to the FCC, CLECs and advanced data-service providers now have a bill of rights with respect to obtaining local-loop access connections from incumbent telephone companies. This bill of rights includes being able to colocate equipment within the central office and interconnect with other service providers economically and expeditiously. The objective of colocation is to speed up the rollout of new services and to encourage competition in local markets. Of course, CLECs and other types of service providers are free to use alternative exchange points, which offer bandwidth provisioning and interconnection as commercial services. The colocation rules for telephone companies and the emergence of alternative exchange points promise to change the telecommunications industry in the United States— perhaps finally achieving the level of competition that is promised by the Telecommunications Act of 1996.

See Also

Alternative Exchange Points

Interconnection Agreements

Local Loop

Unbundled Access

Communications Assistance for Law Enforcement Act

Telephone calls that are transmitted in the computer code of ones and zeros are more difficult to tap than conventional analog phone technology. For this reason, the 1994 Communications Assistance for Law Enforcement Act requires the telecommunications industry to adopt its technologies in order to permit court-ordered wiretaps of digital phones. This functionality is deemed necessary in order to combat terrorism, organized crime, and illegal drug activity.

Compliance with CALEA requires one of the most complicated sets of features ever developed by manufacturers. The industry and the *Federal Bureau of Investigation* (FBI) have been feuding for years over how specific requirements of the law should be met and who will pay for the required modifications to equipment, facilities, and services in order to permit authorized electronic surveillance.

Requested Capabilities

Under CALEA, the Department of Justice and the FBI requested that telecommunications equipment be modified to provide nine essential capabilities:

- *Content of subject-initiated conference calls*   This capability would enable a *Law-Enforcement Agency* (LEA) to access the content of conference calls that are supported by the subject's service, including the call content of parties who are on hold.

- *Party hold, join, or drop*   Messages would be sent to law-enforcement personnel who would identify the active parties of a call. Specifically, for a conference call, these messages would indicate whether a party is on hold and whether a party has joined or has been dropped from the conference call.

- *Subject-initiated dialing and signaling information*   Provides a LEA access to all dialing and signaling information that is available from the subject, informing law-enforcement personnel of a subject's use of features, such as the use of flash-hook and other feature keys

- *In-band and out-of-band signaling (notification message)*   A message would be sent to a LEA whenever a subject's service sends a tone or other network message to the subject or associate (e.g., notification that a line is ringing or busy)

- *Timing information* Information that is necessary for correlating call-identifying information with the call content of a communications interception would be sent to a LEA.

- *Surveillance status* A message that would verify that an interception is still functioning on the appropriate subject would be sent to an LEA.

- *Continuity check tone (c-tone)* An electronic signal would alert an LEA if the facility that is used for delivery of call-content interception has failed or has lost continuity.

- *Feature status* A message would affirmatively notify a LEA of any changes in features to which a subject subscribes.

- *Dialed digit extraction* Information that is sent to an LEA would include those digits that are dialed by a subject after the initial call setup is completed.

After years of negotiations between the United States Justice Department, the FBI, and the telecommunications industry regarding a plan to provide these capabilities, the FCC stepped in to sort through the issues and to devise rules for implementation. In the process, the Justice Department and the FBI got much of what they sought. The FCC, however, decided against requiring companies to give law-enforcement officials automatic notice when a wiretapped person changes features of the telephone service, such as call-waiting or voice mail. The FBI's request for a continuity check tone and feature-status messages also were denied, because they were not considered by the FCC to be necessary for meeting the mandates of CALEA.

Summary

CALEA is intended to preserve the ability of law-enforcement officials to conduct electronic surveillance effectively and efficiently in the face of rapid advances in telecommunications technology. Under the FCC's order (FCC 99-230), which has the force of law, telecommunications companies have until September 2001 to implement equipment standards that integrate the CALEA requirements. To assist the telecommunication industry with meeting its CALEA compliance obligations, Congress authorized $500 million to upgrade systems and equipment that was installed before 1995.

See Also

Federal Communications Commission (FCC)

Competitive Local Exchange Carriers (CLECs)

Competitive Local Exchange Carriers (CLECs) offer voice and data services at significantly lower prices than the ILEC, enabling residential and business users to save money on local access lines and usage charges. In addition to CLECs, regional teleports, metropolitan fiber carriers, CATV operators, electric utility companies, and *Interexchange Carriers* (IXCs) are among the types of companies that are permitted to provide local telecommunications services. Typically, these alternative-access carriers offer services in major cities where traffic volumes are greatest—and, consequently, users are the hardest hit with high local-exchange charges.

CLECs might compete in the market for local services by setting up their own networks or by reselling lines and services that are purchased from the ILEC. Most CLECs prefer to have their own networks, because the profit margins are higher than for resale. Many CLECs start out in new markets as resellers, however, which enables them to establish brand awareness and to begin building a customer base until they can assemble their own facilities-based network.

CLECs employ different technologies for competing in the local services market. Some set up their own Class 5 central office switches, enabling them to offer dial tone, and the usual voice services (including ISDN, as well as features such as caller ID and voice messaging). The larger CLECs build fiber rings to serve their metropolitan customers with high-speed data services. Some CLECs have chosen to specialize in broad-band data services by leveraging existing copper-based local loops, offering DSL services for Internet access. Others bypass the local loop entirely through the use of broad-band wireless technology, such as *Local Multipoint Distribution Service* (LMDS)—enabling them to feed customer traffic into their nationwide fiber backbone networks without the incumbent carrier's involvement.

Unlike the ILECs, which currently might not offer services outside of their *Local Access and Transport Areas* (LATAs), the CLECs are not bound by this restriction. In fact, some CLECs have nationwide fiber backbones that enable them to offer integrated voice-data services end-to-end. These backbones are used to interconnect metropolitan fiber rings and to consolidate traffic from DSL and LMDS for long-haul transport. CLECs that do not have their own fiber networks or that cannot reach all locations with their own fiber nets usually hand off this traffic to other carriers under a prearranged agreement.

Summary

More than 1,000 CLECs exist in the United States, 400 of which started operations in 1999. With the Telecommunications Act of 1996, CLECs and other types of carriers can compete in the offering of local exchange services and must have the capability to obtain the same service and feature connections as the ILECs have for themselves—on an unbundled basis. If the ILEC does not meet the requirements of a 14-point checklist to open its network in this way and in other ways, it cannot receive permission from the FCC to compete in the market for long-distance services. In the three years since the Telecommunications Act of 1996 was passed by Congress and was signed into law by President Bill Clinton, no ILEC qualified for entry into the long-distance market. This situation will undoubtedly change starting in the year 2000.

See Also

Dominant Carrier Status

Federal Communications Commission (FCC)

Interexchange Carriers (ICs)

Local Access and Transport Areas (LATAs)

Telecommunications Act of 1996

Unbundled Access

Communications Services Management (CSM)

Communications Services Management (CSM) involves the provision and management of telecommunications facilities and services within a multi-tenant office building, campus, or office park. The CSM concept began in the early 1980s when it was known as *Shared Tenant Service* (STS), intelligent buildings, smart buildings, or multi-tenant telecommunications.

Under this arrangement, a provider installs a high-capacity central PBX in an office building and offers a suite of telecommunications services to tenants on a shared basis. The PBX is usually owned by the provider or jointly by the provider and the property owner. To ensure success, a large anchor tenant usually must be persuaded to subscribe to the service.

The cost of equipment, lines, services, and maintenance are distributed among the tenants who subscribe to the service. The tenants benefit by obtaining a level of service and support that they would not normally be able to afford by themselves. Organizations find CSM appealing for a variety of other reasons, including the following:

- The convenience of one-stop shopping for all of their telecommunications needs
- Cost savings from discounted services, based on the cumulative volume of traffic from all tenants
- Time savings from having the CSM provider do all of the service and support planning and installation, as well as daily moves, adds, and changes
- Reduced operating costs, because the need for technical expertise, diagnostic tools, and a spare inventory is eliminated

Types of Services

CSM firms provide a wide variety of telecommunications services through a complex web of partnerships that include local and long-distance carriers, computer companies, integration firms, and consultants. Thus, CSM providers can offer a variety of standard and custom services that meet the needs of many different constituencies. These services typically include the following:

- Basic, local telephone service
- Discounted intrastate, interstate, and international telephone service
- Cable, DSL, and broad-band wireless services for high-speed Internet access
- Telecommunications system rental, maintenance, and upgrades
- Custom billing and management reports

- Help desk and message-center services
- Cable and wiring installation and maintenance
- Consulting, customization, and systems integration
- Directory publishing
- Network design and consulting
- LAN/WAN service and support
- Disaster planning, hot-site management, and recovery

Summary

Many companies are finding that managing their own telecommunications services is too great of a drain on resources and that it diverts their attention from core business issues, especially considering the fast-changing nature of telecommunications technologies. Businesses are drawn to office buildings that provide CSM primarily because of the discounted rates on services, the elimination of capital investments in hardware, and the freedom from day-to-day management hassles. In addition, the arrangement offers a wider range of services, features, and discounts than would normally be available to a small or mid-sized business.

See Also

Outsourcing
Smart Buildings

Computer-Telephony Integration (CTI)

In recent years, applications that rely on the integration of voice and data technologies have increased productivity and customer response. *Computer-Telephony Integration* (CTI) makes use of the advanced call-processing capabilities of digital telephone systems and the open application environment of PCs and LANs via intelligent computer-to-telephone system interfaces. The resulting benefits are especially suited to telephone-intensive environments such as customer service, technical support, and telemarketing.

In rudimentary form, CTI has been around for more than a decade. The original CTI approach was to link the PBX with a host, such as a mainframe or a minicomputer. With today's LAN capabilities, a server that has a database management system can act as the host, bringing integrated voice-data applications to every agent desktop. Not only can database records be matched with a customer call for delivery to the next available agent, but e-mail, text chat, and voice calls can be matched from Web sessions, as well.

Many of the attributes of CTI trace their lineage from the call center environment; specifically, from the following elements:

- *Automatic Call Distributor* (ACD)   Manages incoming calls in a variety of ways, including holding them in queues and parceling them out to available agents
- *Automatic Number Identification* (ANI)   Provides the system with the telephone number of the incoming call, in order to identify the caller
- *Database matching*   Provides the means to look up customer data, based on ANI, for delivery to the agent by the time the call is answered
- *Call accounting*   Entails the collection of call-related information for cost containment, internal billing, trend analysis, and agent performance

Today's concept of CTI takes this integration further, treating voice as simply another data type that can be manipulated by the user. In this integrated environment, voice is a messaging format on par with e-mail, facsimile, and even paper. Once a link to a voice-processing system or PBX is in place, you can work with voice mail and with other telephone functions much like simple desktop applications, tying them into e-mail messaging systems and creating entirely new categories of applications that are telephone-enabled.

Applications

The CTI architecture enables one or more computer applications to communicate with the telephone switch (either an ACD or PBX). Among the possible applications of CTI are the following:

- *Inbound call information* Information that is passed from the telephone network to the telephone switch, such as the caller's telephone number and the number dialed, is passed to computer applications. Applications can then identify the caller (by the calling number) and the purpose of the call (from the number dialed), which enables the application to automatically deliver caller information and data that is specific to the purpose of the call to a workstation as the telephone rings.

- *Computerized call processing* Commands that are passed from computer applications instruct the telephone system to perform call-processing functions such as making a call, answering a call, etc., which enables application-controlled call routing based on inbound call information and numbers in a computer database.

- *Outbound calling* CTI increases productivity in outbound calling environments. With automated dialing applications, agents proceed from one active call to the next. No time is wasted listening to busy signals, listening to unanswered ringing, or manually dialing numbers.

In these and other cases, computer applications use a call-processing server and APIs to originate, answer, and manipulate calls. The call-processing server interfaces with the telephone switch and invokes the required function, as requested by the client applications. The server keeps track of call-status information on the telephone-switch side and session status on the application side, making the logical association between the two.

The Role of APIs

The challenge for CTI vendors has been to come up with a standardized way of enabling developers to build and to implement integrated voice-data solutions that work across vendor domains. Their strategy has been to establish standardized *Application Programming Interfaces* (APIs) that work across vendor boundaries.

Intel and Microsoft have developed the *Telephony Application Programming Interface* (TAPI), which is intended to create a single specification for Windows application developers to use when connecting their products to the telephone network. At the LAN server level, AT&T and Novell offer their jointly developed NetWare Telephony Services API, which is intended to make NetWare the platform for this kind of integration. Sun Microsystems offers a Java-based API for developing CTI applications.

Telephone Application Programming Interface **(TAPI)**   TAPI enables custom applications to be built around inexpensive personal computers; specifically, the Windows Telephony API provides a standard development interface between PCs and myriad telephone-network APIs. TAPI is intended to insulate software developers from the underlying complexity of the telephone network and enables developers to focus entirely on the application without having to take into account the type of telephone connection: PBX, ISDN, Centrex, cellular, or *Plain-Old Telephone Service* (POTS). They can specify the features that they want to use without worrying about how the hardware is ultimately linked.

Application Classes   TAPI facilitates the development of three classes of Windows applications. The first class of applications are telephone-enabled versions of existing applications. TAPI creates standard access to telephone functions such as call initiation, call answering, call hold, and call transfer for Windows applications. TAPI addresses only the control of the call, not its content; however, the specification can be applied to any type of call, whether the call is voice, data, fax, or even video.

The second class of telephone-centric applications might embrace visual call control or telephone-based conferencing and collaborative computing. Although such applications have long been available, they previously relied on incompatible APIs. The third class of applications enables the telephone to act as an input/output device for audio data, including voice across data networks.

Application Components   An actual TAPI product implementation is comprised of three distinct components:

- The TAPI-aware application
- A TAPI *Dynamic Link Library* (DLL)
- One or more Windows drivers that interface with the telephone hardware

A TAPI application is any piece of software that makes use of the telephone system. An obvious example might be a *Personal Information Manager* (PIM), which could dial phone numbers automatically. An application becomes TAPI compliant by writing to the APIs that are defined in the TAPI specification. The TAPI DLL is another major component. The application talks to the DLL via the standard APIs. The DLL translates those API calls and controls the telephone system via the device driver. The final component is the *Service Provider Interface* (SPI), which is a driver that is unique to each TAPI hardware product. The TAPI specification supports more than one type of telephone adapter. In turn, the adapters can support more than one line.

TAPI-Enabled Features   TAPI facilitates the development of applications that enable the user to control the telephone from a Windows PC. A number of possible control features are available, including the following:

- *Visual call control*   Provides a Windows interface to common PBX functions such as call hold, call transfer, and call conferencing. Replacing difficult-to-remember dialing codes with Windows icons will make even the most complicated telephone system functions easy to implement.

- *Call filtering*   In conjunction with ANI, this function enables the user to specify the telephone numbers that can reach them. All other calls will be routed to an attendant, a message center, or a voice mailbox. Or, the call can be automatically forwarded to another extension while the user is out of the office.

- *Custom menu systems*   Enables users to build menu systems in order to help callers find the right information, agent, or department. Using the drag-and-drop technique, the menu system can be revised daily to suit changing business needs. The menu system can be interactive, enabling the caller to respond to voice prompts by dialing different numbers. A different voice message can be associated with each response. Voice messages can be created instantly via the PC's microphone.

NetWare *Telephony Services API* (TSAPI)   While TAPI defines the connection between a single phone and a PC, the NetWare TSAPI defines the connection between a networked file server and a PBX. TSAPI is the result of a joint effort by Novell and AT&T to integrate computer and telephone functions at the desktop by using a logical connection that is established over the LAN.

In connecting a NetWare server to the PBX, individual PCs are given control over telephone-system functions. TSAPI is implemented with *NetWare Loadable Modules* (NLMs) that run on Novell servers, along with another NLM that contains a PBX driver. No special hardware is required at the desktop; the PBX supports its own physical connection and uses its own software. The physical link is an ISDN *Basic Rate Interface* (BRI) card in the server that enables the connection between the NetWare server and the PBX.

NetWare Telephony Services consists of a Telephony Server NLM, a set of DLLs for the client, and a sample server application (a simple point-and-click telephone listing that is integrated with directory services). Novell also offers a driver for every major PBX. Alternatively, users can obtain a driver from their PBX vendor.

The NLM's features include drag-and-drop conference calling, the capability to put voice, facsimile, and e-mail messages in one mailbox, third-party call control, and integration between telephones and computer databases. Noteworthy among these is third-party call control, which provides the capability to control a call without being a part of the call. This feature would be used for setting up a conference call, for example.

Unlike Microsoft's TAPI, which enables only first-party call control, NetWare Telephony Services uses third-party constructs as an integral part of its system. The `Make Call` command, for example, has two parameters: one for addressing the originating party, and the other for addressing the destination party. An application that is using this command would therefore enable users to designate an address that is different from their own as the originating party and to establish a connection without becoming a participant in the call. This third-party call control also enables users to set up automatic routing schemes.

Java Telephony Application Programming Interface (JTAPI)

The platform independence of Java is being exploited for CTI applications; specifically, the *Java Telephony Application Programming Interface* (JTAPI) offers the means to build applications that will run on a variety of operating systems and hardware platforms over a variety of telephony networks.

JTAPI defines a reusable set of telephone call control objects that enables application portability across computer platforms. The scaleability of JTAPI enables it to be implemented on devices ranging from hand-held phones to desktop computers to large servers. This functionality enables enterprises to blend together Internet and telephony technology components within a single application environment as they design and deploy new business strategies for improving customer service levels, including launching their presence on the Web.

JTAPI consists of a set of Java language packages, and each package provides a specific piece of functionality for a certain aspect of computer-telephony applications. Implementations of telephony servers choose the packages they support, depending on the capabilities of their underlying platform and hardware. Applications might query for the packages that are supported by the implementation that they are currently using. Additionally, application developers might concern themselves with only the supported packages that the application needs in order to accomplish a task.

At the center of the Java Telephony API is the core package, which provides the basic framework for modeling telephone calls and utilizing rudimentary telephony features. These features include placing a telephone call, answering a telephone call, and disconnecting a telephone call. Simple telephony applications require the core package to accomplish their tasks and do not need to concern themselves with the details of other packages. For example, the core package permits applet designers to easily add telephone capabilities to a Web page.

A number of standard packages extend the JTAPI core package. Among the extension packages are those for call control, call center, media, phone, private data, and capabilities:

- *Call control* Extends the core package by providing more advanced call-control features such as placing calls on hold, transferring telephone calls, and conferencing telephone calls
- *Call center* Enables applications to perform advanced features that are necessary for managing large call centers (such as routing, automated call distribution, predictive calling, and associating application data with telephony objects)
- *Media* Enables applications to access the media streams that are associated with a telephone call. They have the capability to read and write data from these media streams. DTMF (touch-tone) and non-DTMF tone detection and generation is also provided in this package.
- *Phone* Permits applications to control the physical features of telephone sets
- *Capabilities* Enables applications to query whether certain actions can be performed. Capabilities take two forms. Static capabilities indicate whether an implementation supports a feature, whereas dynamic capabilities indicate whether a certain action is permitted— given the current state of the call model.
- *Private data* Enables applications to communicate data directly with the underlying hardware switch. This data can be used to instruct the hardware to perform a switch-specific action.

JTAPI also defines call-model objects that work together to describe telephone calls and defines the endpoints that are involved in a telephone call:

- *Provider* This object might manage a PBX that is connected to a server, a telephony/fax card in a desktop machine, or a computer-networking technology such as IP. The provider hides the service-specific aspects of the telephony subsystem and enables Java applications and applets to interact with the telephony subsystem in a device-independent manner.

- *Call* This object represents a telephone call. In a two-party call, a telephone call has one Call object and two connections. A conference call is three or more connections that are associated with one Call object.

- *Address* This object represents a telephone number.

- *Connection* This object models the communication link, which is the relationship between a Call object and an Address object.

- *Terminal* This object represents a physical device, such as a telephone and its associated properties. Each Terminal object can have one or more Address objects (telephone numbers) associated with it, as in the case of some office phones that are capable of managing multiple line appearances.

- *Terminal Connection* This object models the relationship between a Connection and the physical endpoint of a call, which is represented by the Terminal object.

Applications that are built with JTAPI use the Java sandbox model for controlling access to sensitive operations. Callers of JTAPI methods are categorized as trusted or untrusted by using criteria that is determined by the run-time system. Trusted callers are allowed full access to JTAPI functionality. Untrusted callers are limited to operations that cannot compromise the system's integrity. In addition, JTAPI can be used to access telephony servers or implementations that provide their own security mechanisms, such as a username and password.

Summary

CTI removes the barriers between telephony and other information and productivity tools, providing users with substantial gains in efficiency and information management in an easy-to-use environment. Under the CTI concept, the most appropriate pieces of technology are combined in practical applications for a more productive workplace.

See Also

> *Automatic Call Distributors* (ACDs)
>
> *Automatic Number Identification* (ANI)
>
> *Private Branch Exchange* (PBX)

Cordless Telecommunications

The familiar cordless telephone, which was introduced in the early 1980s, has become a key factor in reshaping voice communications. Because people cannot be tied to their desks, as many as 70 percent of business calls do not reach the right person on the first attempt. This situation has seen a dramatic improvement with cordless technology, which makes phones as mobile as their users. Now, almost 30 percent of business calls reach the right person on the first attempt.

Cordless versus Cellular

Although cellular phones and cordless phones are both wireless, they have assumed distinct and separate meanings based on their areas of use and the differing technologies that were developed in order to meet user requirements.

Briefly, cellular telephones are intended for off-site use. The systems are designed for a relatively low density of users. Here, macrocellular technology provides wide-area coverage and the capability to make calls while traveling at high speeds. Cordless telephones, on the other hand, are designed for users whose movements are within a well-defined area. The cordless user makes calls from a portable handset that is linked by radio signals to a fixed base station (refer to Figure C-13). The base station is connected either directly or indirectly to the public network. Cellular and cordless are implemented with their own standards-based technologies.

Cordless Standards

The cordless system standards are referred to as CT0, CT1, CT2, CT3, and DECT, with CT standing for Cordless Telecommunications.

Figure C-13
The familiar cordless
telephone that is
found in many
homes

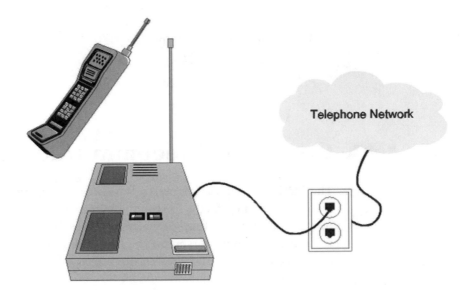

CT0 and CT1 were the technologies for first-generation analog cordless telephones. Comprising base station, charger, and handset and primarily intended for residential use, they had a range of 100 to 200 meters. They used analog radio transmission on two separate channels: one to transmit, and one to receive. The potential disadvantage of CT0 and CT1 systems is that the limited number of frequencies can result in interference between handsets, even with the relatively low density of residential subscribers.

Also targeted at the residential user, CT2 represented an improved version of CT0 and CT1. Using *Frequency Division Multiple Access* (FDMA), the CT2 system creates capacity by splitting bandwidth into radio channels in the assigned frequency domain. In the initial call setup, the handset scans the available channels and locks onto an unoccupied channel for the duration of the call. Using *Time-Division Duplexing* (TDD), the call is split into time blocks that alternate between transmitting and receiving.

The *Digital Enhanced Cordless Telecommunications* (DECT) standard started as a European standard for cordless communications, with applications that included residential telephones, the wireless PBX, and *Wireless Local Loop* (WLL) access to the public network. Primarily, DECT was designed to solve the problem of providing cordless telephones in high-density, high-traffic offices and other business environments.

CT3, on the other hand, is a technology developed by Ericsson in advance of the final agreement on the DECT standard and is designed specifically for the wireless PBX application. Because DECT is essentially based on CT3 technology, the two standards are similar. Both enable the user to make and receive calls when the user is within the range of a base station. Depending on the specific conditions, this distance amounts to a radius of between 50 meters and 250 meters from the station. To provide service throughout the site, multiple base stations are set up in order to create a picocellular network. Signal handoff between the cells is supported by one or more radio exchange units that are ultimately connected to the host PBX.

Both DECT and CT3 have been designed to cope with the highest-density telephone environments, such as city office districts in which user densities can reach 50,000 per square kilometer. A feature called *Continuous Dynamic Channel Selection* (CDCS) ensures seamless handoff between cells, which is particularly important in a picocellular environment where several handoffs might be necessary (even during a short call). The digital radio links are encrypted in order to provide absolute call privacy. The two standards are based on a multi-carrier *Time Division Multiple Access / Time Division Duplexing* (TMDA/TDD). They do not use the same operating frequencies, though, and consequently have different overall bit rates and call-carrying capacities.

The difference in frequencies governs the commercial availability of DECT and CT3 around the world. Europe is committed to implementing the DECT standard with the frequency range of 1.8GHz to 1.9GHz. Other countries, however, have made frequencies in the 800MHz to 1,000MHz band available for wireless PBXs, thereby paving the way for the introduction of CT3.

Summary

Many of the problems that arise, from the non-availability of staff to a wired PBX, can be avoided with cordless telephones. They are ideal for people who by the nature of their work can be difficult to locate (maintenance engineers, warehouse staff, messengers, etc.) and for locations on a company's premises that cannot be effectively covered by a wired PBX (warehouses, factories, refineries, exhibition halls, dispatch points, etc.).

A key advantage of cordless telecommunications is that it can simply be integrated into the corporate telecommunications system with add-on products and without the need to replace existing equipment. Another advantage of cordless telecommunications is that the amount of telephone wiring is dramatically reduced. Because companies typically spend between 10 percent and 20 percent of the original cost of their PBX on wiring the system, the use of cordless technology can have a significant impact on cost.

Considerable benefit also exists in terms of administration. For example, when moving offices, employees do not need to change their extension numbers, nor does the PBX need to be reprogrammed in order to reflect the change.

See Also

>   Cellular Communications
>   Digital Enhanced Cordless Telecommunications

Cross-Talk

In general, cross-talk is the unwanted coupling of communications paths. Cross-talk is caused when two telephone lines run physically close and parallel for a considerable distance. The result is that a conversation on one line can be heard on the other line. Shielding diminishes the effect of cross-talk, as does the installation of a load coil on the wire that is experiencing the problem. A load coil isolates the frequency band that is used for voice—300Hz to 3,300Hz—so that noise, including cross-talk, cannot disrupt communication.[1] When twisted-pair wires are used as a conduit for electrical voice and data signals, one wire in the pair carries positive voltage and the other carries negative voltage. Ideally, the voltage that is carried by the two wires should be the same above and below zero. For instance, if one wire carries +5 volts, the other should carry -5 volts. If both signals are affected by a voltage spike from another pair of wires, they should have the capability to withstand the hit if the two wires are affected equally. If the two wires are unbalanced, however, such as if one of the wires is better protected from outside voltages than the other wire, the differential will change—resulting in a cross-talk condition.

As frequencies increase, so does the potential for cross-talk. Today, the biggest cause of cross-talk in the local loop is residential DSL service, which brings high-power digital signaling to the local loop. Although voice calls are relatively unaffected by the degraded connections that are created by cross-talk—the parties at each end merely talk louder over the noise—frag-

---

[1]The installation of load coils also prevents the transmission of high-speed data that is available through DSL technologies, which use the frequency range above 3,300Hz.

ile data packets are easily corrupted. Techniques such as error detection and retransmission can mitigate some of the damage, but at a certain point, the connection might drop.

Several methods exist for dealing with cross-talk. For companies that are in control of their own cable plant, they can buy newer cables that are of much better quality than those that were made only five years ago. The newer cables are much better equipped to prevent cross-talk from occurring. An operational strategy is to measure transmitted cross-talk and to subtract it at the *Data Communications Equipment* (DCE). While effective, this method is an expensive process if applied to each line. Improvements in installation procedures and technician training also can cut down on the escaping signals that cause cross-talk.

Several types of cross-talk exist:

- *Near-End Cross Talk* (NEXT)   Cross-talk that appears at the output of a wire pair at the transmitter (near) end of the cable
- *Far-End Cross Talk* (FEXT)   A measure of the unwanted signal coupling from a transmitter at the near-end into a neighboring pair that is measured at the far-end
- *Equal-Level Far-End Cross-Talk* (ELFEXT)   A measure of the unwanted signal coupling from a transmitter at the near-end into a neighboring pair that is measured at the far-end, relative to the received signal level that is measured on that same pair
- *Power Sum Equal-Level Far-End Cross-Talk* (PSELFEXT)   A computation of the unwanted signal coupling from multiple transmitters at the near-end into a pair that is measured at the far-end, relative to the received signal level on that same pair.

Summary

Opportunities for cross-talk will always exist. To deal with this situation, many carriers are working to keep high-powered signals separated in the copper distribution network and in the copper binder groups. Other carriers have focused on a fiber deployment strategy, which has diverted their attention from the proper maintenance of their copper loops. That lack of maintenance has now resulted in a host of cross-talk problems that are associated with the deployment of DSL services.

See Also

Analog Line Impairment Testing

Line Conditioning

Custom Local Area Signaling Services (CLASS)

Custom Local Area Signaling Services (CLASS) are a collection of services that are offered to local-area residential and business customers on a pre-subscription basis. These features support both local and inter-office applications where *Signaling System 7* (SS7) is deployed. The most popular CLASS features include the following:

- *Call Return* Redials the last caller, whether the call was completed or not. The user does not have to know the number in order to return the call, because the service provides it. The service also enables the user to know whether the number was blocked, private, or out of the area. If the line is busy, the service checks for a free line for up to 30 minutes.

- *Caller ID* Displays the seven-digit or 10-digit telephone number of an incoming call. This service requires the subscriber to have a CLASS-compatible display phone or stand-alone display unit.

- *Calling Number Delivery Blocking* Enables users to block their number from being displayed on a display telephone or on a display unit

- *Call Trace* Enables the user to trace harassing or life-threatening telephone calls on demand. In extreme cases, the telephone might release traced information to appropriate law-enforcement officials. The user must file a complaint with the proper authorities and fill out a complaint form at the telephone company office. This feature usually works only with calls that are placed within the local service area.

- *Repeat Dialing* Automatically calls the last number that the user dialed, whether it was answered, unanswered, or busy. The service can redial a busy number for up to 30 minutes. When the line is free, the subscriber hears a distinctive ring, indicating that the call can now be established.

- *Preferred Call Forwarding* Enables the user to forward only selected calls to a special number and enables the user to store up to six numbers. When incoming calls originate from one of those numbers, those calls are forwarded to the desired number.

- *Call Block* Enables the user to reject calls from selected numbers and enables the user to store up to six numbers. When incoming calls originate from one of those numbers, the user's phone will not even ring. Callers hear a recording that indicates that the user is not accepting calls.

- *Call Selector* Enables the user to choose numbers that will ring distinctively. The user can store up to six numbers. When incoming calls originate from one of those numbers, the user's phone rings distinctively.

- *Wake-Up Service* Enables the user to program the phone to ring at a certain time

Summary

Other custom calling features are available, and more features are continually being added. In most cases, CLASS services entail an extra charge of between $2.50 and $5 per month and are itemized on the monthly phone bill. Periodically, many telephone companies run promotions for new calling features, giving their customers 30 days to try the services at no charge. Unless the customer specifically cancels the service, however, the customer will be billed automatically after the trial period.

See Also

> *Advanced Intelligent Network* (AIN)
> *Signaling System 7* (SS7)

CHAPTER **D**

Data Compression

Data compression has become a standard feature of most bridges and routers, as well as modems. In its simplest implementation, compression capitalizes on the redundancies that are found in the data. The algorithm detects repeating characters or strings of characters and represents them as a symbol or token. At the receiving end, the process works in reverse in order to restore the original data.

The compression ratio tends to differ by application. The compression ratio can be as high as 6-to-1 when the traffic consists of heavy-duty file transfers. The compression ratio is less than 4-to-1 when the traffic consists of mostly database queries. When there are only keep-alive signals or sporadic query traffic on a T1 line, the compression ratio can dip below 2-to-1.

Encrypted data exhibits little or no compression because the encryption process expands the data and uses more bandwidth. If data expansion is detected and compression is withheld until the encrypted data is completely transmitted, however, the need for more bandwidth can be avoided. The use of data compression is particularly advantageous in the following situations:

- When data traffic is increasing due to the addition or expansion of LANs and associated data-intensive, bursty traffic
- When LAN and legacy traffic are contending for the same limited bandwidth
- When reducing or limiting the number of 56Kbps/64Kbps lines is desirable in order to reduce operational costs
- When lowering the *Committed Information Rate* (CIR) for Frame Relay services or sending fewer packets over an X.25 network can result in substantial cost savings

The greatest cost savings from data compression most often occurs at remote sites, where bandwidth is typically in short supply. Data compression can extend the life of 56Kbps/64Kbps leased lines, thus avoiding the need for more expensive fractional T1 lines or Nx64 services. Depending on the application, a 56Kbps/64Kbps leased line can deliver 112Kbps to 256Kbps or higher throughput when data compression is applied.

Types of Data Compression

Several different data compression methods are in use today over WANs—among which are TCP/IP header compression, link compression, and multi-channel payload compression. Depending on the method used, there can be a significant tradeoff between lower bandwidth consumption and increased packet delay.

TCP/IP Header Compression   With TCP/IP header compression, the packet headers are compressed—but the data payload remains unchanged. Because the TCP/IP header must be replaced at each node for IP routing to be possible, this compression method requires hop-by-hop compression and decompression processing. This method adds delay to each compressed/decompressed packet and puts an added burden on the router's CPU.

Link Compression   With link compression, the entire frame—both protocol header and payload—are compressed. This form of compression is typically used in LAN-only or legacy-only environments. This method, however, requires error-correction and packet-sequencing software, which adds to the processing overhead that is already introduced by link compression, and results in increased packet delays. Also, similar to TCP/IP header compression, link compression requires hop-to-hop compression and decompression, so processor loading and packet delays occur at each router node that the data traverses.

With link compression, a single data-compression vocabulary dictionary or history buffer is maintained for all virtual circuits that are compressed over the WAN link. This buffer holds a running history of the data that has been transmitted in order to help make future transmissions more efficient. To obtain optimal compression ratios, the history buffer must be large, requiring a significant amount of memory. The vocabulary dictionary resets at the end of each frame. This technique offers lower compression ratios than multi-channel, multi-history buffer (vocabularies) data compression methods. This situation is particularly true when transmitting mixed LAN and serial protocol traffic over the WAN link and when frame sizes are 2Kb or less. This situation translates into higher costs, but if more memory is added in order to achieve better ratios, this method increases the up-front cost of the solution.

Mixed-Channel Payload Data Compression   By using separate history buffers or vocabularies for each virtual circuit, multi-channel payload data compression can yield higher compression ratios that require much less memory than other data compression methods. This statement is particularly true in cases where mixed LAN and serial protocol traffic traverses the network. Higher compression ratios translate into lower WAN bandwidth requirements and greater cost savings.

Performance varies, however, because different vendors define payload data compression differently. Some consider it to be the compression of everything that follows the IP header; however, the IP header can be a significant number of bytes. For overall compression to be effective, header compression must be applied, which adds to the processing burden of the CPU and increases packet delays.

External Data Compression Solutions

Although bridges and routers can perform data compression, external compression devices are often required to connect to higher-speed links. The reason is because data compression is extremely processor intensive, with multi-channel payload data compression being the most burdensome. The faster the packets must move through the router, the more difficult it is for the router's processor to keep up.

The advantages of internal data-compression engines are that they can provide multi-channel compression, a lower cost, and simplified management. By using a separate internal *Digital Signal Processor* (DSP) for data compression, however, instead of the software-only approach, all of the other basic functions within the router can continue to be processed simultaneously. This parallel processing approach minimizes the packet delay that can occur when the router's CPU is forced to handle all of these tasks by itself.

Summary

Data compression will become increasingly important to most organizations as the volume of data traffic at branch locations begins to exceed the capacity of the wide-area links. Multi-channel payload solutions provide the highest compression ratios and reduce the number of packets that are transmitted across the network. Reducing packet latency can be effectively achieved via a dedicated processor such as a DSP and by employing end-to-end compression techniques, rather than node-to-node compression/decompression. All of these factors contribute to reducing WAN circuit and equipment costs, as well as to improving the network response time and availability for user applications.

See Also

Voice Compression

Data Service Units (DSU)

The *Data Service Unit* (DSU) is a type of *Data Communications Equipment* (DCE) that connects various *Data Terminal Equipment* (DTE) with carrier-provided digital services. The DSU comes in stand-alone, rack-mounted, or router-integrated versions that connect various DTE via RS-232, RS-449, or V.35 interfaces with widely available digital services that offer 56Kbps/64Kbps access, including *Digital Data Service* (DDS) and cell-based/frame-based services such as ATM, SMDS, and Frame Relay. Typical applications for the DSU include LAN interconnection, dedicated Internet access, and dedicated remote PC access to local hosts.

DDS operates at speeds of 1.2Kbps to 56Kbps/64Kbps and supports point-to-point or multi-point applications. Most DSUs have a built-in asynchronous-to-synchronous converter, which accommodates asynchronous input devices that operate at speeds of 1.2Kbps to 57.6Kbps—as well as synchronous input devices that operate at either 56Kbps or 64Kbps. When packaged with *Channel Service Unit* (CSU) functions, the CSU/DSU device interfaces with T1 services at 64Kbps and with N × 64Kbps up to 1.536Kbps.

The CSU component protects the T1 line and the user equipment from lightning strikes and from other types of electrical interference. The CSU also provides a keep-alive signal during periods in which no user data traverses the line. The CSU component also offers capabilities for carrier-initiated loop-back testing (refer to Figure D-1) and provides storage for accumulating performance statistics.

Basic Functions

The DSU supplies timing to each user port, takes the incoming user data signals (e.g., RS-449, RS-232, or V.35), and converts them into the form that

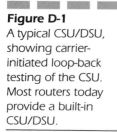

Figure D-1
A typical CSU/DSU, showing carrier-initiated loop-back testing of the CSU. Most routers today provide a built-in CSU/DSU.

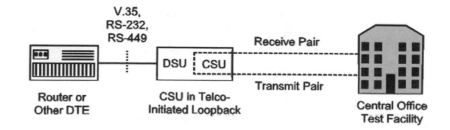

V.35,
RS-232,
RS-449

Receive Pair

DSU : CSU

Transmit Pair

Router or
Other DTE

CSU in Telco-
Initiated Loopback

Central Office
Test Facility

is needed for transmission over the carrier-provided line. This conversion transforms the input signal into the specified line code and framing format. Any device that connects directly to a digital line (via an external or internal CSU) must perform these functions; otherwise, the device needs a DSU. Any piece of network equipment that does not have a bipolar port needs a DSU in order to connect to a CSU. The most common type of access device is the combination unit, which offers DSU and CSU functionality and eliminates these concerns (in addition to reducing the number of devices that must be managed).

The DSU's front panel provides a set of LED indicators that show the status of the DTE interface, various test modes, and the loop status. These LEDs respond to standard loop-back commands from the service provider or from the user side. Included with the remote loop-back capability are selectable bit-error rate test patterns, the results of which are displayed on the front panel.

For users who do not want to perform testing through the diagnostic and loop-back test capabilities that are built into many of today's CSU/DSUs, carrier-provided test services are available. The test procedure is initiated by placing a call to the local carrier's STC to report a problem. Test center technicians troubleshoot the customer's entire network, all the way down to the CSU/DSUs. Such tests are disruptive, however, because they are conducted on an in-band basis; that is, the test signals replace the user's production data. To lessen the impact, users can schedule circuit testing for off-peak hours. Despite this situation, many smaller organizations that have neither the diagnostic test equipment nor the technical expertise that is required to perform their own tests have come to rely on the local carrier's test service for maintaining their networks.

Management

Similar to CSUs, DSUs can be managed by either the vendor's proprietary network-management system or by *Simple Network Management Protocol* (SNMP) tools. The DSU provides a non-disruptive, in-band SNMP management channel over a DDS-leased circuit. For Frame Relay connections, SNMP management is provided with a connection from the DSU's management port to a router port or to a external LAN adapter at the remote site. In this case, management information is sent across the network as framed data and is then delivered to the DSU by the remote router. The router forwards the management information to the DSU through an external connection to the management port on the DSU.

Simple versions of DSUs that are not designed to be managed are now available. They will automatically install on any standard 56Kbps/64Kbps facility and will provide the correct data rate to the DTE port. Remote control and testing of the distant DSU is accomplished via an external modem connection to the ASCII interface.

Modular routers, such as Cisco Systems' 1600 and 3600 series routers, can be equipped with a WAN interface card that incorporates a fully managed CSU/DSU to facilitate the deployment and management of Internet and intranet connectivity. This integrated solution eliminates the need for external CSU/DSUs and enables all of these components, including the router, to be managed both locally and remotely as a single entity via SNMP or as a Telnet session that uses the Cisco command-line interface.

Summary

DSUs, as well as CSUs, not only provide an interface between DTE and the carrier's network, but such devices also help network managers fine-tune their networks for performance and cost savings. For example, when using the diagnostic capabilities of a CSU/DSU that is connected to a Frame Relay network, the network manager can monitor traffic on each *Permanent Virtual Connection* (PVC) in order to set an appropriate *Committed Information Rate* (CIR) and to set a permitted burst rate for each circuit. In addition, the delay between network nodes can be measured—as well as the performance of a line between the user and local carrier—to determine whether the carrier is actually delivering the level of service that has been promised.

See Also

  Channel Service Units (CSUs)
  Digital Data Services (DDS)

Data Switches

Data switches have been in mainframe computer environments since their introduction in 1972 and have been the forerunners of today's hubs and LAN switches. Through port selection or port contention, data switches enable a larger number of users to share a limited number of host ports. Data switches also perform the necessary protocol conversions that enable

Figure D-2
PC-to-mainframe
connectivity (with
partitioning) that is
implemented by a
data switch

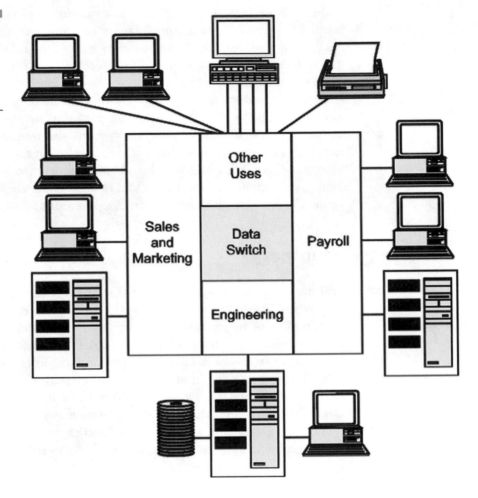

PCs to communicate with the mainframe. They have evolved to become an economical means of controlling access to the mainframe via partitioning (refer to Figure D-2) and via other security features. Typically, these switches support data transmission speeds of up to 19.2Kbps for asynchronous data and 64Kbps for synchronous data.

Connectivity

As the architecture of data switches became more modular, world-wide computer ports grew incrementally (into the thousands). Today, the data switch can also serve as a LAN server or as a gateway to packet and T1 networks for remote host access.

PCs and other devices can be connected to the data switch in a variety of ways, including direct connection with EIA-232C cabling for distances of up to 50 feet. For longer distances of one to two miles (between buildings, for example), PCs and peripherals can be connected to the data switch via line drivers or local multiplexers. Some data switch vendors have integrated these functions into their data switches in the form of optional plug-in cards. Connections between PCs and the data switch are usually accomplished via the extra twisted pairs of telephone wire that are already in place in most office environments. Remote data switches can be accessed via dedicated or dialup phone lines.

When a user enters the connect command and personal password, the data switch will attempt to complete the requested connection. If the requested port is busy, the user is placed into the queue and is notified of changes in position with a screen message. Some users might be assigned a higher priority in the queue than others. When a high-priority user attempts to access a busy port, other waiting users are bumped and the priority user assumes the first place in the queue. All other contenders for that port are then notified of their new status.

Features

Not only do some data switches permit different configurations to be loaded, but they can also automatically implement one or the other on a scheduled basis. A late-night crew of CAD/CAM professionals can plod along with cumbersome design work on high-performance workstations, for example, and can be restricted to accessing the mainframe from 6 p.m. to 8 a.m. The class-of-service feature keeps these users from being able to access sensitive financial information that might also be stored in the mainframe. When the workday begins at 8 a.m., the time-of-day clock changes the data switch back to the primary configuration for general access (but without interrupting existing connections). This setup enhances overall network performance in that available resources can be reallocated during the day for optimal usage, based on the varying needs of different classes of users. Various alternative configurations can be stored on disk and can be implemented by an authorized PC with only a few keystrokes.

A session-toggling feature enhances operator productivity by permitting two connections: primary and secondary. The operator can toggle back and forth between two host sessions on different links in order to perform multiple tasks simultaneously. For example, a batch file transfer can be in progress over the secondary link while a real-time database search is performed over the primary link. When the batch file transfer is complete, the primary link can be put on hold while another file transfer is initiated over the secondary link.

The data switch automatically adapts to different transmission rates. The network administrator does not have to match terminals with computer ports; rather, each computer port can be set at its highest rate. The data switch uses a buffer to perform the rate conversion for any device that communicates with another device at a faster or slower rate. In other words, users do not have to be concerned about speed, and network administrators do not have to waste time changing the transmission speeds of computer ports in order to accommodate lower-speed devices. Thus, a computer port that is set at 19.2Kbps can send data to a much slower device. For reliable data-rate conversion, however, the connecting devices must be capable of flow control; otherwise, data could be lost.

When XON/XOFF is used for flow control, the switch buffer will be prevented from overflowing. When the buffer is in danger of overflowing, an XOFF signal is sent to the computer, telling it to suspend transmission. When the buffer clears, an XON signal is sent to the computer, telling it to resume transmission. These settings are also used for reformatting character structures, enabling devices of different manufacturers to communicate with each other through the data switch.

Administration

Instead of being confined to one terminal, many data switches enable the network administrator to log on to the computer from any terminal. Once a connection is established, a variety of functions can be invoked in order to enhance operating efficiency.

A broadcast feature enables the network administrator to transmit messages to individual users, with delivery controlled by the time-of-day clock. The same message can be sent every hour, or a different message can be sent to different users simultaneously. A special link feature enables the network administrator to make permanent connections between a terminal and a port. Sometimes called nail-up, this feature enables continuous access to certain devices, such as printers.

With the force disconnect feature, the network administrator can disconnect any port at any time, for any reason. Open files are even closed automatically with the proper disconnect sequences, including any required control characters. This capability is also found in a timeout feature, which enhances system efficiency by automatically disconnecting idle ports after a predefined period of inactivity.

When the data switch is equipped with a logging port, the network administrator can obtain a complete record of port connects and disconnects to aid in maintaining security. When used with the optional security call-back, this feature provides a precise audit trail of connections as well as a description of the users who made them. In addition, all alarms can be logged for output to a PC or to a designated port on the host, in order to aid in network analysis and management.

The network administrator might decide to group ports with similar characteristics and capabilities under a class name to permit a more efficient utilization of shared resources. Password-protected classes can also be used to restrict access and to enhance system security. The class of service might designate the use of particular host ports for general access, while other classes might designate a high-speed printer, a modem pool, or a gateway to a packet network.

Summary

Data switches were advocated as an economical alternative to LANs when host connectivity needs were relatively simple. They also made a suitable LAN server when networking needs grew more complex. Gateway functions could be added to data switches through plug-in cards that provided the appropriate protocol converters. This functionality permitted asynchronous or synchronous terminals from one vendor to communicate with the mainframes and network nodes that were provided by other vendors, in addition to permitting the data switch to connect to various types of LANs. Although data switches are still available, advances in hub technology (and more recently, LAN switches) have overtaken the switches to the point that they are rarely considered for new data centers.

See Also

Hubs

LAN Switches

Data Warehouses

A data warehouse is an extension of the *Database Management System* (DBMS), which consolidates information from various sources into a high-level, integrated form that is used to identify trends and make business decisions. For a large company, the amount of information in a data warehouse could be several trillion bytes or terabytes (TB). The technologies that are used to build data warehouses include relational databases, powerful, scaleable processors, and sophisticated tools to manipulate and analyze large volumes of data for the purpose of identifying previously undetectable patterns and relationships. Benefits include increased revenue and decreased costs due to the more effective handling of great volumes of data.

In addition, the more effective handling of corporate data (identifying patterns of repeat purchases, the frequency and manner with which customers use a company's products, and their propensity to switch vendors when they are offered better prices or more targeted features) can improve customer satisfaction and cement customer loyalty. A change of only a few percentage points of customer retention can equate to hundreds of millions of dollars to a large company.

System Components

Data-warehousing frameworks typically consist of four elements: data that resides in one or more database systems, software to translate data, connectivity software to transfer data between databases and platforms, and end-user query tools.

Key system components include an information store of historical events (the data warehouse), warehouse administration tools, data manipulation tools, and DSSs that enable the strategic analysis of the information. The effectiveness of the data warehouse architecture depends on how well it addresses the issues that are related to each of the following major components:

■ *Warehouse Population*   A central repository of warehouse data is obtained by downloading and consolidating data from various operational systems. The data consists of historical (transaction-based) events and related information that is needed in order to isolate and aggregate those events. Volume tends to be high, so performance and cost are key considerations for warehouse and operational data sources.

- *Warehouse Volume* The data in a large warehouse can be made more accessible by arranging the information into data marts, or specialized subsets of the data warehouse. Days might be necessary for a query to run through a multi-terabyte data warehouse. Data marts emerged to improve system and network performance, because it is not always necessary for everyone in the organization to have direct access to all of the data in the warehouse.

- *Warehouse Administration* This component focuses on maintaining the metadata (data about data) that provides analytical derivation, exception recognition, integrity, controls, and security. Metadata, which resides above the warehouse data, defines the rules and content of the views that are provided from the entire domain of available information. Metadata also maps user queries to the operational data sources that are needed to satisfy the request.

- *Operational Data Store* The ODS draws its data from the various operational systems in the corporation, but it also can add information that is derived from data keys in a data mart. As such, it can not only add the value of consolidating a common view of enterprise data, but it can also add data derived from trend analyses that are performed on the data marts.

Data Types

A comprehensive warehouse contains all or most of the following data types:

- Historical events downloaded from operational systems, such as invoice transactions, claims, and payments. The level of granularity is determined during the design phase of the data-warehouse architecture.

- Master entities referenced by event, such as customer, product, vendor, and patient

- Master entity hierarchical rollups, such as customer, store, district, region, and country

- Summarization of historical events, such as monthly sales by store, district, and region. These elements constitute preemptive queries. Data is not real-time; rather, it aggregates when periodically added to the warehouse. In this way, data that is requested by a user is always accurate. In many systems, critical data is maintained in both a granular state and with one or more levels of summarization for rapid access.

- Miscellaneous domain data, such as codes, flags, validation and translation data. Domain data is important for assigning the correct attributes to data fields and for correlating different codes that are used on separate operational systems.

Another data type is metadata, which contains information that is needed to map the warehouse to DSS views, operational sources, and related data in the warehouse. Metadata determines the analytical strength of the warehouse in terms of overall flexibility, extensibility, and adaptability. Examples of metadata include the source of warehouse information, data-aggregation methods and rules, purge and retention periods, replication and distribution rules, the history of extracts, and event data from outside sources (such as subscription databases).

Decision Support System (DSS)

The DSS supports managerial decisions and provides a user interface and tools for the heuristic analysis of large amounts of data. The DSS works with metadata in order to offer a flexible, responsive, interactive, intuitive, and easy-to-use method of constructing and executing warehouse queries. DSS enables users to develop a hypothesis, to test its validity, and to create queries based on the validated hypothesis. Among the key characteristics of a DSS are the following:

- Automatic monitoring capability, to control runaway queries
- Flexibility to transparently request data from a central, local, or desktop database
- Data-staging capabilities for temporary data stores, which improves system performance for conversational access
- Drill-down capabilities to access exception data at lower levels
- Import capabilities to translators, filters, and other desktop tools
- A scrubber, to merge redundant data, resolve conflicting data, and integrate data from other systems
- Usage statistics, including response times

A key capability of the DSS is data mining, which uses sophisticated tools to detect trends, patterns, and correlations that are hidden in vast amounts of data. Information discoveries are presented to the user and provide the basis for strategic decisions and action plans that can improve corporate financial performance.

Multi-Dimensional Analysis

Online Analytical Processing (OLAP) is a sophisticated form of DSS that provides business intelligence through the multi-dimensional analysis of information that is stored in relational and other tabular (two-dimensional) databases. Users can interrogate data warehouses dynamically and intuitively by slicing, dicing, and otherwise carving up the cube in order to answer complex, real-world business questions such as, "Which products sell best in each region?", or "How can inventory be reduced to free working capital?". The cubes themselves are optimized for efficient calculation via built-in algorithms and programmable business rules.

Summary

Data warehouses are reaching the terabyte level because businesses and government agencies are not only collecting more data, but they are keeping the data longer for the purpose of analyzing trends. Raw data that is pulled from transaction-processing systems and other sources, however, generally accounts for a fraction of the overall size of warehouses. Indexes, temporary files, backup capacity, mirrored data, and workspace make up the rest. Furthermore, as databases grow, modeling the database, loading it with data, scrubbing the data, and creating indexes becomes more complicated and time consuming. Performance can bog down as ad hoc queries plow through the data warehouse.

Moving to a distributed architecture by deploying data marts can make large data warehouses more manageable, but the disadvantage of this approach is that it becomes harder to build a central repository of corporate information (if necessary for the business). A distributed, logical warehouse is another option. The database can be segmented in order to achieve greater efficiency—and, in the process, to provide redundancy and load balancing to guard against data loss and to improve overall performance.

See Also

Storage Media

Decibel

The decibel (dB) is a unit of measurement expressing the relationship of some reference point and another point that is above or below that reference point. The base reference point is 0dB, and subsequent measurements are relative to that reference point. When used as an absolute measure, it must be either a ratio comparing it to another measured or measurable value (e.g. dBi, a ratio with isotropic radiation). The decibel can also be related to some fixed value (e.g. dBm, which is related to the fixed standard, the milliwatt). Some of the more useful dB variations are as follows:

- dBi is the antenna gain in dB, relative to an isotropic source.
- dBm is the power in dB, relative to one milliwatt.
- dBW is the power in dB, relative to one watt.
- dBmV is referenced to one millivolt and is often used as a measure of signal levels (or noise) on a network.

When converting millivolts to dBmV, every 6dBmV doubles the voltage. For example, one millivolt is equal to 0dBmV. Doubling that to 2 millivolts is +6dBmV, and doubling it again to 4 millivolts is +12dBmV. Converting fractions of a millivolt to decibels uses the same principle of notation. For example, 0.5 millivolts converts to -6dBmV, while 0.25 millivolts converts to -12dBmV. The dB scale related to power is different from the dB scale relating to voltage. In power measurements, the power level doubles every 3dB, instead of every 6dB as in voltage. The dB scale related to audio output is different from the dB scales relating to voltage and power.

Human Hearing

Because the range of audio intensities that the human ear can detect is so large, the scale that is frequently used to measure intensity is based on multiples of 10. This type of scale is sometimes referred to as a logarithmic scale. The threshold of hearing is assigned a sound level of zero decibels. A sound that is 10 times more intense is assigned a sound level of 10dB. A sound that is 10 times more intense (10×10) is assigned a sound level of 20dB. A sound that is 10 times more intense ($10 \times 10 \times 10$) is assigned a sound level of 30dB. A sound that is 10 times more intense ($10 \times 10 \times 10 \times 10$) is assigned a sound level of 40dB. The range of frequencies that the human ear can detect is between 20 Hertz and 20,000 Hertz (20KHz). The following table lists some common and some not-so-common sounds, with an estimate of their intensity and decibel level.

Source	Number of Times Greater Than the Threshold of Hearing (TOH)	Decibels
Threshold of Hearing (TOH)	10^0	0dB
Rustling leaves	10^1	10dB
Whisper	10^2	20dB
Normal conversation	10^6	60dB
Busy street traffic	10^7	70dB
Vacuum cleaner	10^8	80dB
Heavy truck traffic	10^9	90dB
A walkman at maximum level	10^{10}	100dB
Power tools	10^{11}	110dB
Threshold of pain to the ear	10^{12}	120dB
Airport runway	10^{13}	130dB
Sonic boom	10^{14}	140dB
Perforation of eardrum	10^{16}	160dB
12 feet from a battleship cannon muzzle	10^{20}	220dB

Summary

A variety of test instruments are available to handle virtually any noise measurement requirement, including analog impulse meters to measure quick bursts of sound. These devices typically have output jacks for connections to charting devices, which plot continuous noise levels across a roll of paper. Digital devices output measurements to *Light-Emitting Diode* (LED) screens. Band filters enable the selection of narrow frequency ranges in order to isolate specific noises for measurement. Optional calibrators are available for in-field adjustments.

See Also

>   Attenuation
>
>   Hertz
>
>   Jitter

Dialing Parity

In today's competitive telecommunications environment, it is common for customers to switch service providers in order to get the best deal on pricing. In the past, many customers were discouraged from changing their carrier because it meant having to dial extra digits. This situation was advantageous for the *Incumbent Local Exchange Carriers* (ILECs), which did not require customers to dial an access code. Today, however, the ILECs must provide dialing parity to competing providers of local toll and long-distance services. In other words, subscribers no longer have to dial an access code (e.g., 10-ATT) in order to place a local toll call.

Instead of using the local telephone company for local toll calls, customers can presubscribe to an alternative carrier to handle them. With presubscription, there is no access code to dial or special equipment to buy. Phone numbers are dialed in the ordinary manner. Presubscription is actually the result of deregulation at the state level. In states where local toll presubscription has not been ordered by state authorities, it will occur simultaneously with the ILECs' eventual entry into the long-distance market.

Summary

The Telecommunications Act of 1996, among other things, lists the interconnection obligations (Section 251) of all LECs. Among these obligations, each LEC "has the duty to provide dialing parity to competing providers of telephone exchange service and telephone toll service, and the duty to permit all such providers to have nondiscriminatory access to telephone numbers, operator services, directory assistance, and directory listing, with no unreasonable dialing delays."

See Also

>   Access Charges
>
>   Dialing Parity

Digital Cross-Connect Systems (DCS)

In 1981, AT&T developed a T-carrier device called the *Digital Cross-Connect System* (DCS) in order to automate the entire process of circuit provisioning. Instead of having a technician manually patch access lines to long-haul transport facilities, for example, the DCS enabled customer-ordered circuits to be set up between two points from a remote location via a keyboard command, thereby expediting circuit setup.

AT&T started offering these capabilities to its Accunet T1.5 subscribers in 1985. Instead of waiting for service orders to be processed before they could rearrange their networks, users were given the capability via a tariffed service called *Customer-Controlled Reconfiguration* (CCR).

With a terminal that is connected to AT&T's central control system, customers can reconfigure their networks without carrier involvement. Because the reconfigurations are software-defined, they can be implemented in a matter of minutes. Most long-distance and local telephone companies now offer this capability as a service. Corporations also can implement this capability via cross-connect systems that are installed on their private networks.

Central Control System

The reconfiguration capabilities of cross-connect systems are useful for instituting disaster-recovery procedures, bypassing critical systems so that scheduled preventive maintenance can be performed, meeting peak traffic demand, and implementing temporary applications such as videoconferencing or collaborative computing.

The network manager issues reconfiguration instructions via a dialup or dedicated connection from an on-premises terminal to the carrier's central control system. The central control facility holds network routing maps for each subscriber. With the maps held in memory, each subscriber can invoke predefined, alternate configurations with only a few keystrokes from the on-premises control terminal. At periodic intervals, the carrier's control system distributes these configurations to the various cross-connect systems that will be involved with implementing the routing changes.

Several customers can share the cross-connect system, enabling the telephone company to offer VPNs to small companies—many of whom appreciate the benefits of private networks but still cannot afford to implement multi-node configurations of their own. Hubbing off the carrier's cross-connect system makes a viable alternative to setting up a separate network, because the arrangement can provide all of the control and flexibility of a private network but without the huge capital investment in equipment and the risk of obsolescence. Access to management features is password protected in order to prevent one customer from interfering with the network of another customer. The carrier is responsible for circuit testing and system maintenance.

Drop and Insert

Aside from customer-controlled reconfiguration, the key feature of DCS is drop and insert. This term refers to the capability of the DCS to exchange channels from one facility to another, with the purpose of routing the traffic appropriately, rerouting traffic around failed facilities, or increasing the efficiency of all of the available digital facilities. This drop-and-insert capability is also offered by some T1 and T3 multiplexers.

The most sophisticated cross-connects provide three levels of switching: DS3, DS1, and DS0. At the DS3 level, a battery of 28 T1 facilities can be switched. At the DS1 level, the entire composite of 24 channels (DS0s) can be switched from one T1 facility to another. At the DS0 level, individual 64Kbps channels can be switched from one DS1 stream to another, and other channels can be inserted in their place. In other words, while one or more DS0s can be dropped at an intermediate location, others can be inserted into the bit stream at that time for transmission to another location (refer to Figure D-3).

Some cross-connect systems can even perform switching at the sub-DS0 level, enabling individual subchannels operating at 2.4Kbps, 4.8Kbps, or 9.6Kbps to be bundled into one 56Kbps/64Kbps channel and then to be separated at another cross-connect for individual routing to their respective destinations.

Next Generation

A more flexible type of DCS has been available from Lucent Technologies since 1995. The company's DACS II is a software-based, multiprocessor-controlled, fully non-blocking cross-connect and test-access system that

Figure D-3
Rerouting channels
in a DCS-based
network via a
management
terminal under CCR

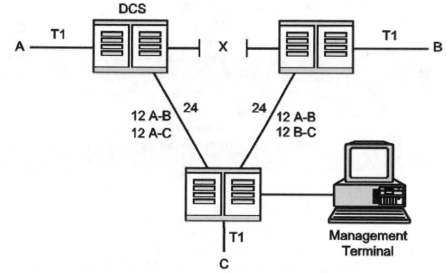

Figure D-3
Rerouting channels in a DCS-based network via a management terminal under CCR

supports a number of wireline and wireless service provider applications for business overlay, transport, and private networks.

When installed at the network's edge, DCS remotely grooms traffic before it reaches a *Point of Presence* (POP) or hub central office. This setup enables network service providers to expedite service to their customers while making the most efficient use of outside plant facilities. When deployed in wireless networks, the DACS II provides high density in wireless hubs or cell sites and end offices, meeting the demands of wireless providers for a compact, cost-effective, and reliable cross-connect solution.

The DACS II can scale from DS-0 to DS-3, offering the capabilities of a much larger system at a lower cost (and with a smaller footprint). The product also integrates test functionality and supports up to 88 DS1 or E1 interfaces. A software-based add-on computer system can be provisioned for remote monitoring. Placed at either the core of the network or at the edge, the DACS II is a flexible solution for POP, colocated, and customer-premise environments.

Summary

Although the DCS is more accurate and flexible than a manual patch panel, it is not as well suited for setting up calls or handling real-time disaster recovery because of the time that it takes to establish the desired pathways. For these functions, a channel bank or a T1 multiplexer is required. Also, unlike the central office switch that sets up, supervises, and tears down communication paths every time a call is placed, a cross-connect system keeps communication paths in place for continuous use over a period of months or even years. But for networks that have constantly changing needs, circuits can be added, deleted, or rearranged as demands warrant. This capability can result in tremendous cost savings, because it eliminates the tendency to over-order facilities to meet any contingency and eliminates the delay in going through the carrier for ordering circuit reconfigurations.

See Also

Channel Banks

Multiplexers

Digital Data Services (DDS)

AT&T pioneered digital communications with the introduction of its *Dataphone Digital Service* (DDS) in the early 1970s. The acronym DDS has come to refer to either the transport method Digital Data Service or the brand name of the AT&T service itself. Until 1984, when T1 facilities were first tariffed, 56Kbps DDS facilities were the fastest digital systems that were commercially available. DDS is suitable for the simultaneous two-way transmission of data at synchronous speeds of 2.4Kbps, 4.8Kbps, 9.6Kbps, or 56Kbps. Later, 19.2Kbps and 64Kbps speeds were added to DDS, along with an optional secondary channel for low-speed data applications (such as diagnostics).

DDS is available from the AT&T central offices that are listed in Tariff FCC No. 10; however, DDS might not be available in every LATA or in every wire center area within a LATA. DDS is configured by combining components to connect two or more AT&T central offices. Data-terminal equipment interconnects via DDS by means of standard RS-232C or V.35 interfaces. Both the CSU and DSU are required on the front end of the digital line. Most new DDS circuits are terminated in an integrated CSU/DSU device. DDS can be connected to a switched digital service, in which case the appropriate office connection is also required.

Service Provisioning

DDS is a private-line, inter-office service. As such, DDS can be provisioned as a hub-based service and is available in point-to-point and multi-point synchronous configurations. With hub-oriented services, the user's traffic must be routed through a series of special hub offices. Because there are fewer DDS hubs nationwide (AT&T has about 100 hubs), longer circuit mileage is typical, which inflates the cost of the service. The average back-haul distance with DDS is 60 miles, compared with more economical *Generic Digital Services* (GDS) that are implemented from *Serving Wire Centers* (SWCs). Because there are about 20,000 SWCs among all carriers nationwide, the average back-haul distance for GDS is only six miles. The advantage of DDS is that it offers a higher-quality transmission and less down time than GDS.

AT&T issues credits to customers as compensation for service interruptions. DDS is considered to be inoperative when there has been a loss of continuity or when the error performance is below the design objective of 99.5 percent error-free seconds (measured over a continuous 24-hour period).

Secondary Channel (SC)

Optionally, *DDS with a Secondary Channel* (DDS/SC) can be provided to the customer. Because this secondary channel operates at a relatively low bit rate, it is typically used by customers to measure the end-to-end error-rate performance of the primary channel. Vendors of DDS equipment often use the secondary channel to provide remote unit configuration and monitoring capabilities. These capabilities include host control over remote DSU optioning, surveillance of the remote DSU/terminal interface, remote alarming, performance testing of the network and equipment, and reporting of reference information that resides in the firmware of each remote CSU/DSU. The following table summarizes the primary and secondary channel bit rates:

Primary Channel	Secondary Channel
2.4Kbps	133 bps
4.8Kbps	266 bps
9.6Kbps	533 bps
19.2Kbps	1,066 bps
56Kbps	2,132 bps
64Kbps	Not available

User-Controlled Diagnostics

In providing a completely independent, low-speed (auxiliary) data channel, DDS/SC enables network surveillance to be performed on a continuous and non-disruptive basis over the secondary channel (refer to Figure D-4)—without interfering with the production data, which continues to flow over the primary channel.

While there are CSU/DSUs that provide a virtual secondary channel on conventional DDS circuits (by multiplexing test data along with user data), these devices do not provide the testing flexibility of DDS/SC. Because the primary and secondary channels that are derived with this approach are not truly independent, continuous, real-time tests cannot be performed without impacting production data.

Summary

Initially, DDS met the emerging need for a high-speed, high-quality private-line service. The addition of a secondary channel gave users the means to

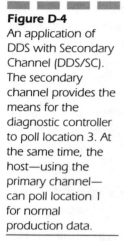

Figure D-4
An application of DDS with Secondary Channel (DDS/SC). The secondary channel provides the means for the diagnostic controller to poll location 3. At the same time, the host—using the primary channel—can poll location 1 for normal production data.

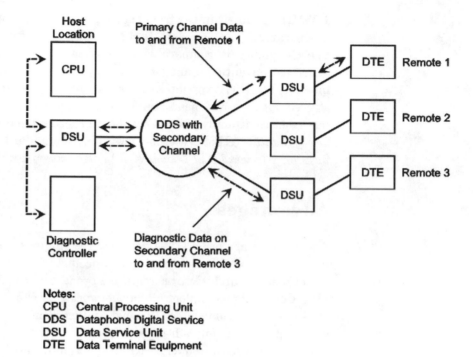

Notes:
CPU Central Processing Unit
DDS Dataphone Digital Service
DSU Data Service Unit
DTE Data Terminal Equipment

control and diagnose problems without interfering with normal production data on the primary channel. GDS made its debut in the late 1980s as a more economical alternative to DDS. The lower cost of GDS is mainly attributable to its implementation through carrier-serving wire centers, which eliminates the need to back-haul traffic. Because traffic does not have to travel as far with GDS, it has become a popular alternative to DDS.

See Also

  Generic Digital Services (GDS)

Digital Enhanced Cordless Telecommunications

The *Digital Enhanced Cordless Telecommunication* (DECT) standard defines a protocol for secure digital telecommunications and is intended to offer an economical alternative to existing cordless and wireless solutions. DECT uses *Time-Division Multiple Access* (TDMA) technology to provide 10

1.75MHz channels in the frequency band between 1.88Ghz and 1.90GHz. Each channel can carry up to 12 simultaneous two-way conversations. Speech quality is comparable to conventional land-based phone lines. Whereas conventional analog cordless phones have a range of about 100 meters, the DECT version can operate reliably up to 300 meters. What started out as a European standard for replacing analog cordless phones has been continually refined by the *European Telecommunication Standards Institute* (ETSI) to become a worldwide standard that provides a platform for *Wireless Local Loops* (WLL) and wireless LANs, as well.

Advantages

A key advantage of DECT is dynamic reconfiguration, which means that implementation does not require advance load, frequency, or cell planning. Other wireless architectures require a predetermined frequency-allocation plan. Conventional analog cellular networks, for example, are organized as cells in honeycomb fashion. To avoid conflict from adjacent cells, each base station is allotted only a fraction of the permitted frequencies. Changing a particular station's frequency band to accommodate the addition of more base stations to increase network capacity involves an often difficult and expensive hardware upgrade. However sparsely the base stations are constructed at the start of an installation, all possible base stations must be assigned frequencies before any physical systems are put into place.

In a DECT system, planning for uncertain future growth is unnecessary, because a DECT base station can dynamically assign a call to any available frequency channel in its band. The 12 conversations that are occurring at any one time can take place on any of the 10 channels in any combination. The handset that initiates a call identifies an open frequency and time slot on the nearest base station and grabs it. DECT systems also can reconfigure themselves on the fly, in order to cope with changing traffic patterns. Therefore, adding a base station requires no modification of existing base stations and no prior planning of channel allocations.

Compared to conventional analog systems, DECT systems do not suffer from interference or cross-talk. Neither different mobile units nor adjacent DECT cells can pose interference problems, because DECT dynamically manages the availability of frequencies and time slots. This dynamic reconfiguration capability makes DECT useful also as a platform for WLLs. DECT enables the deployment of a few base stations to meet initial service demands, with the easy addition of more base stations as traffic levels grow.

Voice compression (i.e., ADPCM) and the higher levels of the DECT protocol are not implemented at the base stations but are handled separately by a concentrator. The concentrator routes calls between the WLL network and the PSTN. This distributed architecture frees base station processing power so that it can better handle the up to 12 concurrent transmission and reception activities.

For high-end residential and small-business users, DECT permits wireless versions of conventional PBX equipment, supporting standard functions such as incoming and outgoing calls, call hold, call forwarding, and voice-mail—without having to install new wiring. In this application, DECT dynamic reconfiguration means that implementation does not require advance load, frequency, or cell planning. Users can begin with a small system, then simply add components as needs change.

The DECT/GSM Interworking Profile enables a single handset to address both DECT systems and conventional cellular networks. This functionality enables users to take advantage of the virtually free wireless PBX service within a corporate facility and then to seamlessly switch over to GSM when the handset passes out of range of the PBX base station. When the call is handled by GSM, appropriate cellular charges accrue to the user.

Wireless Local Loops (WLLs)

Although residential cordless communication represents the largest current market for DECT-based products, other applications look promising for the future. In developing countries, where the lack of a universal wired telecommunications infrastructure can limit economic growth, DECT permits the creation of a WLL—thereby avoiding the considerable time and expense that is required to lay wire lines. Wireless local loops can be implemented in several ways, which are summarized in Figure D-5.

In a small cell installation in densely populated urban areas, the existing telephone network can be used as a backbone that connects the base stations for each DECT cell. These DECT base stations can be installed on telephone poles or on other facilities. Customer boxes (i.e., transceivers) that are installed on the outside of houses and office buildings connect common phone, fax, and modem jacks inside. Through the transceivers, customers use their telephone, fax, and modem equipment to communicate with the base stations outside. In addition, customers can use DECT-compliant mobile phones, which can receive and transmit calls to the same base station.

In larger cell installations, such as in suburban or rural areas, fiber-optic lines might provide the backbone that connects local relay stations to the nearest base station. These relay stations transmit and receive data to and

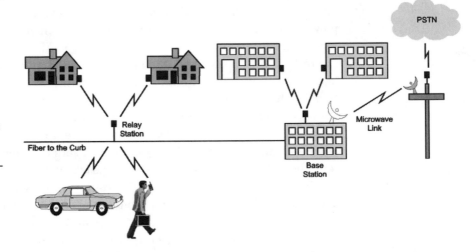

from customer boxes. In these installations, the customer box must have a direct line of sight to the relay station.

Network feeds over long distances can be accomplished via microwave links, which are more economical than having to install new copper or fiber-optic lines. Large cells can be converted easily into smaller cells by installing additional base units or relay stations. Because DECT has a self-organizing air interface, no top-down frequency planning is necessary—as is the case with other wireless connection techniques such as GSM or DCS 1800.

While most WLL installations focus on regular telephone and fax services, DECT paves the way for enhanced services. Multiple channels can be bundled to provide wider bandwidth (which can be tailored for each customer and billed accordingly). This functionality enables the mapping of ISDN services all the way through the network to the mobile unit.

Wireless LANs

In many data applications that have low bit-rate requirements, DECT can be a cost-effective solution. One example is remote wireless access to corporate LANs. By bundling channels, full duplex transmission of up to 480Kbps per frequency carrier is theoretically possible. For multiple data links, a DECT base station can be complemented by additional DECT base stations that are controlled by a DECT server, which forms a multi-cell system for higher traffic requirements. With a transparent interface to ISDN, data access and videoconferencing through wireless links can be utilized.

Such installations can also include services such as voice mail, automatic call back, answering and messaging services, data on demand, and Internet access.

Summary

As DECT becomes a true commodity technology, we might see DECT modules incorporated into building control and security systems—providing intelligent systems that enable automatic control and alerting in order to augment or replace today's customized telemetry and wired systems. DECT might also find its way into the home, providing automatic security alerting in the event of unauthorized entry, fire, or flood or for the remote telephone control of appliances and return channels for interactive television.

See Also

Global System for Mobile (GSM) Telecommunications
PCS 1900

Digital Loop Carrier Systems (DLCS)

Digital Loop Carrier Systems (DLCS) enable telephone companies to expand their networks in an economical manner to accommodate additional subscribers and to efficiently handle the higher traffic load. Expanding a network with such systems provides telephone companies with the flexibility to accommodate growth, expands efficient use of digital or analog central office facilities, and provides a cost-effective alternative to replacing or upgrading older offices or wire centers.

Operating Environment

Within the LATA are numerous wire centers that serve specific communities of interest. A wire center can be a 50,000-line central office, or it can be a *Community Dial Office* (CDO) that serves only a few hundred subscribers. Service can also be provided by a *Remote Terminal* (RT) which, for example, gives tenants of a distant office park access to the telephone network and to all of the capabilities and features of the nearest central office switch.

The central office typically serves subscribers via copper pairs (24-26 AWG) within a 2.5-mile (12,000 feet) radius. To reach subscribers who are farther away, two conventional strategies have been employed: setting up a CDO with trunks linking it to the central office, or setting up a remote terminal to link a fairly limited number of subscribers to the network. The remote terminal can be linked to a CDO via a secondary feeder, or a series of secondary feeders can link RTs directly to the central office via the primary feeder. Subscribers who are connected through the CDO or through the remote terminals have access to all the features of the central office to which they are ultimately linked.

Territories that are served by remote terminals are called *Carrier Serving Areas* (CSAs). CSAs are simply planning entities that consist of a distinct geographical area that is served by a carrier remote terminal site. The CSA concept (refer to Figure D-6) enables the telephone company to plan its

Figure D-6
Carrier serving areas are territories that are served by remote terminals.

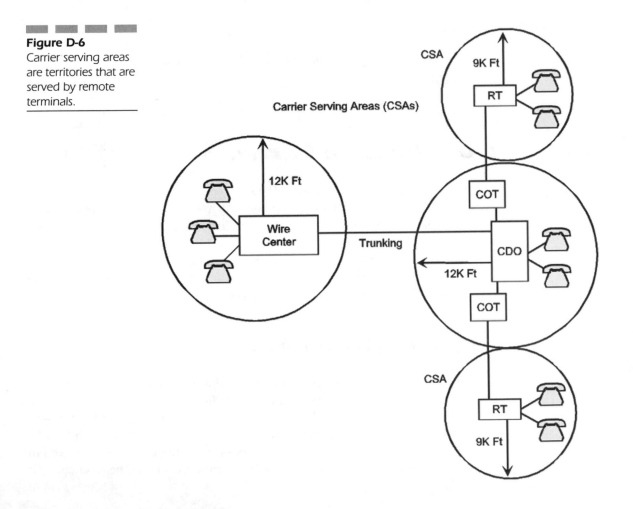

network in order to accommodate incremental growth. The telecommunications provider can provide service to an ever-increasing number of subscribers in any number of remote locations.

The primary feeder from the central office to the CDO might consist of as many as 3,600 copper wire pairs. A secondary feeder from a CDO to a remote terminal can consist of as many as 1,800 copper wire pairs. From the secondary feeder, the pairs branch off to individual subscriber locations. A problem occurs when all of the wire pairs in a primary or secondary feeder are used up. The telecommunications provider must find a way to continue accommodating additional subscribers and efficiently handle the higher traffic load.

Before digital-loop carrier systems were introduced in the mid-1970s, the only viable solution was to segment each CSA and to add more wire and equipment. But that was an extreme solution that involved increased installation and maintenance costs, particularly in metropolitan areas where costs for conduit/duct expansion were prohibitive. Today, with demand often outpacing feeder capacity, telecommunications providers require a simpler, more cost-effective way of coming to grips with the growth problem. One attractive solution is to use T1 lines to free copper wire pairs and to increase the traffic capacity of the network—and, in the process, retain a high degree of flexibility in network planning. The following scenario illustrates an application for digital-loop carrier systems that yields three benefits: CDO replacement, feeder relief (which also eliminates the need for conduit/duct expansion in major metropolitan areas), and efficient distribution.

Applications

The XYZ Company, a nation-wide electronics firm, has just announced plans for a new manufacturing plant in an undeveloped location outside city limits. While county officials are extolling the virtues of free enterprise —and the 200 new jobs that the business would create for their constituents —city officials are already drawing up plans to annex the site of the planned facility in order to broaden the tax base. The XYZ Company, which cannot afford the additional tax burden, abruptly changes its location plans to a small town across the county's border that is well beyond the reach of the city. Unfortunately, the local telephone company cannot serve the new location with its present facilities.

Anticipating growth around the city, the telecommunications provider installed community dial offices. The provider had planned to serve the XYZ Company's new plant with an underutilized CDO nearby, but it was caught completely off guard by the company's last minute decision to build the facility in a small town 10 miles away, where 50 residents were served by a remote terminal.

With potentially hundreds of new subscribers to serve, the telecommunications provider would quickly run out of wire pairs at its remote terminal site, yet the increase in subscribers was not enough to justify adding a CDO. In this case, the telecommunications provider decided to replace an existing CDO with a DLCS (refer to Figure D-7). With a DLCS, the telecommunications provider achieved several objectives.

First, the provider retired wire pairs from the remote terminal to the CDO through the use of T1 span lines, which in turn handled the higher concentration of traffic from the small town. Because the T1 span lines

Figure D-7
Digital-loop carrier systems enable carriers to extend emergency services and other services to outlying areas while freeing wire pairs for possible future growth.

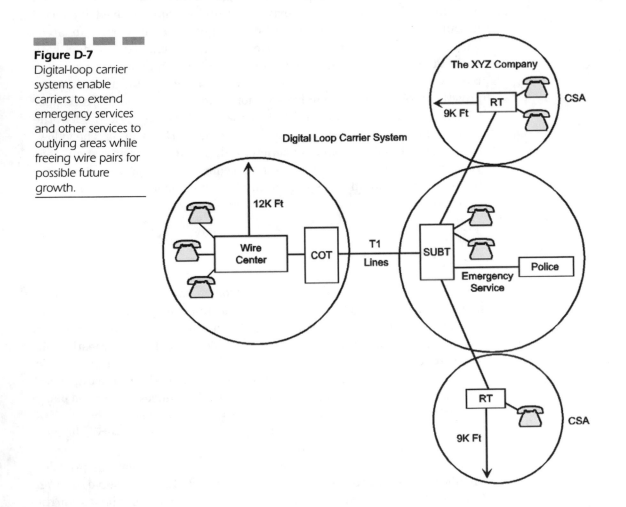

carry traffic in digital format, the DLCS can be readily integrated into the telecommunications provider's digital switch. Second, the telecommunications provider could offer emergency services and other special services, such as data, coin, and *Foreign Exchange* (FX) lines, simply by adding plug-in modules to the DLCS's *Subscriber Terminal* (SUBT) and by linking the subscriber terminal to the *Central Office Terminal* (COT) via T1 lines. Also, the architecture of the DLCS permits deferral of a significant portion of the terminal costs (plug-ins) until forecasts are clearer or until specific services are actually needed. Third, with the T1 span lines between the DLCS subsystems (COT and SUBT), enough wire pairs in the feeder plant could be retired or used for distribution in order to accommodate future subscriber growth. The telecommunications provider also achieved a degree of planning flexibility that it did not have before, which saved on future installation and maintenance.

An additional but not so apparent benefit of the DLCS is that it eliminates the need for load coils. Present wire-pair media are limited to the voice-frequency bandwidth of approximately 4KHz, which is sufficient for *Plain-Old Telephone Service* (POTS) but not for the high-speed data services. Load coils are used to improve base-band (4KHz) transmission for distances longer than 18,000 feet. Wire pairs without load coils (non-loaded) are capable of supporting wider bandwidths than the 4KHz base-band that is used for voice, enabling broad-band *Digital Subscriber Loop* (DSL) data services to be supported in the higher frequency range.

Summary

Digital-loop carrier systems were introduced in the mid-1970s as a solution for economical network expansion. Savings have accrued not only from reducing analog facilities by up to 80 percent but also from building and real-estate liquidation and maintenance efficiencies. Today, the DLCS has evolved into an economical, efficient vehicle for generating new revenues for telephone companies. These systems will continue to evolve in order to meet increasingly sophisticated network needs.

See Also

Central Office Switches

Digital Subscriber Line Technologies

Local Access and Transport Areas (LATAs)

Digital Subscriber Line (DSL) Technologies

DSL is a category of local-loop technologies that turns an existing twisted-pair line (which is normally used for POTS) into a high-speed digital line that can support multimedia applications, Web surfing, news feeds, VPN access, and in some cases, telephone calls. The electronics at both ends of the local loop compensate for impairments that would normally impede high-speed data transmission over ordinary twisted-pair copper lines. This functionality enables ILECs and their competitors to offer high-speed connections over which a variety of advanced broad-band data services and value-added applications can be offered.

DSL provides a fast, economical way to satisfy surging demand among businesses and consumers for huge amounts of bandwidth (which is especially needed for Internet access). In leveraging existing copper local loops, DSL obviates the need for a huge capital investment to bring fiber to the customer premises or to the curb in order to offer broad-band services. Instead, idle twisted pairs at the customer premises can be provisioned to support high-speed data services. Users are added by simply installing DSL access products at the customer premises and by connecting the DSL line to the appropriate voice or data network switch via a DSL concentrator at the *Central Office* (CO) or at a *Serving Wire Center* (SWC), where the data and voice are split for distribution to the appropriate network (refer to Figure D-8).

A dozen DSL technologies are currently available—each of which is optimized for a given level of performance that is relative to the distance of the customer premises from the CO or SWC. The farther away the customer is from the CO or SWC, the lower the speed of the DSL in both the upstream (toward the network) and downstream (toward the user) directions. The closer the customer location is from the CO or SWC, the greater the speed in both directions.

Common Characteristics of DSL

Regardless of the specific type of technology that is used to implement DSL, the different varieties share some common characteristics:

■ All DSL services are provisioned over the same unshielded twisted-pair copper wiring that is commonly used for POTS. Thus, no special

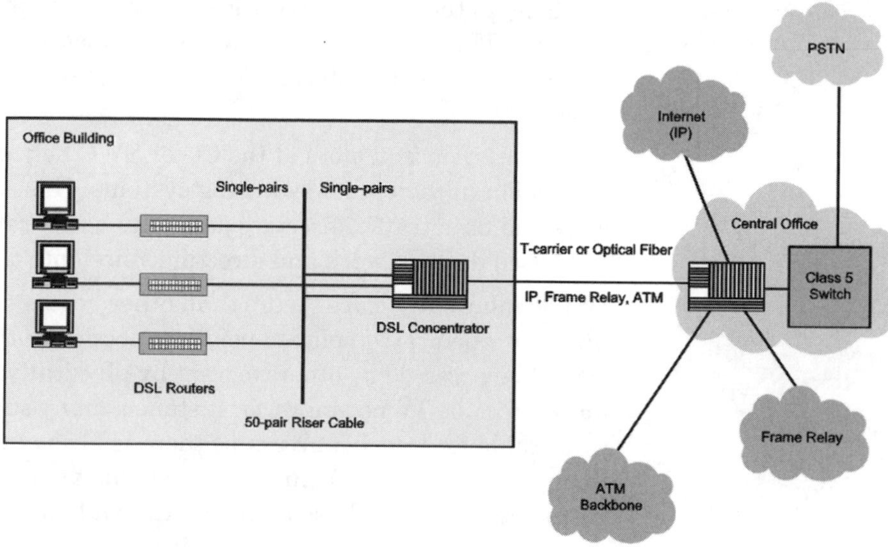

■■■ ■■■ ■■■ ■■■

Figure D-8
Each customer location is equipped with one or more DSL modems. DSL lines from multiple customers are concentrated at a DSL Access Multiplexer (DSLAM) at the central office, where voice and data are split for delivery to appropriate networks.

connection must be installed in the customer premises in order to obtain this type of service.

■ All of the DSL varieties offer a means to turn a low-quality, voice-grade POTS line into a high-quality, broad-band data line. The use of electronic equipment at both ends of the POTS lines adapts or compensates for impairments that would normally corrupt high-speed data transmission over ordinary twisted-pair copper lines. The method by which this line conditioning is achieved differs by equipment vendor, which helps account for the slight variance in maximum line speeds for the same DSL service at comparable distances. Other factors that account for these variances include the specific gauge of the wire (e.g., 24 AWG gauge versus 26 AWG gauge) over which the DSL service is provisioned.

■ All DSL technologies conform to a similar configuration of user and service-provider equipment. At a telecommuter's home, branch office, or other corporate facility, DSL access requires a copper phone line that is connected to a DSL-compatible modem, which is actually a router. At the service provider's central office or serving wire center, concentration/server equipment connects multiple users and passes the transmissions to their respective voice and data networks.

■ Once the DSL modem or router is connected to the digital subscriber line and the service is activated by the carrier, it remains continuously available to the user without the need to dial up every time access is required. The service is always available, just like a LAN. To use the service to access the Internet, for example, users merely open their Web browser.

■ DSL access concentrators at the CO or SWC help carriers relieve congestion in their voice-switching systems. This equipment partitions voice and data traffic, directing data onto a separate packet, frame, or cell-based data network and directing voice onto the PSTN.

■ DSL is inherently more secure than other access technologies. DSL provides a dedicated point-to-point connection to the network that cannot be accessed by others (except by physically tapping into the line). On cable TV networks, for instance, many subscribers share the same cable for Internet access, so there is always the possibility of an intrusion by hackers. To guard against this breach of security, cable customers must purchase their own firewall and have the technical expertise to configure and manage it.

■ DSL is also flexible in the types of traffic data formats that it can accommodate. In addition to voice, DSL can be used to transport IP, Frame Relay, and ATM traffic from the customer premises through the local loop and to the appropriate network, including the PSTN, the Internet, or a corporate VPN.

When it comes to price, DSL is far more economical than other digital technologies such as T1. Similar to T1, DSL is priced at a flat rate per month with unlimited hours of access. But because of the line conditioning that is required, T1 access circuits cost anywhere from $800 to $1,800 per month, depending on the market. By comparison, bandwidth that is provisioned over DSL can cost much less. Bell Atlantic's Infospeed DSL Plus, for example, costs $65 a month for 640Kbps; $115 for 1.6Mbps; and $205 for 7.1Mbps. The service supports voice and data, and Bell Atlantic even provides customers with wiring assistance, DSL modem installation, 24-hour technical support, and a 30-day money-back guarantee.

Asymmetrical versus Symmetrical

The bandwidth that is available over the DSL is carved up in a variety of ways to meet the needs of particular applications. When the upstream and downstream speeds are different, the DSL is referred to as asymmetrical; that is, much greater bandwidth is available in the downstream direction than in the upstream direction, as in the case of *Asymmetrical Digital Subscriber Line* (ASDL).

ADSL runs at up to 8Mbps in the downstream direction and up to 640Kbps in the upstream direction, with the actual speeds depending upon the distance of the customer location to the CO or SWC. This setup would meet the needs of Internet users who want to retrieve multimedia Web content quickly, without waiting indefinitely for the pages to be loaded to their computer—as is the case with dialup modem connections. The lower upstream speed of ADSL is more than adequate for issuing information requests from the computer to the network, because simple queries usually do not require more than 16 Kbps. This situation leaves enough bandwidth capacity to handle multiple voice channels, as well as data.

Asymmetrical operation is fine for activities such as Web surfing, but it might not be appropriate for applications such as server mirroring, where huge amounts of data must have the capability to flow in both directions. *Symmetrical Digital Subscriber Line* (SDSL) service, on the other hand, offers the same amount of bandwidth in both the upstream and downstream directions—160Kbps to about 2Mbps are available in each direction, depending on the distance of the user's location to the CO or SWC. SDSL and other symmetrical DSL services, such as *High Bit-Rate DSL* (HDSL), are more suited to applications that once required a T1/E1 line. Among the popular symmetrical applications are video conferencing, interactive distance learning, and telecommuting.

Two-Wire versus Four-Wire

Traditionally, voice has been handled in the local loop via two wires (i.e., a twisted pair). This method continues to be an economical way to provide millions of residential customers with POTS. Businesses, however, require better-quality local loops for high-speed digital communication, which is achieved by providing them with four-wire connections for services such as T1. Businesses pay a premium price for these connections.

With the growing popularity of the Internet, however, even residential customers have a need for more bandwidth than can be obtained with 56Kbps modems and Basic-Rate ISDN. Where the infrastructure is almost exclusively two-wire, new DSL technologies have been developed that are capable of providing bandwidth in the multi-megabit-per-second range—without requiring expensive upgrades of the local loop.

The improvements in two-wire technologies for residential customers have had a corresponding effect on four-wire technologies that are used by businesses. Increasingly, two-wire DSL solutions are moving into the business sector, and their performance now exceeds the previous performance levels of traditional four-wire solutions. In other words, carriers can offer services to twice the number of businesses or double the transmission speed to the same businesses—without incurring major local-loop upgrade costs.

Other advantages exist to offering two-wire DSL solutions to businesses. Provisioning T1 service requires the installation of repeaters every 4,000 to 6,000 feet in order to boost signal strength. This addition is an expensive, time-consuming task for the carrier and inflates the costs of T1 service. Some two-wire DSL technologies, such as *High-Speed Digital Subscriber Line Two-Wire* (HDSL2), match or exceed the performance of T1 without the need for repeaters.

Rate Adaption

A version of ADSL is available that adjusts dynamically to varying lengths and qualities of twisted-pair local-access lines. Similar to ADSL, *Rate-Adaptive DSL* (RADSL) delivers a high-capacity downstream channel and a lower-speed upstream channel while simultaneously providing POTS over standard copper loops. Unlike ADSL, which does not tune itself to changing line conditions, RADSL adjusts data rates up or down in much the same way that ordinary modems do. With RADSL, it is also possible to connect over different lines at varying speeds. Connection speed can be determined when the line synchs up, while the connection is active, or when a signal is given from the central office.

Inverse Multiplexing

Multiple DSL lines can be bonded together to provide users with higher-speed services. For example, two-wire SDSL, which tops out at 2Mbps, can be bonded with another two-wire SDSL to offer an access speed of up to 4Mbps. With this much bandwidth, small to medium-size companies can obtain the transmission capacity they need without resorting to more expensive four-wire T1 access lines.

The bonding arrangement for DSL requires the user to have multiple telephone lines over which the higher speed service can be run. No additional hardware is required. A simple software change to the inverse multiplexers in the service provider's network implements the bonding process. Once the DSL lines are bonded together in the service provider's network, the user's data load is balanced across the active lines.

Service Provisioning

Most DSL service providers offer customers the means to check for local DSL availability via forms that are posted on their Web sites that prompt for the address and phone number of the DSL location (refer to Figure D-9). If DSL service is available, the database application notifies the user of the type of service that is available as well as the speed of the connection. Because this type of service is not accurate, a quirk of DSL services is that the actual speed of the connection will not be positively known until the service is actually provisioned. Customers in high-rise office buildings have to factor in the vertical distance as part of the total circuit distance. Now, online service users can call and conduct live tests of their line to determine whether the setup is suitable for DSL and at what speed. These services can even determine the existence of a load coil on the line, which would preclude DSL operation.

Once DSL service is ordered, the service provider will arrange with the LEC to connect a line to the network interface outside the customer premises, or to share the existing POTS line for data. Either way, the DSL service provider (or a local agent) visits the customer premises and connects the line to a DSL modem.[1] The DSL modem is typically leased from the service provider, but users can now purchase them from a retail source and configure them through a Web browser (refer to Figure D-10). By connecting the DSL modem to a hub, multiple LAN users can share the available bandwidth to access the Internet for applications such as e-mail, Web browsing, and newsgroup discussions.

---

[1]So-called DSL modems are actually routers. They are called modems because most consumers have become familiar with them for dialing into the Internet or for sending faxes from their home computers. Manufacturers stayed with this familiar term, rather than risking the confusion of consumers with the unfamiliar term *router*.

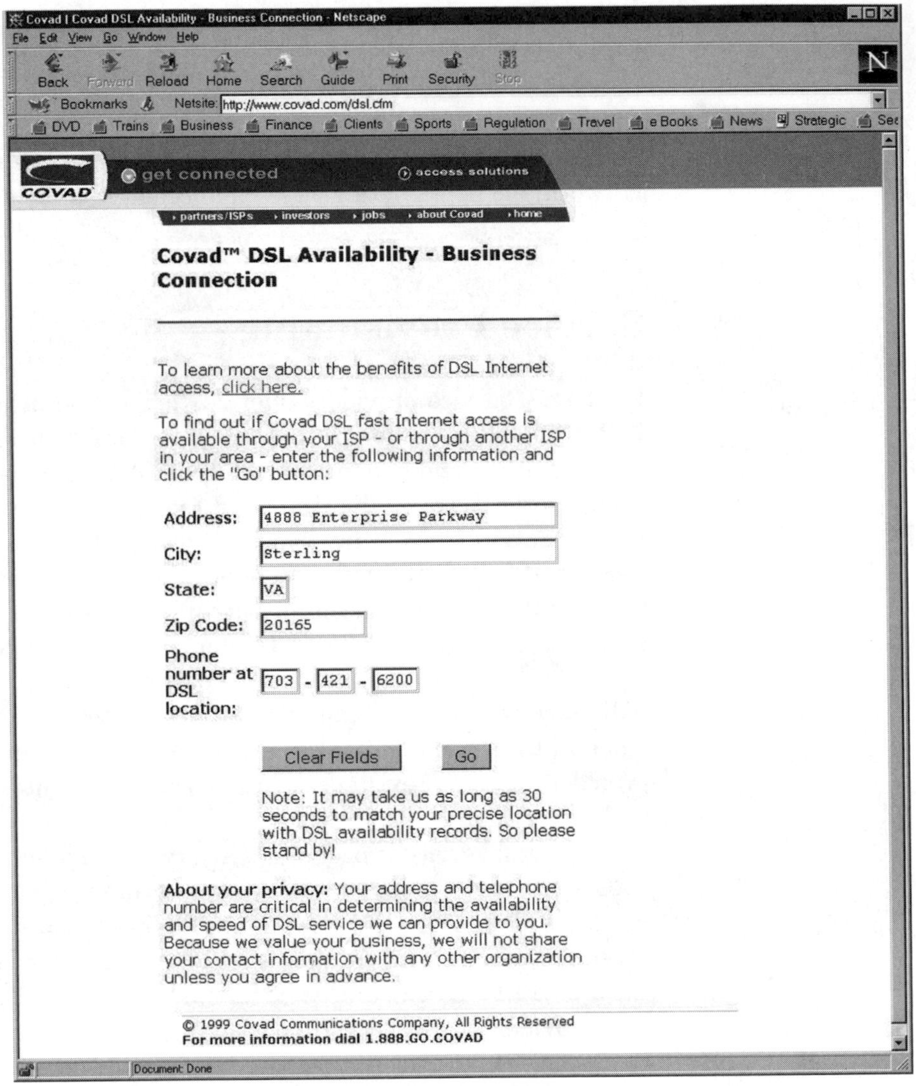

Figure D-9
Covad
Communications is
among the many
providers that
provide a Web form
to enable consumers
to check on the
availability and speed
of DSL in their service
area.

For large installations, like a multi-tenant building, an appropriately sized DSL concentrator will usually be installed on the premises in order to aggregate the wires from individual DSL modems in each office. The building manager can have his own technician install the modems at each workstation and connect the wires to the concentrator or can have the service provider do it for a fee. The concentrator will usually be scaleable in terms of port density, in order to accommodate future growth and to support multiple types of DSL to meet the varying needs of users.

Figure D-10

The Linksys Instant Broadband EtherFast Cable/DSL Router is one of a growing number of off-the-shelf products that is easily configured by the user through a Web browser interface.

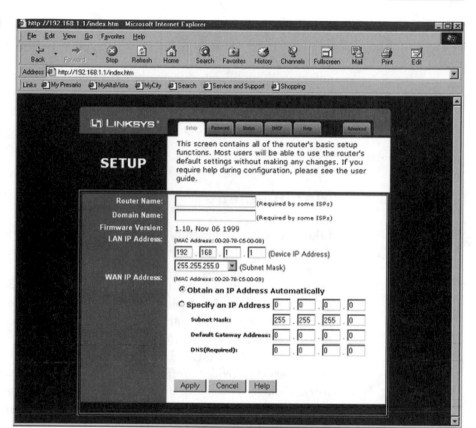

Several methods can be used to send aggregated traffic from the on-premises DSL concentrator to the service provider's location. Depending on the amount of traffic involved, leased T1 lines, Frame Relay over DS3, and ATM over SONET can be used. At the service provider's location, another concentrator splits individual user sessions and tunnels for distribution to other networks.

Depending on the switching platform used, value-added services can be offered over DSL. Cisco's BPX 8650 switching platform, for example, enables service providers to offer DSL on an IP-over-ATM backbone. This platform will enable the carrier to offer QoS, either through ATM or through *Multi-Protocol Label Switching* (MPLS)—thus permitting the delivery of value-added services such as voice-over-DSL.

As noted, DSL service cannot be provisioned over lines that have load coils. These are used on some voice lines to improve signal quality and are typically inserted at 3,000-foot intervals along the line. While load coils improve voice service, they do not enable any signals above the 4KHz voice band to pass through them. Because all types of DSL use frequencies that are well above the voice region, the load coils prevent the use of DSL technology. Although load coils can be removed, carriers are reluctant to do so for fear of impacting the QoS of other users on adjacent lines. For users who have lines that are outfitted with load coils, the carrier might have to steal an idle pair from another customer in order to provision DSL.

Management

Like any other service, managing DSL involves carrier and customer components. Initially, DSL services were hard for the carriers to provision, which stalled service delivery. But new tools have become available to enable carriers to streamline the rollout, setup, and ongoing management of DSL. There are now tools that can be used by customers, enabling them to change services, add bandwidth, and monitor carrier performance for compliance with *Service Level Agreements* (SLAs).

Paradyne's Service-Level Agreement Reporter, for example, can be used by both a service provider to offer SLAs on its DSL service and by corporate network managers to ensure that such SLAs are being met. The SLA Reporter supports Paradyne's multi-service DSLAM that carriers use to offer a variety of DSL services. The SLA Reporter enables service providers to obtain factual, network-based statistics and operational information in order to verify the quality of services that are being provided to their customers.

Graphical charts provide the performance information that is delivered to customers via a secure Web site. Using familiar Web browser software, customers can view and even interact with the data that is provided in the SLA Reporter output. This functionality enables corporate network planners and architects to obtain their own local views of network performance, throughput, capacity planning, and line quality.

Another company, Syndesis, offers service-provisioning software for DSL. Its NetProvision Creator streamlines the carrier setup of DSL lines and extends service ordering functions to the customer. The company's NetProvision Activator issues the commands that configure the individual network devices, including the customer modem, the DSL access multiplexer in the carrier switching office, and core switches and routers. By eliminating paper-based provisioning requests, the process is not only faster, but the number of order errors is reduced.

NetProvision Creator enables DSL service profiles to be defined in plain language, such as the speed of a line. This feature makes it easy for sales representatives to use the software to take customer orders for DSL services, and it enables customers to modify their DSL-based services via the Web. For example, a customer with a 256Kbps DSL service to an ISP could increase that bandwidth to 384Kbps via NetProvision Creator, which is installed on a Web site. Requested changes are then carried out by NetProvision Activator, which speaks the language of the technology and enables single-step activation of unlimited subscriber accounts across a diverse network infrastructure.

Technology Innovation

DSL technology is changing rapidly, and the trend toward voice-data convergence is taking hold here as well as in other advanced services. One of the most compelling developments comes from Copper Mountain Networks. The company offers technology that provides distance and efficiency breakthroughs in the delivery of DSL that will enable carriers to more effectively address the voice and data needs of the growing telecommuter and small-business markets.

In addition to combining voice and data on the same local loop, the solution extends the reach of DSL from 18,000 feet to more than 100,000 feet. By extending the distance so dramatically, carriers can reach many more customers. An estimated 23 million local loops currently extend well beyond 18,000 feet. To achieve longer loops of up to 100,000 feet (depending on the wire gauge), line-powered repeaters and *Remote Terminal Units* (RTUs) are used.

The solution is implemented with the company's CopperEdge 200 Concentrator, which consists of an inverse multiplexer and a DSLAM. The inverse multiplexer sits on the customer premises, while the DSLAM goes in the CO or SWC of the service provider. The DSLAM handles multiple types of DSL connections. A couple of pairs of wires are used to link the inverse multiplexer with the DSLAM. The two devices rely on *Multi-Link Frame Relay* (MFR) technology to perform the actual bonding. MFR is the Frame Relay Forum's proposed standard that defines the bonding of any physical connection to a higher-speed virtual connection.

Copper Mountain's decision to use MFR instead of *Multi-Link Point-to-Point Protocol* (MLPPP) enables carriers to provision end-to-end *Permanent*

Virtual Circuits (PVCs) by using RFC 1490, which defines the use of multiple protocols over Frame Relay. This functionality makes for interoperability between Frame Relay devices from different vendors, which is important because it enables customers who are using traditional T1 lines to more easily migrate to DSL-based Frame Relay services for substantial cost savings. This flexible new capability for DSL provides carriers with the means to support several functions, including the following:

- *Bandwidth bonding*   If a small business requires bandwidth that is greater than 144Kbps, the carrier can inverse-multiplex up to three DSL circuits from the concentrator for 432Kbps on a single pair of wires, at distances of up to 100,000 feet. Future product releases will boost bandwidth to 12Mbps, making possible DSL services at Fractional T3 rates.

- *Adding/dropping DSL*   Once a carrier has extended DSL to its farthest point, the carrier can add more subscribers to the same circuit anywhere between the CO and the last drop by using an *Add/Drop Repeater* (ADR) and an RTU at the customer premises.

- *Adding new DSL services*   With the DSL Concentrator, carriers can introduce other DSL services in the future by using the same chassis.

Although DSL is only a few years old, Copper Mountain's pair-gain solution indicates how much flexibility there really is with DSL technologies and how far continued innovation can take the industry, both in terms of services to customers and competitiveness in the market for local services.

Summary

The first wave of DSL offerings targeted small and medium-size businesses, the branch offices, and telecommuters of larger firms—as well as Internet power users in the residential market. The current wave is aimed at large enterprises and multi-tenant buildings. The next major wave of DSL deployment will involve service providers mining their installed base in order to offer technology enhancements that turn DSL into an economical platform that is available for multi-service networking (capable of supporting virtually any protocol and traffic type, including voice-over packets, frames, and cells). In addition, the reach of DSL is being extended to remote areas that were never before served by broad-band technology.

See Also

Local Loop

T-Carrier

Direct Broadcast Satellite (DBS)

Direct Broadcast Satellite (DBS) meets consumer demand for entertainment programming, Internet connectivity, and multimedia applications. DBS offers more programming choices for consumers and a platform for the development of future services. Much of the growing popularity of DBS services is attributable to the picture quality that is provided by digital technology.

One of the most popular DBS services is DirecTV. First introduced in the United States in 1994 by Hughes Electronics and Thomson Consumer Electronics (now Thomson MultiMedia), DirecTV is now marketed worldwide. DirecTV was the first DBS service to deliver up to 175 channels of digital-quality programming. The number of channels that DirecTV offers is now 210. The satellite service requires the user to have an 18-inch dish, a digital set-top decoder box, and a remote control. The system features an on-screen guide that enables users to scan and select programming choices by using the remote. Customers can also use the remote control to instantly order pay-per-view movies as well as set parental controls and spending limits.

The DirecTV installation includes an access card that provides security and encryption information and that enables customers to control the use of the system. The access card also enables DirecTV to capture billing information. A standard telephone connection is also used to download billing information from the decoder box to the DirecTV billing center. This telephone-line link enables DirecTV subscribers to order pay-per-view transmission as desired.

DirecTV enables users to integrate local broadcast channels with satellite-based transmissions. In markets where broadcast or cable systems are in place, users can maintain a basic cable subscription or connect a broadcast antenna to the DirecTV digital receiver to receive local and network broadcasts. A switch that is built into the remote control enables consumers to instantly switch between DirecTV and local stations.

Internet access is provided through DirecPC, a product that uses DirecTV technology in conjunction with a PC in order to deliver high-bandwidth, satellite-based access to the Internet. The DirecPC package includes a satellite dish and an expansion card that is designed for a PC's I/O bus. This receiver card transmits data from the Internet to the computer at 400Kbps—a rate that is 14 times faster than that of a 28.8Kbps modem connection.

Users connect to the ISP through a modem connection, but the ISP is responsible for routing data through the satellite uplink and for transmitting the data to the receiver card and into the computer (refer to Figure D-11). The service also provides users with the option to narrowcast software from the head end of a network, in order to branch users during off-peak hours. Additionally, DirecPC transmits television broadcasts from major networks, such as CNN and ESPN, to the user's computer system.

Operation

DBS operates in the Ku band, which is the group of frequencies from 12GHz to 18GHz. TV shows and movies are stored on tape or in digital form

Figure D-11
Typical DBS configuration for Internet access

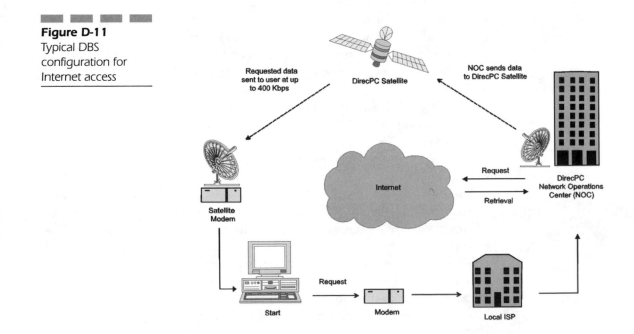

Figure D-12
Typical DBS
configuration for
television
programming

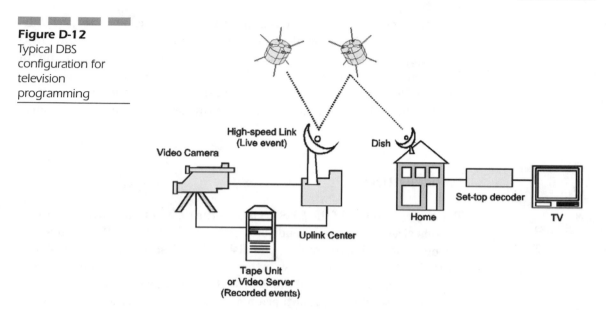

at a video server, while live events are broadcast directly to a satellite (refer to Figure D-12). Stored programs are sent to the uplink center (ground-to-satellite center) manually, via tape, or are electronically sent from the video server over fiber-optic cable. Live events also pass through the uplink center. At that location, all programs—whether live or stored—are digitized (or redigitized) and compressed before they are uplinked to the satellites. All DBS systems use the MPEG-2 compression scheme because it delivers a clean, high-resolution video signal and CD-quality sound. The satellites broadcast up to 200 channels simultaneously via the downlink. The home satellite dish picks up all of the channels and sends them via a cable to a set-top decoder. The set-top decoder tunes one channel, decodes the video, and sends an analog signal to the TV.

Service Providers

More than one million U.S. residents have installed small TV satellite dishes in order to receive programming via satellite services. At this writing, there are four direct-broadcast satellite systems in operation: PrimeStar, EchoStar, *Digital Satellite Service* (DSS), and AlphStar. DirectTV uses DSS and PrimeStar.

Ordering PrimeStar service is similar to receiving cable. After the order is placed, a technician installs the dish and activates programming. DSS, EchoStar, and AlphaStar services also give users the option of installing the dish themselves. The dish must be placed so that it can capture a clean signal from the nearest satellite—usually on the roof, facing south. To activate service, the user calls the programming provider to obtain a unique satellite dish address.

Equipment

The key component of the DBS system is the dish antenna, which comes in various sizes. Dish size depends on the strength of the satellite signal: the stronger the signal, the smaller the dish can be. Users select the dish based on their geographic proximity to the satellite source, which also explains why it is necessary to install the dish so that it points in a specific direction. If the satellite sits on the southern horizon, the dish must be pointed south.

The user also needs a receiver-decoder unit, which tunes in one channel from the multitude of channels that it receives from the dish. The decoder then decompresses and decodes the video signal in real time so that the programs can be watched on the television set. These set-top units might also include a phone-line connection for pay-per-view ordering and Internet access. Taping DBS programs requires the set-top unit to be tuned to the correct channel. To make recording easier, some receiver-decoders include an event scheduler and an on-screen programming guide.

As with most audio-video components, DBS units come with a remote control. Some manufacturers offer a universal remote that can also be used to operate the TV and VCR. The accessories that are available for DBS systems deal with secondary and tertiary installations. Users can buy additional receiver-decoder units or multi-room distribution kits, which use either cable or radio frequencies to transmit the signals from the original set-top unit to other rooms. Some kits enable the VCR to be plugged into the distributor.

Programming

Each of the four DBS systems that are currently available provides similar core services. The differences lie in the availability of premium movie channels, audio channels, and pay-per-view events. With more than 200 channels to choose from, the on-screen programming guide can become an important factor when selecting a service. Most guides enable users to sort the available programming based on content area, such as sports, movies, or comedies, or to list favorite channels at the top of the menu. Depending on the equipment selected, users can even store the favorite channel profiles of multiple family members.

Parental lockout enables adults to block specific channels or programming with a specific content rating or to set a maximum pay-per-view spending limit. Channel-blocking options are protected by passwords; with multi-profile units, parents can customize the system for each child.

Regulation

Despite increases in the number of subscribers to DBS systems in recent years (DirecTV alone has eight million subscribers), CATV systems remain the dominant supplier in what is called the *Multi-Channel Video Program Distribution* (MVPD) market. The FCC has regulatory authority over DBS and in mid-1999 cleared the way for DBS to compete more effectively with cable. Specifically, the FCC approved the transfer of an authorization to construct, launch, and operate a high-powered DBS service on 28 channels from MCI WorldCom to EchoStar.

The FCC found that the transfer of this additional satellite capacity to EchoStar is in the public interest because it enables EchoStar to increase substantially its number of program offerings and to compete more effectively with cable in the MVPD market. The EchoStar transaction enables DBS to become a stronger competitor. As a result, consumers should benefit from the increased competition that should lead to increased program offerings and options and/or lower prices in the marketplace for multi-channel video programming. EchoStar had already been providing DBS service, but as a result of this transaction, it will now have the capability to operate DBS satellites from two orbital locations that are capable of transmitting DBS signals to all portions of the continental United States.

Summary

Industry experts say that there could be as many as 15 million DBS households by 2001, giving DBS 20 percent of the market—up from its current 6 percent market share in 1999. This factor has attracted AT&T and MCI WorldCom to the market. AT&T, for example, resells the DBS services of DirecTV, offering one-stop shopping ranging from ordering equipment and service to home installation, customer service, and financing.

The future of DBS service might include specialized markets for information services. DBS is already being used for Internet access and could be deployed in ways to offer sales-force automation, inter-enterprise communications, home shopping, and demographically targeted programming. DBS services also are positioned to take advantage of emerging technologies such as interactive multimedia and HDTV.

See Also

Local Multi-Point Distribution Service

Satellite Communications

Directory Assistance—411

Two types of directory-assistance service are available to customers throughout the United States: local directory-assistance service and non-local directory-assistance service. Directory-assistance service is considered local whenever a customer requests the telephone number of a subscriber who is located within the same LATA or area code. Local directory assistance typically is provided by a customer's LEC. Under the *Modified Final Judgment* (MFJ) of 1982, the BOCs were permitted to use their *Official Services Networks* (OSN), which cross LATA boundaries, for the provision of local directory-assistance service to their own local-exchange customers. Currently, most customers dial 411, 1-411, or 555-1212 to access their LEC's local directory-assistance service and pay a nominal charge of 50 cents per call (after two free calls a month).[2]

Types of Directory Assistance

Directory-assistance service is considered non-local whenever a customer requests the telephone number of a subscriber who is located outside the LATA or area code. The BOCs were prohibited under the MFJ from providing non-local directory-assistance service. In view of this restriction, non-local directory-assistance service has traditionally been provided by IXCs. To access non-local directory assistance, most customers can dial the

---

[2]While there are charges for local and non-local directory-assistance services, directory assistance for 800, 888, and 877 numbers is free and available by calling 1-800-555-1212.

National Numbering Plan at 1-NPA-555-1212. Thus, in order to use this service, customers must know the area code of the customer or entity that they wish to call. When a customer does not know the area code, he or she dials the number for local directory-assistance service in order to obtain such information from the local directory-assistance operator. Customers who are using non-local directory-assistance service typically can obtain up to two directory listings from the same area code per call. In order to obtain a listing from a different area code, the customer has to dial the number for directory-assistance service in that area.

When a customer dials 411 or 1-411, the local central office switch will route the call to an operator services switch that adds the *Automatic Directory Assistance Services* (ADAS) component to the call. The ADAS system, created by Nortel, delivers a script requesting the city, state, and listing that is desired. If the requested number is local, the call is routed to an operator who has access to local directory-listing information. The LEC uses its *Official Services Network* (OSN) or leased common-carrier lines to connect end users to local directory-assistance operators and to connect local directory-assistance operators to their local directory-assistance databases.

If the requested number is non-local, the call is routed to an operator who handles non-local directory-listing requests. The non-local request could be for either a telephone number in the LEC's region or a telephone number outside its region. If the request is for a telephone number within the carrier's region, the number will be retrieved from its own directory-assistance database. If the request is for a telephone number outside the carrier's region, the number will be retrieved from another database that might be owned and operated by another carrier or vendor. After identifying the telephone number that matches the customer's request, the non-local directory-assistance operator will either quote the number verbally or will cause the number to be automatically quoted by an audio response system.

Case Study

Nortel is one of the vendors that provides a national directory-assistance service to telephone companies. In addition to ADAS, Nortel offers the Quest411 system—which is used by US WEST and enables its customers to obtain listed telephone numbers from anywhere in the United States, simply by making a local telephone call. Customers who want phone listings from other parts of the country no longer must go through the trouble of searching for the right area code and then placing a long-distance call to the appropriate directory-assistance service (many times, without success).

With the Quest411 system, callers in Colorado, for example, can simply dial their current local directory-assistance number—1 + 411—to obtain listings from any part of the country. The Quest411 system integrates with ADAS and uses Nortel's *Flexible Vocabulary Recognition* (FVR), which is an advanced speech-recognition system that enables callers to simply respond to a voice prompt with the name, city, and state of the person, business, or government agency that they are seeking.

Quest411 provides US WEST operators with the capability to search both its ADAS data and a national database of 120 million up-to-date listings from a host of telephone companies. These listings are compiled into a single database by Metromail Corporation's On-Line Services division. Depending on the type of request, ADAS automatically recognizes the response and routes directory searches either to a local directory-assistance operator within US WEST's 14-state region or to the US WEST National Directory Assistance bureau. In each case, operators hear a recording of the customer request, and locality listings for particular cities and states automatically appear on the operator's computer screen.

Regulatory Issues

Section 271 of the Telecommunications Act of 1996 prohibits BOCs from providing long-distance services originating in the states where they currently provide local exchange and exchange access services before they satisfy certain statutory requirements. BOCs, however, are not required to receive regulatory approval in order to provide long-distance services that are either previously authorized services under section 271(f) or incidental interLATA services as defined in section 271(g).

In mid-1999, the FCC issued a ruling that exempted in-region directory assistance from treatment as a long-distance service, concluding that directory assistance falls within the scope of the exception that is provided in section 271(g)(4). Accordingly, the FCC found that BOCs can continue their provision of region-wide directory-assistance service. The FCC also concluded, however, that the nation-wide component of BOCs' non-local directory assistance does not qualify for this same exception if the BOC does not own the database that is used to provide the national directory-assistance information.

The BOCs must provide region-wide directory-assistance service on a structurally separate basis. They are also required to make available to all unaffiliated entities all of the telephone numbers that they use to provide region-wide directory-assistance service at the same rates, terms, and conditions that they give to themselves. Because of their historic monopoly control over these telephone numbers, the FCC found it necessary to impose a non-discrimination requirement with respect to the telephone numbers that the BOCs use to provide region-wide directory-assistance service. This requirement is intended to promote the development of a fully competitive market for this service by ensuring that no one competitor will have an undue advantage in the region-wide directory-assistance services market.

National Directory Assistance

In 1999, Bell Atlantic introduced its National 411 directory-assistance service, which enables customers from Maine to Virginia to dial 411 for local directory assistance that can get them numbers in all 50 states. The service provides the caller with a local listing at regular low prices and 95 cents for up to two national listings. Competing directory-assistance services charge up to $1.40 for all lookups.

AT&T and MCI WorldCom offer their own non-local directory-assistance services in which callers can use one telephone number to obtain directory listings from anywhere in the United States. AT&T's 00 INFO service, for example, permits customers to obtain telephone numbers from anywhere in the United States by dialing 00. Customers do not need to know the area code. Similarly, by dialing MCI WorldCom's 10-10-9000 number, customers can obtain telephone listings from across the country.

Enhanced Directory Assistance

Enhanced directory-assistance service is aimed at mobile phone users and includes many features that are currently ignored by traditional directory-assistance providers. Some of the features that are included in enhanced directory-assistance service are as follows:

- Yellow Pages searches
- Restaurant guide
- Movie listings
- Local event information
- Personal phone book

- Emergency road service
- Weather reports

Internet Directory Assistance

The latest innovation in the provision of national directory assistance is Internet-based service. MasterFiles' offering of its Reach Directory Assistance enables users to access national directory assistance interactively on the Web at only 33 cents per search for both local and national numbers— a savings of 65 percent when compared to MCI WorldCom's 10-10-9000 and AT&T's 00 INFO. Volume pricing plans are available, as well.

MasterFiles guarantees the most accurate information at the lowest rates on the Internet. Whereas the RBOCs update their listings daily, reflecting changes from the previous business day, MasterFiles offers the same information in real time. MasterFiles keeps the subscriber listings physically stored at each individual RBOC and has built the necessary hardware, software, and networking components that link these databases in order to provide the most current information available.

Access to the service is achieved with special client software that enables users to access the telephone company directory information directly from their Windows-based computer and dialup/dedicated Internet connection. Users can receive up to 20 listings per screen, instead of one or two listings at a time via the phone. In batch-processing mode, the client software can process hundreds of requests per hour.

Summary

Local and interexchange carriers and their competitors offer accurate and up-to-date directory information. Recorded announcements provide the caller with the option of having the call completed automatically. In some locations, the directory-assistance operator can perform reverse lookups, providing the customer's name and address when the caller gives the phone number. In some cases, ZIP code information is also provided. The trend in directory-assistance services is to offer simple dialing and universal information.

See Also

Emergency Service—911

Distance Learning

Distance learning involves the use of digital networks and videoconferencing technologies to deliver a variety of education programs to remote locations. Through the use of various wireline or wireless technologies, courses can be delivered to rural areas, for example, or to link schools in a district in order to provide a common curriculum. Distance-learning programs are even available through the Internet.

With a distance-learning program, a university can extend its reach to suburbs and offer the same quality of instruction that is available on campus. A corporation or government agency can use distance learning to educate and inform staff while cutting the cost of travel and the amount of lost productivity time. In the years ahead, this method of education is expected to have an increasingly significant impact on learning in the home, school, and workplace.

Numerous reasons exist for this optimistic outlook on distance learning. According to the *United States Distance Learning Association* (USDLA), the following factors will expand the importance of distance learning:

- Mounting frustration with school systems nationwide, driven by unacceptable student performance and under-achievement
- Parents' interest in additional learning services for their children outside of the traditional school environment
- Business interest in the capability of information technology and networks to deliver training and learning services in a timely and cost-effective manner
- The need of employees in both the public and private sector to continually update their job skills and knowledge

All 50 states have distance-learning projects underway, most of which involve the cooperation of state and federal government agencies, businesses, and one or more telephone companies.

Applications

Although a variety of high-speed networks and videoconferencing technologies are integral components of distance-learning programs, any successful program must focus on the instructional needs of the participants, rather than on the technology itself. An important factor for successful distance learning is a competent instructor who is experienced, who is at ease with the equipment, and who uses the media creatively. Often, the effectiveness of distance-learning programs also hinges on the level of interactivity between the instructor and students and among the students themselves.

Public Schools   The chronic budgetary problems of the nation's public school systems and the widespread anti-tax sentiment that prevents an easy remedy are other factors that are increasing the popularity of distance-learning programs. Distance-learning networks can go a long way toward making up budgetary shortfalls by permitting schools in the same town or city to make use of the same teacher, in order to simultaneously deliver live instruction to multiple classrooms.

Tying local schools together via distance-learning networks can also go a long way toward balancing the quality of education between rich and poor schools in the same district. For example, the best math, science, and language teachers at rich schools can deliver courses to students at poor schools at the same time via interactive videoconferencing systems. Also, courses that might not be available at one school can be accessed from another school without requiring students or teachers to change locations during school hours.

Colleges   Colleges have been offering distance-learning programs since the 1980s. Now, they are seeking to leverage their distance-learning networks in order to enhance the educational opportunities of geographically dispersed students. One way to accomplish this goal is by tying in high-speed automated library services. Via the distance-learning network, remote users have access to the college's CD-ROM-based library services, including books, journals, videos, government documents, and databases.

Additionally, such networks permit users to share and retrieve information that is available at other academic and public libraries via the college's library consortium membership—and at the state, national, and international level—via the Internet. Through Internet access, the distance-learning network can deliver such information as bulletin board data and graphics, database and archival resources, and computer programs, as well as file transfer and e-mail services.

Corporations   Increasingly, corporations are delivering training programs to remote locations via their private networks. A properly designed distance-learning program can reduce training time by 30 percent to 50 percent and can eliminate costs that are associated with travel. Added benefits include the following: consistent quality of delivery, consistent content, and the consistent application of professional standards to each location.

Associations   Associations are also offering distance-learning programs for their members. Medical and hospital associations, for example, operate videoconferencing networks that (among other things) offer instruction on new surgical procedures, new drugs, and new medical technologies. Associations that represent the automotive and retail industries, as well as the travel and hospitality industries, are among the many other providers of distance-learning programs for their members.

Military   The military is also among the largest users of distance-learning programs. In addition to providing instruction on a variety of military topics, these programs include college-level and graduate-level courses to help members of the armed forces obtain their degrees and upgrade their management and communications skills.

Government Agencies   Government agencies are continually subject to public criticism for wasting money and consequently face continued attempts by Congress to cut their budgets. Often, significant amounts of money can be saved by cutting back on travel, without negatively impacting staff operations. For many agencies, the answer is distance learning, which can be used to provide instruction on new regulations, compliance matters, and administrative policies and procedures and to improve supervisory and management skills.

Summary

Distance learning has demonstrated its value in a variety of environments, including public schools and universities, corporations, and government agencies. Through the application of appropriate wireline and wireless networking technologies, learning can occur independently of time and distance constraints. Distance learning can afford learners in rural areas the same educational opportunities as those in urban areas, and it can also help spread the cost of technology, computer courseware, and multimedia references and instruction to a broader base of users.

See Also

Telemedicine

Videoconferencing

Dominant Carrier Status

To encourage competition in the long-distance market, the FCC in 1980 devised the dominant/non-dominant regulatory scheme for rate and entry regulation. The FCC defined a dominant carrier as one that "possesses market power" and noted that control of bottleneck facilities was " . . . evidence of market power requiring detailed regulatory scrutiny." The Commission also determined that if a common carrier is determined to be "non-dominant," regulatory requirements would be "streamlined." Specifically, tariffs that were filed by non-dominant carriers would be presumed lawful and would be subject to reduced notice periods.

Because AT&T held 90 percent of the long-distance market in 1980 and could use its position to control prices, it was classified by the FCC as "dominant." MCI and Sprint, on the other hand, were classified as "non-dominant." The FCC also found that AT&T was dominant in the provision of international service, as well. AT&T controlled the overwhelming share of that market, had exclusive operating agreements with the carriers in most major foreign markets, and had few rivals in the provision of essential United States international submarine cable facilities. The FCC had ample reason to conclude that AT&T exercised market power and should be regulated as dominant for its provision of international calling services. Applying dominant carrier regulatory safeguards to AT&T enabled the FCC to monitor changes in AT&T's circuit capacity, which could indicate anti-competitive activity.

Over the past decade, competitive conditions have changed significantly. AT&T's competitors now hold operating agreements and international facilities for all major markets. They share ownership of all major international facilities with AT&T, and the new, state-of-the-art submarine cable facilities have reserve capacity available to all owners that exceeds AT&T's own capacity on the facilities. Moreover, there are three facilities-based networks for domestic long-distance services that compete with AT&T's network in order to link international facilities to United States customers. This domestic competition prevents AT&T from leveraging control over its domestic network in order to shut out competition on the international segment. In short, it is no longer plausible to view AT&T as controlling bottle-neck facilities.

Changes in market share and consumer behavior also reflect significant shifts in the market structure. AT&T's share of the overall domestic market has declined to less than 60 percent and is now below 70 percent in all of the top 50 international markets. Demand elasticity is substantial, as demonstrated by great volatility in household choice of long distance and international carriers in response to specialized pricing and marketing plans. These developments collectively reveal a market in which the FCC now believes that AT&T cannot unilaterally exercise market power to the detriment of its competitors.

In 1995, the FCC determined that AT&T was no longer a dominant carrier, because it no longer possessed individual market power. Accordingly, AT&T was relieved of the regulatory burdens that were imposed by the FCC's dominance standard. Specifically, AT&T no longer had to wait up to 45 days to introduce a new service or to change pricing. The company could now file for new services on one day's notice, just like other long-distance companies.

Summary

With the change in AT&T's status from that of dominant carrier to non-dominant carrier, AT&T became free to quickly bring its full range of capabilities to the marketplace. The change ended a vestige of regulation that was rooted in the predivestiture Bell System. This change does not end the FCC's oversight of AT&T; rather, it merely puts AT&T on the same regulatory footing as its competitors, who for years have used AT&T's dominant carrier status to try to stifle the company's capability to offer innovative new services and pricing plans. The former BOCs, on the other hand, are still classified by the FCC as dominant carriers. For this reason, they have been consistently denied permission by the FCC (from 1996 to 1999) to enter the long-distance market until they open their local loops to competition.[3] The FCC has always presumed that the *Competitive Local Exchange Carriers* (CLECs) are non-dominant.

---

[3] In December 1999, Bell Atlantic became the first ILEC to receive authorization from the FCC to enter the long-distance market. The carrier commenced long-distance service in New York in January 2000.

See Also

> *Competitive Local Exchange Carriers* (CLECs)
>
> *Federal Communications Commission* (FCC)
>
> *Interexchange Carriers* (ICs)
>
> Telecommunications Act of 1996
>
> Telecommunications Industry Mergers

Downsizing

From a technical perspective, downsizing refers to the process through which companies ease the load of the central mainframe by distributing appropriate processing and information resources to LANs; specifically, the servers and clients that are located throughout the organization.[4] This arrangement can yield optimal results for all users—providing PC users with ready access to the information that they need in a friendly format, while permitting mission-critical applications and databases to remain on the mainframe where security and access privileges can best be applied.

Benefits of Downsizing

In moving applications and information closer to departments and individuals, knowledge workers of all levels in the organization are empowered to improve the quality and timeliness of decision-making and to enable a more effective corporate response to customer issues and market dynamics. In addition, the following benefits can occur:

- Downsized applications might run as fast or faster on PCs and servers than on mainframes, at only a fraction of the cost.
- Even if downsizing does not improve response time, it can improve response-time consistency.
- In the process of rewriting applications to run in the downsized client-server environment, there is the opportunity to realize greater functionality and efficiency from the software.

---

[4] Downsizing has also been used to describe how organizations restructure themselves to become more competitive by trimming the number of employees, outsourcing specialized tasks, and spinning off operations that are unrelated to the core business.

- With greater reliance on off-the-shelf applications, the number of programmers and analysts that are needed to support the organization is minimized.

- The lag time between applications development and business processes can be substantially shortened, enabling significant competitive advantages to be realized.

- The organization does not need to be locked into a particular vendor, as is typically the case with centralized host-based architectures.

Distributed Environment

The distributed computing environment that is brought about by downsizing can take many forms. For example, dedicated file servers can be used on the LAN in order to control access to application software and to prevent users from modifying or deleting certain types of files. Several file servers can be deployed throughout the network, each supporting a single application (e.g., e-mail, facsimile, graphics, or specific types of databases). Metering tools can be included on the server in order to monitor usage and to prevent unauthorized copying, which might violate various software licenses.

When connected to a LAN, an appropriately equipped PC can act as a server. With the client/server approach, an application program is divided into two parts on the network. The client portion of the program, or the front end, executes on the user's workstation—enabling tasks such as querying databases, producing printed reports, or entering new records. The server portion of the program, or the back end, resides on a computer (i.e., server) that is configured to support multiple clients. This setup offers users shared access to numerous application programs as well as to printers, file storage, database management, communications, and other capabilities. Automated tools ensure data concurrency, implement security, and ease software-maintenance tasks.

Mainframe as Server

The mainframe is still valued for its unequaled processing and security capabilities. The mainframe can continue supporting traditional applications that require processing power and access to legacy applications, but it can also act as a server in the distributed computing environment. In the case of IBM, for example, the addition of the following special software enables the mainframe to act as a server:

- *LAN Resource Extension and Services* (LANCES)   A server-based software product that is used with NetWare LANs
- *Data Facility Storage Management Subsystem* (DFSMS)   A suite of software programs that automates storage management
- *Network File Server* (NFS)   Originally designed to operate on LANs, this software can now operate with MVS on the mainframe.
- *File Transfer Protocol* (FTP)   The FTP server application can be used as part of the native TCP/IP stack under VTAM or as a single, third-party FTP server application that also runs as a VTAM application.

The mainframe can also play the role of the master server in a hierarchical arrangement, backing up, restoring, and archiving data from multiple LAN servers.

Summary

Determining the benefits of downsizing, the extent to which downsizing will be implemented, who will do the downsizing, and the justification of downsizing's up-front costs are highly subjective activities. The answers will not appear in the form of cookie-cutter solutions from market-savvy vendors, either. In fact, the likelihood of success will be improved greatly if the downsizing effort is approached within the context of business-process re-engineering.

See Also

   Business Process Re-Engineering
   Client/Server Networks

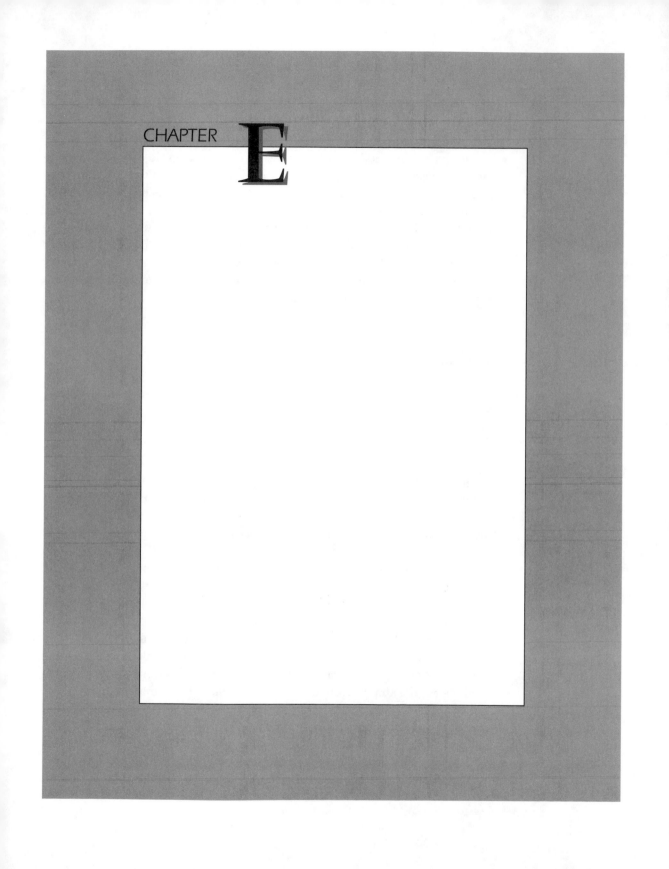

CHAPTER E

Electronic Commerce

The increasing popularity of the Internet has awakened companies to the numerous commercial opportunities that involve the sale of goods and services to a vast, global marketplace. According to various industry estimates, about 50,000 merchants have set up shop on the Web as of year-end 1999. More businesses are also engaged in e-commerce with each other. The investment firm Goldman Sachs estimates that the value of business-to-business e-commerce conducted in the United States will be approximately $1.5 trillion by 2004, up from an estimated $114 billion in 1999. Better security, simplified payment systems, and easy-to-use e-commerce development tools have contributed to the success of e-commerce in recent years.

Payment Systems

A number of tools are available for creating and managing secure electronic payment systems. Such tools typically offer a shopping cart that can be accessed by a Web browser (refer to Figure E-1). With these tools, retail merchants of any size can set up and manage virtual storefronts on the Internet with templates that are included in the software. With the storefront in place, companies can then process credit card payments over the Internet. The application takes payment information from Internet clients who are protected by 128-bit *Secure Sockets Layer* (SSL) encryption. This system also enables retailers to add product-related searching, sales tax and audit reports, and discounts and coupons.

Other types of payment systems also make online shopping faster and more convenient. One of the newest payment systems is the virtual credit card. Trintech Group, for example, offers the ezCard, which is a virtual version of a consumer's plastic credit card or debit card. The card resides on the cardholder's PC and auto-populates online merchant payment forms.

The ezCards are issued by the cardholder's issuing bank and contain the cardholder's basic card information. The ezCard application sits on the cardholder's PC desktop or toolbar until it is invoked for online payment. Once the card is activated by dragging-and-dropping onto the merchant payment form, ezCard automatically fills in the payment fields on the form. The cardholder information resides on the issuer's secure server and is only transmitted from the secure server when the customer makes an online purchase.

The ezCards can be distributed by issuing institutions such as banks, card organizations, digital service providers, and financial portals. These

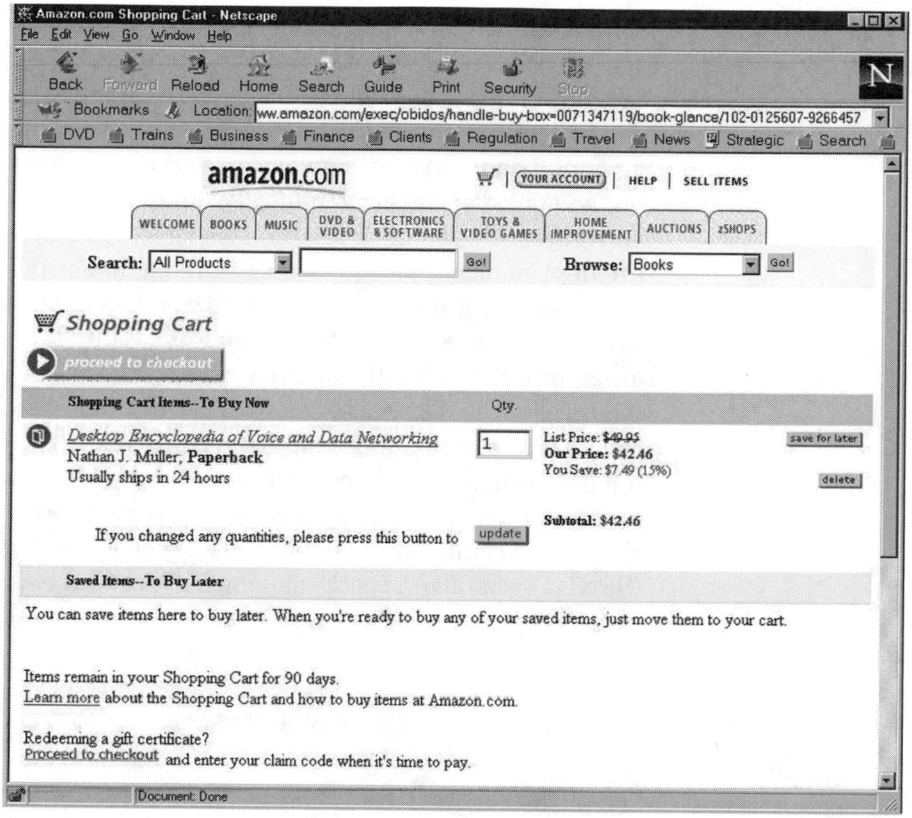

Figure E-1

Once users find an item of interest, they simply click the Add to Shopping Cart button. In this example from Amazon.com, the shopping cart page lists the item, quantity, and shipping schedule. When users have finished adding items to their shopping cart, they click the Proceed to Checkout icon, which takes them through the rest of the ordering process.

organizations need to have PayGate NetIssuer installed from Trintech, which distributes the ezCards. To activate their ezCards, cardholders visit the card issuer's secure Web site and access their account by entering unique personal information, such as their date of birth or their PIN. With Trintech's NetIssuer installed on the issuer's secure server, the issuer's Web site automatically downloads, installs, and launches the ezCard application on the cardholders's PC. The cardholder then completes ezCard activation by choosing a personal password, shipping address, telephone number, and e-mail address.

Every time cardholders make a purchase by dragging ezCard from the desktop to the merchant payment form, they are prompted to type their personal password for verification. The ezCard automatically fills in the form, which the cardholder reviews and then submits in order to complete the purchase. Safe and secure shopping is ensured by NetIssuer's SSL encryption of payment card and address information.

Online Banking

Not only is the Internet suited for buying and selling, but it also offers online banking, as well. Via the Internet, bank customers can monitor account balances, transfer money from their savings account to their checking account, pay bills electronically, apply for car loans, and prequalify for mortgages. Hundreds of financial service organizations currently offer (or are implementing) remote-access banking and/or bill payment services.

Among the growing number of virtual banks on the Internet is Atlanta Internet Bank, a service of Carolina First Bank. The *Federal Deposit Insurance Corporation* (FDIC)-insured virtual bank offers a comprehensive portfolio of services that can be accessed online via its Web page. Customers can choose any service that a traditional bank offers (refer to Figure E-2). Clicking the Transfer Funds icon, for example, enables the user to open a secure connection for transferring money from a money market fund to his or her checking account. Some virtual banks, such as *Security First Network Bank* (SFNB), even offer a check-imaging feature that enables users to view and print a copy of each check that has cleared in the past 90 days. Customers

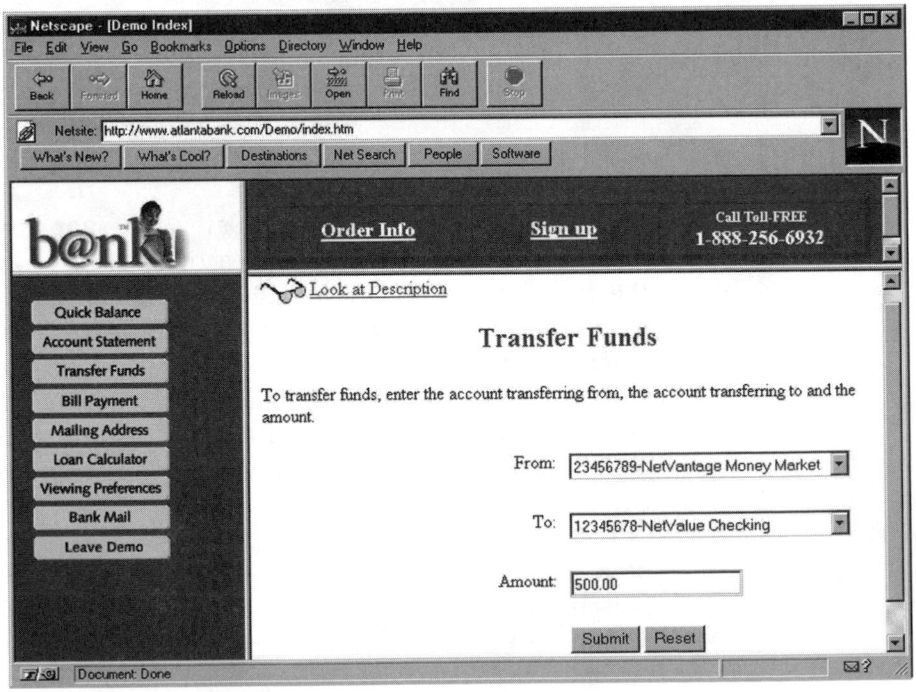

Figure E-2
The Web page of Atlanta Internet Bank, an FDIC-protected bank that operates on the Internet

can also transfer funds back and forth between accounts through point-and-click commands.

Internet checking accounts at SFNB have no monthly fees or minimum balance requirements. Customers receive 20 free electronic payments each month, unlimited check-writing privileges, and a free ATM or debit card. SFNB is even insured by the FDIC, just like brick-and-mortar banks. SFNB expects to offer customers online discount stock brokerage, insurance purchasing, and consumer loans. A single balance sheet will display all assets and liabilities, brokerage activity, and account information. Extending these and other financial services over the Internet requires retail companies, credit card issuers, and financial institutions to implement secure transaction systems that guard against tampering by hackers and to protect sensitive financial information from falling into the wrong hands.

Standards

Several major industry groups have developed standards that secure electronic transactions over the Internet.

Secure Electronic Transaction (SET) *Secure Electronic Transaction* (SET) is a technical specification for securing payment-card transactions over open networks, such as the Internet. SET helps the buyer and seller complete a transaction and have the transaction authorized by a bank. SET also secures payment-card transactions over the Internet by using a combination of data encryption, user-authentication services, and digital certificates in order to safeguard a buyer's card number and to ensure that the transaction data is immune from tampering.

Encryption is tamper-proof because of data partitioning, which entails separating the credit card number from the payment information that is traveling across the Internet. Today's *Electronic Funds Transfer* (EFT) systems are based on the same concept: *Electronic Data Interchange* (EDI) information acts as a wrapper for the EFT payment message. Corporations have settled financial transactions this way for years; SET merely extends the technique to consumers.

The major advantage of SET over other online security systems is the addition of digital certificates that associate the cardholder and merchant with a financial institution and the Visa or MasterCard payment system. Digital certificates prevent a level of fraud that other systems do not address. The certificates also provide cardholders and merchants with a higher degree of confidence that the transaction will be processed in the same high-quality manner as conventional Visa and MasterCard transactions in retail stores. Once the transaction is handed off to the credit card company's Web site, the SET protocol is shed—and the data rides as it usually would across MasterCard, Visa, and other private transaction-processing networks.

The specification is open and free to anyone who wishes to use it in order to develop SET-compliant software. ETSs based on SET have been in operation since 1998, but acceptance of SET has been slow. The standard has taken three years to develop, and there have been a number of well-publicized conflicts among vendors' offerings. SET's fraud-blocking features are seen as helpful but not critical for most merchants who are already online and who have purchased software or developed systems to perform the same function in a less-cumbersome fashion.

Open Financial Exchange **(OFX)**   Microsoft, Intuit (a financial software publisher), and CheckFree (an e-commerce service provider) have developed a single, unified technical specification that enables financial institutions to exchange data over the Internet with other customers. *Open Financial Exchange* (OFX) consolidates three proprietary standards: Microsoft's Open Financial Connectivity, Intuit's OpenExchange, and CheckFree's electronic banking and payment protocols. OFX is implemented via Internet security standards; specifically, SSL and *Private Communication Technology* (PCT).

OFX enables financial institutions to exchange financial data over the Internet with users of desktop and Web-based financial software by streamlining the process that financial institutions undertake in order to connect to multiple customer interfaces and systems. By making it more compelling for financial institutions to implement online banking, OFX opens the door to online transaction services for a growing number of consumers, enabling them to manage finances online with the institution of their choice.

OFX supports a wide range of financial activities, including consumer and small-business banking, consumer and small-business bill payment, and investments (including stocks, bonds, and mutual funds). The companies plan to add other financial services, including financial planning, insurance and tax preparation, and filing. Many financial institutions, including Visa Interactive, Fidelity, Schwab, and Royal Bank of Scotland, have endorsed the OFX standard.

Joint Electronic Payments Initiative **(JEPI)**   The *Joint Electronic Payments Initiative* (JEPI), sponsored by CommerceNet and the *World Wide Web Consortium* (W3C), provides a universal payment platform to enable merchants and consumers to transact business over the Internet by using different forms of payment. JEPI enables different payment instruments and protocols to exchange information. Because not all merchants accept all forms of payment and transport mechanisms, a payment-negotiation protocol could provide a standard way in which applications could negotiate the appropriate payment methods.

JEPI enables clients and servers to negotiate payment instruments, protocols, and transport between one another. JEPI consists of two parts: an extension layer that rides on the *Hypertext Transfer Protocol* (HTTP), and a negotiation protocol that identifies the appropriate payment method. The protocols make payment negotiations automatic for end users (this negotiation occurs at the moment of purchase, based on configurations within the browser).

The protocol can be used to build a type of financial services middleware that helps merchants and buyers identify whether specific transactions will be handled by e-mail or by file transfer, for example. The protocol also helps users identify whether the transaction supports such protocols as the payment card industry's SET specification.

Among the companies that are involved in JEPI are Microsoft, IBM, CyberCash, Xerox, British Telecom, and Digital Equipment. These companies are incorporating the payment-negotiation protocol into their e-commerce systems.

Summary

From the consumer's perspective, certain online transactions finally make economic sense. On a transaction-by-transaction basis, for example, online banking can actually be cheaper than postage stamps. With point-and-click ease, users can now launch Quicken, Microsoft Money, or a proprietary banking-software package and pay bills, open accounts, close accounts, transfer money between accounts, monitor stock portfolios, make adjustments to investments, and even create monthly ledgers for tax purposes. These online services offer automation, immediacy, and flexibility—key selling points, especially among harried urban consumers. Even the smallest businesses can now execute monthly financial transactions automatically and save money on bookkeeping —all over the Internet through participating financial institutions.

See Also

Electronic Data Interchange (EDI)

Electronic Data Interchange (EDI)

Since the 1970s, *Electronic Data Interchange* (EDI) has been promoted as the way to enhance business transactions with speed, accuracy, and cost savings. Essentially, EDI is the electronic transfer of business documents between companies in a structured, computer-processed data format. Because the information is transmitted over a *Value-Added Network* (VAN) and does not have to be rekeyed at the other end, the speed and accuracy of business transactions is greatly improved. Because business documents are in electronic form, cost savings can range from $3 to $10 per transaction over manual, paper-based procedures.

The idea behind EDI is simple. Instead of processing a purchase order with multiple paper forms and mailing it to the supplier, for example, the data is passed through an application link, where software maps the data into a standard, machine-readable data format. That data is then transmitted to the supplier's computer, where it is passed to an application link that maps the data into the internal format that is expected by the supplier's order-entry system.

While e-mail also transfers business data electronically, it generally uses a free format rather than a structured format. Because the sender can choose any format, it would be difficult to design an application program that would directly accept e-mail input from many different sources without significant manual editing. EDI standards exist in order to overcome this situation.

Benefits of EDI

EDI offers many benefits to participants. Data flow within the organization is streamlined, making it easier to develop and maintain complete audit trails for all transactions. Having all of this information online also provides a means of tracking vendor performance, performing cost-benefit analyses, improving project management, and enhancing overall corporate financial control.

A reduced order-to-pay cycle is the natural result of eliminating the use of regular mail and cutting the time that is required to process paper at both ends of the transaction. In other words, buyers can wait longer before replenishing stock that is ordered from suppliers, thereby reducing inventory and associated costs. The inherent efficiency of EDI means that buyers can pay for goods sooner—perhaps qualifying for further discounts. At the same time, sellers can improve their cash position through timely payments from buyers.

With more accurate and timely data, planning and forecasting can be greatly improved. Companies can better plan the receipt of materials to coincide with assembly-line schedules or with continuous control-processing operations, thus eliminating unnecessary down time caused by material shortages. Finally, dramatic cost savings that are associated with daily business transactions can accrue to users of EDI, because the manual tasks of sorting, matching, filing, reconciling, and mailing documents are eliminated.

For many small to mid-size companies, the pressure for an EDI implementation is increasingly coming from larger customers who indicate that EDI is the preferred method of transacting business. In some cases, EDI implementation is becoming a prerequisite for continuation of the business relationship. Some larger companies are even subsidizing EDI implementation for their smaller suppliers, recovering the cost by realizing even greater efficiencies and economies in transaction processing.

Role of VANs

The main service that a VAN offers is a reliable, secure clearinghouse for EDI transmissions. The VAN operates similarly to an e-mail system. The user sends the VAN a single transmission containing several interchanges that are destined for various trading partners. The VAN provides assurances that the data was received intact and that it was distributed to each recipient's e-mail mailbox. In turn, the service provider sends the contents of each mailbox to the appropriate subscriber upon request. The VAN also provides security in order to guarantee that no one but the addressee can access the data. The VAN performs other services, including compliance checking and translation of data from one protocol standard to another. VANs also provide software installation and troubleshooting assistance to their subscribers.

The VAN supplies EDI translation software to its subscribers, and this software maps the data from an application (order entry) and translates it into the EDI format. This software runs on a modem-equipped PC with communications software to provide dialup access to the VAN. If the volume of transactions justifies, a dedicated connection to the VAN can be used instead.

Another approach, which represents a more thorough adoption of EDI, entails a company integrating several business applications into the EDI

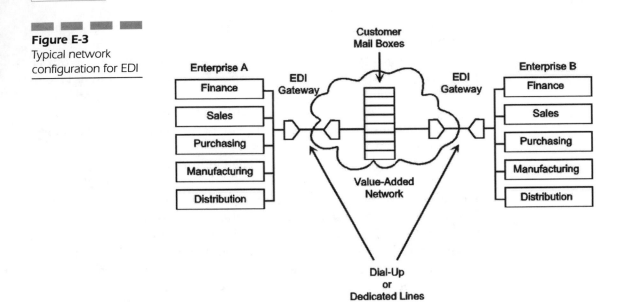

system and using its own communication facilities and services to carry the transaction messages to an EDI gateway that connects to a VAN (refer to Figure E-3). Each application is re-engineered so that the output is already in the prescribed EDI format, making further translations unnecessary.

Interactive versus Batch Processing

The problem with traditional batch processing for EDI transactions is that as a company adds trading partners and increases the volume of messages, batch jobs begin to contend with one another for access to computer and communications resources. EDI messages, such as ship notices, might be delayed because the EDI translator is busy processing a large batch job or because all available communications channels are being used. Aside from rescheduling jobs, these problems can be solved through the purchase of additional equipment or the leasing of more lines.

To improve this situation, companies are adopting fast-batch EDI and interactive EDI. Fast-batch EDI essentially speeds up the store-and-forward process by sending messages through e-mail mailboxes on VANs directly into a recipient's computer system. Interactive EDI typically involves establishing a two-way link that enables trading partners to rapidly exchange records or fields within an EDI message, rather than the entire message.

Web-Based EDI

The latest innovation in EDI entails using the Web to send business documents. This functionality enables smaller firms that have not traditionally used EDI software and that have subscribed to the services of a VAN to participate in the process at a low cost of entry. The common format for Web-based document exchanges is the *Hypertext Markup Language* (HTML), which can work in several ways. In one scenario, when a company fills out HTML-based forms, these forms are converted behind a VAN's Web site (i.e., gateway) into EDI-formatted messages before being passed on to the recipient. In another scenario, two trading partners might exchange documents directly over the Web, without going through a VAN. A back-end program puts the information in the right format so that it can be stored in a database. Security is maintained during transmission over the Web through the *Secure Hypertext Transfer Protocol* (SHTTP) or SSL protocol.

Summary

In the future, businesses that have not (or will not) implement EDI will be at a competitive disadvantage. Companies that derive considerable benefit from EDI will be reluctant to deal with those companies that force them to become bogged down with manually processing paper transactions. This situation goes against the trend toward flatter, decentralized organizational infrastructures, which many companies believe is necessary to compete more effectively on a global scale. The traditionally high cost of EDI is plummeting as more vendors offer inexpensive software that enables EDI to be run over the Web. Even traditional VAN providers see the writing on the wall and are starting to offer Web-based EDI service. This situation will widen the appeal of EDI to even the smallest companies and will further stimulate the growth of e-commerce.

See Also

Electronic Commerce

Value-Added Networks (VANs)

World Wide Web (WWW)

Electronic Mail

The popularity of *Electronic Mail* (E-Mail) in recent years has paralleled the growth of the Internet. More mail is now delivered in electronic form over the Internet than is delivered by the United States Postal Service. That amounts to tens of billions of messages per month. Message delivery usually takes only minutes over the Internet, instead of days or weeks (as is typical with postal services in many countries).

Advantages of E-Mail

Automating information delivery and processing with e-mail can dramatically reduce the cost of doing business, because the manual tasks of sorting, matching, filing, reconciling, and mailing paper files are virtually eliminated. Also, businesses can take advantage of attendant cost savings on overnight delivery services, supplies, file storage, and clerical personnel. Users can even send e-mail from within any application that supports Microsoft's *Messaging Applications Programming Interface* (MAPI) specification. Thus, a file that is created in a MAPI-compliant word processor or spreadsheet application, for example, can be e-mailed as an attachment without having to leave that application.

Through mail gateways, e-mail can be sent to people who might subscribe to some other type of service, such as *America Online* (AOL) or the *Microsoft Network* (MSN). Some gateway services, such as RadioMail, can even transport messages from the Internet to wireless devices such as *Personal Digital Assistants* (PDAs) and portable computers that are equipped with radio modems. Also, with more paging services supporting short-text messaging, e-mail can even be sent over the Internet to alphanumeric pagers.

The servers and gateways on the Internet take care of message routing and delivery. The e-mail address contains all of the information that is necessary to route the message. If a message cannot be delivered because of some problem on the Internet, an error message is returned that explains the reason and estimates the time of delivery.

Internet Protocols

Currently, the dominant mail protocols used on the Internet are the *Simple Mail Transfer Protocol* (SMTP) and *Post Office Protocol 3* (POP3). When installed on a server, SMTP gives the server the capability to send and route messages over the Internet. POP3 is also installed on a server, giving the server the capability to hold incoming e-mail until the recipient is ready to download the mail onto his or her own computer. Once downloaded, the e-mail message can be opened, edited for reply, cut and pasted into another application, filed for future reference, forwarded, or deleted (refer to Figure E-4).

POP was designed to support offline message access. Messages must be downloaded before they can be read and then deleted from the mail server. This mode of access is not compatible with access from multiple computers, because it tends to distribute messages across all of the computers that are used for mail access.

A new protocol is now available called the *Internet Mail Access Protocol* (IMAP4), which enables users to access messages on the server as if the messages were local. For example, e-mail that is stored on an IMAP server can be manipulated from a desktop computer at home, from a workstation at the office, and from a notebook computer while traveling—without the need to transfer messages or files back and forth between these computers. IMAP's capability to access messages (both new and saved) from more than one computer has become extremely important as reliance on electronic

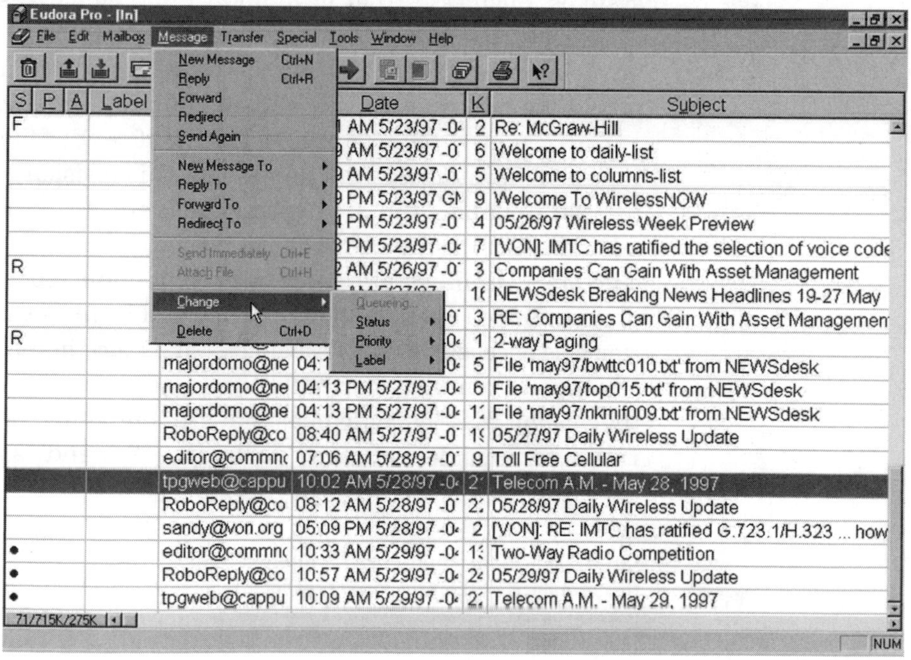

Figure E-4

A typical e-mail interface; in this case, QualComm's Eudora Pro messaging and the use of multiple computers among mobile professionals increases. The key advantages of IMAP include the following:

- Full compatibility with other Internet messaging standards, such as MIME[1]
- Message access and management from more than one computer
- Message access without reliance on less-efficient file-access protocols
- Support for online, offline, and disconnected access modes
- Support for concurrent access to shared mailboxes
- No required knowledge about the server's file-store format

The IMAP protocol includes operations for creating, deleting, and renaming mailboxes; checking for new messages; permanently removing messages; setting and clearing flags; server-based MIME parsing (relieving clients of this burden) and searching; and selective fetching of message attributes and text.

Outsourcing Arrangements

Many businesses find it difficult and time-consuming to run their own e-mail systems. Businesses that have up to 1,000 employees might find that outsourcing their e-mail is a more attractive option. By subscribing to a carrier-provided e-mail service, these companies can save from 50 percent to 75 percent on the cost of buying and maintaining an in-house system.

Carriers even offer guaranteed levels of service that will minimize down time and maintain 24-hour support for this mission-critical application. An interface enables the subscribing company to partition and administer accounts. Companies can even create multiple domains for divisions or contractors. Administration is entirely conducted through the service provider's Web site. Changes are easy and intuitive; no programming experience is necessary, and the changes take place immediately. Many services offer spam controls to virtually eliminate junk mail. Outsourcing can level the playing field for companies that are competing in the Internet economy. Businesses no longer need to budget for endless rounds of software

---

[1]MIME stands for Multi-Purpose Internet Mail Extensions, which is a technique for encoding text, graphics, audio, and video files as attachments to SMTP-compatible Internet mail messages.

upgrades. A carrier-provided service can also include value-added applications such as collaboration, calendaring, and unified messaging services.

For companies that are not ready to entrust their entire e-mail operation to a third-party service provider, customized solutions are available that enable them to selectively outsource certain aspects and functions. For example, a company can choose to have its e-mail system hosted on the service provider's server. Such midsourcing can reduce the costs of administration and support and can provide improved performance. In addition, the company can easily add new features without incurring a costly upgrade.

Summary

Only a decade ago, e-mail was considered by most companies to be just a fad. Now, there is a great appreciation for e-mail and its role in supporting daily business operations. In fact, the popularity of e-mail is so great now that many companies are being forced to upgrade the capacity of their communications links and systems in order to accommodate the growing traffic load. Steps are also being taken to minimize unnecessary traffic, such as storing only one copy of e-mail attachments on a server, rather than enabling the same attachment to be duplicated to all recipients of the message. Other companies are considering outsourcing their e-mail operations to carriers and to *Internet Service Providers* (ISPs), which is part of the new trend toward applications outsourcing.

See Also

> *Application Service Provider* (ASP)
>
> Electronic Commerce
>
> Facsimile
>
> Paging
>
> Unified Messaging

Electronic Software Distribution

With a growing population of PC and workstation users deployed across widely dispersed geographical locations—each potentially using different combinations of operating systems, applications, databases, and network protocols—software has become infinitely more complex and difficult to install, maintain, and meter. The capability to perform these tasks over a network from a central administration point can leverage investments in software, enforce vendor license agreements, and greatly reduce network support costs.

Industry experts estimate that the average five-year cost of managing a single desktop PC exceeds $45,000 and that the five-year cost of deployment and managing changes to new client/server applications will average nearly an additional $45,000 to $55,000 per user. These support costs can be cut by as much as 50 percent by automating the distribution and maintenance of software from a central administration facility. *Electronic Software Distribution* (ESD) tools can also provide useful reports that can aid in problem resolution.

Automating File Distribution

The complexity of managing the distribution and implementation of software at the desktop requires that network administrators make use of automated file-distribution tools. By assisting a network administrator with tasks such as packaging applications, checking for dependencies, and offering links to event and fault-management platforms, these tools reduce installation time, lower costs, and speed problem resolution.

One of these tools is a programmable file-distribution agent that is used to automate the process of distributing files to particular groups or workstations. A file distribution job can be defined as software installations and upgrades, startup file updates, or file deletions. Using a file-distribution agent, these types of changes can be automatically applied to each workstation or group.

The agent can be set up to collect file distribution status information. The network administrator can view this information at the console to determine whether files were distributed successfully. The console provides view formats and types that enable the administrator to review status data, such as which workstations are set up for file distributions, the stations to which files have been distributed, and the number of stations that are waiting for distributions.

Because users can be authorized to log in at one or more workstations, the file-distribution agent determines where to distribute files based on the primary user (owner) of the workstation. The owner is established the first time someone takes a hardware/software inventory of the workstation. Before automated file distributions are run, the hardware inventory agent is usually run in order to check for resource availability, including memory and hard-disk space. Distributions are made only if the required resources are available to run the software.

Via scripts, the network administrator can define distribution criteria, including the group or station to receive files and the day or days on which the files are to be distributed. Scripts usually identify the files for distribution and the hardware requirements that are needed to run the file distribution job successfully. Many vendors provide templates in order to ease script creation. The templates are displayed as a preset list of common file tasks. Using the templates, the network administrator can outline a file-distribution script, then use the outline to actually generate the script.

To help network administrators prepare for a major software distribution, some products offer routines called wizards that walk administrators through the steps that are required to assemble a package. A package is a complete set of scripts, files, and recipients that are necessary to successfully complete a distribution and to install the software. To reduce network traffic associated with software distributions, some products automatically compress packages before they are sent to another server or workstation. At the destination, the package is automatically decompressed when accessed. When a file distribution job is about to run, target users receive a message indicating that files are about to be sent and requesting that they choose to either continue or cancel the job.

Managing Installed Software

Maintaining a software inventory enables the network administrator to quickly determine which operating systems and applications are installed on various servers and clients. In addition to knowing which software packages are installed and where they are located, the network administrator can track application usage in order to ensure compliance with vendor license agreements (refer to Figure E-5).

The ESD tool creates and maintains a software inventory by scanning all of the disk drives on the network. Usually, software-management tools come with preset lists of software packages that they can identify during a scan of all disk drives on the network. Some tools can recognize several

Application	License File	Total	Used	Inactive	Free	Queued
WordPerfect 6.0a	WP	19	19	4	0	0
Microsoft Office 4.2	MSOFFICE	5	3	1	2	0
Microsoft Excel for Windows 5.0a	EXCEL	5	3	1	2	0
WordPerfect 5.1+	WPDOS	3	3	1	0	1
Lotus SmartSuite	SMART	5	2	1	3	0
Microsoft Project for Windows 4.0	WINPROJ	3	1	1	2	0
Corel CorelDRAW 5.0	CORELDRW	5	5	2	0	0
Borland Paradox 4.5	PDOXDOS	2	1	0	1	0
Fifth Generation FastBack Plus 6.0	FB	1	1	0	0	1
Computer Select 3.0	COMPSEL	2	0	0	2	0

thousand software packages. Software that cannot be identified during a scan is tagged for further inquiry and manual data entry. The next time a scan is performed, the added software packages will be properly identified.

Whether the initial software inventory is established by drive scanning or by manual entry, the following information about the various software packages usually can be added or updated at any time:

■ *Package information*   The number of available software licenses, product manufacturers, and project codes to log application use to a particular project

■ *Software availability*   Describes when the software can be accessed by users (open or closed). The software package can be closed during the upgrade process, and a message can be displayed to indicate when the application becomes available. A startup message can also be displayed when users open particular applications. For example, the network administrator can post a message telling users that this application is a beta copy of an application or that a new release of the software will be installed on a specific date. Being able to relay such information to users helps prevent unnecessary support calls to the help desk.

■ *Files in package*   Executable files that are associated with the software and files that are used to verify integrity (for example, the presence of a computer virus or unauthorized file access)

- *Additional information about the application*   Which departments in the organization have access to the application or to the vendor's technical support phone number

- *Optional information*   Whether the software is a Windows or OS/2 application, for example, and the directory to which the software should be added

In addition, access to applications can be restricted by user, workgroup, department, or project. Certain financial software, for example, can be restricted to accounting personnel. Personnel-management software can be restricted to the human resources department. CAD/CAM software can be restricted to an engineering workgroup. Locking out unauthorized users prevents inadvertent data loss or malicious damage to files.

Metering Software Usage

The capability to track software usage helps the network administrator ensure that the organization complies with software license agreements while also making sure that users have access to required applications. Tracking software usage also helps reduce software acquisition costs, because accurate usage information can be used to determine which applications are run most before deciding on upgrades and choosing how many copies to buy.

Metering enables the network administrator to control the number of concurrent users of each application. The network administrator can also choose to be notified of the times when users are denied access to particular applications because all available copies are in use. This option might identify the need to purchase additional copies of the software or the need to pay an additional license charge to the vendor so that more users can access the application.

Before users are granted access to metered software, the software inventory is checked to determine whether any copies are available. If no copies are available, a status message is issued indicating that there are no copies available. The user waits in a queue until a copy becomes available.

Metering software can save money on software purchases and ensure compliance with software copyright laws. For example, if there are 100 Microsoft Word users on an enterprise subnetwork and only half of them use Word concurrently, the software metering tool's load-balancing feature automatically handles the transfer of available software licenses to another subnetwork on a temporary or permanent basis. Load balancing helps network administrators purchase licenses based on need, rather than on the number of potential users at a given location.

License-management capabilities are important, because it is a felony under U.S. federal law to copy and use (or sell) software. Companies that are found guilty of copyright infringement face financial penalties of up to $100,000 per violation. The *Software Publisher's Association* (SPA) runs a toll-free hotline and receives about 40 calls a day from whistle-blowers. The SPA sponsors an average of 250 lawsuits a year against companies that are suspected of software-copyright violations. Since 1988, every case in which the SPA has been involved has been settled successfully. Having a license-management capability can help the network administrator track down illegal copies of software and eliminate a company's exposure to litigation and financial risk.

Some software-metering packages enable application usage to be tracked by department, project, workgroup, and individual for charge-back purposes. Charges can be assigned on the basis of general network use, such as time spent logged onto the network or disk space consumed. Reports and graphs of user groups or department charges can be printed or exported to other programs, such as to an accounting application.

Although companies might not require departments or divisions to pay for application or network usage, charge-back capabilities can still be a valuable tool for breaking down operations costs and planning for budget increases.

Distribution over the Internet

Since 1996, corporations and ASPs have had the option of subscribing to Internet-based software-distribution services. Such services are usually hosted on a Web server and provide managed Internet-based delivery and tracking of business-critical software applications and documents. Available via a link from any Web or extranet site, such services offer all of the functionalities of an enterprise-class software-management system with the convenience and flexibility of a hosted Web service.

Such services are designed for large and mid-size companies that are using the Web to extend their businesses and to lower the cost of customer support and product distribution. As these companies make broader use of their extranets, there is an increasing need to distribute proprietary digital files in order to support online business services. On an outsourcing basis, service providers offer confirmed delivery and tracking of these software and document packages in a secure and cost-effective manner.

The entire service is available from a standard browser, offering users the flexibility and ease-of-use of an intuitive Web interface. Businesses that seek to outsource their software distribution operations simply create authorized user lists and packages of software, documents, and digital files that will be available for downloading. The service provider's management server authenticates users and maintains a database in order to track distribution transactions, package versions, and customer accounts—providing comprehensive reports to support business operations.

Once the system is set up, users are assigned an authorization code and are directed to a secure, fully customized download portal site that is hosted by the service provider. Here, they can browse through available software, check their download histories, and select packages by name, version, or category. The download manager is a Java applet that loads dynamically when a customer requests an update package. The applet executes a managed download, confirming receipt of the files on the customer's computer. As additional packages are made available, the service provides automatic e-mail notification of the updates, providing a way to instantly notify authorized users and groups.

The service might also include a block-level restart feature, which automatically resumes the file transfer if a download is interrupted and synchronizes from the point of failure. All download activities are secure, using multi-level encryption and authentication features.

In addition to a completely managed service, some service providers also offer a lower-cost self-drive option, enabling IT and Web site managers to run the service from a Web browser—with all distribution and tracking activity operated from the service provider's distribution center.

Summary

ESD tools are essential for managing software assets. They can help trim support costs by permitting software to be distributed and installed from a central administration point, in addition to ensuring compliance with vendor license agreements, thereby eliminating exposure to lawsuits for copyright infringement. They can also help companies manage software in order to minimize their investments while meeting the needs of all users.

See Also

> *Applications Service Providers* (ASPs)
>
> Asset Management

Emergency Service—911

The number 911 is the designated universal emergency number in North America for both wireline and wireless telephone service. Dialing 911 puts the caller in immediate contact with a *Public Safety Answering Point* (PSAP) operator who arranges for the dispatch of appropriate emergency services—ambulance, fire, police, and/or rescue—based on the nature of the reported problem. Since its inception in 1968, this concept has amply demonstrated its value by saving countless lives in thousands of cities and towns across the United States and Canada.

Although 911 represents a major advance in the seven-digit, multiple-agency emergency-service concept of years past, the systems that implemented the abbreviated dialing plan in early years had serious limitations. The PSAP operators had to rely on their notetaking abilities during crisis situations that were called in by near-hysterical people who had to be calmed down before a meaningful exchange of information could occur. This procedure wasted valuable time, risked the omission of vital information, and left too much room for error.

If the caller hung up before providing location information, there was no way for the PSAP operator to know who called without involving the telephone company, whose call-trace procedures might or might not have been successful in obtaining that information. Without the originating telephone number, the PSAP operator could not determine whether calls were really emergencies or just hoaxes. Even when the phone number was provided by the telephone company, without additional information from the caller, the PSAP operator often did not know what type of emergency service to dispatch to the location.

Advances in technology led to the introduction of *Enhanced 911* (E911) service in the mid-1980s. The primary enhancement consisted of automating the entire sequence of events leading to the dispatch of appropriate emergency services. Automation eliminated human error and provided precise location information, which resulted in faster response times, fewer wasted efforts on false alarms, and a substantial cost savings on the delivery of emergency services. Automation also resulted in a dramatic increase in the number of lives saved.

System Components and Operation

While a number of sophisticated management tools are now available to increase the operating efficiency of E911 systems, there are only four standard capabilities required to provide E911 service (refer to Figure E-6): *Automatic Number Identification* (ANI), *Automatic Location Information* (ALI), selective routing, and call transfer.

The selective routing system located at the local central office identifies 911 calls and matches the caller's directory number with its assigned primary and secondary PSAPs. The ANI control system is typically located at the PSAP. The device answers incoming calls from the central office and then requests the ANI for decoding.

Next, the ANI control system forwards call ringing to the PSAP's phone system and sends a ring back to the caller via the central office. (The caller hears this ring as a distinct second ring; the first ring comes from central office ring circuits, while the second comes from the PSAP's ANI control system.) The calling number goes to the ANI display at the PSAP attendant's

Figure E-6

Schematic showing the main hardware components of an E911 system

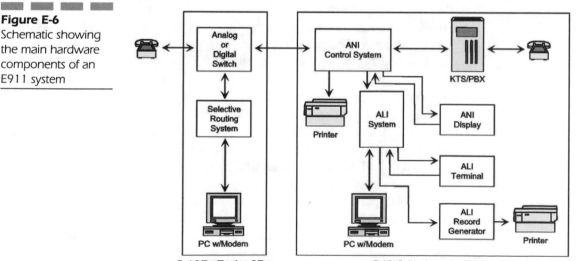

station. Metropolitan PSAPs might be equipped with *Automatic Call Distributors* (ACDs), which place calls in a queue for the next available operator. After a timeout period, the ACD will divert the call to a recorded announcement or will send the call back to the tandem office for rerouting to the designated secondary PSAP. Some manufacturers offer ACD as an integral part of E911 systems.

At the same time the calling number goes to the ANI display, the call is forwarded to the ALI system, where it is cross-referenced with the ALI database that provides detailed location and identification information. This database includes a file for each directory number. Each file contains such standard information as street address, occupant name(s), and the nearest facility for police, fire, medical service, and poison control.

This database can be customized to include the unique attributes of commercial and residential structures, hazardous conditions (explosives, chemicals, and radioactive materials), and specific information about the occupants (disabled person, children, elderly, etc.). Expanded screen formats enable the input of even more detailed information, which can be accumulated on a continuous basis. Database maintenance is accomplished via a PC, with updates loaded to the main computer on demand or automatically according to a predefined schedule. The data-entry system includes several security features, including passwords, in order to prevent unauthorized access.

Some E911 systems can also call up maps that can zero in to any level of detail, showing the nearest locations of such objects as power lines, gas and water mains, hydrant positions, and buried cable. With this information, firefighters and rescue teams can arrive at the scene mentally prepared and appropriately equipped to deal with virtually any emergency.

The ALI system is located at the PSAP. In addition to interfacing with the ANI control system and matching ANI information with the appropriate database file for the caller, the ALI system automatically forwards location and identification information to the proper agency's attendant-position terminal. These terminals can be located at the police station or fire station or at a centralized emergency-response center.

An optional ALI record-generator system interfaces with the telecommunications company's change-order system to automatically assign the nearest PSAP, police and fire departments, emergency medical service, and poison control center to street addresses. When used with an ALI system, the record-generator system automatically updates all on-site PSAP databases.

The selective routing system is a multi-functional device that can be located at either a tandem office or an end office. As its name implies, the system selectively routes 911 calls by directory number to primary/secondary

PSAPs. If the primary PSAP is busy or if there is no response due to a common equipment failure, the selective routing system will reroute the call to a predefined secondary PSAP.

Most E911 systems include an automatic callback capability that is used to verify the legitimacy of incoming calls. Optional single-button call transfer enables PSAP operators to transfer calls on a discretionary basis to appropriate secondary PSAPs or emergency service providers. Also, a call-sharing feature enables two or more PSAPs or emergency service providers to share information, which is displayed simultaneously on operator terminals. Local police, sheriffs, and state troopers, for example, can share information and immediately agree on appropriate jurisdiction before dispatching services and personnel.

Based on the processing options selected, the E911 system also can perform the following actions:

- Print a copy of the information that is displayed on the screen
- Accept input from the PSAP operator regarding the type of response to the call and its outcome
- Write a record to the daily call file, showing the incoming telephone number, date, and time, for later report generation

Additionally, the system can optionally produce a variety of reports (such as an online log report, which shows all incoming calls) and other information in chronological order. Also available are monthly and quarterly reports showing the number of calls by operator, the number of calls by response, and other statistical information.

Wireless E911

Enhanced wireless E911 services help ensure that in emergencies, wireless phones provide vital information in order to assist 911 call centers, or PSAPs, with locating the caller. Adequate funding for PSAPs remains critical to successful E911 implementation. Therefore, a mechanism for PSAP cost recovery must be in place before a wireless carrier is obligated to provide E911 services. To help ensure that carriers are not required to make unnecessary expenditures before a PSAP is ready to use E911 information —and to encourage and support local and state authorities in the funding of wireless E911—the FCC requires a mechanism to be in place for PSAP cost recovery.

The FCC prefers that negotiations between wireless carriers and PSAPs should remain the primary means of ensuring the expeditious selection of transmission methods that meet the individual requirements of the PSAP and carrier in each situation. In the event that an impasse arises, the FCC will be available to help resolve these disagreements.

The *Incumbent Local Exchange Carriers* (ILECs) play a key role in the implementation of wireless E911 service by transmitting calls from the wireless carrier to the PSAP. Although the FCC does not have specific rules that apply to the ILECs with regard to supporting wireless E911 calls, parties could bring complaints before the state public-service commissions or the FCC if an ILEC fails to perform of any of its obligations.

Wireless carriers must have the capability to recover the costs that they incur for implementing E911. The responsibility for setting up a cost-recovery mechanism is the responsibility of the states. Many states have passed legislation that applies a subscriber surcharge to wireless customers, much like the surcharge that has traditionally been applied to wireline service in order to pay for E911 services. In most areas, wireline customers currently pay a surcharge for all services, but this charge does not cover the wireless E911 costs.

At the same time, PSAPs have reported that the current wireline surcharge is not sufficient to cover their expenses for provisioning 911 services to wireless subscribers, who make nearly 25 percent of all 911 calls. Moreover, wireless carriers will incur a hefty expense in retro-fitting existing systems to meet the new FCC requirements. For these reasons, a feasible cost-recovery mechanism is needed to offset the expected high costs of implementing wireless E911.

Summary

Emergency 911 service has become a valuable tool in rendering prompt and appropriate assistance to people who are in critical need. Most states have laws that mandate prompt action on all calls that are received by a PSAP operator. Unfortunately, the 911 system is so taken for granted that many calls are not for emergencies at all, and expensive resources end up being needlessly expended on trivial pursuits. PSAP operators now receive calls on such matters as garbage-collection dates, late mail delivery, a leaky faucet or a heater that the landlord will not fix, directions to stores and restaurants, and whether or not to see a lawyer for a particular problem. The 911 systems in some communities are so bogged down with non-emergency calls that public-awareness campaigns are launched periodically by telephone companies and emergency service providers with the aid of the local news media.

See Also

> *Automatic Number Identification* (ANI)
> Cellular Voice Communication

Ethernet

The Ethernet LAN originated as a result of the experimental work that was done by the Xerox Corporation at its *Palo Alto Research Center* (PARC) in the mid-1970s. Once developed, Ethernet quickly became a de facto standard with the backing of DEC and Intel. Xerox licensed Ethernet to other companies that developed products based on the formal specification issued by Xerox, Intel, and DEC in 1980. Because Ethernet enabled interoperability among the equipment of different vendors, a number of independent vendors embraced Ethernet as the industry standard for LANs.

The pure Ethernet specification that was jointly developed by Xerox, DEC, and Intel defined the functions that were performed by the lowest levels of the *Open Systems Interconnection* (OSI) Reference Model—the physical and data link layers. Essentially, Ethernet is a multi-access, packet-switched network that uses a contention-based method of broadcast. Much of the original Ethernet design was incorporated into the 802.3 standard that was adopted in 1980 by the *Institute of Electrical and Electronic Engineers* (IEEE).

Ethernet is based on a bus topology that is contention-based, meaning that stations compete with each other for access to the network—a process that is controlled by a statistical arbitration scheme. Each station listens to the network to determine whether the network is idle. Upon sensing that no traffic is currently on the line, the station is free to transmit. If the network is already in use, the station backs off and tries again.

Frame Format

The IEEE 802.3 standard defines a multi-field frame format, which differs only slightly from that of pure Ethernet (refer to Figure E-7).

Preamble   The frame begins with an eight-byte field called a preamble, which consists of 56 bits having alternating 1 and 0 values. These bits are

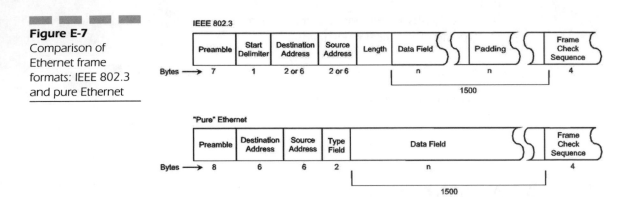

used for synchronization and to mark the start of the frame. The same bit pattern that is used in the pure Ethernet preamble is used in the IEEE 802.3 preamble, which includes the one-byte start frame delimiter field.

Start Frame Delimiter   The IEEE 802.3 standard specifies a start frame delimiter field, which is really a part of the preamble. This field is used to indicate the start of a frame.

Address Fields   The destination address field identifies the station(s) that are to receive the frame. The source address field identifies the station that sent the frame. If addresses are locally assigned, the address field can be either two bytes (16 bits) or six bytes (48 bits) in length. A destination address can refer to one station, to a group of stations, or to all stations. The original Ethernet specifies the use of 48-bit addresses, while IEEE 802.3 permits either 16-bit or 48-bit addresses.

Length Count   The length of the data field is indicated by the two-byte count field. This IEEE 802.3-specified field is used to determine the length of the information field when a pad field is included in the frame.

Pad Field   To detect collisions properly, the frame that is transmitted must contain a certain number of bytes. The IEEE 802.3 standard specifies that if a frame being assembled for transmission does not meet this minimum length, a pad field must be added in order to bring the frame up to that length.

Type Field   Pure Ethernet does not support length and pad fields, as does IEEE 802.3. Instead, two bytes are used for a type field. The value that is specified in the type field is only meaningful to the higher network layers and is not defined in the original Ethernet specification.

Data Field This portion of the frame is passed by the client layer to the data link layer in the form of eight-bit bytes. The minimum frame size is 72 bytes, while the maximum frame size is 1,526 bytes (including the pre-amble). If the data to be sent uses a frame that is smaller than 72 bytes, the pad field is used to stuff the frame with extra bytes. Fewer problems occur during collision handling than when you define a minimum frame size. If the data to be sent uses a frame that is larger than 1,526 bytes, the higher layers must break the data into individual packets in a procedure called fragmentation. The maximum frame size reflects practical considerations related to adapter card buffer sizes and the need to limit the length of time that the medium is tied up in transmitting a single frame.

Frame Check Sequence A properly formatted frame ends with a frame check sequence, which provides the means to check for errors. When the sending station assembles a frame, it performs a *Cyclical Redundancy Check* (CRC) calculation on the bits in the frame. The sending station stores the results of the calculation in the four-byte frame check sequence field before sending the frame. At the receiving station, an identical CRC calculation is performed, and a comparison is made with the original value in the frame check sequence field. If the two values do not match, the receiving station assumes that a transmission error has occurred and requests that the frame be retransmitted. In pure Ethernet, there is no provision for error correction. If the two values do not match, the client layer receives notification that an error has occurred.

Media Access Control (MAC)

Several key processes are involved in transmitting data across the network. Among these processes are data encapsulation/decapsulation and media-access management, both of which are performed by the MAC sublayer of OSI's data link layer.

Data Encapsulation/Decapsulation Data encapsulation is performed at the sending station. This process entails adding information to the beginning and end of the data unit to be transmitted. The data unit is received by the MAC sublayer from the LLC sublayer. The added information is used to perform the following tasks:

- Synchronize the receiving station with the signal
- Indicate the start and end of the frame
- Identify the addresses of sending and receiving stations
- Detect transmission errors

The data encapsulation function is responsible for constructing a transmission frame in the proper format. The destination address, source address, and type and information fields are passed to the data link layer by the client layer in the form of a packet. Control information that is necessary for transmission is encapsulated into the offered packet. The CRC value for the frame check sequence field is calculated, and the frame is constructed.

When a frame is received, the data decapsulation function that is performed at the receiving station is responsible for recognizing the destination address, performing error checking, and then removing the control information that was added by the data encapsulation function at the sending station. If no errors are detected, the frame is passed to the LLC sublayer.

Specific types of errors are checked in the decapsulation process, including whether the frame is a multiple of eight bits or whether it exceeds the maximum packet length. The address is also checked to determine whether the frame should be accepted and processed further. If so, a CRC value is calculated and checked against the value in the frame check sequence field. If the values match, the destination address, source address, and type and data fields are passed to the client layer. The packet in its original form is passed to the station.

Media-Access Management   The method that is used to control access to the transmission medium is known as media access management in IEEE terms, but this concept is called link management in Ethernet parlance. Link management is responsible for several functions, starting with collision avoidance and collision handling (which are defined by the IEEE 802.3 standard for contention networks).

Collision avoidance   Collision avoidance entails monitoring the line for the presence or absence of a signal (carrier). This portion is the carrier sense portion of *Carrier-Sense Multiple Access with Collision Detection* (CSMA/CD). The absence of a signal indicates that the channel is not being used and that it is safe to begin transmission. Detection of a signal indicates that the channel is already in use and that transmission must be withheld. If no collision is detected during the period of time known as the collision window, the station acquires the channel and can complete the transmission without risking a collision.

Collision handling   When two or more frames are offered for transmission at the same time, a collision occurs that triggers the transmission of a sequence of bits called a jam. When a jam occurs, all stations on the network recognize that a collision has occurred. At that point, all transmissions in progress are terminated, and retransmissions are attempted at calculated intervals. If there are repeated collisions, link management uses a backing-off process that involves increasing the retransmission wait time following each successive collision.

On the receiving side, link management is responsible for recognizing and filtering fragments of frames that resulted from a transmission that was interrupted by a collision. Any frame that is less than the minimum size is assumed to be a collision fragment and is not reported to the client layer as an error.

New methods have been developed to improve the performance of Ethernet by reducing or totally eliminating the chance for collisions, without having to segment the LAN into smaller subnetworks. Special algorithms sense when frames are on a collision course and will temporarily block one frame while enabling the other frame to pass.

Summary

Ethernet is the most popular type of LAN, and its success has spawned some interesting innovations. 10BaseT Ethernet, for example, enables the LAN to operate over ubiquitous *Unshielded Twisted-Pair* (UTP) wiring instead of thick or thin coaxial cable. Also, Isochronous Ethernet brings ISDN to the desktop to support multimedia applications, including voice. For those who find 10Mbps inadequate for supporting large file transfers and graphic-intensive applications, higher-speed versions of Ethernet are available—including Fast Ethernet at 100Mbps and Gigabit Ethernet at 1Gbps. At the time of this writing, a standard for a 10Gbps version of Ethernet is in progress.

See Also

Ethernet (10BaseT)

Ethernet (100BaseT)

Ethernet (1000BaseT)

Ethernet (Isochronous)

Ethernet (VG-AnyLAN)

Simple Network Management Protocol (SNMP)

Ethernet (10BaseT)

10BaseT is an IEEE standard for providing 10Mbps Ethernet performance and functionality over ubiquitously available UTP wiring. This standard is noteworthy in that it specifies a star topology, unlike traditional 10Base2 and 10Base5, which use coaxial cabling laid out in a bus or ring topology. The star topology permits centralized network monitoring, which enhances fault isolation and bandwidth management. For new installations, twisted-pair wire is substantially less expensive as well as easier to install and maintain than Ethernet's original thick coaxial cable (10Base5) or the thin coaxial alternative (10Base2).

Performance

Traditionally, Ethernet has relied upon coaxial cable with multi-drop connections attached to a LAN segment. Ethernet can be extended by joining multiple cable segments with repeaters, which boost the signals. The key disadvantage of this topology is that any disturbance to the continuity of the cable at any point along the bus will isolate the segments.

In contrast, the star topology that is inherent in twisted-pair hub systems dedicates a link to each user. Therefore, if an individual link were to exhibit an impairment, only the user on that link would be affected. 10BaseT promises this exact benefit. Because the network can tolerate a malfunction of any end-user device or its physical link, the rest of the network will not be affected—resulting in improved network availability. This situation is made possible by the link test and auto-partition logic that is inherent in the port-level circuitry of a 10BaseT hub.

A network-management system can make problems even easier to identify. When a pre-defined error threshold is exceeded, for example, the network management system will alert the LAN administrator, enabling a technician to be dispatched before the user even realizes that a problem exists. Alarm notification is usually by a screen message or visual indicator; some systems can even notify the LAN administrator via pager.

10BaseT and traditional bus Ethernet LANs have different distance limitations: 10BaseT operates reliably over cable segments not exceeding 100 meters, whereas traditional Ethernet LANs using thick coaxial cable operate reliably over segments of up to 500 meters in length. 10Base2 Ethernets are effective at 200 meters using thin coax. The cable limitation of 10BaseT aside, its bit error-rate performance is at least as good as 10Base2 and 10Base5 systems. The 10BaseT specification enables bit-error rates of no more than one in 100 million bits. The data-encoding scheme that is used for 10BaseT systems is the same as the scheme that is used for coaxial-based Ethernets: self-clocking Manchester Encoding.

10BaseT LANs are designed to support the same applications as traditional Ethernet LANs. The relatively short distance over which 10BaseT LANs operate is rarely a factor in its capability to support these applications. During the standardization process, a survey by AT&T revealed that more than 99 percent of employee desktops are located within 100 meters of a telephone wire closet in which all of the connections meet at a patch panel.

System Components

Because the 10BaseT standard specifies a star configuration, installation entails the use of some equipment that is not always found on traditional networks. The most obvious component of 10BaseT LANs is the type of media used: twisted-pair wiring.

The Media   Most installed telephone wiring is known as 24 *American Wire Gauge* (AWG). Even if the existing telephone wiring follows the required star configuration and is within the roughly 100-meter (330 feet) distance limitation, it might still not be suitable to handle the 10Mbps data rate. No particular wire gauge or type is specified in the 10BaseT standard, although 24 AWG is what most equipment vendors have used in conformance tests to confirm that their products transmit reliably at up to 100 meters.

More important than gauge, however, are the attenuation (signal loss), impedance (resistance), delay, and cross-talk characteristics of the wiring. Minimally acceptable levels for all of these factors are detailed in the 10BaseT specifications. Generally, the inside wiring that has been installed in the past 20 years for telephone connections meets the 10BaseT specifi-cations at cabling runs of up to 100 meters. Older wiring might not meet the standards, and as a result, poor or erratic LAN performance could occur—and maximum transmission distances could be considerably less than 100 meters.

The 10BaseT standard was designed to eliminate the requirement for shielded wiring. This standard relies on the twists in twisted-pair wire to hold down frequency loss, which in turn improves the integrity of the signal. The twists minimize the effects of this loss by preventing high-frequency sig-nal energy from radiating to and corrupting the signal that is being carried over nearby twisted pairs (cross-talk). Even the individual wire pairs in 25-pair telephone cable are twisted. This cable is widely used in large installa-tions; it enables as many as 12 10BaseT segments to be neatly carried, separated, and patched to a panel for distribution to the hub.

Adapter Cards Adapter cards, also known as *Network Interface Cards* (NICs), are boards that insert into a PC's or server's expansion slot. The adapter card connects the device's bus directly to the LAN segment, elimi-nating the need for a separate transceiver. Depending on the type of adapter, the card permits connections to thick or thin coaxial cable as well as to unshielded twisted pairs. An adapter with an *Attachment Unit Inter-face* (AUI) port will permit connection to a transceiver, enabling the card to be used with thick and thin coaxial cables and fiber-optic cables.

10BaseT adapters vary considerably in their capabilities and features for reporting link status. Some adapters notify the user of a miswired connec-tion to the LAN. Many adapters have LED displays that indicate link sta-tus after the connection is made to the hub, such as link, collision, transmit, and receive. Others provide minimal information (indicating that the link is correctly wired and working, for example), while some only light up when something is wrong. Still other cards have no LEDs to display link infor-mation. Some cards feature a menu-driven diagnostic program to help the user isolate problems with the cards, such as finding out whether the adapter is capable of responding to commands from the software.

Transceivers Transceivers connect PCs and peripherals that are already equipped with legacy Ethernet cards to 10BaseT wiring. Typically, a transceiver consists of a small external box with an RJ45 jack at one end for connecting to the unshielded twisted-pair LAN segment and an AUI port at the other end for connecting to the Ethernet adapter card.

***Medium-Access Unit* (MAU)**   MAUs terminate each end of the 10BaseT link. As such, the MAU accommodates two wire pairs: one pair for transmitting the Ethernet signal, and the other pair for receiving the signal. The 10BaseT standard describes seven basic functions that the MAU performs. The transmit, receive, collision detection, and loop-back functions direct data transfer through the MAU. The jabber detect, signal-quality error test, and link integrity functions define ancillary services that the MAU provides.

The jabber function removes equipment from the network whenever it continuously transmits for periods that are significantly longer than required for a maximum-length packet (indicating a possible problem with the NIC). The signal-quality error test detects silent failures in the circuitry, while the link-integrity signal detects breaks in the wire pairs. Both components assist with fault isolation.

The Hub   All 10BaseT stations are connected to the hub via two twisted-pair wire pairs—one two-wire path for transmitting, and the other two-wire path for receiving—over a point-to-point link. In essence, the hub acts as a multi-port repeater and consists of the necessary circuitry to retime and regenerate the signal that is received from any of the wire segments that connect at the hub to each of the other segments. A 10BaseT hub, however, is more than a simple repeater: it also serves as an active filter that rejects damaged packets.

The multi-port repeater provides packet steering, fragment extension, and automatic partitioning. The packet-steering function broadcasts copies of packets that are received at one repeater port to all of its other ports. Fragment extension ensures that partially filled packets are sent to their proper destination. Automatic partitioning isolates a faulty or misconnected 10BaseT link to prevent the link from disrupting traffic on the rest of the network.

Several types of 10BaseT hubs exist—the most common being chassis-based solutions and stackable 10BaseT hubs that can be cascaded with appropriate cable connections. At the low end, hubs come in managed or unmanaged versions. The chassis models are high-end devices that provide a variety of connections to the WAN, support multiple LAN topologies and media, and offer advanced network management with SNMP typically included, as well. This type of 10BaseT hub is also the most expensive. Aside from enhancing their products with more interfaces and network-management features to further differentiate them from competitive offerings, vendors are continually increasing the port capacity of these devices to bring down the price per port, making them more appealing to larger users.

Depending on port capacity, stackable hubs are used to link members of a workgroup or department. With a pass-through or crossover cable, additional units can be added in a cascading arrangement in order to meet growth requirements or to connect to a high-end hub.

Management

The promise of 10BaseT networks was not only that Ethernet would run over economical unshielded twisted-pair wiring and offer unprecedented configuration flexibility, but that it would offer a superior approach to LAN management. Some of the most important management capabilities that are available through the hub include the following:

- Support for the IEEE 802.3 Repeater Management standard and the Internet's Repeater *Management Information Base* (MIB)
- Remote site manageability from a central management station
- Provision of performance statistics (not only at the port level, where such information has been traditionally available, but at the module and hub level, as well)
- Auto-partitioning, which involves the capability to automatically remove disruptive ports from the network
- The capability to set performance thresholds that notify managers of a problem or automatically take action to address the problem
- Port address-association features that connect the Ethernet MAC address of a device with the port to which it is attached
- Source/destination address information, which aids network redesign, traffic redistribution, troubleshooting, and security

Summary

Overall, 10BaseT LANs offer a sound technical solution for most applications that would normally make use of Ethernet. In the office environment, packet throughput and error rates over twisted-pair wiring are the same as with coaxial systems, but the former is limited to shorter distances. The standard offers protection against equipment and media faults that can potentially disrupt the network, and the signaling method used is reasonably immune from most sources of electromagnetic interference that are commonly found in the office environment. With the increasing demand for bandwidth to support today's data-intensive applications, 10BaseT is rapidly being displaced by 100Mbps (100BaseT) and 1000Mbps (1000BaseT) LANs. In recent years, Ethernet NICs and hubs have supported both 10Mbps and 100Mbps segments.

See Also

> Ethernet
>
> Ethernet (100BaseT)
>
> Ethernet (1000BaseT)
>
> Ethernet (Isochronous)
>
> Ethernet (VG-AnyLAN)
>
> *Simple Network Management Protocol* (SNMP)

Ethernet (100BaseT)

The increasing complexity of desktop-computing applications is fueling the need for high-speed networks. Emerging data-intensive applications and technologies (such as multimedia, groupware, imaging, and a continuation of the explosive growth in the use of high-performance database software packages on PC platforms) all tax today's client-server environments and demand even greater bandwidth and improved client-server response times.

A number of high-speed LAN technologies are available to address the needs for more bandwidth and improved response times. Among them is Fast Ethernet, or 100BaseT, which is a technology that is designed to provide a smooth migration from current 10BaseT Ethernet—the dominant 10Mbps network type in use today—to high-speed 100Mbps performance.

Compatibility

Fast Ethernet uses the same contention-based MAC method: *Carrier-Sense Multiple Access with Collision Detection* (CSMA/CD), which is at the core of IEEE 802.3 Ethernet. The Fast Ethernet MAC specification simply reduces the bit time (the time duration of each bit transmitted) by a factor of 10, enabling a 10 times greater boost in speed over 10BaseT. Fast Ethernet's scaled CSMA/CD MAC leaves the remainder of the MAC unchanged. The packet format, packet length, error control, and management information in 100BaseT are all identical to those in 10BaseT.

Fast Ethernet enables the use of the same UTP cabling that is already installed for 10BaseT networks and is implemented in the same star topology as a 10BaseT network. In addition, all of the 100BaseT Fast Ethernet products (as well as the 10BaseT products) can be managed from a single SNMP management application.

Because no protocol translation is required, data can pass between 10BaseT and 100BaseT stations via a hub that is equipped with a 10Mbps/100Mbps bridge module. Both technologies are also full-duplex capable. This compatibility enables existing LANs to be inexpensively upgraded to the higher speed as demand warrants.

Media Choices

To ease the migration from 10BaseT to 100BaseT, Fast Ethernet can run over Category 3, 4, or 5 UTP cable while preserving the critical 100-meter segment length between hubs and end stations. The use of fiber enables even more flexibility with regard to distance. For example, the maximum distance from a 100BaseT repeater to a fiber-optic bridge, router, or switch using fiber-optic cable is 225 meters (742 feet). The maximum fiber distance between bridges, routers, or switches is 450 meters (1,485 feet). The maximum distance between a fiber bridge, router, or switch (when the network is configured for half-duplex) is 2 kilometers (1.2 miles). By connecting repeaters and internetworking devices, large, well-structured networks can be easily created with 100BaseT.

The types of media that are used to implement 100BaseT networks are summarized as follows:

- *100BaseTX*   A two-pair system for data grade (EIA 568 Category 5) UTP and STP cabling
- *100BaseT4*   A four-pair system for both voice and data grade (Category 3, 4, or 5) UTP cabling
- *100BaseFX*   A multi-mode, two-strand fiber system

Together, the 100BaseTX and 100BaseT4 media specifications cover all cable types that are currently in use in 10BaseT networks. Because 100BaseTX, 100BaseT4, and 100BaseFX systems can be mixed and interconnected through a hub, users can retain their existing cabling infrastructure while migrating to Fast Ethernet. 100BaseT also includes an MII specification that is similar to the 10Mbps AUI. The MII provides a single interface that can support external transceivers for any of the 100BaseT media specifications.

Summary

Unlike other high-speed technologies, Ethernet has been installed for more than 20 years in business, government, and educational networks. The migration to 100Mbps Ethernet is made easier by the compatibility of 10BaseT and 100BaseT technologies, making it unnecessary to alter existing applications for transport at a higher speed. This compatibility enables 10BaseT and 100BaseT segments to be combined in both shared and switched architectures, enabling network administrators to apply the right amount of bandwidth in an easy, precise, and cost-effective manner. Fast Ethernet is managed with the same tools as 10BaseT networks, and no changes to current applications are required to run them over the higher-speed 100BaseT network.

See Also

Ethernet

Ethernet (10BaseT)

Ethernet (1000BaseT)

Ethernet (Isochronous)

Ethernet (VG-AnyLAN)

Simple Network Management Protocol (SNMP)

Ethernet (1000BaseT)

Ethernet is a highly scaleable LAN technology. Long available in two versions, 10Mbps Ethernet and 100Mbps Fast Ethernet, a third version has now become standardized—offering another capability for bandwidth increase. Offering a raw data rate of 1000Mbps or 1Gbps, so-called Gigabit Ethernet uses the same frame format and size as previous Ethernet technologies. Gigabit Ethernet also maintains full compatibility with the huge installed base of Ethernet nodes through the use of LAN hubs, switches, and routers.

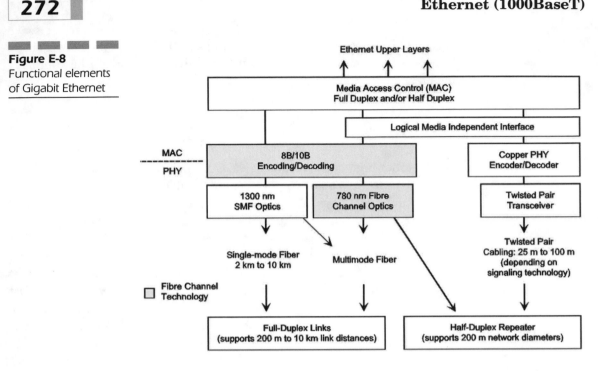

Gigabit Ethernet supports full-duplex operating modes for switch-to-switch and switch-to-end station connections and half-duplex operating modes for shared connections using repeaters and the CSMA/CD access method. Figure E-8 illustrates the functional elements of Gigabit Ethernet.

The initial efforts in the IEEE 802.3z standards process drew heavily on the use of Fibre Channel and other high-speed networking components. Fibre Channel encoding/decoding integrated circuits and optical components are readily available and are specified and optimized for high performance at relatively low costs. The first implementations of Gigabit Ethernet employed Fibre Channel's high-speed, 780-nm (short wavelength) optical components for signaling over fiber optics and 8B/10B encoding/decoding schemes for serialization and deserialization. Fibre Channel technology, operating at 1.063Gbps, was enhanced to run at 1.250Gbps—thus providing the full 1000-Mbps data rate for Gigabit Ethernet. Link distances—up to 2 km over single-mode fiber and up to 550 meters over 62.5-micron multi-mode fiber—were specified as well.

In mid-1999, the IEEE and the Gigabit Ethernet Alliance formally ratified the standard for Gigabit Ethernet over copper. The IEEE 802.3ab standard defines Gigabit Ethernet operation over distances of up to 100 meters, using four pairs of Category 5 balanced copper cabling. The standard adds a Gigabit Ethernet physical layer (1000Base-T) to the original 802.3 standard, enabling a higher speed over the existing base of Category 5 unshielded twisted-pair wiring. The standard also enables auto-negotiation between 100-Mbps and 1000-Mbps equipment.

The following table summarizes the outcome of Gigabit Ethernet standards over various media:

Specification	Transmission Facility	Purpose
1000BaseLX	Long-wavelength laser transceivers	Support links of up to 550 meters of multi-mode fiber or 3,000 meters of single-mode fiber
1000BaseSX	Short-wavelength laser transceivers operating on multi-mode fiber	Support links of up to 300 meters using 62.5-micron multi-mode fiber or links of up to 550 meters using 50-micron multi-mode fiber
1000BaseCX	STP cable spanning no more than 25 meters	Support links among devices that are located within a single room or equipment rack
1000BaseT	UTP cable	Support links of up to 100 meters using four-pair Category 5 UTP

Source: IEEE 802.3z Gigabit Task Force

For the first few years of product rollouts, Gigabit Ethernet is not expected to be deployed on the desktop. The initial applications for Gigabit Ethernet will be campuses or buildings that require greater bandwidth between routers, switches, hubs and repeaters, and servers. Examples include switch-to-router, switch-to-switch, switch-to-server, and repeater-to-switch links.

At this writing, a draft specification for Ethernet at the SONET OC-192 rate of 10Gbps is in the works. This technology will not be used for building high-speed campus and building backbones; rather, it will be used as a service offering from carriers for *Metropolitan-Area Networks* (MANs) or as a fat pipe to access the Internet. Service providers are looking for a familiar, low-cost, high-speed technology to support networked applications, such as VPNs, IP telephony, transparent LAN services, and e-commerce. Ethernet running at 10Gbps offers all of those functions and more, in addition to reducing operational costs (because it is the one technology that can be used from LAN to MAN to WAN). Because there is no need for ancillary equipment, such as protocol translators, equipment and operational costs can be greatly reduced.

Summary

The seamless connectivity to the installed base of 10-Mbps and 100-Mbps equipment, combined with Ethernet's scaleability and flexibility to handle new applications and data types over a variety of media, makes Gigabit Ethernet a practical choice for high-speed, high-bandwidth networking. 1000BaseT enables the deployment of Gigabit Ethernet into the large installed base of Category 5 cabling, preserving the investment that organizations have made in existing cable infrastructure.

See Also

> Ethernet
>
> Ethernet (10BaseT)
>
> Ethernet (100BaseT)
>
> Ethernet (Isochronous)
>
> Ethernet (VG-AnyLAN)
>
> Fibre Channel

Ethernet (Isochronous)

With billions of dollars already invested in Ethernet LANs, companies are finding that it is impractical to either replace or install parallel networks in order to deliver high-bandwidth applications to individual desktops—especially multimedia applications that might also have to traverse the WAN. Economical methods are available for improving and extending current LAN infrastructures so that they can accommodate such applications. This idea drives the IEEE 802.9a standard, which is also referred to as ISLAN-16T or isochronous Ethernet. Isochronous Ethernet was originally developed by National Semiconductor.

As its name implies, isochronous (which literally means "same time") Ethernet simply adds the isochronous component, in the form of multi-rate ISDN, to standard Ethernet. The marriage of Ethernet and ISDN provides for the guaranteed QoS for voice in networked multimedia applications. In addition, isochronous Ethernet is interoperable with H.320, T.120, and MPEG standards for videoconferencing, document conferencing, and video distribution.

The premise behind isochronous Ethernet is that voice—not data or video —is the critical component in multimedia communications. During a conversation, even the slightest delay creates major disruption. Echoes caused by delays can confuse even the most articulate speakers. A split-second delay can disrupt a simple two-way conversation by causing one participant to ask a question again, only to interrupt the other's attempt to answer. Anyone who has experienced a telephone conversation via a round-trip satellite link is acquainted with the potentially disruptive effects of delay.

Guarding against this situation requires the use of isochronous protocols (in this case, ISDN). A network is isochronous when it operates in real time, with time being defined by the worldwide standard 8KHz clock for voice communications. Today's LANs are asynchronous and do not provide the 8KHz clocking signal that the voice component of a multimedia application requires (delivered through an H.320 codec, for instance). Because isochronous Ethernet inherently provides this synchronization, it offers the means to operate over the PSTN and delivers the same clocking to every desktop on the LAN. The IEEE 802.9a standard is the only LAN technology that extends 8KHz clocking to the desktop. All others, including *Asynchronous Transfer Mode* (ATM) networks, require expensive buffering to smooth out traffic bursts and adaptation techniques to transmit and receive synchronous traffic, such as H.320 voice and video. This functionality makes real-time multimedia conversations possible with isoEthernet.

Isochronous Ethernet is a hybrid network that integrates standard 10Mbps Ethernet (IEEE 802.3's 10BaseT) LAN technology with 6.144Mbps of isochronous bandwidth, for a total of 16Mbps that is available to any user; hence the IEEE designation of Integrated Service Local Area Network, or ISLAN 16-T (refer to Figure E-9). Through an encoding scheme called 4B:5B, the total bandwidth is increased to 16Mbps by using the same 20MHz clock that provides only 10Mbps using the Manchester encoding scheme of traditional Ethernet.

Over the same ubiquitous Category 5 unshielded twisted-pair wire that is used for10BaseT, isochronous Ethernet provides 96 56Kbps/64Kbps ISDN B channels to bring packet plus wide-band, circuit-switched multimedia services to the desktop. This amount represents at least 30 times more bandwidth than is needed to support a multi-point video conference with six participants, each using 384Kbps, plus interactive whiteboarding and presentation graphics. In the background, the user can even receive a facsimile as well as e-mail and voice mail messages over separate channels.

Isochronous Ethernet can be integrated into an existing 10BaseT environment simply by putting an isochronous Ethernet hub into the wiring closet

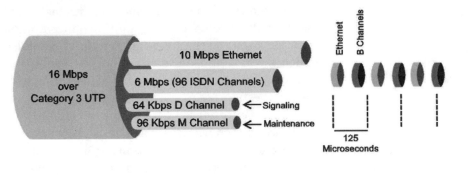

with the 10BaseT hub and connecting the two hubs with an *Attachment Unit Interface* (AUI). This link provides basic connectivity among all users.

Multimedia workstations, outfitted with isochronous Ethernet adapter cards, connect to the isochronous Ethernet hub. Because an isoEthernet hub and isoEthernet cards are only needed to connect a workgroup of multimedia users, existing investments in Ethernet 10BaseT infrastructure are protected, including bridges/routers, servers, applications, test equipment, and management tools. Furthermore, isoEthernet can be added incrementally on a user-by-user basis without changing the existing LAN applications, networking operating systems, or core network infrastructure.

If multimedia applications must traverse the WAN, ISDN links are used. The isoEthernet hub provides synchronization between WAN and LAN services. Because all of the major LECs already support ISDN, they do not have to perform any special circuit provisioning in order to accommodate multimedia applications.

Summary

At one time, isochronous Ethernet represented one of the most attractive and affordable network solutions available for the delivery of multimedia. Although there are proprietary solutions from numerous vendors, isochronous Ethernet is the only solution that is capable of integrating LAN and WAN environments without carriers and companies having to make extensive changes to their networks. Despite its apparent advantages, isochronous Ethernet never really caught on in the marketplace—and today, few vendors offer products that support this functionality.

See Also

>*Asynchronous Transfer Mode* (ATM)
>
>Ethernet
>
>*Integrated Services Digital Network* (ISDN)

Ethernet (100VG-AnyLAN)

An alternative to 100BaseT Ethernet is 100VG-AnyLAN. Both operate at 100Mbps. The 100VG-AnyLAN standard was developed by the IEEE 802.12 committee with the backing of AT&T and Hewlett-Packard. Similar to 100BaseT, 100VG-AnyLAN can be deployed in both shared-media and switched LAN environments. Also, similar to 100BaseT, 100VG supports current network topologies without significant reconfiguration, including 10Mbps Ethernet and 4Mbps and 16Mbps Token Ring structures.

100VG-AnyLAN runs up to 100 meters (node-to-node) over four-pair Category 3, 4, and 5 UTP wiring, 150 meters over two-pair Category 5 STP wiring, and up to 2,000 meters over fiber-optic cable (both single-mode and multi-mode).

Advantages

100VG offers several advantages over 100BaseT. As its name implies, 100VG-AnyLAN can support both Ethernet and Token Ring legacy applications. Each hub in a 100VG network can be configured to support either 802.3 10BaseT Ethernet frames or 802.5 Token Ring frames. A single hub cannot support both frame formats at the same time, and all hubs on the network must be configured in order to use the same frame format. Because 100VG supports both Ethernet and Token Ring, a router is used to move traffic between a 100VG network that uses Ethernet frames and a network that transports Token Ring frames. Going from either type of 100VG network and ATM, FDDI, or other topology also requires a router or a translating bridge (refer to Figure E-10).

100VG eliminates packet collisions and permits more efficient use of network bandwidth by using a demand-priority access scheme instead of the

Figure E-10
100VG-AnyLAN
configuration,
supporting various
other transport
technologies

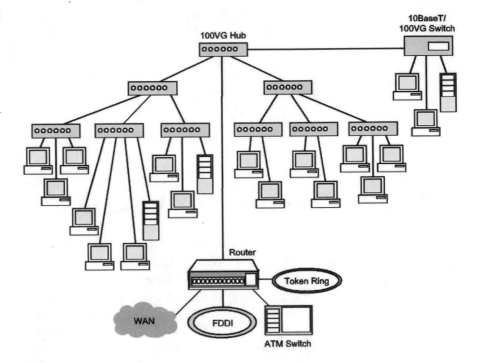

traditional CSMA/CD scheme that is used in 10BaseT and 100BaseT Ethernet LANs. Demand priority also permits rudimentary prioritization of time-sensitive traffic, such as real-time voice and video, making 100VG better suited for multimedia applications.

Media Access Control (MAC)

100VG-AnyLAN uses a unique MAC layer in which a collisionless demand-priority access scheme replaces pure Ethernet's CSMA/CD scheme. Instead of collision detection or circulating tokens, demand priority uses a round-robin polling scheme that is implemented inside a hub or switch. Round-robin polling makes it possible for every single-port node on a 100VG network to send one packet. If a multi-port hub or switch serves as a node on the network, it must request access from the root hub at the next-higher level. Multi-port nodes can transmit one packet for each of their ports.

Using the demand-priority scheme, the 100VG hub or switch permits only one node to access the network segment at a time, thus eliminating the opportunity for packet collisions. When a node needs to send data over the network, it places a request with the hub or switch. The hub or switch then services each node on the segment in sequence. If the node has data to send, the hub or switch permits access to the segment. If the node does not have anything to send, the hub moves on to the next node. Because all nodes requesting access are served during each polling round, all nodes are assured fair access to the network.

In 100VG networks that use Token Ring frames, demand priority follows the same procedure, with the 100VG hub or switch essentially acting as the circulating token. Instead of waiting to catch the token before transmitting, a Token Ring node waits for permission from the hub or switch. As in conventional Token Ring architecture, only one node is permitted to transmit across a network segment at any given time. Multi-port 100VG nodes adhere to the same procedures when using Token Ring packets as they do when using Ethernet. When granted access to the root hub in a 100VG Token Ring, the multi-port 100VG hub or switch might transmit one packet per port.

Demand priority can overcome some inherent limitations of CSMA/CD, such as restricted network size. The more devices the network supports, the more collisions and contention for bandwidth. Although adding users can adversely affect network performance with 100VG, access to the network is always fair. The orderly operating procedure of demand priority makes it easier to add users to the network. Also, its capability to distinguish between high-priority and normal traffic ensures that bursty data applications in high-volume networks will not overwhelm time-sensitive multimedia traffic.

Multimedia

The most important requirement of multimedia transmission is that the receiver should have the capability to receive a long stream of information coherently. Ultimately, the application determines the degree to which this process occurs. Because demand priority is a deterministic protocol, it enables multimedia application developers to model a worst-case scenario and to use that information to determine when compression and timing routines should be invoked. By contrast, CSMA/CD is a first-come, first-served access method that offers no guarantee for data delivery (and it can result in poor performance).

With 100VG-AnyLAN's demand-priority access scheme, multimedia applications can request high-priority service from the 100VG hub. Ultimately, the combination of 100Mbps, deterministic operation, and prioritization enhances the performance of multimedia applications.

Summary

In 1998, there was talk in IEEE circles of extending 100VG-AnyLAN to a 1Gbps version that would compete with the emerging standard for Gigabit Ethernet. The idea was never pursued, and today, few vendors even support 100VG-AnyLAN in their products. Similar to isochronous Ethernet, 100VG-AnyLAN was simply overtaken by other more economical versions of Ethernet.

See Also

Ethernet

Ethernet (10BaseT)

Ethernet (100BaseT)

Ethernet (Gigabit)

Ethernet (Isochronous)

Facsimile

More than 50 million facsimile (fax) machines are in use in the United States, demonstrating the fact that fax transmission is still a critical channel for a wide variety of business communications, ranging from office correspondence and narrative documents of all kinds to computer-generated invoices, purchase orders, and other business forms.

Facsimile transmissions have become an indispensable part of everyday business life. In the United States, faxing accounts for 30 percent of corporate telecommunications bills. Fortune 500 companies alone spend an average of $15 million annually on fax-related transmission charges. In other countries where telecommunications costs more, faxing accounts for 40 percent to 65 percent of the bill.

Fax transmission can be implemented with a stand-alone fax machine, from desktop computers that are equipped with a modem/fax card, or through a fax server on a LAN. Stand-alone fax machines are still popular, despite the heavy reliance on LANs for all forms of communication. While they offer an economical solution in that they can be shared among many users, these stand-alone devices require the user to go to the machine whenever he or she needs to send or receive faxes. This situation wastes time and exposes documents to prying eyes. The big advantage of stand-alone devices, however, is that they accept printed documents that are not already in electronic form on someone's computer.

Many desktop computers now come equipped with a modem/fax card, enabling the computer to act like a fax machine. Large organizations find this solution expensive, however, because each computer must be equipped with an extra telephone line in order to handle incoming and outgoing faxes. This solution is also limited because the fax card cannot handle documents that are not already in electronic form, unless the user has access to a scanner. A fax server has the same limitations but has key advantages for certain applications. Because the fax server is attached to a LAN, anyone can use the fax server to send and receive documents.

Fax Servers

Fax servers consist of hardware and software components. The hardware is a server that is equipped with multiple modem/fax boards, communications interfaces, an appropriate amount of memory and hard-disk storage, and a battery backup subsystem. The software handles the formatting and conversion of faxes as well as the sending and receiving of faxes. The software also offers various inbound/outbound routing features.

Several automated, inbound routing techniques are in use today. With *Direct Inward Dialing* (DID), for example, each user on the network is assigned a personal fax number. Faxes are sent over a dedicated trunk line, received by the fax server, and then are routed according to the fax number used. Users are notified of an incoming fax in the same way that they are notified of e-mail, and they can then retrieve their fax from the server.

With *Dual-Tone Multifrequency* (DTMF) dialing, the sender dials the recipient's special extension number after the fax connection is made. The receiving machine then identifies the recipient by the tones and routes the fax accordingly. Tones sometimes signal the sender to enter the extension number, or an automated voice prompt might request the extension number.

With channel-based or line-based routing, a separate fax line is assigned to each recipient or department within a company. These lines are connected to the fax server, which receives all incoming faxes and routes them to the appropriate person or group. Source ID routing routes incoming faxes according to where they originated. A list of fax numbers is maintained at the fax server, so incoming faxes can be forwarded to the recipient who has been designated to receive all faxes from a particular fax number.

Fax servers that are equipped with *Optical Character Recognition* (OCR) can read the recipient's name or identification number on the fax cover sheet and then route the fax accordingly. This technology only works if the information is typed on the cover sheet, because handwriting cannot yet be read reliably.

Once an incoming alert has been delivered, a fax recipient can view the incoming fax on screen, store the fax, print the fax, or forward the fax. Voice notification of incoming faxes is an emerging trend that enables the fax server to place a call to a telephone, pager, or e-mail address in order to alert a user to an incoming fax. This feature is essential to universal messaging.

Role of *Call Subscriber ID* (CSID)

Most fax machines transmit a *Call Subscriber ID* (CSID) calling string with the documents that they send. The CSID value can be programmed into the fax machine or entered into a field from the fax software. This subscriber ID is intended to label the specific fax machine, but it also can be used to support different methods of automatic routing.

For CSID routing, the fax software builds a table that associates CSID strings to network nodes or e-mail addresses. The software captures the string, then routes the fax to the associated network station or sends the fax via e-mail. The table can also associate a group of recipients to a single CSID. CSIDs can be used to implement two basic types of automatic routing. If a particular user is the only one in the organization who receives faxes from a specific fax machine—perhaps from a branch office, a supplier, or a client—then the administrator could associate the CSID of the remote fax to a network or e-mail address (refer to Figure F-1). The fax software at the server would then automatically route any faxes from that remote fax machine to that person.

The administrator also could assign CSID strings to specific network addresses or to a group of addresses. When someone wants to send a fax, that person can tell the administrator which CSID code to input. The sender then would change the CSID field (if sending via software) or reconfigure the hardware to transmit the proper CSID. The routing table in the fax server would capture the designated string and would send the fax directly to the right person or workgroup. This setup works well, but only if the sender knows how to change the CSID at his or her end.

Figure F-1

Most fax machines transmit a Call Subscriber ID (CSID), which aids routing by associating the sender with a fixed e-mail address.

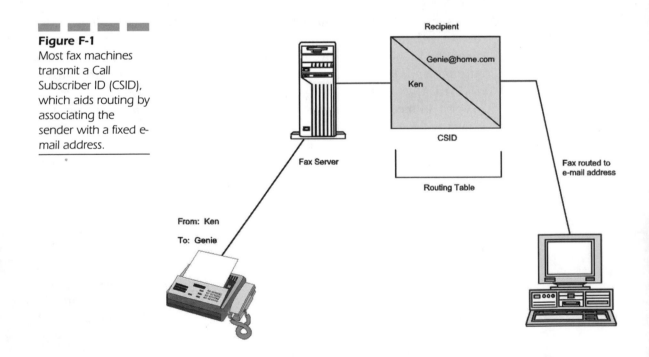

Other Features

Among the outbound routing features, fax servers offer broadcasting and delayed transmission. Broadcasting enables the same document to be faxed automatically to multiple recipients. Users can choose to send a fax to an entire group or only to certain individuals who are listed in the directory. Delayed transmission permits outgoing faxes to be collected for transmission after normal business hours, when long-distance telephone rates are lower. This capability is especially useful for fax transmissions to international locations.

Many fax servers also provide image-processing capabilities. Logos, signatures, and other graphics can often be incorporated into a fax document. Through the use of a print-capture utility that redirects print output from applications to the fax server, fax documents can be transmitted directly from within an application. In most cases, a cover sheet is automatically added to the fax file by the fax server software. The software also implements periodic cleanup procedures, which help prevent the server's hard disk from running out of space.

Many businesses offer important constituencies—customers, prospects, members, press, staff, or vendors—the capability to request documents by calling into their fax server 24 hours a day. These documents might be price lists, forms, sales literature, instruction sheets, or press releases. This type of service is called fax-on-demand, and it enables users to help themselves to documents, rather than tying up office staff in order to fulfill document requests manually. The user typically dials the telephone number of the fax machine and enters an extra number, indicating the document to fax back on the machine of the caller's choice.

Centralized Management

Like any other network device, the fax server itself can be monitored and controlled via familiar SNMP commands that are entered at the management station. Via an SNMP-compliant management module, network line failures, internal errors, and fax port/board problems can be reported. The administrator can be warned through a series of notification methods that include a local beep, an SNMP trap that is sent to a remote management station, a hard copy report, an e-mail message, or a signal to a pager device.

In addition, the fax server usually comes with tools that enable administrators to obtain current and long-term status information regarding both inbound and outbound faxes. The types of data that are provided include which faxes were transmitted successfully, when they were sent, how many pages were sent, and whether any retries were necessary. These logs help users track their individual faxes and enable administrators to keep track of overall usage. Some models offer an account-billing module that tracks fax usage and generates chargeback and billing reports. The administration tools of some fax servers also permit the manipulation of jobs in the queue, changes to priority settings, and cover-page creation.

Fax Service Bureaus

Transmitting thousands of faxes not only presents technical and support challenges, but it also creates a need for different types of reporting, accounting, and user interfaces. A standard fax board that sends fax after fax and generates thousands of confirmation reports is not up to the task. Consequently, organizations that have a heavy volume of faxes might use a service bureau for their document distribution needs. Some of these service bureaus rely extensively on the Internet to distribute imaged documents.

Summary

With the high cost of faxing, many users have turned to e-mail instead; after all, it is virtually free over the Internet. Many documents are not in electronic form. In fact, they might contain annotations, signatures, drawings, or clippings from other sources that are important to retain in their original form—even if it they can easily be put into electronic form for transmission as e-mail. Another consideration is that not everyone has an Internet connection in order to receive e-mail. All they might have is a stand-alone fax machine. Thus, there continues to be heavy reliance on traditional facsimile machines not only among large companies, but also among telecommuters and home-based businesses and individuals, as well. While many industry pundits predicted that the Internet would render fax machines obsolete, fax traffic is actually growing at 20 percent annually and is expected to exceed 500 billion pages in the year 2000—demonstrating that faxing continues to be the easiest and most widely-used form of document communication.

See Also

Electronic Mail

Internet Facsimile

Universal Messaging

Family Radio Service (FRS)

Family Radio Service (FRS) is one of the *Citizens Band* (CB) radio services. FRS is for family, friends, and associates to communicate among themselves within the neighborhood and while they are on group outings. Users can select any of the 14 FRS channels on a take-turns basis. No FRS channel is assigned to any specific individual or organization.

Although manufacturers advertise a range of up to two miles, users can expect a communication range of less than one mile. Although FRS can be used for business-related communications, it cannot be used for telephone calls.

License documents are neither needed nor issued. FRS Rule 1 provides all of the authority that is necessary to operate an FRS unit (refer to Figure F-2) in places where the FCC regulates radio communications, as long as only an unmodified, FCC-certified FRS unit is used. An FCC-certified FRS

Figure F-2
Motorola's TalkAbout 280 SLK is a palm-sized FRS unit that is small enough to carry in a shirt pocket. The six-ounce radio runs on three AA batteries.

unit has an identifying label placed on it by the manufacturer. No age or citizenship requirement is in place.

FRS units can be operated within the territorial limits of the 50 United States, the District of Columbia, and the Caribbean and Pacific insular areas. Such units can also be operated on or over any other area of the world, except within the territorial limits of areas in which radio communications are regulated by another agency of the United States or within the territorial limits of any foreign government.

Users cannot make any internal modification to an FRS unit. Any internal modification cancels the FCC certification and voids the user's authority to operate the unit over the FRS. In addition, users cannot attach any antenna, power amplifier, or other apparatus to an FRS unit that has not been FCC-certified as part of that FRS unit. No exceptions to this rule exist, and attaching any such apparatus to an FRS unit cancels the FCC certification and voids everyone's authority to operate the unit over the FRS.

Summary

FRS is used for conducting two-way voice communications with another person. One-way transmission can be used only to establish communications with another person, to send an emergency message, to provide traveler assistance, to make a voice page, or to conduct a brief test. At all times and on all channels, operators must give priority to emergency communication messages concerning the immediate safety of life or the immediate protection of property. No FRS unit can be interconnected to the PSTN.

See Also

> *Citizens Band* (CB) Radio Service
> General Mobile Radio Service
> Low-Power Radio Service

Federal Communications Commission (FCC)

The *Federal Communications Commission* (FCC) is an independent federal agency in the United States that reports directly to Congress. Established by the Communications Act of 1934, the FCC is charged with regulating interstate and international communications by radio, television, wire, satellite, and cable. Its jurisdiction covers the 50 states and territories, the District of Columbia, and U.S. possessions.

The FCC is directed by five commissioners who are appointed by the President and are confirmed by the Senate for five-year terms, except when filling an unexpired term. The President designates one of the commissioners to serve as chairman, and this chairman presides over all FCC meetings. The commissioners hold regular open and closed agenda meetings and special meetings. By law, the commission must hold at least one open meeting per month. They also can act between meetings by circulation—a procedure in which a document is submitted to each commissioner individually for consideration and official action.

Certain other functions are delegated to staff units and bureaus and to commissioners' committees. The chairman coordinates and organizes the work of the FCC and represents the agency in legislative matters and in relations with other government departments and agencies.

Operating Bureaus

At the staff level, the FCC is divided into operating bureaus and offices. Most issues that the FCC considers are developed by one of seven operating bureaus and offices that are organized by substantive area (refer to Figure F-3):

- The Common Carrier Bureau handles domestic wireline telephony.
- The Mass Media Bureau regulates television and radio broadcasts.
- The Wireless Bureau oversees wireless services such as private radio, cellular telephone, *Personal Communications Service* (PCS), and pagers.
- The Cable Services Bureau regulates cable television and related services.
- The International Bureau regulates international and satellite communications.
- The Compliance and Information Bureau investigates violations and answers questions.
- The Office of Engineering and Technology evaluates technologies and equipment.

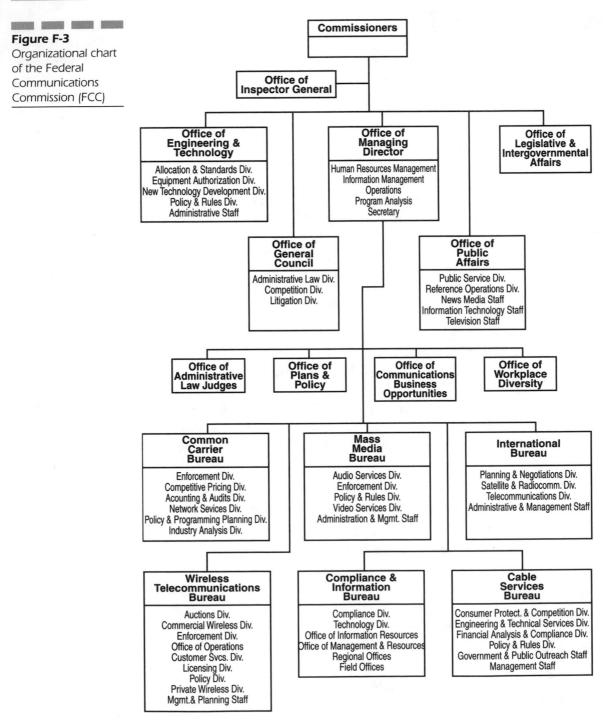

Figure F-3

Organizational chart of the Federal Communications Commission (FCC)

Other Offices

In addition, the FCC includes other offices:

- The Office of Plans and Policy develops and analyzes policy proposals.
- The Office of the General Counsel reviews legal issues and defends FCC actions in court.
- The Office of the Secretary oversees the filing of documents in FCC proceedings.
- The Office of Public Affairs distributes information to the public and to the media.
- The Office of the Managing Director manages the internal administration of the FCC.
- The Office of Legislative and Intergovernmental Affairs coordinates FCC activities with other branches of government.
- The Office of the Inspector General reviews FCC activities.
- The Office of Communications Business Opportunities provides assistance to small businesses in the communications industry.
- The Office of Administrative Law Judges adjudicates disputes.
- The Office of Workplace Diversity ensures equal employment opportunities within the FCC.

Reorganization Plan

In mid-1999, the FCC unveiled a five-year restructuring plan that will enable it to better meet the fast-changing needs of the communications business. The plan seeks to eliminate unnecessary rules in areas where competition has emerged and to reorganize the FCC along functional (rather than technological) lines. The plan hinges on thriving competition in five years, which is enough to reduce the need for the FCC to regulate directly. Under the plan, the FCC will be different both in structure and mission, with the commission managing the transition from an industry regulator to a market facilitator. The plan addresses the disintegration of boundaries between wire, wireless, satellite, broadcast, and cable communications—categories around which the FCC is currently grouped.

The new plan reorganizes the agency so that its internal bureaus are grouped around functions such as licensing and competition, rather than by technology. The plan consolidates enforcement and consumer information into two separate bureaus of the FCC, rather than having those functions spread across the agency. The plan also aims to streamline and speed up the FCC's services, for example by instituting agency-wide electronic filing and automated licensing systems. Other goals include reducing backlog and making greater use of alternative dispute-resolution mechanisms. The plan also makes it faster and easier for consumers to interact with the agency. Under the reorganization plan, the FCC keeps many of its existing priorities, such as protecting consumers from fraud and keeping phone rates affordable for the poor.

Summary

The top-down regulatory model of the Industrial Age is as out of place in today's digital economy as the rotary telephone. As competition and convergence develop, the FCC must streamline its operations and continue to eliminate regulatory burdens. Despite the FCC's planned reorganization from regulator to facilitator, however, the FCC will still have to contend with a set of core functions that are not normally addressed by market forces. These core functions include the following elements: universal services, consumer protection and information, enforcement and promotion of pro-competition goals domestically and worldwide, and spectrum management.

See Also

Public Utility Commissions

Regulatory Process

Telecommunications Act of 1996

Telecommunications Industry Mergers

Universal Service

Federal Telecommunications System 2000 (FTS2000)

The *Federal Telecommunications System 2000* (FTS2000) is an integrated communications network that is used by all agencies of the U.S. government. As such, it is the world's largest private network. AT&T, the primary service provider, completed its portion of the FTS2000 network in June

1990. FTS2000 serves 1.3 million users who work for 92 government agencies in some 4,200 locations that are scattered among the 50 states, including Alaska, Puerto Rico, Guam, and the Virgin Islands. Sprint also provides a portion of this network.

In addition to flexible and customized management, control, administration, and billing, FTS2000 provides the following services:

- Switched voice service
- Switched data service
- ISDN
- Packet-switched services
- Video transmission services
- Dedicated transmission service

All of these services support an extensive array of features, and most are accessible by individual users.

FTS2000 Architecture

The FTS2000 network enables users to select any or all of a wide range of FTS2000 services through *Service Delivery Points* (SDPs). An SDP is the combined physical and service interface between the network and the government's premises equipment, off-premises switching and transmission equipment, and other facilities. For Centrex service, for example, the SDP is the *Local Exchange Carrier* (LEC) central office. SDPs are used for billing purposes and are also the monitoring points from which volume discounts are calculated, based on monthly network usage.

The architecture of the FTS2000 network consists of the following basic elements (refer to Figure F-4):

- Service nodes
- Network access
- Distributed network intelligence
- Transport

Service Nodes The service node contains components that are engineered to provide users easy and efficient access to each of the FTS2000 services. These components are as follows:

Figure F-4
FTS2000 network
architecture

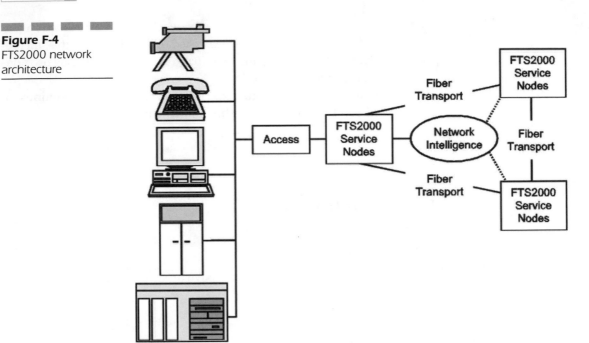

- Class 5 central office switch (5ESS)
- X.25 packet switch
- *Digital Access and Cross-Connect System* (DACS)

To accommodate the circuit-switching requirements of both Switched Voice Service and Switched Data Service in a single system, AT&T uses its 5ESS Switch in the service node. To meet the need for X.25 packet switching, AT&T uses its 1 PSS packet switch. Access to the 1 PSS switch is either direct or switched via the 5ESS switch. Integrated access to the various switches and digital transmission channels is accomplished with AT&T's DACS in the service node. DACS is an intelligent facilities-management system with channel drop-and-insert capabilities that increase the efficiency of digital circuit utilization and reduce facility administration and maintenance costs.

Network Access Network access provides connectivity between SDPs and the service nodes. The initial segment of access, from the government location to an FTS2000-access facility junction, is generally provided through the transmission facilities and serving wire centers of the *Regional Holding Companies* (RHCs) and/or LECs. The access-facility junction is the

AT&T-provided interface between the *Regional Bell Operating Company* (RBOC)/LEC and AT&T. This access-facility junction provides standard functions that are required between an LEC and an *Interexchange Carrier* (IXC). As the prime contractor, AT&T is responsible for all access to the FTS2000 network.

The connection to the FTS2000 network from the SDP to the AT&T access-facility junction is provided by the LECs, using dedicated facilities to ensure the isolation of FTS2000 access lines from the public network (except in those cases where the small amount of voice traffic proves virtual, on-net treatment in order to be more economical). The final segment of the access network is a dedicated, AT&T-provided circuit from the access-facility junction to an FTS2000 node.

Off-net access to public network facilities is provided at all service nodes. Calls from an on-net location to an off-net location remain on the FTS2000 network until those calls reach the service node that is nearest to the call destination (referred to as tail-end hop-off). This functionality maximizes the network utilization and high-performance transmission within the network.

Distributed Intelligence   The service nodes are complemented by distributed intelligence that consists of *Network Control Points* (NCPs), *No. 2 Signal Transfer Points* (#2STPs), and signaling links. The NCPs share call-processing functions with the 5ESS switches, enabling the network to handle peak user demands. The NCP databases also store user information that provides customized services. The #2STPs are elements of the AT&T *Common Channel Signaling* (CCS) network and are also used to provide SS7 signaling, which provides the network with ISDN capabilities.

Transport   Transport facilities are provisioned from existing AT&T fiber-optic facilities. FTS2000 is assigned exclusive use of selected 45Mbps (DS3) segments of AT&T's nationwide fiber network.

Summary

The competition to provide the federal government's next-generation long-distance network began with the May 1997 release of a *Request for Proposals* (RFP) from the United States General Service Administration's *Federal Telecommunications Service 2001* (FTS2001) program. GSA has awarded multiple contracts for FTS2001, which are worth more than $5 billion over eight years. GSA awarded the FTS2001 contracts to Sprint and to MCI WorldCom in December 1998 and January 1999, respectively. They replace the two FTS2000 contracts that were held by AT&T and Sprint that expired in December 1998.

Fiber-Distributed Data Interface (FDDI)

The *Fiber-Distributed Data Interface* (FDDI) is a 100Mbps token-passing network that employs a dual counter-rotating ring topology for fault tolerance. Originally conceived to operate over multi-mode fiber-optic cable, the FDDI standard has evolved to embrace single-mode fiber-optic cable, shielded twisted-pair copper wiring, and even unshielded twisted-pair copper wiring. FDDI is designed to provide high-bandwidth, general-purpose interconnection between computers and peripherals, including the interconnection of LANs (refer to Figure F-5) and other networks within a building or campus environment.

FDDI Operation

A timed token-passing access protocol is used to pass frames of up to 4,500 bytes in size, supporting up to 1,000 connections over a maximum multi-mode fiber path of 200 km (124 miles) in length. Each station along the path serves as the means for attaching and identifying devices on the network, regenerating and repeating frames that are sent to the station. Unlike other types of LANs, FDDI enables both asynchronous (time-insensitive) and synchronous (time-sensitive) devices to share the network. Synchronous services (e.g., voice and video) are intolerant of delays and must be guaranteed a fixed bandwidth or time slot. Synchronous traffic is therefore given priority over asynchronous traffic, which is better capable to withstand delay. FDDI stresses reliability, and its architecture includes integral manage-

Figure F-5
FDDI can carry Ethernet and Token Ring frames as data, providing a multi-protocol backbone network.

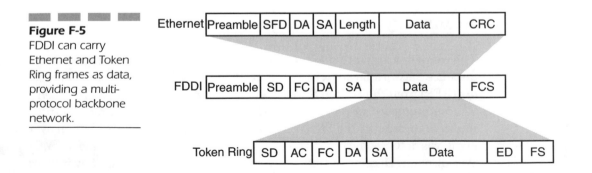

ment capabilities (including automatic failure detection and network reconfiguration).

Any change in the network status—such as power-up or the addition of a new station—leads to a claim process during which all stations on the network bid for the right to initialize the network. Every station indicates how often it must see the token in order to support its synchronous service. The lowest bid represents the station that must see the token most frequently. That request is stored as the *Target Token Rotation Time* (TTRT). Every station is guaranteed to see the token within $2 \times$ TTRT seconds of its last appearance.

This process is completed when a station receives its own claim token. The winning station issues the first unrestricted token, initializing the network on the first rotation. On the second rotation, synchronous devices can start transmitting. On the third and subsequent rotations, asynchronous devices can transmit if there is available bandwidth. Errors are corrected automatically via a beacon-and-recovery process, during which the individual stations seek to correct the situation.

FDDI Architecture

These processes are defined in a set of standards that are sanctioned by the *American National Standards Institute* (ANSI). The standards address four functional areas of the FDDI architecture (refer to Figure F-6).

Physical Media-Dependent **(PMD)**   Data is transmitted between stations after converting the data bits into a series of optical pulses. The pulses are then transmitted over the cable that links the various stations. The *Physical Media-Dependent* (PMD) sublayer describes the optical transceivers; specifically, the minimum optical power and sensitivity levels over the optical data link. This layer also defines the connectors and media characteristics for point-to-point communications between stations on the FDDI network. The PMD sublayer is a subset of the physical layer of the OSI Reference Model, defining all of the services that are needed to transport a bit stream from station to station. The PMD sublayer also specifies the cabling requirements for FDDI-compliant cable plant, including worst-case jitter and variations in cable attenuation.

Physical Layer **(PHY)**   The *Physical Layer* (PHY) protocol defines those portions of the physical layer that are media independent, describing data encoding/decoding, establishing clock synchronization, and defining the

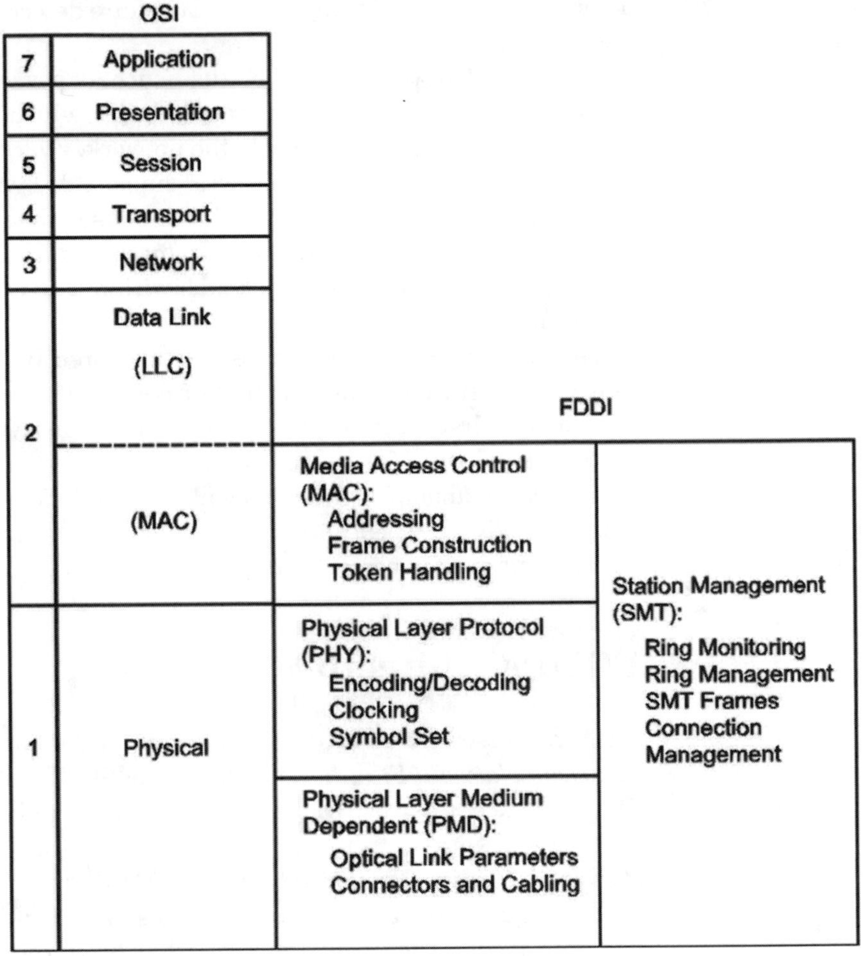

Figure F-6
FDDI layers and their relationship to the seven-layer OSI Reference Model

OSI

Layer	OSI	FDDI	
7	Application		
6	Presentation		
5	Session		
4	Transport		
3	Network		
2	Data Link (LLC)		

2	(MAC)	Media Access Control (MAC):
		    Addressing
		    Frame Construction
		    Token Handling
			Station Management (SMT):
			    Ring Monitoring
			    Ring Management
			    SMT Frames
			    Connection
			    Management

1	Physical	Physical Layer Protocol (PHY):
		    Encoding/Decoding
		    Clocking
		    Symbol Set	

1	Physical	Physical Layer Medium Dependent (PMD):
		    Optical Link Parameters
		    Connectors and Cabling	

handshaking sequence that is used between adjacent stations in order to test link integrity. The PHY also provides the synchronization of incoming and outgoing code-bit clocks and delineates octet boundaries as required for the transmission of information to or from higher layers. These processes enable the receiving station to synchronize its clock to the transmitting station.

***Media Access Control* (MAC)**   FDDI's data link layer is divided into two sublayers. The *Media Access Control* (MAC) sublayer governs access to the medium and describes the frame format, interprets frame content, generates and repeats frames, issues and captures tokens, controls timers, monitors the ring, and interfaces with station management.

The *Logical Link Control* (LLC) sublayer is required for proper ring operation and is part of the IEEE 802.2 standard. In keeping with the IEEE model, the FDDI MAC is fully compatible with the IEEE 802.2 LLC standard. Applications that interface to the LLC and operate over existing LANs, such as IEEE 802.3 CSMA/CD or 802.5 Token Ring, have the capacity to operate over an FDDI network.

The FDDI MAC, similar to the 802.5 Token Ring MAC, has two types of protocol data units: a frame and a token. Frames are used to carry data (such as LLC frames), while tokens are used to control a station's access to the network. At the MAC layer, data is transmitted in four-bit blocks called 4B:5B symbols. The symbol coding is such that four bits of data are converted to a five-bit pattern; thus, the 100Mbps FDDI rate is provided at 125 million signals per second on the medium. This signaling type is employed in order to maintain signal synchronization on the fiber.

***Station Management* (SMT)**   The *Station Management* (SMT) facility provides the system-management services for the FDDI protocol suite, detailing control requirements for the proper operation and interoperability of stations on the FDDI ring. SMT acts in concert with the PMD, PHY, and MAC layers. The SMT facility is used to manage connections, configurations, and interfaces and defines such services as ring and station initialization, fault isolation and recovery, and error control. SMT is also used for statistics gathering, address administration, and ring partitioning.

FDDI Topology

FDDI is a token-passing ring network. Similar to all rings, FDDI consists of a set of stations that are connected by point-to-point links in order to form a closed loop. Each station receives signals on its input side and regenerates them for transmission on the output side. Theoretically, any number of stations can be attached to the network, although default values in the FDDI standard assume no more than 1,000 physical attachments and a 200-km path.

FDDI uses two counter-rotating rings: a primary ring and a secondary ring. Data traffic usually travels on the primary ring. The secondary ring operates in the opposite direction and is available for fault tolerance. If the rings are appropriately configured, stations can transmit simultaneously on both rings—thereby doubling the bandwidth of the network to 200Mbps.

Three classes of equipment are used in the FDDI environment: *Single Attached Stations* (SASs), *Dual Attached Stations* (DASs), and *Concentrators* (CONs). A DAS physically connects to both rings, while an SAS connects only to the primary ring via a wiring concentrator. In the case of a link failure, the internal circuitry of a DAS can heal the network by using a combination of the primary and secondary rings. If a link failure occurs between a concentrator and an SAS, the SAS becomes isolated from the network.

These equipment types can be arranged in any of three topologies: dual ring, tree, and dual ring of trees (refer to Figure F-7). In the dual-ring topology, DASs form a physical loop in which all of the stations are dual attached. In a tree topology, remote SASs are linked to a concentrator, which is connected to another concentrator on the main ring.

Any DAS that is connected to a concentrator performs as an SAS. Concentrators can be used to create a network hierarchy, which is known as a dual ring of trees. This topology offers a flexible, hierarchical system design that is efficient and economical. Devices that require highly reliable communications attach to the main ring, while those that are less critical attach to branches off the main ring. Thus, SAS devices can communicate with the main ring, but without the added cost of equipping them with a dual-ring interface or a loop-around capability that would otherwise be required in order to ensure the reliability of the ring in the event of a station failure.

Failure Protection

FDDI provides an optional bypass switch at each node in order to overcome a failure anywhere on the network. In the event of a node failure, the node is bypassed optically—removing it from the network. Up to three nodes in sequence can be bypassed; enough optical power will remain to support the operable portions of the network.

In the event of a cable break, the dual counter-rotating ring topology of FDDI enables the use of redundant cable in order to handle normal 100Mbps traffic. If both the primary and secondary cables fail, the stations that are adjacent to the failures automatically loop the data around and between rings (refer to Figure F-8), thus forming a new C-shaped ring from the operational portions of the original two rings. When the fault is healed, the network will reconfigure itself again.

Figure F-7
FDDI dual-ring
topology with three
types of
interconnecting
devices

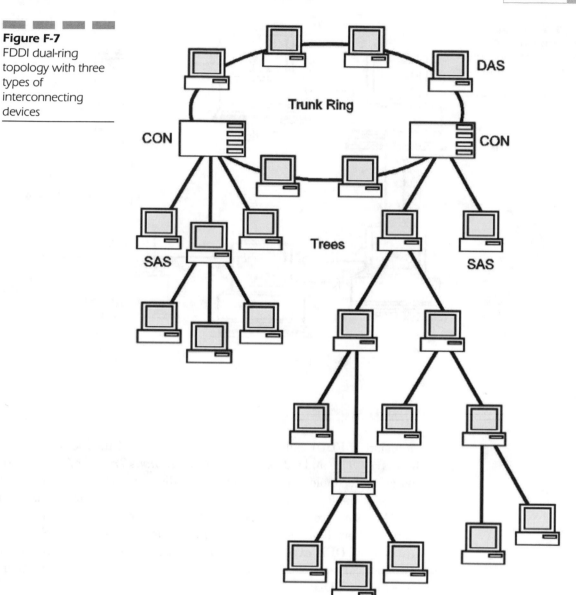

Normally, FDDI concentrators offer two buses that correspond to the two FDDI backbone rings. Fault tolerance is also provided for stations that are connected to the ring via a concentrator, because the concentrator provides the loop-around function for attached stations.

Figure F-8
Self-healing capability
of FDDI's dual-ring
topology

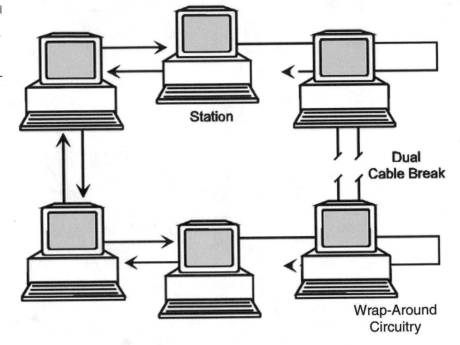

Station

Dual
Cable Break

Wrap-Around
Circuitry

Summary

An extension of FDDI called FDDI-2 uses portions of its 100Mbps band-width to carry voice and video, just like *Asynchronous Transfer Mode* (ATM) cell-switching technology. FDDI is limited by distance, however, while ATM is a highly scalable, broad-band networking technology that spans both LAN and WAN environments. FDDI has fallen out of favor in recent years. Although it offers high reliability and fault tolerance for mission-critical applications, FDDI is expensive to implement and has been overtaken by more economical technologies (such as Faster Ethernet and Gigabit Ethernet). In addition, FDDI has no upward migration path to higher speeds, and it tends to crash under sustained heavy loads.

See Also

Asynchronous Transfer Mode (ATM)
Ethernet (100BaseT)
Ethernet (1000BaseT)

Fiber in the Loop (FITL)

Fiber in the Loop (FITL) is a system in which services to contiguous groupings of residential and business customers are delivered via fiber-optic media in either all or a portion of the loop distribution network. FITL is an umbrella term that encompasses various systems that depend to a greater or lesser degree on fiber optics, including the following:

- *Fiber to the Neighborhood* (FTTN)   A technology that involves bringing fiber into the neighborhood. From there, signals would be carried to businesses and residences via the existing copper wiring.

- *Fiber to the Curb* (FTTC)   A technology that involves bringing fiber into the neighborhood and up to the curb. From there, signals would be carried to businesses and residences via the existing copper wiring.

- *Fiber to the Home* (FTTH)   A technology that involves bringing fiber all the way to the home. Inside the home, signals would be carried over the existing copper wire or coaxial cable.

- *Fiber to the Building* (FTTB)   A technology that involves bringing fiber all the way to the building. Inside the building, signals would be carried over the existing copper wire or coaxial cable or over optical fiber.

FTTC systems have the capacity to offer a dedicated rate of 52Mbps on the downstream path and a dedicated rate of 1.6Mbps on the upstream path. With this much bandwidth available, the network operator would have the capability to offer multiple services over the FTTC system, in which the bandwidth will be shared among different applications (including voice and video) within a home or a building.

Summary

Many services providers have concluded that the cost of offering FTTC on a mass-market basis is currently out of reach, due to the high cost of installing extra fiber and electronics in the local loop. Instead, *Hybrid Fiber/Coax* (HFC) systems and broad-band wireless technologies, such as *Local Multi-Point Distribution Service* (LMDS), offer a more economical approach. Nevertheless, FTTC might yet emerge as the ultimate architecture—at least, in high-density metropolitan areas.

See Also

Fiber-Optic Technology

Hybrid Fiber / Coax (HFC)

Local Multi-Point Distribution Service (LMDS)

Fiber-Optic Technology

Fiber-optic transmission systems have been in commercial use for almost 25 years. Optical fiber provides many performance advantages over copper-based wire. These advantages make optical fiber the most advanced transmission medium available today, offering more bandwidth for the growing volume of voice and data traffic, more security for today's mission-critical business applications, and a low-cost alternative to satellites for international communications.

Bandwidth Capacity

The laser components at each end of the fiber-optic link enable high encoding and decoding frequencies. For this reason, optical fiber offers much more bandwidth capacity than copper-pair wires. Data can be transmitted over optical fiber at multi-gigabit-per-second speeds. Today's commercial optical systems use *Dense Wavelength-Division Multiplexing* (DWDM) technology to provide up to 100 wavelengths on a single optical fiber, with each wavelength constituting a channel that can move data at 5Gbps for an aggregate capacity of 500Gbps. By comparison, early *Wavelength-Division Multiplexing* (WDM) systems provided only four, eight, or 16 optical channels.

While DWDM offers unprecedented transmission capacity for next-generation data networks, continued developments in fiber-optic technology promise even more capacity. With today's experimental techniques, a single strand of fiber can have more than 1,000 channels, and each channel can operate at up to 160Gbps. These technologies, referred to as *Ultra-Dense WDM* (UDWDM), were announced by Bell Labs in November 1999. In its capability to multiply 160Gbps over additional wavelengths, Bell Labs expects to have the power to scale up the capacity of optical fiber to many trillions of bits a second in the near future.

The rapid growth of data networking is prompting the need for fiber-optic transmission systems that can deliver hundreds of megabits-per-second to a large number of ports at a low cost. Transmitters with high channel counts and minimal hardware are attractive candidates for such applications.

Signal Attenuation

Signal attenuation, which is measured in *decibels* (dB), refers to signal loss during transmission (when the signal that is received is not as strong as the signal that is transmitted, for example). Signal attenuation is attributed to the inherent resistance of the transmission medium. For transmissions over metallic cable, loss increases with frequency as the signal radiates outward. The characteristics of optical fiber, however, are such that little or no inherent resistance exists.

The low resistance of fiber enables the use of higher frequencies to derive enough bandwidth in order to accommodate thousands of voice channels. An analog line on the local loop has a frequency range of up to 4KHz for voice transmission, whereas a single optical fiber has a frequency range of 3GHz and higher. Fiber's lower resistance also results in a constant signal level that can be maintained over much longer distances than copper lines, without the need for repeaters to regenerate the signals at various intervals along the span.

Data Integrity

Data integrity refers to a performance rating based on the number of undetected errors in a transmission. Once again, optical fiber surpasses metallic cabling. A typical fiber-optic transmission system produces a bit-error rate of less than 10^{-9}, while metallic cabling typically produces a bit-error rate of 10^{-6}.

Because of their high data integrity, fiber-optic systems do not require an extensive use of error-checking protocols that are common in metallic cable systems. Without the need for error-checking functions, data transmission rates are enhanced. In addition, because the required number of retransmissions is reduced with fiber, overall system performance is greatly improved.

Immunity to Interference

Optical fibers are immune to *Electromagnetic Interference* (EMI) and *Radio-Frequency Interference* (RFI), which are the principal sources of data errors in transmissions over metallic-cable systems. This immunity facilitates fiber installation, because fiber-optic cables do not need to be rerouted around elevators, machinery, auxiliary power generators, fluorescent lighting, and other potential sources of interference.

Fiber's immunity to interference makes it more economical to install, not only because less time is required to route the physical cables, but because there is no need to build special conduits in order to shield fiber from the external environment. In addition, because optical fibers do not generate the electromagnetic radiation that often causes cross-talk on metallic cables, multiple fibers can be safely bundled into a single cable in order to simplify installation.

Security

Fiber is a more secure transmission medium than unshielded metallic cable, because optical fibers do not radiate electromagnetic or radio-frequency energy. In order to intercept a fiber-optic transmission, the wire core must be physically broken, and a connection must be fused to the wire. This procedure is routinely used to add nodes to the fiber cable, but it prohibits the transmission of light beyond the point of the break until the splice is completed. This setup makes unauthorized access easily detectable, because an alarm is raised at a network management station.

Durability

Optical fiber is not a delicate material; in fact, fiber's pull strength (the maximum pressure that can be exerted on the cable before damage occurs) is 200 pounds—eight times that of Category 5 UTP copper wire. Fiber-optic cables are reinforced with a strengthening member inside the cable and a protective jacket around the outside of the cable. These reinforcements produce the same tensile strength as steel wire of an equal diameter.

The inherent strength of fiber, combined with the added reinforcement of being bundled into cable form, gives fiber-optic cables the durability that is necessary to withstand being pulled through walls, floors, and underground conduits without being damaged. In addition, fiber cables are designed to withstand higher temperatures than copper, which makes fiber networks better capable to survive potentially disastrous fires. In typical operating environments, fiber is also more resistant to corrosion than copper wire (and consequently has a longer useful life when it must be installed on telephone poles).

Types of Fiber

Two types of optical fiber exist: single-mode and multi-mode. Single-mode fibers transmit only one light wave along the core, while multi-mode fibers transmit many light waves. Single-mode fibers entail lower signal loss and support higher transmission rates than multi-mode fibers (and therefore are most often selected by carriers for use on the public network). More than 90 percent of fiber cable that is installed by carriers is single-mode.

An example of single-mode fiber is Lucent's TrueWave, which is specifically designed for DWDM-enhanced long-distance networks. The key characteristic of Lucent's high-performance fiber is that it provides consistent amounts of dispersion throughout a broader operating range. Dispersion is the property of optical fiber that causes signals to spread out and to interfere with each other. Lucent's low-dispersion fiber improves network performance by reducing the need for complex dispersion-management equipment.

Multi-mode fibers have relatively large cores. Light pulses that simultaneously enter a multi-mode fiber can take many paths and can exit at slightly different times. This phenomenon, called intermodal pulse dispersion, creates minor signal distortion and thereby limits both the data rate of the optical signal and the distance that the optical signal can be sent without repeaters. For this reason, multi-mode fiber is most often used for short distances and for applications in which slower data rates are acceptable.

Multi-mode fiber can be further categorized as step-index or graded-index. Step-index fiber has a silica core that is encased with plastic cladding. The silica is more dense than the plastic cladding, and the result is a sharp, step-like difference in the refractive index between the two substances. This difference prevents light pulses from escaping as they pass through the optical fiber. Graded-index fiber contains multiple layers of silica at its core, with lower refractive indices toward the outer layers. The graded core increases the speed of the light pulses in the outer layers in order to match the rate of the pulses that traverse the shorter path directly down the center of the fiber.

The fibers in most of today's fiber-optic cable have an outside (or cladding) diameter of 125 microns. (A micron is one millionth of a meter.) The core diameter depends on the type of cable. The cores of multi-mode fibers comprise many concentric cylinders of glass, with each cylinder having a different index of refraction. The layers are arranged so that light that is introduced to the fiber at an angle will be bent back toward the center. The bending results in light that travels in a sine-wave pattern down the fiber core and that enables an inexpensive, non-coherent light source to be used at the transmitter.

Almost all multi-mode fibers have a core diameter of 62.5 microns. Bandwidth restrictions of 200MHz/km to 300MHz/km limit the maximum length of multi-mode segments to a few kilometers. Wavelengths of 850 nm to 1300 nm are used with multi-mode fiber-optic cable. Single-mode fiber consists of a single 8-micron to 10-micron core. In other words, a carefully focused, coherent light source (such as a laser) must be used in order to ensure that light is sent directly down the small aperture. Single-mode fiber is normally operated with light at a wavelength of 1300 nm to 1620 nm.

Summary

Fiber-optic cable is installed in carrier networks because of its capability to reliably and securely carry voluminous amounts of data and large numbers of voice channels over great distances. Fiber is used for transoceanic links between international locations because it has more capacity than satellite links—which also makes fiber more economical—and because it offers fewer delays during voice conversations. Fiber is installed on much of the long-distance network in the United States and is installed in the ring topology around major cities. Reliability is enhanced with dual rings that provide additional capacity for disaster recovery. Businesses use fiber to connect high-speed workstations, computers, and LANs in campus environments and office buildings. In some areas of the country, fiber is even being installed to the curb in order to deliver entertainment services (such as video on demand) to the home and to support interactive multimedia applications. In other places, hybrid network architectures that combine coaxial cable with fiber-optic backbones are being installed in order to provide advanced communications and entertainment services to the home in a more economical manner than fiber-only networks.

See Also

Fiber in the Loop (FITL)

Fibre Channel

Synchronous Optical Network (SONET)

Wavelength-Division Multiplexing (WDM)

Fibre Channel

Fibre Channel is a high-performance interconnect standard that is designed for bidirectional, point-to-point serial data channels between desktop workstations, mass storage subsystems, peripherals, and host systems. Serialization of the data permits much greater distances to be covered than with parallel communications. Unlike networks in which each node must share the bandwidth capacity of the media, Fibre Channel devices are connected through a flexible circuit/packet switch that is capable of simultaneously providing the full bandwidth to all connections.

Advantages

The key advantage of Fibre Channel is speed. Fibre Channel is 10 to 250 times faster than typical LAN speeds. Measured in megabytes per second (MB/s), Fibre Channel offers a transmission rate of 100MB/s (200MB/s in full-duplex mode), which is the equivalent of 60,000 pages of text per second. Such speeds are achieved simply by transferring data between one buffer at the source device and another buffer at the destination device, without regard for how the data is formatted (cells, packets, or frames). What the individual protocols do with the data before or after the data is in the buffer is inconsequential. Fibre Channel provides complete control over the transfer only and offers simple error checking.

Unlike many of today's interfaces—including the *Small Computer Systems Interface* (SCSI)—Fibre Channel is bidirectional, achieving 100MB/s in both directions simultaneously. Thus, it provides a 200MB/s channel if usage is balanced in both directions. Fibre Channel also overcomes the restrictions on the number of devices that can be connected, so that any number of SCSI devices can be accessed instead of the normal limit of 15, for example.

Fibre Channel overcomes the distance limitations of today's interfaces. A fast SCSI parallel link from a disk drive to a workstation, for example, can transmit data at 20MB/s, but it is restricted in length to about 20 meters. In contrast, a quarter-speed Fibre Channel link transmits information at 25MB/s over a single, compact optical cable pair at up to 10 kilometers in length, which enables disk drives to be placed almost anywhere and enables more flexible site planning.

Applications

The high-speed, low-latency connections that can be established via Fibre Channel make it ideal for a variety of data-intensive applications, including the following:

- *Backbones* Fibre Channel provides the parallelism, high bandwidth, and fault tolerance that is needed for high-speed backbones. Fibre Channel is the ideal solution for mission-critical internetworking. The scaleability of Fibre Channel also makes it practical to create backbones that grow as one's needs grow—from a few servers to an entire enterprise network.

- *Workstation clusters* Fibre Channel is a natural choice to enable supercomputer-power processing at workstation costs.

- *Imaging* Fibre Channel provides the bandwidth on demand that is needed for high-resolution medical, scientific, and prepress imaging applications, among others.

- *Scientific / engineering* Fibre Channel delivers the needed throughput for today's new breed of visualization, simulation, CAD/CAM, and other scientific, engineering, and manufacturing applications that demand megabytes of bandwidth per node.

- *Mass storage* Current mass-storage access is limited in rate, distance, and addressability. Fibre Channel provides mass storage attachments at distances of up to several kilometers. Fibre Channel also interfaces with current SCSI, *High Performance Peripheral Interface* (HIPPI), and *Intelligent Peripheral Interface* (IPI) connections, among others.

- *Multimedia* Fibre Channel's bandwidth supports real-time videoconferencing and document collaboration between several workstation users and is capable of delivering multimedia applications that consist of voice, music, animation, and video.

Topology

Fibre Channel uses a flexible-circuit/packet-switched topology to connect devices. Through the switch, Fibre Channel has the capacity to establish multiple, simultaneous point-to-point connections. Devices that are attached to the switch do not have to contend for the transmission medium as they do in a network. Through its intrinsic flow control and acknowledgment capabilities, Fibre Channel also supports connectionless traffic without suffering the congestion of the shared transmission media that is used in traditional networks.

The fabric relieves each Fibre Channel port of the responsibility for station management. All that a Fibre Channel port has to do is manage a simple point-to-point connection between itself and the fabric. If an invalid connection is attempted, the fabric rejects it. If there is a congestion problem en route, the fabric responds with a busy signal, and the calling port tries again.

Fiber Channel Layers

Fibre Channel employs a five-layer stack that defines the physical media and transmission rates, encoding scheme, framing protocol and flow control, common services, and the upper-level applications interfaces.

- *FC-0* The lowest layer, which specifies the physical characteristics of the media, transmitters, receivers and connectors that can be used with Fibre Channel, including electrical and optical characteristics, transmission rates, and other physical components of the standard

- *FC-l* Defines the 8B/10B encoding/decoding scheme that is used to integrate the data with the clock information that is required by serial transmission techniques. Fibre Channel uses 10 bits to represent each eight bits of real data, requiring it to operate at a speed that is sufficient to accommodate this 25 percent overhead. The two extra bits are used for error detection and correction (known as disparity control).

- *FC-2* Defines the rules for framing the data to be transferred between ports, the different mechanisms for using Fibre Channel's circuit and packet-switched service classes (discussed as follows), and the means of managing the sequence of data transfer. All frames belonging to a single transfer are uniquely identified by sequential numbering from zero through n, enabling the receiver to determine whether a frame is missing (and if so, which one is missing).

- *FC-3* Provides the common services that are required for advanced features such as striping (to multiply bandwidth) and hunt groups (the capability for more than one port to respond to the same alias address). A hunt group can be likened to a business that has 10 phone lines but that requires only a single number to be dialed. Whichever line is free will ring.

- *FC-4* Provides seamless integration of legacy standards, including FDDI, HIPPI, IPI, SCSI, and IP, as well as IBM's *Single Byte Command Code Set* (SBCCS) of the *Block Multiplexer Channel* (BMC), Ethernet, Token Ring, and ATM.

Classes of Service

To accommodate a wide range of communications needs, Fibre Channel provides different classes of service at the FC-2 layer:

- *Class 1* This class of service provides exclusive use of the connection for the duration of a session, much like a dedicated physical channel. This class is used for time-critical, non-bursty traffic such as a link between two supercomputers.

- *Class 2* A connectionless, frame-switched link that provides guaranteed delivery and confirms receipt of traffic. No dedicated connection is established between ports as in Class 1; instead, each frame is sent to its destination over any available route. This service is typically used for data transfers to and from a shared mass-storage system that is physically located at some distance from several individual workstations.

- *Class 3* A one-to-many connectionless service that enables data to be sent rapidly to multiple devices that are attached to the fabric. This service is used for real-time broadcasts and for any other application that can tolerate lost packets. Because no confirmation of receipt is given, this service is faster than Class 2 service.

- *Class 4* A connection-based service that offers guaranteed fractional bandwidth and guaranteed latency levels. This goal is achieved by enabling users to lock down a physical path through the Fibre Channel switch fabric.

The same switching matrix can support multiple classes of service simultaneously, per the requirements of each application (refer to Figure F-9).

Fibre Connection (FICON)

Fibre Connection (FICON) is a high-speed I/O interface for connecting mainframes to storage devices. The channel-link speed of FICON is 100MB/s, which is more than five times faster than the 17MB/s of IBM's *Enterprise*

Figure F-9
A switching matrix supporting Fibre Channel's dedicated (Class 1) and switched (Class 2) connections

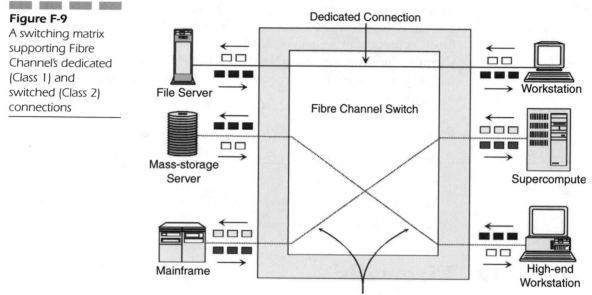

Systems Connection (ESCON), which was the previous standard for transferring data to and from a mainframe over an optical channel connection.

FICON supports full-duplex data transfers (meaning that data can be read and written over the same link at the same time), while ESCON operates in half-duplex mode. FICON uses a mapping layer that is based on technology developed for Fibre Channel. Because it uses Fibre Channel's multiplexing capabilities, small data transfers are multiplexed on the link alongside larger transfers. That way, small data transfers (typical for transactions) do not have to wait for large data transfers to complete. For users who want to mix and match FICON and ESCON, however, IBM offers a FICON director that supports both technologies.

Summary

Fibre Channel is not a telecommunications solution for the wide area; rather, it is a cost-effective, high-speed technology for transporting large volumes of data in the local area (where link distances do not exceed 10 km). Fibre Channel's capability to transfer data at speeds of up to 100MB/s securely and bidirectionally make it an effective, high-performance communications option for distributed computing environments—particularly those that involve mass storage and server clusters. While there is some overlap of capability between ATM and Fibre Channel, each can do some function that the other cannot. ATM can span the local and wide area. Fibre Channel, however, can attach CPU and peripheral devices directly to a high-speed network infrastructure. ATM and Fibre Channel are complementary technologies.

See Also

> *Asynchronous Transfer Mode* (ATM)
>
> Fiber-Optic Technology
>
> *Synchronous Optical Network* (SONET)

Firewalls

Firewalls occupy a strategic position between a trusted corporate network and an untrusted network, such as the Internet (refer to Figure F-10). Firewalls implement perimeter security by monitoring all traffic to and from the enterprise network, in order to determine which packets can pass and which cannot. A firewall can identify suspected break-in events and issue appropriate alarms to a management station, then invoke a predefined action to head off the attack. Firewalls can also be used to trace attempted intrusions from the Internet through logging and auditing functions.

Operation

A packet-filtering firewall examines all of the packets that it receives, then forwards them or drops them based on predefined rules (refer to Figure F-11). The network administrator can control how packet filtering is performed, as far as permitting or denying connections, by using criteria that are based upon the source and destination host or network and the type of network service (refer to Figure F-12).

In addition to packet filtering, a firewall offers other useful security features, such as the following:

■ *Stateful packet inspection* State information is derived from past communications and other applications in order to make the control decision for new communication attempts. With this method of security, the packet is intercepted by an inspection engine that

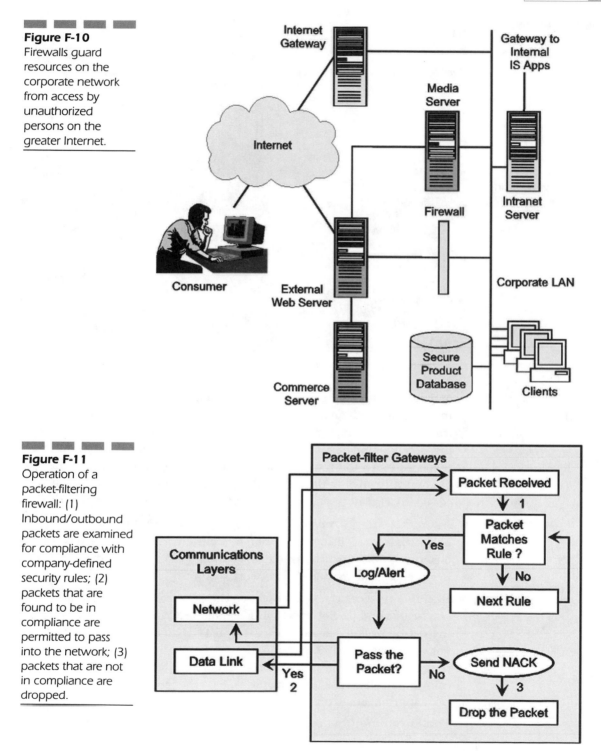

Figure F-10
Firewalls guard resources on the corporate network from access by unauthorized persons on the greater Internet.

Figure F-11
Operation of a packet-filtering firewall: (1) Inbound/outbound packets are examined for compliance with company-defined security rules; (2) packets that are found to be in compliance are permitted to pass into the network; (3) packets that are not in compliance are dropped.

Like other products, CyberGuard Corporation's Firewall 3 enables the network administrator to control packet filtering, which permits or denies connections by using criteria that are based upon the source and destination host or network and the type of network service.

extracts state-related information. The engine maintains this information in dynamic state tables for evaluating subsequent connection attempts. Packets are only permitted to pass when the inspection engine examines the list and verifies that the attempt is in response to a valid request. The list of connections is maintained dynamically, so only the required ports are opened. As soon as the session is closed, the ports are locked—ensuring maximum security.

■ *Network Address Translation* (NAT)   Hides internal IP addresses from public view, preventing them from being used for spoofing (a technique for impersonating authorized users by using a valid IP address in order to gain access to an internal network)

■ *Denial-of-service detection*   Defends the network against syn flooding, port scans, and packet injection. This goal is accomplished by inspecting packet sequence numbers in TCP connections. If they are not within expected ranges, the firewall drops them as suspicious packets. When the firewall detects unusually high rates of new connections, it issues an alert message so that the administrator can take appropriate action.

■ *Virus scanning*   In addition to hiding viruses, Java applets and ActiveX controls can be used to hide intelligent agents that can give intruders access to corporate resources once they get inside the enterprise network. With the increasing use of Java applets and ActiveX controls on Web sites, more firewalls offer the means to either deny corporate users access to Web pages that contain these elements or to filter such content from the Web pages when it is encountered.

■ *Probe detection*   Firewalls offer alarms that are activated when probing is detected. The alarm system can be configured to watch for TCP or UDP probes from either external or internal networks. Alarms can be configured to trigger e-mail, popup windows, or output messages to a local printer.

■ *Event logging*   Automatically logs system error messages to a console terminal or syslog server, enabling administrators to track potential security breaches or other non-standard activities on a real-time basis (refer to Figure F-13)

Automated Intrusion Detection

Many vendors now offer automated intrusion-detection tools for their fire-walls, but one tool in particular is notable for its ease of use. AXENT Tech-

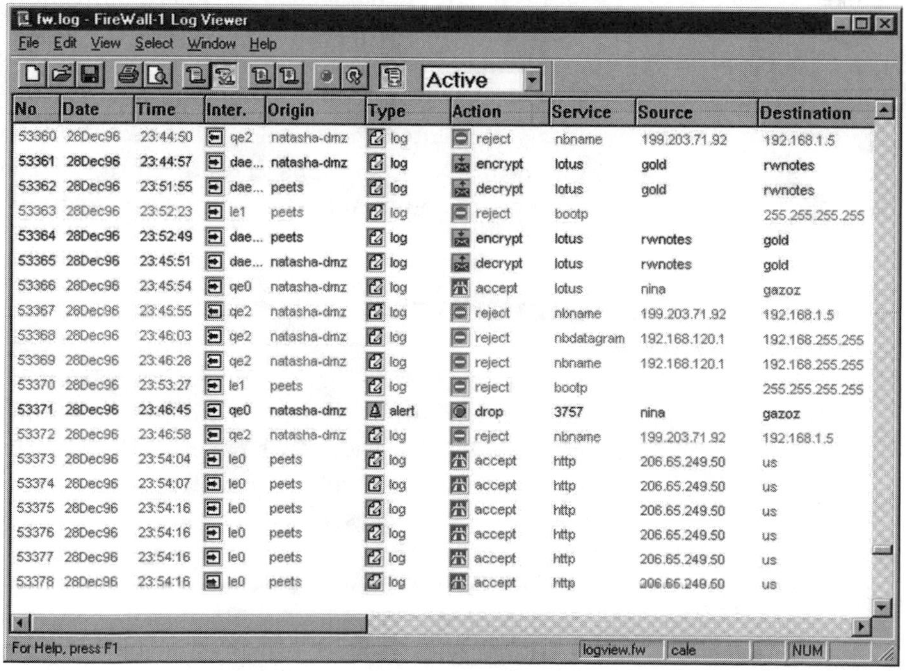

Figure F-13
Check Point Software Technologies' FireWall-1 includes the Live Connections Monitor, which gives network administrators the capability to view all currently active connections. The live connections are stored and handled in the same way as ordinary log records, but they are stored in a special file that is continuously updated as connections start and end.

No	Date	Time	Inter.	Origin	Type	Action	Service	Source	Destination
53360	28Dec96	23:44:50	qe2	natasha-dmz	log	reject	nbname	199.203.71.92	192.168.1.5
53361	28Dec96	23:44:57	dae...	natasha-dmz	log	encrypt	lotus	gold	rwnotes
53362	28Dec96	23:51:55	dae...	peets	log	decrypt	lotus	gold	rwnotes
53363	28Dec96	23:52:23	le1	peets	log	reject	bootp		255.255.255.255
53364	28Dec96	23:52:49	dae...	peets	log	encrypt	lotus	rwnotes	gold
53365	28Dec96	23:45:51	dae...	natasha-dmz	log	decrypt	lotus	rwnotes	gold
53366	28Dec96	23:45:54	qe0	natasha-dmz	log	accept	lotus	nina	gazoz
53367	28Dec96	23:45:55	qe2	natasha-dmz	log	reject	nbname	199.203.71.92	192.168.1.5
53368	28Dec96	23:46:03	qe2	natasha-dmz	log	reject	nbdatagram	192.168.120.1	192.168.255.255
53369	28Dec96	23:46:28	qe2	natasha-dmz	log	reject	nbname	192.168.120.1	192.168.255.255
53370	28Dec96	23:53:27	le1	peets	log	reject	bootp		255.255.255.255
53371	28Dec96	23:46:45	qe0	natasha-dmz	alert	drop	3757	nina	gazoz
53372	28Dec96	23:46:58	qe2	natasha-dmz	log	reject	nbname	199.203.71.92	192.168.1.5
53373	28Dec96	23:54:04	le0	peets	log	accept	http	206.65.249.50	us
53374	28Dec96	23:54:07	le0	peets	log	accept	http	206.65.249.50	us
53375	28Dec96	23:54:16	le0	peets	log	accept	http	206.65.249.50	us
53376	28Dec96	23:54:16	le0	peets	log	accept	http	206.65.249.50	us
53377	28Dec96	23:54:16	le0	peets	log	accept	http	206.65.249.50	us
53378	28Dec96	23:54:16	le0	peets	log	accept	http	206.65.249.60	us

For Help, press F1   logview.fw   cale   NUM

nologies offers 200 so-called drop-and-detect security scenarios in its Omni-Guard/Intruder Alert. These preconfigured scenarios enable organizations to install Intruder Alert to instantly protect systems against hundreds of the most common and dangerous security threats to Windows NT and other key enterprise systems.

Intruder Alert uses a real-time, manager/agent architecture to monitor the audit trails of distributed systems for footprints that signal suspicious or unauthorized activity on all major operating systems, Web servers, firewalls, routers, applications, databases, and SNMP traps from other network devices. Unlike other intrusion-detection tools, which typically report suspicious activity hours or even days after it occurs, Intruder Alert instantly takes action in order to alert network administrators, shut down systems, terminate offending sessions, and execute commands in order to stop intrusions before they damage critical systems.

As new security threats emerge, network administrators can quickly protect their systems by loading new drop-and-detect scenarios that have been researched and developed by AXENT's Information Security SWAT Team, a group of computer security professionals that focused on hacking techniques and the latest computer security threats. These new scenarios, which can be downloaded from the SWAT Team Web site and can be installed enterprise-wide, make it easy for network administrators to keep systems safe from evolving threats.

From a single management workstation, network administrators can quickly drag new security policies and attack scenarios to different enterprise domains, implementing additional protection for hundreds or thousands of systems in a matter of minutes. The enterprise console also provides a correlated, graphical view of security trends, enabling network administrators to view graphs that illustrate real-time security trends and to drill down to additional details concerning activity.

Security Appliances

For small remote offices or remote users who cannot justify the expense of an enterprise firewall, there are appliances that combine firewall and VPN security capabilities. Security is particularly important when *Digital Subscriber Line* (DSL) or cable is used for Internet access. One drawback of these always-on services is that they are vulnerable to security attacks.

NetScreen Technologies, for example, offers its NetScreen-5 security appliance for use in small offices and telecommuter sites with DSL or cable modem access. Its stateful-inspection firewall prevents hacker attacks, and the VPN delivers secure remote access to a corporate network through encrypted tunnels. The NetScreen-5 unit is installed between the PC and cable/DSL modem. Because small remote offices typically do not have staff that has technical expertise, the box can be configured by a network administrator and shipped to a user for plug-and-play installation. The network administrator can then centrally manage and reconfigure dispersed units through the NetScreen Global Manager or through a Web browser.

Administrators can check the status of multiple NetScreen appliances, monitor performance, troubleshoot existing configurations, or add remote sites to the network from one location. All such activities are conducted via VPN tunnels for the highest level of security.

Risk Assessment

Even after you implement a firewall solution, you should periodically conduct a comprehensive security-risk assessment. This assessment helps network administrators identify and resolve security breaches before hackers discover and exploit them, causing serious problems later.

A number of risk-assessment tools are available, including Hacker Shield, which was developed by BindView, a provider of network scanning and response software. Hacker Shield scans and detects networks for potential security holes and offers the user patches or corrective actions to fix the breaches before they become a threat. Hacker Shield identifies and resolves security vulnerabilities at both the operating-system level and the network level, protecting against both internal and external threats. Hacker Shield also monitors key system files for unauthorized changes—and, by referencing a one-million-word dictionary, identifies vulnerable user passwords through a variety of password-cracking techniques.

A detailed report provides network administrators with a description of each vulnerability and corrective action, as well as a ranking of vulnerabilities by the risk(s) that they pose to a site's security. Network administrators also have a high-level overview of the vulnerability and its solution, with an option to link to more detailed explanation and reference materials.

Employing an implementation model similar to anti-virus products, BindView provides ongoing security updates via the Internet in order to ensure that users are protected from the latest threats. BindView uses secure push technology to broadcast the vulnerability updates. Users are not required to reinstall the software in order to integrate the updates.

Load Balancing

While the value of firewalls is undisputed, they can degrade performance and create a single point of network failure. Tasks such as stateful packet inspection, encryption, and virus scanning require significant amounts of processing. As traffic load increases, the firewall can become bogged down. Also, because the firewall sits directly in the data path, it constitutes a single point of failure. If the firewall cannot keep up with filtering all of the packets that are coming through, the firewall will go down and isolate the whole network behind it.

A solution to this problem is to add a firewall for redundancy and to put dynamic load-balancing switches on each side. That way, the switches can distribute incoming requests to the firewalls based on their availability. This configuration eliminates the firewalls as single points of failure, dramatically simplifies configuration, and increases end-user performance by balancing traffic among multiple firewalls.

The switches monitor the health of attached firewalls through automatic, periodic health checks. They also monitor the physical link status of the switch ports that are connected to each firewall. Because the firewalls are no longer directly inline and traffic is evenly distributed among them, the end-user experience is improved. The switches automatically recognize failed firewalls and redirect entire sessions through other available firewalls while maintaining the state of each session.

Summary

Although firewalls can provide a formidable defense against many kinds of attacks, they are not a panacea for all network security problems—particularly those that originate from the corporate side. For example, even if virus scanning is provided at the firewall, this scanning will only protect against viruses that come from the Internet. This scanning does nothing to guard against more likely sources, such as floppy disks that are brought into the company by employees who upload the contents to a desktop computer and inadvertently (or deliberately) spread the virus throughout the network. Companies must take appropriate internal security measures to safeguard mission-critical resources.

See Also

Network Security

Fixed Wireless Access

Fixed wireless-access technology provides a wireless link to the PSTN as an alternative to traditional wire-based local telephone service. Because calls and other information (e.g., data and images) are transmitted through the air, rather than through conventional cables and wires, companies can avoid the labor-intensive cost of providing and maintaining telephone poles and cables.

Fixed wireless-access systems come in two varieties: narrow-band and broad-band. A narrow-band fixed wireless-access service can provide bandwidth up to 128Kbps, which can support one voice conversation and a data session (such as Internet access or fax transmission). A broad-band fixed wireless-access service can provide bandwidth in the multi-megabits-per-second range, which is enough to support telephone calls, television programming, and broad-band Internet access.

A narrow-band fixed wireless service requires a wireless access unit, which is installed on the exterior of a home or business (refer to Figure F-14) in order to enable customers to originate and receive calls with no change to their existing analog telephones. This transceiver is positioned to provide an unobstructed view to the nearest base-station receiver. Voice and data calls are transmitted from the transceiver at the customer's location to the base-station equipment, which relays the call through the carrier's existing network facilities to the appropriate destination. No investment in special phones or facsimile machines is required; customers use all of their existing equipment.

Figure F-14
Fixed wireless access
configuration

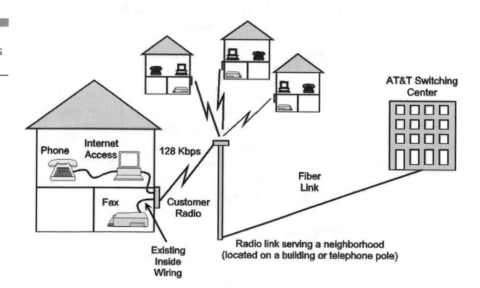

Depending on the vendor, a variety of technologies are employed for the narrow-band wireless access units and network infrastructure:

- *GSM*   Available at 900MHz, 1800MHz, and 1900MHz frequencies, the wireless access units use GSM's *Enhanced Full-Rate Codec* (EFRC) for wireline-equivalent voice quality. These units also provide a variety of popular, revenue-generating GSM subscriber services, including calling-line ID, call forwarding, call waiting, conferencing, and voice mail.

- *TDMA*   800MHz wireless access units operate in analog or TDMA digital mode and can be used by the carrier to provide local phone service as an adjunct to existing mobile cellular services.

- *CDMA*   800MHz and 1900MHz wireless access units operate in digital mode and can also be used by the carrier to provide local phone service as an adjunct to existing mobile cellular services.

Broad-band fixed wireless-access systems are relatively new. They are based on microwave technology operating in the licensed 28GHz to 31GHz frequency range and are referred to as *Local Multi-Point Distribution Service* (LMDS). The FCC completed spectrum auctions in early 1999 for LMDS. At this writing, only about a dozen CLECs offer LMDS, mostly for offering broad-band Internet access at speeds of up to 48Mbps. Although LMDS can be used for feeding voice and data traffic to the fiber-optic networks of IXCs and nation-wide CLECs, bypassing the local loops of the ILECs, few carriers (to date) have deployed LMDS for this purpose. This situation will change as the build-out of LMDS progresses.

Summary

Fixed wireless-access technology originated from the need to contain carriers' operating costs in rural areas, where pole and cable installation and maintenance are more expensive than in urban and suburban areas. Wireless access technology can also be used in urban areas, however, to bypass the LEC for long-distance calls. Because the IXC or CLEC avoids having to pay the ILEC's local loop interconnection charges, the savings can be passed back to the customer.

See Also

Cellular Voice Communications

Local Multi-Point Distribution Service (LMDS)

Frame Relay

In the 1970s, technicians built WANs that used low-speed analog facilities and that were used primarily for voice traffic. For reliable data transmission, however, private and public packet-switched networks had to employ the X.25 suite of protocols. These protocols were conceived in the 1970s networking environment of copper lines and electromechanical switches. At the time, analog lines and equipment were noisy and were subject to a variety of other impairments that made the transmission of data difficult. To deal with this environment, packet switches were deployed via the X.25 protocols. X.25 was endowed with substantial error-correction capabilities, so any node on the network could request a retransmission of errored data from the node that sent it. Errors had to be detected and corrected within the network, because the user's equipment typically did not have the intelligence and spare processing power to devote to this task.

The error correction and flow-control capabilities of the X.25 protocol, however, plus its many other functions, entail an overhead burden that limits network throughput. This situation, in turn, limits X.25 to niche applications (such as point-of-sale transaction processing, where the reliable transmission of credit card numbers and other financial information, not speed, is the overriding concern).

With the increasing use of digital facilities, there is less need for error protection. At the same time, users' end devices have a high level of intelligence, processing power, and storage, making them better at handling error control and diverse protocols. Consequently, the communications protocol that is used over the network might be scaled down to its bare essentials in order to greatly increase throughput. This idea drives the concept of Frame Relay, which can support voice traffic as well as data—packaged in variable-sized frames of up to 4,000 bytes in length.

Frame Relay was introduced commercially in May 1992 and initially gained acceptance as a method for providing end users with a solution for data connectivity requirements, such as LAN-to-LAN connections. Frame Relay provided both an efficient and flexible data-transport mechanism and also enabled a cheaper bandwidth cost associated with connecting multi-protocol networks and other devices.

Whereas X.25 operates at the bottom-three layers of the OSI Reference Model, Frame Relay operates at the first layer and at the lower half of the second layer—which cuts the amount of processing by as much as 50 percent, improving network throughput by a factor estimated at between three and 10 times. Although the Frame Relay network can detect errors, it does not correct them. Bad frames are simply discarded. When the receiving device detects corrupt or missing frames, it can request a retransmission from the originating device, whereupon the appropriate frames are resent.

Advantages

The most compelling advantages of a carrier-provided Frame Relay service include the following:

- *Any-to-any connectivity*   Any node that is connected to the Frame Relay service can communicate with any other node via predefined PVCs or dynamically via SVCs. The need for a highly meshed private-line network is eliminated for substantial cost savings.
- *Higher speeds*   Frame Relay service supports transmission speeds up to 44.736Mbps.
- *Improved throughput / low delay*   Frame Relay service uses high-quality digital circuits end to end, making it possible to eliminate the multiple levels of error checking and error control. The result is higher throughput and fewer delays, compared to legacy packet-switched networks.
- *Cost savings*   Multiple PVCs can share one access link, eliminating the cost of multiple private-line circuits and their associated *Customer-Premises Equipment* (CPE) for substantial cost savings.
- *Flat-rate charges*   Once a customer location is connected to the Frame Relay network, there are no distance-sensitive charges for traffic within the cloud.
- *Simplified network management*   Customers have fewer circuits and less equipment to monitor. In addition, the carrier provides proactive monitoring and network maintenance 24 hours a day.
- *Inter-carrier compatibility*   Frame Relay service is compatible between the networks of various carriers, enabling data to reach locations that are not served by the primary service provider.
- *Customer-controlled network management*   Enables customers to obtain network-management information via in-band SNMP queries and pings that are launched from their own network-management stations
- *Performance reports*   Enables customers to manage their Frame Relay service to its maximum advantage. Available network reports, which are accessible on the carrier's secure Web site, include reports for utilization, errors, health, trends, and exceptions.

Types of Circuits

Packet networks make use of virtual circuits (sometimes referred to as logical channels). The two primary types of virtual circuits that Frame Relay supports are *Switched Virtual Circuits* (SVCs) and *Permanent Virtual Circuits* (PVCs). SVCs are analogous to dialup connections, which require path setup and tear-down. A key advantage of SVCs is that they permit any-to-any connectivity between devices that are connected to the Frame Relay network. PVCs are more similar to dedicated private lines. Once they are set up, the predefined logical connections between sites that are attached to the Frame Relay network stay in place. This feature enables logical channels to be dedicated to specific terminals. The SVC requires fewer logical channels at the host, because the terminals contend for a lesser number of logical channels. Of course, we assume that not everyone will require access to the host at the same time.

Another type of virtual circuit is the *Multicast Virtual Circuit* (MVC), which is used to broadcast the same data to a group of users over a reserved data-link connection in the Frame Relay network. This type of virtual circuit might be useful for expediting communications among members of a single workgroup who are dispersed over multiple locations, or to facilitate interdepartmental collaboration on a major project. MVC can also be used for broadcast faxing, news feeds, and push applications.

The same Frame Relay interface can be used to set up SVCs, PVCs, and MVCs. All three can share the same digital facility. In supporting multiple types of virtual circuits, Frame Relay networks provide a high degree of configuration flexibility as well as more efficient utilization of the available bandwidth.

Congestion Control

Real-time congestion control must accomplish the following critical objectives in a Frame Relay network:

- Maintaining high throughput by minimizing timeouts and out-of-sequence frame deliveries

- Preventing session disconnects, unless they are required for congestion control

- Protecting against unfair users who attempt to hog the available network resources by exceeding their *Committed Information Rate* (CIR) or established burst size

■ Preventing the spread of congestion to other parts of the network

■ Providing delays that are consistent with application requirements and service objectives

In the Frame Relay network, congestion can be avoided through control mechanisms that provide *Backward Explicit Congestion Notification* (BECN) and *Forward Explicit Congestion Notification* (FECN), which are depicted in Figure F-15.

BECN is indicated by a bit set in the data frame by the network to notify the user's equipment that congestion avoidance procedures should be initiated for traffic in the opposite direction of the received frame. FECN is indicated by a bit set in the data frame by the network to notify the user that congestion avoidance procedures should be initiated for traffic in the direction of the received frame. Upon receiving either indication, the end point (i.e., bridge, router or other internetworking device) takes the appropriate action to ease congestion.

The response to congestion notification depends upon the protocols and flow-control mechanism that are employed by the end point. The BECN bit would typically be used by protocols that are capable of controlling traffic flow at the source. The FECN bit would typically be used by protocols that implement flow control at the destination.

Figure F-15
Congestion notification on the Frame Relay network

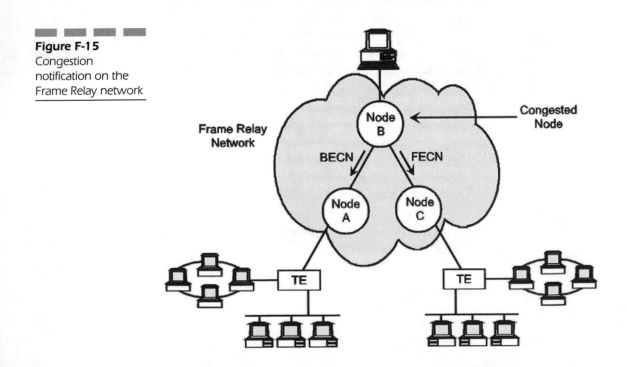

Upon receipt of a frame with the BECN bit set, the end point must reduce its offered rate to the CIR for that Frame Relay connection. If consecutive data frames are received with the BECN bit set, the end point must reduce its rate to the next step rate below the current offered rate. The step rates are 0.675, 0.50, and 0.25 of the current rate. After the end point has reduced its offered rate in response to receiving the BECN, it might increase its rate by a factor of 0.125 times the current rate after receiving two consecutive frames with the BECN bit cleared.

If the end point does not respond to the congestion notification; or if the user's data flow into the network is not significantly reduced as a result of the response to the congestion notification; or if an end point is experiencing a problem that exacerbates the congestion problem, then the network invokes congestion-recovery procedures. These procedures include discarding frames (in which case, the end-to-end protocols that the end points employ are responsible for detecting and requesting the retransmission of missing frames).

Frame discard can be conducted on a priority basis; that is, a decision is made as to whether certain frames should be discarded in preference to other frames in a congestion situation, based on predetermined criteria. Frames are discarded based on their discard-eligibility setting of 1 or 0, as specified in the data frame. A setting of 1 indicates that the frame should be discarded during congestion, while a setting of 0 indicates that the frame should not be discarded unless there are no alternatives.

The discard eligibility can be determined in several ways. The user can declare whether the frames are eligible for discard by setting the discard-eligibility bit in the data frame to 1. Or, the network-access interface can be configured to set the discard eligibility bit to 1 when the user's data has exceeded the CIR; in which case, the data is considered excess and subject to discard. For users who subscribe to CIR=0, which moves data through the Frame Relay network on a best-efforts basis (subject to bandwidth availability), all traffic is discard eligible.

Frame Relay Charges

Frame Relay service providers can invoice users for a wide range of items, including the following:

- Port charge for access to the nearest Frame Relay switch, which is applied to every user location that is attached to the Frame Relay network

- Local loop charge, which is the cost of the facility providing access to the Frame Relay network

- Charges for the number of PVCs and SVCs that are required

- Burst characteristics, usually determined by a committed burst size and a burst excess size

- CIR, which is the anticipated average rate of data flow through the access point

- Customer-premises equipment, which includes the Frame Relay or internetworking access equipment that is optionally leased from the service provider and bundled into the cost of the service

- Distance between locations. Usually, there is one price for local Frame Relay service and another price for national Frame Relay service.

Voice-Over Frame Relay (VoFR)

Voice-Over-Frame Relay (VoFR) is receiving growing attention. Most data-oriented *Frame Relay Access Devices* (FRADs) and routers use the *First-In, First-Out* (FIFO) method of handling traffic. In order to achieve the best voice quality, however, voice frames cannot be permitted to accumulate behind a long queue of data frames. Voice FRADs and routers, therefore, employ traffic-prioritization schemes to minimize delay for voice traffic.

Traffic-prioritization schemes ensure that voice packets have preference over data. During times of network congestion, one of the easiest prioritization methods is to simply discard frames. In such cases, data frames (rather than voice frames) will be discarded first, giving voice a better chance of making it through the network.

Some service providers offer prioritization of PVCs within the Frame Relay network. Prioritization features on both the CPE and the Frame Relay network can result in better voice-application performance. The CPE ensures that higher-priority traffic is sent to the network first, while PVC prioritization within the network ensures that higher-priority traffic is delivered to its destination first.

VoFR equipment compresses the voice signal from 64Kbps to at least 32Kbps. In most cases, compression to 16Kbps or even 8Kbps is possible. Some equipment vendors support dynamic compression options. When bandwidth is available, a higher voice quality is achieved by using 32Kbps, but as other calls are placed or as other traffic requires bandwidth, a 16Kbps or 8Kbps compression algorithm is implemented. Most voice FRADs also support fax traffic. A fax can take up as little as 9.6Kbps of bandwidth for each active line. VoFR usually enables a company to use its existing phones and numbering plan. In most cases, an internal dialing plan can be set up that enables users to dial fewer digits to connect to internal locations.

A persistent myth about VoFR is that voice calls can be carried free on an existing Frame Relay network. In fact, VoFR requires special CPE, entails an increase in the port speed, and requires a boost in the CIR of the PVCs —all of which carry a price tag.

Summary

The need for Frame Relay arose partly from the emergence of digital networks, which are faster and less prone to transmission errors than older analog lines. Although the X.25 protocol overcomes the limitations of analog lines, it does so with a significant performance penalty—due mainly to its error checking and flow-control capabilities. In having the capability to do without these and other functions, plus support different types of virtual circuits, Frame Relay offers more efficient utilization of available bandwidth—and, as a result, offers more configuration flexibility.

See Also

Asynchronous Transfer Mode (ATM)

Packet Switched Network—X.25

Switched Multi-Megabit Data Service (SMDS)

Frequency Division Multiple Access (FDMA)

Three multiple-access schemes are in use today that provide the foundation for mobile communications systems:

- *Frequency Division Multiple Access* (FDMA) Serves the calls with different frequency channels
- *Time Division Multiple Access* (TDMA) Serves the calls with different time slots
- *Code Division Multiple Access* (CDMA) Serves the calls with different code sequences

Of the three, FDMA is the simplest and still the most widespread technology in use today for mobile communications. For example, FDMA is used in the CT2 system for cordless telecommunications. The familiar cordless phone that is used in the home is representative of this type of system. FDMA creates capacity by splitting bandwidth into radio channels in the frequency domain. In the initial call setup, the handset scans the available channels and locks onto an unoccupied channel for the duration of the call. Based on *Time-Division Duplexing* (TDD), the call will be split into time blocks that alternate between transmitting and receiving.

The traditional analog cellular systems, such as those that are based on the *Advanced Mobile Phone Service* (AMPS) standards, also use FDMA to derive the channel. In the case of AMPS, the channel is a 30KHz slice of the spectrum. Only one subscriber at a time is assigned to the channel. No other conversations can access the channel until the subscriber's call is finished or until the call is handed off to a different channel in an adjacent cell.

The analog operating environment poses several problems. One is that the wireless devices are often in motion. Current analog technology does not deal with call hand-offs well, as evidenced by the high incidence of dropped calls. This environment is particularly harsh for data and is less tolerant of transmission problems than voice. Whereas momentary signal fade is a nuisance in voice communications, for instance, it might cause a data connection to drop.

Another problem with analog systems is their limited capacity. To increase the capacity of analog cellular systems, the 30KHz channel can be divided into three narrower channels of 10KHz each. This setup is the basis of the *Narrow-Band AMPS* (N-AMPS) standard. This band-splitting technique, however, incurs significant base-station costs, and its limited growth potential makes it suitable only as a short-term solution.

While cell subdivision is often used to increase capacity, this solution has its limits. Because adjacent cells cannot use the same frequencies without risking interference, a limited number of frequencies are being reused at closer distances, which makes it increasingly difficult to maintain the quality of communications. Subdividing cells also increases the amount of overhead signaling that must be used in order to set up and manage the calls, which can overburden switch resources. In addition, property or rights-of-way for cell sites is difficult to obtain in metropolitan areas where traffic volume is highest and future substantial growth is anticipated.

These and other limitations of analog FM radio technology have led to the development of second-generation cellular systems based on digital radio technology and advanced networking principles. Providing reliable service in this dynamic environment requires digital radio systems that employ advanced signal-processing technologies for modulation, error correction, and diversity. TDMA and CDMA provide these capabilities.

Summary

FM systems have supported cellular service for nearly 15 years, during which the demand has finally caught up with the available capacity. Now, first-generation cellular systems based on analog FM radio technology are rapidly being phased out in favor of digital systems that offer higher capacity, better voice quality, and advanced call-handling features. TDMA-based and CDMA-based systems are contending for acceptance among analog cellular carriers worldwide.

See Also

  Code Division Multiple Access (CDMA)

  Cordless Telecommunications

  Time Division Multiple Access (TDMA)

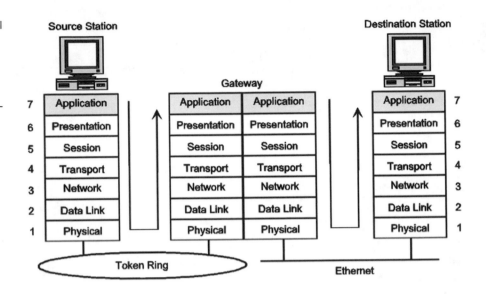

Gateways

Gateways interconnect dissimilar networks or applications and operate at the highest layer of the *Open Systems Interconnection* (OSI) reference model: the application layer (refer to Figure G-1). A gateway consists of protocol-conversion software that usually resides in a server, minicomputer, mainframe, or front-end device. One application of gateways is interconnecting disparate networks or media by processing the various protocols that each network uses, so that information from the sender is intelligible to the receiver (despite differences in network protocols or computing platforms).

For example, when you use an SNA gateway to connect an asynchronous PC to a synchronous IBM SNA mainframe, the gateway acts as both a conduit through which the computers communicate and as a translator between the various protocol layers. The translation process consumes considerable processing power, resulting in relatively slow transmission rates when compared to other interconnection methods (hundreds of packets per second for a gateway, versus tens of thousands of packets per second for a bridge).

In addition to its translation capabilities, a gateway can check on the various protocols that are in use, ensuring that there is enough protocol processing power available for any given application. A gateway also helps ensure that the network links maintain a level of reliability for handling applications in conformance with predefined error-rate thresholds.

Gateway Applications

Gateways have a variety of applications. In addition to facilitating LAN workstation connections to various host environments, such as IBM's SNA systems and mid-range systems, they facilitate connections to X.25 packet-switching networks. Other applications of gateways include the interconnection of various e-mail systems, enabling the exchange of mail between normally incompatible formats. Servers that are equipped with the X.400 international messaging protocol actually provide the gateway function.

In some cases, you can use gateways to consolidate hardware and software. For example, you can use multiple, networked PCs that share an SNA 3270 gateway in place of IBM's 3270 Information Display System or in place of many individual 3270 emulation products. Although the IBM systems offer a standard means of achieving the PC-host connection, it is expensive when you use a gateway to attach a large number of stand-alone PCs. The relatively high connection cost per computer discourages host access for occasional users and limits the central control of information.

If the PCs are attached to a LAN, however, one gateway can emulate a cluster controller and thereby provide all workstations with host access at a low cost. Cluster-controller emulators use an RS-232C or compatible serial interface to a host adapter or communications controller, such as an IBM 3720 or 3745. They can support up to 254 simultaneous sessions.

IP-PSTN Gateways

A relatively new type of gateway provides connections between the Internet and the PSTN, enabling users to place phone calls from their multimedia PCs or conventional telephones over the Internet, or from a carrier's managed IP data network (and vice versa). This arrangement enables users to save on long-distance and international call charges.

The IP-PSTN gateways perform the translations between the two types of networks. When a standard voice call is received at a near-end gateway, the analog voice signal is digitized, compressed, and packetized for transmission over an IP network. At the far-end gateway, the process is reversed and the packets are decompressed and returned to analog form before the call is delivered to its intended destination on the PSTN.

The gateways support one or more of the internationally recognized G.7xx voice codec specifications for toll-quality voice. The most commonly supported codec specifications are as follows:

- G.711   Describes the requirements for a codec using *Pulse-Code Modulation* (PCM) of voice frequencies to achieve 64Kbps, providing toll-quality voice on managed IP networks with sufficient available bandwidth

- G.723.1   Describes the requirements for a dual-rate speech codec for multimedia communications (e.g., videoconferencing) transmitting at 5.3Kbps and 6.3Kbps. This codec provides near toll-quality voice on managed IP networks.[1]

- G.729A   Describes the requirements for a low-complexity codec that transmits digitized and compressed voice at 8Kbps. This codec provides toll-quality voice on managed IP networks.

The specific codec to be used is negotiated on a call-by-call basis between the gateways by using the H.245 control protocol. Among other features, the H.245 protocol provides for capability exchange, enabling the gateways to implement the same codec at the time the call is placed. The gateways can be configured to implement a specific codec at the time the call is established, based on predefined criteria such as the following:

- Use G.711 only   In which case, the G.711 codec will be used for all calls
- Use G.729 (A) only   In which case, the G.729 (A) codec will be used for all calls
- Use highest common bit rate codec   In which case, the codec that will provide the best voice quality is selected
- Use lowest common bit rate codec   In which case, the codec that will provide the lowest packet bandwidth requirement is selected

This capability-exchange feature provides carriers and ISPs with the flexibility to offer different quality voice services at different price points, in addition to enabling corporate customers to specify a preferred proprietary codec to support voice or a voice-enabled application through an intranet or IP-based VPN.

Summary

Gateways are available as software or can be dedicated hardware systems that are equipped with appropriate software to make the translations

---

[1] The *Mean Opinion Score* (MOS) that is used to rate the quality of speech codecs measures toll-quality voice as having a top score of 4.0. With G.723.1, voice quality is rated at 3.98, which is only 2 percent less than that of analog telephone.

between different applications, networked devices, or different types of networks. Gateways can even be used to reconcile the differences between network-management systems or *Operations Support Systems* (OSSs), enabling them to interoperate with the systems of other vendors.

See Also

Bridges

Open Systems Interconnection (OSI)

Repeaters

Routers

Voice-Over IP (VoIP)

General Mobile Radio Service (GMRS)

General Mobile Radio Service (GMRS) is one of several personal radio services; specifically, it is a personal two-way UHF voice-communication service that can be used to facilitate the activities of the individual's immediate family members. This FCC-licensed service has a communications range of five to 25 miles and cannot be used to make telephone calls.

A GMRS system consists of station operators, a mobile station (which often consists of several mobile units) and sometimes one or more land stations. The classes of land stations are as follows: base station, mobile relay station (also known as a repeater), and control stations. A small base station is one that has an antenna no more than 20 feet above the ground or above the tree on which it is mounted and that transmits with no more than 5 watts.

Users communicate with a GMRS radio unit (refer to Figure G-2) over the general area of their residence in an urban or rural area. This area must be within the territorial limits of the 50 United States, the District of Columbia, and the Caribbean and Pacific insular areas. In transient use, mobile station units from one GMRS system can communicate through a mobile relay station in another GMRS system with the permission of its licensee.

Twenty-three GMRS channels are available. None of the GMRS channels are assigned for the exclusive use of any system, however. License applicants and licensees must cooperate with the selection and use of the

Figure G-2
Motorola's TalkAbout
Distance DPS is a
hand-held GMRS
radio that weighs
11.7 ounces and
runs on a
rechargeable NiCad
battery or six AA
batteries. A GMRS
license, which is
issued by the FCC,
and a fee are
required to use this
radio.

channels in order to make the most effective use of them and to reduce the possibility of interference.

Any mobile station or small base station in a GMRS system that operates in the simplex mode can transmit voice-type emissions with no more than 5 watts on the following 462MHz channels: 462.5625, 462.5875, 462.6125, 462.6375, 462.6625, 462.6875, and 462.7125MHz. The *Family Radio Service* (FRS) shares these channels.

Any mobile station in a GMRS system can transmit on the 467.675MHz channel to communicate through a mobile relay station that is transmitting on the 462.675MHz channel. The communications must be for the purpose of soliciting or rendering assistance to a traveler, or for communicating in an emergency that pertains to the immediate safety of life or the immediate protection of property.

Each GMRS system license assigns one or two of eight possible channels or channel pairs (one 462MHz channel and one 467MHz channel that are spaced 5MHz apart), as requested by the applicant. Applicants for GMRS system licenses are advised to investigate or monitor to determine the best available channel(s) before making their selection. Each applicant must select the channel(s) or channel pair(s) for the stations in the proposed system from the following list:

- For a base station, mobile relay station, fixed station, or mobile station: 462.550, 462.575, 462.600, 462.625, 462.650, 462.675, 462.700, and 462.725MHz

- For a mobile station, control station, or fixed station in a duplex system: 467.550, 467.575, 467.600, 467.625, 467.650, 467.675, 467.700, and 467.725MHz

GMRS system station operators must cooperate and share the assigned channel with station operators in other GMRS systems by monitoring the channel before initiating transmissions, waiting until communications in progress are completed before initiating transmissions, engaging in only permissible communications, and limiting transmissions to the minimum practical transmission time.

Other Services

Other private, personal radio services are available for short-distance two-way voice communications. You can use the *Citizens Band* (CB) radio service, over which users are authorized to operate an FCC type-accepted CB unit for communications covering a distance of one to five miles. No license document is issued. Also, the *Family Radio Service* (FRS) is available, over which users are authorized to operate an FCC-certified FRS unit for communications covering a distance of less than one mile. No license document is issued.

Summary

Any individual 18 years of age or older who is not a representative of a foreign government is eligible to apply for a GMRS system license. Application for a GMRS system license is made on FCC Form 574. A booklet titled, "Instructions for Completion of FCC Form 574" is helpful for preparing the application. You can obtain this booklet from the FCC's Consumer Assistance Branch by calling 1-800-322-1117 (a filing fee applies).

See Also

Citizens Band (CB) Radio Service

Family Radio Service (FRS)

Low-Power Radio Service

Generic Digital Services (GDS)

Generic Digital Services (GDS) are dedicated, digital private-line services that are offered at transmission speeds of 2.4, 4.8, 9.6, 19.2, 56, and 64Kbps. They are aimed at both analog service users who seek higher-quality transmission and at AT&T's *Dataphone Digital Service* (DDS) customers who seek a more affordable digital service. With the exception of 64Kbps, all transmission speeds include a secondary channel for low-speed data-transport applications, including diagnostic signaling.

With *Customer Network Reconfiguration* (CNR), users have the flexibility to alter their network from an on-premises network-management terminal according to time-of-day volume needs, as well as to perform disaster recovery quickly. Alternatively, the user can opt to have the carrier reconfigure the network. These features are implemented at the carrier's *Digital Cross-Connect System* (DCS). Through the DCS, multi-point and point-to-point topologies are supported.

The following table provides the brand names under which generic digital services are provided by their respective carriers:

Carrier	Service
Ameritech	Optinet
Bell Atlantic	Digital Connect Service and DigiPath
BellSouth	Synchronet
Pacific Bell	Advanced Digital Network
SBC	MegaLink
US WEST	DigiCom

Applications

GDS can be used to support the routine information management tasks that are essential to the daily operations on all businesses. The following table summarizes the common applications of GDS in the education, banking/finance, and retail sectors.

Application	Education	Banking/Finance	Retail
File Transfer	■	■	■
Network Management	■	■	■
Disaster Recovery	■	■	■
Time of Day Routing	■	■	■
Electronic Funds Transfer		■	■
Automatic Teller Machines		■	
Order Entry			■
Inventory Management			■
Information Retrieval	■	■	■
CAD/CAM	■		
LAN-to-LAN Interconnection	■	■	■

DCS nodes located in the carrier's *Serving Wire Centers* (SWCs) are used to support GDS. The cross-connect nodes facilitate network administration, disaster recovery, and remote testing. Through the DCS, users can make or order changes to their network in order to meet daily, occasional, or seasonal data traffic needs. The carrier automatically takes steps to correct any errors that occur on the transmission links, with the goal of 99.96 percent error-free seconds, which equates to 3.5 hours of down time per year.

GDS can connect to interLATA services that are provided by the user's long-distance carrier of choice, making connectivity to every central office that is served by a digital T-carrier facility possible. GDS also provides a migration path to more sophisticated technologies as data needs change, including T1 and ISDN. The carrier can provide all of the equipment that is necessary to use GDS, such as the CSU/DSUs that are required for the front-end of the link(s), as well as bridges/routers for LAN-to-LAN interconnection. Alternatively, users can purchase or use their own equipment.

The multiplexing capability of the DSU makes it particularly compatible with multiple stand-alone terminals or applications that different vendors supply. The banking industry is representative of this environment. All banks essentially run the same applications at each location, with each application being served with separate multi-point lines:

- A teller application, which runs at 4.8Kbps or 9.6Kbps
- *Automated Teller Machines* (ATMs), which run at 1.2Kbps or 24Kbps
- A platform application, which runs at 4.8Kbps or 9.6Kbps
- A security application, which runs at 75bps to 1.2Kbps

Costs for GDS at 56Kbps are similar to the costs of a voice-grade private line in many areas. In such cases, a multi-port, multi-point DSU affords even greater cost savings. A network of three multi-point lines with six drops each can be replaced with one 56Kbps multi-point line, using a DSU at each location to multiplex the applications. Many DSUs have software-selectable port rates of 75bps to 64Kbps. These speeds can be used in any combination, as long as they do not exceed the total bandwidth of 64Kbps.

Multi-drop networks require a *Multi-Point Junction Unit* (MJU) to bridge or connect the individual circuit segments. The unit is composed of hardware and software that is integral to the generic service. Use of the MJU requires the customer to designate a control station, with the other stations on the multi-point circuit designated as remote stations. One unit can support one control station and four remote stations, but these can be cascaded in order to support additional remote stations.

Much of the cost savings associated with GDS is attributable to the elimination of traffic back-hauling. With hub-oriented services (e.g., DDS), the user's traffic must be routed through a special hub office. Because there are fewer DDS hubs nationwide (AT&T has about 100), longer circuit mileage is typical, which inflates the service cost. With GDS, the service is available from the serving wire centers of the LECs (about 20,000 among all carriers), so there is less of a need to back-haul traffic—six miles with GDS, versus 60 miles with DDS.

Service Components

Generic digital services are supported by the following network elements (refer to Figure G-3): the local loop, a serving wire center, inter-office facilities, a digital cross-connect system, an MJU, and a D4 channel bank that is equipped with an office channel unit data port.

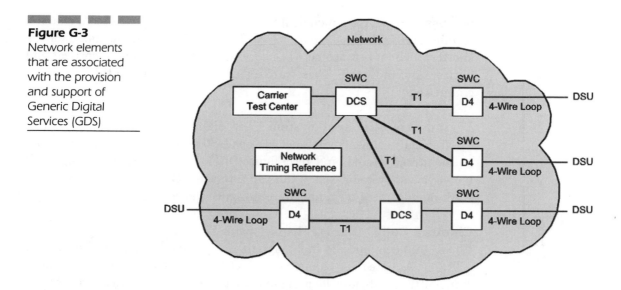

Figure G-3
Network elements
that are associated
with the provision
and support of
Generic Digital
Services (GDS)

Local Loop The local loop uses copper wire pairs that connect the customer premises to the LEC's serving wire center. The local loop is terminated at the customer premises at the network interface. On the customer side of the network interface is network channel terminating equipment, which for GDS is a *Digital Service Unit* (DSU). The functions of the DSU include the generation and reconstruction of the digital signal; signal encoding and formatting; timing extraction from the incoming signal, including sampling and loop timing; and generation and recognition of control signals.

Serving Wire Center At the serving wire center, the local loop is terminated on an *Office Channel Unit Data Port* (OCUDP) of a D4 digital channel bank. The functions that are provided by this port include transmitting outgoing loop signals to the customer station; reshaping, retiming, and regenerating the incoming loop signal; assembling the data into a format suitable for network and loop transmission; providing electrical current to the loop; and passing network-originated commands (e.g., loop-backs) to the DSU in order to implement diagnostics.

Inter-Office Facilities Inter-office facilities carry 24 individual 64Kbps (DS0) channels on the 1.544Mbps (DS1) transport facilities between the digital nodes. In addition to the cost of the local loop, the customer is charged for inter-office facilities between serving wire centers and the customer premises. The charges include a fixed element (generally called a channel termination) and a mileage-sensitive element. Mileage is calculated for the shortest airline distance between serving wire centers, using vertical and horizontal coordinates. These rate elements consist of monthly recurring charges only.

Digital Cross-Connect System   The digital cross-connect system can electronically connect individual 64Kbps DS0 channels. These channels can be monitored, tested, and reconfigured from the local exchange carrier's network test systems. Users can tap into the DCS with on-premises network-management terminals to reconfigure the channels.

Channel Bank   The D4 channel bank is a *Time-Division Multiplexer (TDM)* that combines 24 input signals for transport to a digital node over a T1 link. As noted, the channel bank's OCUDP performs several functions: transmitting outgoing loop signals to the customer station; reshaping, retiming, and regenerating the incoming loop signal; assembling the data into a format suitable for network and loop transmission; providing current to the loop; and performing the diagnostic and functional commands that are received from the network test center.

Secondary Channel   The secondary channel gives customers an independent low-speed, derived data channel that operates in parallel with the primary data channel. The secondary channel can be used to perform network surveillance on a continuous and non-disruptive basis, meaning that production data can continue to flow in an unaffected manner over the primary channel.

Although the secondary channel is used primarily for the exchange of test, diagnostic, measurement, and control messages between a central site and its remote terminals, it can also be used as a low-speed, general-purpose data channel. To take advantage of the secondary channel, the user must have DSUs that specifically support this capability. The table on the following page summarizes the primary and secondary channels that are available with GDS.

Summary

A significant advantage of generic digital services is that bandwidth can be used more efficiently than with other types of digital services. The flexible multiplexing scheme improves the integration of different applications on

Primary Channel	Secondary Channel
2.4Kbps	133 bps
4.8Kbps	266 bps
9.6Kbps	533 bps
19.2Kbps	1,066 bps
56Kbps	2,132 bps
64Kbps	Not available

one network, and the use of multi-port, multi-point DSUs improves band-width efficiency even more. If the somewhat lower service quality and avail-ability of GDS (compared to DDS) is acceptable, they are an excellent way to improve a network's efficiency and reduce costs.

See Also

  Digital Data Services (DDS)

Global Positioning System (GPS)

The *Global Positioning System* (GPS) is a network of 24 Navstar satellites orbiting Earth at 11,000 miles. Originally established by the United States Department of Defense at a cost of about $13 billion, access to GPS is free to all users, including those in other countries. The system's positioning and timing data are used for a variety of applications, including air, land, and sea navigation, vehicle and vessel tracking, surveying and mapping, and asset and natural resource management. With military accuracy restric-tions lifted in March 1996, the GPS can now pinpoint the location of objects as small as a dime anywhere on the earth's surface.

The first GPS satellite was launched in 1978. The first 10 satellites were developmental satellites, called Block I. From 1989 to 1993, 23 production satellites, called Block II, were launched. The launch of the 24th satellite in 1994 completed the system. The satellites are positioned so that signals from six of them can be received nearly 100 percent of the time at any point on earth.

GPS Components

The GPS consists of satellites, receivers, and ground control systems. The satellites transmit signals (1575.42MHz) that can be detected by GPS receivers on the ground. These receivers can be portable or mounted in ships, planes, and cars to provide exact position information, regardless of weather conditions. They detect, decode, and process GPS satellite signals to give the precise position of the user.

The GPS control or ground segment consists of five unmanned monitor stations located in Hawaii, Kwajalein in the Pacific Ocean, Diego Garcia in the Indian Ocean, Ascension Island in the Atlantic Ocean, and Colorado Springs, Colorado. There is also a master ground station at Falcon Air Force Base in Colorado Springs and four large ground antenna stations that broadcast signals to the satellites. The stations also track and monitor the GPS satellites.

System Operation

With GPS, signals from several satellites are triangulated to identify the exact position of the user. To triangulate, GPS measures distance by using the travel time of a radio message from the satellite to a ground receiver. To measure travel time, GPS uses accurate clocks in the satellites. Once the distance to a satellite is known, knowledge of the satellite's location in space is used to complete the calculation. GPS receivers on the ground have an almanac stored in their computer memory that indicates where each satellite will be in the sky at any given time. GPS receivers calculate for ionosphere and atmosphere delays in order to further tune the position measurement.

To make sure that both satellite and receiver are synchronized, each satellite has four atomic clocks that keep time to within three nanoseconds, or three billionths of a second. For cost savings, the clocks in the ground receivers are not that accurate. To compensate, an extra satellite range measurement is taken. Trigonometry says that if three perfect measurements locate a point in three-dimensional space, then a fourth measurement can eliminate any timing offset. This fourth measurement compensates for the receiver's imperfect synchronization.

The ground unit receives the satellite signals, which travels at the speed of light. Even at this speed, the signals take a measurable amount of time to reach the receiver. The difference between the time the signals are sent and the time they are received, multiplied by the speed of light, enables the

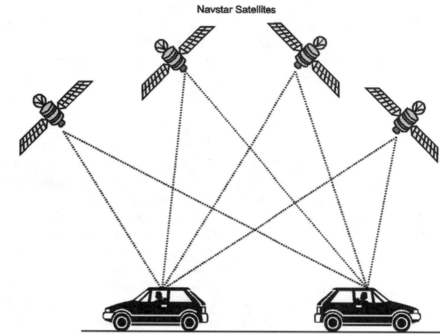

Figure G-4
Signals from four satellites, captured by a vehicle's onboard GPS receiver, are used to determine precise location information.

Navstar Satellites

GPS-equipped Vehicles

receiver to calculate the distance to the satellite. To measure precise latitude, longitude, and altitude, the receiver measures the time that it took for the signals from several satellites to reach the receiver (refer to Figure G-4).

GPS uses a system of coordinates called the *Worldwide Geodetic System 1984* (WGS-84). This system is similar to the latitude and longitude lines that are commonly seen on large wall maps used in schools. The WGS-84 system provides a built-in, standardized frame of reference, enabling receivers from any vendor to provide exactly the same positioning information.

GPS Applications

Although the GPS system was only completed in 1994, it has already proven itself in military applications, most notably in Operation Desert Storm where United States troops and their allies faced a vast, featureless desert. Without a reliable navigation system, sophisticated troop maneuvers could not have been performed. This situation could have prolonged the operation well beyond the 100 hours that it actually took.

With GPS, troops had the capability to go places and maneuver in sand-storms or at night when even the troops who were native to the area could not. Initially, more than 1,000 portable commercial receivers were purchased for their use. The demand was so great that before the end of the conflict, more than 9,000 commercial receivers were in use in the Persian Gulf region. They were carried by ground troops and attached to vehicles, helicopters, and aircraft instrument panels. GPS receivers were used in several aircraft, including F-16 fighters, KC-135 tankers, and B-52s. Navy ships used GPS receivers for rendezvous, mine-sweeping, and aircraft operations. Today, GPS has become an important element of military operations and weapons systems. In addition, GPS is used by satellites to obtain highly accurate orbit data and to control spacecraft orientation.

While the GPS system was originally developed to meet the needs of the military community, new ways to use its capabilities are continually being found, from the exotic to the mundane. Among the former is the use of GPS for wildlife management. Endangered species such as Montana elk and Mojave Desert tortoises have been fitted with tiny GPS transceivers to help determine population distribution patterns and possible sources of disease. In Africa, GPS receivers are used to monitor the migration patterns of large herds for a variety of research purposes.

Hand-held GPS receivers are now routinely used in field applications that require precise information gathering, including field surveying by utility companies, mapping by oil and gas explorers, and resource planning by timber companies. GPS-equipped balloons are used to monitor holes in the ozone layer over the polar ice caps. Air quality is being monitored by using GPS receivers. Buoys that track major oil spills transmit data via GPS. Archaeologists and explorers use the system to mark remote land and ocean sites until they can return with proper equipment and funding.

Vehicle tracking is one of the fastest-growing GPS applications. GPS-equipped fleet vehicles, public transportation systems, delivery trucks, and courier services use receivers to monitor their locations at all times. GPS is also helping save lives. Many police, fire, and emergency medical service units are using GPS receivers to determine the police car, fire truck, or ambulance that is nearest to an emergency—enabling the quickest possible response in life-or-death situations.

GPS data will become more useful to consumers when it is linked with digital mapping. Accordingly, some automobile manufacturers are offering moving-map displays guided by GPS receivers as an option on new vehicles. The displays can even be removed and taken into a home to plan a trip. Some GPS-equipped vehicles give directions to drivers on display screens and through synthesized voice instructions. These features enable drivers to get where they want to go more rapidly and safely than has ever been possible before.

When GPS data is used in conjunction with geographic data collection systems, it is possible to instantaneously arrive at submeter positions together with feature descriptions to compile highly accurate *Geographic Information Systems* (GIS). When used by cities and towns, for example, GPS can help in the management of the following kinds of assets:

Point Features	Line Features	Area Features
Signs	Streets	Parks
Manhole covers	Sidewalks	Landfills
Fire hydrants	Fitness trails	Wetlands
Light poles	Sewer lines	Planning zones
Storm drains	Water lines	Subdivisions
Driveways	Bus routes	Recycling centers

Some government agencies, academic institutions, and private companies are using GPS to determine the location of a multitude of features, including point features such as pollutant discharges and water supply wells, line features such as roads and streams, and area features such as waste lagoons and property boundaries. Before GPS, such features had to be located with surveying equipment, aerial photographs, or satellite imagery. The precise location of such features can now be determined by one individual who uses hand-held GPS equipment.

GPS and Cellular

GPS technology is even being used in conjunction with cellular technology to provide value-added services. With the push of a button on a cellular telephone, automobile drivers and operators of commercial vehicles in some areas can talk to a service provider and simultaneously signal their position, emergency status, or equipment failure information to auto clubs, security services, or central dispatch services.

This functionality is possible with Motorola's Cellular Positioning and Emergency Messaging Unit, for example, which offers mobile security and tracking to those who drive automobiles and/or operate fleets. The system is designed for sale to systems integrators that configure consumer and commercial systems that operate via cellular telephony. The Cellular Positioning and Emergency Messaging Unit communicates GPS-determined vehicle position and status, making it suited for use in systems that support road-side assistance providers, home security monitoring firms, cellular carriers, rental car companies, commercial fleet operators, and auto manufacturers who are seeking a competitive advantage.

Skytel, the provider of wireless messaging services, is marketing its AutoLink system for automobiles (which uses a GPS receiver). The AutoLink system provides automatic emergency response, theft deterrence, vehicle tracking and immobilization, two-way personal paging, remote vehicle unlocking, driver personalization, navigational guidance, and location-based information service.

Summary

Due to its accuracy, GPS is rapidly becoming the locational data collection method of choice for a variety of commercial, government, and military applications. GPS has certainly become an important and cost effective method for locating terrestrial features too numerous or too dynamic to be mapped by traditional methods. Although originally funded by the United States Departmentof Defense, access to the GPS network is free to all users in any country. This availability has encouraged applications development and has created an entirely new consumer market, particularly in the area of vehicular location and highway navigation.

See Also

Cellular Voice Communications

Global System for Mobile (GSM) Telecommunications

Global System for Mobile (GSM) Telecommunications—formerly known as Groupe Spéciale Mobile for the group that started developing the standard in 1982—was designed from the beginning as an international digital cellular service. Initially, developers intended for GSM subscribers to be able to cross national borders and find that their mobile services crossed along with them. Today, GSM is well established in most countries, with the highest concentration of service providers and users in Europe.

Originally, the 900MHz band was reserved for GSM services. Since GSM first entered commercial service in 1992, it has been adapted to work at 1800MHz for the PCNs in Europe and at 1900MHz for PCS in the United States.

GSM Services

GSM telecommunication services are divided into teleservices, bearer services, and supplementary services.

Teleservices   The most basic teleservice supported by GSM is telephony, which includes an emergency service in which the nearest emergency service provider is notified by dialing three digits (similar to 911). Group 3 fax, an analog method described in ITU-T recommendation T.30, is also supported by GSM through the use of an appropriate fax adapter.

Bearer Services   A unique feature of GSM compared to older analog systems is the *Short Message Service* (SMS). SMS is a bi-directional service for sending short alphanumeric messages (up to 160 bytes) in a store-and-forward manner. For point-to-point SMS, a message can be sent to another subscriber to the service, and an acknowledgment of receipt is provided to the sender. SMS can also be used in cell-broadcast mode for sending messages such as traffic updates or news updates. Messages can be stored in a smart card called the *Subscriber Identity Module* (SIM) for later retrieval.

Because GSM is based on digital technology, it enables synchronous and asynchronous data to be transported as a bearer service to or from an ISDN terminal. The data rates supported by GSM are 300 bps, 600 bps, 1200 bps, 2400 bps, and 9600 bps. Data can use either the transparent service, which has a fixed delay but no guarantee of data integrity, or a non-transparent service, which guarantees data integrity through an *Automatic Repeat Request* (ARQ) mechanism—but with variable delay.

GSM has much more potential in terms of supporting data. Finnish telecommunications operator Sonera offers the world's fastest digital GSM mobile phone service by using the *High Speed Circuit Switched Data* (HSCSD) system. The system offers up to 38.4Kbps, compared with 9.6Kbps for regular GSM networks.

Transmission speeds of up to 171.2Kbps are already becoming available with *General Packet Radio Service* (GPRS). The high bandwidth is achieved by using all eight time slots at the same time. GPRS facilitates several new applications that have not previously been available over GSM networks, due to the limitations in speed of circuit switched data (9.6Kbps) and message length of the Short Message Service (160 characters). GPRS will fully enable the Internet applications that are normally accessed from a desktop computer, from Web browsing to chatting over the mobile network. Other new applications for GPRS include file transfer and home automation—the capability to remotely access and control in-house appliances and machines.

GPRS will require new GSM phones that specifically support the new service. Both HSCSD and GPRS are steps toward the third-generation mobile technology, called *Universal Mobile Telecommunications System* (UMTS), which will be launched to work beside GSM and other digital second-generation systems in advanced markets. UMTS will enable wireless networks such as GSM to work at up to 2Mbps.

Supplementary Services   Supplementary services are provided on top of teleservices or bearer services and include such features as caller identification, call forwarding, call waiting, and multi-party conversations. A lockout feature also exists that prevents the dialing of certain types of calls, such as international calls.

Network Architecture

A GSM network consists of the following elements: mobile station, base station subsystem, and *Mobile Switching Center* (MSC). Each GSM network also has an operations and maintenance center, which oversees the proper operation and setup of the network. Two Air Interfaces exist: the Um interface is a radio link over which the mobile station and the base station subsystem communicate; the A interface is a radio link over which the base station subsystem communicates with the MSC.

***Mobile Station* (MS)**   The *Mobile Station* (MS) consists of the radio transceiver, display and digital signal processors, and the *Subscriber Identity Module* (SIM). The SIM provides personal mobility so that the subscriber can have access to all services, regardless of the terminal's location or the specific terminal used. By removing the SIM from one GSM cellular phone and inserting it into another GSM cellular phone, the user can receive calls at that phone, make calls from that phone, or receive other subscribed services. The SIM card can be protected against unauthorized use by a password or *Personal Identification Number* (PIN).

An *International Mobile Equipment Identity* (IMEI) number uniquely identifies each mobile station. The SIM card contains an *International Mobile Subscriber Identity* (IMSI) number, which identifies the subscriber, a secret key for authentication, and other user information. Because the IMEI and IMSI are independent, this arrangement provides users with a high degree of security.

Base Station Subsystem The base station subsystem consists of two parts: the *Base Transceiver Station* (BTS) and the *Base Station Controller* (BSC). These communicate across the air interface, enabling operation between components that different suppliers produce.

The base transceiver station contains the radio transceivers that define a cell and handles the radio link protocols with the mobile stations. In a large urban area, there will typically be a number of BTSs to support a large subscriber base of mobile service users.

The base station controller provides the connection between the mobile stations and the MSC, in addition to managing the radio resources for the BTSs, handling such functions as radio channel setup, frequency hopping, and handoffs. The BSC also translates the 13Kbps voice channel used over the radio link to the standard 64Kbps channel used by the land-based PSTN or ISDN.

Mobile Services Switching Center (MSC) The MSC acts similar to an ordinary switching node on the PSTN or ISDN and provides all of the functionalities that are needed to handle a mobile subscriber, such as registration, authentication, location updating, handoffs, and call routing to a roaming subscriber. These services are provided in conjunction with several other components, which together form the network subsystem. The MSC provides the connection to the public network (PSTN or ISDN) and signaling between various network elements that use *Signaling System 7* (SS7).

The MSC contains no information about particular mobile stations. This information is stored in two location registers, which are essentially databases. The *Home Location Register* (HLR) and *Visitor Location Register* (VLR), together with the MSC, provide the call routing and roaming (national and international) capabilities of GSM.

The HLR contains administrative information for each subscriber who is registered in the corresponding GSM network, along with the current location of the mobile device. The current location of the mobile device is in the form of a *Mobile Station Roaming Number* (MSRN), which is a regular ISDN number used to route a call to the MSC where the mobile device is currently located. Only one HLR is needed per GSM network, although it might be implemented as a distributed database. The VLR contains selected administrative information from the HLR, which is necessary for call control and provision of the subscribed services, for each mobile device currently located in the geographical area controlled by the VLR.

Two other registers are used for authentication and security purposes. The *Equipment Identity Register* (EIR) is a database that contains a list of all valid mobile equipment on the network, where each mobile station is identified by its IMEI. An IMEI is marked as invalid if it has been reported stolen or is not type approved. The authentication center is a protected database that stores a copy of the secret key that is stored in each subscriber's SIM card, which is used for authentication.

Channel Derivation and Types

Because the radio spectrum is a limited resource that is shared by all users, a method must be devised in order to divide up the bandwidth among as many users as possible. The method that GSM uses is a combination of *Time Division Multiple Access* (TDMA) and *Frequency Division Multiple Access* (TDMA/FDMA).

The FDMA part involves the division by frequency of the total 25MHz bandwidth into 124 carrier frequencies of 200KHz bandwidth. One or more carrier frequencies are then assigned to each base station. Each of these carrier frequencies is then divided in time, using a TDMA scheme, into eight time slots. One time slot is used for transmission by the mobile device and one for reception. They are separated in time so that the mobile unit does not receive and transmit at the same time.

Within the framework of TDMA, two types of channels are provided: traffic channels and control channels. Traffic channels carry voice and data between users, while the control channels carry information that is used by the network for supervision and management. Among the control channels are the following:

- *Fast Associated Control Channel* (FACCH)   This channel is created by robbing slots from a traffic channel in order to transmit power control and call hand-off messages.

- *Broadcast Control Channel* (BCCH)   Continually broadcasts (on the downlink) information including base station identity, frequency allocations, and frequency hopping sequences

- *Standalone Dedicated Control Channel* (SDCCH)   Used for registration, authentication, call setup, and location updating
- *Common Control Channel* (CCCH)   Comprised of three control channels that are used during call origination and call paging
- *Random Access Channel* (RACH)   Used to request access to the network
- *Paging Channel* (PCH)   Used to alert the mobile station of an incoming call

Authentication and Security

Because radio signals can be accessed by virtually anyone, authenticating users in order to prove that they are who they claim to be is an important feature of a mobile network.

Authentication involves two functional entities: the SIM card in the mobile unit and the *Authentication Center* (AC). Each subscriber is given a secret key, one copy of which is stored in the SIM card and the other copy is stored in the AC. During authentication, the AC generates a random number that it sends to the mobile unit. Both the mobile unit and the AC then use the random number, in conjunction with the subscriber's secret key and an encryption algorithm called A3, to generate a number that is sent back to the AC. If the number that the mobile unit sends is the same as the one that the AC calculated, the subscriber is authenticated.

The calculated number is also used, together with a TDMA frame number and another encryption algorithm called A5, to encrypt the data sent over the radio link, preventing others from eavesdropping. Encryption provides an added measure of security, because the signal is already coded, interleaved, and transmitted in a TDMA manner—thus providing protection from all but the most technically astute eavesdroppers.

Another level of security is performed on the mobile equipment, as opposed to the mobile subscriber. As noted, each GSM terminal is identified by a unique *International Mobile Equipment Identity* (IMEI) number. A list of IMEIs in the network is stored in the *Equipment Identity Register* (EIR). The status that is returned in response to an IMEI query to the EIR is one of the following:

- *White-listed*   Indicates that the terminal can connect to the network
- *Gray-listed*   Indicates that the terminal is under observation from the network for possible problems

■ *Black-listed*   Indicates that the terminal has either been reported as stolen, or it is not type approved (i.e., it is not the correct type of terminal for a GSM network). Such terminals cannot connect to the network.

Summary

GSM customers now equal about 35 percent of the entire world market for wireless services. One new subscriber signs up for service every second of the day and night. At year-end 1999, GSM in North America had some four million customers across the United States and Canada. By mid-1999, seventeen operators offered commercial service in about 3,500 cities in 45 states, the District of Columbia and four Canadian provinces. According to the North American GSM Alliance, GSM coverage now reaches 162 million North Americans covering more than 52 percent of the Canadian population and nearly 62 percent of the United States population.

See Also

     Code Division Multiple Access (CDMA)

     Digital Enhanced Cordless Telecommunications (DECT)

     Frequency Division Multiple Access (FDMA)

     General Packet Radio Service (GPRS)

     PCS 1900

     Time Division Multiple Access (TDMA)

CHAPTER H

Help Desks

With the proliferation of hardware and software throughout corporations, *Information Technology* (IT) and telecommunications managers are faced with the task of providing troubleshooting assistance to users who are scattered throughout the organization. The consequences of not providing adequate levels of assistance are too compelling to ignore: lost corporate productivity, slow responses to competitive pressures, and the eventual loss of market share. One way to efficiently and economically service the needs of a growing population of computer and communications users is to set up a help desk.

Briefly, the help desk acts as a central clearinghouse for support issues and is manned by a technical staff that addresses support problems and attempts to solve them in-house before calling in vendors or carriers. The help-desk operator logs every call—and, if possible, attempts to isolate the cause of the problem. If the problem cannot be solved over the phone, the operator dispatches a technician and monitors progress to a satisfactory conclusion before ending the transaction.

Help-desk operators are usually able to answer from 50 percent to 70 percent of all calls without having to pass the calls along to another authority. Aside from handling calls from users, help desks can provide services such as order and delivery tracking, asset and inventory tracking, preventive maintenance, and vendor-performance monitoring.

Although the concept of the help desk originated in the mainframe environment, the role of the help desk has expanded into realms that have not been ventured into by many mainframe professionals—due to the increased corporate reliance on LANs. Fortunately, there are now software packages available that assist with help-desk administration. Problem-determination tools that are based on expert systems can also bring untrained help-desk personnel up to competency in a quick manner.

A recent trend has been the extension of help-desk support for the Internet. With the capability to update transactions via the Web, field technicians now can update information that pertains to outstanding requests from any location in which they have Web browser access. Users can perform a range of actions, such as authorizing a change request and adding new information to an existing trouble ticket. Help-desk applications and data generate Web forms (schemas) and hyperlinks dynamically (refer to Figure H-1), which enables help-desk managers focus on serving customers and improving business processes, instead of constantly maintaining static Web pages.

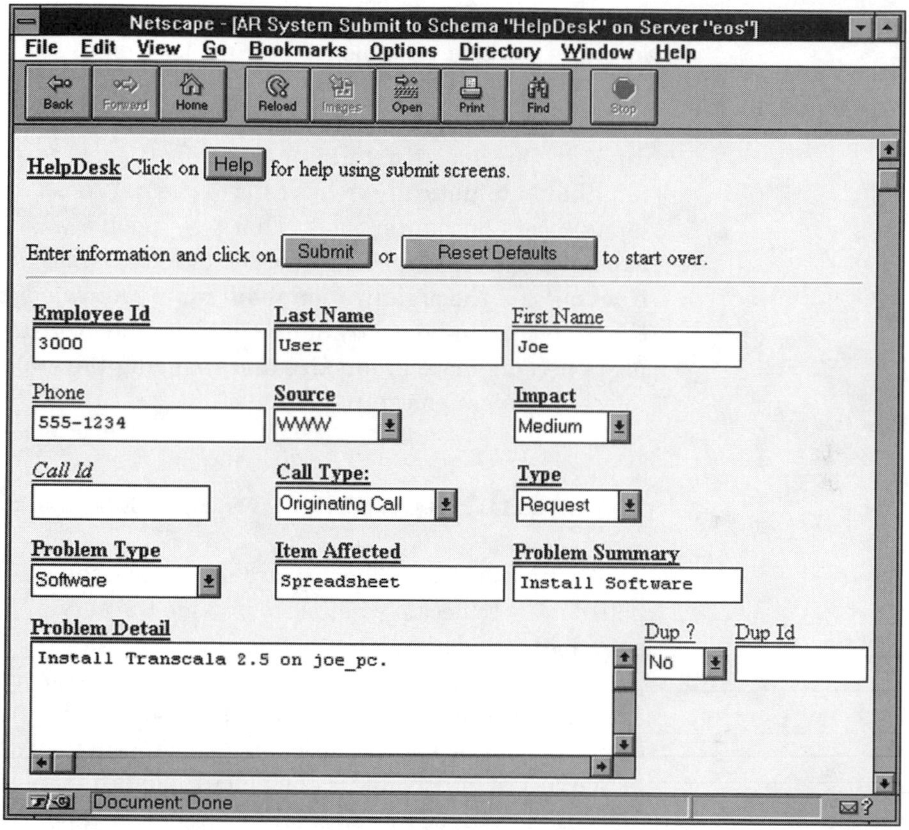

◼◼ ◼◼ ◼◼ ◼◼

Figure H-1

Remedy Corporation's
Action Request
System has
extensions to the
Web. The company's
ARWeb client enables
organizations to
create a Web-based
help desk that can be
accessed by anyone
who has a Web
browser. In this case,
a user can access a
Web form to report a
problem to the help
desk.

Benefits

Establishing a centralized help desk to coordinate the resolution of system
problems offers a number of benefits. Users have a single number to
remember; support personnel are assured of an orderly, controlled flow of
tasks and assignments; and managers are provided with an effective means
of tracking problems and solutions. The help desk provides users with a
warm and fuzzy level of support. Knowing that someone is available to
solve any problem—or even to help them find their way through unfriendly
documentation—adds to an individual's confidence and willingness to learn
new applications and office technologies.

Although a help desk costs money, it can pay for itself in many ways that
unfortunately can be hard to quantify. The fact is that most companies
invest millions of dollars in computer and communication systems, and
they also have millions of dollars invested in people. To ensure that both

resources are utilized to their optimal advantage, there should be an entity in place that is capable of solving the many different problems that inevitably arise.

Determining which level of support the help desk should provide can be difficult. One way to determine the proper level of support is to have a system that automatically tracks calls to the help desk and records what problem(s) users encounter most. That way, people who have the appropriate expertise can be identified in order to lend assistance or can be recruited from outside the organization to fill in any knowledge gaps. Depending on the problems, in-house training might provide the dual benefits of helping users become more productive and reducing the support burden of IT and telecommunications staff.

Internal Support versus External Support

Assuming that these benefits are attractive, the next step in setting up a help desk is to decide whether to provide the service internally or to rely on an outside vendor. Another option is to offer the help desk as a billable corporate service, which can prevent the help desk from becoming deluged with calls from users who can just as easily look up the information in manuals or use the online help facilities that come with many application packages.

While it is important for the help-desk staff to have technical skills, people skills are much more useful. When a user calls with a problem, the support person must extract information from someone who does not know the technical jargon that is needed to reach a solution or from someone who does not know how to use seemingly arcane procedures in order to isolate and solve problems.

The help-desk operator can perform all of the maintenance activities that once required the dispatch of an on-site technician to accomplish. Previously, when users had a problem, the help staff or technician had to go to users' desks and physically look over their shoulder. With a remote-control product, the help-desk operator can stay put, hit a few keys, and instantly see a user's screen and control the keyboard. By taking control of the user's computer, the help-desk operator can often assess the problem immediately.

An alternative to internally staffed help desks is to subscribe to a commercial service. Subscribers typically call an 800 number to obtain help for Windows-based PCs and software programs and peripherals, for example. A variety of pricing schemes are available. An annual subscription, for instance, might entitle the company to unlimited advice and consultation. Per-call pricing is also available. The cost of basic services is between $10 to $15 per call, depending on call volume. Depending on the service provider, calls are usually limited to 15 minutes or 20 minutes. If the problem cannot be resolved within that time, the call might even be free.

Resolution can involve anything from talking the caller through a system reboot to helping the user ascertain that a PC-to-mainframe link is out of order and deciding what must be done in order to get the link back into service. Callers can also obtain solutions to common software problems, ranging from how to recover an erased file to importing files from one package to another, to modifying system-configuration files when adding new application packages or hardware.

Some hotline services even offer advice concerning the selection of software and hardware products, provide software installation support, and guide users through maintenance and troubleshooting procedures. Other service providers specialize in LANs of one type or another.

Because many callers are not familiar with the configuration details of their hardware and software, this information is compiled into a subscriber profile for easy access by online technicians. With this information readily available, the time that is spent with any single caller is greatly reduced (which helps keep the cost per call low and helps attract new customers). The cost of a customer profile varies greatly, depending on the size of the database that must be compiled. The cost might be based on the number of potential callers, or it might entail a flat yearly charge for the entire organization.

The operations of some service providers are becoming quite sophisticated. To service their clients, staff might access a shared knowledge base that consists of tens of thousands of questions and answers. Sometimes called expert systems, the initial knowledge base is compiled by technical experts in their respective fields. As online technicians encounter new problems and devise solutions, this information is added to the knowledge base for immediate access when the same problem is encountered at a later time.

Even users who have access to vendors' free help-line services can benefit from a third-party help line. The third-party service might cover situations in which more than one application or hardware platform is being used, whereas vendor help-line services only provide assistance that covers their products. Moreover, most users rarely bother to phone the vendors because of constantly busy lines. So, rather than supplanting vendor services and in-house support desks (which might be overburdened), third-party services can be used to complement them.

In addition to delivering a variety of help services, for an extra charge, third-party service providers can issue call-tracking and accounting reports to help clients keep a lid on expenses for this kind of service and to allocate expenses appropriately among departments and other internal cost centers.

Summary

The help desk can improve the QoS while decreasing service costs. In addition, the help desk frees up experts to work in other areas and provides consistent answers to questions. Some expert systems use the database to generate graphics and text reports concerning which types of hardware and software cause the most problems. This information can be used to guide product purchases and to ascertain the response times of vendors. The continued growth of distributed computing over LANs and WANs and the increasing complexity of hardware and software will expand the role of the corporate help desk.

See Also

    Asset Management

    Network Management

Hertz

The frequency of electromagnetic waves generated by radio transmitters is measured in cycles per second, or hertz (Hz). An electromagnetic wave is composed of complete cycles. The number of cycles that occur each second gives radio waves their frequency, while the peak-to-peak distance of the waveform gives the amplitude of the signal (refer to Figure H-2).

The frequency of standard speech is about 3,000 cycles per second, or 3 kilohertz (KHz). Some radio waves might have frequencies of many millions of hertz (megahertz, or MHz), and even billions of hertz (gigahertz, or GHz). The following table provides the range of frequencies and their band classification.

Frequency	Band Classification
Less than 30KHz	*Very Low Frequency* (VLF)
30KHz to 300KHz	*Low Frequency* (LF)
300KHz to 3MHz	*Medium Frequency* (MF)
3MHz to 30MHz	*High Frequency* (HF)
30MHz to 300MHz	*Very High Frequency* (VHF)
300MHz to 3GHz	*Ultra High Frequency* (UHF)
3GHz to 30GHz	*Super High Frequency* (SHF)
More than 30GHz	*Extremely High Frequency* (EHF)

The term *hertz* was adopted in 1960 by an international group of scientists and engineers at the General Conference of Weights and Measures, in honor of Heinrich R. Hertz (1857-1894), a German physicist (refer to Figure H-3). Hertz is best known for proving the existence of electromagnetic waves, which had been predicted by British scientist James Clerk Maxwell in 1864. Hertz used a rapidly oscillating electric spark to produce ultra-

Figure H-2

Each cycle per second equates to 1 hertz (Hz). In this case, three cycles occur in one second, which equates to 3Hz.

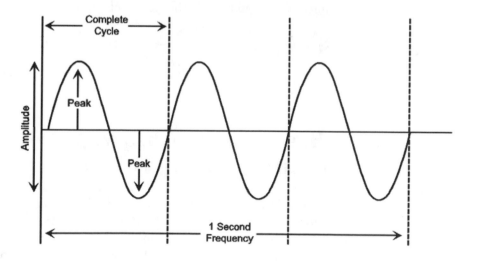

Figure H-3
German physicist
Heinrich R. Hertz
(1857-1894) proved
the existence of
electromagnetic
waves, which led to
the development of
radio, microwave,
radar, and other
forms of wireless
communication.

high-frequency waves. These waves caused similar electrical oscillations in a distant wire loop. The discovery of electromagnetic waves and how they could be manipulated paved the way for the development of radio, microwave, radar, and other forms of wireless communication.

See Also

Decibel (dB)

Hierarchical Storage Management (HSM)

Hierarchical Storage Management (HSM) uses two or more levels of storage (three is typical: online, offline, and near-line) to provide a cost-effective and efficient solution in order to meet the demand for increased storage space and appropriate data-retrieval response time. HSM came about because of the need to move low-volume and infrequently accessed files from disk, thus freeing up valuable online storage space.

Although more disks can be added to file servers in order to keep up with storage needs, budget constraints often limit the long-term viability of this solution. In an HSM scheme, data can be categorized according to its frequency of usage and can therefore be stored appropriately: online, near-line or offline. Different storage media come into play for each of these categories, and migration operations are under the control of an HSM management system (refer to Figure H-4).

Frequently used files are stored online on local disk drives that are installed in a server or workstation. Occasionally used files are stored near-line on secondary storage devices such as optical disks, *Re-Recordable/ Rewriteable Compact Discs* (CD-RWs), or *Recordable Compact Discs* (CD-Rs), which are installed in a server-like device called an autochanger or jukebox. Infrequently used files are usually migrated offline to tape cartridges that are stored in a tape jukebox or to a library facility that is capable of holding hundreds or thousands of tapes. The library facility uses sophisticated robotics to retrieve individual bar-coded cartridges and to insert them into a tape drive so that the data can be moved to local storage media. The exchange time can be several minutes.

For organizations that have mixed needs for online and offline storage, a near-time automated tape library offers the best compromise between price and performance. These systems bridge the gap between fast, expensive

Figure H-4

Hierarchical data storage spans magnetic disk, optical disk, and tape options.

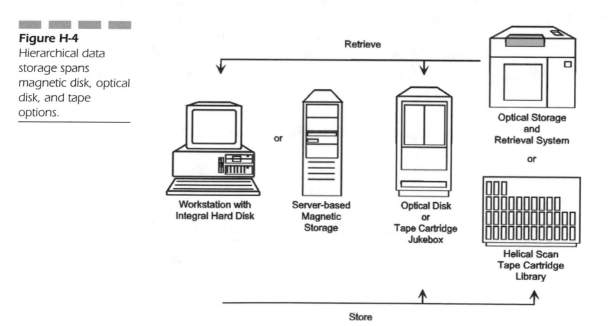

online disk storage and slow, high-capacity, offline tape libraries. The exchange time is about 30 seconds.

A management system determines when a file should be transferred or retrieved, initiates the transfer, and keeps track of its new location. As files are moved from one type of media to another, they are put into the proper directory for user access. The management system automatically optimizes storage utilization across different media types by removing files from one to the other until they are permanently archived in the most economical way, usually via a tape library. At the same time, individual files or whole directories can be excluded from migration.

Data migration can be controlled according to criteria such as file size and last date of access. Files can also be migrated when the hard disk reaches a specified capacity threshold. For example, when magnetic disk storage reaches the established threshold of 80 percent full, files are migrated automatically to optical storage, freeing up magnetic storage until it reaches another specified threshold (say, 60 percent).

Summary

Huge amounts of data—hundreds of gigabytes or even terabytes—cannot be managed efficiently without the help of HSM solutions. HSM balances the cost, capacity, and efficiency of different storage media according to the frequency of data usage, which provides all clients and servers with expanded online storage through the migration of files between magnetic disks, optical disks, and tape, according to predefined rules.

See Also

> *Redundant Array of Inexpensive Disks* (RAID)
>
> Storage Media

High-Definition Television (HDTV)

High-Definition Television (HDTV) uses digital technology to improve both the production and reception of television programs. HDTV will eventually replace all existing analog television systems that are in use today. During the transition, which will be completed in 2006, analog television systems will continue to be supported.

HDTV offers several improvements on the current television standard that was developed by the *National Television Standards Committee* (NTSC). The NTSC standard consists of a picture with 525 lines of resolution, a 4:3 horizontal-to-vertical aspect ratio, and analog sound. HDTV offers 1,124 lines of resolution, a panoramic 16:9 horizontal-to-vertical aspect ratio, and digital sound. HDTV also permits multiple digital audio tracks that enable stereo, CD-quality sound and audio in several languages.

HDTV offers a temporal resolution of 60 frames per second, which is twice the temporal resolution that is offered under the NTSC standard. This feature enables smooth motion and high picture clarity. The picture will be displayed in a way that is more like movies, adding the feeling of realism to TV.

Development Efforts

Research into HDTV was initiated in 1968 by *Nippon Hoso and Kyokai* (NHK), the Japan Broadcasting Corporation. This original system, called MUSE, was analog and required 36MHz channels—six times wider than the 6MHz standard that the United States uses. Japan and Europe further developed the system, and each launched experimental services. In 1985, the quest for a single world-wide production standard was thwarted in a political struggle between European and Japanese competitors. But the development of HDTV continued, driven mostly by the fear that Japan would gain commercial domination of key technologies in the next century (particularly in the field of communications).

Many U.S. companies responded to this perceived threat and began developing their own versions of HDTV. The development of HDTV standards has been underway in the United States since 1987. Initially, all of the proposals were for analog transmissions (until 1990, when the FCC was confronted with four digital proposals—among which it could not choose). Instead of selecting one of these proposals, the FCC suggested the idea of a grand alliance of companies and research organizations that would all work together to come up with one standard for HDTV.

Members of the so-called Grand Alliance included AT&T, the General Instrument Corporation, the *Massachusetts Institute of Technology* (MIT), Philips Consumer Electronics, the David Sarnoff Research Center, Thomson Consumer Electronics, and Zenith Electronics. The Grand Alliance issued a proposal based on recommendations from the *Advanced Television Systems Committee* (ATSC), which included support for 18 different digital formats of information and MPEG-2 for compression. The result was a standard called *Digital Television* (DTV), which was proposed to the FCC in 1996.

FCC Approval

In the Telecommunications Act of 1996, Congress directed the FCC to issue licenses for digital television to incumbent television broadcasters. Accordingly, in April 1997, the FCC issued its plan for the commercial implementation of DTV. The goal of the plan was to ensure the success of free, local, digital-broadcast television.

To bolster DTV's chance for success, the FCC decided to permit broadcasters to use their channels according to their best business judgment, as long as they continued to offer free programming upon which the public relied. Specifically, broadcasters must provide a free digital video-programming service that is at least comparable in resolution to today's service and is aired during the same time periods as today's analog service. The FCC does not require broadcasters to air high-definition programming or initially to simulcast their analog programming on the digital channel.

Broadcasters will be able to put together whatever package of digital product they believe will best attract customers and to develop partnerships with others, in order to help make the most productive and efficient use of their channels. These services could include data transfer, subscription video, interactive materials, audio signals, and whatever other innovations that could be promoted and used to gain profit. Giving broadcasters flexibility in how they use their digital channel will enable them to put together the best mix of services and programming in order to stimulate consumer acceptance of digital technology and the purchase of new digital receivers.

Summary

At this writing, HDTV service is available in the top 10 markets of Atlanta, Boston, Chicago, Dallas-Fort Worth, Detroit, Los Angeles, New York, Philadelphia, San Francisco-Oakland, and Washington, D.C. HDTV service will become available in the top 11 to 30 markets by year-end 2000. The FCC has set a target of 2006 as a reasonable end date for NTSC service. After that, there will be no analog TV. Toward that end, the FCC will review the progress of DTV implementation every two years and make adjustments in the timetables if necessary. Stations that cannot meet the target dates for HDTV service due to unforeseen circumstances will be given deadline extensions.

See Also

> *Direct Broadcast Satellite* (DBS)
> Switched Digital Video
> Video on Demand

Home Phone-Line Networking

The number of multiple-PC homes is growing faster than the number of single-PC homes. Currently, there are 35 million homes in the United States that have at least two PCs. The need for a network in the home to share printers, modems for Internet access, CD-ROMs, hard drives, and other equipment becomes essential—if only to save money on expensive resources, including multiple phone lines and Internet access accounts. Beyond cost savings, a home network offers convenience. Users no longer have to waste time looking all over the house to check each computer for the documents, spreadsheets, images, or software that they need for office work, school work, tax preparation, or to play their favorite games.

Furthermore, a network in the home provides the opportunity to implement special features such as a family message board and voice intercom. The cost of installing and configuring a basic two-node network in the home is less than $100 (U.S. dollars), and several vendors offer kits that include all of the necessary components. Third-party software is usually required for features such as internal messaging and voice intercom. More computers can be added to the network by purchasing extra adapter cards in order to make the phone line connections. Such networks can be built and configured at a leisurely pace, with professional results achievable within a Saturday afternoon.

The first home phone-line products were introduced in 1998 but offered a top transmission speed of only 1Mbps. Today's phone line networking products offer a top transmission speed of 10Mbps. In either case, the existing phone wiring within the home provides the connectivity among computers, including shared Internet access, without the need for a hub device.

In addition, the newer 10Mbps products are backward compatible with the older 1Mbps products.

Data runs over the home's existing phone wiring without disrupting normal phone service. This goal is accomplished through the use of *Frequency-Division Multiplexing* (FDM), which essentially divides the data that travels over the phone lines into separate frequencies: one for voice, one for high-bandwidth Internet access (such as DSL), and one for the network data. These frequencies can coexist on the same telephone line without interfering.

Components

For each computer that will be networked in the home, a *Network Interface Card* (NIC) is required—just like at the office. The NIC is an adapter that plugs into a vacant slot in the computer. For phone-line networking, the NIC will have two RJ11 ports into which a segment of phone wire will be connected: one to the phone, and the other to the wall jack. Many NICs are optimized to work in Windows environments—specifically, PCs that run Windows 95, Windows 98, or Windows NT. Windows provides the software drivers for most brands of NICs, making installation and configuration essentially a plug-and-play affair.

With the computer turned off, the user opens the case and inserts the NIC into the appropriate type of card slot. The short white slots are for *Peripheral Component Interconnect* (PCI) cards, and the longer black slots are for the older *Industry-Standard Architecture* (ISA) cards. Due to the limited number of vacant card slots in each computer—usually only three or four of each type—it is best to take an inventory of each computer before actually buying the NICs or a network kit. Whenever possible, PCI-type cards should be used for their higher performance.

Once the card is installed, the computer can be restarted. Windows will recognize the new hardware and configure it automatically. If the driver or a driver component is not found, however, it might be necessary to get out the Windows CD-ROM or the NIC vendor's 3.5-inch installation disk so that Windows can find the components it needs.

After Windows has recognized and configured the NICs, you can connect each computer. One segment of phone line is inserted into the wall jack, and the other end is inserted into an RJ11 port on the computer's NIC. Another phone-line segment plugs into the phone and terminates at the other RJ11 port on the NIC, which enables the phone to be used (even while a file transfer is in progress between computers).

Instead of using one of the two RJ11 ports for connection to a telephone, the extra RJ11 port can be used for daisy-chaining computers over the same phone line. Some vendors also include an RJ45 port on the NIC, which enables users to migrate to 100Mbps by using Category-5 LAN cabling.

To actually share files and peripherals over the phone line connections, each computer must be configured with the right protocols and should be set up for file and printer sharing. Once this configuration is complete, a modem can also be shared. Up to 25 users can share that modem for Internet access, eliminating the need for separate telephone lines, modems, and accounts with an *Internet Service Provider* (ISP). If Internet access is provided over a broad-band service such as cable or DSL, the computers that are connected over the phone-line network can share these resources, as well.

Configuring the Network

As noted, after installing an NIC and powering up the computer, Windows will recognize the new hardware and automatically install the appropriate network-card drivers. If the drivers are not already available on the system, Windows will prompt the user to insert the manufacturer's disk that contains the drivers, and they will be installed automatically.

Next, the user must select the client type. Because we are creating a Microsoft peer-to-peer network, the user must add Client for Microsoft Networks as the primary network logon (refer to Figure H-5). Because the main advantage of networking is resource sharing, it is important to enable the sharing of both printers and files. This task is performed by clicking the File and Print Sharing button and choosing one or both of these capabilities (again, refer to Figure H-5).

Identification and security are the next steps in the configuration process. From the Identification tab of the dialog box, the user must select a unique name for the computer and the workgroup to which it belongs, as well as a brief description of the computer (refer to Figure H-6). This information will be visible to others when they use Network Neighborhood to browse the network.

From the Access Control tab of the dialog box, the user selects the security type. For a small peer-to-peer network, share-level access is adequate

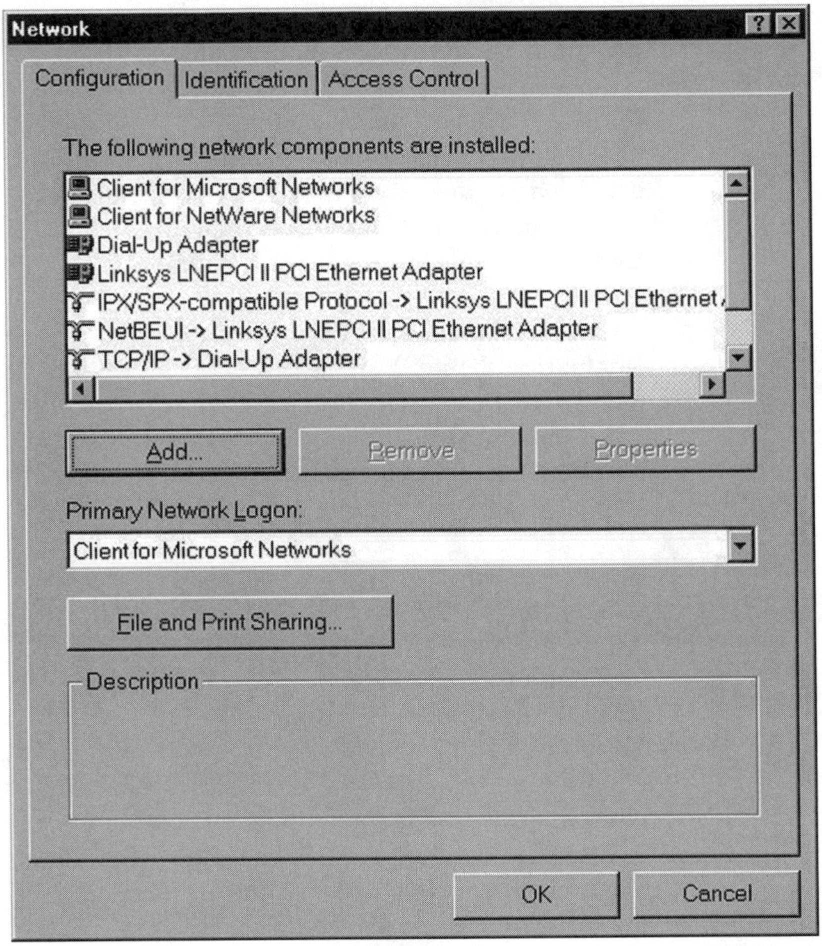

Figure H-5
By accessing the
Windows control
panel and double-
clicking the network
icon, this dialog box
opens. Here, the user
can configure the
computer for the
primary network
logon, plus file and
printer sharing.

(refer to Figure H-7). This level of access enables printers, drives, directories, and CD-ROM drives to be shared and enables the user to establish password access for each of these resources. In addition, read-only access enables users to view (not modify) a file or directory.

To permit or deny disk drives to be shared, the user double clicks the My Computer icon on the desktop, then right-clicks the drive to be shared. Next, the user selects Sharing from the pick list (refer to Figure H-8). The type of access can be set as Read-Only, Full, or Depends on Password. This action is performed for each drive that the user wants to share, including the CD-ROM, CD-RW, CD-R, and *Digital Video Disk* (DVD) optical disk drives. Individual directories can be set for no sharing in the same way.

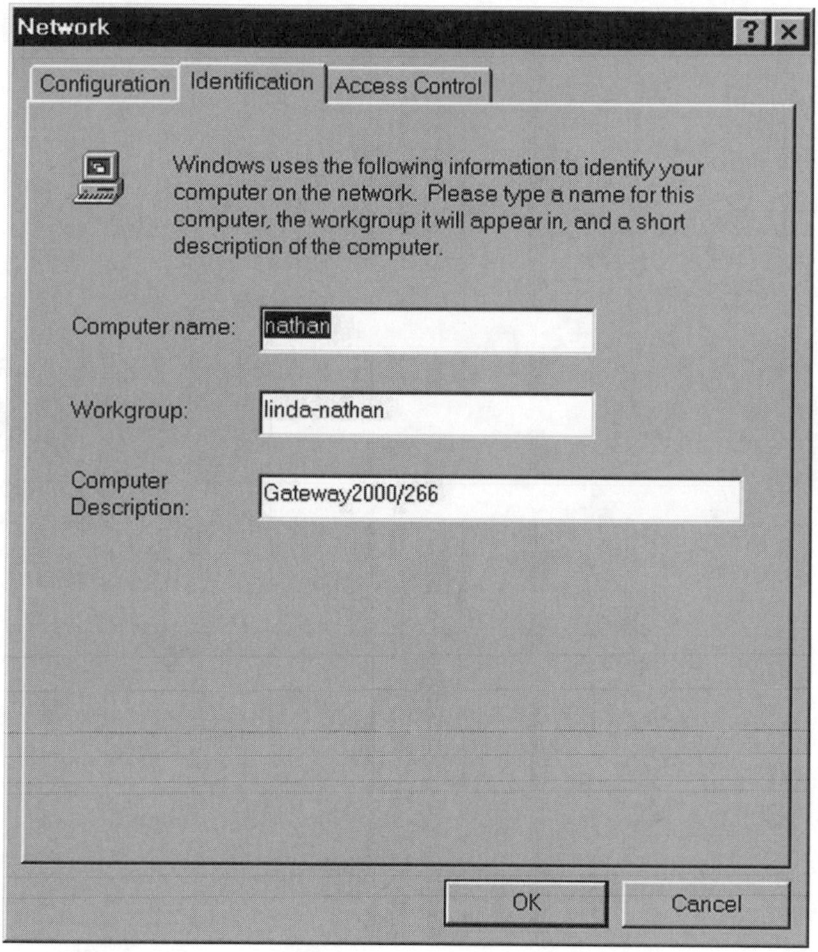

Figure H-6
A unique name for the computer, the workgroup to which it belongs, and a brief description of the computer all identify the computer to other users when they access Network Neighborhood to browse the network.

To enable a printer to be shared, the user right-clicks the printer icon in the Control Panel and selects Sharing from the pick list (refer to Figure H-9). Next, the user clicks the Shared As radio button and enters a unique name for the printer (refer to Figure H-10). If desired, you can give this resource a password, as well. When another computer tries to access the printer, the user will be prompted to enter a password. If a password is not necessary, the password field is left blank.

Another security option on the Access Control tab is user-level access, which is used to limit resource access by username. This function eliminates the need to remember passwords for each shared resource. Each user simply logs on to the network with a unique name and password, and the

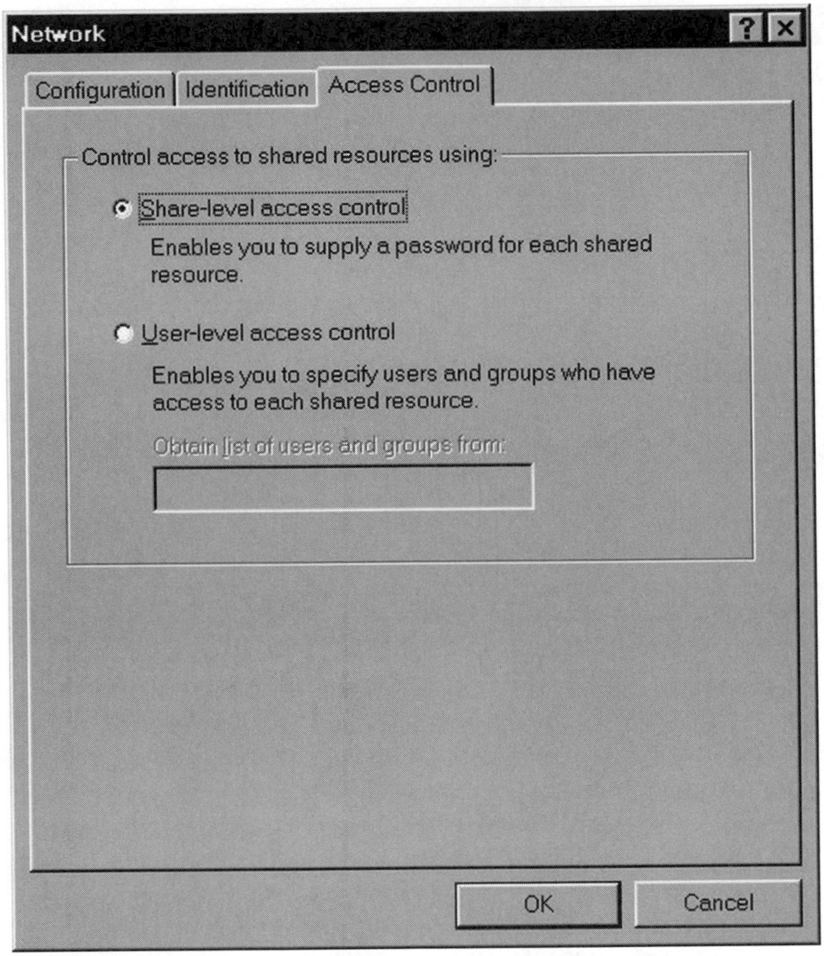

Figure H-7
Choosing share-level
access enables the
user to password-
protect each shared
resource.

network administrator governs who can do what on the network. This
approach, however, requires the computers to be part of a larger network
with a central server—one running Windows NT Server—that maintains
the *Access Control List* (ACL) for the whole network. Because Windows 95
and Windows NT workstations support the same protocols, Windows 95
computers can participate in a Windows NT Server domain.

Figure H-8
Any type of hard
drive or optical drive
on any computer can
be shared. Access
privileges can be
associated with each
drive, such as Read-
Only, Full, or
Depends on
Password.

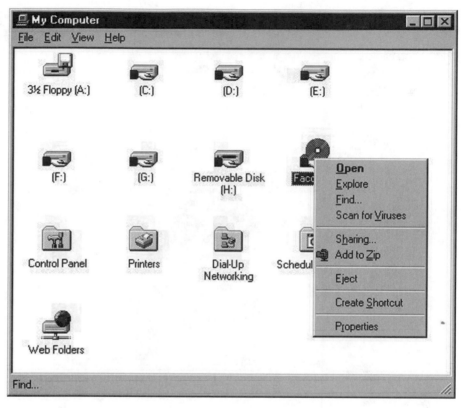

Peer services can be combined with standard client/server networking. For example, if a Windows 95 computer is a member of a Windows NT network and has a color printer to share, the resource owner can share that printer with other computers on the network. The server's ACL determines who is eligible to share resources.

Once the computers are properly configured and connected through the existing phone lines, the network is operational. This peer-to-peer network is an inexpensive way for small companies within a household to share resources among a small group of computers. This type of network provides many of the same functions as the traditional client/server network, including the capability to run network versions of popular software packages.

Figure H-9
By right clicking a
printer, you can set
up a printer for
sharing.

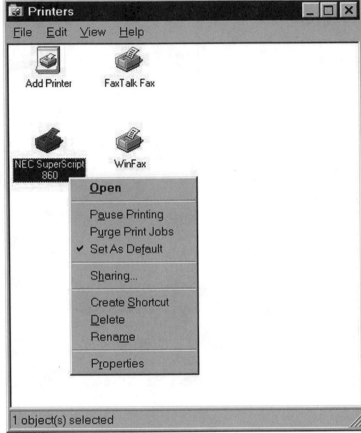

Standards

To standardize the products that interconnect computers over standard
telephone wiring in the home, the vendor-oriented Home Phone-Line Networking Alliance was established in 1998. Products that adhere to the
HomePNA standard permit the creation of simple, cost-effective home networks using existing phone wiring. The use of phone wiring for this purpose
eliminates the need for cable installation and a hub, yet enables shared
Internet access with one ISP account without interfering with the regular
phone service. The HomePNA Specification 2.0 delivers a 10Mbps data rate
for home phone-line networking, while maintaining full backward compatibility and interoperability with the previous specification (which offered a
data rate of 1Mbps).

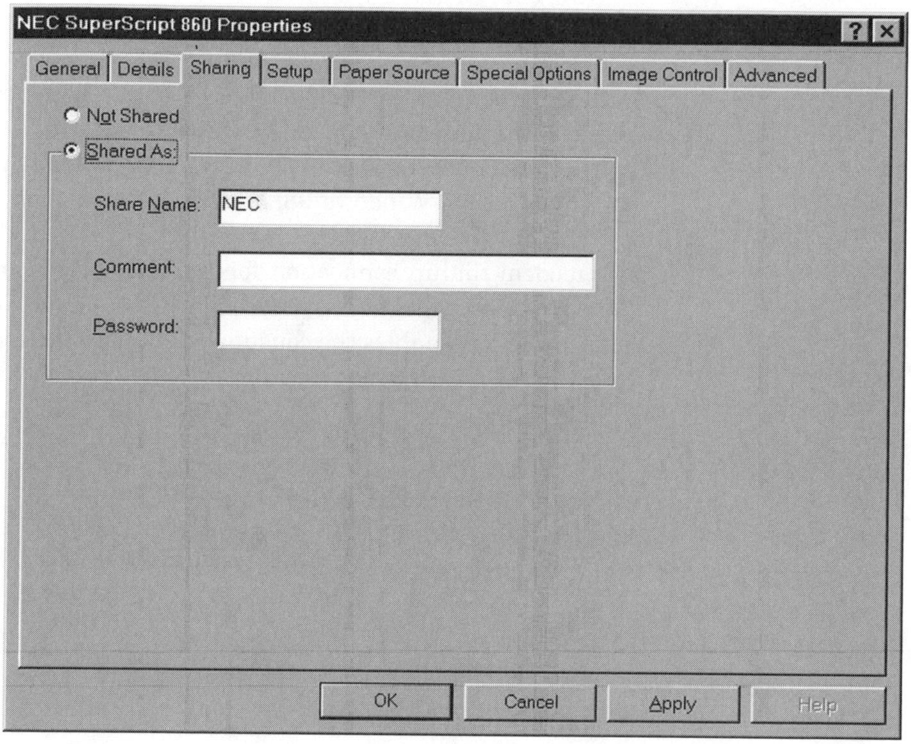

Figure H-10
A shared printer is named and can be password protected if necessary.

Adding data to voice over existing phone wiring does not pose interference problems, because different frequencies are used for each component. Standard voice occupies the range from 20Hz to 3.4KHz in the United States (slightly higher internationally), while phone-line networking operates in a frequency range that is higher than 2MHz. By comparison, DSL services such as ADSL occupy the frequency range from 25KHz to 1.1MHz. The frequencies used are far enough apart that the same wiring can support all three.

Interoperability is the key to enabling mass consumer adoption of home networks. Accordingly, the HomePNA has partnered with *CEBus Industry Council*'s (CIC) PlugLab to be its official testing facility. The CIC PlugLab conducts an extensive certification program in order to ensure that member companies' home phone-line networking products meet HomePNA's exacting interoperability and performance standards and that the products merit its seal of approval. This certification program helps ensure consumers that the products that they buy from different manufacturers will work together on the same network.

Summary

As the number of PCs and peripherals in the home continues to increase, the need for a network to leverage these assets and to provide shared access to the Internet becomes more apparent. PCs on home phone networks must run Windows 95, Windows 98, Windows NT, or some other software that supports file sharing. Once networked, the PCs can share a printer as well as a modem dialup connection for Internet or corporate network access. Users can also work high-speed DSL and cable modems into the mix. In fact, HomePNA enables the consumer to choose the method of WAN access, which can also include ISDN and wireless services.

See Also

Cable Television Networks

Digital Subscriber Line Technologies

Ethernet

Hubs

Internet

Modems

Hubs

With today's networks becoming increasingly more complex, the conventional bus and ring LAN topologies have exhibited shortcomings, especially with regard to cable maintenance. In the past, a fault anywhere in the cabling often brought down the entire network or a significant portion of the network. This weakness was compounded by the inability of technicians to readily identify the point of failure from a central administration point, which tended to prolong network down time. This situation led to the development of the wiring hub in the mid-1980s.

Hubs provide a central point at which all wires meet. They are at the center of the star configuration, with the wires (i.e., segments) radiating outward to connect the various network devices, which can include bridge/routers that connect to remote LANs via the WAN (refer to Figure H-11). Wiring hubs physically convert the networks from a bus or ring topology to a star topology while logically maintaining their Ethernet or Token Ring characteristics. The advantage of this configuration is that the failure of one

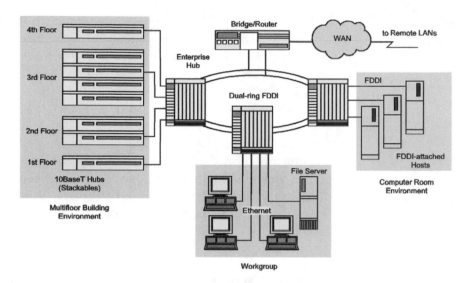

Some connectivity options are available through various types of hubs that serve multi-floor building, workgroup, and computer-room environments.

segment—which might be shared among several devices or might be dedicated to just one device—does not necessarily impact the performance of other segments.

Not only do hubs limit the impact of cabling faults to a particular segment, but they also provide a centralized administration point for the entire network. If the wiring hub also employs some *Central Processing Units* (CPUs) and management software to automate fault isolation, network reconfiguration, and statistics-gathering operations, it is no longer just a hub; rather, it is an intelligent hub that is capable of solving a wide range of connectivity problems efficiently and economically.

Types of Hubs

High-end hubs are modular in design, enabling the addition of ports, network interfaces, and special features as needed by the organization. These enterprise-level hubs can support networks that combine different LAN topologies and media types in a single chassis. Ethernet, Token Ring, and FDDI networks can coexist in a single hub. LAN segments that use twisted-pair wiring, coaxial cable, and optical fiber can also be interconnected through the hub.

Other hubs are available in fixed configurations for departments or workgroups that do not anticipate future growth. A variation of the fixed-configuration hub is the stackable hub. A unique feature of stackable hubs is that they can be interconnected through a modular backplane interface, which offers managers the capability to economically expand workgroup and departmental networks as needed. Whereas the high-end modular systems are used to build large-scale enterprise networks, stackables are designed for small to medium-size networks.

A relatively new category of hub is the superhub. These are modular units that provide at least an uplink to a stand-alone ATM switch, if not some level of integral ATM switching, in addition to 100Mbps LAN support, and integrated LAN switching and routing. Fully populated superhubs support in excess of 500 ports of mixed-media and shared and switched connectivity over a gigabit-per-second backplane in a software-manageable, fault-tolerant, hot-swappable modular chassis that can cost more than $100,000.

Hub Components

Enterprise-level intelligent hubs contain four basic components: chassis, backplane, plug-in modules, and a network-management system.

Chassis   The chassis is the hub's most visible component, containing an integral power supply and/or a primary controller unit and varying in the number of available module slots. The modules insert into the chassis and are connected by a series of buses, each of which can constitute a separate network or can integrate with one or more backbone networks. The chassis holds the individual modules. In fitting into the chassis, each module is instantly connected to other modules via the hub's high-speed backplane.

Backplane   The main artery of the hub is its backplane, which is a board that contains one or more buses that carry all communications between LAN segments. The hub's backplane is analogous to a PC bus through which various interface cards can be interconnected. The data path that carries traffic from card to card is often called a channel; however, unlike the PC, the hub's backplane typically consists of multiple physical or logical channels. Minimally, the hub accommodates one LAN segment for each channel on the backplane.

Segmenting the backplane in this way enables multiple independent LANs or LAN segments to coexist within the same chassis. Usually, a separate backplane channel exists in order to carry management information. The segmented backplane typically has dedicated channels for Ethernet, Token Ring, and FDDI. Some hubs employ a multiplexing technique across the backplane in order to divide the available bandwidth into multiple logical channels. Other hubs support load sharing that enables network modules to select the backplane channel that will transport the traffic. Still other hubs are designed to enable backplanes to be added or upgraded, in order to accommodate network expansion and new technologies.

The potential bandwidth capacity of newer backplane designs that support ATM switching is quite impressive, reaching well into the gigabits-per-second range—more than enough to accommodate several Ethernet, Token Ring, and FDDI networks simultaneously.

Modules The functionality of hubs is determined by individual modules. The types of modules that are available depends on the hub vendor. Typically, the vendor will provide multi-user Ethernet and Token Ring cards, LAN management, and LAN bridge and router cards. The use of bridge and router modules in hubs overcomes the distance limitations that are imposed by the LAN cabling and facilitates communication between LANs and WANs.

Plug-in modules for terminal servers, communications servers, file/application servers, and SNA gateways also exist. Hub vendors also offer a variety of WAN interfaces, including those for X.25, Frame Relay, ISDN, fractional T1, T1, T3, SMDS, and ATM. As many as 60 different types of modules might be available from a single hub vendor, many of which are provided under third-party OEM, technology-swap, and other vendor-partnering arrangements.

Modules plug into vacant chassis slots. Depending on the vendor, the modules can plug into any vacant slot(s) that is specifically devoted to their function. Hubs that support any-slot insertion automatically detect the type of module that is inserted into the chassis and establish the connections to other compatible modules. In addition, many vendors offer a hot-swap capability that permits modules to be removed or inserted without powering down the hub.

Management System

Hubs occupy a strategic position on the network, providing the central point of connection for workstations, servers, hosts, bridges, and routers on the LAN and over the WAN. The hub's management system is used to view and control all devices that connect to it, providing information that can greatly aid in troubleshooting, fault isolation, and administration. The management tools typically fall into five categories: accounting management, configuration management, performance management, fault management, and security management.

Hub vendors typically provide proprietary management systems that offer value-added features that can make it easier to track down problem-causing workstations or servers. Most of these management systems support the *Simple Network Management Protocol* (SNMP), enabling them to be controlled and managed through an existing management platform such as IBM's SystemView for AIX, Hewlett-Packard's OpenView, and Sun's Solstice SunNet Manager. Some hubs have *Remote Monitoring* (RMON) embedded in the hub, making possible more advanced network monitoring and analysis up to the application layer.

Summary

Hubs are now the central point of control and management for the elements that make up departmental and enterprise networks. Hubs, which were developed to simplify the management of structured wiring as networks became bigger and more complex, enable the wiring infrastructure to expand in a cost-effective manner as the organization's computer systems grow and move—and as interconnectivity requirements become more sophisticated.

See Also

  Bridges

  Routers

  LAN Switches

Hybrid Fiber/Coax (HFC)

As its name implies, *Hybrid Fiber/Coax* (HFC) is the combination of optical fiber and coaxial cable on the same network. Such networks are used to provide high-speed digital services to homes and businesses. An HFC system could deliver the following benefits to each customer:

- Multiple telephone lines
- 25 to 40 broadcast analog TV channels
- 200 broadcast digital TV channels

- 275 to 475 digital point-cast channels (which deliver programming at a time that the customer selects)
- High-speed, two-way digital links for Internet and corporate LAN access

HFC divides the total bandwidth into a downstream band (to the home or business) and an upstream band (to the network). The downstream band typically occupies the 50MHz to 750MHz frequency range, while the upstream band typically occupies the 5MHz to 40MHz frequency range. The higher-bandwidth downstream band is necessary for delivering cable television broadcasts and multimedia Internet content to the user.

Originally, cable operators envisioned a coaxial tree-and-branch architecture in order to bring advanced services to the home. The capacity of fiber-optic transmission technology, however, led many cable operators to shift to an approach that combined fiber and coax networks for an optimal advantage. Transmission over fiber has two key advantages over coaxial cable:

- A wider range of frequencies can be sent over the fiber, increasing the bandwidth that is available for transmission.
- Signals can be transmitted over greater distances without amplification.

Fiber to the Neighborhood (FTTN)

A key disadvantage of fiber is that the optical components that are required to send and receive data are still too expensive to deploy at each subscriber location. Therefore, cable operators have adopted an intermediate approach that is known as *Fiber to the Neighborhood* (FTTN). In this approach, fiber reaches into the neighborhood, and coaxial cable branches out to each subscriber location. This arrangement increases the bandwidth that the plant is capable of carrying, while reducing the number of amplifiers that are needed and the number of amplifiers that are in cascade between the cable operator's head-end office and each subscriber.

The total number of amplifiers is an important economic consideration, because each amplifier must be upgraded (or more typically, replaced) in order to pass the larger bandwidth that the fiber and shorter coaxial cable runs enable. The number of amplifiers in cascade is important for ensuring signal quality. Because each amplifier is an active component that can fail, the fewer the amplifiers in cascade, the lower the chance of failure. Fewer amplifiers and shorter trees also introduce less noise into the cable signal.

These improvements translate into higher bandwidth, better-quality service, and reduced maintenance and operating expenses for the cable operator.

HFC Advantages

In HFC networks, fiber is run from a service distribution hub (i.e., a head-end office) to an optical feeder node in the neighborhood, with tree-and-branch coax distribution in the local loop (refer to Figure H-12). Two overriding goals in an HFC architecture are to minimize the fiber investment by distributing the fiber over the maximum number of subscribers and to use the upstream bandwidth efficiently for the highest subscriber fan-out.

Approximately 200 million TV sets and 100 million VCRs are deployed in the United States, and none of them are digital. This situation casts doubt about the immediate viability of all-digital *Fiber to the Curb* (FTTC) networks. Any serious attempt to provide video services must deal with this embedded base. Services that provide digital television signals require a separate digital decoder for each TV tuner (including the VCR), making FTTC an expensive proposition for subscribers—especially because the digital set-top converter would be required even to receive basic television channels. The advantage of HFC networks, however, is that they carry *Radio Frequency* (RF) signals, delivering video signals directly to the home in the exact format that television sets and VCRs were designed to receive.

Figure H-12
Typical topology of
an HFC network

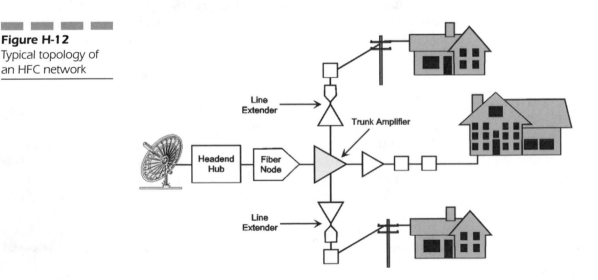

HFC networks also have the capability to evolve over time from a basic broadcast plant to a two-way network with interactive bandwidth. This evolution is achieved as needed by activating unused dark fiber and subdividing existing nodes to serve fewer homes per node.

During the initial stages of interactive service development in each market, it is anticipated that consumer demand will be low, with early subscribers scattered throughout the service area. To provide services, FTTC network operators must adhere to standard network-design concepts, which entail bringing fiber to every curb (whether or not a particular household or business wants to subscribe to the service). With HFC networks, scattered demand can be accommodated by providing cable modems as needed. Using cable modems, digital bandwidth can be allocated flexibly on demand, with virtually arbitrary bit rates. This situation enables the service provider to inexpensively add bandwidth capacity as demand builds, spreading out capital expenditures in order to meet subscriber growth in any given service area.

Modulation Technologies

For example, if the HFC network operator wants to supply digital HDTV service, today's HFC network designs can already accommodate individual users of digital HDTV at 20Mbps—even if the demand is scattered. With a technique called *Quadrature Amplitude Modulation* (QAM), a single HDTV channel could be supplied in a standard 6MHz channel slot within the digital band between 550MHz and 750MHz. HFC network equipment operates equally well with *National Television Standards Committee* (NTSC) signals, which occupy the 6MHz channels, or *Phase Alternating by Line* (PAL) standard broadcast analog signals—all on the same network. Channel assignments are completely arbitrary within the 750MHz spectrum.

One of the major difficulties with HFC is that the cable system was never intended for reliable, high-speed data transmission. Cable was installed with the aim of providing a low-cost conventional (analog) broadcast TV service. As a result of the original limited performance requirements, CATV networks are noisy, and communications channels are subject to degradation. In particular, they can suffer badly from strong narrow-band interference called ingress noise. Ingress noise plays havoc with conventional single-carrier modulation schemes such as QAM, which cannot avoid the noisy region.

A new modulation technique—*Discrete Wavelet Multi Tone* (DWMT)—is a wavelet transform-based, multi-carrier modulation scheme that provides better channel isolation, thereby increasing bandwidth efficiency and noise immunity. DWMT divides the channel bandwidth into a large number of narrow-band subchannels and adaptively optimizes the number of bits per second that can be transmitted over each subchannel. DWMT provides throughput of 32 DS0s (64Kbps each) over each megahertz of bandwidth on the coaxial cable. When a subchannel is too noisy, DWMT does not use the subchannel—thereby avoiding channel impairments and maintaining a reliable, high bit-rate throughput.

Using DWMT, higher data rates can be achieved over longer distances, which translates to more channels. DWMT uses the cable bandwidth and communications infrastructure more efficiently, and that enables cable operators to offer more services.

Summary

The growth of video and interactive communications services, coupled with developments in digital compression, have driven both CATV and telephone operators to seek effective ways to integrate interactive video and data services with traditional communications networks. With two-way capability, HFC enables cable operators to offer telephone service over the cable as well as hundreds of channels of interactive TV, digital services, and more. By using fiber links from the central site to a neighborhood hub (and coax cable from there to a few hundred homes and businesses), HFC provides an efficient and economical way to deliver the next generation of communications services while supporting current services.

See Also

Cable Telephony

Fiber in the Loop (FITL)

Digital Subscriber-Line (DSL) *Technologies*

CHAPTER **I**

Incumbent Local Exchange Carriers (ILECs)

Incumbent Local Exchange Carriers (ILECs) is a term that refers to the 22 former *Bell Operating Companies* (BOCs) that divested from AT&T in 1984, as well as Cincinnati Bell, *Southern New England Telephone* (SNET), and the larger independent telephone companies of GTE and United Telecommunications. In addition, some 1,300 smaller telephone companies are also in operation, serving mostly rural areas.

Originally, the former BOCs were owned by seven regional holding companies: Ameritech, Bell Atlantic, Bell South, NYNEX, Pacific Telesis, *Southwestern Bell Communications* (SBC), and US WEST. Over the years, some of these regional companies merged to the point that today, only four are left. Bell Atlantic and NYNEX were the first to merge in 1994. Bell Atlantic also completed a $53 billion merger with GTE in mid-1999. SBC Communications merged with Pacific Telesis in 1997 and then with Ameritech in 1999.

Regulatory Approval

All mergers must pass regulatory approval at the state and national level. The FCC approves mergers with input from the *Department of Justice* (DoJ). In the case of the SBC-Ameritech merger, the FCC imposed 28 conditions on SBC in exchange for its approval of the transaction.

The approval package contains a sweeping array of conditions that will make SBC-Ameritech's markets the most open in the nation, boosting local competition by providing competitors with the nation's steepest discounts for resold local service and full access to *Operating Support Systems* (OSSs). The package also requires SBC to accelerate its entry into new markets by six months, so that the company will compete in 30 new markets within 30 months after the merger closes. In addition, SBC will establish a separate subsidiary in order to provide advanced services [such as high-speed Internet access via *Asymmetrical Digital Subscriber Line* (ADSL)]. Also, SBC will not charge its residential customers minimum monthly long-distance fees for at least three years after it enters the long-distance market. The conditions also require stringent performance monitoring, reporting, and enforcement provisions that could trigger more than $2 billion in payments by the company if certain goals are not met.

Summary

The monopoly status of the ILECs officially ended with passage of the Telecommunications Act of 1996. Not only can other types of carriers enter the market for local services in competition with the telecommunications companies, but the regional parent companies of the telecommunications companies can also compete in each other's territories. Through mergers, the combined companies can enter out-of-region local markets on a broad scale quickly and efficiently enough to become effective national competitors. The merger of SBC and Ameritech, for example, enables the company to enter 30 new U.S. local markets beyond its combined 13-state region. After the merger, the new SBC will offer an additional 70 million people a new competitor for their telecommunications services dollars, bringing the industry to new levels of consumer choice. The combined company will serve customers in all of the top 50 U.S. population markets. Connecting these 50 markets will help SBC build a national network (and eventually, an international network).

See Also

Competitive Local Exchange Carriers (CLECs)

Interexchange Carriers (IXCs)

Local Exchange Carriers (LECs)

Telecommunications Industry Mergers

Interexchange Carriers (IXCs)

Interexchange Carriers (IXCs), otherwise known as long-distance carriers, include the big three: AT&T, MCI WorldCom, and Sprint. Since 1984, the ILECs have been limited to providing local calling services within their own *Local Access and Transport Areas* (LATAs). Long-distance calls between LATAs must be handed off to the IXCs that have established *Points of Presence* (POPs) within each LATA, for the purpose of receiving and terminating interLATA traffic.

In addition to providing long-distance telephone service, the IXCs offer business services such as ISDN, Frame Relay, leased lines, and a variety of other digital services. Many IXCs are also *Internet Service Providers* (ISPs), which offer Internet access services, *Virtual Private Networks* (VPNs), e-mail, Web hosting, and other Internet-related services.

Bypass

Traditionally limited to providing service between LATAs, the Telecommunications Act of 1996 enables IXCs to offer local exchange services in competition with the ILECs. But because the ILECs charge too much for local loop connections and services (and do not deliver them in a consistent, timely manner), the larger IXCs have implemented technologies that enable them to bypass the local exchange. Among the methods that IXCs are using to bypass the local exchange are CATV networks and broad-band wireless technologies.

In the former case, AT&T, for example, has acquired the nation's two largest cable companies, TCI and MediaOne, to bring local telephone services to consumers—in addition to television programming and broad-band Internet access. As these bundled services are introduced in each market, they are provided to consumers at an attractive price with the added convenience of a single monthly bill.

Long-Distance Market

Today, there are more than 600 long-distance providers—20 of which have revenues of more than $100 million, and eight of which have annual revenues exceeding $1 billion. Competition in the long-distance market is fueled in large part by the massive deployment of long-distance capacity in the hands of emerging carriers. Since 1996, more than 62 million miles of new fiber-optic cable has been laid—71 percent of which was added by emerging carriers. For example, Qwest has reported that the currently lit portion of its network has sufficient capacity to handle the current combined traffic of AT&T, MCI WorldCom, and Sprint.

Competition in long distance and any distance will increase dramatically as the BOCs enter the long-distance market in their regions. Bell Atlantic has predicted that it will take a 26 percent share of the long-distance market in New York when it is permitted to enter that business.

For the past seven years, there has been a decline in long-distance rates. In addition to downward price pressure from competition, the growth of fiber-optic links and lower access fees have reduced the service-provisioning costs of the long-distance carriers. Local access fees have dropped from 5.8 cents per minute in 1992 to 3.3 cents per minute today, and proposals are floating around to bring the costs down to 1.1 cents per minute.

The downward price pressure is likely to continue for the foreseeable future, especially as *Voice-Over IP* (VoIP) becomes more popular among businesses as well as consumers. As more voice continues to migrate to high-speed data networks, some analysts claim that carriers will virtually give away voice services in order to capture more of the bundled-services market (long distance, Internet, wireless, and entertainment), where the profit margin is higher.

Merger Activity

Similar to the ILECs, the IXCs have been merging in recent years. In July 1998, MCI and WorldCom received final approval to complete their planned merger. As a condition of the merger, MCI had to sell its Internet backbone facilities and wholesale and retail Internet businesses. The feeling among regulators was that WorldCom's ownership of UUNET and MCI's extensive Internet holdings have potential anti-competitive impacts. Accordingly, MCI sold its Internet businesses to Cable & Wireless for $1.75 billion.

In November 1999, MCI WorldCom and Sprint filed for FCC approval of their planned merger. The two companies want to create a strong, third force to compete with AT&T and with the so-called mega-Bells. By combining the companies' local assets and technical and entrepreneurial skills, the new WorldCom hopes to be far-better positioned to crack open the BOCs' lock on access to customers—in order to compete in all markets nationwide, most especially the market for all-distance service.

The merger will create a company that can effectively compete in advanced broad-band and wireless services, as well. One way in which the merged MCI WorldCom and Sprint will have the capability to more effectively deploy advanced services is by combining their complementary *Multi-Channel Multi-Point Distribution Service* (MMDS) wireless assets in order to provide a national broad-band service. When those assets are combined with *Digital Subscriber Line* (DSL) and MCI World-Com's fiber rings in 102 cities, the joint assets will create an independent broad-band network that reaches residences and businesses across the country.

Summary

The domestic market for long-distance services is estimated at about $90 billion (U.S. dollars). Growing competition in long-distance services has eroded AT&T's market share from its former monopoly level to about 50 percent. With this competition has come increasing availability of low-cost calling plans for a broad range of consumers. As a result, average revenue per minute earned by carriers has been declining steadily for several years, while long-distance usage has increased substantially in order to make up for that revenue shortfall.

See Also

Federal Communications Commission (FCC)

Incumbent Local Exchange Carriers (ILECs)

Telecommunications Act of 1996

Voice-Over IP (VoIP)

Infrared Networking

Infrared technology is used to implement wireless LANs (as well as the wireless interface) in order to connect laptops and other portable machines to the desktop computer that is equipped with an infrared transceiver. Infrared LANs are proprietary in nature, so users must rely on a single vendor for all of the equipment. The infrared interface for connecting portable devices with the desktop computer, however, is standardized by the *Infrared Data Association* (IrDA).

Infrared LANs

Infrared LANs typically use the wavelength band between 780 nm and 950 nm, primarily because of the ready availability of inexpensive, reliable system components.

There are two categories of infrared systems that are commonly used for wireless LANs. One is directed infrared, which uses narrow laser beams to transmit data over one to three miles. This approach can be used to connect LANs in different buildings. Although transmissions over laser beam are virtually immune to electromechanical interference and would be extremely difficult to intercept, such systems are not widely used because their performance can be impaired by atmospheric conditions, which can vary daily. Effects such as absorption, scattering, and shimmer can reduce the amount of light energy that is picked up by the receiver, causing the data to be lost or corrupted.

The other category of infrared systems is non-directed, which uses a less-focused approach. Instead of a narrow beam to convey the signal, the light energy is spread out and is bounced off narrowly defined target areas or larger surfaces, such as office walls and ceilings. Non-directed infrared links can be further categorized as either line-of-sight or diffuse (refer to Figure I-1). Line-of-sight links require a clear path between transmitter and receiver but generally offer higher performance.

The line-of-sight limitation can be overcome by incorporating a recovery mechanism in the infrared LAN, which is managed and implemented by a separate device called a *Multiple Access Unit* (MAU), to which the workstations are connected. When a line-of-sight signal between two stations is temporarily blocked, the MAU's internal optical link-control circuitry automatically changes the link's path to get around the obstruction. When the original path is cleared, the MAU restores the link over that path. No data is lost during this recovery process.

Diffuse links rely on light that is bounced off reflective surfaces. Because it is difficult to block all of the light that is reflected from large surface areas, diffuse links are generally more robust than line-of-sight links. The disadvantage of diffused infrared is that a great deal of energy

Figure I-1

Line-of-sight versus diffuse configurations for infrared links

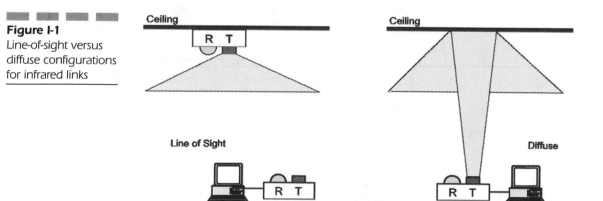

is lost—and, consequently, the data rates and operating distances are much lower.

System Components *Light-Emitting Diodes* (LEDs) or *Laser Diodes* (LDs) are used for transmitters. LEDs are less efficient than LDs, typically exhibiting only 10 percent to 20 percent electro-optical power-conversion efficiency, while LDs offer an electro-optical conversion efficiency of 30 percent to 70 percent. LEDs, however, are much less expensive than LDs, which is why most commercial systems use them.

Two types of low-capacity silicon photodiodes are used for receivers: *Positive-Intrinsic-Negative* (PIN) and avalanche. The simpler and less expensive PIN photodiode is typically used in receivers that operate in environments that have bright illumination, whereas the more complex and more expensive avalanche photodiode is used in receivers that must operate in environments where background illumination is weak. The difference in the two types of photodiodes is their sensitivity.

The PIN photodiode produces an electrical current in proportion to the amount of light energy that is projected onto it. Although the avalanche photodiode requires more complex receiver circuitry, it operates in much the same way as the PIN diode (except that when light is projected onto it, there is a slight amplification of the light energy). This feature makes it more appropriate for weakly illuminated environments. The avalanche photodiode also offers a faster response time than the PIN photodiode.

Operating Performance Current applications of infrared technology yield performance that matches or exceeds the data rate of wire-based LANs: 10Mbps for Ethernet, and 16Mbps for Token Ring. Infrared technology has a much higher performance potential, however (transmission systems operating at 50Mbps and 100Mbps have already been demonstrated).

Because of its limited range and inability to penetrate walls, non-directed infrared can be easily secured against eavesdropping. Even signals that go out windows are useless to eavesdroppers, because they do not travel far and might be distorted by impurities in the glass (as well as by the glass's placement angle).

Infrared offers high immunity from electromagnetic interference, which makes it suitable for operation in harsh environments such as factory floors. Because of its limited range and inability to penetrate walls, several infrared LANs might operate in different areas of the same building without interfering with each other. Because there is less chance of multi-path fading (large fluctuations in received signal amplitude and phase), infrared links are highly robust. Many indoor environments have incandescent or fluorescent lighting, which induces noise in infrared receivers. This problem is overcome by using directional infrared transceivers with special filters to reject background light.

Media Access Control **(MAC)**   Infrared supports both contention-based and deterministic *Media Access Control* (MAC) techniques, making it suitable for Ethernet as well as for Token Ring LANs. To implement Ethernet's contention protocol—*Carrier Sense Multiple Access* (CSMA)—each computer's infrared transceiver is typically aimed at the ceiling. Light bounces off the reflector in all directions to enable each user to receive data from other users (refer to Figure I-2). CSMA ensures that only one station can transmit data at a time. Only the station(s) to which packets are addressed can actually receive the packets.

Deterministic MAC relies on token passing to ensure that all stations in turn get an equal chance to transmit data. This technique is used in Token Ring LANs, where each station uses a pair of highly directive (line-of-sight) infrared transceivers. The outgoing transducer is pointed at the incoming transducer of a station down line, thus forming a closed ring with the wireless-infrared links among the computers (refer to Figure I-3). With this configuration, much higher data rates can be achieved because of the gain that is associated with the directive infrared signals. This approach

Figure I-2
Ethernet
implementation
using diffuse infrared

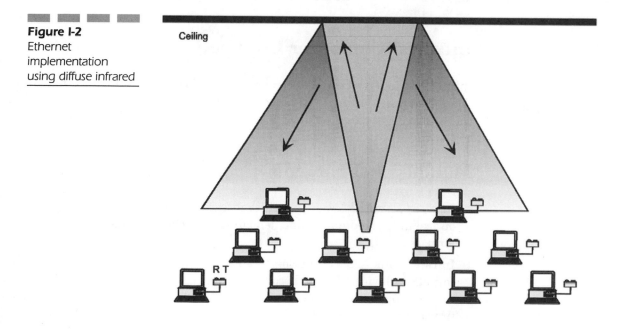

Figure I-3
Token Ring
implementation
using line-of-sight
infrared

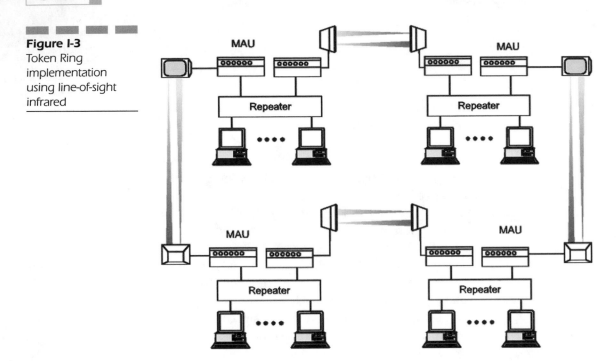

improves overall throughput, because fewer bit errors will occur (minimizing the need for retransmissions).

Infrared Computer Connectivity

Infrared products for computer connectivity conform to the standards that are developed by the *Infrared Data Association* (IrDA). The complete IrDA protocol suite contains three mandatory protocols and seven optional protocols.

Mandatory Protocols

- *Infrared Physical Layer*   Specifies infrared transmitter and receiver optical links, modulation and demodulation schemes, and frame formats
- Infrared Link Access Protocol (*IrLAP*)   Has the responsibility for link initiation, device address discovery, address conflict resolution, and connection startup, in addition to ensuring reliable data delivery and providing disconnect services

■ Infrared Link Management Protocol (*IrLMP*)   Enables multiple software applications to operate independently and concurrently, sharing a single IrLAP session between a portable PC and network-access device

Optional Protocols

■ Infrared Transport Protocol (*IrTTP*), or Tiny TP   Has responsibility for data-flow control, packet segmentation, and reassembly

■ *IrLAN*   Defines how a network connection is established over an IrDA link

■ *IrCOM*   Provides COM (serial and parallel) port emulation for legacy COM applications, printing, and modem devices

■ *IrOBEX*   Provides object exchange services that are similar to the *Hypertext Transfer Protocol* (HTTP) that is used to move information around the Web

■ *IrDA Lite*   Provides methods of reducing the size of IrDA code while maintaining compatibility with full implementations

■ *IrTran-P*   Provides image exchange for digital image-capture devices/cameras

■ *IrMC*   Specifies how mobile telephony and communication devices can exchange information, including phone book, calendar, and message data

In addition, there is a protocol called IrDA Control that enables cordless peripherals such as keyboards, mice, game pads, joysticks, and pointing devices to interact with many types of intelligent host devices. Host devices include PCs, home appliances, game machines, and television/Web set top boxes.

An extension called Very Fast IR (VFIR) provides a maximum transfer rate of 16Mbps, a four-fold increase in speed from the previous maximum data rate of 4Mbps. The extension brings end-users faster throughput without an increase in cost and is backward compatible with equipment employing the previous data rate.

In addition to directly connecting portable and desktop computers, infrared links can be used to connect portable computers to the corporate LAN via a network-access device. Although this setup can also be accomplished by inserting the portable computer into a docking station that is connected to the LAN, these devices are typically based on proprietary technology—and, in most cases, do not work with portable PCs from other vendors.

Summary

Infrared's primary impact will take the form of benefits for mobile professional users. Infrared enables simple, point-and-shoot connectivity to standard networks, which streamlines users' workflow and enables them to reap more of the productivity gains that are promised by portable computing. IrDA technology is supported in more than 100 million electronic devices, including desktops, notebooks, palm PCs, printers, digital cameras, public phones/kiosks, cellular phones, pagers, PDAs, electronic books, electronic wallets, and other mobile devices.

When used on a LAN, infrared technology also confers substantial benefits to network administrators. Infrared is easy to install and configure, requires no maintenance, and imposes no remote-access tracking hassles. Infrared does not disrupt other network operations, and it provides data security. Also, because it makes connectivity so easy, infrared encourages the use of high-productivity network and groupware applications on portables, thus helping administrators amortize the costs of these packages across a larger user base.

See Also

Spread-Spectrum Radio

Inside Telephone Wiring

Part 68 of the FCC's rules governs the terms and conditions under which *Customer-Premises Equipment* (CPE) and wiring can be connected to the telephone network. Although not put into effect in until 1975, Part 68 traces its origins to the FCC's Carterfone decision of 1968.

The purpose of Part 68 is to ensure that terminal equipment and wiring can be connected to the PSTN without causing harm to the network. Part 68 restrictions should be no greater than necessary in order to ensure network protection. Carriers have the responsibility to show that any particular Part 68 restriction is necessary.

In 1984, the FCC refined Part 68 to enable customers to connect one-line and two-line business and residential telephone wiring to the network. The

FCC established a demarcation point to mark the end of the carrier network and the beginning of customer-controlled wiring. The demarcation point would be located on the subscriber's side of the network. The FCC also issued orders detariffing the installation and maintenance of telephone inside wiring (the Commission first detariffed the installation of complex wiring). In 1986, the Commission extended detariffing to the installation of simple inside wiring and the maintenance of all inside wiring. The FCC enabled carriers to retain ownership of telephone inside wiring but prohibited carriers from performing the following actions:

- Using their ownership to restrict the removal, replacement, rearrangement, or maintenance of telephone inside wiring
- Requiring customers to purchase telephone inside wiring
- Imposing a charge for the use of such wiring

In 1990, the Commission amended the definition of the demarcation point for both simple and complex wiring, in order to ensure that the demarcation point would be near the point where the wiring entered the customer's premises. The revised definition required that the demarcation point generally be no farther than 12 inches inside the customer's premises. For existing multi-unit installations, the demarcation point is determined in accordance with the carrier's reasonable and nondiscriminatory standard operating practices. For new wiring installations in multi-unit premises, including additions, modifications, and rearrangements of existing wiring, the carrier can establish a reasonable and nondiscriminatory practice of placing the demarcation point at the minimum point of entry. When the carrier does not have such a practice, the multi-unit premises owner determines the location of the demarcation point(s). If there are multiple demarcation points for either existing or new multi-unit installations, the demarcation point for any particular customer cannot be farther inside the customer's premises than 12 inches from the point at which the wiring enters the customer's premises.

In June 1997, the FCC clarified its position on inside telephone wiring, stating that the carrier's standard operating practices for determining the demarcation point for multi-unit installations are those practices that were in effect on August 13, 1990. Its subsequent rulings do not authorize changing the demarcation point for an existing building to the minimum point of entry. Reiterating that carriers cannot require the customer or building owner to purchase or pay for the use of carrier-installed wiring that is now on the customer's side of the demarcation point, the FCC concluded that the carrier cannot remove such wiring.

The FCC also amended its previous definition of the telephone demarcation point as follows:

- The demarcation point can be located within 12 inches of the point at which the wiring enters the customer's premises "or as near thereto as practicable."

- Only major additions or rearrangements of existing wiring can be treated as new installations.

- Multi-unit building owners can restrict customer access to only that wiring that is located in the customers' individual unit.

- Local telephone companies are required to provide building owners with all available information regarding carrier-installed wiring on the customer's side of the demarcation point, in order to facilitate the servicing and maintenance of that wiring.

In addition, building owners have a right to access wiring on their side of the demarcation point and have a responsibility to maintain that wiring. Accordingly, the FCC has restricted the capability of carriers to interfere with customer access and maintenance activities. Carriers cannot use claims of ownership as a basis for imposing restrictions on the building owner's removal, rearrangement, replacement, or maintenance of the wiring. Because there are already procedures under which carriers recover the costs of inside wiring that was originally installed or maintained under tariff, carriers are not entitled to additional compensation for wiring—nor can carriers require that the wiring be purchased or impose a charge for its use.

Finally, because carrier removal of installed wiring would prevent customer access and maintenance of that wiring—and because the threat of removal could be used to force the customer to purchase the wiring—the FCC has ruled that carriers cannot remove installed inside wiring.

Quality Standards

In January 2000, the FCC issued quality guidelines for newly installed inside telephone wiring. At a minimum, inside telephone wiring must be solid, 24 gauge or larger, twisted pairs that comply with the electrical specifications for Category 3, as defined in the ANSI EIA/TIA Building Wiring Standards.

Conductors must have insulation with a 1500-volt minimum breakdown rating. This rating is established by covering the jacket or sheath with at least 6 inches (15 cm) of conductive foil. A potential difference between the foil and all of the individual conductors that are connected together must be established, such that the potential difference is gradually increased over a 30-second time period to 1500 volts (60Hz) and then is applied continuously for 60 seconds more. At no time during this 90-second time interval should the current between these points exceed 10 milliamperes peak.

All wire and connectors that meet these requirements must be marked, in a manner visible to the consumer, with the symbol CAT 3 or a symbol consisting of a C with a 3 contained within the C character, at intervals not to exceed 12 inches along the length of the wire.

In establishing minimum standards for simple inside wiring that eventually connects to the PSTN, consumers and small businesses will benefit by having wiring that is capable of accommodating clear telecommunications and digital transmissions. Consumers and small businesses will also benefit from the decreased necessity for the expensive replacement of poor-quality, simple inside wiring, as might be required in order to accommodate extra lines for additional telephones, personal computers, fax machines, and ISDN or DSL services. The quality guidelines will also stem the increasing incidence of cross-talk and the risk of network harm that is associated with the installation of poor-quality inside wiring.

Summary

The FCC has consistently issued rulings that are aimed at encouraging competition in the inside wiring installation and maintenance markets, promoting new entry into those markets, and fostering the development of an unregulated, competitive telecommunications marketplace. These rulings include setting quality standards for inside telephone wiring that will enable consumers and small businesses to obtain access to broad-band communications services, such as DSL.

See Also

Carterfone Decision
Inside Cable Wiring

Inside Cable Wiring

Inside cable wiring is located within a subscriber's home or apartment and is usually installed by a cable operator or contractor. This wiring does not include items such as amplifiers, converters, decoder boxes, or remote-control units. Extensions to the initially installed cable can be performed by the subscriber by using materials purchased from retail sources and without the permission of the cable operator.

In accordance with FCC rules, after a subscriber voluntarily terminates cable service, the cable operator can leave the home wiring in place (or notify the consumer that the cable company will remove the wiring) unless the consumer purchases the wire from the cable operator on a per-foot replacement-cost basis. If the wire was previously transferred or sold to the subscriber, the subscriber owns the wiring, and the cable operator cannot remove it or restrict its use—regardless of the reason for service termination. In either case, the cable operator cannot be held responsible for any signal leakage that occurs from the home wiring once the operator ceases providing service over that wiring.

If the subscriber does not already own the wiring and declines to purchase it from the cable operator, the cable operator can remove the home wiring within 30 days. The cable operator must do so at no charge to the subscriber and must pay the cost of any damage that is caused by its removal.

To leave the wiring inside and to remove the wiring outside the subscriber's home, the cable operator might, for single-unit dwellings, sever the cable approximately 12 inches outside the point where the cable wire enters the outside wall of the subscriber's home. For *Multiple-Unit Dwellings* (MUD), the cable operator might sever the wire approximately 12 inches outside the point where the cable wire enters the subscriber's individual dwelling unit, except in cases of loop-through or other similar series wire configurations that are not covered by the home wiring rules.

Where an operator fails to adhere to these procedures, the operator is deemed to have immediately relinquished any and all ownership interests in the home wiring and is not entitled to compensation for the wiring. Furthermore, the cable operator cannot subsequently attempt to remove the home wiring or restrict its use.

If the cable operator informs the subscriber of his or her rights and the subscriber agrees to purchase the wiring, constructive ownership over the home wiring will transfer immediately to the subscriber, who can authorize a competing service provider to connect with and use the home wiring. If the subscriber declines to purchase the home wiring, the operator has seven business days to remove the wiring; otherwise, the subscriber cannot make a subsequent attempt to remove it or to restrict its use.

If the cable operator owns the inside wiring, the operator cannot charge its subscribers a separate wire-maintenance fee—because the lease rate for operator-owned wiring would already include a component for maintenance and repair. If the subscriber owns the inside wiring, the cable operator cannot charge a lease rate but could offer a separate, optional wire-maintenance service and charge hourly rates for service as necessary.

Multiple Dwelling Units (MDUs)

For *Multiple Dwelling Units* (MDUs), the cable demarcation point is set at (or about) 12 inches outside of where the cable wire enters the subscriber's individual dwelling unit. Generally, each subscriber in an MDU has a dedicated line (often called a home run) running to his or her premises from a common feeder line or riser cable that serves as the source of video-programming signals for the entire MDU. The riser cable typically runs vertically in a multi-story building and connects to the dedicated home-run wiring at a tap or a multi-tap, which extracts portions of the signal strength from the riser and distributes individual signals to subscribers.

Depending on the size of the building, the taps are usually located in a security box (often called a lock box) or utility closet that is located on each floor, or at a single point in the basement. Each time the riser cable encounters a tap, its signal strength decreases. In addition, the strength of a signal diminishes as the signal passes through the coaxial cable. As a result, cable wiring often requires periodic amplification within an MDU in order to maintain picture quality. Amplifiers are installed at periodic intervals along the riser, based upon the number of taps and the length of coaxial cable within the MDU.

Certain other video service providers typically employ a similar inside-wiring scheme. Among them are providers of MMDS, LMDS, *Satellite Master-Antenna Services* (SMATV), and *Direct Broadcast Satellite* (DBS) providers. These providers use wireless technologies to deliver their signal to an antenna on the roof of an MDU, then they run their riser cable down from the roof to the taps and dedicated home-run wires. These types of providers are not always given the same access to customers by the MDU owners.

At this writing, a debate exists as to who actually owns access to multi-tenant buildings for the purpose of installing the inside cable wiring. There is contention among two groups of users: the millions of Americans who live in apartment buildings and other MDUs and the many businesses, including small businesses, that are located in office buildings that they do not control.

Ownership and Access

The special difficulty with offering competitive facilities-based services to these customers arises from the need to transport signals across the building owner's premises to the individual customer's unit. For a telecommunications reseller or a user of the ILEC's unbundled local loops, this transport is typically accomplished by piggybacking on the incumbent's existing facilities as part of the resale or unbundled access agreement. A carrier that uses its own wireline or wireless facilities to reach the building owner's premises, however, must then either install its own equipment or obtain access to existing in-building facilities in order to reach individual customers.

Depending on state law and local practices, some or all of the locations and facilities to which competing carriers might require access can be controlled by the incumbent service provider, the building owner, or both. The rules governing ownership and control of existing facilities also differ depending on whether the facilities are used for telecommunications or video-programming services. In order for facilities-based competition to be fully available to all customers, however, reasonable and nondiscriminatory access to competing providers must be provided by whomever controls these facilities.

At this writing, there appears to be no industry consensus on who owns or controls the right of access to MDUs. The multi-tenant building industry has put up strong resistance to attempts at depriving building owners of their property rights. Even assuming that the FCC ultimately determines that it has the authority to take action under existing law, the arguments in opposition might well form the basis for protracted litigation.

At this writing, it is unclear whether congressional legislation is appropriate to facilitate competitive telecommunications carriers' access to MDUs. Legislation could clarify the authority of the FCC to take action in the public interest, in order to promote reasonable and non-discriminatory access to MDUs. Legislation could also prevent the imposition of restrictions that might discriminate or inhibit the capability of competitive providers to install the facilities necessary to offer their services in MDUs, including wireless equipment such as antennas on the rooftops of apartments and office buildings. Legislation could also provide guidance to the FCC, and to reviewing courts, concerning the proper scope of agency action in this area and the principles that should apply (while still leaving implementation details to be determined by the FCC).

Summary

The FCC enables subscribers to purchase the cable home wiring inside the premises up to the demarcation point. As in the telephone context, a demarcation point generally is the point where a service provider's system wiring ends and the customer-controlled wiring begins. From the customers' point of view, this point is significant because it defines the wiring that they can own or control. For purposes of competition, the demarcation point is significant because it defines the point where an alternative service provider can attach its wiring to the customer's wiring in order to provide service. Access to the inside cable wiring in MDUs is a contentious issue that involves the property rights of building owners and the rights of individual subscribers to access services of their own choosing.

See Also

Inside Telephone Wiring

Institute of Electrical and Electronics Engineers (IEEE)

The *Institute of Electrical and Electronics Engineers* (IEEE) is the world's largest technical professional society with more than 320,000 members who conduct and participate in its activities through 1,400 chapters in 147 countries. The IEEE and its predecessors—the *American Institute of Electrical Engineers* (AIEE) and the *Institute of Radio Engineers* (IRE)—date to 1884. In 1961, the AIEE and IRE merged to form the IEEE.

The technical objectives of the IEEE focus on advancing the theory and practice of electrical, electronic, and computer engineering and computer science—publishing nearly 25 percent of the world's technical papers in these fields. The IEEE sponsors technical conferences, symposia, and local meetings worldwide, in addition to providing educational programs for members.

Technical Societies

As of year-end 1999, the IEEE had 36 technical societies that provided publications, conferences, and other benefits to members within specialized areas—from aerospace and electronic systems to vehicular technology. Each of these societies has technical committees that define and implement the technical directions of the society. For example, there are 19 technical committees within the communications society:

- Cable-Based Delivery and Access Systems
- Communications Software
- Communication Switching
- Communications Systems Integration and Modeling
- Communication Theory
- Computer Communications
- Enterprise Networking
- Gigabit Networking
- Information Infrastructure
- Interconnections in High-Speed Digital Systems
- Internet
- Multimedia Communications
- Network Operations & Management
- Optical Communications
- Personal Communications
- Quality Assurance Management
- Radio Communications
- Satellite and Space Communications
- Signal Processing and Communications Electronics
- Signal Processing for Storage
- Transmission and Access and Optical Systems

Standards Board

The IEEE Standards Board is responsible for all matters regarding standards in the fields of electrical engineering, electronics, radio, and the allied branches of engineering. Currently, there are 10 standing committees of the IEEE Standards Board.

Procedures Committee (ProCom) This committee is responsible for recommending to the IEEE Standards Board improvements and procedural changes in order to promote efficient discharge of responsibilities by the IEEE Standards Board, its committees, and other committees of the Institute that are engaged in standards activities.

New Standards Committee (NesCom)   This committee is responsible for ensuring that proposed standards projects are within the scope and purpose of the IEEE, that standards projects are assigned to the proper society or other organizational body, and that interested parties are appropriately represented in the development of IEEE standards. This committee examines project-authorization requests and makes recommendations to the IEEE Standards Board regarding their approval.

Standards Review Committee (RevCom)   This committee is responsible for reviewing submittals for the approval of new and revised standards and for the reaffirmation or withdrawal of existing standards, in order to ensure that the submittals represent a consensus of the parties that have a significant interest in the covered subjects. The committee makes recommendations to the IEEE Standards Board regarding the approval of these submittals.

Awards and Recognition Committee (ArCom)   This committee is responsible for the administration of all awards that are presented by the IEEE Standards Board. The ArCom acts on behalf of the Board to approve nominations for IEEE Standards Awards, in addition to submitting nominations for standards awards that other organizations sponsor.

New Opportunities in Standards Committee (NosCom)   This committee is responsible for identifying and exploring avenues for enhancing IEEE leadership in areas of new technological growth and for recommending to the IEEE Standards Board actions to achieve this purpose.

Procedures Audit Committee (AudCom)   This committee provides oversight of the standards-development activities of the societies, their standards-developing entities, and the *Standards Coordinating Committees* (SCCs)[1] of the IEEE Standards Board.

Seminars Committee (SemCom)   This committee provides oversight for the operation of the seminar program by providing technical expertise and support. The committee reviews and proposes new seminars to ensure that the topics covered are appropriate.

---

[1]When a proposed standard does not fall into the subject area that is covered by one of the technical societies—or if a technical society cannot handle the workload—a Standards Coordinating Committee is established and coordinated by the IEEE Standards Board.

International Committee (IntCom)   This committee is responsible for coordinating IEEE Standards activities with non-IEEE standards organizations. The committee also assists in the adoption by IEEE of non-IEEE standards when appropriate.

Administrative Committee (AdCom)   This committee acts for the Standards Board between meetings and makes recommendations to the Standards Board for its disposition at regular meetings.

Patent Committee (PatCom)   This committee provides oversight on the use of any patents and patent information in IEEE standards. PatCom also reviews any patent information submitted to the IEEE Standards Board to determine conformity with the patent procedures and guidelines.

Summary

The IEEE is not a member of the *International Electrotechnical Commission* (IEC) or the *International Organization for Standardization* (ISO)—since only countries, not standards bodies, can be members of IEC and ISO. However, when IEEE working groups need global participation in their projects, they can go through any IEC or ISO member country to make submissions to their IEC or ISO technical committees.

See Also

  American National Standards Institute (ANSI)

  International Electrotechnical Commission (IEC)

  International Organization for Standardization (ISO)

  International Telecommunication Union (ITU)

Integrated Access Devices

Integrated access devices (IADs) support voice, data and video services over the same T1/E1 access line. These multi-service access devices typically support *Asynchronous Transfer Mode* (ATM) technology (Figure I-4), which turns the different traffic types into fixed-length 53-byte cells for transmission to the carrier's ATM network. The consolidation of multiple traffic types over the same access facility eliminates the need for separate lines

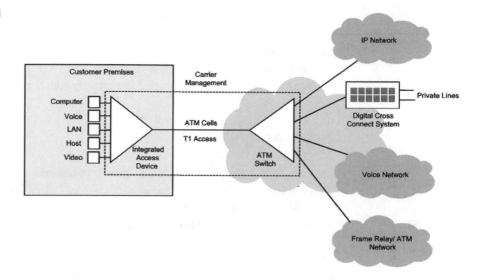

Figure I-4
An integrated access device (IAD) support voice, data and video services over the same access line. Traffic is split out on the carrier side of the network and routed to its appropriate destination. The carrier usually extends management to the IAD on the customer premises.

and having to subscribe to separate services. Management of the IAD is usually the responsibility of the service provider.

Carriers benefit from this arrangement as well. Today's IADs are feature-rich. They offer end-to-end support for *ATM Adaptation Layer 2* (AAL2), the industry standard for transporting compressed voice over ATM. This enables service providers to deliver eight-fold the call traffic of traditional T1/E1 lines without degradation in the quality of voice. IADs enable carriers to offer not just bandwidth, but also a way to deliver integrated services to their customers with quality, reliability and flexibility over that bandwidth. IADs also provide carriers with the means to support any service—ATM, frame relay or circuit emulation—over any channel within the same T1/E1 line simultaneously

In addition, such products provide queuing algorithms, which provide traffic prioritization capabilities for both IP and circuit-based traffic. With multiple classes of service available, carriers can guarantee differentiated levels of service for both data and voice traffic at the enterprise, enabling them to deliver value-added applications such as *virtual private networks* (VPNs).

A like device in the central office delivers voice compression, while reducing costs by nearly 30 percent per channel. Quality of voice features such as silence suppression and echo cancellation, as well as support for fax/modem call detection and international code conversion, are also available. The common platform—customer premises and carrier *point of presence* (POP)—ensures interoperability and simplifies operations and management for carriers, while delivering cost efficiency.

Scaleable multi-service access solutions can start with six T1/E1 interfaces and voice compression options, and be incrementally upgraded to support *Inverse Multiplexing over ATM* (IMA), DS3/E3, and OC3 WAN connections. This scaleability enables service providers to meet the demands of a growing enterprise customer with a single edge solution.

Summary

IADs enable users to aggregate a variety of lower speed services into a high-speed ATM infrastructure. A *competitive local exchange carrier* (CLEC), for example, can deploy IADs to provide integrated voice, data and video services to business customers, letting them take advantage of applications necessary to succeed in a highly competitive marketplace. At the same time, IADs enable CLECs and other types of carriers to meet their business needs for cost efficiency and high performance.

See Also

Asynchronous Transfer Mode (ATM)

Multi-Service Networking

Integrated Services Digital Network

The Integrated Services Digital Network (ISDN) made its debut in 1980 with the promise of providing a high-quality, ubiquitous switched digital service for multimedia applications. Although ISDN was intended to become a worldwide standard to facilitate global communications, this was not to be the case for many years. In the United States, non-standard carrier implementations, incompatibilities between customer and carrier equipment, the initial high cost of special adapters and telephones, spotty coverage, and configuration complexity hampered user acceptance of ISDN through the mid-1990s.

Despite these implementation problems, ISDN offers tangible benefits over the existing public telephone network, including:

- Faster call setup and network response times

- Increased network management and control facilities

- Support for such advanced applications as videoconferencing

- Improved configuration flexibility and additional restore options
- The elimination of delays in switching over new lines and services
- The capability to streamline networks through integration, thus reducing the complexity and cost of cabling and equipment

By 1996 many of the problems with ISDN had been solved. The increasing popularity of the Internet sparked consumer demand for ISDN as a means of accessing the Web and improving the response time for navigating Web sites and viewing multimedia documents.

Applications

ISDN can be used for numerous applications, including videoconferencing, medical image transmission, the delivery of multimedia training sessions, and Internet access. Other applications of ISDN include temporarily rerouting traffic around failed leased lines and handling peak traffic loads (Figure I-5). ISDN also can play a key role in various *computer-telephony integration* (CTI), telecommuting, and remote access (i.e., remote control and remote node) applications.

For certain applications, ISDN can even be more economical than leased lines, which entail fixed monthly charges based on distance, whether or not they are fully used. For example, high-bandwidth applications that are used infrequently (e.g., videoconferencing and medical image transmission) may

Figure I-5
Among other applications, ISDN can be used to backup T1 leased lines in case of failure, provide an additional source of temporary bandwidth to handle peak traffic loads, or support special applications such as videoconferencing or medical imaging on an as-needed basis.

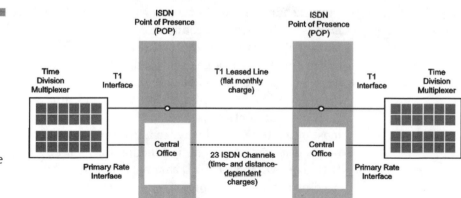

not justify the cost of purchasing leased lines that will be underutilized most of the time. Since users are typically billed for ISDN channels only when used, significant savings can accrue when they are used for infrequent applications.

ISDN Channels

ISDN is a circuit-switched digital service that comes in two varieties. The *basic rate interface* (BRI) provides two bearer channels of 64Kbps each, plus a 16Kbps signaling channel. The *primary rate interface* (PRI) provides 23 bearer channels of 64Kbps each, plus a 64Kbps signaling channel. Any combination of voice and data can be carried over the B channels. Additional channels can be created within the B channels through the use of compression.

ISDN was designed to be compatible with existing digital transmission infrastructures; specifically, T1 in North America and Japan (1.544Mbps) and E1 in Europe (2.048Mbps). Because ISDN can evenly reduce both T1 and E1 into 64Kbps increments, the 64Kbps channels became the worldwide standard. The use of 64Kbps channels also enables users to migrate more easily from private T1/E1 networks to ISDN and build hybrid networks consisting of both public and private facilities.

In both BRI and PRI, ISDN's separate D channel is used for signaling. As such, it has access to the control functions of the various digital switches on the network. It interfaces with *Signaling System 7* (SS7) to provide message exchange between the user's equipment and the network to set up, modify, and clear the B channels. Via SS7, the D channel also gathers information about other devices on the network, such as whether they are idle, busy or off. In being able to check ahead to see if calls can be completed, network bandwidth can be conserved. If the called party is busy, for example, the network can be notified before resources are committed.

The D channel's task is carried out very quickly, so it remains unused most of the time. For this reason, PRI users can assign the 64Kbps D channel to perform the signaling function for as many as eight PRI lines.[2] For BRI users, whenever the D channel is not being used for signaling, it can be used as a bearer channel (if the carrier offers this capability as a service)

---

[2]This is rarely done because if the PRI line with the D channel goes out of service, the other PRI lines that depend on it for signaling also go out of service.

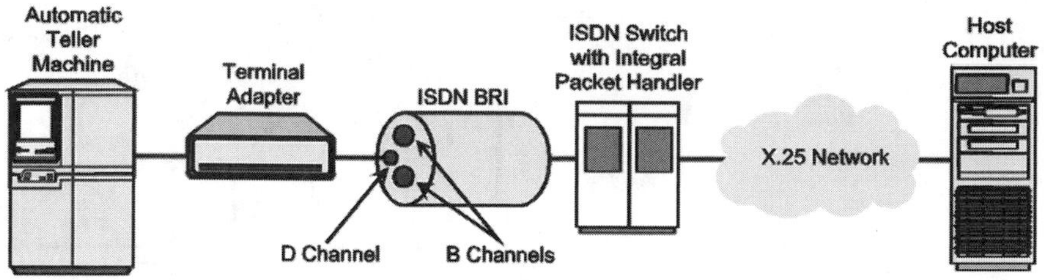

Figure I-6
When ISDN's D
channel is not used
for signaling, it can
be used for low-
speed applications,
while the B channels
carry other voice and
data calls.

Figure I-6
When ISDN's D channel is not used for signaling, it can be used for low-speed applications, while the B channels carry other voice and data calls.

over X.25 networks for point-of-sale applications such as *automatic teller machines* (ATMs), lottery terminals, and cash registers (Figure I-6).

With regard to ISDN PRI, there are two higher-speed transport channels called H channels. The H0 channel operates at 384Kbps, while the H11 operates at 1.536Mbps. These channels are used to carry multiplexed data, data and voice, or video at higher rates than that provided by the 64Kbps B channel. The H channels also are ideally suited for backing up FT1 and T1 leased lines. Eventually other high-speed transport channels will be added in support of broad-band ISDN.

Multi-rate ISDN lets users select appropriate increments of switched digital bandwidth on a per-call basis. Speeds, in increments of 64Kbps, are available up to 1.536Mbps. Multi-rate ISDN is used mostly for multimedia applications and videoconferencing.

ISDN Architectural Elements

The architectural elements of ISDN include several reference points that define network demarcations between the telephone company and the customer premises:

- *R* The reference point separating non-ISDN (TE2) equipment and the *Terminal Adapter* (TA), which provides TE2 with ISDN compatibility.
- *S* The reference point separating terminals (TE1 or TA) from the network terminal (NT2).
- *T* The reference point separating NT2 from NT1 (not required if NT2 and NT1 functionality is provided by the same device).
- *U* The reference point separating the subscriber's portion of the network (NT1) from the carrier's portion of the network (LT).

Figure I-7
ISDN architectural
elements and
reference points.

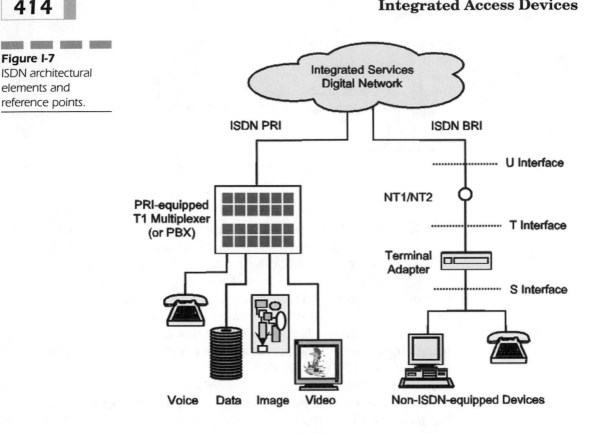

There are also interfaces between various types of customer premises equipment. Network terminators provide network control and management functions, while terminal equipment devices implement user functions. Figure I-7 illustrates the reference points and architectural elements of ISDN.

Although ISDN BRI and PRI services consist of different configurations of communications channels, both services require the use of distinct functional elements to provide network connectivity. One of these elements is the *Network Terminator 1* (NT1), which resides at the user's premises and performs the four-wire to two-wire conversion required by the local loop. Aside from terminating the transmission line from the central office, the NT1 device is used by the telephone company for line maintenance and performance monitoring.

Network Terminator 2 (NT2) devices include all NT1 functions in addition to protocol handling, multiplexing, and switching. These devices are usually integrated with PBX and key systems.

ISDN *terminal equipment* (TE) provides user-to-network digital connectivity. TE1 provides protocol handling, maintenance, and interfacing functions and supports such devices as digital telephones, data terminal equipment, and integrated workstations—all of which comply with the ISDN user-network interface. The large installed base of non-ISDN TE2 devices (e.g., telephones and PCs) can communicate with ISDN-compatible devices when users attach or install a *Terminal Adapter* (TA) to/in the non-ISDN device. A TA takes the place of a modem. Users can connect a maximum of eight TE/TA devices to a single NT2 in a multi-drop configuration.

Summary

Over the years, ISDN has been touted as a breakthrough in the evolution of worldwide telecommunications networks—the single most important technological achievement since the advent of the telephone network itself in the 19th century. Others disagree, pointing out that ISDN implementation has stalled and that other, more powerful telecommunications technologies such as frame relay and ATM are progressing so rapidly that ISDN is now obsolete. The debate over ISDN's relevance to today's telecommunications needs must be put into the context of specific applications. Just like any other service, ISDN will be adequate to serve the needs of some users, but not others. Meanwhile, ISDN will continue to evolve to include broadband services, eventually ending the debate about its obsolescence.

See Also

> *Computer-Telephony Integration* (CTI)
>
> *Digital Subscriber Line* (DSL) Technologies
>
> *Signaling System 7* (SS7)
>
> Time Division Multiplexing
>
> Videoconferencing

Interactive Video and Data Service

Interactive Video and Data Service (IVDS) was devised in 1992 by the Federal Communications Commission (FCC) as a point-to-multi-point, multi-point-to-point, short distance communication service that operates in the 218–218.5 MHz and 218.5–219 MHz frequency bands. In September 1993, the FCC assigned the licenses for IVDS by lottery in the top ten markets.

The following year, two licenses per market were offered for auction at the same time, with the highest bidder given a choice between the two available licenses, and the second highest bidder winning the remaining license. More than 95% of all IVDS licenses were won by small businesses or businesses owned by members of minority groups or women. Additional auctions, ending in 1997, were held for the available spectrum in other markets.

Applications

As envisioned, an IVDS licensee would be able to use IVDS to transmit information, product and service offerings to its subscribers and receive interactive responses. Potential applications included ordering goods or services offered by television services, viewer polling, remote meter reading, vending inventory control, and cable television theft deterrence. An IVDS licensee was able to develop other applications without specific approval from the FCC. An IVDS channel, however, was insufficient for the transmission of conventional full motion video.

Initially, mobile operation of IVDS was not permitted. In 1996, however, the FCC amended its rules to permit IVDS licensees to provide mobile service to subscribers. This action authorized mobile operation of response transmitter units (subscriber units) operated with an effective radiated power of 100 milliwatts or less. The FCC found that these amendments would provide additional flexibility for IVDS licensees to meet the communications needs of the public without increasing the likelihood of interference with TV channel 13.

According to the FCC's rules at the time IVDS spectrum was awarded, licenses cancelled automatically if a licensee did not make its service available to at least 30% of the population or land area within the service area within three years of grant of the system license. Each IVDS system licensee had to file a progress report at the conclusion of three- and five-year benchmark periods to inform the FCC of the construction status of the system. This arrangement was intended to reduce the attractiveness of licenses to parties interested in them only as a speculative vehicle.

Summary

The spectrum for IVDS was initially allocated in 1992 to provide interactive capabilities to television viewers. This did not occur, largely because the enabling technology was too expensive and market demand too low or non-existent. IVDS is now known as 218–219 MHz Service. In September 1999, the FCC revised its rules for the 218–219 MHz Service to maximize the efficient and effective use of this frequency band. The FCC simply redesignated the 218–219 MHz Service from a strictly private radio service—one that is used to support the internal communications requirements of the licensee —to a service that can be used in both common carrier and private operations. Licensees are now free to design any service offering that meets market demand. In addition, licensees now have up to ten years from the date of the license grant to build out their service, without meeting the three- and five-year construction benchmarks.

See Also

Local Multi-Point Distribution Service

Multi-Channel Multi-Point Distribution Service

Interactive Voice Response

Interactive Voice Response (IVR) is a technology that enables callers to obtain requested information stored in a corporate database. IVR technology uses the familiar telephone keypad as an information retrieval and data gathering conduit. Recorded voice messages prompt and respond to the callers' inquiries or commands. Examples of IVR range from simply selecting announcements from a list of options stored in the computer (also known as audiotext) to more complex interactive exchanges that rely on database lookups, such as a bank account balance.

A basic IVR system takes the call, asks questions with multiple choice answers, and responds to the digit entered by the caller with the proper pre-recorded announcement. This self-service solution provides requested information to callers 7 days a week, 24 hours a day. This enables companies to meet the information needs of various constituents efficiently and economically—without having to devote staff to handle routine requests for information.

IVR can be an integral function of a PBX or ACD, or it can be provided as a service by a telecommunications carrier. With Sprint's InterVoice multi-application platform, for example, businesses can offer their customers a variety of call processing functions, including interactive voice response.

Applications

IVR systems can be used for a variety of applications. A brokerage firm can use an IVR system to take routine orders from investors who want to order a corporate prospectus. An investment fund can take routine requests for new account applications. An employer can take routine requests from employees about their benefit plans. A help desk can use an IVR system to take routine questions from computer users, and step them through a preliminary troubleshooting process that reveals the most common hardware or software problems that they can correct themselves. Colleges can use IVR systems to answer routine questions about degree programs, the registration process, and fees.

Just about any organization can use an IVR system to meet the informational needs of callers. Businesses can improve customer service by offering 24-hour access to information, with the added benefit of providing consistent information and transaction capabilities to their customers through the simple use of a telephone. Large inventories of literature can be eliminated, not to mention the costs of storage and the staff to maintain it. Mailing expenses can also drop dramatically.

IVR systems even enable companies to bill callers for services. They can capture credit card information from callers with touch-tone phones, or voice capture callers' names and addresses. For more complicated transactions, the IVR system can also provide the caller with the option of accessing a live operator.

Although IVR systems can deliver requested information via recorded voice announcements, sometimes the information is best delivered in printed form. Some IVR systems include a Fax On Demand capability—also known as fax back—that enables callers to select documents from a menu of available items that are described to them (Figure I-8). Callers can receive information at their fax machine instantly, or they have the option of scheduling delivery at a time more convenient for them.

Process Flow and Navigation

Many IVR systems are designed for a particular application. The process flow and navigation of an IVR system can be illustrated by examining its role in supporting a typical employee benefit plan application.

Employees are provided with an enrollment worksheet, which shows them available options for benefits for the upcoming enrollment period. This worksheet also has a PIN number printed on it for access into the IVR.

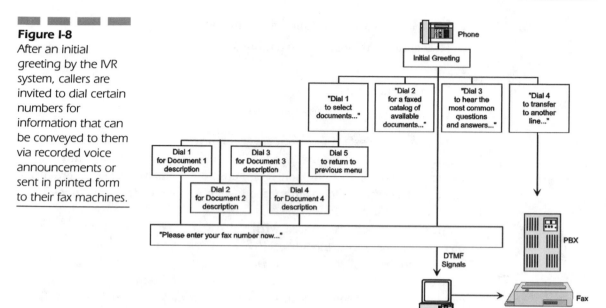

Figure I-8
After an initial greeting by the IVR system, callers are invited to dial certain numbers for information that can be conveyed to them via recorded voice announcements or sent in printed form to their fax machines.

Employees call an 800 number to access the system. The system will prompt them to enter their social security number and PIN. Any incorrect entries will be voiced back as incorrect, along with a message inviting the user to "Try again."

Built-in editing features are used to eliminate errors typically made by employees using a paper-based enrollment. During the call employees are only presented with the options for which they are eligible. The IVR system validates each entry to ensure that the entry conforms to the plan's provisions.

If the employee is a first time caller (no selections have been recorded yet), the system will guide them through the enrollment process. Included in the script are benefit option eligibility prompts that use a dynamic menu for the following typical benefit plans:

- Medical options (only gives the eligible options and HMOs)
- Medical coverage category (Single, Family, etc.)
- Dental options (only gives the available options and DMOs)
- Dental coverage categories
- Vision options
- Vision coverage categories

- Life insurance options (only gives the available coverage)
- LTD options (only gives the available coverage)
- Dependent life
- Spending accounts
- Other plans

If the employees' total benefit coverage price is in excess of company provided flex dollars or credits, the IVR system can communicate any salary reduction or deduction impact. If the employee has an excess of credits, a choice of the credit allocation method, such as cash or transfer to a spending account, can be offered.

Many functions are offered with voice response systems, such as total pay period costs and modeling, PIN changes, current enrollment status inquiry, ordering of materials, *Primary Care Physician* (PCP) data collection, dependent data collection, and new hire enrollments. IVR voice scripts can be designed to communicate this information with ease and simplicity. Additional features—such as custom messages about proof of insurability, waiver of coverage forms, special instructions, and fax back services—can be added to the voice response script.

Summary

Easily accessible and accurate information is important in this increasingly competitive age, marked by corporate staff reductions, streamlined business processes, and the need for cost containment. Consequently, IVR is becoming a vital communication tool that alleviates the time and resource burdens currently facing many businesses. IVR technology uses the familiar telephone keypad for information retrieval and data gathering, essentially offering corporate constituents the means to help themselves. Through a single toll free number, the IVR system can bring together information retrieval, directory services, and transaction capabilities.

See Also

Automatic Call Distributors (ACD)

Call Centers

Facsimile

International Callback Service

Callback is a service that is provided by U.S. international long-distance resellers as a means for their customers, located outside of the United States, to access U.S-based international lines. Typically, a caller places a call to the callback provider's switch located in the United States. If uncompleted call signaling is used, the caller dials the provider's switch in the United States, waits a predetermined number of rings, and hangs up without the switch answering. The switch then automatically returns the call, and upon completion, provides the caller with a U.S. dial tone. All traffic is thus originated at the U.S. switch. The calls are billed at U.S. tariffed rates, which are often much lower than those of the originating country.

Legal Challenge

In 1995, AT&T complained to the FCC that international callback services were illegal, arguing that uncompleted call signaling is an unreasonable practice because it violates the federal wire fraud statute and constitutes theft of service. In response, the service providers offering international callback contended that wire fraud cannot occur without a completed call and, further, that uncompleted call signaling does not constitute wire fraud because there is no wrongful appropriation of money or property, since international facilities-based carriers do not charge for uncompleted calls.

The FCC rejected AT&T's charge that uncompleted call signaling constitutes federal wire fraud in violation of Section 201(b) of the Communications Act. On reconsideration of the matter, the FCC even sought the opinion of the Department of Justice as to whether uncompleted call signaling violates the federal wire fraud statute. The Justice Department agreed with the FCC in its determination that uncompleted call signaling does not constitute wire fraud, adding: "It appears . . . that the U.S. carriers have found and are legitimately exploiting a loophole in AT&T's tariff structure."

With regard to AT&T's contention that uncompleted call signaling is an unreasonable practice under Section 201(b) because it constitutes theft of service, the FCC recognized that uncompleted call signaling constitutes an uncompensated use of the network. However, in the system as currently structured by facilities-based carriers, customers do not expect to pay for an uncompleted call. Nor do carriers expect to be compensated. Because there is no expectation of payment for uncompleted calls, the failure to pay for those calls does not deprive carriers of anything they are otherwise due. Thus, the FCC did not find that any "property" had been "taken" from AT&T.

While AT&T initially opposed international callback, AT&T now offers an international callback service of its own. Customers of its *Software Defined Network* (SDN) service can obtain Alternate Network Access. Under this offering, employees of SDN user companies traveling overseas call a number in the United States to obtain U.S. dial tone for a call to another country. In this way, employees can avoid expensive international direct-dial rates on calls that would otherwise originate overseas.

Summary

International callback is a popular service and has been expanded to include fax transmissions, cellular calls, and calls placed through calling cards as well. Typically, the caller dials a number in the United States and hangs up after the first ring. The service provider's switch in the United States immediately calls back with a dial tone, upon which the caller dials the destination number. The arrangement provides the means for users to save on international calls to the United States because, typically, calls are billed at a higher rate in other countries.

See Also

Cellular Voice Communications

Facsimile

International Common Carriers

Any carrier that provides an international communications service is an *international common carrier* (ICC). For purposes of regulatory status, an ICC may be classified as either dominant or non-dominant for the provision of particular communications services on particular routes. The relevant markets on the foreign end of a U.S. international route include: international transport facilities or services, including cable landing station access and back-haul facilities; inter-city facilities or services; and local access facilities or services on the foreign end of a particular route.

Dominant vs. Non-Dominant

A U.S. carrier that has no affiliation with, and that itself is not, a foreign carrier in a particular country to which it provides service (i.e., a destination country) is presumed by the FCC to be non-dominant for the provision of international communications services on that route. However, a U.S. carrier that has or acquires an affiliation with a foreign carrier that is a monopoly provider of communications services in a destination country is presumed by the FCC to be dominant for the provision of international communications services on that route.

A U.S. carrier that has or acquires an affiliation with a foreign carrier that is not a monopoly provider of communications services in a destination country may seek to be regulated as non-dominant on that route. In such cases, the carrier bears the burden of submitting information to the FCC that is sufficient to demonstrate that its foreign affiliate lacks sufficient market power on the foreign end of the route to adversely affect competition in the U.S. market. If the U.S. carrier demonstrates that the foreign affiliate lacks 50 percent market share in the international transport and the local access markets on the foreign end of the route, the U.S. carrier is presumed by the FCC to be non-dominant.

A carrier that is authorized to provide an international switched service and which provides the service solely through the resale of an unaffiliated U.S. facilities-based carrier's international switched services is presumed by the FCC to be non-dominant for the provision of that service. A carrier regulated as non-dominant must notify the FCC at any time that it begins to provide such service through the resale of an affiliated U.S. facilities-based carrier's international switched services. The carrier will be classified as a dominant carrier on that route, unless the FCC finds that the carrier otherwise qualifies for non-dominant status.

Dominant Carrier Regulation

Any carrier classified as dominant for the provision of particular services on particular routes must comply with certain regulatory requirements in its provision of such services on each route. Among these requirements are:

- File international service tariffs on one day's notice without cost support.
- Provide services as an entity that are separate from its foreign carrier affiliate, and in compliance with the following requirements:

- The authorized carrier must maintain separate books of account from its affiliated foreign carrier.
- The authorized carrier must not jointly own transmission or switching facilities with its affiliated foreign carrier.
- File quarterly reports on traffic and revenue within 90 days from the end of each calendar quarter.

- File quarterly reports summarizing the provisioning and maintenance of all basic network facilities and services procured from the foreign carrier affiliate, or from an allied foreign carrier. These reports should contain the types of circuits and services provided; the average time intervals between order and delivery; the number of outages and intervals between fault report and service restoration; and for circuits used to provide international switched service, the percentage of "peak hour" calls that failed to complete.
- In the case of an authorized facilities-based carrier, file quarterly circuit status reports within 90 days from the end of each calendar quarter.

Foreign Carrier Affiliation

U.S. international carriers that propose to become affiliated with a foreign carrier must notify the FCC and obtain prior approval. In such cases, the FCC must be notified 60 days prior to the consummation of either of the following acquisitions of direct or indirect interests in or by foreign carriers:

- Acquisition of a controlling interest in a foreign carrier by the authorized carrier, or by any entity that controls the authorized carrier, or that directly or indirectly owns more than 25 percent of the capital stock of the authorized carrier.
- Acquisition of a direct or indirect interest greater than 25 percent, or a controlling interest, in the capital stock of the authorized carrier by a foreign carrier or by an entity that controls a foreign carrier.

Any carrier authorized to provide international communications service that becomes affiliated with a foreign carrier that has not previously notified the FCC must provide notification within 30 days after acquiring the affiliation. In order to retain non-dominant status on the affiliated route, the carrier notifying the FCC of a foreign carrier affiliation must provide information to demonstrate that it qualifies for non-dominant classification.

After the FCC issues a public notice of the affiliation, interested parties may file comments within 14 days of the public notice. However, at any time it deems necessary, the FCC can impose dominant carrier regulation on the carrier for the affiliated routes before or after the deadline for submission of public comments.

The FCC will presume the investment to be in the public interest unless it notifies the carrier that the investment raises questions concerning the public interest, convenience and necessity. If notified that the investment raises a substantial and material question, then the carrier is prohibited from consummating the planned investment until it has filed a complete application and the FCC has approved it by formal written order.

Summary

All international common carriers are responsible for the continuing accuracy of their certifications with regard to affiliations with foreign carriers. Whenever the substance of any ICC's certification is no longer accurate, the carrier has 30 days to file a corrected certification. This information may be used by the Commission to determine whether a change in regulatory status may be warranted.

See Also

Dominant Carrier Status
Federal Communications Commission (FCC)

International Electrotechnical Commission

The *International Electrotechnical Commission* (IEC) was founded in 1906 as a result of a resolution passed at the International Electrical Congress held in St. Louis, Missouri (United States) in 1904. The membership currently consists of 52 National Committees, representing the electrotechnical interests in each country, from manufacturing and service industries to government, scientific, research and development, academic and consumer bodies. Membership includes all the world's major trading nations and a growing number of industrializing countries.

The IEC promotes international cooperation on all questions of standardization and related matters—such as the assessment of conformity to standards—in the fields of electricity, electronics and related technologies. It provides a forum for the preparation and implementation of consensus-based voluntary international standards, which serve as a basis for national standardization and as references when drafting international tenders and contracts. To fulfill its mission, the IEC publishes international standards and technical reports.

Organization

The governing authority of the IEC is the Council, which is the general assembly of the National Committees, who are the Commission's members. The IEC also comprises executive and advisory bodies.

Council   The Council deals mainly with administrative matters and is assisted by the *General Policy Committee* (GPC). The decisions and policy of the Council are implemented under the supervision of the Management Board. The Council receives reports from the GPC, the Management Board, the Committee of Action, and the Conformity Assessment Board.

In addition, the *President's Advisory Committee on future Technologies* (PACT) provides a link with private and public research and development activities, keeping the IEC informed of accelerating technological changes and the accompanying demand for new standards.

Committee of Action   The Council delegates the management of standards work to the *Committee of Action* (CA), the membership of which consists of representatives of 12 National Committees.

Among the main tasks of the CA is to set up *Technical Committees* (TCs), follow up and coordinate the work of the TCs, and examine the need to undertake work in new fields.

Conformity Assessment Board   The Council delegates the overall management of the IEC's conformity assessment activities to the *Conformity Assessment Board* (CAB). The CAB provides a single coordinated contact point for high-level negotiations with other conformity assessment bodies at international and regional levels. Among the main tasks of the CAB are setting the IEC's conformity assessment policy so as to serve the present and future needs of international trade. It monitors the operation of IEC conformity assessment schemes by examining their continued relevance. It also coordinates and interfaces with international and regional bodies on CA matters.

Central Office   The Central Office supports the TCs and their *subcommittees* (SCs), as well as the National Committees, by ensuring the reproduction and circulation of working documents and the final texts of standards.

Technical Committees and Subcommittees   As noted, the technical work of the IEC is carried out by TCs and SCs. The TCs prepare technical documents on specific subjects within their respective scopes, which are then sent to the National Committees for voting their approval as international standards.

If a technical committee finds that its scope is too wide to enable all the items on its work program to be dealt with, it may set up SCs, defining in each case a scope covering part of the subjects dealt with by the main committee. The SCs report on their work to the parent TC.

In order to draft documents, a TC or SC may set up *working groups* (WGs) composed of a limited number of experts. An ad hoc group may be set up to examine a particular point and report on it to the TC or SC.

Summary

The IEC cooperates with numerous international organizations, particularly with the *International Organization for Standardization* (ISO) and the *International Telecommunication Union* (ITU). At the regional level, there is a joint working agreement with the *European Committee for Electrotechnical Standardization* (CENELEC), comprising 18 national committees, most of which are also IEC members, and a cooperation agreement with COPANT, the Pan American Standards Commission. Close links are also in place with other bodies in non-electrotechnical areas, such as the World Health Organization, the International Labour Office, the International Organization of Legal Metrology, and the International Atomic Energy Agency.

See Also

> *Institute of Electrical and Electronics Engineers* (IEEE)
>
> *International Organization for Standardization* (ISO)
>
> *International Telecommunication Union* (ITU)

International Organization for Standardization

Established in 1947, the *International Organization for Standardization* (ISO)[3] is a non-governmental, worldwide federation of national standards bodies from 100 countries. A member body of ISO is the national body "most representative of standardization in its country." Only one such body for each country is accepted for membership. The member for the U.S. is the *American National Standards Institute* (ANSI).

ISO promotes the development of standardization and related activities to facilitate the international exchange of goods and services, and develops cooperation in the spheres of intellectual, scientific, technological and economic activity. The ISO's work results in international agreements, which are published as International Standards. The importance of standards is that they provide the end-user with a criterion for judgment, a measurement of quality, and a certain guarantee of compatibility and interoperability.

ISO is best known for its standards work in worldwide communications systems, particularly the development of the seven-layer *Open Systems Interconnection* (OSI) reference model. However, ISO is active in many other fields, including advanced materials, the environment, life sciences, urbanization and construction, and quality assurance. ISO covers all standardization fields except electrical and electronic engineering, which is the responsibility of the *International Electrotechnical Committee* (IEC). The ISO's work in the field of information technology is carried out by a *joint ISO/IEC technical committee* (JTC 1).

---

[3]ISO is a word, not an acronym for the organization. It is derived from the Greek *isos*, meaning "equal," which is the root of the prefix "iso-" that occurs in many other terms, such as "isometric" (of equal measure or dimensions) and "isonomy" (equality of laws, or of people before the law). The line of thinking from "equal" to "standard" led to the choice of ISO as the name of the organization.

The technical work of ISO is highly decentralized, carried out in a hierarchy of about 2,700 technical committees, subcommittees and working groups. In these committees, qualified representatives of industry, research institutes, government authorities, consumer bodies, and international organizations come together as equal partners in the resolution of global standardization problems.

Standards Development Process

There are three main phases in the ISO standards development process. The need for a standard is usually expressed by an industry sector, which communicates this need to a national member body. The latter proposes the new work item to ISO as a whole. Once the need for an International Standard has been recognized and formally agreed upon, the first phase involves definition of the technical scope of the future standard. This phase is usually carried out in working groups, which comprise technical experts from countries interested in the subject matter. Once agreement has been reached on which technical aspects are to be covered in the standard, a second phase is entered during which countries negotiate the detailed specifications within the standard. This is the consensus-building phase.

The final phase comprises the formal approval of the resulting draft International Standard. The acceptance criteria stipulate approval by two-thirds of the ISO members that have participated actively in the standards development process, and approval by 75 percent of all members that vote. Upon approval, the text is published as an ISO International Standard.

Most standards require periodic revision. Several factors combine to render a standard out of date: technological evolution, new methods and materials, and new quality and safety requirements. To take account of these factors, ISO has established the general rule that all ISO standards should be reviewed at intervals of not more than five years. On occasion, it is necessary to revise a standard earlier.

To accelerate the standards process (handling of proposals, drafts, comment reviews, voting, publishing, etc.), ISO makes use of information technology and program management methods. To date, ISO's work has resulted in 9,300 International Standards, representing some 170,700 pages. In 1998 alone, ISO published 1,058 International Standards solving problems in business sectors ranging from traditional activities, such as agriculture and construction, through mechanical engineering, quality and environmental management, to the newest information technology developments, such as the digital coding of audio-visual signals for multimedia applications.

ISO Structure

Like all standards bodies, ISO has an organizational structure that enables it to carry out its mission in the most effective way possible (Figure I-9).

General Assembly   The General Assembly meets once a year. Its agenda includes a multi-year strategic plan and financial status report.

Policy Development Committees   The General Assembly establishes advisory committees, called policy development committees, which are open to all member bodies and correspondent members.

■ Committee on Conformity Assessment (*CASCO*)   Studies the means of assessing the conformity of products, processes, services and quality systems to appropriate standards or other technical specifications.

Figure I-9
Structure of the
International
Organization for
Standardization.

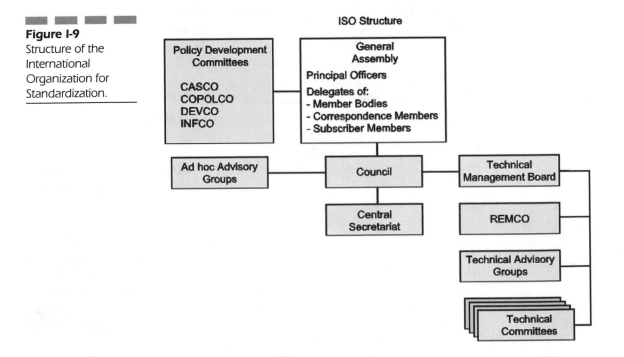

Prepares international guides relating to the testing, inspection and certification of products, processes and services, and the assessment of quality systems, testing laboratories, inspection bodies, certification bodies, and their operation and acceptance.

- Committee on Consumer Policy (*COPOLCO*) Studies the means of assisting consumers to benefit from standardization, and the means of improving their participation in national and international standardization efforts. Promotes from the standardization point of view the information, training and protection of consumers.

- Committee on Developing Country Matters (*DEVCO*) Identifies the needs and requirements of the developing countries in the fields of standardization and assists those countries, as necessary, in defining these needs and requirements. Having done so, it recommends measures to assist the developing countries in meeting them.

- Committee on Information Systems and Services (*INFCO*) Coordinates the activities of ISO and its members in relation to information services, databases, marketing and sales of standards, technical regulations and related matters, including these services and products in electronic form.

Council The operations of ISO are governed by the Council, which consists of the principal officers and 18 elected member bodies. The Council appoints the Treasurer, the 12 members of the Technical Management Board, and the Chairmen of the policy development committees. It also decides on the annual budget of the Central Secretariat.

Central Secretariat The Central Secretariat in Geneva acts to ensure the flow of documentation in all directions, clarifies technical points with secretariats and chairmen, and ensures that the agreements approved by the technical committees are edited, printed, submitted as draft International Standards to ISO member bodies for voting, and then published. Although the greater part of the ISO technical work is done by correspondence, there are, on average, a dozen ISO meetings taking place somewhere in the world every workday of the year.

Ad Hoc Advisory Groups These groups advance the goals and strategic objectives of the organization. Members of such groups may be invited to participate as individuals rather than as representatives of member bodies. Recommendations of such groups are made to the Council for any subsequent action.

Technical Management Board   Reports to and, when relevant, advises the Council on all matters concerning the organization, coordination, strategic planning, and programming of the technical work of ISO. Examines proposals for new fields of ISO technical activity, and decides on all matters concerning the establishment and dissolution of technical committees.

REMCO - Committee on Reference Materials   Establishes definitions, categories, levels and classification of reference materials for use by ISO.

Technical Advisory Groups   *Technical advisory groups* (TAG) are established, when necessary, by the Technical Management Board (and the IEC Committee of Action in cases of Joint ISO/IEC TAGs) to advise the Board (and the IEC Committee of Action when relevant) on matters of basic, sectoral and cross-sectoral coordination, coherent planning and the needs for new work.

Technical Committees   The ISO has over 2,700 technical committees that propose standards in many diverse areas from screw threads and fasteners to in vitro diagnostic test systems and elevating work platforms. The technical committee on Information Technology alone has issued 1,216 ISO standards.

Summary

ISO does not work alone in international standardization. It collaborates very closely with its partner, the IEC. An agreement reached in 1976 defines their respective responsibilities: the IEC covers the field of electrical and electronic engineering, all other subject areas fall under the purview of ISO. When necessary, attribution of responsibility for work programs to ISO or IEC is made by mutual agreement. In specific cases of mutual interest, joint technical bodies or working groups are set up. Common working procedures ensure efficient coordination and the widest possible global application.

Although ISO and the IEC are not part of the United Nations, they have many technical liaisons with various specialized UN agencies. Several UN agencies are actively involved in international standardization, among them, the International Telecommunication Union, the World Health Organization, the Food and Agriculture Organization, and the International Atomic Energy Agency.

See Also

> *American National Standards Institute* (ANSI)
>
> *International Electrotechnical Commission* (IEC)
>
> *International Telecommunication Union* (ITU)

International Telecommunication Union

The *International Telecommunication Union* (ITU) is an international organization within which governments and the private sector coordinate standards for global telecommunications networks and services. ITU activities include the coordination, development, regulation and standardization of telecommunications and the organization of regional and world telecommunications events. Founded in Paris in 1865 as the International Telegraph Union, the International Telecommunication Union took its present name in 1934 and became a specialized agency of the United Nations in 1947. Currently, the ITU is headquartered in Geneva, Switzerland.

The ITU adopts international regulations and treaties governing all terrestrial and space uses of the frequency spectrum as well as the use of geostationary satellite orbits. It also develops standards to facilitate the interconnection of telecommunication systems on a worldwide scale, regardless of the type of technology used. It encourages the development of telecommunications in developing countries and provides technical assistance in the areas of implementing, managing and financing technologies. In essence, the ITU's mission covers the following domains:

- *Technical* to promote the development and efficient operation of telecommunication facilities to improve the efficiency of telecommunication services, their usefulness, and their general availability to the public.

- *Development* to promote and offer technical assistance to developing countries in the field of telecommunications, to promote the mobilization of the human and financial resources needed to develop telecommunications, and to promote the extension of the benefits of new telecommunications technologies to people everywhere.

- *Policy*   to promote, at the international level, the adoption of a broader approach to the issues of telecommunications in the global information economy and society.

Organization

The ITU comprises 187 member countries and 363 members—scientific and industrial companies, public and private operators, broadcasters, regional/international organizations—which participate in the three sectors.

The ITU comprises a plenipotentiary conference, the Council which acts on behalf of the Plenipotentiary Conference, world conferences on international telecommunications, a General Secretariat, the Radiocommunication Sector, a Telecommunication Standardization Sector, and a Telecommunications Development Sector.

Plenipotentiary Conferences   The Plenipotentiary Conference, the supreme authority of the ITU, adopts the fundamental policies of the organization and decides on the organization and activities of the Union in a treaty known as the International Telecommunication Constitution and Convention. The Plenipotentiary Conference is composed of delegations representing all members and is convened every four years. Conferences normally are limited to four weeks and focus on long-term policy issues. In this respect, plenipotentiary conferences take decisions on draft Strategic Plans submitted by the Council outlining the objectives, work programs and expected outcome for each constituent of the Union until the following Conference.

The ITU Council   The ITU Council is composed of forty-six members elected by the Plenipotentiary Conference with due regard to the need for equitable distribution of the seats on the Council among all regions of the world: the Americas, Western Europe, Eastern Europe and Northern Asia, Africa, Asia and Australia.

The role of the Council is to consider, in the interval between two plenipotentiary conferences, broad telecommunication policy issues in order to ensure that the Union's policies and strategy fully respond to the constantly changing telecommunication environment. In addition, the Council is responsible for ensuring the efficient coordination of the work of the Union and for exercising effective financial control over the General Secretariat and the three Sectors.

World Conferences on International Telecommunications World Conferences on International Telecommunications are empowered to revise Telecommunications Regulations. They establish the general principles which relate to the provision and operation of international telecommunications services offered to the public as well as the underlying international telecommunication transport means used to provide such services. They also set the rules applicable to administrations and operators with respect to international telecommunications.

World Conferences on International Telecommunications are open to all ITU Member Administrations and to the United Nations and its specialized agencies, regional telecommunication organizations, intergovernmental organizations operating satellite systems and the International Atomic Energy Agency.

General Secretariat In addition to handling all the administrative and financial aspects of the Union's activities, including provision of computer services, the work of the General Secretariat essentially covers:

- Publication and distribution of information on telecommunication matters.
- Organization and provision of logistical support to the Union's conferences.
- Coordination of the work of the Union with the United Nations and other international organizations.
- Public relations, including relations with members, industry, users, press, and academia.
- Organization of the World and Regional TELECOM Exhibitions and Forums.
- Electronic information exchange and access to ITU documents, publications and databases

Radiocommunication Sector The role of the Radiocommunication Sector is to ensure the rational, equitable, efficient and economical use of the radio-frequency spectrum by all radiocommunication services, including those using the geostationary-satellite orbit, and carry out studies without limit of frequency range on the basis of which recommendations are adopted. Subjects covered include:

- Spectrum utilization and monitoring
- Inter-service sharing and compatibility
- Science services
- Radio wave propagation
- Fixed-satellite service
- Fixed service
- Mobile services
- Sound broadcasting
- Television broadcasting

The Radiocommunication Sector operates through Radio Conferences held every two years along with a Radiocommunication Assembly supported by study groups (legislative functions), an Advisory Group (strategic advice), and a Bureau headed by a Director (administrative functions).

Telecommunication Standardization Sector The Telecommunication Standardization Sector studies technical, operating, and tariff questions and issues recommendations on them with a view to standardizing telecommunications on a world-wide basis, including recommendations on interconnection of radio systems in public telecommunication networks and on the performance required for these interconnections. Technical or operating questions specifically related to radiocommunication come within the purview of the Radiocommunication Sector.

The Telecommunication Standardization Sector operates through World Telecommunication Standardization Conferences supported by study groups (legislative), an Advisory Group on Standardization (strategic advice), and a Standardization Bureau headed by a Director (administrative).

Telecommunications Standardization Study Groups are groups of experts in which administrations and public/private sector entities participate. Their focus of work is on standardization of telecommunication services, operation, performance and maintenance of equipment, systems, networks and services, tariffs principles and accounting methods.

Although they are not binding, ITU Recommendations are generally complied with because they guarantee the interconnectivity of networks and technically enable services to be provided on a worldwide scale. Activities of the telecommunication standardization sector cover:

- Telecommunication services and network operation
- Telecommunication tariffs and accounting principles

- Maintenance
- Protection of outside plant
- Data communication
- Terminal for telematic services
- Switching, signaling and man-machine language
- Transmission performance, systems and equipment
- ISDN

Telecommunications standardization conferences are held every four years. An additional conference may be held at the request of one quarter of the membership, provided a majority of the Members agree. Telecommunications Standardization Conferences approve, modify or reject draft standards (called Recommendations because of their voluntary character) and approve the program of work. On that basis, they also decide which study groups to maintain, set up or abolish.

Telecommunication Development Sector   The role of the Telecommunication Development Sector is to discharge the Union's dual responsibility as a specialized agency of the United Nations and executing agency for implementing projects under the United Nations development system or other funding arrangements. The aim is to facilitate and enhance telecommunications development by offering, organizing, and coordinating technical cooperation and assistance activities. The objectives of the Telecommunication Development Sector are to:

- Raise the level of awareness of decision-makers concerning the important role of telecommunications in the national economic and social development program, and provide information and advice on possible policy and structural options.
- Promote the development, expansion and operation of telecommunication networks and services, particularly in developing countries.
- Enhance the growth of telecommunications through cooperation with regional telecommunications organizations and with global and regional development financing institutions.
- Activate the mobilization of resources to provide assistance in the field of telecommunications to developing countries by promoting the establishment of preferential and favorable lines of credit, and cooperating with international and regional financial and development institutions.

- Promote and coordinate programs to accelerate the transfer of appropriate technologies to the developing countries in the light of changes and developments in the networks of the developed countries.

- Encourage participation by industry in telecommunication development in developing countries, and offer advice on the choice and transfer of appropriate technology.

- Offer advice, carry out or sponsor studies, as necessary, on technical, economic, financial, managerial, regulatory and policy issues, including studies of specific projects in the field of telecommunications.

Summary

The ITU endeavors to respond quickly to the requirements of emerging services and market expectations. By serving as the focal point for coordination with other organizations, forums and consortia worldwide, consumers are eventually provided with access to an increasing range of interoperable products and services. At the same time, the risk of market chaos is greatly reduced, which benefits the economies of all countries.

See Also

American National Standards Institute (ANSI)

Bellcore

Federal Communications Commission (FCC)

Institute of Electrical and Electronics Engineers (IEEE)

Internet

The Internet consists of tens of thousands of interconnected data networks worldwide, all of which use the *Internet protocol* (IP). This "network of networks" has developed since 1969 largely without any central plan, and no single entity can control or speak for the entire system. The technology of the Internet enables new types of services to be layered on top of existing protocols. Numerous users can share physical facilities, the mix of traffic through any point changes constantly, and there are no firm borders that cleanly separate "local" from "long distance" as there are with the *public switched telephone network* (PSTN).

Figure I-10
Conceptual overview
of the Internet.

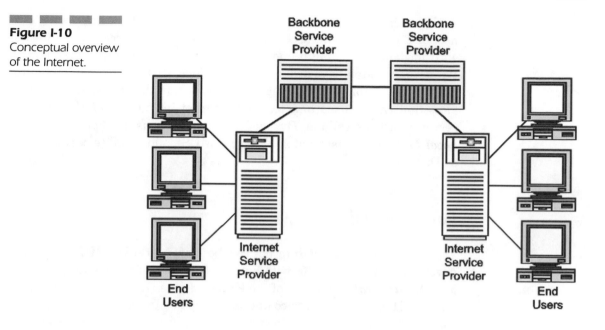

For purposes of understanding how the Internet works, three basic types of entities can be identified: end users, Internet service providers, and backbone providers. Figure I-10 shows the general relationships of these entities.

End users access and send information either through individual connections or through organizations such as universities and businesses. End users include both those who use the Internet primarily to receive information, and content creators who rely on the Internet to distribute information and applications. *Internet service providers* (ISPs), such as AT&T, MCI WorldCom's UUNET, and America Online, connect their subscribers to the Internet over their own backbone networks. The backbones of these providers meet at *Network Access Points* (NAPs), where they exchange traffic with each other as well as with many other smaller backbone providers.

Even the biggest backbone providers also serve as ISPs; for example, MCI WorldCom offers dial-up and dedicated Internet access to business and residential users, but also connects other ISPs and telecommunications carriers to its nationwide backbone. ISPs and backbone providers typically have multiple points of interconnection, and the interrelationships between these providers are changing over time. Since the Internet has no central access point, individual transmissions may be routed through multiple service providers to get data to its destination.

End users may access the Internet though several different types of connections. Most residential and small business users access the Internet with modems over dialup lines, cable, or *Digital Subscriber Line* (DSL). Larger users often have dedicated connections, ISDN or Frame Relay, providing their employees with high-speed access to the Internet. Other services that can be used to access the Internet include terrestrial wireless services such as cellular *Personal Communications Services* (PCS) and *Local Multipoint Distribution Service* (LMDS), and satellite services such as DirecPC.

Internet History

The roots of the current Internet can be traced to ARPANET, a network developed in the late 1960s with funding from the *Advanced Research Projects Administration* (ARPA) of the United States Department of Defense.[4] ARPANET linked together computers at major universities and defense contractors, enabling researchers at those institutions to exchange data. As ARPANET grew during the 1970s and early 1980s, several similar networks were established, primarily between universities. The TCP/IP protocol was adopted as a standard to enable these networks, comprising of many different types of computers, to interconnect.

In the mid-1980s, the *National Science Foundation* (NSF) funded the establishment of NSFNET, a TCP/IP network that initially connected six NSF-funded national supercomputing centers at a data rate of 56Kbps. NSF subsequently awarded a contract to a partnership of Merit (one of the existing research networks), IBM, MCI, and the State of Michigan to upgrade NSFNET to T1 speed (1.544Mbps) and to interconnect additional research networks. The NSFNET backbone, completed in 1988, initially connected thirteen regional networks. Individual sites such as universities could connect to one of these regional networks, which then connected to NSFNET, so that the entire network was linked together in a hierarchical structure. Connections to the federally-subsidized NSFNET were generally free for the regional networks, but the regional networks typically charged smaller networks a flat monthly fee for their connections.

---

[4]Even before the involvement of the Department of Defense, the individuals most credited with the development of the protocols and architecture on which today's Internet is based are Vinton Cerf, Robert Kahn, Leonard Kleinrock and Lawrence Roberts.

The military portion of ARPANET was integrated into the Defense Data Network in the early 1980s, and the civilian ARPANET was taken out of service in 1990. But by that time NSFNET had supplanted ARPANET as a national backbone for an Internet of worldwide interconnected networks. In the late 1980s and early 1990s, NSFNET usage grew dramatically, jumping from 85 million packets in January 1988 to 37 billion packets in September 1993. The capacity of the NSFNET backbone was upgraded to handle this additional demand with the addition of T3 (45Mbps) lines.

In 1992, the NSF announced its intention to phase out federal support for the Internet backbone, and encouraged commercial entities to set up private backbones. Alternative backbones had already begun to develop because NSFNET's acceptable use policy, rooted in its academic and military background, did not permit the transport of commercial data. In the 1990s, the Internet expanded beyond universities and scientific sites to include businesses and individual users connecting through commercial ISPs and other types of online services.

Federal support for the NSFNET backbone ended in April 1995. The NSF handed over management of the four original NAPs to Ameritech (Chicago), Pacific Bell (Northern California), Sprint (New York) and MCI (Washington, D.C.). The NSF continues to fund the VBNS (Very High-Speed Backbone Network Service), a non-commercial research-oriented backbone that is sometimes referred to as Internet II.

Since termination of federal funding for the NSFNET backbone, the Internet has continued to evolve. The number of backbone providers has continued to grow, offering ever-greater bandwidth capacity. In addition, other exchange points have been created to handle the increasing traffic load on the Internet. Despite these developments, Internet usage has increased at an even faster pace.

Operating Characteristics

The fundamental operational characteristics of the Internet are that it is a distributed, interoperable, packet data network. A distributed network has no one central repository of information or control, but comprises an interconnected web of host computers, each of which can be accessed from virtually any point on the network. Routers throughout the network regulate the flow of data at each connection point and reroute data around points of congestion or failure.

An interoperable network uses open protocols so that many different types of networks and facilities can be transparently linked together, and enables multiple services to be provided to different users over the same network. The Internet can run over virtually any type of facility that can transmit data, including copper and fiber optic circuits of telephone companies, coaxial cable of cable companies, and various types of wireless connections. The Internet also interconnects users of thousands of different local and regional networks, using many different types of computers. The interoperability of the Internet is made possible by the TCP/IP protocol, which defines a common structure for Internet data and for the routing of that data through the network.

The data transmitted over the Internet is split up into small chunks, or packets. Unlike circuit-switched networks, such as the *public switched telephone network* (PSTN), a packet-switched network is connectionless. In other words, a dedicated end-to-end transmission path does not need to be opened for each transmission. Rather, each router calculates the best routing for a packet at a particular moment in time, given current traffic patterns, and sends the packet to the next router. Thus, even two packets from the same message may not travel the same physical path through the network. This mechanism is referred to as dynamic routing. When packets arrive at the destination point, they must be reassembled, and packets that do not arrive for whatever reason must generally be re-sent. This system enables network resources to be used more efficiently, as many different communications can be routed simultaneously over the same transmission facilities.

Addressing

When an end user sends information over the Internet, the data is first broken up into packets. Each of these packets includes a header which indicates the point from which the data originates and the point to which it is being sent, as well as other information. TCP/IP defines locations on the Internet through the use of IP numbers. These numbers include four address blocks consisting of numbers between 0 and 256, separated by periods (e.g. 160.130.0.252). Internet users generally do not need to specify the IP number of the destination site because IP numbers can be represented by alphanumeric domain names such as fcc.gov or ibm.com. Domain name servers throughout the network contain tables that cross-reference these domain names with their underlying IP numbers.

Some top-level domains (such as .uk for Britain) are country-specific; others (such as .com) are generic and have no geographical designation. The *domain name system* (DNS) was originally run by the United States Department of Defense, through private contractors. In 1993, responsibility for non-governmental registration of generic domains was handed over to the NSF. The NSF established an exclusive agreement with *Network Solutions Inc.* (NSI), under which NSI handles registration. NSI currently charges $70 for a two-year domain name registration (Figure I-11). The exclusive agreement ended in 1998. Today, there is competition and NSI is one of several domain name registries. Country-specific domains outside the United States are generally handled by registration entities within those countries.

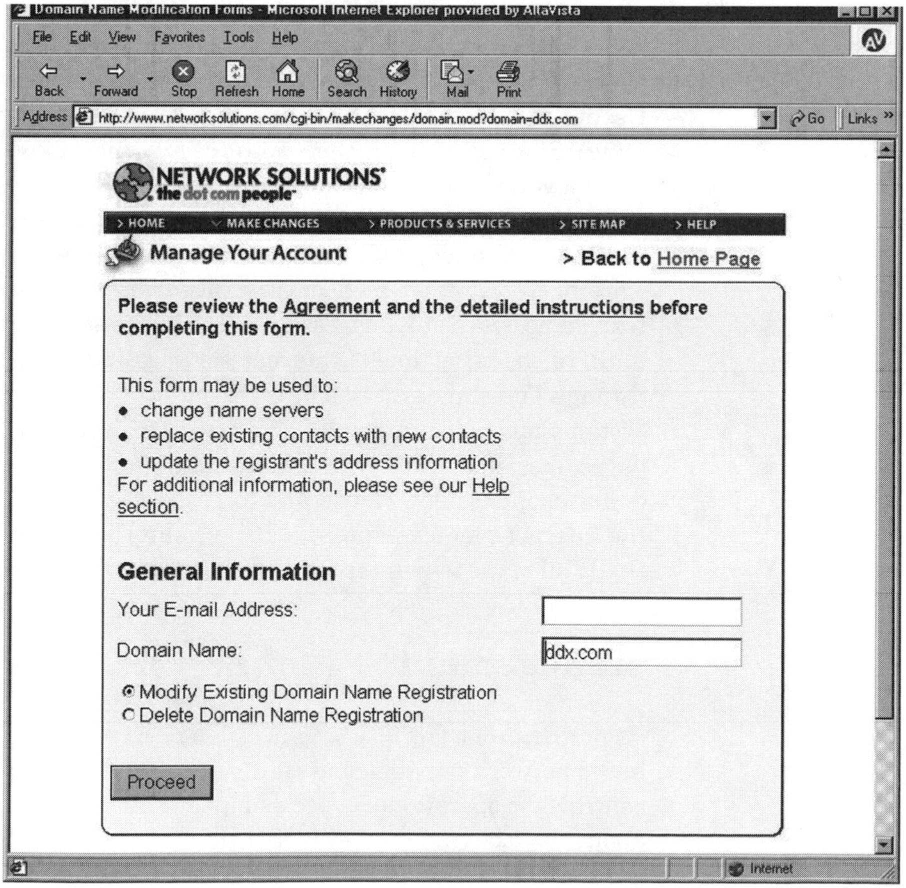

Figure I-11
NSI offers users the ability to manage their domain name account on the Web. This form may be used to: change name servers, replace existing contacts with new contacts, and update the registrant's address information.

Services on the Internet

The actual services provided to end users through the Internet are defined not through the routing mechanisms of TCP/IP, but depend instead on higher-level application protocols, such as *hypertext transport protocol* (HTTP); *file transfer protocol* (FTP); network news transport protocol (NNTP), and *simple mail transfer protocol* (SMTP). Because these protocols are not embedded in the Internet itself, a new application-layer protocol can be operated over the Internet through as little as one server that transmits the data in the proper format. The utility of a service to users, however, increases as the number of servers that provide that service increases.

By the late 1980s, the primary Internet services included e-mail, Telnet, FTP, and Usenet news. E-mail, which is still the most popular Internet service, enables users to send text-based messages to each other using a common addressing system. Telnet enables Internet users to log into a host and access information and applications from a remote location. FTP enables users to download files from a remote host computer onto their own system. Usenet newsgroups enable users to post and review messages on specific topics. Since 1995, with the advent of graphical browsers, the *World Wide Web* (WWW) has become one of the most utilized services on the Internet.

The Web has two primary features that make it a powerful, full service method of accessing information through the Internet. First, the client software, or Web browsers, can access multimedia information—a combination of text, audio, video and images embedded in the same file—plus provide access to all of the other major Internet services such as FTP, e-mail, and news through one standard interface. Second, the Web incorporates a hypertext system that enables individual Web pages to provide direct links to other Web pages, files, and other types of information. Thus, complex services such as online shopping, news feeds, and interactive games can be provided through the Internet over a non-proprietary system. The Web is the foundation for virtually all of the new Internet-based services currently being developed.

Management

As noted, no one entity or organization governs the Internet. Each facilities-based network provider that is interconnected with the global Internet only controls the operational aspects of its own network. No one can even be sure about the exact amount of traffic that passes across the Internet, because each backbone provider can only account for their own traffic and there is no central mechanism for these providers to aggregate their data.

Despite all this, the Internet does not operate in an environment of pure chaos. Certain functions, such as domain name routing, the issuing of IP addresses, and the definition of the TCP/IP protocol, must be coordinated, or traffic would never be able to pass seamlessly between different networks. With tens of thousands of different networks involved, it would be impossible to ensure technical and administrative compatibility if each network provider had to separately coordinate such issues with all other network providers.

These coordinating functions have traditionally been performed by an array of quasi-governmental, intergovernmental, and non-governmental bodies. The United States government, in many cases, has handed over responsibilities to these bodies through contractual or other arrangements. In other cases, entities have simply emerged to address areas of need.

The broadest of these organizations is the *Internet Society* (ISOC), a non-profit professional organization founded in 1992. ISOC organizes working groups and conferences, and coordinates some of the efforts of other Internet administrative bodies. Internet standards and protocols are developed primarily by the *Internet Engineering Task Force* (IETF), an open international body mostly comprising volunteers. The work of the IETF is coordinated by the *Internet Engineering Steering Group* (IESG), and the *Internet Architecture Board* (IAB), which are affiliated with ISOC. The *Internet Assigned Numbers Authority* (IANA) manages the root of the DNS to promote stability and robustness.

Internet Trends

Estimates from numerous sources suggest that as many as half a billion people worldwide will be regular users of the Internet by year-end 2001. As the Internet grows, methods of accessing the Internet will also expand and fuel further growth. Today, most users access the Internet through universities, corporate sites, dedicated ISPs, or consumer online services. Telephone companies, whose financial resources and network facilities dwarf those of most existing ISPs, have also entered the Internet access market and are serving businesses and residential customers.

At the same time as these new access technologies are being developed, new Internet clients are also entering the marketplace. Low-cost Internet devices such as WebTV and its competitors enable users to access Internet services through an ordinary television at far less cost than a personal computer. Various other devices, including *network computers* (NCs) for business users, and Internet-capable video game stations for consumers, are now routinely connected to the Internet. There are now services on the

Internet for mobile users, enabling them to access content through their cellular phones or *personal digital assistants* (PDAs). In the near future, household appliances and environmental systems will be connected to the Internet, reporting problems via e-mail to their owners or to service firms.

An important trend in recent years has been the growth of intranets and other corporate applications. Intranets are internal corporate networks that use the TCP/IP protocol of the Internet. These networks are typically connected to the public Internet through firewalls that enable corporate users to access the Internet, but prevent outside users from accessing information on the corporate network.

Summary

Limited government intervention is a major reason the Internet has grown so rapidly in the United States. The Telecommunications Act of 1996 adopts such a position. The 1996 Act states that it is the policy of the United States "to preserve the vibrant and competitive free market that presently exists for the Internet and other interactive computer services, unfettered by Federal or State regulation," and the FCC has a responsibility to implement that statute. To date, electronic commerce transactions over the Internet have been exempt from taxation in order to enable enough time for the full potential of the Internet to be realized.

See Also

  Electronic Commerce

  Electronic Mail

  Firewalls

  Internet-Enabled Mobile Phones

  Internet Facsimile

  Intranets

  TCP/IP

  World Wide Web (WWW)

Internet-Enabled Mobile Phones

Internet-enabled mobile phones potentially represent an important communications milestone, providing users with access to Web content and applications, including the ability to participate in electronic commerce transactions, delivered through the *Wireless Application Protocol* (WAP).

WAP is an open, global specification that empowers mobile users with wireless devices to easily access and interact with information and services instantly. The first beneficiary of WAP-enabled business applications is expected to be the financial services industry, an early adopter of e-commerce. Because of its dynamic nature, WAP is expected to become a key delivery channel for the secure execution of transactions on the move and the retrieval of ever-changing information, such as bank account details. Banks will be able to provide mobile financial services, including balance inquiries and fund transfers. Future e-commerce services include ticket purchasing and stock trading.

WAP also enables mobile phone users to access news services to retrieve the latest general news, sports news and financial news. Other types of information that will be immediately accessible to mobile phone users include traffic reports, weather forecasts, and airline schedules. Users can also personalize these services by creating a profile, which might request updated stock quotes every half-hour, or specify tastes in music and food. A user could also set up predefined locations, such as home, main office or transit so that the information is relevant for that time and location.

With access to real-time traffic information, for example, users can obtain route guidance on their cell phone screens via the Internet using WAP. Up-to-the-minute road conditions are displayed directly on the cell phone screen. Street-by-street guidance is provided for navigating by car, subway or simply walking—taking into account traffic congestion to work out the best itinerary. Such services can even locate, and guide users to, the nearest facilities such as free parking lots or open gas stations using either an address entered on the phone keypad or automatic positioning techniques.

One vendor that has been particularly active in developing WAP-compliant Internet-enabled mobile phones is Nokia, the world's biggest maker of mobile phones. The company's model 7110 works only on GSM 900 and GSM 1800 in Europe and Asia, but is indicative of the types of new mobile phones that about 70 other manufacturers are targeting at the world's 200 million cellular subscribers. It displays Internet-based information on the same screen used for voice functions. It also supports SMS and e-mail, and includes a calendar and phone book, as well.

The phone's memory can also save up to 500 messages—SMS or e-mail —sorted in various folders such as the inbox, outbox, or specially defined folders. The phone book has enough memory for up to 1000 names, with up to five phone and fax numbers and two addresses for each entry. The user can mark each number and name with a different icon to signify home or office phone, fax number or e-mail address, for example. The phone's built-in calendar can be viewed by day, week or month, showing details of the user's schedule and calendar notes for the day. The week view shows icons for the jobs the user has to do each day. Up to 660 notes in the calendar can be stored in the phone's memory.

Nokia has developed several innovative features to make it faster and easier to access Internet information using a mobile phone:

- *Large display*   the screen has 65 rows of 96 pixels (Figure I-12), enabling it to show large and small fonts, bold or regular, as well as full graphics.

- *Microbrowser*   like a browser on the Internet, enables the user to find information on the Internet by entering a few words to launch a search. When a site of interest is found, its address can be saved in a "favorites" folder, or input using the keypad.

- *Navi Roller*   this built-in mouse looks like a roller (Figure I-13) that is manipulated up and down with a finger to scroll and select items from

Figure I-12
Display screen of the Nokia 7110.

Figure I-13
Close-up of the Navi Roller on the Nokia 7110.

an application menu. In each situation, the Navi Roller knows what to do when it is clicked—select, save, or send.

■ *Predictive text input*   as the user presses various keys to spell words, a built-in dictionary continually compares the word in progress with the words in the database. It selects the most likely word to minimize the need to continue spelling out the word. If there are several word possibilities, the user selects the right one using the Navi Roller. New names and words can be input into the phone's dictionary.

However, the Nokia phone cannot be used to access just any Web site. It can only access Web sites that have been developed using WAP-compliant tools. The WAP standard includes its own *Wireless Markup Language* (WML), which is a simple version of the *Hypertext Markup Language* (HTML) that is widely used for developing Web content. The strength of WAP is that it spans multiple airlink standards and, in the true Internet tradition, enables content publishers and application developers to be unconcerned about the specific delivery mechanism.

Summary

The idea behind Internet-enabled mobile phones is not Web browsing, but accessing services with content that is specifically tailored to the needs of mobile users. This is where WAP comes in; it provides the means to convey this kind of information wirelessly by streamlining the content for proper rendering on cell phone displays. The information services cost between $5 and $10 a month for most users, but carriers could offer them free to customers on high-usage plans.

See Also

Cellular Data Communications

Cellular Voice Communications

Personal Communications Services (PCS)

Internet Facsimile

To help contain telecommunications costs, companies and individuals are leveraging their existing Internet connections to fax documents. There are several ways to send faxes over the Internet: subscribe to a commercial mail-to-fax gateway service, use a technique called remote printing, buy Windows 95 application software, devise a do-it-yourself method, or sign up for a free service.

Mail-to-Fax Gateways

With a mail-to-fax gateway service, subscribers send faxes as they would e-mail. This type of service is of particular value to users in other countries because it enables them to bypass the often costly PTT networks to send faxes to the United States using their existing Internet connections. The service can save users in other countries as much as 80 percent on faxes to the United States.

The service provider's gateway accepts e-mail from the sender with an attachment that contains an image of the document. Attachment support varies among service providers. Most support PostScript, HTML, TIF, JPG and GIF files as attachments. The gateway then routes the document to an Internet server closest to the recipient. The server dials the local number of the recipient's fax machine to deliver the document.

Internet delays are minimized by the architecture of the service provider's network (Figure I-14). After the initial transmission from the sender's desktop to the nearest gateway on the Internet, the service provider uses multiple types of networks as well as the Internet to dynamically select the best overall route for delivering each fax. The use of multiple networks gives the service provider broad geographical coverage.

To send e-mail to the service provider, the sender enters the fax number of the remote device in the "To:" field of the e-mail program using the following format:

```
faxnumber@faxnet.com
```

where `faxnumber` is the telephone number of the remote fax machine (including country code and area code, if any) and `faxnet` is the name of the service provider.

As documents traverse the Internet, confidentiality is assured by encryption supplied by the service provider. At the receiving fax machine, the document arrives just as it would from any other fax machine. When the fax is successfully delivered, the service provider sends back a delivery notice stating the time the fax was received, the e-mail address of the sender and receiver, the subject, number of pages delivered, delivery attempts, called

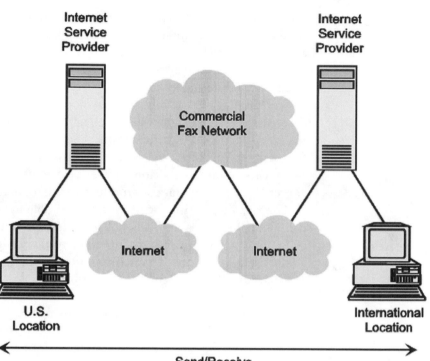

Figure I-14
Architecture of a
typical fax network.
The network links to
the Internet for
broader reach and
cost savings for users,
avoiding expensive
PTT service charges
normally associated
with fax
transmissions.

fax machine identifier, and a fax document reference number. If delivery cannot be completed, an error message is returned to the sender via e-mail explaining the reason.

Some service providers offer a resend capability, in case the remote fax machine is busy; off-line fax queuing, enabling the sender to assemble multiple faxes for transmission at a convenient time; a log of sent, delivered and undelivered faxes; and the ability to import address book information from other applications. A broadcast capability enables subscribers to send the same document to multiple recipients listed in a group name on file with the service provider. The document merges with information from the broadcast fax group before the service provider routes it to all of the listed fax numbers.

Remote Printing

Mail-to-fax gateway services are enhanced versions of the original experiment in remote printing, which has been available to Internet users on a limited basis since mid-1993.[5] The purpose of the experiment is to integrate the e-mail and facsimile communities, providing a way for e-mail users to send documents to the fax machines of people who do not have e-mail. The arrangement is called "remote printing" because the remote fax machine prints out the document.

Working together, many servers cooperatively provide remote printing access to the international telephone network, enabling people to send faxes via e-mail to the calling areas of participating servers equipped with special fax spooling software. The general-purpose Internet e-mail infrastructure takes care of all the routing, delivering the document to the appropriate server (i.e., gateway) for distribution within a predefined calling area.

The remote printing facility is not available in as many locations as the commercial services and it is not intended for heavy usage. In fact, a site administrator may choose to impose a usage limit on a daily or monthly basis. Such limits are intended to balance the desire to encourage legitimate users with the need to prevent abuse.

Commercial services have no usage restrictions; they simply bill subscribers accordingly. But the remote printing facility is free and requires no software other than an e-mail program. For a list of locations, by country code and area code, served by the remote printing facility, address e-mail to:

```
tpc-coverage@town.hall.org
```

To take advantage of the remote printing facility, users send electronic mail to an address which includes the phone number associated with the target facsimile device. Using the *Domain Name System* (DNS), the Internet message-handling infrastructure routes the message to a remote printer server, which provides access to facsimile devices within a specified range. The message is imaged on the target remote printer (i.e., fax machine) and an acknowledgment is sent back to the sender of the message via e-mail.

To send a document via the remote printer facility, the following format is used on the "To:" line of the message:

```
remote-printer.Joe_Smith@14157772525.iddd.tpc.int
```

where `remote-printer` identifies the kind of access and `Joe_Smith` is the name of the recipient. Next comes the familiar @ sign, followed by the

---

[5]The two people who started the remote printing experiment are: Carl Malamud of the Internet Multicasting Service, a non-profit organization, and Marshall Rose of Dover Beach Consulting, Inc.

recipient's fax number. A dot separates the fax number from the acronym `iddd`, which stands for International Direct Dialing Designator. After the next dot comes `tpc.int`, which is the Internet subdomain.

With this information, the message will be routed to a remote printer server, which will transmit it as a fax to the recipient. After delivery, an acknowledgment message is sent back to the originator via e-mail.

Windows Fax Software

There are some inexpensive standalone alternatives to faxing over the Internet. Typically, these are installed as print drivers on desktop computers, enabling documents to be faxed from any Windows application. The fax software itself usually supports directories, off-line queuing, status messages, and the creation of cover sheets. Some enable received faxes to be redirected to another location. At the destination end, the fax is received via e-mail as an attachment.

Attachment support is usually limited with standalone products, and some of these products are not capable of reaching conventional fax machines unless the document goes through a mail-to-fax gateway service. The recipient opens his or her e-mail as usual, and uses an appropriate image viewer to read or print out the attachment.

Mainstream fax software, such as Symantec Corp.'s (formerly, Delrina) WinFax PRO (starting with version 7.5), are beginning to support fax transmissions over the Internet. After preparing the document for faxing, the user is given a choice of delivery methods—through a normal phone line or through the Internet. To the user, transmitting a fax appears to be the same for either method. However, behind the scenes the process differs. Once a user has chosen to send a fax over the Internet, WinFax PRO 7.5 compresses and encrypts the fax and sends it through a service provider called Net-Centric whose server uses intelligent routing algorithms to route the fax over the Internet at the lowest cost possible to the recipient. Throughout the entire process, WinFax PRO 7.5 provides real-time status to the user.

Do-It-Yourself

Anyone with an Internet connection, e-mail software, and a scanner can send faxes over the Internet without the aid of extra-cost mail-to-fax gateway services or special fax software. Documents are simply scanned, saved into a graphic format that can be opened by the recipient, and sent as e-mail attachments. This homegrown approach does not provide the bells and whistles of professional services and products, but the quality of the received faxes is the same.

Free Service

A relatively new way to send faxes over the Internet is to subscribe to a free service. For example, a free consumer service called eFax.com unites the 500 million global fax users with 100 million e-mail users. The company offers a fax number and free usage of that number to individuals, so they can receive private faxes at their own e-mail account. eFax.com targets millions of mobile professionals, small business users, and home office workers, who are seeking faster, more efficient, and more reliable ways to overcome the frustrations and hassles of trying to receive paper documents remotely.

Users go to www.efax.com to set up their personal fax account with minimal registration information. They get an eFax.com number instantly. That number is like any other fax number with an area code and seven digits. Once users give that number to associates, family or friends, they can receive paper faxes from any standard fax machine into their e-mail account. Each paper fax sent to an eFax.com number is automatically routed via the eFax.com Service Center. There, eFaxes are converted to digital form, compressed and password protected, then forwarded to e-mail addresses for fast downloads. Users have a confidential digital copy of the document without needing to wait next to a fax machine, paying lofty hotel charges, or worrying about late or incomplete deliveries, paper jams, or major backlogs at traditional fax machines. Small and home office users can eliminate costs and installation headaches of dedicated fax lines.

eFaxes look and work just like regular e-mails with attachments. Attachments are opened with an eFax viewer so users can read them on-screen, print them, or forward them to other e-mail addresses. The small, proprietary viewer is included with the first fax each customer receives to be installed on the customer's system. Most faxes sent using eFax.com are compressed two to three times more than standard digital documents, making the download process extremely fast. Like regular faxes, eFaxes maintain all original formatting, including text, graphics and handwritten notes or signatures.

Users also have the option to receive confidential voice-mail messages at their eFax number. They can listen to voice-mail messages on their phone or receive it by e-mail as an attachment and play it by opening the eFax Messenger software. Users can forward voice-mail messages to other eFax users. eFax Voice is bundled free with all eFax Services.

eFax hopes to get the bulk of its future revenue from upgrades. The company's eFax-Plus service costs subscribers $2.95 a month. In addition, the premium service enables users to convert faxes to text using OCR software, store up to 10MB of fax messages and to preview faxes before opening them.

Summary

Cost savings is the principle reason for turning to Internet-based fax solutions. Various methods are available to accomplish this. The choice will depend on several factors, including geographical reach, feature requirements, the number of faxes sent per month, cost per fax (if any), and ease of use.

See Also

Electronic Mail

Facsimile

Internet

Voice Mail

Intranets

An intranet is a private data network that enables an organization to leverage the benefits of the public Internet, simply by using the same protocols and services as the public Internet, including those for e-mail, news, and the Web. Companies and other organizations build intranets for improving internal communication, distributing information among employees, and enabling more employees to access critical applications.

The foremost benefit that a company can derive from an intranet is more cost-effective communications. Attaining cost-effective communications entails making information directly accessible to the people who need it. Intranets provide direct access to information, so people can easily find what they need without involving anyone else, either for permission or direction on how to navigate through the information. At the same time, companies can protect their information from people not entitled to access it.

From the perspective of managing information, an intranet can extend the reach of distribution and simplify logistics. For example, it can become quite cumbersome to maintain the distribution list for a typical quarterly status report sent to a large mailing list. If anything changes before the next report is due, an update can easily be developed and posted on an intranet, giving everyone access to the new information.

Publishing information on an intranet is quite simple, especially since intranets use the same protocols as the greater Internet, including the *Hypertext Transport Protocol* (HTTP), which is used for information retrieval on the Web. There are a number of Web publishing tools available that quickly turn individual documents into the *Hypertext Markup Language* (HTML) format.

Not only has Web publishing become much easier, but users are finding it easier to search for the right document just by using key words. It is no longer necessary for employees to ask someone for copies of a document or request that their names be put on a distribution list. There is also more control over what is seen. If the big picture is all a person wants, then only that level of information is delivered. However, if detailed information is desired, such search mechanisms as Boolean parameters, context-sensitivity, and fuzzy logic may be employed. Of course, the user also has the option of accessing greater levels of detail by following the hypertext links embedded in documents.

Of note is that intranets may introduce new chores in managing information. For example, ensuring all departments have the same updated versions of information requires synchronization across separate departmental servers, including directories and security mechanisms. Providing varying levels of information access to different audiences—engineering, manufacturing, marketing, human resources, suppliers, customers—is also an issue. And although hypertext links facilitate information search and retrieval, the links must be maintained to ensure integrity as information changes or is added to the database. Fortunately, there are tools that database administrators can use to identify broken links so appropriate corrective steps can be taken.

Enhanced, timely information exchange within the organization is another benefit of intranet adoption. As people from various organizations, functions and geographic locations increasingly work together, the need for real-time collaboration becomes paramount. Teams need to share information, review and edit documents, incorporate feedback, as well as reuse and consolidate prior work efforts. Intranets that enable collaboration to occur without paper or copies of files can save hours and even days in a project schedule.

Electronic collaboration eliminates many hurdles such problems as distance between co-workers, multiple versions and paper copies of information, and the need to integrate different work efforts. The intranet becomes a unifying communications infrastructure that greatly simplifies system management tasks and makes it easy to switch between internal and external communications.

Companies are also finding that, by extending their intranets beyond their immediate boundaries, they are able to communicate more directly and efficiently with the communities with which they do business. Establishing "shared space" via these extranets with suppliers and other strategic partners, for example, can result in key savings in time and money in communicating inventory levels, tracking orders, announcing new products, and providing ongoing support. Intranets are being used in a range of application areas to leverage access to existing information and extend a company's reach to employees, partners, suppliers, and customers.

As companies progress in such methods of interaction, more sophisticated intra-networking can result in increased responsiveness and shortened order fulfillment time to customers. Suppliers track inventory levels directly, reduce delays in order fulfillment, and save costs of maintaining inventory. More companies are developing this form of information exchange, where the electronic capability actually drives the process.

Security

Increasing the number of people who have access to important data or systems can make a company's information technology infrastructure vulnerable to attack if precautions are not taken to protect it. Integrating security mechanisms into an intranet minimizes exposure to misuse of corporate data and to overall system integrity. A secure intranet solution implies seamless and consistent security function integrated between desktop clients, application servers, and distributed networks. It should include policies and procedures, the ability to monitor and enforce them, as well as robust software security tools that work well together and do not leave any gaps in protection.

The following basic security measures are typically taken to ensure broad coverage, while protecting corporate resources from potential harm:

- Access control via user IDs and passwords enables varying degrees of access to applications and data.
- Secure transmission mechanisms like encryption impede outside parties from eavesdropping or changing data sent over a network.
- Authentication software validates that the information appearing to have been originated and sent by a particular individual was actually sent by that person.

■ Repudiation software prevents people who have bought merchandise or services over the network from claiming they never ordered what they received.

■ Disaster recovery software and procedures assist in recovering data from a server that experiences a major fault.

■ Anti-virus software detects and removes hostile code before it can cause damage.

Intranets that extend beyond organizational or company boundaries, may require integration among various security systems. In addition, special firewall software may be required to prevent attacks from malicious hackers on the Internet.

Development

Costs are an important consideration when developing an intranet. Beyond the list prices for hardware and software components lie the less obvious costs of administration, maintenance, and additional applications development.

The skill sets that are required for developing an intranet are varied and quite specialized. They include technical people with a knowledge of system and network architectures, an understanding of the *Internet Protocol* (IP), and experience in developing applications with such tools as Java, ActiveX, and Perl. There is also a need for creative people, particularly graphic artists and HTML coders who excel at making the content visually compelling through the integration of images, audio, video and text.

The cumulative efforts of many people can go into the initial development and implementation of a corporate intranet. However, many of these people may be only peripherally involved. For example, the same network managers and technical staff that keep the division's network up and running by default keep the intranets up and running, since they may all run off the same server.

The daily maintenance of the two intranets may require only the part-time efforts of a few people from the marketing and technical groups. The caliber of skills these individuals bring to the task that makes all the difference, rather than the number of people.

While it takes people with specialized skills to develop an intranet, it takes a different set of skills to sustain it. Companies usually deal with this situation by recruiting multi-functional people—those who can apply what they normally do on the job to the medium of the intranet.

Extranets

A variation of the corporate intranet is the "extranet." This is a collaborative TCP/IP network that brings together suppliers, distributors, application developers, and customers to achieve common goals via the Web. The concept is totally different from a public Web site or intranet, which are focused around an individual organization's objectives. Among the activities conducted over an extranet are the delivery of product availability and pricing information, custom product configuration and price checking, real-time order entry, and order status inquiry.

Summary

Corporate intranets are becoming as significant to the telecommunications industry as the PC has become to the computer industry. It fundamentally changes the way people in large organizations interact with each other. In the process, intranets can improve employee productivity and customer response. Intranets are also being used to connect companies with their business partners, enabling them to collaborate in such vital areas as research and development, manufacturing, distribution, sales and service. A variety of tools are used for these purposes, including interactive text, audio and video conferencing, file sharing, and whiteboarding. In fact, anything that can be done on the public Internet can also be done on a private intranet—easily, economically, and securely.

See Also

> Firewalls
>
> Internet
>
> TCP/IP
>
> *World Wide Web* (WWW)

Inverse Multiplexers

Inverse multiplexers enable users to dial up appropriate increments of bandwidth to support a given application and pay for the number of local access channels only when they are set up and ready to send voice, data, or video traffic. On completion of the transmission, the channels are taken down and carrier billing stops. This method of access obviates the need for over-provisioning the corporate network to support temporary applications.

Inverse multiplexing can be implemented in *customer premises equipment* (CPE) or as a carrier-provided service. Either way, the advantages of inverse multiplexing include the immediate availability of extra bandwidth when needed, which eliminates the need for standby links that are billed to the user whether fully used or not. This adds up to significant cost savings.

The inverse multiplexer gathers data from a bandwidth-intensive application. The information is then divided among multiple 56/64Kbps or 384Kbps channels that are dialed up as needed and aggregated to achieve what is, in effect, a higher-speed link. The inverse multiplexer synchronizes the information across the channels and transmits it via switched public network services to a similar device at the remote location. There, the data is received as a single data stream (Figure I-15).

Some inverse multiplexers can be configured to support multiple applications simultaneously. For example, an inverse multiplexer can be used to link multiple applications at a single site to the public network via a T1 or ISDN access facility, link a PBX to a *virtual private network* (VPN), a router to a Fractional T1 network, or a video codec to a switched digital service. This capability appeals to users who want to spread the cost of a T1 access line across multiple applications. Some products enable users to switch multiple applications on a call-by-call basis over different carriers' services simultaneously. While some inverse multiplexers interface only to switched services, others can access both switched and dedicated communications facilities.

Another capability of some inverse multiplexers is the transport of bandwidth-intensive data across multiple T1 circuits to achieve a Fractional T3 circuit. T3-level inverse multiplexers are intended for applications that require transport between the T1 and T3 rates of 1.544Mbps and

Figure I-15

A simple inverse multiplexer configuration for a point-to-point videoconference or image transfer.

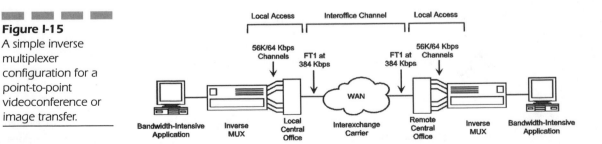

44.736Mbps. As many as 28 T1 circuits can be aggregated to achieve the desired bandwidth. Dynamic bandwidth capabilities enable users to add or delete T1 circuits based on application needs and traffic priorities. A user could send token-ring or Ethernet transmissions across the wide-area network at native speeds, for example.

System Management

The system management interface usually consists of a microcomputer equipped with software that enables the network manager to define and monitor traffic flow, bandwidth requirements, access line quality, and various configuration parameters.

Through this interface administrative functions are also performed, such as the creation of call profiles. A call profile is a file that contains the parameters of a particular data call so that a similar call can be quickly reestablished at another time simply by loading the call profile. Usually the call profile function includes a factory-loaded profile that acts as the template for creating and storing user-defined call profiles. Because each data call may involve as many as 25 separately configurable parameters, the use of call profiles can save a lot of time. Users typically load or edit a call profile using keyboard commands to the management software on the microcomputer.

Inverse multiplexer management interfaces often support remote devices. This capability enables a network administrator at a central location to configure, test, and otherwise manage other inverse multiplexers at remote locations in much the same way as is currently offered by the in-band management systems of some T1 multiplexers. This is accomplished by the management interface reserving a certain amount of the network bandwidth, usually not more than 2 percent, as a subchannel to implement remote management.

Most inverse multiplexers can be remotely monitored and controlled via SNMP. This is usually accomplished with SNMP agent software included with the product. The agent collects detailed error statistics, utilization ratios, and performance histories that can be retrieved for analysis.

Standards

The *Bandwidth On Demand Interoperability Group* (BONDING), formed in late 1991, has defined interoperability standards for inverse multiplexers.

There is also a set of international standards for bandwidth-on-demand services called *Global Bandwidth on Demand* (GloBanD).

The BONDING specification describes four modes of inverse multiplexer operation:

- *Mode 0*   Enables inverse multiplexers to receive two 56Kbps calls from a video codec and initiate dual 56Kbps calls to support a video conference.

- *Mode 1*   Enables inverse multiplexers to spread a high-speed data stream over multiple switched 56K/64Kbps circuits. Because this mode does not provide error checking, the inverse multiplexers operating in this mode have no way of knowing if one of the circuits in a multi-circuit call has failed. In this case, it is up to the receiving node to detect that it has not received the full amount of data and must request more bandwidth.

- *Mode 2*   Adds error checking to each 56K/64Kbps circuit by stealing 1.6 percent of the bandwidth from each circuit for the passage of information that detects circuit failures and re-establishes links.

- *Mode 3*   Uses out-of-band signaling for error checking, which may be derived from a separate dial-up circuit or the unused bandwidth of an existing circuit.

When establishing calls, inverse multiplexers at both ends first determine whether they can interoperate using the vendor's proprietary protocol. If not, this means that the inverse multiplexers of different vendors are being used and that they should use the BONDING protocol to support the transmission.

Summary

The inverse multiplexer enables network managers to match bandwidth to the application. These devices (or a carrier-provided service) provide a degree of configuration flexibility that cannot be matched in efficiency or economy using any other technology. With inverse multiplexers, organizations no longer have to over-provision their networks to handle peak traffic or run occasional high-bandwidth applications. Instead, they can order bandwidth only when it is needed and, in the process, save on line costs.

See Also

Multiplexers

Java

Java is a network programming language that was originally developed by Sun Microsystems. Currently, it is the fastest growing programming language for cross-platform networks, particularly those that call for the thin client model of computing. Java is actually a scaled down version of the C++ programming language that omits many seldom used features, while adding an object orientation. Java provides a cleaner, simpler language that can be processed faster and more efficiently than C or C++ on nearly any microprocessor.

Whereas C or C++ source code is optimized for a particular model of processor, Java source code is compiled into a universal format. It writes for a virtual machine in the form of simple binary instructions. Compiled byte-code is executed by a Java run-time interpreter, performing all the usual activities of a real processor, but within a safe, virtual environment instead of a particular computer platform. This allows the same Java applications to run on all platforms and networks, eliminating the need to "port" an application to different client platforms. In fact, Java applications can run anywhere the virtual machine software is installed, including any Java-enabled browser, such as Microsoft's Internet Explorer and Netscape Communications' Netscape Navigator.

The use of Java enables remote users, mobile professionals, and network managers to access corporate networks, systems and legacy data through Java applets that are downloaded from the server to the remote computer only when needed. An applet is a piece of a larger application that resides on the server. The function of the applet is to extend the capabilities of the larger application to the remote user. This is the fundamental principle of "network computing."

In most cases, the applets are stored in cache on a hard disk at the client location and in others, they are stored in cache memory. Either way, the applet does not take up permanent residence on the client machine. Since applets are delivered to the client only as needed and all software maintenance tasks are performed at the server, users are assured of access to the latest application release level. This not only saves on the cost of software, it permits companies to get away with cheaper computers, since every computer need not be equipped with the resources necessary to handle every conceivable application. At the same time, there is no sacrifice in the capabilities of users to do their work while away from the office.

Rapid Applications Development

The acceptance of Java has spawned a steady stream of visual development tools that aid in rapid application development (RAD). Among the second-generation RAD tools is Borland International's JBuilder. When opened, the tool displays the main window and AppBrowser from where the user can access all the usual development functions through three major panes: the Navigation pane, Content pane, and Structure pane (Figure J-1).

The navigation pane shows a list of projects with associated files, which may include Java, HTML, text or image files. The content and structure panes display information about the selected file. For example, if a Java file is selected, the structure pane shows such information as imported packages, the classes and/or interfaces in the file, any ancestor classes and/or interfaces, and variables and methods. With the AppBrowser in Project Browser mode, the user can manipulate the files in a project.

JBuilder includes an Object Gallery which contains shortcuts that create skeletal instances of many objects, letting the user quickly manufacture

Figure J-1
JBuilder's multi-paned AppBrowser, showing Navigation pane (upper left), Structure pane (bottom left) and Content pane (right).

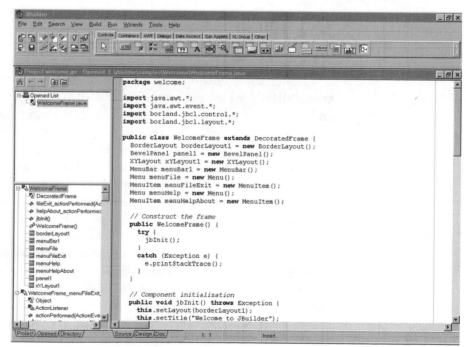

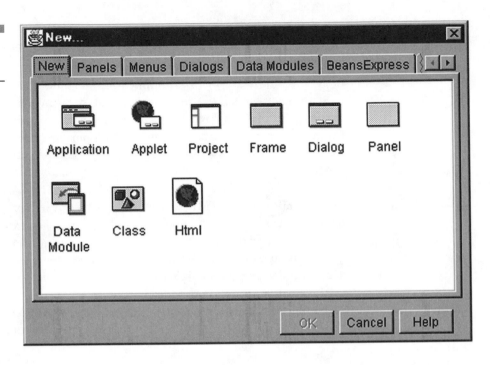

such things as applets, applications, frames, dialogs, panels, data modules, classes and HTML files (Figure J-2).

To guide the user through the major tasks necessary to create Java programs, JBuilder also comes equipped with wizards to create new projects and files and modify existing ones. Among the nine wizards is an application wizard that creates a new Java application shell containing a frame and an applet wizard which creates a minimal applet and HTML file containing the applet. There is a wizard that wraps an existing applet in a JavaBean class, making it look like a JavaBean to other applets and applications. The wizards can even work with existing projects to convert an application to an applet or vice versa. Other wizards can be added to JBuilder, as they become available.

JBuilder was designed with business application development needs in mind. In addition to Java BeansExpress for easy JavaBean creation and deployment, it includes drag-and-drop database components and tools, complete JDBC connectivity, more than 100 JavaBeans—including grid with source code, charting, numerous wizards, and command-line tools—and Local InterBase for offline SQL database development.

Network Management Applications

Many interconnect vendors are using Java for building network management applications that can be accessed through Web browsers. Through hypertext-linked Web pages set up by the vendor, network managers can use their Java-enabled Web browsers to access embedded applets to launch various network management applications. Routers, switches, hubs, multiplexers, CSU/DSUs—virtually any network device—can be configured and monitored in real-time from any location. Applets that provide trend analysis and network reports, access to the vendor's technical support and on-line documentation are also integrated through the Web browser so that configuration changes and network planning can be accomplished using real data instead of guesswork.

One such network management framework—NetDirector@Web from Newbridge Networks—integrates core services such as discovery, topology, and event management offered by open platforms, such as HP OpenView and IBM NetView, and provides distributed network directory services that can be exploited by applications for policy-based management.

The NetDirector Home Page provides a directory for the network, hyperlinking all of the company's VIVID family devices to simplify network navigation. The home page reflects the status of all discovered VIVID devices to show, at a glance, the health of the devices and other useful information such as firmware version and events. The network manager can manage the network from home or on the road by hot linking to the devices. The home page also provides a method for the administrator to specify management policies, such as upgrading firmware and software across multiple devices throughout the network, or defining network behavior in the event of a broadcast storm.

The Web-based applications are bundled with NetDirector, Newbridge Networks' enterprise management solution that integrates with HP Open-View on Solaris, HP-UX, Windows NT platforms.

Because of Java's real-time capabilities, changes in the network status are reflected immediately, without requiring the network manager to reload Web pages. Other Web management offerings only provide static HTML-based configuration reports for various network devices. In addition, Java applets are loaded dynamically from NetDirector@Web servers so that the user does not have to pre-install or continually update the network management software on the system being used to manage the network.

Among the Java-based applications that run under the NetDirector@Web framework is VitalStat, a network diagnostic and analysis tool. VitalStat analyzes baseline response times and other performance characteristics.

When deviations are detected, VitalStat diagnoses the problem, attempts to isolate whether the cause is application, server or network related, and recommends or initiates appropriate corrective actions. VitalStat has a Java-powered Web interface and provides anytime, anywhere management access via a standard Web browser. VitalStat uses intelligent agents that run in the network elements. As a result, it can follow the same path an end-user station uses to access a server in order to detect and diagnose problems. This enables more accurate problem determination for intelligent reporting back to the network administrator. While policies can be configured centrally and reports viewed from the VitalStat graphical user interface, the actual event detection, analysis and response can be addressed seamlessly by the agents themselves without requiring user intervention. The following Figures, J-3 through J-6, illustrate some of the reports available.

NetDirector@Web uses several levels of security to access the network from any location at any time. The first level of security is a firewall. As Net-Director@Web is targeted for management of an intranet within a firewall, the firewall prevents unauthorized users outside the intranet from accessing the network.

In addition, through an access list, administrators can define a list of allowed users/hosts that have access to various management functions. This type of security provides greater control than offered by SNMP community strings. Telnet, FTP and Web connections are secured through host access security and by user name and password authentication. Only the "root"

Figure J-3
Vital Stat shows the number of errors for selected servers at each node by time in minutes after the event.

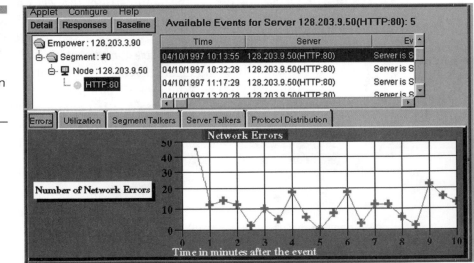

Figure J-4
VitalStat shows percent utilization for selected servers at each node. This information can be used for performance baselining.

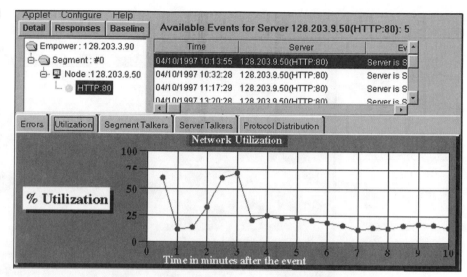

Figure J-5
VitalStat uses SNMP's RMON (remote monitoring) standard to identify a server's top talkers by MAC address in terms of kilobytes sent.

user with a valid password is given access to certain administrative functions. Java provides still another level of security, limiting the operating system and resources that the application can access. Java also has virus protection to prevent viruses from attaching themselves to the applications.

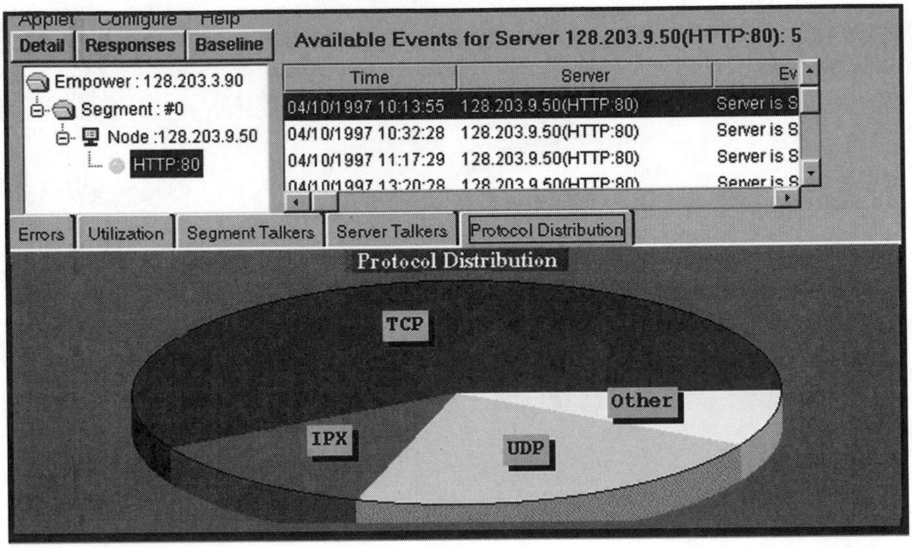

Figure J-6

VitalStat identifies how the traffic of a selected server is distributed by protocol. Among the protocols identified are IP, IPX, TCP, and UDP.

Summary

The early success of the C++ programming language owes a great deal to its ability to access legacy code written in C. Similarly, Java preserves much of C++ and offers a number of compelling benefits: it is object-oriented, portable, and relatively easy to master and maintain. Once written, Java applications and applets can run unchanged on any operating system that has a Java interpreter. These and other benefits of Java can greatly speed the development cycle for Web-based applications, including those for integrated computer telephony and remote network management. The applications themselves are accessed only when needed, with the most updated version downloaded to the client's cache as an applet. When the client disconnects from the network, the applet is flushed from the cache, conserving limited system resources. This is the basis for network computing, a new paradigm that, in essence, treats the Internet as the computer. Many network management vendors have already issued Java versions of their products, making management data readily accessible from any Java-enabled workstation in the enterprise.

See Also

Computer-Telephony Integration

Network Computing

Jitter

While delay is the time it takes to get a unit of information from source to destination through a network, jitter is the variance of the delay. Both can have potentially disrupting effects on applications running over the network, particularly if they are time-sensitive. Examples of time-sensitive applications are telephone calls and videoconferences. Some data applications are also time-sensitive, such as pages and text chat sessions.

In the past, delay and jitter were not important aspects of computer networks. For example, it did not matter if file transfers or e-mail took half a second longer, independent of the total transfer time (delay). Similarly, it did not matter if—on a particular file transfer—70 percent of the data was sent during the first half of the transfer and 30 percent in the last half (jitter).

But jitter and delay matter when it comes to two-way or multi-way conversations and conferences—they must have low delay and jitter to support the natural interaction among participants, since long pauses can be potentially disruptive to a conversation.

Multimedia applications that combine audio and video content are even more sensitive to delay and jitter. To prevent dropouts in an audio stream or jerkiness in video, jitter must be low. For one-way broadcasts, buffering can be used in the end-stations to decrease the effect of jitter, but only at the cost of increased delay. While this delay is acceptable for one-way broadcasts, it is not acceptable for two-way conversation.

The electrical pulses, which are sent through a network as indications of 0s and 1s, are normally sent at very specific intervals of time. The repeaters, bridges, and switches on the network contain buffers to accept the signal from the input side and send it to the output side at a tightly controlled speed. This results in a clean output signal, which can be received by the next piece of equipment on the network.

However, every piece of equipment has a narrow tolerance within which it operates. If a signal passes through several pieces of equipment or cable segments, these tolerances can add up and cause the resulting signal to shift in phase compared to what was originally sent. This makes it difficult or impossible for the next device on the network to lock on to the signal, which causes errors (Figure J-7).

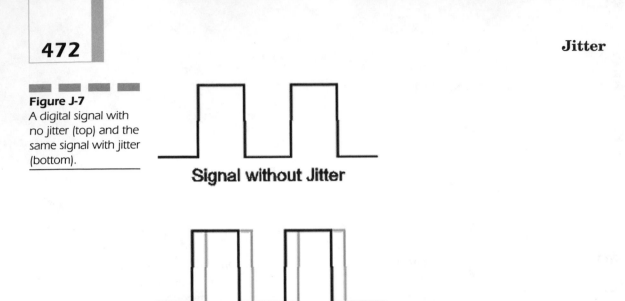

Figure J-7
A digital signal with
no jitter (top) and the
same signal with jitter
(bottom).

Summary

As clock speeds in computers and data rates in communications systems increase, timing budgets become tighter and the need to measure and characterize jitter becomes more critical. Oscilloscopes offer various tools to measure jitter. On new networks with no performance history, a device called a jitter generator can be used to add controlled jitter to digital signals for the purpose of stressing the connections to PSTN, cellular and PCS base stations, satellite modems, and microwave links. Not only does this ensure that the connections are error free before being turned over to user traffic, but once the jitter parameters are known, lower cost oscilloscopes can be used for periodic quality checks.

See Also

Attenuation

Latency

Key Telephone Systems

Traditionally, key telephone systems functioned like Private Branch Exchanges (PBXs), but offered only a subset of the available features. Being more economical than PBXs, small and mid-size businesses found them very appealing, especially for the key system's configuration flexibility, which allowed multiple lines to be terminated at each telephone. Within certain limits, each line can be answered from any multi-line telephone in the system by pressing the appropriate line key.

Since the introduction of the first key telephone system by Bell Telephone in 1938, key systems have evolved from cumbersome units with toggle switches to the sophisticated, electronic systems of today. Key systems have advanced to the point where they can be incrementally equipped with a wealth of call processing and system management features, rivaling those of higher-cost PBXs.

Evolution of Key Systems

The first standard key systems were electromechanical devices (designated 1A key telephones) that used adjunct key units and toggle switches for line control. These 1A systems had clearly defined system components and wiring schemes, and offered an attractive alternative to prior, custom-built telephone systems that lacked standardization.

In 1953, Bell replaced its 1A systems with 1A1 systems that packaged line control equipment into the telephone set. Line status indicators on the 1A1 telephones lit steadily to show a line in use, and flashed to indicate held calls. Current key system standards evolved from early 1A and 1A1 system designs.

In 1963, further technological improvements led to the introduction of 1A2 systems, which used plug-in modules for line control, signaling, and intercom functions. Inter-Tel offered the first solid state enhanced electromechanical 1A2 key system in 1975.

Electromechanical 1A2 key systems were a major improvement over the earlier 1A and 1A1 systems. The older systems were hardwired, but 1A2 systems were built around packaged components that made installation and maintenance much easier. These components included:

- *Key Service Unit*   a wall-mounted or freestanding cabinet that contains the key telephone units and circuit packs that control access to central office (CO) lines, Centrex lines, or PBX lines. The KSU also manages signaling units for private line, intercom, paging, and music-on-hold connections.

- *Key Telephone Units* support CO line connections and provide connections for individual stations and intercom paths, as well as paging and music-on-hold equipment within the system. KTUs also provide tone and ringing generation, lamp indicator control, and dialed digit detectors and decoders.

- *Main Distribution Frame* provides the primary connection points for the KSU and is the termination point for all cables and wires within the key system, including external CO lines and internal station lines.

- *Power Supply* converts commercial voltages to meet the system's unique operating requirements. The power supply can be housed within the KSU or in a separate cabinet near the commercial AC power source.

- *Station Equipment* consists of desktop phones and attendant stations. Each line key on an electromechanical key telephone requires a minimum of three wires for operation—one for the talk path, one for hold, and one for the line lamp.

Stored Program Control

While technical enhancements were continually added to 1A2 key systems, they were rapidly replaced by stored program control (SPC) electronic key systems, which were introduced in the mid-1970s. Electronic key systems are similar to 1A2 systems, but offer several improvements, including reduced cabling requirements, support for more features, and greater configuration flexibility.

The most significant difference is that electronic systems replace the relay-based KSU and KTUs used in electromechanical systems with an electronic KSU and printed circuit boards (PCBs). The electronic KSU architecture typically consists of a card-cage with a prewired backplane. Specialized PCBs plug into the backplane to control various system functions. The modular design of the electronic KSU facilitates the installation and removal of PCBs, allowing new features to be added to the system without wiring changes.

Electronic station sets also use specialized circuits for control. Whereas 1A2 station equipment requires several wire pairs for each key, electronic key systems require only two or three wire pairs per station, regardless of the number of keys. Some key systems also offer T1 and ISDN interfaces and support wireless technology for in-building mobile communication.

Station sets have increased in intelligence to the point that individual telephones can be programmed by users for personal features, such as speed dialing. Many station sets also include liquid crystal displays (LCDs) to identify the calling party, provide message notification, and guide users through feature implementation procedures.

Basic key system call handling features—including call forward, conference, speed-dial, and last number redial—are accessible from electronic station sets. Many sophisticated system features and functions are also accessible from the station keys, including station message detail recording (SMDR), automatic number identification (ANI), and least cost routing (LCR). Auxiliary jacks provide connections to answering machines, facsimile machines, and other peripheral devices.

Some key systems can be connected to an Automatic Call Distributor (ACD). A Windows-based application that runs on a PC displays a range of information necessary to help manage the workload and resources of a small call center. Such systems organize and display the ACD information reported by the telephone system's MIS port. Displays include the number of calls in queue, longest call waiting, number of agents logged-in, number of agents busy, number of agents available, number of agents unavailable, individual agent status and status times.

Attendant Console

Today's KTS attendant consoles operate as a Windows application on a PC, providing advanced easy-to-use features that enable the operator to accurately process calls quickly. Some key systems seamlessly integrate up to four attendant consoles, automatically distributing the load of incoming calls. Among the features that can be implemented at the console are:

- Answer button with answer priority and queuing
- Emergency call priority
- Programmable buttons that allow one-touch operation of the most frequently used features
- Automatic day/night mode switching
- Message center/voice mail transfer
- Call waiting count
- Loop hold display with timer

- Internal name/extension directory
- Outgoing speed dial directory
- Busy lamp field for checking station status at a glance
- Color-coded indicators that simplify monitoring of various status conditions

More key systems are now being programmed by PCs via a computer telephony interface (CTI). With open application programming interfaces (APIs), users can connect the key system with voice mail, call processing software, and other products that come from third-party manufacturers, including fax machines.

More key systems are being equipped to address the growing problem of toll fraud, giving the administrator the ability to program the system to automatically turn off whenever invalid attempts are made to access the system. The system can also be programmed to notify designated individuals via phone or pager when a toll fraud attempt is made.

Summary

Key systems have advanced to the point where they are now "KSU-less." Instead of a KSU with a central processor, KSU-less systems incorporate system and feature control directly within the telephone sets. The use of very large scale integration (VSLI) circuit technology eliminates the need for a separate equipment cabinet. Not only do non-KSU systems retain the multi-line capability, they can also be used with Centrex, allowing users to program one or two keys for Centrex features.

See Also

Automatic Call Distributors

Centrex

Private Branch Exchanges

Kiosks

Kiosks were introduced for a variety of applications in the mid-1980s, but failed to catch on in the U.S. until the recession of the late '80s and early

'90s. At that time, kiosks became a convenient and economical way for businesses to disseminate information about their products and services without burdening overworked staff.

However, the earliest kiosks merely deluged users with facts that were not necessarily relevant to their informational needs. There was very little interaction that let users specify the kind of information they wanted. There was no provision for graphics, let alone video, to hold the attention of users. And if the kiosk broke down, the operator would not know about it until a technician made a scheduled visit.

Today multimedia is being used not only to attract users to the kiosk, but once there, to keep their attention until the entire message is delivered. There is also more opportunity for user interaction, which is important from an entertainment perspective as well as for getting the message across. Kiosks now are a common fixture in hotels, airports, bus stations, convention centers, shopping malls, stores, travel stops, and hospitals.

Applications

At airports and train terminals, for example, kiosks are used to issue tickets. At the kiosk, travelers select an appropriate destination and time of departure and pay for the ticket by passing their credit card through the magnetic stripe reader. The ticket and boarding pass are issued at the kiosk via the integral printer. With ticket and boarding pass in hand, the passenger can immediately board the plane or train without having to wait on a long line at the ticket counter.

In a typical financial application, a bank would set up a kiosk in a convenient lobby location or other public access point to allow customers to see if they qualify for a car loan and how they might manage the payments over three, four or five years. Rather than take up a bank officer's time, customers would approach the kiosk and score themselves on various factors that lead to loan qualification based on answers they provide to a set of standard questions. Also available are the current interest rates and payment schedules. The user can then approach the bank officer with the completed loan application or comparison shop elsewhere.

Many banks are using kiosks for issuing statements and for allowing customers to make account inquiries. A bank cannot provide these types of services with an automatic teller machine (ATM) because the dozen or so people standing in line will not tolerate waiting that long.

Government agencies are also experiencing similar concerns. For example, just about every motor vehicle department in the country is being hamstrung by budget cuts and staff reductions, while the demand for its services continues to rise. This is forcing motor vehicle departments, for instance, to look at ways to streamline workflow and automate such routine services as license renewal and vehicle registration.

One way to do this is by locating kiosks at shopping malls and other public access points, where people can renew their licenses. The user selects "renewal" from the terminal display, allows a photo to be taken, pays the renewal fee with a credit card, and walks away with a temporary license in a matter of minutes. Via a modem connection, the motor vehicle department's host computer can collect information from the kiosks at the end of the day for processing. Within 10 days a permanent license, with the photo, is mailed to the driver.

Other government applications that lend themselves to the use of multimedia kiosks are the handling of court information and the payment of fines via credit card, as well as agency procedural and directory information.

System Components

The multimedia terminals combine video capabilities and touch-screen technology in ergonomic, kiosk packaging. Bundled into the kiosks are such components as central processor, credit card reader, optical disk, printer, loudspeaker, and proximity detector that automatically starts the application when someone comes within a few feet of the system. A custom-designed run-time program presents information in text, graphic or video formats for photo quality images or full-motion video displays.

Retrieving accumulated information input by users is a function of the application. Typically, a dialup connection will transfer all accumulated information stored on the kiosk's hard disk for further processing by the host (Figure K-1). Alternatively, the application could be designed to sort information into separate files based on inputs from users. That way, the host can be more selective in its retrieval request. For example, a file may contain the number of users who made a specific touch selection during the course of the day. This touch selection may indicate a product preference or a request for specific information by mail.

Network Capabilities

The kiosk can be equipped with a leased line modem, giving it the ability to be connected to the host for polling over a multidrop leased line. That way, accumulated information from each unit can be retrieved automatically upon request by the host.

At the same time, dialup or leased line connections enable the kiosks to be updated with new information from a central management facility. Instead of dispatching a technician to exchange CD-ROMs containing a new video presentation, for example, the master station can download the information to the hard disk in each kiosk. The video and images are compressed for transmission and storage using such standards as JPEG or MPEG.

Kiosks can be integrated into just about any type of host environment, including IBM 3270, SNA/SDLC, and UNIX. The system can be networked in the X.25 as well as in TCP/IP environment, or run over Ethernet and token ring LANs.

Remote diagnostics can be implemented over dialup lines via RS-232C interfaces. A dialup connection can be used to gather alarms that report an out-of-paper condition, a full hard disk, no touch screen activity, or that the central processor is down. The system's integral diagnostic functions tell an onsite technician or remote operator what components are not functioning properly.

Internet Kiosks

One of the newest innovations in kiosk technology is its integration with the Internet. Such kiosks are used to offer Web browsing, entertainment and e-commerce, making Internet services accessible to millions of people worldwide. Business travelers, vacationers, students and patrons of a wide variety of businesses are now able to get on-line with Internet kiosks located in restaurants, arcades, cafes, colleges, taverns, airports, bus stations, military bases, and hotels.

Such kiosks permit business owners to provide Web access on a pay-per-use basis to the public in a wide variety of commercial and public venues. Special software protects the unattended systems from hackers and computer viruses. A number of services are available through the kiosk, including:

- Sending and receiving email
- Internet shopping
- Financial services
- Web browsing and gaming
- Travel services and reservations
- Chat lines
- Sports statistics and scores
- Internet gambling
- Up-to-the-minute national and world news
- Stock market quotes and trading
- Internet telephony and teleconferencing

The features of the Internet kiosk can be mixed and matched, enabling businesses to build their own systems that offer specific services. As needs change, services can be added to the system by purchasing appropriate software modules.

Summary

Kiosks are used to assist businesses and government agencies with a variety of product and service support requirements, while offering consumers a convenient and private method of access. At the same time, kiosks can alleviate the workload of overburden staff, especially with regard to handling routine requests. When networked together through a central management facility, kiosks make an efficient and economical vehicle for the delivery of information and services. When connected to the Internet, kiosks can provide a multitude of services, including voice and video calls, at convenient public locations.

See Also

Multimedia Networking

Voice over IP

CHAPTER **L**

Local-Area Network (LAN) Telephony

Local-Area Network (LAN) telephony integrates voice and data over the same medium, enabling automated call distribution, voice mail, and interactive voice response, as well as voice calls and teleconferencing between workstations on a LAN. The benefit of LAN-based telephony is eliminating the costly, proprietary nature of a PBX and replacing it with a standards-based Ethernet/IP solution. By carrying voice conversations in the form of IP packets, local calls can traverse the Ethernet LAN while long-distance calls can go out to the wide-area, IP-based intranet. Through the use of IP/PSTN gateways, calls can even reach conventional telephones off the IP network.

With LAN telephony, users who are working away from their offices—at home or in a hotel, for example—can use a single phone line to carry both data and voice traffic. The user dials to access the corporate intranet, which would be equipped and engineered to carry real-time voice traffic. Such a system provides an integrated directory view, enabling remote users to locate individuals within the corporation for voice connection or e-mail connection in a unified way. Likewise, phone callers (internal or external to the corporation) can locate the mobile workers who are connected to any part of the intranet. Thus, LAN telephony enables users to work seamlessly from any location.

By using the LAN-based conferencing standards, transparent connectivity of different terminal equipment can be achieved. The media that is used by any conference participant would be limited only by what is supported by his or her terminal equipment. Connectivity to room-based conference systems or analog telephones can be achieved by gateways, which would perform the required protocol and media translations.

IP PBX

The PBX is a circuit switch that provides organizations with access to communications services and call-handling features. PBX sets up a communication path between the calling and called party, supervises the circuit for various events (e.g. answer, busy, and disconnect), and tears down the path when it is no longer required. In many ways, the PBX mimics a telephone company's central office switch, except that it is smaller in scale and is privately owned or leased.

PBXs that are based on IP can transport intra-office voice over an Ethernet LAN and wide-area voice over the PSTN or over a managed IP network. Full-featured digital phone sets link directly to the Ethernet LAN via a 10BaseT interface, without requiring connection to a desktop computer. Phone features can be configured by using a Web browser. Existing analog devices, such as phones and fax machines, can be linked to the LAN via a gateway. In addition to IP networks, calls can be placed or received by using T1, PRI ISDN, or traditional analog telephone lines.

All of the desktop devices have access to the calling features that are offered through the IP PBX management software that runs on a LAN server. The call-management software enables client devices on the network, such as phones and computers, to perform functions such as call hold, call transfer, call forward, call park, and calling-party ID. Even advanced PBX functions, such as multiple lines per phone or multiple phones per line, can be performed by the management software. The software also offers directory services. Unified messaging capabilities enable voice-mail messages to be sent to an existing electronic mailbox.

Standards

The building block of LAN telephony is the international H.323 standard, which specifies the visual telephone system and equipment for packet-switched networks. H.323 is an umbrella standard that covers a number of audio and video encoding standards. Among these standards is H.225 for formatting voice into packets. H.225 is based on the *Internet Engineering Task Force*'s (IETF) *Real-Time Protocol* (RTP) specification and the H.245 protocol for capability exchange between workstations.

On the sending side, uncompressed audio/video information is passed to the encoders by the drivers, then is given to the audio/video application program. For transmission, the information is passed to the terminal-management application, which might be the same as the audio/video application. The media streams are carried over RTP/UDP, and call control is performed by using H.225-H.245/TCP.

Gateways provide the interoperability between H.323 and the *Public-Switched Telephone Network* (PSTN), as well as between networks that are running other teleconferencing standards such as H.320 for ISDN, H.324 for voice, and H.310/H.321 for ATM. An example H.323 deployment scenario involves H.323 terminals that are interconnected in the same local area by a switched LAN. Access to remote sites is provided by gateways, routers, or integrated gateway/router devices. The gateways provide communication with H.320 and H.324 terminals that are remotely connected to the ISDN and PSTN, respectively. H.323-to-H.323 communication between two remote sites can be achieved by using routers that directly carry IP traffic over the *Point-to-Point* Protocol (PPP) that runs on ISDN. For better channel efficiency, gateways can translate H.323 streams into H.320 to be carried over ISDN lines (and vice versa).

Vendor Implementation

NBX Corporation, now a unit of 3Com, is among the growing number of vendors that offer telephony systems that operate over LANs. The company's NBX 100 Communication System leverages the ubiquity of Ethernet. The NBX 100 connects to the LAN infrastructure in the same manner as PCs, printers, servers, and other network devices. The NBX 100 employs packet technology that delivers high-quality voice reliably over the Ethernet. The system includes a network call processor for call control, voice mail, auto attendant, browser administration, and connectivity to the PSTN and WAN/Internet.

With connectivity to the WAN/Internet, the NBX 100 also supports the growing demand for voice communication from telecommuters and branch offices. Using a standard multi-protocol router, remote users can gain access to the full suite of NBX 100 voice communication features—just as if they were attached to the LAN at the office headquarters. Simultaneously, remote users also gain access to all of the resources on their company's LAN, including e-mail, file servers, and intranet/Internet access.

At this writing, 3Com is bringing LAN telephony to large enterprises. Its NBX technology is being extended to its workgroup and core network switches. The company's SuperStack II switches support up to 750 users per switch, enabling multiple SuperStack IIs to handle thousands of users in hundreds of locations via a distributed architecture. Eventually, 3Com will enhance its CoreBuilder 9000 with voice capabilities, as well. On the WAN, 3Com plans to help enterprises replace expensive voice-only trunking with converged voice/data networks. Its PathBuilder switches—which already support voice-over ATM, voice-over Frame Relay, and Voice-Over IP —will also provide voice-over *Digital Subscriber Line* (DSL).

Other vendors are aggressively pursuing the market for LAN telephony. For example, 3Com's nemesis, Cisco Systems, acquired Selsius Systems, which offers an IP-based call-processing system. The Selsius family of IP phones includes a 12-button and 30-button phone, a 12-button and 30-button model with speakerphone, and a 12-button and 30-button model with display and speakerphone. All of the phones support G.711 mu-law or G.723 voice encoding. Automated device installation enables a LAN administrator to plug a phone into the IP network and to have the device automatically acquire an IP address, register with the system's CallManager, and download a configuration template and available directory number. The IP PBX can be accessed through a Web browser from anywhere in the world for remote management and diagnostics. The system comes with an interface to the Windows NT Event Viewer, enabling network managers to view system events. Call-detail records are available in a flat-file, comma-delimited format that can be imported into other software programs for reporting purposes.

Summary

A LAN-based PBX eliminates the need for IP telephony software to be loaded on each client PC, in addition to enabling organizations to avoid having to set up and manage separate LAN and PBX infrastructures. A unified backbone to the desktop enables common delivery of voice and data for reduced wiring and maintenance costs. Using a switched 100Mbps Ethernet, network engineers can design telephone networks with essentially unlimited capacity. When the need arises for more workstations (i.e., extensions), another Ethernet switch is added. Administering these systems is done locally through a Windows *Graphical User Interface* (GUI) or remotely through a Web browser. The use of ITU-standard G.711/G.723 audio compression means that less bandwidth is required to transmit voice, so the available bandwidth can be used more efficiently.

See Also

Call Centers

Ethernet

LAN Switching

Transfer Control Protocol / Internet Protocol (TCP/IP)

Voice-Over IP (VoIP)

LAN Switching

When shared LANs become too slow, performance can be improved by creating segments that are linked by bridges. Bridges keep local traffic on a particular LAN segment while enabling packets that are destined for other segments to pass through in a process called filtering. But no matter how many segments are created, LAN performance tends to diminish—if only because more users are continually added to the network. The greater the number of workstations that are simultaneously accessing the LAN, the smaller each workstation's available bandwidth becomes.

With a switched LAN (Ethernet or Token Ring), each user can have access to the network at full native speed instead of having to share it with multiple users. Dedicated LAN links improve network performance for all users, allowing them to be more productive and making the network easier to manage. In some cases, switched LANs provide enough performance improvements to hold off purchases of more expensive ATM or FDDI networks

LAN switching is implemented in conjunction with an intelligent wiring hub or a dedicated LAN switch. Bandwidth can be controlled by restricting access to any logical segment to only authorized members of the workgroup (refer to Figure L-1). Creation of these virtual workgroups also provides security, since packets—broadcast and multi-cast—will only be seen by authorized users within that virtual workgroup. While users of one logical network cannot access another logical network (thus enforcing security), multiple virtual workgroups can still share centralized resources, such as

Figure L-1
A standard hub versus a switching hub. A standard hub (left) broadcasts packets to all ports, while a switching hub (right) supports virtual connections to limit traffic to only specific addressees.

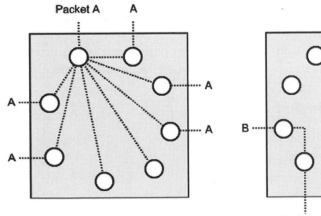

e-mail servers. Highly secure logical segments can be established and modified from an SNMP-based network management station and can encompass any device on the network.

Port versus Segment Switching

Two basic types of switching hubs exist: port switching and segment switching. Port-switching hubs enable administrators to assign ports to segments via network-management software. In effect, these hubs act as software-controlled patch panels. Segment-switching hubs treat ports as separate segments and forward packets from port to port. How they accomplish this task varies by vendor and has important implications, depending on the application. What all segment-switching hubs have in common is that they can substantially increase available network bandwidth. Acting like high-speed multi-port bridges, segment-switching Ethernet hubs offer 10Mbps to each port, for example.

Ethernet Switching

In the traditional Ethernet environment, stations contend with each other for access to the network—a process that is controlled by a statistical arbitration scheme. Each station listens to the network to determine whether the network is idle. Upon sensing that no traffic is currently on the network, the station is free to transmit. If the channel is already in use, the station backs off and tries again later. If multiple stations sense that the network is idle and send out packets simultaneously, a collision occurs (which corrupts the data). When a collision is detected, the stations back off and try again at staggered intervals. This process is called *Carrier-Sense Multiple Access with Collision Detection* (CSMA/CD).

The problem with CSMA/CD is that collisions force retransmissions, which causes the network to slow down. In turn, this situation results in less bandwidth availability for all users. A related problem with Ethernet is that a station's packets are automatically broadcast to all other stations. In other words, all other stations are aware of the packets, although only one station can actually read them. The broadcast nature of Ethernet increases the likelihood of collisions.

In Ethernet switching, the MAC-layer address determines to which hub or switch port the packet will travel. Because no other ports are aware of the packet's existence, the stations do not have to be concerned about whether their packets will collide with data from other stations as they transmit toward the hub or switch. In Ethernet switching, a virtual connection is created between the sending and receiving ports. This dedicated connection remains in place only long enough to pass a packet between the sending and receiving stations.

If a station has packets for a busy port, that station's port momentarily holds them in its buffer. When the busy port is free, a virtual connection is established—and the packet is released from the buffer and is sent to the newly freed port. This mechanism works well unless the buffer becomes filled, in which case packets are lost. To avoid this situation, some vendors offer a throttling capability. When a port's buffer begins to fill up, that port begins to send packets back to the workstation. This process slows the workstation's transmission speed and evens out the pressure at the port. Because no packets are lost, many more packets make it through the hub than would otherwise be possible under heavy traffic conditions.

Some LAN switching products offer a choice of different packet-switching modes: cut-through and store-and-forward. Cut-through speeds frame processing by beginning the transmission on the destination LAN before the entire frame arrives in the input buffer. Store-and-forward works like a traditional LAN bridge, where frame forwarding begins only after the entire frame arrives in the input buffer. Store-and-forward mode results in more of a performance hit but ensures error-free delivery. Some vendors support automatic switching between cut-through and store-and-forward modes. With this technique, called error-free cut-through switching, the switch changes from cut-through to store-and-forward mode if the percentage of bad packets flowing through the switch exceeds a predefined threshold.

Some LAN switching devices support only one address per port, while others support 1,500 or more. Some devices are capable of dynamically learning port addresses and enabling or prohibiting new port addresses. Disallowing new port addresses enhances hub security (when ignoring new port addresses, the corresponding port is disabled, thereby preventing unauthorized access).

Some vendors offer full-duplex Ethernet connections, providing each user with 20Mbps of dedicated bandwidth (send and receive) over unshielded twisted-pair wiring. With half-duplex signaling and collision detection disabled, one pair of wires can transmit at 10Mbps while the other pair receives at 10Mbps. This situation creates a collision-free connection that can ease bottlenecks between similarly equipped switching hubs or servers.

Token-Ring Switching

LAN switching is also available for the Token Ring environment. Token-Ring switching is a technology for dedicating 4Mbps or 16Mbps of bandwidth to each user on a LAN. Software-controlled switching of individual ports into any one of a number of Token Rings eliminates the need to patch network cables physically to change LAN configurations and provides centralized, per-port control over Token Ring LAN configurations.

Similar to switched Ethernet LANs, switched Token Ring LANs can be configured to operate in full-duplex in order to send and receive data simultaneously. The total per-port capacity of each switch module would then be 32Mbps.

Associated with Token Ring switching is the capability to automatically protect the ring from potential disruptions. Although some vendors offer the capability to mix 4Mbps and 16Mbps on the same module, the ports can detect and deny entry to any device attempting to connect to a port with a transmission frequency different from the port's predefined configuration. This feature prevents a device that is configured for 16Mbps transmission from accessing a port configured for 4Mbps transmission, for example.

Summary

Enterprise-level LAN switches are available that provide seamless switching between fast LAN technologies including FDDI, Fast Ethernet (100Base-T), and 100VG-AnyLAN while providing a safe migration from these and existing Ethernet and Token Ring LANs to ATM. There are even gigabit switches for supporting large departments in enterprise LANs. Combined with LAN bandwidth-management intelligence, LAN switches provide users with the capability to build and manage reliable, high-performance, switched intra-networks.

See Also

Ethernet

Hubs

Token Ring

Latency

Latency is the amount of delay that affects all types of communications links. Delay on telecommunications networks is usually measured in milliseconds (ms) or thousandths of a second.

A rule of thumb that is used by the telephone industry is that the round-trip delay for a telephone call should be less than 100ms. If the delay is much more than 100ms, participants think they hear a slight pause in the conversation and use that pause as an opportunity to begin speaking. But by the time their words arrive at the other end, the other speaker has already begun the next sentence and feels that he or she is being interrupted. When telephone calls take place over satellite links, the round-trip delay is typically about 250ms, and conversations become full of awkward pauses and accidental interruptions.

Latency affects the performance of applications on data networks, as well. On the Internet, for example, excessive delay can cause packets to arrive at their destination out of order, especially during busy hours. The reason why packets might arrive out of sequence is that they can take different routes on the network. The packets are held in a buffer at the receiving device until all packets arrive and are put in the right order. While this situation does not affect e-mail and file transfers, which are not real-time applications, excess latency does affect multimedia applications in that it causes voice and video components to arrive out of synchronization.

If the packets that contain voice or video do not arrive within a reasonable time, they are dropped. When packets that contain voice are dropped, a condition known as clipping occurs—the cutting off of the first or final syllables in a conversation. Dropped packets of video cause the image to be jerky. Excessive latency also causes the voice and video components to arrive out of synchronization with each other, causing the video component to run more slowly than the voice component (for example, in a video conference, a person's lips will not match what he or she is really saying).

The problem of latency can be addressed by assigning a class of service or *Quality of Service* (QoS) to a multimedia application, which identifies it as being time sensitive and requiring priority over other less time-sensitive data types. QoS can be handled by the network (i.e., routers, hubs, and switches), by the operating system, or by the hardware and operating system working together.

See Also

Jitter

Leasing

Network managers today are not only responsible for selecting equipment that best satisfies corporate communications requirements, but they are increasingly being called upon to recommend the most cost-effective way to procure that equipment. Although the methods for technology procurement have become more creative in recent years, mostly in the structuring of contract terms and conditions, the fundamental decision still boils down to one of leasing or purchasing. Leasing can often provide organizations with many financial and non-financial benefits.

Financial Incentives

A number of financial reasons exist for considering leasing over purchasing, such as when a company cannot secure credit for an installment loan on the system and has no available line of credit, or cannot make the down payment that is required for an installment loan. Leasing can improve a company's cash position, because costs are spread over a period of years. Leasing can free capital for other uses and can even cost-justify technology acquisitions that would normally prove too expensive to purchase. Leasing also makes it possible to procure equipment on short notice that has not been planned or budgeted.

A purchase, on the other hand, increases the debt relative to equity and worsens the company's financial ratios in the eyes of investors, creditors, and potential customers. An operating lease on rental equipment can reduce balance-sheet debt, because the lease or rental obligation is not reported as a liability.

Therefore, at the least, an operating lease represents an additional source of capital and preserves credit lines. Beyond that, leasing can help companies comply with the covenants in loan agreements that restrict the amount of new debt that can be incurred during the loan period. The purpose of such provisions is to make sure that the company does not jeopardize its capacity to pay back the loan. But in providing additional capacity for acquiring equipment without violating loan agreements or hurting debt-to-equity ratios, leasing enables companies to have their cake and eat it, too.

With major improvements in technology becoming available every 12 to 18 months, leasing can prevent a company from becoming saddled with obsolete equipment. In other words, the potential for losses when replacing equipment that has not been fully depreciated can be minimized by leasing, rather than by purchasing. Furthermore, with rapid advancements in technology and consequent shortened product life cycles, it is becoming more difficult to sell used equipment. Leasing eliminates this problem as well, because the leasing company owns the equipment.

For organizations that are concerned with controlling staff size, leasing also minimizes the amount of time and resources that are spent to cost-justify capital expenditures, evaluate new equipment and dispose of old equipment, negotiate trade-ins, compare the capabilities of vendors, perform reference checks on vendors, and review contractual options. Also, no need exists for additional administrative staff to keep track of configuration details, spare parts, service records, and equipment warranties. Also, because the leasing firm is usually responsible for installing and servicing the equipment, there is no need to spend money on skilled technicians or on outside consulting services. Lease agreements might be structured to include ongoing technical support and even a help hotline for end users.

This situation brings up another advantage to leasing: it can minimize maintenance and repair costs. Because the lessor has a stake in keeping the equipment functioning properly, the lessor usually offers on-site repair and the immediate replacement of defective components and subsystems. In extreme cases, the lessor might even swap the entire system for a properly functioning unit.

Although contracts vary, maintenance and repair services that are bundled into the lease can eliminate the hidden costs that are often associated with an outright purchase. When purchasing equipment from multiple vendors, the user will often become bogged down with processing, tracking, and reconciling multiple vendor purchase orders and invoices in order to obtain a complete system or network. Under a lease agreement, the leasing firm provides a single-source purchase order and invoicing, which cuts down on the user's administration, personnel, and paperwork costs.

Finally, leasing usually enables more flexibility in customizing contract terms and conditions than normal purchasing arrangements, due simply to the fact that there are no set rates and contracts when leasing. Unlike many purchase agreements, each lease is negotiated on an individual case basis. The items that are typically negotiated in a lease are the equipment specifications, the schedule for upgrades, the maintenance and repair services, and training.

Another negotiable item has to do with the end-of-lease options, which can include signing another lease for the same equipment, signing another lease for more advanced equipment, or buying the equipment. Many lessors will allow customers to end a lease ahead of schedule without penalty if the customer agrees to a new lease of upgraded equipment.

Non-Financial Incentives

Some compelling non-financial reasons also exist for considering leasing over purchasing. In some cases, leasing can make it easier to try new technologies or to try the offerings of vendors that would not normally be considered. After all, leases always expire or can be canceled (a penalty usually applies), but few vendors are willing to take back purchased equipment.

Leasing permits users to take full advantage of the most up-to-date products at the least risk—and often under attractive financial terms. This kind of arrangement is particularly attractive for companies that use technology for competitive advantages, because it means that they can continually upgrade by renegotiating the lease (often with little or no penalty for terminating the existing lease early). Similarly, if the company grows faster than anticipated, it can swap the leased equipment for an upgrade.

You should note, however, that many computer and communications systems are now modular in design, so the fear of early obsolescence might not be as great as it once was. Nevertheless, leasing offers an inducement to try vendor implementations on a limited basis without committing to a particular platform or architecture, with minimal disruption of mainstream business operations.

Companies that lease the equipment can avoid a problem that invariably affects companies that purchase equipment: how to get rid of outdated equipment. Generally, no used equipment is worth more on a price/performance basis than new equipment, even if it is functionally identical. Also, as new equipment is introduced, it erodes the value of older equipment. These byproducts of improved technology make it difficult for users to unload older, purchased equipment.

With equipment coming off lease, the leasing company assumes the responsibility of finding a buyer. Typically, the leasing company is staffed with marketers who know how and where to sell used equipment. They know how to prospect for customers for whom state-of-the-art technology is more than they need, but a second-hand system might be a step up from the seven-year-old hardware that they are currently using.

Also, a convenience factor is associated with leasing, because the lessee does not have to maintain detailed depreciation schedules for accounting and tax purposes. Budget planning is also easier, because the lease involves fixed monthly payments. Leasing locks in pricing over the term of the lease, enabling the company to know in advance what the equipment costs will be over a particular planning period.

With leasing, there is also less of an overhead burden with which to contend. For example, there is no need to stockpile equipment spares, subassemblies, repair parts, and cabling. The leasing firm has the responsibility to keep inventories up to date. Their technicians (usually third-party service firms) make on-site visits in order to swap boards and to arrange for overnight shipping of larger components when necessary.

Leasing can also shorten the delivery lead time on desired equipment. Eight weeks or longer might be necessary to obtain the equipment that is purchased from a manufacturer. In contrast, it might take from one to 10 working days to obtain the same equipment from a leasing firm. Often, the equipment is immediately available from the leasing firm's lease/rental pool. For businesses that need equipment that is not readily available, some leasing firms will make a special procurement and have the equipment in a matter of two or three days if the lease term is long enough to make the effort worthwhile.

Many leasing companies offer a master agreement, giving the customer a preassigned credit limit. All of the equipment that the customer wants goes on the master lease and is automatically covered by its terms and conditions. In essence, the master lease works similarly to a credit card.

Summary

Leasing is another form of buying on credit. Payments are made monthly, and the total price includes interest on the principal amount. Equipment leasing is done routinely by computer vendors, telecommunications carriers, disaster recovery providers, systems and network integrators, third-party service firms, and leasing companies that specialize in financing computer systems and communication networks. For network managers, commercial software packages are available that automate the lease-versus-purchase decision process. Often, the company that is leasing the equipment will also provide asset-management software in order to help customers keep track of all of the items under lease.

See Also

Asset Management

Lifeline Service

Under the concept of universal service, all households that want access to the telecommunications network are capable of affording it. The *Federal Communications Commission* (FCC) offers universal service support to the following:

■ Telecommunications carriers in rural, insular, and high-cost areas in which telecommunications services are often more expensive to provide

■ Low-income consumers through the Lifeline program, which provides monthly reductions in service charges, and the Link Up program, which provides reductions in initial connection charges

■ Schools, libraries, and rural health-care providers

The Lifeline and Link Up programs directly benefit eligible low-income consumers. Although the universal service program provides support to telecommunications carriers in rural, insular, and high cost areas and produces lower rates in those areas, it is not targeted to low-income customers.

Lifeline Assistance

The federal Lifeline Assistance program provides between $3.50 and $7 per month to telecommunications service providers, in order to enable them to reduce eligible consumers' monthly charges. The amount of federal support will vary, depending on decisions that are made by the state commission (such as whether to provide state support). But eligible low-income consumers in every state, territory, and possession will receive at least a $3.50 reduction on their telephone bill in the form of a credit against their $3.50 per month subscriber-line charge as a result of the federal universal service support program. The reduction applies to a single telephone line at the qualifying consumer's principal place of residence.

To qualify for Lifeline Assistance in states that provide state support, a consumer must meet the criteria that is established by the state commission. The state commission is required to establish narrowly targeted qualification criteria based on income or factors that are directly related to income. In states that do not provide state support, a consumer must participate in one of the following programs: Medicaid, food stamps, *Social Security Income* (SSI), federal public housing assistance, or *Low-Income*

Home Energy Assistance Program (LIHEAP). The named subscriber to the local telecommunications service must participate in one of these assistance programs in order for that household to receive Lifeline support.

All qualifying low-income consumers receive the following services: voice-grade access to the public-switched network, *Dual-Tone Multi-Frequency* (DTMF) signaling, single-party service or its functional equivalent, access to emergency services, access to operator services, access to interexchange service, access to directory assistance, and toll limitation free of charge (provided that the carrier is technically capable of providing toll limitation). Toll limitation includes both toll blocking, which prevents the placement of any long-distance calls, and toll control, which limits the amount of long-distance calls to a preset amount that the consumer selects. Carriers that provide Lifeline Assistance cannot collect a service deposit in order to initiate Lifeline service if the qualifying low-income consumer voluntarily elects toll blocking.

Link Up

Link Up, also known as Lifeline Connection Assistance, offers eligible low-income consumers a reduction in the local telephone company's charges for starting telephone service. The reduction is one-half of the telephone company's charge or $30, whichever is less. Link Up offers a deferred payment plan for charges that are assessed for starting service (for which eligible consumers do not have to pay interest). Eligible consumers are relieved of paying interest charges of up to $200 that are deferred for a period not to exceed one year.

Link Up does not reduce or eliminate any permissible security deposits. The Link Up reduction applies to a single telephone line at an eligible consumer's principal place of residence. Qualifying consumers can receive the reduction in connection charges more than once, only if they change residences. The eligibility standard for Link Up is the same as for Lifeline Assistance.

Summary

The federal universal service program is funded by all providers of interstate telecommunications services. The objective of Lifeline and Link Up services is to increase subscriptions among low-income consumers. All carriers that are designated by their state commission as eligible telecommunications carriers must offer Lifeline and Link Up to qualifying consumers.

See Also

Universal Service

Line Conditioning

Line conditioning refers to the process of compensating for various line impairments in order to improve data transmission on analog or digital leased lines. Most modems perform elaborate handshaking processes before beginning transmission. As part of these routines, they test for various line impairments and apply appropriate correction techniques. During the transmission, the modems at each end constantly test the line for changes in quality. The bit rate will increase or decrease based on the quality of the line. A *Channel Service Unit* (CSU) is an interface for digital leased lines. In addition to performing loopback testing, the CSU performs line conditioning to enable leased lines to perform at their rated speed.

By adding equalizers to analog leased lines, circuit quality can be improved in order to permit fast and reliable data transmission. Equalizers mainly correct for variances in amplitude and delay. Amplitude equalization minimizes the effects of amplitude (loudness) or frequency (phase or time-delay) distortion by compensating for the variations in data-channel quality. Among the common types of equalization are the following:

- Frequency equalization, which restores the amplitude of different frequencies that have suffered non-linear distortion
- Dynamic equalization, which counteracts delay distortion in analog circuits
- Timing equalization, which corrects for jitter

Circuit Parameters

Among the circuit parameters that can be controlled or corrected by equalizers are attenuation distortion, envelope-delay distortion, inadequate S/N ratios, non-linear distortion, impulse noise, and phase-jitter characteristics.

Attenuation distortion   The loss of signal strength, measured as the difference in the power level between the transmitted signal and the received signal. Attenuation is usually expressed in *decibels* (dB) per kilometer.

Envelope-delay distortion   Also known as phase distortion, this type of line impairment involves the transmission of different frequencies that travel at different speeds in communications circuits. This situation can cause signal distortion when the signals arrive out of phase. The middle-range frequencies tend to arrive first. In analog copper circuits, this problem is generally not noticeable for voice, because the main effect of such delays is to change the phase of the signal slightly. For high-speed modems where symbols are coded by phase changes, however, this problem is serious. Equalizers can be used to compensate for this distortion by artificially delaying the faster frequencies.

Signal/Noise ratio   For the transmission of data, the ratio of signal to noise must be high; that is, there must be little noise on the line compared to the signal level. As the signal-to-noise ratio diminishes, the more likely the data transmission will be disrupted, necessitating the retransmission of corrupted or missing data. Among the ways to condition the line in order to eliminate noise is to splice a load coil into the problem circuit. This action isolates the usable frequency, such that only signals below the 4000Hz range can pass through the circuit. While the use of load coils improves voice quality, it prevents the passage of high-speed data that uses the frequencies above 4000Hz.

Non-linear distortion   This impairment is a form of signal-processing error that creates signals at frequencies that are not necessarily present in the input signal. The new frequencies might be harmonics, named so because they exist at frequencies that are integer multiples of the input signal. For example, the harmonics of 1KHz are at 2KHz, 3KHz, 4KHz, and so on.

Impulse noise   A random occurrence of energy spikes having random amplitude and spectral content. Impulse noise in a data channel can be a definitive cause of data-transmission errors.

Phase jitter   Just as the signal is affected by noise, the phase of the signal is affected by phase noise, which can be generated by additive amplitude noise as well as by true phase modulation of the signal. The resulting phase noise, or jitter characteristic, is much like amplitude noise in that it has nominal background jitter levels that are similar to the nominal amplitude noise level in the channel. Also, occasional abrupt changes or hits, which are similar to amplitude impulse noise, can occur. This phase jitter results in the displacement in the phase of the signals that are transmitted through the voice-frequency channel. The amount of phase displacement increases with the magnitude of the phase noise and varies at a rate that is equal to the frequencies at which the phase noise occurs. A common cause of these deviations is the modulation of carrier in frequency-division multiplex systems by the power-line frequency, or some harmonic of the power-line frequency. Other causes of phase jitter are interfering tones on the channel or variations in the medium.

Summary

Line conditioning is a service that can be applied by the carrier (but not to switched circuits, because the path through the network is constantly changing). If the circuit is leased and is dedicated to one connection only, however, line impairments can then be measured, and corrections can be applied. Generally, this form of leased-line conditioning results in data rates that can be many times more than those that are possible through switched circuits.

See Also

>   Analog Line Impairment Testing
>   Attenuation
>   Cross-Talk
>   *Decibel* (dB)
>   Hertz
>   Jitter
>   Latency

Local Access and Transport Areas (LATAs)

Under the 1982 Consent Decree that was handed down by the United States Federal District Court in Washington, D.C., AT&T had to divest its ownership of the *Bell Operating Companies* (BOCs). Accordingly, all Bell territory in the continental United States was divided into geographic areas called *Local Access and Transport Areas* (LATAs). The BOCs were permitted to provide telephone service within a LATA (intraLATA service) but were not permitted to carry traffic across LATA boundaries (refer to Figure L-2).

Originally there were 161 LATAs, but according to the FCC, today the number of LATAs stands at 193. Each LATA is identified regionally by a three-digit number that bears no relation to area codes. A state can have several LATAs or just one. In a few cases, LATAs can cross state lines.

According to the Consent Decree, interLATA traffic was to be carried by IXCs. The LATAs did not cover territory that was served by the independent telephone companies, and they were not subject to the restrictions that were imposed by the Consent Decree. While a BOC could not carry traffic across a LATA boundary, an independent telephone company could carry traffic—regardless of whether that traffic crossed LATA boundaries.

When the LATAs were created, most independent exchanges were classified as associated with a particular LATA. The BOCs were only permitted to provide service within a LATA and the associated exchanges. This restriction was intended to restrain the former monopoly phone companies from engaging in anti-competitive behavior.

The Court, however, noted that there were often joint operating arrangements between independent exchanges and neighboring BOC facilities. For

Figure L-2
Representation of LATA boundaries and call traffic handled by BOCs and Interexchange Carriers (IXCs)

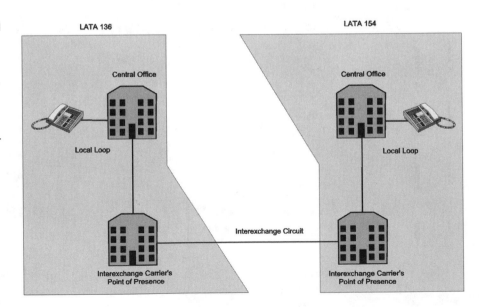

example, BOCs often switched traffic between their end offices and the end offices of the independents, which then carried the traffic to its final destination. If all of this traffic were considered interLATA, BOCs could not participate in these arrangements, and significant and costly network rearrangements would have been necessary. To prevent the need for such rearrangements, the Court classified most independent exchanges as associated with a particular BOC LATA. Traffic between a BOC LATA and an associated exchange was treated as intraLATA and could be carried by the BOC; traffic between a BOC LATA and an unassociated exchange was treated as interLATA and thus could not be carried by the BOC.

Also in establishing the LATAs, the Court recognized that there were existing local calling areas that would cross the newly created LATA boundaries. The Court did not intend for LATA boundaries to interfere with local calling areas that had been established by state regulators; thus, the Court granted exceptions to permit BOCs to carry interLATA traffic if necessary, in order to preserve existing LATA arrangements. In granting such exceptions, known as *Expanded Local Calling Service* (ELCS) requests, the Court and *Department of Justice* (DoJ) considered factors such as the number of customers or access lines involved; the degree of community interest; the degree of calling between the exchanges; and the competitive effects.

Similarly, the Court received more than 100 additional requests involving LATA associations, including requests for new associations, disassociations, and changes in association from one LATA to another. LATA association requests were generally granted if the changes in associations would avoid the need for expensive reconfiguration and would not endanger competition.

With the Telecommunications Act of 1996, the local exchange carriers can qualify to handle interLATA calls by meeting the requirements of a 14-point competitive checklist and by receiving approval of the state *Public-Utility Commission* (PUC) and FCC. The Telecommunications Act of 1996 not only preserves the LATA concept but also gives the FCC exclusive authority over LATA boundaries.

LATA Modifications

Passage of the Telecommunications Act of 1996 changed the procedure by which a LATA boundary is modified. Consequently, BOCs can modify LATA boundaries if such modifications are approved by the FCC.

The BOCs and PUCs often petition the FCC for changes in LATA boundaries. The FCC issues a *Notice of Proposed Rule Making* (NPRM), asking for comments on the merits of such proposals. If there is no serious objection on public-interest grounds, certain types of requests are usually granted. Among the reasons for requesting LATA changes are as follows:

- To provide expanded local calling service
- To offer services such as ISDN and Internet access to communities that could not obtain these services by any other means
- To better serve the needs of subscribers that straddle LATA boundaries
- To enable exchanges that are purchased from an independent carrier to be included in a BOC's existing LATA
- To permit independent telephone companies to change the LATA association of specific exchanges
- To redefine the boundaries of multiple LATAs within the state to become a one-LATA state in order to simplify regulation

While the first five requests stand a reasonable chance of approval, the last request is always turned down on the grounds that the FCC does not delegate its authority over the definition of LATA boundaries to the states.

Summary

Subscribers can choose a preferred long-distance carrier for intraLATA service and reap several advantages. By combining all toll calls (intraLATA and interLATA) into one bill, corporate and residential subscribers can qualify for greater volume discounts and can realize such savings more quickly. In addition, the features that are used on long-distance calls, such as call accounting, account codes, and geographic restrictions, can be applied to the intraLATA local toll calls.

See Also

Federal Communications Commission (FCC)

Local Telecommunications Charges

Long-Distance Telecommunications Charges

Telecommunications Act of 1996

Local Exchange Carriers (LECs)

Local Exchange Carriers (LECs) provide residential, business, and interexchange access services. In addition to Centrex, many of the larger LECs are developing and/or offering value-added services such as voice and data messaging via cellular and *Personal Communications Services* (PCS) networks. The LECs, which are commonly called telephone companies (or telcos), include the 22 former BOCs that divested from AT&T in 1984, as well as Cincinnati Bell, *Southern New England Telephone* (SNET), and the telephone companies GTE and Sprint. These companies are now referred to as *Incumbent Local Exchange Carriers* (ILECs) in order to distinguish them from competitors in the local market and from the hundreds of smaller telephone companies that serve largely rural areas.

In addition to providing local phone service and providing IXCs with access to the local loop, the LECs provide billing services. Phone bills that come from a LEC can actually represent charges from a number of services and providers. A telephone bill can comprise many basic elements, including charges for local *Message Units* (MSUs), special service offerings, directory service, 911 emergency service, cellular calls, and Internet access. These and other charges are totaled in a consolidated monthly invoice from the LEC. In addition, charges for long-distance calls that are carried by IXCs and cellular service providers might appear on the monthly invoice, as well.

Another type of LEC is the *Competitive Local Exchange Carrier* (CLEC). This type of service provider offers business and residential users lines and services on a resale basis or from its own facilities-based network, enabling customers to save money. Regional teleports, metropolitan fiber carriers, and CATV operators are among the types of companies that are now involved with providing competitive local-exchange services. Typically, these alternative access carriers offer service in major markets where traffic volumes are greatest—and, consequently, users are hardest hit with high local service charges.

Summary

With passage of the Telecommunications Act of 1996, new entities are allowed in the market for local telephone service, including cable operators, electric utilities, *Internet Service Providers* (ISPs), and entertainment companies. Until late 1999, the ILECs were restricted to providing local service within their assigned serving areas (called LATAs). The first ILEC to obtain FCC permission to offer long-distance service in its own territory is Bell Atlantic. At this writing, the company is rolling out long-distance service, starting with its subscribers in New York.

See Also

Competitive Local Exchange Carriers (CLECs)

Incumbent Local Exchange Carriers (ILECs)

Interexchange Carriers (IECs)

International Common Carriers

Local Access and Transport Areas (LATAs)

Local Loop

The local loop is an *Unbundled Network Element* (UNE) that the *Federal Communications Commission* (FCC) defines as a transmission facility between a distribution frame, or its equivalent, in a central office and the network interface device at the customer premises. This definition includes, for example, two-wire and four-wire analog voice-grade loops and two-wire and four-wire loops that are conditioned to transmit the digital signals that are needed to provide services such as ISDN, DSL, and DS1 signals.[1]

ILECs are required to provide access to these transmission facilities only to an extent that is technically feasible. If it is not technically feasible to condition a loop facility to support a particular service, the ILEC does not need to provide unbundled access to that loop. For example, a local loop that exceeds the maximum length permitted for the provision of *High Bit-Rate Digital Subscriber Line* (HDSL) service could not feasibly be conditioned for such service.

---

[1]Carriers have traditionally defined local loops in more detailed terms than are discussed here. Likewise, the definition of the term differs among state *Public-Utility Commissions* (PUCs). The FCC has taken a general approach in an effort to minimize complex and resource-intensive disputes between ILECs and competitors over whether a particular function qualifies as a loop.

The FCC's definition of loops in some instances requires the ILEC to take affirmative steps in order to condition existing loop facilities to enable requesting carriers to provide services that are not currently provided over such facilities. For example, if a competitor seeks to provide a digital service such as DSL and the loop is not currently conditioned to carry digital signals (but it is technically feasible to do so), the ILEC must condition the loop to permit the transmission of digital signals. Thus, the FCC rejects the arguments of some ILECs that competitors take the networks as they find them with respect to unbundled network elements. The requesting carrier must, however, bear the cost of compensating the ILEC for such conditioning.

The FCC further requires that ILECs provide competitors with access to unbundled loops, regardless of whether the incumbent LEC uses *Integrated Digital-Loop Carrier* (IDLC) technology, or similar remote concentration devices, for the particular loop that is sought by the competitor. IDLC technology enables a carrier to aggregate and multiplex loop traffic at a remote concentration point and to deliver that multiplexed traffic directly into the switch, without first demultiplexing the individual loops. If the FCC did not require ILECs to unbundle IDLC-delivered loops, end users who are served by such technologies would not have the same choice of competing providers as end users who are served by other loop types. Furthermore, such an exception would encourage ILECs to hide loops from competitors through the use of IDLC technology.

In most cases, it is technically feasible to unbundle IDLC-delivered loops. One way to unbundle an individual loop from an IDLC is to use a demultiplexer to separate the unbundled loop(s) prior to connecting the remaining loops to the switch. Other ways exist to separate individual loops from IDLC facilities, including methods that do not require demultiplexing. For example, IDLC loops can be moved onto other loop carrier links or alternatively can be removed from the multiplexed signal through a grooming process. Again, the costs that are associated with these mechanisms must be borne by requesting carriers.

The FCC also requires ILECs to offer unbundled access to the *Network Interface Device* (NID) at the customer premises. When a competitor deploys its own loops, the competitor must be able to connect its loops to customers' inside wiring in order to provide competing service, especially in multi-tenant buildings. In many cases, inside wiring is connected to the ILEC's loop plant at the NID. In order to provide service, a competitor must have access to this facility. Therefore, a requesting carrier is entitled to connect its loops, via its own NID, to the ILEC's NID. The new entrant bears the cost of connecting its NID to the incumbent LEC's NID.

Summary

The purpose of requiring incumbent LECs to make unbundled local loops available is to facilitate market entry and to improve consumer welfare. Without access to unbundled local loops, new entrants would need to invest immediately in duplicate facilities in order to compete for customers. Such investment and building would likely delay market entry and postpone the benefits of local telephone competition for consumers. Moreover, without access to unbundled loops, new entrants would be required to make a large initial investment in loop facilities before they had a customer base that was large enough to justify such an expenditure. This situation would increase the risk of entry and would raise the new entrant's cost of capital. By contrast, the ability of a new entrant to purchase unbundled loops from an ILEC enables the new entrant to build facilities gradually and to deploy loops for its customers where it is efficient to do so.

See Also

> *Digital Subscriber-Line Technologies* (DSLT)
>
> *Integrated Service Digital Network* (ISDN)
>
> Unbundled Access

Local Multi-Point Distribution Service (**LMDS**)

Local Multi-Point Distribution Service (LMDS) is a two-way microwave technology that operates in the 27GHz to 31GHz range. This broad-band service enables communications providers to offer a variety of high-value, quality services to homes and businesses, including broad-band Internet access. LMDS offers greater bandwidth capabilities than a previous technology called *Multi-Channel Multi-Point Distribution Service* (MMDS) but has a maximum range of only 7.5 miles from the carrier's hub to the customer premises.

Applications

LMDS provides enormous bandwidth—enough to support 16,000 voice conversations plus 200 channels of television programming. The following

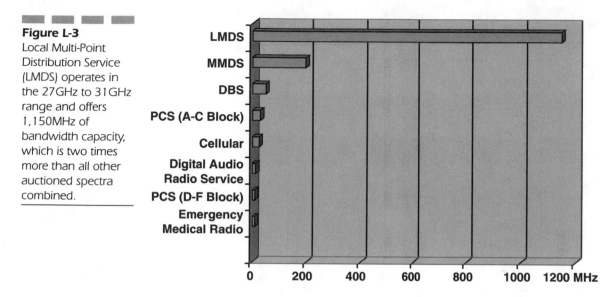

Figure L-3
Local Multi-Point
Distribution Service
(LMDS) operates in
the 27GHz to 31GHz
range and offers
1,150MHz of
bandwidth capacity,
which is two times
more than all other
auctioned spectra
combined.

chart (refer to Figure L-3) contrasts LMDS with the bandwidth that is available over other wireless services.

CLECs can deploy LMDS to completely bypass the local loops of the ILECs, eliminating access charges and avoiding long service-provisioning delays. LMDS costs far less to deploy than installing new fiber, which enables CLECs to economically bring customer traffic onto their existing fiber networks. The strategy among many CLECs is to offer LMDS to owners of multi-tenant buildings. LMDS offers a single, wireless pipe for multiple voice, data, and video applications. Subscribers can use LMDS for a variety of high-bandwidth applications, including television broadcast, video conferencing, LAN interconnection, broad-band Internet access, and telemedicine.

Operation

LMDS operation requires a clear line of sight between the carrier's hub station antenna and the antenna at each customer location. The maximum range between the two is 7.5 miles; however, LMDS is also capable of operating without having a direct line-of-sight with the receiver. This feature, which is highly desirable in built-up urban areas, can be achieved by bouncing signals off buildings so that the signals get around obstructions. A potential problem for LMDS users is that the microwave signals can be disrupted by heavy rainfall and dense fog.

At the carrier's hub location, there is a roof-mounted multi-sectored antenna (refer to Figure L-4). Each sector of the antenna receives/transmits signals between itself and a specific customer location. This antenna is small (some measure only 12 inches in diameter). The hub antenna brings the multiplexed traffic to an indoor switch (refer to Figure L-5), which processes the data into 53-byte *Asynchronous Transfer Mode* (ATM) cells for transmission over the carrier's fiber network. These individually addressed cells are converted back to their native format before going off the carrier's network to their proper destinations: the Internet, the PSTN, or the customer's remote location.

At each customer's location, there is a rooftop antenna that sends/receives multiplexed traffic. This traffic passes through an indoor *Network Interface Unit* (NIU) that provides the gateway between the RF components and the in-building equipment, such as a LAN hub, PBX, or video-conferencing system. The NIU includes an up/down converter that changes the frequency of the microwave signals to a lower *Intermediate Frequency* (IF) that the electronics in the office equipment can more easily (and inexpensively) manipulate.

Spectrum Auctions

In May 1999, the FCC completed the last auction for LMDS spectra. More than 100 companies qualified for the auctions, bidding against each other

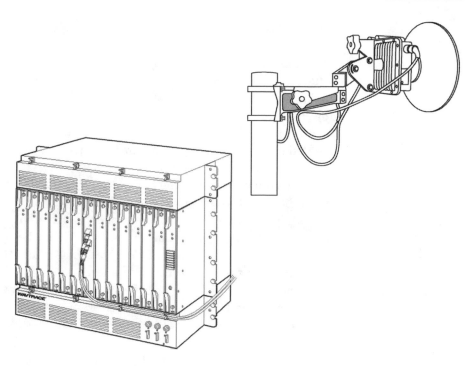

Figure L-5

A microwave transceiver (top right) handles multiple point-to-point downstream and upstream channels to customers. The transceiver is connected via coaxial cables to an indoor switch (bottom left), which provides the connectivity to the carrier's fiber network. The traffic is conveyed over the fiber network in the form of 53-byte ATM cells. Source: Wavtrace, Inc.

for licenses in select *Basic Trading Areas* (BTAs).[2] The FCC auctioned two types of licenses in each market. An A-block license permits the holder to provision 1,150MHz of spectrum for distribution among its customers, while a B-block license permits the holder to provision 150MHz. Most of the A-block licenses in the largest BTAs were snapped up by the large CLECs, while the B-block licenses were taken by smaller companies, ISPs, universities, and government agencies.[3] The licenses are granted for a 10-year period, after which the FCC can take them back if the holder does not have service up and running.

---

[2]Based on data compiled by Rand McNally. Most large cities are *Metropolitan Trading Areas* (MTAs), and most of the larger U.S. towns are classified as BTAs. These are not the same as LATAs, which have defined the local service boundaries of the former BOCs since their divestiture from AT&T in 1984.

[3]The ILECs, such as the RBOCs, are forbidden to enter the LMDS market for three years. By then, they probably will have obtained clearance from the FCC to enter the long-distance market. When they do, they can use LMDS to bypass each other's local loops.

History

Bernard Bossard is generally recognized as the inventor of LMDS. Bossard, who had worked with microwaves for the military, believed he could make point-to-multi-point video work in the 28GHz band. Not interested in sending high-powered, low-frequency signals over long distances, Bossard focused instead on sending low-powered, high-frequency signals over a short distance. The result was LMDS. In 1986, he received funding and formed CellularVision with his financial backers. CellularVision then spun off the technical rights to the technology into a separate subsidiary, CT&T, which licenses the technology to other companies.

CellularVision was awarded a pioneer's preference license by the FCC for its role in developing LMDS. CellularVision began operating a commercial LMDS in metropolitan New York, providing video programming to subscribers in the Brighton Beach area. In 1998, CellularVision changed its name to SPEEDUS.COM. The company has a network operations center and recently has been expanding the number of operating cells in the New York area (and now claims more than 12,000 residential and business subscribers).

SPEED is delivered via 14 fully functional Internet broadcast stations in operation under SPEEDUS.COM's FCC license, which covers metropolitan New York. SPEED subscribers can browse the Web by using the company's SPEED modem, which is capable of downstream speeds of up to 48Mbps (31 times faster than a full T1 line).

In the SPEED.COM system, cable programming is downlinked from geosynchronous satellites to the company's head-end facility, where local broadcast transmissions are also received. At the company's master control room, the programming signals are then amplified, sequenced, scrambled, and converted to 28GHz. The SPEED.COM transmitters and repeaters then broadcast a polarized FM signal in the 28GHz spectrum over a radius of up to three miles to subscribers and to adjacent cells for transmission. A 6-inch square, highly directional flat plate, window, roof, or wall-mounted antenna receives the scrambled signal and delivers it to the addressable set-top converter, which decodes the signals. The subscriber receives 49 channels of high-quality video and audio programming, including pay-per-view and premium channels.

More than 100 companies own licenses for LMDS. Nextlink is one of the largest single holders of LMDS licenses in the United States, having invested more than $800 million in such systems (largely through the acquisition of other companies that held LMDS licenses). Nextlink is a CLEC and is using LMDS to feed traffic to its fiber networks. Its approach to building a city is to install fiber. In areas where that process will take too long or where permits are too hard to come by, Nextlink will use the following components (in this order): LMDS, DSL, and ILEC. In addition to Nextlink, other companies that offer LMDS include Teligent and Winstar Communications.

Summary

Fiber optics is the primary transmission medium for broad-band connectivity today. Of the estimated 4.6 million commercial buildings in the United States, however, 99 percent are not served by fiber. Businesses are at a competitive disadvantage in today's information-intensive world unless they have access to broad-band access services, including high-speed Internet access. These businesses, including many data-intensive, high-technology companies, can be adequately served with LMDS. In fact, the LMDS market has the potential to become a significant portion of the global access market, which will include a mix of many technologies (including DSL, cable modems, broad-band satellite, and fiber-optic systems).

See Also

Asynchronous Transfer Mode (ATM)

Cable Television Networks

Digital Subscriber-Line Technologies (DSLT)

Fiber-Optic Technology

Microwave Communications

Multi-Channel Multi-Point Distribution Service (MMDS)

T-Carrier Facilities

Local Telecommunications Charges

The telecommunications bill from the *Local Exchange Carrier* (LEC) actually represents charges for a number of services, some of which might come from other service providers and vendors. Most of these charges are totaled and are consolidated into a single phone bill, making it easier for customers to reconcile and providing a convenient method of payment.

Basic Charges

A number of basic charges are itemized in the monthly statement from the LEC.

Local Service   LECs operate central office facilities that carry all local inbound and outbound traffic—and, in some instances, long-distance, Internet, and telemetry traffic. Typically, companies receive an itemized statement of outbound calls for each month. As *Automatic Number Identification* (ANI) and caller-identification services are enhanced, some types of inbound traffic information are also being made available.

With passage of the Telecommunications Act of 1996, CLECs can apply to the state PUCs and to the FCC in order to provide local exchange services. Upon approval of the application and subsequent interconnection arrangements, these services providers might do their own billing for local service. Whatever the source of local service, the outbound call-detail statement is a valuable analytical tool that enables anomalies to be noted and investigated. Unverified charges can then be brought to the attention of the LEC, and customers can seek credit adjustments.

Special Central-Office Services

With the increasing implementation of *Advanced Information Network* (AIN) services, the LECs have begun to offer a variety of enhanced services to customers. These services can include general business solutions, voice services, data services, and value-added services. All can be priced on a usage basis or a flat fee, depending on the type of service.

General business solutions   Examples of business solutions include contingency planning, disaster recovery through bandwidth on demand, video conferencing, and LAN interconnection for client/server computing. Some LECs are offering ATM-based services in order to extend voice, video, and data to the desktop for a wide range of business applications.

Voice services   Voice services options range from premises-based to central office-based offerings. These services can include Centrex, cellular communications, and digital transport services. Some LECs also offer selected, integrated services that are targeted at the small business market. These services include the following:

- *Centrex*   Offers the features of a PBX/ACD, but without the customer having to invest in his or her own equipment. Centrex charges are based on line usage and choice of feature package. Various types of Centrex service are available, including basic, key/hybrid Centrex, and ISDN. Enhanced Centrex offerings include voice mail, e-mail, message-center support, modem pooling, and packet-switched data transmission.

- *Cellular communications*   This option enables users to add wireless business communications to existing Centrex, PBX, or key systems, converting a corporate facility into an intra-company cellular system. Wireless services can include services ranging from personal communications for individuals to a complete network of wireless and data communications for an enterprise.

- *Digital transport*   This option provides switched exchange-access services such as direct inward dialing, one-way incoming or outgoing PBX trunks, two-way incoming/outgoing PBX trunks, and Custom 800 services.

Voice enhancements   These services include features such as automatic callback, call forwarding, remote call forwarding, voice mail, call waiting, call redirection, caller ID, caller ID with name, speed calling, three-way calling, automated voice messaging, and the many other services that have become an integral part of conducting business. Pricing usually consists of a one-time setup charge and a fixed monthly charge thereafter.

Voice equipment   The LECs also provide equipment that can be purchased or leased. The equipment can include items such as Centrex telephone sets, PBX systems, key systems, single-line and multi-line telephone sets, and cellular telephones. If the equipment is leased, the monthly bill will include these charges.

Data services   The LECs have become increasingly involved in providing data services, both switched and dedicated line. These usually cover the entire data transmission bandwidth and can range from less than 9600bps switched services to SONET and ESCON-based dedicated services.

Value-added services Value-added services can include options such as telemetry (for remote security and environmental monitoring), telemanagement, asset-management products and services, and Centrex management services. Some LECs provide tools that are needed for on-site control over an organization's voice and data communications, as well as for equipment and facilities management. These tools can include features such as automated-call attendant services to handle incoming calls with the purpose of improving customer service, or interactive voice-response service to automate routines such as confirming an account balance, checking the status of an order, or registering for classes. Centrex management features can include options such as automatic route selection, uniform call distribution, Centrex call management, and wireless Centrex services.

Internet Services

Now, many LECs also offer access to the Internet. Typically, costs depend on the type of access that is chosen. Most carriers provide both switched and dedicated services for Internet access. Switched services include X.25, 56Kbps, and ISDN. Dedicated services include T1, *Digital Subscriber Line* (DSL), and Frame Relay. Costs include a monthly charge that increases as transmission speed increases and a one-time set-up charge. Charges to access Usenet newsgroups are incurred based on the number of users, while e-mail pricing can include a setup charge per mailbox and a monthly fee per mailbox. Most LECs provide free Internet browsers.

Operator and System-Assisted Calls

The LEC also provides separate line-item billing for operator and system-assisted calls.

Directory Assistance Directory-assistance charges are typically summarized and reflected as a single line item. Most businesses are given a fixed number of directory-assistance calls per month, with each subsequent call being billed at a fixed rate (usually between 60 cents and 75 cents per call).

To reduce directory-assistance charges, telecommunications managers can employ a variety of cost-control methods, including the following:

- Providing telephone directories for frequently called areas
- Setting up system speed-dial numbers or individual speed-dial lists for users through the telephone system
- Programming the telephone system to block access to directory assistance

Directory Advertising The telephone company typically charges for two different types of directory listings: Yellow Pages listings and White Pages listings. Directory charges are usually summarized and reflected as a line item on the LEC bill.

Yellow Pages advertising tends to be more elaborate and expensive than White Pages advertising. Many LECs now also provide interactive Yellow Pages, which offer enhanced search functions by a variety of criteria. Often, the telecommunications manager has no authority in the advertising area, because the company's marketing department makes deals directly with the LEC customer-service representative. The telecommunications manager's job is to simply review the ad charges and to note any variations in the billing.

White Pages listings are often provided by the LEC at no charge (if the listings are relatively simple). Listings of entire departments, individual branches, and the like, however, usually result in a charge on the billing statement.

System-Assisted Calls System-assisted calls include *69 calls, which automatically establish a connection to the party who called last. A nominal charge exists for such services, usually 50 cents to 75 cents per call.

Government-Mandated Services

One or two line items on the local exchange bill might represent legislatively mandated services such as emergency services and relay-center services. Most LEC bills include a fixed monthly charge for Emergency (or Enhanced Emergency) 911 services. This issue is generally parochial, determined and authorized by the municipal, county, or state government. In most cases, the charge should be no more than pennies per month per line.

The second such line item is for support of relay-center services for the hearing-impaired or for other Americans with Disabilities Act-related services (including a small surcharge on every business line). The purpose of this surcharge is to offset the costs for providing communications services for the disabled. As with the 911 surcharge, the ADA surcharge should be no more than pennies per month per line.

Long-Distance Services

Because almost every business in the United States is located in an equal-access area, virtually every company can designate the long-distance carrier(s) to appear on its 1+ routes. The LEC routes all of the company's long-distance calls to the designated carrier. The LEC also acts as the billing agent for the designated 1+ long-distance carrier. Typically, the LEC does nothing more than post the long-distance carrier's billing information on the local exchange bill and act as a collection agent for the long-distance service provider.

With passage of the Telecommunications Act of 1996, LECs can apply to the state PUCs and to the FCC in order to provide conventional long-distance service. Upon approval of the application and subsequent interconnection arrangements with other carriers, long-distance services will be billed by the carrier that provides the local service.

Some LECs provide long-distance cellular service as well as local cellular service. Businesses can have their cellular calls appear on the same telephone bill as other types of charges, including Internet access.

Special Calling Services

A wide variety of special calling services (976-type or 900-type services, for example) are currently available. Calling these information services, which offer details about everything from Dow-Jones Reports to psychic readings, is rarely useful in the business environment, so charges for these services should be monitored carefully.

The LEC typically acts as a billing agent for special calling-service providers and posts charges that are received from the service provider to the local exchange bill. These charges can be eliminated by programming the telephone system to block access to numbers with 976 and 900 prefixes.

Nearly all LECs now provide a telephone calling card to which all calls can be billed. Calling-card charges appear on the customer's monthly telephone bill and vary according to usage.

Professional Services

All LECs also typically offer a broad range of technical support services that are designed for companies that outsource many technical project components. These professional services can range from program management to temporary staffing consultants and training. Usually, professional services are billed on an as-needed basis and are typically used when a new service or a new technology is being implemented.

Other Charges

The invoice from the LEC might also include some or all of the following charges:

- *Inside wiring maintenance charges* Many state PUCs have passed rules that transfer ownership of inside wire from the local-exchange service provider to the owner of the building. As an option, customers can elect to maintain inside wire themselves, or they can opt for the LEC's maintenance service for a nominal monthly charge (which is reflected as a fixed cost on a single line item on the billing statement).

- *Installation and engineering charges* LECs typically charge for the installation, setup, and testing of new lines and equipment, as well as for engineering services. Engineering services can include everything from site preparation—including *Heating, Ventilation, and Air Conditioning* (HVAC), if necessary—to traffic analysis, network modeling, and periodic fine tuning of the communication system.

- *Federal subscriber-line charge* The LEC collects a network-access charge that is mandated by the FCC. This fee is not a tax. The subscriber-line charge is part of a program for local telephone companies to recover the cost of connecting individual phone lines to interstate long-distance networks. These costs include telephone wires, poles, and other facilities that link each telephone subscriber or customer to the telephone network. The maximum subscriber-line charge for single-line business customers will remain capped at a maximum charge of $3.50 per line per month. The maximum subscriber-line charge for multi-line business customers is the local company's average interstate cost of providing a line in that state or $9 per line per month, whichever is lower. The maximum subscriber-line charge for each non-primary residential line is capped at a maximum charge of $5 per line per month.

- *Presubscribed interexchange carrier charge* This charge is an access charge that long-distance companies pay to local telephone companies as of January 1, 1998. Instead of paying a higher charge per minute to the local telephone companies (as was required under the old rules), the long-distance companies now pay to local telephone companies a flat-rate, per-telephone line charge plus a lower charge per minute. As of July 1999, the maximum PICC that is paid by the long-distance companies for primary residential lines and single-line business lines is $1.04 per line per month. For non-primary residential lines, the maximum PICC that is paid by the long-distance companies is $2.53 per line per month. (Local telephone companies treat a line as non-primary when it serves the same address as the primary line, even if the bill is in a different name at the same address.) The maximum PICC that is paid by the long-distance companies for each multi-line business line is $4.31. Each year, the maximum multi-line business PICC will increase by $1.50, as adjusted by inflation. A long-distance company pays this charge for each residential and business telephone line that is presubscribed to that long-distance company. If a consumer or business has not selected a long-distance company for its telephone lines, the local telephone company can bill the consumer or business for the PICC.

Other Charges and Credits The billing statement might have a catch-all category that includes charges and/or credits for service orders and setup fees for new features that the LEC provides. All of these services tend to have one-time, non-recurring fees that can be waived during special promotions.

Summary

The Telecommunications Act of 1996 clears the way for competition in the provision of local services. There is already significant competition in the major telecommunications markets, which has resulted in lower costs for services. As competition expands, customers will have an expanded menu of choices available and will have the charges appear on the same telephone bill, making it easier to reconcile and providing a convenient method of payment.

See Also

Long-Distance Telecommunications Charges

Local Telephone Number Portability

The Telecommunications Act of 1996 requires the *Federal Communications Commission* (FCC) to implement the mandate of Congress to perform the following tasks:

- Ensure that all Americans have the ability to keep their existing telephone number at the same location when changing local telephone-service providers
- Promote competition in the local telephone market
- Implement number-portability service without degradation in the quality of telecommunications service

Congress enacted the Telecommunications Act of 1996 to establish a national framework that promotes competition and reduces regulation in all telecommunications markets. Congress realized that bringing competition to the local telephone marketplace was the best way to reduce prices, to motivate telephone companies to provide high-quality service, to deploy advanced services, and to provide more overall choices for customers.

To increase competition in the local telephone market service, Congress recognized that certain barriers to competition must be eliminated. One of the major barriers to competition was the inability of customers to switch from one telephone company to another and retain the same telephone number. Congress realized that customers would be reluctant to switch to new telephone-service providers if they were unable to keep their existing telephone numbers. Congress directed local telephone companies to offer telephone-number portability in accordance with requirements prescribed by the FCC.

In order to provide the kind of telephone-number portability envisioned by Congress, telephone companies have had to invest in upgrades to their networks. In May 1998, the FCC determined what types of costs that local telephone companies will be permitted to recover through separate charges for establishing and providing telephone-number portability service and which costs they must treat as part of the overall cost of doing business. The

FCC determined that incumbent local telephone companies were allowed, but not required, to recover the costs of implementing and providing telephone-number portability through two kinds of charges: charges paid by other telephone companies that use a telephone company's number-portability facilities to process their own calls, and a small, fixed monthly charge that is assessed to customers. Telephone companies started to assess customers charges in February 1999.

Long-Term Telephone-Number Portability Charge

Long-term telephone-number portability is a service that provides residential and business telephone customers with the capability to retain, at the same location, their existing local telephone numbers when switching from one local telephone-service provider to another.

This service can entail a fixed monthly charge through which local telephone companies can recover certain costs of providing long-term number-portability service. Recoverable costs include the costs of creating new facilities, physically upgrading or improving the existing PSTN, and performing the ongoing functions that are associated with providing long-term number portability.

Customer Charges

The FCC determined that ILECs can, but are not required to, recover certain costs of providing number portability through a monthly charge that is billed to their customers. The local telephone companies can only charge customers in areas in which local telephone-number portability is available. The FCC's rules prohibit local telephone companies from passing on to their customers any of the costs that they have incurred in establishing telephone number portability until telephone-number portability service is actually available in the customer's service area.

Telephone-number portability is being implemented first in major metropolitan areas and must be made available in other areas within six months after a new telephone company requests that the incumbent local telephone company offer telephone-number portability in that service area. New entrants to the local telephone market, wireless telephone-service providers, and long-distance companies also incur additional costs of handling calls to numbers that are portable. Because the FCC does not regulate the rates of these carriers, which are subject to substantial competition, these carriers might decide to recover their costs of providing long-term telephone-number portability in any lawful manner that is consistent with their obligations under the Telecommunications Act of 1996.

Local telephone companies can continue to assess this charge on customers' telephone bills for a period of five years from the date that the local telephone company first began collecting the charge. The FCC required the telephone companies to spread out the costs over a five-year period and to charge the same amount each month, so that customers would know what to expect on their monthly bills. At the end of the five-year period, the telephone company must stop assessing this charge.

The amount that is charged for this service might differ by region. Different telephone companies have various types of network equipment. These companies will incur different costs as they prepare their local telephone networks to provide number portability. The charges that appear on customers' monthly telephone bills are being collected by the local telephone company to cover certain costs of implementing and providing number-portability service.

The FCC ensures that the amounts that are charged by local telephone companies are reasonable. The telephone companies are required to file tariffs, or schedules, establishing the rates that they will charge customers before the charges are actually applied to the telephone bill. The FCC reviews the tariffs to determine whether the charges are reasonable and whether they are in accordance with current rules and orders.

Customers are required to pay the long-term number portability charge even if they still have the same telephone-service provider. All customers receive benefits from telephone-number portability any time they call someone who has changed local service providers. For example, because number portability enables a telephone customer to change carriers without changing numbers, callers will not have to contact directory assistance for the new telephone number. In this way, all consumers benefit from number portability.

Multi-Line Customers

Customers who have more than one telephone line pay more for local telephone-number portability service than those who have only one line. The FCC's long-term telephone-number portability cost-recovery system enables incumbent local telephone companies to assess one monthly telephone-number portability charge for each telephone line. Customers who have more than one line will have the opportunity to port each telephone number that they have to another carrier—and therefore are required to pay for number portability in connection with each number they have.

Local telephone companies can charge business customers who have different kinds of lines more than one long-term telephone-number portability charge per line. Business customers can be charged nine, five, or one monthly long-term telephone-number portability charge per line, depending upon their line arrangement. For example, a business customer who has a PBX will be assessed nine monthly number-portability charges.

Exemptions

Cellular and other wireless carriers are not required to provide telephone number portability at this time. Their customers cannot retain the same local telephone number if they change their local service from a wireline local telephone company to a wireless carrier, such as a cellular or PCS service provider. Likewise, customers cannot switch from a cellular or PCS service provider to a local wireline service provider and keep the same cellular or PCS telephone number. Such customers are not charged for local telephone-number portability service. In addition, carriers cannot impose the monthly long-term telephone-number portability charge on customers of the Lifeline Assistance program.

Summary

Customers might have little recourse but to pay the local telephone-portability charge. If the local telephone company has placed this charge on the telephone bill, customers can call the company to ask questions about the service and to make certain that long-term number portability is available in their area. Customers also can look for competition in the local telephone market. New competitors that are offering local telephone service might take different approaches to how they charge customers to cover their costs of providing long-term number-portability service. When deciding whether to change local service providers, customers can factor into their decision the cost savings from lower local telephone-portability charges or the absence of such charges.

See Also

Local Telecommunications Charges

Long-Distance Telecommunications Charges

Long-Distance Telecommunications Charges

Long-distance carriers, also known as *Interexchange Carriers* (IXCs), offer a variety of network services that can be tailored to meet the specific needs of small to large businesses, telecommuters, mobile professionals, and self-employed people who work from offices in their homes. Many of these offerings can be aggregated for call-volume discounts and are consolidated on a single monthly billing statement.

Long-Distance Service Charges

The continued pace of innovation has produced an enormous choice of long-distance services. They are delivered over dedicated private lines, the switched public network, cellular facilities, satellite, or a combination of these elements. Each service has various features and options, with the choice influencing costs and volume discounts. Billing statements consolidate all call activity, making it convenient to track and control long-distance telecommunications costs.

Outbound Services   The major long-distance carriers offer outbound calling services, which are traditionally known as *Wide-Area Telecommunications Service* (WATS). This type of service is a bulk-rate toll service that is priced according to call distance or rate band. Introduced 38 years ago by AT&T, traditional WATS has been replaced by more flexible and manageable virtual WATS-like services in which billing is based on time of day and call duration (as well as distance). Discount plans based on call volume and the number of corporate locations that are enrolled in the plan are available.

Most carriers can provide detailed billing reports that are available on customer request and in a choice of media. Such reports provide specific call detail—and, in some instances, exception reporting, which flags unusual usage patterns, anomalies in billing, and/or significant changes in calling patterns.

Inbound Services   Most IXCs also offer 800 service and the newer 888 service, both of which are also known as inbound WATS. These services are bulk-rate inbound toll service available in various areas. Carriers offer either switched-access or dedicated-access 800 and 888 service. Switched service enables 800 and 888 calls to be received over regular telephone lines; dedicated service provides a private line that is digitally connected from the carrier network to the business network. Each service provides several options. For example, users can geographically screen their calls or block calls from certain parts of the country; other services automatically route calls to specific locations based on customer-specified requirements.

The larger IXCs also offer 800 service for international calls. Callers from international locations use country-specific numbers in order to route calls to a subscriber's access line in the United States. Personal 800 service also exists for individuals who work from home.

Billing for 800 services is based on several factors: destination of the call, time of day when the call is placed, duration of the call, and selection of switched versus dedicated access. Most carriers have discount plans based on the time of day and the day of the week, in addition to offering volume usage discounts; similar discount plans and feature options are available for international 800 services. Detailed billing reports for 800 services, available on customer request, provide specific call detail and exception reporting. The carriers also offer the option to receive call detail information in real time or on a daily or monthly basis via PCs. Such services can be used to measure marketing responses, track lost calls, and gauge the effectiveness of call-center operations.

Virtual Network Services   The three major long-distance carriers—AT&T, MCI WorldCom, and Sprint—all offer voice-oriented *Virtual Private Network* (VPN) services. While each offering has its own name, the three services work in essentially the same way. Each carrier provides feature and option packages, aimed at various-size companies, which combine outbound and inbound services, and offer varying levels of service and discounts based on combined usage.

Basically, VPN services provide capabilities similar to private lines (including line conditioning, error testing, and high-speed, full-duplex transmission with line quality adequate for data) over the public-switched network. VPNs provide on-demand dialup circuits or bandwidths that can be dynamically allocated, reducing or eliminating the need for fixed point-to-point private lines. Most VPNs support analog data transmission speeds at 56Kbps. Most carriers also provide international VPN capabilities with their partners in foreign countries. The international VPNs provide point-to-point, two-way calling capabilities for voice, fax, and data.

Billing is based on the call destination, time of day, and call duration. All of the VPN carriers also have discount plans for volume usage and plans based on the day-of-week usage. There is even the option for cellular access, with all cellular calls added to the total volume discount. Detailed billing reports are provided for VPN service. VPN call detail reports provide specific information about each call—and, in some instances, report exceptions, as well. The bill consolidates call activity and charges for all corporate locations on the VPN.

Special Line Charges   Long-distance carriers offer special access facilities, including (but not limited to) tie lines, *Foreign Exchange* (FX) lines, *Off-Premises Extensions* (OPX), T-carrier facilities, Frame Relay, SONET, ATM, international facilities, and other such services. In addition to the provision of the interexchange portion of the facility, the long-distance carriers also provide coordination for the installation of these facilities with the LECs and the local carriers at international locations. When digital service to a foreign country is not available, the carriers also offer voice-grade private line service. This type of service supports lower-speed applications such as facsimile and e-mail.

Charges for special access facilities usually appear as a single line item under the special access-line heading on the long-distance billing statement. Itemization of these charges is available on a separate facilities record, which can be obtained from the long-distance carrier upon request. This record lists each individual facility by circuit number and describes all associated components and related equipment. Typically, it also identifies the circuit end points by location address and serving central office.

Cellular Services   Similar to the LECs, the IXCs offer cellular services for voice and data. The charges for cellular services can be included on the same billing statement as other types of services and, in many cases, can be included with other types of calls in the volume discount plan.

Some cellular services are data-oriented and are billed according to the number of packets sent and received, as well as call duration and distance of the call. Voice calls are billed according to call duration and distance of the call. The bill shows total air-time charges and call details, and might also contain charges for such optional services as call waiting, roaming, and voice mail. In some cases, there might be the option for unlimited weekend calling for a flat fee, or a monthly package of minutes billed for a flat fee. Also, if the individual or company has elected to ensure the phone against theft or loss, the monthly insurance charge will also appear on the billing statement.

Network-Management Services   As networks have become more complex, the larger carriers have begun offering end-to-end network-management services for multi-vendor environments. These services include management of private line and switched services, WANs, LANs, mainframe connections, and networking equipment such as routers, switches, PBXs, and modems. Managed SNA services even involve the carrier managing and maintaining a Frame Relay network that has been specifically designed and performance-tuned to carry delay-sensitive SNA traffic. These management services are provided 24 hours a day, seven days a week.

These services are essentially custom services, because each company will have its own connectivity, performance, and management requirements. Accordingly, the network management service is priced on an individual company basis, with costs varying by the size of the network, the services included, and the equipment that is to be managed. Once the total cost is determined, the customer is billed a fixed monthly amount. As changes are requested by the customer, these changes are added to the billing statement.

Satellite Services   A few long-distance carriers also provide satellite-based end-to-end network services. The satellite service is usually available on a dedicated or shared basis in order to receive and transmit data, voice, and video signals, with transmission speeds ranging from 56Kbps to 1.544Mbps. Costs vary by the type of service chosen but typically include a monthly service charge and transmission cost. If equipment is purchased, such as *Very Small Aperture Terminals* (VSATs), there is a one-time equipment charge. Often, this type of equipment can be leased as well. In this case, fixed monthly payments will be shown on the bill, along with the transmission charges.

Calling-Card Services   All of the IXCs also offer long-distance calling cards—and, like all of the other services, a wide variety of options are available that affect billing. Standard calling cards enable long-distance calls to be made by dialing an account number and *Personal Identification Number* (PIN). Prepaid calling cards are available that enable users to control long-distance calling time, thereby containing service charges. Charges for this service are based on long-distance usage and options that are chosen with the calling card.

Conferencing Services The IXCs offer conferencing services that include both video conferencing and telephone conferencing. Switched video services provide video conferencing and related options, such as video applications help desks, multi-point conferencing, digital video broadcast services, and connections to other carriers' networks. Some carriers also provide equipment options that are either sold or leased. Most carriers also provide an operator as needed to assist with setting up and connecting all parties to the conference. Telephone conferencing can support communication among hundreds of locations simultaneously. Monthly billing options for both types of conferencing include call detail by department, division, or username; a management report that summarizes conference usage company-wide; and usage-analysis reports, including a sort by cost center for charge-back purposes.

Messaging Services IXC-provided messaging services enable either caller-paid or sponsor-paid live or recorded one-way messages, two-way conversations, and electronic counting for opinion polls, with a wide range of options that are available for each type of service. Other messaging services might include electronic delivery of news clippings, e-mail, fax delivery, and interactive voice services. All carriers are also now providing *Electronic Data Interchange* (EDI) services, which enable companies to electronically exchange business documents with trading partners. The billing charges for these messaging services are based on the options that are chosen, bundling with other service, and usage.

Other Charges and Credits The catch-all category on the long-distance billing statement, Other, includes charges and/or credits for installation, service orders, engineering charges, and other such services that are provided by the long-distance carrier. These service fees tend to be one-time, non-recurring service fees and will show up on the billing statement only once—unless arrangements have been made to spread the costs over multiple billing periods.

Summary

In the current competitive environment, telecommunications companies are learning to differentiate themselves in a variety of ways. In addition to services, features and options, they are seeking to differentiate themselves by the billing services they can offer. In response to customer demand for more sophisticated billing options, the LECs and long-distance carriers are now viewing their billing systems not only as basic collection tools but also as full-fledged information systems that can be tailored to the needs of their customers. Business customers now have the option of viewing their bills via the Internet, just as some credit card issuers now enable customers to view their monthly statements on the Internet with their Web browsers.

See Also

Local Telecommunications Charges
Truth in Billing

Low-Power Frequency-Modulated (LPFM) Radio Service

In January 2000, the *Federal Communications Commission* (FCC) created two new classes of non-commercial radio stations that were referred to as *Low-Power Frequency-Modulated* (LPFM) radio services. LPFM radio services are designed to serve localized communities or underrepresented groups within communities.

The LP100 service operates in the power range of 50 watts to 100 watts and has a service radius of about 3.5 miles. The LP10 service operates in the power range of 1 watt to 10 watts and has a service radius of about 1 mile to 2 miles. In conjunction with the new radio services, the FCC adopted interference-protection requirements based on distance separation between stations. This functionality is intended to preserve the integrity of existing FM radio stations and to safeguard their capability to transition to digital-transmission capabilities.

The FCC put into place minimum distance separations as the best practical means of preventing interference between low-power radio and full-power FM stations. This separation requires minimum distances between stations that use the same or first adjacent channels. Third-adjacent channel and possibly second-adjacent channel separations, however, might not be necessary in view of the low power levels of LPFM radio.

License Requirements

Eligible LPFM licensees can be non-commercial government or private educational organizations, associations or entities; non-profit entities with educational purposes; or government or non-profit entities providing local public safety or transportation services. LPFM licenses will be awarded throughout the FM radio band, however, and will not be limited to the channels that are reserved for use by non-commercial educational radio stations.

To further its goals of diversity and create opportunities for new voices, no existing broadcaster or other media entity can have an ownership interest or can enter into any program or operating agreement with any LPFM station. In addition, in order to encourage locally originated programming, LPFM stations are prohibited from operating as translators.

To foster local ownership and diversity, during the first two years of LPFM license eligibility, licensees are limited to local entities—certifying that they are physically headquartered, have a campus, or have 75 percent of their board members residing within 10 miles of the station that they seek to operate. During this time, no entity can own more than one LPFM station in any given community. After two years from the date the first applications are accepted, in order to bring into use whatever low-power stations remain available but unapplied for, applications will be accepted from non-local entities. For the first two years, no entity will be permitted to operate more than one LPFM station nationwide. After the second year, eligible entities can own up to five stations nationwide, and after three years, up to 10 stations nationwide. LPFM stations will be licensed for eight-year renewable terms. These licenses will not be transferable. Licensees will receive four-letter call signs with the letters LP appended.

In the event that multiple applications are received for the same LPFM license, the FCC will implement a selection process that awards applicants one point each for the following:

- Certifying an established community presence of at least two years prior to the application
- Pledging to operate at least 12 hours daily
- Pledging to air at least eight hours of locally originated programming daily

If applicants have the same number of points, time-sharing proposals will be used as a tiebreaker. Where ties have not been resolved, a group of up to eight mutually exclusive applicants will be awarded successive license terms of at least one year for a total of eight years. These eight-year licenses will not be renewable.

LPFM stations will be required to broadcast a minimum of 36 hours per week—the same requirement that is imposed on full-power, non-commercial educational licensees. They will be subject to statutory rules, such as sponsorship identification, political programming, not airing obscene or indecent programming, and requirements to provide periodic call-sign announcements. They also will be required to participate in the national *Emergency Alert System* (EBS).

Summary

According to the FCC, the new LPFM service will enhance community-oriented radio broadcasting. During the proceedings leading up to the new classes of radio service, broad national interest in LPFM service was demonstrated by the thousands of comments received from state and local government entities, religious groups, students, labor unions, community organizations, musicians, and others supporting the introduction of a new LPFM service. The FCC expects that the local nature of the LPFM service, coupled with the eligibility and selection criteria, will ensure that LPFM licensees will meet the needs and interests of their communities.

See Also

> *Federal Communications Commission* (FCC)
>
> Spectrum Auctions

Low-Power Radio Service

Low-Power Radio Service (LPRS) is one of the *Citizens Band* (CB) Radio Services. LPRS is a one-way short-distance VHF communication service providing auditory assistance to persons with disabilities, persons who require language translation, and persons in educational settings. LPRS also provides health care assistance to the ill, law enforcement tracking services in cooperation with a law enforcement agency, and point-to-point network control communications for *Automated Marine Telecommunications*

System (AMTS) coast stations. In all applications, two-way voice communications are prohibited.

A license from the FCC is not needed to use most LPRS transmitters. To operate an LPRS transmitter for AMTS purposes, however, the user must hold an AMTS license. Otherwise, provided that the user is not a representative of a foreign government, anyone can operate an FCC type-accepted LPRS transmitter for voice, data, or tracking signals.

An LPRS transmitter can be operated within the territorial limits of the 50 United States, the District of Columbia, and the Caribbean and Pacific insular areas. The transmitter can also be operated on or over any other area of the world, except within the territorial limits of areas where radiocommunications are regulated by another agency of the United States, or within the territorial limits of any foreign government. The transmitting antenna must not exceed 30.5 meters (100 feet) above ground level. This height limitation does not apply, however, to LPRS transmitter units that are located indoors or where the antenna is an integral part of the unit.

260 channels are available for LPRS. These channels are available on a shared basis only and are not assigned for the exclusive use of any entity. Certain channels (19, 20, 50, and 151 through 160) are reserved for law-enforcement tracking purposes. Further, AMTS-related transmissions are limited to the upper portion of the band (216.750MHz to 217.000MHz).

Users must cooperate in the selection and use of channels in order to reduce interference and make the most effective use of the authorized facilities. Channels must be selected in an effort to avoid interference to other LPRS transmissions. In other words, if users are experiencing interference on a particular channel, they should change to another channel until a clear one is found.

Finally, operation is subject to the conditions that no harmful interference is caused to the United States Navy's SPASUR radar system (216.88MHz to 217.08MHz) or to a Channel 13 television station.

Summary

LPRS can operate anywhere CB station operation is permitted. An LPRS station is not required to transmit a station identification announcement. The LPRS transmitting device may not interfere with TV reception or federal government radar, and must accept any interference received, including interference that may cause undesired operation. All LPRS system equipment must be made available for inspection by an authorized FCC representative upon request.

See Also

  Citizens Band (CB) Radio Service
  Family Radio Service (FRS)
  General Mobile Radio Service (GMRS)

CHAPTER M

Managed Systems Network Architecture (SNA) Services

As companies move from IBM's host-centric (i.e., mainframe) to the distributed computing environment (i.e., *Local-Area Networks* [LANs]), the major carriers are offering *Wide-Area Network* (WAN) services that address the specific needs of SNA users. Typically, these managed SNA services operate over Frame Relay connections, but they can also operate over well-managed IP backbones. Such services are especially suited to organizations that have many remote locations that must be tied into one or more hosts. Among the advantages of managed SNA services over Frame Relay are the following:

- Frame Relay's *Permanent Virtual Circuits* (PVCs) replace expensive *Synchronous Data Link Control* (SDLC) and *Binary Synchronous Communications* (BSC) multi-drop networks between the host and branch offices.

- Consolidating connections through Frame Relay eliminates costly *Serial-Line Interface Coupler* (SLIC) ports on *Front End Processors* (FEPs) while increasing performance.

- WAN access extends the useful lives of SDLC/BSC controllers and 3270 terminals.

- The availability and reliability of SNA connections are increased by enabling controllers to take advantage of WAN connections that have multiple host paths.

A managed Frame Relay service includes the *Frame Relay Access Devices* (FRADs)—leased or purchased—that transport SNA traffic over the PVCs. FRADs are more adept than routers at congestion control and prioritization. Some FRADs multiplex multiple SNA/SDLC devices onto the same PVC instead of requiring separate PVCs for each attached device, resulting in even greater cost savings.

The FRADs encapsulate SNA/LLC2 frames with minimal overhead and enable traffic to be selectively prioritized, ensuring that mission-critical SNA data arrives in a timely fashion. Most FRADs perform local polling acknowledgment of keep-alive frames in order to minimize the risk of timing out SNA sessions. Some services permit NetView visibility and control to be extended to the attached SNA devices.

For a legacy SNA shop that does not have the expertise or resources, a managed SNA service can be an economical interconnectivity option. Although Frame Relay networks are much more difficult to configure, administer, and troubleshoot than private lines, the carrier assumes these responsibilities. In addition, the carrier provides design and reconfiguration assistance in order to ensure that mission-critical applications are providing the highest level of performance, reliability, and availability.

A legitimate concern exists that Frame Relay is not yet effective in guarding against the loss of data during congestion conditions. Although warnings of impending congestion are relayed from the carriers' Frame Relay switches to the FRADs on the edge of the network, they cannot force them to adjust the frame rate. If the FRADs do not react to these congestion indicators, the carrier networks can discard frames in order to reduce congestion. This issue is usually not serious, however. The carriers routinely over-provision their Frame Relay networks to guard against congestion and possible frame loss, and they will continue to do so in order to attract new SNA customers.

The Frame Relay services themselves are priced attractively. On average, the Frame Relay service costs about 25 percent less than the equivalent private network. In some cases, discounts of up to 40 percent are possible. Penalty fees are waived if upgrading the network entails having to break contracts for other services or leases on hardware. In some cases, it might be possible to get the first month of the managed SNA service free, so the old services can continue to be used until the changeover to Frame Relay is complete.

Summary

Customer comfort levels with SNA over Frame Relay are rising because of the service's maturity and track record of reliability. In addition, SNA-specific services that prioritize SNA traffic, that guarantee performance with *Service-Level Agreements* (SLAs), and that assist with SNA private line to Frame Relay migration provide additional incentives.

See Also

Frame Relay

Microwave Communications

A microwave is a short radio wave that varies from 1 millimeter to 30 centimeters in length. Because microwaves can pass through the ionosphere, which blocks or reflects longer radio waves, microwaves are well suited for satellite communications. This reliability makes microwaves well suited to terrestrial communications as well, such as those that are delivered by *Local Multi-Point Distribution Service* (LMDS) and *Multi-Channel Multi-Point Distribution Service* (MMDS).

Much of the microwave technology that is in use today for point-to-point communications was derived from radar that was developed during World War II. Initially, microwave systems carried multiplexed speech signals over common carrier and military communications networks, but today they are used to handle all types of information (voice, data, facsimile, and video) in either an analog or digital format.

The first microwave transmission occurred in 1933, when European engineers succeeded in communicating reliably across the English Channel—a distance of about 12 miles (20 km). In 1947, the first commercial microwave network in the United States came online. Built by Bell Laboratories, this network was a New York-to-Boston system consisting of 10 relay stations carrying television signals and multiplexed voice conversations.

A year later, New York was linked to San Francisco via 109 microwave relay stations. By the 1950s, transcontinental microwave networks were routinely handling more than 2,000 voice channels on hops averaging 25 miles (41.5 km) in length. By the 1970s, not a single telephone call, television show, telegram, or data message crossed the country without spending some time on a microwave link.

Over the years, microwave systems have matured to the point that they have become major components of the nation's *Public Switched Telephone Network* (PSTN) and an essential technology with which private organizations can satisfy internal communications requirements. Microwave systems can even exceed the 99.99 percent reliability standard that is set by the telephone companies for their phone lines.

Microwave Applications

Early technology limited the operations of microwave systems to the radio spectrum in the 1GHz range, but due to improvements in solid-state technology, today's government systems are transmitting in the 153GHz region while commercial systems are transmitting in the 40GHz region with FCC approval. The 64GHz to 71GHz band is reserved for inter-satellite links.[1] These frequency bands offer short-range wireless radio systems the means to provide communications capacities approaching those that are now achievable only with coaxial cable and optical fiber.

These spectrum allocations offer a variety of possibilities, such as use in short range, high-capacity wireless systems that support educational and medical applications, and wireless access to libraries or to other information databases. Short-haul microwave communications equipment is also routinely used by hotel chains, CATV service providers, and government agencies. Corporations are making greater use of short-haul microwave, especially for extending the reach of LANs in places where the cost of local T1 lines is prohibitive. Common carriers use microwave systems for backup in the event of fiber cuts and on terrain where laying fiber is not economically feasible. Cellular service providers use microwave to interconnect cell sites as well as to the regular telephone network. Some interexchange carriers (and corporations) even use short-haul microwave to bypass incumbent local-exchange carriers to avoid lengthy service-provisioning delays and to avoid paying hefty local-access charges.

Network Configurations

Now, more than 25,000 microwave networks exist in the United States alone. Basically two microwave network configurations exist: point-to-point and point-to-multi-point. The first type meets a variety of low and medium-density communications requirements, ranging from simple links to more complex extended networks, such as the following:

- Sub-T1/E1 data links
- Ethernet/Token Ring LAN extensions
- Low-density digital backbone for wide-area mobile radio and paging services
- PBX/OPX/FX voice, fax, and data extensions
- Facility-to-facility bulk data transfer

---

[1]The FCC maintains a 60-page *Table of Frequency Allocations* that can be found on its Web page at http://www.fcc.gov/oet/info/database/spectrum/Welcome.html.

Point-to-multi-point microwave systems provide communications between a central command and control site and remote data units. A typical radio communications system provides connections between the master control point and remote data collection and control sites. Repeater configurations are also possible. The basic equipment requirements for a point-to-multi-point system include the following:

- *Antennas* For the master, an omni-directional antenna; for the remotes, a highly directional antenna that is aimed at the master station's location
- *Tower (or other structure, such as a mast)* To support the antenna and transmission line
- *Transmission line* Low-loss coaxial cable that connects the antenna and the radio
- *Master station radio* Interfaces with the central computer; transmits and receives data from the remote radio sites and can request diagnostic information from the remote transceivers. The master radio can also serve as a repeater.
- *Remote radio transceiver* Interfaces to the remote data unit; receives and transmits to the master radio
- *Management station* A computer that can be connected to the master station's diagnostic system, either directly or remotely, for control and collection of diagnostic information from master and remote radios

Wireless Cable

Traditionally, cable-system operators have used microwave transmission systems to link cable networks. These *Cable Antenna Relay Services* (CARS) have experienced declining usage as cable operators have deployed more optical fiber in their transmission systems. Improvements in microwave technology and the opening of new frequencies for commercial use, however, have contributed to the resurgence in short-haul microwave. In the broadcast industry, short-haul microwave is often referred to as wireless cable, which comes in the form of *Local Multi-Point Distribution Service* (LMDS) and *Multi-Channel Multi-Point Distribution Service* (MMDS).

These wireless cable technologies have two key advantages, one of which is availability. With an FCC license, they can be made available in areas of scattered population and in other areas where it is too expensive to build a traditional cable station. The other advantage is affordability. Due to the lower costs of building a wireless cable station, savings can be passed on to subscribers.

Summary

Microwave is now almost exclusively a short-haul transmission medium, while optical fiber and satellite have become the long-haul transmission medium of choice. Short-haul microwave is now one of the most agile and adaptable transmission media available, with the capability of supporting data, voice, and video. Short-haul microwave is also used to back up fiber-optic facilities and to provide communications services in locations where it is not economically feasible to install fiber.

See Also

Local Multi-Point Distribution Service (LMDS)

Multi-Channel Multi-Point Distribution Service (MMDS)

Satellite Communications

Modems

A modem (modulator-demodulator) converts the digital signals that are generated by a computer into analog signals that are suitable for transmission over dialup telephone lines or voice-grade leased lines. Another modem, located at the receiving end of the transmission, converts the analog signals back into digital form for manipulation by the data recipient. Although long-distance lines are digital, most local lines are not—which explains why modems are often required to transfer files, access bulletin boards, send e-mail, and connect to host computers from remote sites.

Modems are packaged as internal, external, or rack-mounted models. Internal modems insert into the vacant slot of a desktop computer or notebook, while external desktop modems connect to a computer's RS-232 serial port. Rack-mounted modems are full or half cards that are housed in an equipment frame located in a wiring closet. From there, cables connect to each PC's serial port. Modem cards can also be installed in a communications server that is attached to the corporate LAN for sharing by many users.

Modem manufacturers are continually redesigning their products to incorporate the latest standards, to enhance existing features, and to add new features. The advancement of modulation techniques, error correction, data compression, and diagnostics are among the continuing activities of modem manufacturers.

Modulation Techniques

Modems use modulation techniques to encode the serial, digital data that is generated by a computer onto the analog carrier signal. The simplest modulation techniques employ two signal manipulations in order to transmit information: *Frequency-Shift Keying* (FSK) and *Phase-Shift Keying* (PSK). FSK is similar to the frequency modulation technique that is used to broadcast FM radio signals. By forcing the signal to shift back and forth between two frequencies, the modem can encode one frequency as a 1 and the other as a 0. This modulation technique was widely used in the early 300bps modems. At higher rates, FSK is too vulnerable to line noise to be effective.

PSK, another early modulation technique, makes use of shifts in a signal's phase to indicate 1s and 0s. The problem with this method is that *phase* refers to the position of a waveform in time; therefore, the data-terminal clocks at both ends of the transmission must be synchronized precisely.

Another method, known as *Differential-Shift Phase Keying* (DPSK), uses the phase transition to indicate the logic level. With this scheme, you do not have to assign a specific binary state to each phase; rather, it only matters that some phase shift has taken place. The telephone bandwidth is limited, however, so it is only possible to have 600 phase transitions per second on each channel, thus limiting transmission speeds to about 600bps. To increase speed, you can expand DPSK from a two-state to a four-state pattern that is represented by four two-bit symbols (known as a *dibit*) as follows:

- Maintain the same state (0,0)
- Shift counterclockwise (0,1)
- Shift clockwise (1,0)
- Shift to the opposite state (1,1)

Other modulation techniques, such as trellis encoding, are much more sophisticated and capable of moving data at much higher speeds. Trellis encoding entails the use of a 32-bit constellation with *quintbits*, in order to pack more information into the carrier signal and to offer more immunity from noise. The use of quintbits offers 16 extra possible state symbols. These extra transition states enable dial-line modems to use transitions between points, rather than specific points, to represent state symbols. The receiving modem uses probability rules to eliminate illegal transitions and to obtain the correct symbol, which gives the transmission greater immunity to line impairments. Additionally, the fifth bit can be used as a redundant bit, or checksum, in order to increase throughput by reducing the probability of errors.

By increasing the number of points in the signal constellation, you can encode greater amounts of information in order to increase the modem's throughput. Ever slighter variations in the phase-modulated signal might be used to represent coded information, which translates into higher throughput. Some modem manufacturers use constellations consisting of 256 or more points.

Transmission Techniques

Modems support two types of transmission techniques: asynchronous or synchronous. The user's operating environment determines whether an asynchronous or synchronous modem is required. During asynchronous transmission, start and stop bits frame each segment of data during transfer in order to distinguish each bit from the one that precedes it. Synchronous transmission transfers data in one continuous stream; therefore, the transmitting and receiving devices must be synchronized precisely in order to distinguish each character in the data stream. Some types of modems support both types of transmission.

Modem Speed

For many users, the most important modem characteristic is data rate. The quality of the connection has a lot to do with the actual speed of the modem. If the connection is noisy, for example, the modem might have to step down to a lower speed in order to continue transmitting data. Some modems can sense improvements in line quality and can automatically step up to higher data rates as line quality improves.

In 1997, a new class of modems became available that offered data transmission rates of up to 56Kbps. The modems are based on technology that exploits the fact that for most of its length, an analog modem connection is

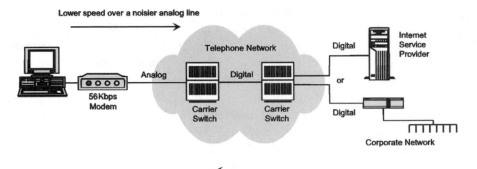

Figure M-1
Today's 56Kbps modems can send data at top speed, but only from a digital source. Because the signal is already in digital form, the traffic is free of impairments from noise that are introduced when an analog modem signal is made digital within the carrier network. From an analog source, the top speed is quite a bit lower than 56Kbps, because the traffic is subject to impairment from noise.

really digital. When an analog signal leaves the user's modem, the signal is carried to a phone company's central office to be digitized. If the signal is destined for a remote analog line, the signal is converted back to analog at the central office that is nearest to the receiving user. The conversion is made at only one place: where the analog line meets the central office. During the conversion, noise is introduced that cuts throughput. But the noise is less in the other direction, from digital to analog, enabling the greater downstream throughput (refer to Figure M-1).

You can also tie together two or three 56Kbps modems in order to achieve a combined data rate of up to 168Kbps (3 x 56 Kbps)[2] over multiple dialup lines (refer to Figure M-2). When two or three modems in a modem-pool device are used for Internet access, for example, they call the *Internet Service Provider* (ISP) simultaneously and share the downloading of Web pages, resulting in the greater throughput rate. Download time can be cut by as much as two-thirds. Although the ISP does not have to do anything

---

[2]Because the quality of each line can differ at any given time, the aggregate speed of three dialup lines will actually be much less than 168Kbps.

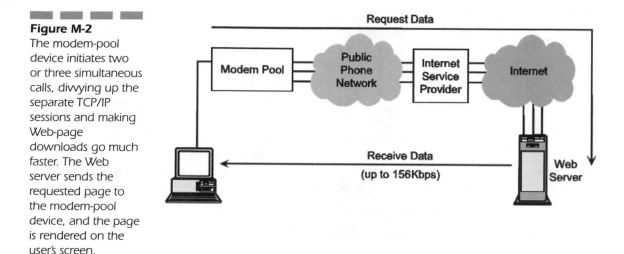

Figure M-2
The modem-pool device initiates two or three simultaneous calls, divvying up the separate TCP/IP sessions and making Web-page downloads go much faster. The Web server sends the requested page to the modem-pool device, and the page is rendered on the user's screen.

different as far as hardware is concerned, except have enough 56Kbps modems, the ISP must enable users to establish multiple sessions with a single user ID and password.

Modem Features

Most modems come equipped with the same basic features, including error correction and data compression. In addition, they have features that are associated with the network interface, such as flow control and diagnostics. Various security features that modems implement are also available.

Error Correction   Networks often contain disturbances that modems must handle (or, in some cases, overcome). These disturbances include attenuation distortion, envelope-delay distortion, phase jitter, impulse noise, background noise, and harmonic distortion—all of which negatively affect data transmission. To alleviate the disturbances that are encountered when transferring data over leased lines (without line conditioning) and dialup lines, most products include an error-correction technique in which a processor puts a bit stream through a series of complex algorithms prior to data transmission.

The most prominent error-correction technique has been the *Microcom Networking Protocol* (MNP), which uses the *Cyclic Redundancy Check* (CRC) method for detecting packet errors and requests retransmissions when necessary.

Link Access Procedure B (LAP-B), a similar technique, is a member of the *High-Level Data-Link Control* (HDLC) protocol family, the error-correcting protocol in X.25 for packet-switched networks. LAP-M is an extension to that standard for modem use and is the core of the ITU error-correcting standard, V.42. This standard also supports MNP Stages 1 through 4. Full conformance with the V.42 standard requires both LAP-M and MNP Stages 1 through 4 to be supported by the modem. Virtually all modems that are currently made by major manufacturers conform to the V.42 standard.

The MNP is divided into nine classes. Only the first four classes deal with error recovery, which is why only those four are referenced in V.42. The other five classes deal with data compression. The MNP error-recovery classes are as follows:

- MNP Classes 1 to 3, which packetize data (and, Microcom claims, ensure 100 percent data integrity)
- MNP Class 4, which achieves up to 120 percent link-throughput efficiency via Microcom's Adaptive Packet Assembly and Data Phase Optimization, which automatically adjusts packet size relative to line conditions and reduces protocol overhead

Data Compression   With the adoption of the V.42bis recommendation by the ITU in 1988, there is a single data-compression standard: Lempel-Ziv. This algorithm compresses most data types, including executable programs, graphics, numerics, ASCII text, or binary data streams. Compression ratios of 4:1 can be achieved, although actual throughput gains from data compression depend on the types of data that are being compressed. Text files are most likely to yield performance gains, followed by spreadsheet and database files. Executable files are most resistant to compression algorithms because of the random nature of the data.

Diagnostics and Other Features   Most modems perform a series of diagnostic tests to identify internal and transmission-line problems. Most modems also offer standard loopback tests, such as local analog, local digital, and remote digital loopback. Once a modem is set in test mode, characters that are entered on the keyboard are looped back to the screen for verification.

Most modems also include call-handling features such as automatic dial, answer, redial, fallback, and call-progress monitoring. Calling features simplify the chore of establishing and maintaining a communications connection by automating the dialing process. Telephone numbers can be stored in non-volatile memory.

Other standard modem features that are commonly offered include fallback capability and remote operation. Fallback enables a modem to automatically drop, or fall back, to a lower speed in the event of line noise and then revert to the original transmission speed after line conditions improve. Remote operation, as the name implies, enables users to activate and configure a modem from a remote terminal.

Security   Modems that offer security features typically provide two levels of protection: password and dial-back. The former requires the user to enter a password, which is verified against an internal security table. The dial-back feature offers an even higher level of protection. Incoming calls are prompted for a password, and the modem either calls back to the originating modem by using a number that is stored in the security table or prompts the user for a telephone number and then calls back.

Security procedures can be implemented before the modem handshaking sequence, rather than after the sequence. This procedure effectively eliminates the access opportunity for potential intruders. In addition to saving connection establishment time, this method uses a precision, high-speed, analog security sequence that is not even detectable by advanced line-monitoring equipment.

Transmission Facilities

Modems support two types of lines: leased or dialup. The primary difference between the two involves the procedure for establishing a connection, as opposed to the line itself.

Dialup Lines   Dialup lines are used for typical telephone service. These lines usually connect to a small, modular wall jack (called an RJ-11) and a companion plug that is inserted into the jack in order to establish a connection to the telephone or modem. When a voice-grade analog line is used with a modem, a short cord with an RJ-11 connector on both ends is inserted into the jack and into the modem.

Although the RJ-11 modular-jack connection is common for low-speed modem connection, problems can arise with regard to the consistency of the signals that are transmitted over the line. To ensure a consistently high signal level, a special data-line jack, such as RJ-41 or RJ-45, can be installed. These data-line jacks are designed specifically to operate with modem circuitry and are often used in leased-line environments to help maintain the quality of the transmitted signal.

Leased Lines   Leased lines are available in two-wire and four-wire versions. Four-wire leased lines differ from their two-wire counterparts in terms of cost (four-wire lines are more expensive) and in the mode of modem operation that is supported. Not all modems can support leased lines. An effective way of determining whether a modem can support a leased-line connection is to examine the way in which the telephone line is connected. Modems that are designed for two-wire and four-wire leased-line operation have two sets of terminal screws with which to attach the two pairs of lines.

To sustain the optimal performance of leased-line modems, the lines might be specifically selected for their desirable characteristics. This service is an extra cost called line conditioning, which is provided by the carriers on a best-effort basis. AT&T's D6 line conditioning, for example, addresses phase jitter, attenuation distortion, and envelope-delay distortion—all of which can impair transmission at data rates approaching 19.2Kbps.

The monthly charges and installation cost of D6 conditioning are higher than for other levels of conditioning, if only because there are fewer wire pairs available that exhibit the higher immunity from generic noise and non-linear distortion in high-density locations. Immunity from noise and non-linear distortion occurs on copper pairs more by chance than by design, because they might be caused externally and might be beyond the control of the carrier. In other words, numerous wire pairs must be tested before those having the desired characteristics can be identified and put into service as conditioned lines.

Wireless Links   Wireless modems are required to transfer data over public wireless services and private wireless networks. These modems come in a variety of hardware configurations: stand-alone, built-in, and removable PC card. The newer modems are programmable and are therefore capable of being used with a variety of wireless services that use different frequencies and protocols. There are even modems that mimic wireline protocols, enabling existing applications to be run over the wireless network without modification.

Single-frequency modems   Private wireless networks operate in a range of unique frequency bands in order to ensure privacy. Using radio modems that operate over dedicated frequencies within these frequency bands also permits the transmission of business-critical information without interference problems. Furthermore, the strategic deployment of radio modems can provide metropolitan-area coverage without the use of expensive antenna arrays.

Such modems are designed to provide a wireless, protocol-independent interface between host computers and remote terminals that are located as far away as 30 miles. Most provide a transmission rate of at least 19.2Kbps point-to-point in either half-duplex or full-duplex mode. Some radio modems even support point-to-multi-point radio network configurations, serving as a virtual multi-drop radio link that replaces the need for expensive, dedicated lines (refer to Figure M-3). In this configuration, one modem is designated as the master, passing polling information and responses between the host and terminals over two different frequencies.

In multi-drop configurations, a given radio network is capable of supporting one type of asynchronous or synchronous polling protocol. Because such modems perform no processing or interpreting of the protocol, the host (or front-end processor) must generate all required protocol framing, line discipline, node addressing, and data encapsulation. Also, depending on the vendor, these modems might be optionally equipped with an integral repeater in order to maintain signal integrity over longer distances.

Multi-frequency modems   Regardless of the transmission technology or hardware configuration that is used, the modem must be tuned to the frequency of the service provider's wireless network in order to operate properly. Until recently, modems were offered in different versions according to the wireless network to which the modem would connect. This situation delayed product development and inflated the cost of manufacturing, and this cost was passed on to users in the form of higher prices for equipment. To overcome these problems, chip manufacturers developed programmable chip sets that are not limited to a specific network's radio frequency. Newer wireless modems are computer configurable; within specified frequency ranges, the transmit and receive frequencies are independently selectable via software.

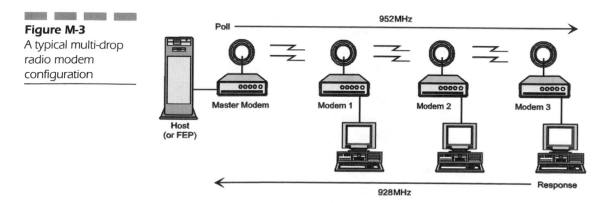

Figure M-3
A typical multi-drop radio modem configuration

Multimedia modems   Not only can modems be programmed for multi-frequency use, but they can also provide seamless integration of multiple media—wireline and wireless—through a common, programmable interface. This goal is accomplished with a chip set that supports both wireline and wireless communications. Special software that is used with the chip set provides a method for connecting cellular phones to modems, which is important because cellular phones lack dial tones and other features that modems use on the wireline phone network. The software makes it appear that those features exist.

PC Card Modems

The advantages of PC Card (formerly known as PCMCIA) technology now extend to modems. These modems are increasingly being used for connecting portable computer users with hosts and LANs at corporate locations. A variety of such modems are available for portable computers, enabling users to access the Internet, e-mail services, and facsimile services. These credit card-sized devices use the Type II slot that now comes standard with most portable computers. Also, cards are available that connect to various wireless messaging services, including those that are based on *Cellular Digital Packet Data* (CDPD) technology. They have a built-in antenna and can even act as stand-alone receivers when the computer is shut off. Some cards are programmable, enabling users to access or receive messages from different wireless services.

CDPD modems are available that work with any DOS-based or Windows-based computer, supporting the V.22bis, V.23, V.23bis, V.42bis, Group 3 fax, and V.17 wireline fax and data protocols, plus Microcom's MNP-10 cellular protocol or Paradyne's *Enhanced Cellular Throughput* (ETC) protocol. Some modems can automatically identify the types of modem protocols that are used at the receiving end and adjust their own operation accordingly.

The latest PC Card trend melds a 10Base-T Ethernet LAN adapter and a fax/modem on a single card. Some manufacturers have even added wireless capabilities to such cards, providing support for one-way and two-way paging as well as direct cellular phone connection. These multi-function cards are an ideal choice for subnotebook computers and *Personal Digital Assistants* (PDAs). Of note, however, is that these multi-function cards do not yet support all of the special features that are typically found in single-function cards.

Multi-Function Modems

Multi-function modems use programmable *Digital Signal Processing* (DSP) technology to turn a computer into a complete desktop message center, enabling the user to control telephone, voice (recording and playback), fax, data transfers, and e-mail. Typical features include multiple mailboxes for voice mail, caller ID support, call forwarding, remote message retrieval, a phone directory, and a contact database.

In some cases, the modem is actually on a full-duplex sound card. By plugging in speakers and a subwoofer, the user can even enjoy a stereo-sound speakerphone. A separate connection to a CD-ROM player enables the user to work at the computer while listening to music. With DSP, the modem can be easily upgraded to the latest communications standards, and new capabilities can be added by simply loading additional software. For example, a 33.6Kbps can be upgraded to 56Kbps by installing new software (instead of having to buy new hardware)—often at no extra charge from the vendor.

Multimedia modems use *Digital Simultaneous Voice and Data* (DSVD), which enables the user to send voice and data at the same time over a single telephone line. The biggest advantage of DSVD is that users no longer need to interrupt telephone conversations or install a separate line in order to transmit data or receive faxes. Multimedia modems typically include full-duplex speakerphone, fax, modem, and 16-bit stereo audio capabilities.

Another new way in which vendors are packaging modems is by integrating them with ISDN terminal adapters, which enables users to communicate with conventional dialup services and to also take advantage of ISDN when possible—all without cluttering the desktop or having to use up scarce slots in the PC.

Soft Modems

The latest step in modem packaging is to eliminate the need for dedicated hardware altogether. Intel's *Native Signal Processing* (NSP) initiative, for example, embeds software emulation of modem and sound-card hardware in the Pentium chip itself. The idea is to enable any Windows application to have access to features that are implemented in a special driver—to send and receive e-mail and faxes or to play sound files—with no modem or specialized hardware, aside from the main CPU.

Motorola takes a different approach, offering a line of *Host Signal Processor* (HSP) modems that rely heavily on software and the processing power of the host PC. But rather than relying on its own digital signal processor, this type of modem is based on a cheaper *Application-Specific Integrated Circuit* (ASIC) that takes advantage of a PC's central Pentium chip for the processing power. HSP technology has been around for several years but has only recently become feasible because of the growing availability of faster desktop computers based on the Pentium chip.

Summary

Once thought to be an outdated technology that would be supplanted by ISDN terminal adapters that are connected to digital lines, modems are not only growing in use but are undergoing a surge in innovation, as well. Higher-speed modems, multi-function modems, and new technologies that rely more on a computer's CPU for carrying out modem functions have all combined in order to breathe new life into this market segment.

See Also

Cable Television Networks

Digital Subscriber-Line Technologies (DSLT)

Integrated Services Digital Network (ISDN)

Multimedia Networking

Multimedia networking involves the delivery of real-time applications over LANs and WANs. Because multimedia applications can consist of several data types (text, voice, images, and video, for example)—all of which can be stored in different locations—the problem involves how to ensure their synchronized delivery. Even when only one data type is involved, such as a video stream or an audio stream, the problem involves how to ensure that its delivery is smooth and is not interrupted whenever the network becomes clogged with other kinds of traffic. For example, if the audio portion of a video conference does not arrive at the same time as the video components, the lip movements of participants will not be synchronized with their actual conversations.

The problem of synchronization can be addressed by assigning a class of service or *Quality of Service* (QoS) to a multimedia application, which identifies it as being time sensitive and requiring priority over other less time-sensitive data types. While standards-based ATM networks are designed to integrally support QoS for multimedia applications, the more popular installed base of Ethernet LANs was not designed for this functionality. Consequently, other techniques must be employed in order to run multimedia applications over legacy LANs, as well as the ubiquitous Internet and the growing number of IP-based intranets.

Multimedia over LANs

To produce integrated, high-quality graphics and sound, multimedia applications need regular, predictable data delivery. But Ethernet delivery timing is not predictable. Not only must workstations contend for network access, but a full-load Ethernet transmission is also subject to excess latency and jitter. Latency is the delay between the time that data is transmitted and the time that it is received at the destination. Jitter is the uncertainty in the arrival time of a packet or the variability of latency. Although video-compression techniques such as MPEG are essential to high-quality multimedia, they are not enough to overcome these problems.

But even the limitations of Ethernet can be surmounted in order to support multimedia applications. A variety of techniques can be used to overcome excess packet delay, including the use of star-wired switching configurations and vendor-specific enhancements to Ethernet that are transparent to each end-system adapter and that are backward-compatible with existing Ethernet adapters.

One of these enhancements is a traffic-control algorithm that is implemented at the hub and enables each Ethernet segment to operate at more than 98 percent efficiency—and, even under full load, to service a mix of real-time and conventional data traffic. The use of such traffic-control algorithms provides predictable LAN transmission by regulating the flow of traffic on the link in order to minimize jitter. The result is increased Ethernet predictability in support of real-time multimedia applications, even at 100Mbps.

A serious limitation of Ethernet for multimedia transmission is that it offers no priority-access scheme. All traffic must contend for access on a best-efforts basis, causing delay in getting data onto the network in the first place. Some vendors offer a method of prioritizing traffic over Ethernet in order to deliver QoS to applications. In one scheme, traffic can be prioritized as either high or low, with high-priority traffic permitted onto the network before low-priority traffic. When a workstation receives a video stream from the server, the data is temporarily held in the workstation's buffer. Through flow-control techniques, the video stream is released for viewing at a consistent bit rate, which has the effect of smoothing out any jerkiness in the video and enabling the audio to better synchronize with the video. The whole process provides the illusion of real-time multimedia transmission. As the buffer empties, it is continually filled as the server contends for priority access to the network.

Similar priority schemes can be applied to Token Ring LANs with even better results, because Token Ring LANs provide each workstation with guaranteed access to the transmission facility. With each workstation given its turn to access the transmission facility for a given length of time, latency is predictable and jitter becomes much less of a problem. Natively, therefore, Token Ring is better at handling multimedia applications than Ethernet. The reason why Ethernet surpasses Token Ring in market share is that Token Ring is far more expensive to implement than Ethernet, and Ethernet has always had a growth path toward higher speeds.

Multimedia over IP

The problem of delay is more serious when one tries to run multimedia applications over IP networks, including corporate intranets and the public Internet. A key disadvantage of the *Internet Protocol* (IP) is that it does not allocate a specific path or amount of bandwidth to a particular session. The resulting delay can vary wildly and unpredictably, disrupting real-time applications. A number of solutions have been proposed that attempt to perform resource-setup functions that are similar to that of the Q.93x signaling protocol that is used in the public circuit-switched network. There, a dedicated path is set up between two parties that remains for the duration of the conversation, which results in excellent speech quality with little or no delay.

A similar mode of operation can be applied to IP networks with the addition of certain protocols to the routers. One of the most promising of these is the *Resource Reservation Protocol* (RSVP), developed by the *Internet Engineering Task Force* (IETF).

As an Internet control protocol, RSVP runs on top of IP in order to provide receiver-initiated setup of resource reservations on behalf of an application data stream. When an application requests a specific quality of service for its data stream, RSVP is used to deliver the request to each router along the path(s) of the data stream and to maintain router and host states in order to support the requested level of service. In this way, RSVP essentially enables a router-based network to mimic a circuit-switched network on a best-efforts basis.

At each node, the RSVP program applies a local decision procedure, called admission control, to determine whether it can supply the requested QoS. If admission control succeeds, the RSVP program in each router passes incoming data packets to a packet classifier that determines the route and the QoS class for each packet. The packets are then queued as necessary in a packet scheduler that allocates resources for transmission on the particular link. If admission control fails at any node, the RSVP program returns an error indication to the application that originated the request.

The advantage of RSVP is that it will work with any physical network architecture. In addition to Ethernet, RSVP will run over other popular networks such as Token Ring and FDDI—as long as IP is the underlying network protocol. This functionality makes RSVP suitable for company-wide networks as well as for the Internet, providing end-to-end service between the two.

RSVP also meshes well with the next generation of IP (version 6 or IPv6), enabling users to set up end-to-end connections with a specified amount of flow control for a given time period. This functionality is made possible by the capability of IPv6 to label packets in traffic patterns, making it easier to identify packets that belong to particular traffic flows for which the sender requests special handling. In other words, time-sensitive services such as real-time video and voice will get the special handling that they require along the route path. IPv6 could take a few years to be widely implemented throughout the Internet. Until IPv6 is widely implemented, many vendors do not see a reason to support RSVP.

Role of Middleware

Middleware can play a key role in implementing multimedia applications over LANs, particularly in enabling Windows applications to run over ATM. The middleware runs invisibly under the application layer and above the network layer (refer to Figure M-4), bridging the gap between ATM standards and the de facto Microsoft Windows standards. In this scenario, middleware brings ATM QoS directly to the application, resulting in high-quality voice, video, text, and images to the desktop while leveraging

Figure M-4
A middleware
solution enables
Windows
applications to run
over ATM. A
predetermined class
of service is assigned
according to the type
of data stream that is
detected by the
middleware (in this
case, LAN data,
MPEG, or a H.320
video conference).

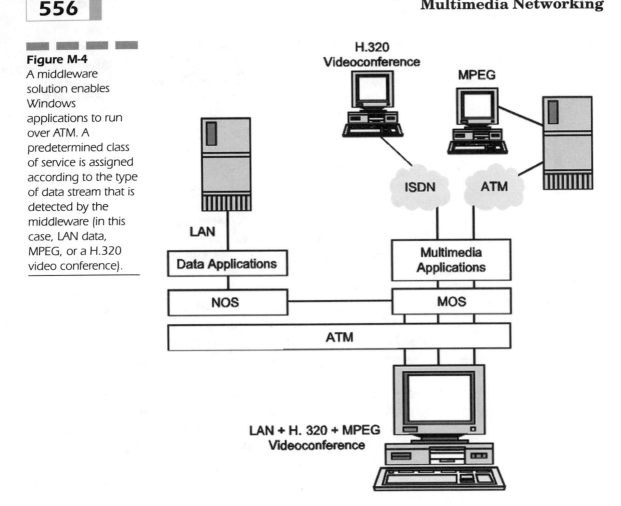

the inherent management, security, and authentication features of the existing *Network Operating System* (NOS).

Users on ATM-attached PCs still run the same applications, and they can access the LAN server as before—except that they can run networked multimedia. QoS is carried out automatically by the middleware, based on the type of traffic. Predetermined delay and bandwidth parameters are allocated to the applications according to their bandwidth and delay needs (measured in milliseconds). For example, MPEG traffic can be assigned 1.5Mbps of bandwidth, while an audio file is granted 5Mbps. For a H.320-based video conference, 384Kbps is assigned with a delay of 200 milliseconds to 300 milliseconds. All of these services—including applications that run over TCP/IP, IPX, and NetBIOS protocols—can be delivered with guaranteed end-to-end quality by using the existing desktop wiring infrastructure. Furthermore, no changes to existing hardware and software are required.

To implement video conferencing, for example, each PC must have an ATM adapter with a direct connection to an H.320-based codec interface board. An H.320-compliant multimedia gateway server connects local ATM workgroups to remote sites through ISDN for video conferencing across the WAN.

The multimedia gateway server is an integrated hardware and software package that upgrades a designated PC to an ATM/ISDN gateway with BRI or PRI connections. An ISDN connection is made via the server, rather than directly at the PC level. The server initiates external calls over the WAN through ISDN's Q.931 call setup and management signaling protocol. Video conferencing PCs communicate with each other or with the gateway server through an ATM connection to a media switch. The ATM adapter and middleware provide ISDN redirection, setting up calls from one PC to another for video conferencing.

The portion of the middleware that resides on the end user's PC sits between the applications and the LAN emulation software, where it sets up real-time calls for voice, video, and audio. The software redirects real-time data streams that are generated by Windows programs from the local disk drive to a destination across the ATM network. In essence, this redirection spoofs or fools applications into thinking that they are executing locally.

When a Windows program requests a file, the middleware consults a map to determine whether the file is stored locally, on the LAN server, or on the media server. If the file is stored on the LAN server, the request is processed normally through the appropriate protocol stack. The LAN-emulation software on the adapter sets up an ATM call to the LAN server and segments the IP or IPX packets into 53-byte cells and sends them. At the other end, the ATM-to-Ethernet adapter at the remote port of the media switch performs cell-to-packet reassembly and passes them on to the local LAN server.

A component of the middleware also resides on a media server, where its primary function is to handle client requests and retrieve data. When the client requests a file that is stored there, the middleware sets up an ATM call to the media server and passes the request along to it. At the core of this middleware component is a real-time kernel that performs call scheduling. This function enables the media server to support multiple users simultaneously.

Summary

These are only a few of the ways in which multimedia networking is being implemented. There is no question that many areas can benefit from multimedia, enhancing collaborative efforts in online engineering, manufacturing, and health care (to name a few). Multimedia can even facilitate the real-time integration of multiple sites into a virtual corporation and can add personalized assistance to e-commerce applications on the Internet.

See Also

Asynchronous Transfer Mode (ATM)

Bandwidth Management Systems

Isochronous Ethernet

Video Streaming

Multiplexers

Multiplexers first appeared in the mid-1980s when private networks became popular. These devices combine voice, data, and video traffic from various input channels so they can be transmitted over a single higher-speed digital link—usually a T1 line. At the other end, another multiplexer separates the individual lower-speed channels, sending the traffic to the appropriate terminals. Multiplexers enable businesses to reduce telecommunications costs by making the most efficient use of the leased line's available bandwidth. Because the line is billed at a flat monthly rate, there is ample incentive to load it with as much traffic as possible. Using different levels of voice and/or data compression, the channel capacity of the leased line can easily be doubled or quadrupled to save even more money.

There are several types of multiplexing in common use today. On private leased lines, the dominant technologies are *Time-Division Multiplexing* (TDM) and *Statistical Time-Division Multiplexing* (STDM). Each lends itself to particular types of applications. TDMs are used when most of the applications must run in real time, including voice, video conferencing, and multimedia. When an input device has nothing to send, however, its assigned channel is wasted. With STDM, if a device has nothing to send, the channel that it would have used is dynamically reassigned to a device that does have something to send. If all channels are busy, input devices wait in a queue until a channel becomes available. STDMs are used in situations where efficient bandwidth usage is valued and the applications are not bothered by delay.

Although used mostly on private networks, both types of multiplexers can interface with the PSTN, as well. For example, if a private T1 line degrades or fails, the multiplexer can be configured to automatically switch traffic to an ISDN link until the private line is restored to service.

Time Division Multiplexing

With TDM, each input device is assigned its own time slot, or channel, into which data or digitized voice is placed for transport over a high-speed link. When a T1 line is used, there are 24 channels—each of which operates at 64Kbps. The link carries the channels from the transmitting multiplexer to the receiving multiplexer, where they are separated and are sent on to assigned output devices. As noted, if an input device has nothing to send, the assigned channel is left empty—and that increment of bandwidth goes unused.

The TDM manages access to the high-speed line and cyclically scans (or polls) the terminal lines, extracts bits or characters, and interleaves them into the assigned time slots (e.g., frames) for output to the high-speed line. The multiplexer includes channel cards for each low-speed channel and its associated device, a scanner/distributor, and common equipment for the high-speed line. The low-speed channel cards handle the data and control signals for the terminal devices. They also provide storage capacity through registers that provide bit or character buffering for placing or receiving data from the time slots in the high-speed data stream.

The TDM's scanner/distributor scans and integrates information that is received from the low-speed devices into the message frame for transmission on the high-speed line and also distributes data received from the high-speed line to the appropriate terminals at the other end. The common equipment provides the logical functions that are used to multiplex and demultiplex incoming and outgoing signals. This equipment contains the necessary logic to communicate with both the low-speed devices and with the high-speed devices, also generating data, control, and clock signals (which ensure that the time slots are perfectly synchronized at both ends of the link).

When digital lines are used on the network side, the TDM is connected to a *Channel-Service Unit/Digital-Service Unit* (CSU/DSU), which is a required network interface for carrier-provided digital facilities (typically provided as a plug-in card by most multiplexer vendors). The CSU is positioned at the front end of a circuit in order to equalize the received signal, to filter both the transmitted and received waveforms, and to interact with the carrier's test facilities. The DSU element transforms the encoded waveform from *Alternate Mark Inversion* (AMI) to a standard business-equipment interface, such as RS-232 or V.35, in addition to performing data regeneration, control signaling, synchronous sampling, and timing.

Operation   TDM technology supports asynchronous, synchronous, and isochronous data transmission. Asynchronous data transfer requires the framing of each character by a start bit and stop bit, which enables the originating terminal to control the timing of each transmitted character. Synchronous data-transfer timing is controlled by the multiplexer. Terminals send synchronous blocks of data that are framed by characters. Bits within a block are synchronized to clock signals that are generated by the TDM. Synchronous terminals operate at higher speeds than their asynchronous counterparts, but both are multiplexed in a similar manner.

Isochronous transmission supports multimedia applications when voice and other data must arrive together. In this type of transmission, the individual terminals generate their own clock signals, with all clocks running at the same nominal rate. Isochronous multiplexers provide some buffering and rate adjustment to compensate for slight variations among the clock rates.

A TDM samples data from each terminal input channel and integrates it into a message frame for transmission over the high-speed line. Message frames consist of time slots, and each time-slot position is allocated to a specific terminal. Interleaving is the technique that multiplexers use to format data from multiple devices for aggregate transmission over the link (refer to Figure M-5).

Most of the market leaders offer multiplexers that will interface with public networks via the byte-interleaving technique. Some vendors support both bit-interleaving and byte-interleaving, enabling their products to readily interconnect with both private facilities for maximum efficiency (i.e., bit) and public-switched services for increased connectivity (i.e., byte). This con-

Figure M-5

Data from multiple
input sources is
interleaved by the
time-division
multiplexer for
transmission over the
high-speed link. Note
the empty time slot. If
a device has nothing
to send, this amount
of bandwidth goes
unused.

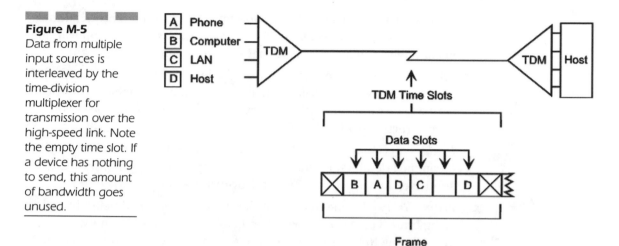

figuration flexibility enables companies to take advantage of both environments, according to shifting economics or application needs.

Features   High-end T1 multiplexers provide numerous standard and
optional features, including the following:

- *Drop-and-insert*   With this feature, a multiplexer can accept a high-
  speed, composite data stream from another multiplexer, demultiplex
  (remove) a portion of the data stream, modify it (e.g., add additional
  data frames), and transmit the altered high-speed data stream to a
  third multiplexer. A complementary feature, called bypass, enables a
  multiplexer to pass through a high-speed data stream without
  modification.

- *Subrate data multiplexing*   Provides programmable data rates at
  56Kbps and below for both synchronous and asynchronous data.
  Subrate channels, together with normal channels (e.g., DS0), are
  carried over the same physical T1 facility. To maintain transparency,
  the subrate multiplexing technique accommodates independent
  clocking of the transmitted and received data.

- *ISDN*   Via plug-in cards, many T1 vendors support primary-rate ISDN
  (23B+D). In addition to the 23 64Kbps B (bearer) channels and one
  64Kbps D channel for out-of-band signaling, there are two high-
  capacity ISDN channels that can be supported by the multiplexer: the

384Kbps H0 channel and the 1.536Mbps H11 channel. These channels are best suited for interconnecting LANs, high-speed data applications, video conferencing, and private network restoring.

■ *LAN Adapters* Optional plug-in LAN adapters are available with most multiplexers. A 10BaseT module, for example, enables Ethernet traffic to be combined with voice, synchronous data, and video traffic over a fractional T1 or full T1 link.

■ *Bridges and routers* To interconnect LANs, multiplexers can accommodate plug-in bridge/router modules. A variety of protocols are supported, including TCP/IP and Novell's IPX protocol.

■ *Frame Relay* Frame Relay interfaces give users added flexibility when tying branch office and workgroup LANs to the backbone network. With Frame Relay, users can create a virtual packet network that can be overlaid onto a high-end T1 or statistical-multiplexer network. The subnetwork can expand and contract in order to use available network resources. Some vendors even support voice traffic over the Frame Relay network. A separate voice-compression module digitizes analog voice for priority transmission.

■ *ATM* Low-speed ATM interfaces for T1 multiplexers provide *Constant Bit-Rate* (CBR) capability, enabling the ATM interface to handle delay-sensitive applications such as voice and video conferencing. When the video conference is over, the CBR capability can be disabled and the bandwidth can be made available to other data applications.

■ *SMDS* Some T1 multiplexers can be equipped with the *Distributed-Queue Dual Bus* (DQDB) *Subscriber-Network Interface* (SNI) and *Data-Exchange Interface* (DXI) for SMDS. In addition, the ports can be individually configured in order to convert SMDS to Frame Relay or ATM, or to convert Frame Relay to ATM.

■ *Testing* TDMs support local loopback tests on both the high-speed and low-speed ports. They also support remote loopback tests for the low-speed ports at the other end of the link. System diagnostic tests can be performed when the loopback tests are combined with character generation and error-detection capabilities. Some multiplexers automatically busy-out individual, remote, low-speed ports when a failure is detected on a low-speed modem or computer port.

System Management With support for the *Simple Network-Management Protocol* (SNMP), the network manager can configure, monitor, and control

T1 multiplexers from the same console that controls the LAN devices. A *Management Information Base* (MIB) for the T1 multiplexer gives administrators at a remote site the same configuration flexibility that they would have if they were at the device's control panel. A special MIB gives network administrators access to, and control of, every configurable element of the device. GET messages enable a network administrator to receive status information from a device, while TRAP messages report alarm conditions and SET messages reconfigure network devices.

A number of advanced system-management features are available with T1 multiplexers. To help network managers effectively control system resources, some multiplexer management systems maintain a detailed physical inventory of every card in the network. When new cards are plugged into a node, the card ID information is automatically reported to the network-management system—ensuring that equipment inventory is always current. This information includes a card slot, card type, serial number, hardware revision level, and firmware revision level.

The network-management systems of some multiplexers provide centralized order entry and order tracking. When a request is received for network service, for example, it is necessary to generate a network order for more bandwidth, which can be put into service immediately or at a specific time.

The alarm filter enables network operators to monitor alarms in real time and to set filters to determine which information should be displayed at the network-management system console. Most network-management systems provide a visual and audible indication of the most recent alarm-summary information and track and update alarm conditions automatically in the database. A level of severity can be assigned to each alarm type: critical, major, minor, alert, and ignore.

The network-management database contains an event log. This information is date/time stamped and describes how the event was created. For example, if a technician pulls a card from a multiplexer, a state change (event) is registered. Another state change is registered when a new card is inserted. Other types of events include the following:

- Operator actions
- Alarms
- Status changes
- Order installation and removal
- System errors
- Capacity

The event log provides a chronological history of all activity within the network, giving the network manager a single source of information in order to track and analyze network performance.

Security   A multi-level password capability enables unique, individual access to the network. On an individual password basis, the system administrator can restrict access of any menu, submenu, or operating screen in the network-management system. For example, a network supervisor might have access to all screens and functions, while an order-entry clerk might have access only to the order-management area in order to add or view orders (not to modify or delete them).

Redundancy   The redundant components of a multiplexer can include CPUs, buses, trunk cards, power supplies and fans, and network-management systems. If the primary CPU/bus pair fails, for example, the secondary CPU/bus pair will take over nodal processing and clocking functions. When the switch takes place, the network-management system is notified, and an alarm is presented to the operator. Each CPU maintains a mirror image of information that is contained in the redundant CPU, so in the event of switchovers, information is not lost. Switchover from an active to a redundant CPU will not corrupt active circuits. The same applies to other redundant system components.

Management Reports   Both standard and user-customized reports are available from most network-management systems. Standard reports can be modified in order to facilitate network maintenance, capacity planning, inventory tracking, cost allocation, and vendor relations. In addition, ad-hoc reports can be created from the management system's database by using *Structured Query Language* (SQL).

Statistical Time-Division Multiplexers (STDMs)

STDM is a more efficient multiplexing technique in that input/output devices are not assigned their own channels. If an input device has nothing to send at a particular time, another input device can use the channel. This process uses the available bandwidth more efficiently. An STDM can be purchased as a stand-alone device, or it can be an add-on feature to a TDM, providing service over one or more assigned channels.

Operation   STDM operation is similar to that of TDMs; the high-speed side appears much like a TDM high-speed side, while the low-speed side is quite different. The STDM allocates high-speed channel capacity based on demand from the devices that are connected to the low-speed side. This allocation by demand (or contention) provides a more efficient use of the available capacity on the high-speed line. In the variable-allocation scheme of an STDM frame, the time slots do not occur in a fixed sequence.

An STDM increases high-speed line usage by supporting input channels whose combined data rates would exceed the maximum rate supported by the high-speed port. When any given channel is idle (not sending or receiving data), input from another active channel is used in the time slot instead. The STDM has the option of turning off the flow of data from a sender if there is insufficient line capacity and then turning the flow back on when the capacity becomes available.

Features   TDMs and STDMs share many of the same operational and management features. One is data compression. Similar to TDMs, STDMs support techniques for compressing data so that they can actually transmit fewer bits per character. Data compression shrinks the time slot for the STDM and enables it to transmit more time slots per frame.

While TDMs detect and flag errors, STDMs have the capability to correct them. The sending STDM stores each transmitted data frame and waits for the receiving STDM or computer to acknowledge receipt of the frame. A *Positive Acknowledgment* (ACK) or *Negative Acknowledgment* (NAK) is returned. If an ACK is received, the STDM discards the stored frame and continues sending the next frame. If a NAK is returned, the STDM retransmits the questionable frame and any subsequent frames. The process is repeated until the problematic frame is accepted, or until a frame retransmission counter reaches a predetermined number of attempts and activates an alarm.

For applications such as asynchronous data transmission, where error detection is not performed as part of the protocol, STDM error detection and correction is a valuable feature. For protocols such as IBM's *Binary Synchronous Communication* (BSC), however, which contains its own error-detection algorithm, STDM-performed error control adds additional delays and redundancies that might not be appropriate for the application.

Several throughput enhancements are available for STDMs, which can be added later when needs change. These features include the following:

- *Per-channel compression*   Each channel has its own compression table.
- *Fast-packet technology*   Increases throughput by sending part of a frame before the entire frame is built

- *Data-prioritization*   Inserts shorter, interactive frames between larger, variable, run-length frames
- *Run-length compression*   Removes redundant characters from the transmission in order to improve performance

Summary

Despite the continuous price reductions on leased lines since the 1980s, businesses are always looking to cut the cost of telecommunications. One of the most effective ways to reach this goal is through the deployment of multiplexers that increase the bandwidth utilization of leased lines. Such lines are billed at a flat monthly rate by the carriers, no matter how little or how much traffic is actually carried over them. Not only can the business save money by using multiplexers, but the cost of the devices themselves can be recovered in a matter of a few months from the money that is saved. Using different levels of voice and/or data compression, the channel capacity of a leased line can be easily doubled or quadrupled to save even more money— enabling the business to recover the cost of the equipment even faster.

See Also

Inverse Multiplexing

Wavelength-Division Multiplexing (WDM)

Multi-Channel Multi-Point Distribution Service (MMDS)

Multi-Channel Multi-Point Distribution Service (MMDS) traces its origins to 1972 when it was introduced as an analog technology called *Multi-Point Distribution Service* (MDS). For many years, MMDS was used for one-way broadcast of television programming, but in early 1999, the *Federal Communications Commission* (FCC) opened this spectrum to enable two-way transmissions, making it useful for delivering telecommunication services, including high-speed Internet access, to homes and businesses. This technology, which has now been updated to digital, operates in the 2GHz to 3GHz range, enabling large amounts of data to be carried over the air from the operator's antenna towers to small receiving dishes that are installed at each customer location.

Operation

With MMDS, a complete package of TV programs can be transmitted on a high frequency (2GHz to 3GHz). On these microwave frequencies, up to 24 stations can be combined for delivery at the same transmitting antenna. Because microwave signals travel only in a straight line of sight, the height of the antenna is important (it has to be as high as possible). Generally, one MMDS head end covers areas between 20 km and 50 km. MMDS signals are normally encrypted to enable the operator to control and bill for services.

Most of the time, the operator receives programming via a satellite downlink. Large satellite antennas that are installed at the head end collect these signals and feed them into encoders that encrypt the programming. The encrypted video and audio signals are modulated via *Amplitude Modulation* (AM) and *Frequency Modulation* (FM) to an *Intermediate Frequency* (IF) signal. These IF signals are up-converted to MMDS frequencies and are then amplified and combined for delivery to a coax or waveguide feed line. This feed line is connected to the transmitting antenna, which can have an omni-directional or sectional pattern.

The small antennas at each subscriber location receive the signals and pass them via a cable to a set-top box that is connected to the television. If the service also supports high-speed Internet access, a cable also goes to a special modem that is connected to the subscriber's PC.

Service Providers

Although MMDS originally supported the broadcast of television programming, companies such as MCI WorldCom and Sprint have embraced MMDS as an alternative to procuring copper phone lines from *Incumbent Local Exchange Carriers* (ILECs) and other rivals in order to serve hard-to-reach customers. The companies see wireless cable systems as an eventual alternative to other high-speed Internet access options such as copper-based *Digital Subscriber Line* (DSL) and cable modems, as well as traditional telephone modems.

The merged company of MCI WorldCom-Sprint, assuming that it passes muster with regulators, has the potential to reach more than 50 million homes through MMDS once its systems are built. That is almost as many as AT&T will have through its cable deals with TCI, MediaOne, and Time Warner, which enable AT&T to reach 60 million homes. MMDS can deliver 1Mbps to 2Mbps (download speeds) at a competitive price, in addition to local and long-distance calling and video channels.

MCI WorldCom and Sprint acquired their wireless cable holdings only recently. In five separate deals that are still pending, Sprint is spending about $1.3 billion for systems that can reach roughly 30 million U.S. homes. MCI WorldCom paid about $350 million for CAI Wireless Systems, gaining access to about 24 million homes through CAI and affiliated operations. That brings spending by MCI WorldCom-Sprint so far to more than $1.6 billion for access to more than 54 million homes.

Summary

To date, only a few U.S. wireless cable systems are operating, and still fewer systems are offering telephone services and Internet access over the air. But one of those systems—now owned by Sprint—has attracted more than 10,000 customers for its Internet access service in Phoenix, Arizona in less than a year.

See Also

  Cable Television Networks

  Digital Subscriber-Line Technologies (DSLT)

  Local Multi-Point Distribution Service (LMDS)

  Microwave Communications

Multi-Service Networking

Under the concept of multi-service networking, voice-enabled routers, LAN and WAN switches, SNA-to-LAN integration solutions, dialup and other access products, Web site management tools, Internet appliances, and network-management software are brought together in an effort to address current and emerging business applications. By definition, multi-service networks inherently support any type of traffic (and therefore, any type of applications-networking requirement). The multi-service network typically consists of many components that all work together in order to ensure consistent QoS for networked applications.

IP Telephony

A key capability of the multi-service network is IP telephony, which can be implemented by adding an IP module to a conventional PBX, connecting an IP/PSTN gateway to the PBX, or implementing an IP PBX to handle all intra-company voice calls. The complete IP telephony system consists of a desktop telephone, call-manager software that is installed on a communications server, and a WAN gateway—all of which are attached to an existing LAN/WAN infrastructure. Locally, the infrastructure is usually a shared or switched Ethernet that provides the bandwidth and connectivity between the attached devices (IP phones or multimedia computers). IP address assignment is provided through the *Dynamic Host Control Protocol* (DHCP) that is installed on a communication server.

The call-management software provides the intelligence that is necessary for PBX-like features. This application is usually installed, along with DHCP, on a Windows NT server and provides basic call processing, signaling, and connection services to IP phones and virtual phones, *Voice-over-IP* (VoIP) gateways, and other local and remote devices. This application includes the management and control of various signaling protocols such as Q.931 for ISDN WAN control and H.225/H.245 for IP packet control.

The call-management software also implements supplementary and enhanced services such as hold, transfer, forward, conference, multiple line appearances, automatic route selection, speed dial, last-number redial, and other features that are extended to IP phones and gateways via parameters that are stored in a configuration database. Microsoft's *Internet Information Server* (IIS), for example, could be installed at the communications server in order to provide a browser interface to the configuration database. With administrator privileges, users can configure their own phones through a Web interface.

IP/PSTN Gateways

Gateways convert voice from the packet domain to the circuit-switched domain. Specifically, this type of device converts the voice packets that have been placed into an Ethernet frame into the format that can be accepted by the PSTN. Gateways—digital or analog—also pass signaling information, including dial tone and network signaling (as well as caller ID).

A digital gateway supports G.711 audio encoding (64Kbps), offers Ethernet access, supports ISDN PRI, and provides integrated *Digital Signal Processor* (DSP) functions. Each T1 interface card supports 24 channels with line echo cancellation and packet-to-circuit conversion for voice or fax calls. In addition to H.323 compliance, which enables interoperability with other H.323 client applications and gateways, the digital gateway also supports supplementary services such as call forward, transfer, and hold. The gateway is configured by using a Web interface.

An analog gateway supports G.711 (64Kbps) or G.723 (dual-rate 5.3Kbps or 6.3Kbps) audio compression and comes with integrated DSPs and modular, analog, circuit-switched interfaces. The analog system not only connects to local analog telephone company lines but also provides connectivity to devices such as fax machines, voice mail, and analog phones.

Multi-Service Access Routers

With companies spending billions of dollars each year on internal phone calls and faxes between their own offices, there is ample incentive to reduce these costs by integrating voice, fax, and data onto a single network infrastructure. With voice and fax over IP, companies can deploy integrated, scaleable networks without sacrificing voice and fax quality. In addition, the deployment of these multi-service capabilities can be done without changing the way in which phone calls are made or faxes are sent.

Routers exist that enable traditional telephony traffic, such as voice and fax, to be integrated with traditional data traffic such as IP, IPX, and SNA. This integration is achieved with traditional telephony interfaces so that PBXs, key systems, traditional phones, IP phones, fax machines, and even the PSTN can physically connect to the router. Once these connections are established, the voice and fax traffic are processed by the DSPs and are placed into IP packets or Frame Relay cells for transfer to other network locations. In keeping as much voice and fax traffic as possible on the data network, toll charges can be eliminated or greatly reduced. These routers also enable users to take advantage of advanced applications such as secure Internet access, managed network services, VPNs, and e-commerce.

Workstations on the remote LAN can be assigned IP addresses dynamically by either a communications server at the corporate location or by the access routers that use DHCP. The access routers also support *Network-Address Translation* (NAT), which effectively creates a private network that is invisible to the outside world. NAT enables network administrators to assign IP addresses that are normally reserved for the Internet for private use over a remote LAN. For businesses that want to permit select access to the network, NAT can be configured to permit only certain types of data requests, such as Web browsing, e-mail, or file transfers.

Security features that are implemented through the router's *Operating System* (OS) protect the privacy of company communications and commerce transactions over the Internet. The OS also provides the means to build custom security solutions, including standard and extended *Access-Control Lists* (ACLs), dynamic ACLs, router and route authentication, and *Generic Routing Encapsulation* (GRE) for tunneling. Perimeter security features control traffic entry and exit between private networks, intranets, or the Internet. To protect the corporate LAN from unauthorized access, the routers can also support token cards, *Password Authentication Protocol* (PAP), *Challenge Handshake Authentication Protocol* (CHAP), and other security features that are available through an optional firewall. Access routers can be managed with a software application that provides configuration and security management, as well as performance and fault monitoring. Centralized administration and management can be applied via the *Simple Network-Management Protocol* (SNMP), Telnet, or local management through the router's console port.

Multi-Service Concentrator

A multi-service concentrator is a wire-speed T1 router and serial data device that has voice, video, and ATM capabilities. This concentrator includes Ethernet LAN and data capabilities, as well as IP and SNA suites. This type of router can be deployed over private or public networks in order to reduce equipment and connection costs, simplify network management, and improve application performance. Through the concentrator's operating system, these systems perform multi-protocol routing and bridging. They can also be tightly integrated with smaller access routers at branch-office locations.

In addition to providing 24 channels of voice through the T1 port, the multi-service concentrator provides echo cancellation for all voice channels and achieves further cost savings through the use of *Voice-Activity Detection* (VAD). This feature halts voice traffic during the silent periods of a conversation, enabling the idle bandwidth to be used for data. Further bandwidth efficiencies can be achieved with voice compression at 8Kbps (G.729, G.729a) or 32Kbps (G.711, ADPCM).

The multi-service concentrator connects to any standard PBX switch, key system, or telephone. At smaller corporate sites, the concentrator can be used as a local voice system that can possibly obviate the need for Centrex, key system, or PBX. This concentrator offers an ISDN BRI voice interface and supports an array of call-handling capabilities for voice connections, and it can also be used in tie line and ring-down modes. This concentrator can also support *Dual-Tone Multi-Frequency* (DTMF), digit-based, per-call switching by using dialed digits to select destination sites and network calls.

The multi-service concentrator supports transparent *Common Channel Signaling* (CCS) and the Q.SIG voice-signaling protocol. Q.SIG is a form of CCS that is based on ISDN Q.931, the signaling method that is used by the D channel for call setup and teardown. Q.SIG provides transparent support for supplementary PBX services so that proprietary PBX features are not lost when connecting PBXs to networks that are comprised of multi-service concentrators.

In addition to voice calls, the multi-service concentrator also supports both circuit and packet-mode video. Circuit video is transported bit-by-bit through circuit emulation over a *Constant Bit-Rate* (CBR) ATM connection. Packet video can be supported over a *Variable Bit-Rate* (VBR) ATM connection or over the LAN, through the router engine, and over an *Unspecified Bit-Rate* (UBR) connection.

The multi-service concentrator is also compatible with the drop-and-insert capability of *Digital Cross-Connect Systems* (DCS) that are used on the PSTN. Drop and insert refers to the software-defined capability of the DCS to exchange channels from one digital facility to another, in order to implement the appropriate routing of the traffic, reroute traffic around failed facilities, or increase the efficiency of all of the available digital facilities. Accordingly, the multi-service concentrator enables some time slots of a T1 facility to be used for on-net traffic and services, while the rest can be dropped/inserted off-net for transport over the PSTN when necessary.

Multi-Service LAN Switches

A multi-service LAN switch implements telephony in the LAN and provides seamless integration with campus and WAN systems. In addition to a range of connectivity options and network services, this type of switch also provides an IP telephone and call manager to enable organizations to build corporate intranets for multicast, mission-critical voice applications. These switches also offer redundancy and topology resiliency for high availability and Gigabit Ethernet and ATM interfacing for high performance. In addition, they reduce complexity with application awareness and policy classification, which eliminates the need for network managers and administrators to engage in detailed configuration of QoS parameters.

Such switches provide Layer 3 routing capabilities over Fast and Gigabit Ethernets, T1 to OC-48 ATM, and *Packet over SONET* (PoS) uplinks. The ATM switching capabilities provide campus backbones with the means to integrate data, voice, and video traffic over ATM *LAN Emulation* (LANE), *Multi-Protocol over ATM* (MPOA), or *Multi-Protocol Labeling Switching* (MPLS)/Tag-Switching networks.

LANE is a Layer 2 bridging protocol that makes a connection-oriented ATM network look like just another connectionless LAN segment to higher-layer protocols and applications. As such, LANE provides a means to migrate today's legacy networks toward ATM networks without requiring the existing protocols and applications to be modified. The scheme supports backbone implementations, directly attached ATM servers and hosts, and high-performance, scaleable computing workgroups. By defining multiple emulated LANs across an ATM network, switched virtual LANs can be created for improved security and greater configuration flexibility. Additional benefits include minimal latency for real-time applications and QoS for emulated LANs.

MPOA preserves the benefits of LAN emulation while enabling intersubnet, internetwork communication over ATM virtual circuits without requiring routers in the data path. This framework synthesizes bridging and routing with ATM in an environment of diverse protocols, network technologies, and virtual LANs. MPOA is capable of using both routing and bridging information in order to locate the optimal exit from the ATM cloud. MPOA enables the physical separation of internetwork-layer route calculation and forwarding—a technique known as virtual routing. This separation enables efficient intersubnet communication and enhances manageability by decreasing the number of devices that must be configured in order to perform internetwork layer route calculation. MPOA also reduces the number of devices that are participating in internetwork-layer route calculation and eliminates the need for edge devices to perform internetwork-layer route calculation.

MPLS, based on Cisco Systems' tag switching, is an *Internet Engineering Task Force* (IETF) standard for IP service delivery. MPLS labels or tags provide the capability to differentiate service classes for individual data flows.

The tags work similarly to address labels on packages in an express delivery system—they expedite packet delivery on large corporate-enterprise networks, enabling the creation of faster, lower-latency intranets that can effectively support data, voice, and video on a common network infrastructure.

Configuring and deploying QoS policies is achieved through a policy manager. The policy manager's *Graphical User Interface* (GUI) enables network administrators to define traffic classification and QoS enforcement policies. This system includes a rules-based policy builder, integrated policy validation and error reporting, and a policy-distribution manager that downloads policies to the various network devices. These features enable network administrators to quickly apply a mix of QoS policy objectives that protect business-critical application performance. In the process, organizations can more easily make the transition from unconstrained bandwidth utilization toward more consistent application performance over currently available bandwidth. This functionality is not only increasingly necessary in order to eliminate unpredictable performance for business-critical applications that run in corporate networks, but it is also required in order to integrate data, voice, and video over a common network infrastructure.

Multi-Service Routers

Multi-service routers provide organizations with the flexibility to meet the constantly changing requirements at the core and distribution points of their networks. Typically, these routers deliver up to 1Gbps of throughput over a mid-plane that provides the capability to switch DS0 time slots between multi-channel T1 interfaces, much like a TDM multiplexer. At the same time, such routers provide digital voice connectivity via an ATM circuit emulation service module. Together, these capabilities enable the router to be connected to an ATM network on one side and to the TDM network on the other side.

Integral networking software provides routing and bridging functions for a wide variety of protocols and network media, including any combination of Ethernet, Fast Ethernet, Token Ring, FDDI, ATM, serial, ISDN, and *High-Speed Serial Interface* (HSSI). Port and service adapters are connected to the router's *Peripheral-Component Interconnect* (PCI) buses, enabling connection to external networks.

Multi-service routers are generally equipped with the most advanced reliability features. For example, software-defined configuration changes take effect without rebooting or interrupting network applications and services. Port adapters and service adapters can be inserted and removed while the system is online. An automatic reconfiguration capability enables seamless upgrades to higher density and new port adapters without the need for rebooting, taking the system offline, or manual intervention. Dual hot-swappable, load-sharing power supplies provide system power redundancy; if one power supply or power source fails or is taken offline, the other power supply maintains system power without interruption. Alerts are issued when potentially problematic system fluctuations occur before they become critical, thereby enabling resolution while the system remains online.

Multi-Service Edge Switches

Multi-service edge switches provide data, voice, and video integration over the wide area. They extend the WAN backbone to branch offices, providing high service levels regardless of location. They internetwork between existing routers and LAN switches in order to provide seamless traffic flow between LAN/campus and WAN domains. By combining ATM's dynamic bandwidth management with queuing techniques, the edge switches minimize recurring WAN bandwidth costs, while ensuring fairness and high QoS for individual applications. The switches support legacy applications, large-scale packet voice, Frame Relay, and ATM and provide integrated LAN interfaces. Internetworking and management functions between the edge switch and LAN switching, routing, and branch-office devices improves end-to-end network performance and QoS.

Port interfaces ranging from 1.2Kbps to 155Mbps are typically supported, as are network or trunk interfaces ranging from T1 to OC-3. Advanced traffic-management features and multiplexing techniques deliver high levels of bandwidth utilization and efficiency. The voice compression, silence suppression, and fax relay capabilities deliver additional bandwidth savings.

In being capable to consolidate multiple WAN infrastructures, organizations have the flexibility to deploy VoIP, voice-over ATM, or voice-over Frame Relay. Full control over network resources is exercised in a variety of ways—with per-*Virtual Circuit* (VC) queuing, per-VC rate scheduling, and multiple classes of service (CoS)—guaranteeing QoS levels for the individual applications. This functionality enables all applications to be supported, according to their specific requirements, by using advanced traffic-management and CoS features.

Access devices can be connected to the backbone through a leased line or though public Frame Relay or ATM WANs. Depending upon the modules that are used to connect to the access device, different levels of internetworking are possible. Data applications can be based on Ethernet or Token Ring LANs, as well as Frame Relay, ATM, and legacy protocols. Voice applications can use different transport technologies. In addition to VoIP, the edge switch supports *Voice-Over Frame Relay* (VoFR) and *Voice Transport over ATM* (VtoA). Dialup frame relay, SNA networking, and frame forwarding are also supported. For locations that need ATM broad-band speeds, the systems offer Frame Relay to ATM service internetworking.

Frame-to-ATM service internetworking is implemented by an edge switch and segments and maps variable-length Frame Relay frames into fixed-length ATM cells. This functionality enables the switch to provide transparent connectivity between large ATM and small Frame Relay locations, independent of the wide-area access protocol. Small sites that do not have enough traffic to justify a dedicated access connection can connect to a Frame Relay network-on-demand basis by using switched (dialup) lines.

Through frame forwarding, the edge switch can transport frame-based protocols, such as SDLC, X.25, or any other HDLC-based protocol, at speeds ranging from 9.6Kbps to 16Mbps. With frame forwarding, all valid HDLC frames are forwarded or tunneled from one frame relay port to another port without frame relay header processing or *Local Management Interface* (LMI) control. This method of transport results in efficient bandwidth usage and low latency on the corporate network.

Circuit data services are provided by the edge switch for transport of asynchronous or synchronous circuit data or video, which is transparently carried through a fixed-delay, fixed-throughput, zero-discard, point-to-point data connection across the WAN. These capabilities enable the transport of both legacy and TDM traffic and facilitate the migration to future ATM networks. The reliability of multi-service edge switches is enhanced by common equipment that can be configured for redundancy. New software releases can be remotely downloaded onto the redundant processor for background installation while traffic continues to run. Advanced distributed-intelligence algorithms enable the network to automatically route new connections—and, if necessary, reroute traffic around failures in network facilities.

Summary

Multi-service networking has emerged as a strategically important issue for both companies and carriers. This networking involves the convergence of all types of communications—data, voice, and video—over a single packet or cell-based infrastructure. The benefits of multi-service networking are reduced operational costs, higher performance, greater flexibility, integration and control, and faster deployment of new applications and services than can otherwise be achieved over traditional voice-oriented PSTNs or even TDM-based, leased-line, private networks. The interest among organizations for data, voice, and video integration is being fueled by short-term cost savings. This integration also meets medium-term requirements for the support of emerging applications and leads to the long-term objectives of reducing complexity and network redundancy through technology convergence.

See Also

IP Telephony

CHAPTER N

Network Agents

Network agents are special programs that accomplish specified tasks by executing commands remotely. Network managers can create and use intelligent agents to execute critical processes, including performance monitoring, fault detection and correction, and asset management. The agent-manager concept is not new. The manager-agent relationship is intrinsic to most standard network-management protocols, including the *Simple Network-Management Protocol* (SNMP) that is used to manage TCP/IP networks. In fact, SNMP agents are widely available for all kinds of network devices, including bridges, routers, hubs, multiplexers, and switches.

In the SNMP world, agents respond to polls from a management station on the operational status of the various devices on the network. Based on the information that is returned, agents can then be directed by the management station in order to obtain more data, set performance variables, or generate traps when specified events occur. To retrieve the collected data, however, the agents must be polled by central management software—a process that increases network traffic. On WANs, which are being increasingly burdened with multimedia and other delay-sensitive applications, traffic from continuous polling and the resulting data transfers can degrade network performance. So-called intelligent agents address this problem.

What makes these agents so smart is the addition of programming code that consists of rules that tell them exactly what to do, how to do it, and when to do it. In essence, the intelligent agent plays the dual role of manager and agent. Under this rules-based scheme, polling is localized, events and alarms are collected and correlated, various tasks are automated, and only the most relevant information is forwarded to the central management station (refer to Figure N-1) for analysis. In the process, network traffic is greatly reduced, and problems are resolved faster.

Network Agent Applications

Agent technology has been available for several years and still represents one of the fastest-growing areas in network management. In a global economy that encourages the expansion of networks to reach new markets and discourages the addition of personnel to minimize operating costs, it makes sense to automate as many management tasks as possible through the use of intelligent agents. In recognition of these new business realities, the list of tasks that are being handled by network agents is continually growing.

Performance management   Network performance monitoring can help determine network service-level objectives by providing measurements to help managers understand typical network behavior and normal periods. The following capabilities of intelligent agents are particularly useful for building a network performance profile:

- *Baselining and network trending*   Identifies the true parameters of the network by defining typical and normal behavior. Baselining also provides long-term measurements to check service-level objectives and show out-of-the-norm conditions (which, if left unchecked, can diminish the productivity of network users).

- *Application usage and analysis*   Identifies the overall load of network traffic, which times in the day that certain applications load the network, which applications are running between critical servers and clients, and what their load is throughout the day, week, and month

- *Client/server performance analysis*   Identifies which servers can be overutilized, which clients are hogging server resources, and which applications or protocols they are running

- *Internetwork performance*   Identifies traffic rates between subnets so that the network manager can determine which nodes are using WAN links to communicate. This information can be used to define typical throughput rates between interconnected devices.

- *Data correlation*   Enables peak network-usage intervals to be selected throughout the day to determine which nodes were contributing most to the network load. Traffic source and associated destinations can be determined with seven-layer protocol identification.

Fault management   When faults occur on the network, problems be resolved quickly to decrease the negative impact on user productivity. The following capabilities of intelligent agents can be used to gather and sort the data that is needed to quickly identify the cause of faults on the network:

- *Packet interrogation*   Isolates the actual session that is causing the network problem, enabling the network manager to assess the nature of the problem quickly

- *Data correlation*   Because managers cannot always be on constant watch for network faults, it is important to have historical data available that provides views of key network metrics at the time of the fault. Such metrics can be used to answer questions such as, "What is the overall error/packet rate, and what are the types of errors that occurred?," "What applications were running at the time of the fault?," "Which servers were most active?," "Which clients were accessing these active servers?," and "Which applications were they running?"

- *Identification of top error generators*   Identifies the network nodes that are generating the faults and contributing to problems (such as bottlenecks caused by errors and network down time)

- *Immediate fault notification*   With immediate notification of network faults, managers can instantly learn when a problem has occurred before users do. Proactive alarms help detect and solve the problem as it is happening.

- *Automated resolution procedures*   The intelligent agents can be configured to automatically fix a problem when it occurs. The agent can even be programmed to automatically e-mail or notify help-desk personnel with on-screen instructions on how to solve the problem.

Capacity planning and reporting   Capacity planning and reporting enables the collection and evaluation of information in order to make informed decisions about how to respond to network growth. For this purpose, the following capabilities of intelligent agents are useful:

- *Baselining*   This capability enables the network manager to determine the true operating parameters of the network, against which future performance can be measured.

- *Load balancing*   Load-balancing capabilities enable the network manager to compare internetwork service objectives from multiple sites at once to determine which subnets are over-utilized or underutilized. Load balancing also helps the network manager discover which subnets can sustain increased growth and which require immediate attention for possible upgrades.

■ *Protocol/application distribution*   Protocol and application-distribution capabilities can help the network manager understand which applications have outgrown which domains or subnets. For example, these capabilities can indicate whether certain applications are continuously taking up more precious bandwidth and resources throughout the enterprise. With this kind of information, the network manager can better plan for the future.

■ *Host load balancing*   Enables the network manager to obtain a list of the top network-wide servers and clients that use mission-critical applications. For example, the information that is collected from intelligent agents might reveal whether specific servers always dominate precious LAN or WAN bandwidth, or it might help spot when a CPU is becoming overloaded. In either case, an agent on the LAN segment, WAN device, or host can initiate load balancing automatically when predefined performance thresholds are met.

■ *Traffic-profile optimization*   To ensure adequate service-level performance, the ability of network managers to compare actual network configurations against proposed configurations is valuable. From the information that is gathered and reported by intelligent agents, traffic profiles can be developed that enable what-if scenarios to be put together and tested before incurring the cost of physically redesigning the network.

A growing number of capacity-planning and reporting tools have become available in recent years. One tool is Bay Networks' (a business unit of Nortel Networks) Optivity Enterprise, a family of network-management products that includes the Optivity Design and Analysis suite of network design and optimization applications for Ethernet and Token Ring environments. Among the tools that are available in this suite is DesignMan, which performs simulation activities by using live traffic information that is gathered by embedded management agents on the network (refer to Figure N-2).

Security management   A properly functioning and secure corporate network plays a key role in maintaining an organization's competitive advantage. Setting up security objectives that are related to network access must be considered before mission-critical applications are run over untrusted networks—particularly, the Internet. The following capabilities of network

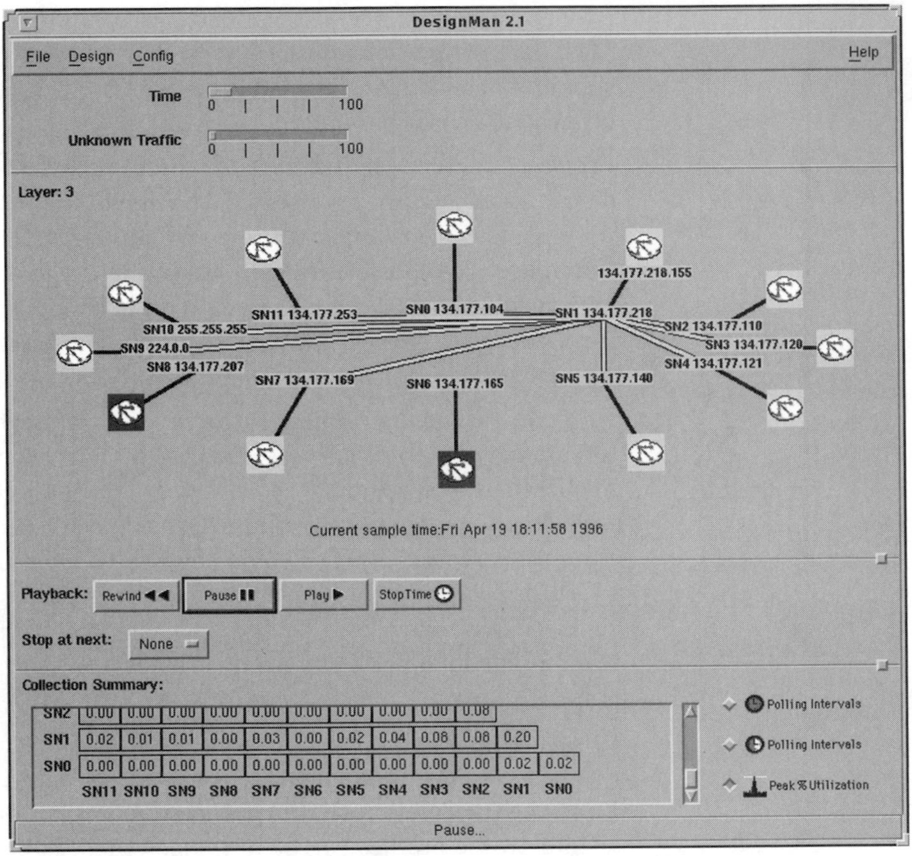

Figure N-2
Operating at Layer 3 (network), Bay Networks' DesignMan application shows the traffic flow between logically connected subnets. The user can apply what-if scenarios to the traffic data that is collected by intelligent agents, in order to see the effect of moving a server, for example, from one subnet to another. The application uses a VCR metaphor, enabling the user to play, pause, stop, and rewind the scenario in order to view its impact on the entire network.

agents can help discover holes in security by continuously monitoring access attempts:

- *Monitor effects of firewall configurations* By monitoring firewall traffic, the network manager can determine whether the firewall is functioning properly. For example, if the firewall was recently programmed to prohibit access to a corporate host via Telnet—but the program's syntax is wrong—the intelligent agent will report this fact immediately.

- *Show access to and from secure subnets* By monitoring access from internal and external sites to secure data centers or subnets, the network manager can set up security service-level objectives and firewall configurations based on the findings. For example, the information that is reported by the intelligent agent can be used to determine which external sites should have access to the company's database servers or legacy hosts.

- *Trigger packet capture of network security signatures* Intelligent agents can be set up in order to issue alarms and to automatically capture packets upon the occurrence of external intrusions or unauthorized application access. This information can be used to track down the source of security breaches. Some network agents even have the capability to initiate a trace procedure to discover a breach's point of origination.

- *Show access to secure servers and nodes with data correlation* This capability reveals which external or internal nodes are accessing potentially secure servers or nodes and identifies which applications they are running.

- *Show applications running on secure nets with application monitoring* This capability evaluates applications and protocol use on secure networks or traffic components to and from secure nodes.

- *Watch protocol and application use throughout the enterprise* This capability enables the network manager to select applications or protocols for monitoring by the intelligent agent so that the flow of information throughout the enterprise can be viewed. This information can identify who is browsing the Web and who is accessing database client/server applications or running Notes, for example.

Intelligent agents can be used for a variety of other tasks, including Internet-related tasks. They can monitor information that is logged by servers on the Web, for example. When the log entries exceed a designated threshold, this situation might indicate a high demand for applications and impending congestion if the logging rate continues. An intelligent agent can act on this information in order to redirect traffic to another server, to balance the load across the available Web servers.

Applications management There are now client-side agents that continuously monitor the performance and availability of applications from the end user's perspective. A just-in-time applications performance-management capability captures detailed diagnostic information at the precise moment when a problem or performance degradation occurs, pinpointing the source of the problem so that it can be resolved immediately. Such agents are installed on clients as well as on application servers. They monitor every transaction that crosses the user's desktop, traversing networks, application servers, and database servers. They monitor all distributed applications and environmental conditions in real time, comparing actual availability and performance with service-level thresholds.

Via a management console, a window is provided into application availability and performance throughout the enterprise. Via the console, IT personnel can identify which users are experiencing problems, then drill down to view successive layers of problem and diagnostic detail. Through the console, the IT administrator can also define service-level performance thresholds, specify automated corrective action plans, and fine-tune data collection and reporting. A repository stores all exception and historical end-user application-usage data in a standard SQL database.

Remote user support   With the increasing number of mobile professionals and telecommuters—most of whom have no permanent connections to a LAN or WAN—IT administrators are faced with the challenge of managing this growing base of unattached computers. They have the task of ensuring that each system is working properly, that the system is configured to corporate standards, that the system is running the right versions of the right software, and that the system is functioning reliably. Typical management solutions, however, are not proactive and often involve long-distance calls to the help desk, user down time, and shipping costs for sending a system in for upgrades or repairs.

To deal with the problems of providing remote users with support, you can install agent software on the client that gives IT administrators a presence on each machine, regardless of its location. Mobile Automation, Inc., for example, offers its RightState agent software that enables IT administrators to define profiles for these out-of-reach PCs and how they should be configured, what software must be on the hard drive, and how often they should run programs such as diagnostics and virus checks.

The RightState server component communicates with each PC, comparing the correct configuration profile with the computer's current profile. This component notes exactly which files and scripts that a particular client system needs, requests only those files, and downloads and installs them. This exchange between the client and server continues until the system exactly fits the defined configuration. RightState even checks periodically to make sure that the configuration stays the same, alerting the IT administrator of any changes or problems.

Communication between the server component and the remote clients takes place through e-mail. Because remote users typically check e-mail several times a day, the agent can report the status of the client several times a day, as well. The agent can even be instructed to log on to an FTP server and download files. If the transmission is interrupted, the agent can pick up the installation where it left off during the next login.

The agent does not hog limited bandwidth to perform its tasks; rather, it breaks the data packets into smaller pieces and sends them encrypted, one by one, in order to optimize bandwidth. The IT administrator can determine the size of the packets before they are compressed and transmitted sequentially. The agent gathers all of the packets at the client side, whether they arrive over one e-mail session or multiple sessions. The agent decrypts and reassembles them into standard sized packets before implementation. Therefore, a large file (such as a word-processing program) can be sent over a period of days before being installed.

This method of providing remote support is non-intrusive. The user is not interrupted and might never even know what is going on in the background. The agent takes remote users out of the loop so that IT administrators will not have to deal with resistance or non-compliance from busy users.

WAN service-level management   Despite the migration from private WANs to public WANs in recent years, network managers are still accountable for overall network performance. Consequently, they must ensure that their carrier is equally as concerned about service quality. An effective approach for ensuring service quality is to implement the relatively new concept of WAN service-level management, which is a collaborative effort between subscriber and service provider in order to manage the service quality of public network services. In this arrangement, both the carrier and the customer work together to plan, monitor, and troubleshoot WAN service quality.

WAN service-level management offers a number of benefits. Subscribers can increase network availability and performance, reduce the need for recurring support, and ensure that business needs are met at the lowest possible cost. Service providers can reduce operational support costs, prioritize responses to alarms, issue trouble tickets, set expectations for service quality, and help justify recommendations to upgrade bandwidth.

Ensuring the success of WAN service-level management requires historical data and the collection of WAN service-quality information to arrive at baseline performance metrics. Normally, these tasks are difficult, but today's network agents automate the collection, interpretation, and presentation of WAN service-level information—making it easier to monitor and verify the performance of carrier-provided services.

Summary

Over the years, intelligent agents have proven to be indispensable tools for providing network-management assistance. Problems can be identified and resolved locally by network agents, rather than by harried operators at a central management console or by sending technicians to remote locations —both of which are expensive and time-consuming. In many cases, intelligent agents can implement restorations automatically in response to specified events. These actions can be as simple as resetting a device by turning it off and then on again. Other times, the restoration might consist of balancing the load across multiple lines or servers to avoid impending congestion. Agents will become even more indispensable as networks continue to expand to international locations. In today's global economy, having the capability to effectively monitor remote systems and networks becomes even more important, especially when organizations are under pressure to minimize staff in order to reduce operating costs.

See Also

Network Management

Simple Network-Management Protocol (SNMP)

Network Backup

Network backup is not a simple matter for most businesses. One reason is because it is difficult to find a backup system that is capable of supporting different NOSs and data, especially if mid-range systems and mainframes are involved. Protecting mission-critical data that is stored on LANs requires backup procedures that are well defined and rigorous. These procedures include backing up data in a proper rotation, using proper media, and testing the data to ensure that it can be easily and quickly restored in an emergency. Enterprise-wide backups are especially problematic, because typically, there are multiple servers and operating systems as well as isolated workstations that often hold mission-critical data. Moreover, the network and client/server environments have special backup needs: back up too often and throughput suffers; back up too infrequently, and data can be lost.

Backup Procedures

Deciding which files to back up can be more complicated than picking the right storage media. The most thorough backup is a full backup in which every file on every server is copied to one or more tapes or disks. The size of most databases, however, makes this process impractical to do more than once a month.

Incremental backups only copy files that have changed since the last backup. Although this method is faster, it requires careful management because each tape might contain different files. To restore a system that is made with incremental backups requires all of the incremental backups (in the right order) that were made since the last full backup.

Differential backups split the difference between full and incremental techniques. Similar to an incremental backup, a differential backup requires a tape with the full set of files. Each differential tape, however, contains all of the files that have changed since the last full backup, so restoration requires just that full set and the most recent differential.

Scheduling and Automation

The scheduling of backups is determined by several factors, including the critical nature of business applications, network availability, and legal requirements. Network backup software with calendar-based planning features enables the system administrator to perform functions such as scheduling the weekly archiving of all files on LAN-attached workstations. The backup can be scheduled for non-business hours in order to avoid disrupting user applications and to avoid network congestion.

Some scheduling tools enable the system administrator to set precise parameters with regard to network backups. For example, the backup can target only files that have not been accessed in the past 60 days, with the objective of freeing at least 100MB of disk space on a particular server. When the backup is complete, a report is generated that lists the files that have been archived to tape, along with their file size and date of last access. The total number of bytes are also provided, enabling the system administrator to confirm that at least 100MB of storage has been freed on the server.

Event-based scheduling enables the system administrator to run predefined workloads when dynamic events occur in the system, such as the close of a specific file or the start or termination of a job. With regard to network backups, the administrator can decide which events to monitor and what the automated response will be to those events. For example, the administrator can decide to archive all files in a directory after the last print job or to perform a database update after closing a particular spreadsheet.

Although most network backup programs can grab files from individual workstations on the LAN, there might be thousands of users who have similar or identical system configurations. Instead of backing up 1,000 copies of Windows 95/Windows 98, for example, the network backup program can be directed to copy only each user's system configuration files. That way, if a workstation experiences a disk crash, a new copy of Windows 95/Windows 98 can be downloaded from the server, along with the user's applications, data, and configuration files.

With the right management tools, network backup can be automated under centralized control. Such tools can go a long way toward lowering operating and resource costs by reducing the time that is spent on backup and recovery. These tools enhance media management by providing over-write protection, log-file analysis, media labeling, and the capability to recycle backup media. In addition, the journaling and scheduling capabilities of some tools relieve the operator of the time-consuming tasks of tracking, logging, and rescheduling network and system backups. Another useful feature of such tools is data compression, which reduces media costs by increasing media capacity. This automated feature also increases backup performance while reducing network traffic.

When these tools are integrated with high-level management platforms, such as Hewlett-Packard's OpenView or IBM's NetView or operating systems such as Sun's Solaris, problems or errors that occur during automated network backup are reported to the central management console. The console operator is notified of any problem or error via a color change of the respective backup application symbol on the network map. By clicking the symbol, the operator can directly access the network backup application to determine the cause of the problem or to correct the error in order to resume the backup operation.

Capabilities and Features

Depending on the particular operating environment, some of the key areas to consider when evaluating network-backup software include the following:

- Storage capacity and data transfer rate of the backup system
- A fast-start capability that enables full network backups to be performed immediately and fine-tuning of the backup parameters to be performed later
- The capacity to back up the NetWare bindery (if applicable), security information, and file and directory attributes

- Support for multiple file systems, including NetWare, the *Apple File Protocol* (AFP), OS/2 High-Performance File System, Sun Microsystems' *Network File System* (NFS), and OSI *File Transfer, Access, and Management* (FTAM)

- Capability to monitor, back up, and log the activities of multiple file servers simultaneously

- Tape labeling, rotation, and script-file schemes for automating the backup and recovery process

- Reporting and audit-log capabilities

- Fast-search capability that enables a system administrator to easily and quickly locate and retrieve files

- File archiving and grooming methods that enable automatic file and directory storage, including the capability to delete data that has not been accessed for a specified period of time

- Integrated network virus protection

- Security features that limit access to backups to only authorized users

Another key feature of network-backup software is the availability of agents that enable such programs to bypass operating-system constraints in order to store files that are still open, even if the application is accessing or updating them while the backup is in progress. This capability eliminates incomplete backups that typically result when files are not closed and is of particular value to organizations that need around-the-clock access to information while performing complete backups.

When evaluating software for LAN backups in the mainframe environment, some of the key areas to consider include the following:

- Whether all or most of the platforms at the server level are supported, such as LAN Manager, NetWare, and UNIX

- Whether all or most workstation platforms are supported, such as DOS/Windows, OS/2, UNIX, and Macintosh

- Whether non-LAN-connected PCs are supported, such as those with 3270 emulation cards with direct connections to controllers or *Front-End Processors* (FEPs)

- Whether other WAN connections are supported or just TCP/IP

- Whether users can set windows of availability to force backups and recoveries to take place during non-peak hours

- Whether the product supports options for restores to be performed by the database administrator, the LAN administrator, or individual workstation users

■ Whether the product supports both a command-line interface for expert use and a *Graphical User Interface* (GUI) for end-user access

■ Whether the product supports automatic archiving of files that have not been accessed for a specified period, thus freeing up server or workstation disk space

■ Whether the product supports the skipping of redundant files in the backup process

■ Whether the product supports other features, such as heterogeneous file transfers, remote command execution, and job submission from PC to host

Summary

As LANs continue to carry increasing volumes of critical data in varying file formats, vendors continue to push the limits of backup technology. On the software side, the trend is toward increasing levels of intelligence. Backup systems must not only ensure that files are backed up but that they are also easily located and restored. Systems intelligence has already progressed to the point where the user does not need to know the tape, the location on the tape, or even the exact name of a lost file in order to restore the file.

See Also

Hierarchical Storage Management
Storage Media

Network Computing

The concept of network computing originated with Oracle Corporation in 1995, when the company articulated its vision of a minimally-equipped computer that would depend mostly on the network for its applications; specifically, on local servers or remote servers that are located on the Internet or a private corporate intranet. With applications deployed, managed, supported, and executed on servers, organizations would not need to go through the greater expense of equipping every desktop with its own resources for independent operation. Instead, desktops could be equipped with cheaper, application-specific, thin clients.

This type of server-based computing model is especially useful in that it enables enterprises to overcome the critical application-deployment challenges of management, access, performance, and security. As a result, organizations can more quickly realize value from the applications and data that are required to run their businesses, receive the greatest return on computing investments, and accommodate both current and future enterprise computing needs economically.

When Oracle first introduced the concept of network computing, the company's motives were questioned. Critics charged that Oracle was merely trying to increase corporate dependence on servers so that it could boost sagging sales. The benefits of the thin-client architecture were compelling, however. Today, there are competing thin-client architectures, each attracting third-party developers to build hardware components, systems, applications, and management tools that plug into the overall framework. In this regard, Oracle's nearest competitor is Citrix Systems, which offers its *Independent Computing Architecture* (ICA). Through its Citrix Business Alliance program, third-party vendors work with Citrix to develop complementary products and markets for the company's WinFrame and MetaFrame thin-client/server software.

Change of Direction

In November 1998, Oracle officially abandoned its original vision of network computing in favor of Internet computing. Instead of requiring a larger number of smaller databases to be placed on every LAN, as called for in the network-computing model, the Internet-computing model relies on a smaller number of larger databases to which users connect over the Internet in order to access data and applications.

The cornerstone of the revised vision is Oracle8i, which is billed as the world's only Internet database that runs in conjunction with prepackaged server software on industry-standard hardware. Oracle8i supports both interpreted and compiled Java. The platform can also consolidate data and Java objects (and Windows files, as well) through its *Internet File System* (IFS). Users can drag and drop application files into IFS and search on the fields, just as they would search and query database data.

The value of Internet computing is its capability to help small, medium, and large businesses lower computing costs without the complexities of general-purpose operating systems. Oracle8i simplifies a company's systems by consolidating business data onto large servers for easy management, global Internet access, and higher-quality business information.

Oracle also offers subscription-based, remote support services to help customers overcome labor costs.

Oracle estimates that the Oracle8i platform could deliver a 10-to-one cost savings over client/server computing. The company is so convinced of the superiority of this new approach to network computing that it will no longer offer client/server versions of its products. According to Oracle, client/server distributes complexity and takes a tremendous amount of work to back up and maintain all of the data and applications on users' desktops. The Internet-computing model combines the best of the mainframe and client/server worlds by centralizing backups and by giving users the benefit of an intuitive graphical interface.

Benefits of Thin Clients

Businesses that have embraced thin clients are using them for a variety of applications. Most thin clients are used to access an office suite such as Microsoft Office, but some are used to run mission-critical applications such as accounting, transaction processing, and order-entry applications. To a lesser extent, thin clients are also running engineering, *Enterprise-Resource Planning* (ERP), and medical applications.

Users of thin clients are usually task-oriented and prefer to do their work without being distracted by technology issues. These users are front-line professionals, such as doctors in HMOs, accountants, engineers, and salespeople of big-ticket items (such as industrial equipment and real estate). Thin clients are also used for back-office operations that are supported by clerical and administrative staff, low-level salespeople, and workers on the shop floor. The human resources department of large companies can use thin clients that are installed solely with a browser, in order to enable job applicants to fill out forms.

Thin-client computing has several compelling benefits that are of key concern to organizations that are concerned about escalating IT costs:

- *Cost of ownership*   Thin-client computing lowers the total cost of ownership for the IT infrastructure.
- *Platform independence*   The use of thin clients enables applications to be written and deployed without regard for the desktop platform on which they will run.
- *Flexibility*   The use of thin clients eases the deployment of new applications, because they are installed and maintained at the server rather than at every desktop.

■ *Security* In terms of administration and overall protection of the network, some security features (such as virus filtering) are best implemented at the server, rather than at every desktop. In addition, centralizing security at the server enables easier control of access to files, applications, and networks.

If there is one disadvantage to thin-client computing, it is that many companies might have to upgrade their servers in order to accommodate the increased load that will inevitably occur when they are forced to support more clients. In addition to adding ports and interfaces to other networks, the servers might have to be upgraded with redundancy features in order to ensure continuous availability. Even with these one-time costs over a span of three to five years, however, companies can achieve significant savings in total cost of ownership. The savings accumulate primarily through the easier administration of the computing environment.

Role of Applets

Applets are Java programs that are embedded in and controlled by a larger application, such as a Web browser. Netscape Navigator and Microsoft's Internet Explorer have built-in Java interpreters.

With a Java-enabled browser, Web users can take advantage of all of the functionalities that applets provide. For example, through a requisition applet, Web users can have easy access to a company's product catalog and order merchandise online via an electronic order form. Once the user has completed a purchase order, the order is automatically routed for approval and processing through a workflow application until it reaches the shipping department, where the order is filled. A copy of the completed order is routed to the customer-service database. The advantage of using Java in this case is that the Web user does not need to download the entire workflow application to his or her desktop; rather, only the applet is required to collect the order information.

Java is being used as the foundation for developing interactive trading, insurance, investment planning, and stock-quote applications that can be accessed over the Web. Java applets are being used for implementing online banking, enabling customers to download their account information and interactively conduct bank business. Java applets are also being used by transportation companies, enabling customers to access shipping documents—bills of lading, container manifests, and shipment routings—from Web browsers.

Fat versus Thin Clients

The terms *fat* and *thin* refer primarily to the amount of processing that is being performed at the client. Terminals are the ultimate thin clients, because they rely exclusively on the server for applications and processing. Stand-alone PCs are the ultimate fat clients, because they have the resources to run all applications locally and to handle the processing themselves. Spanning the continuum from all-server processing to all-client processing is the client/server environment, where there is a distribution of work between the different processors.

Client/server was once thought to be the ideal computing solution. Despite the initial promises that were held out for client/server solutions, today, there is much dissatisfaction with their implementation. Client/server solutions are too complex; desktops are too expensive to administer and upgrade; and the applications still are not secure and reliable enough. Furthermore, client/server applications take too long to develop and deploy, and incompatible desktops prevent universal access. The network-computing paradigm appears to overcome these limitations.

Role of the Virtual Machine

Applets are specifically designed for distribution over the Internet. As such, they are always hosted by another program such as Netscape's Navigator or Microsoft's Internet Explorer, both of which contain a *Virtual Machine* (VM) that runs the Java code. Because the Java code is written for the virtual machine, rather than for a particular computer or operating system, all Java programs are cross-platform applications by default.

Java applications are fast because today's processors can provide efficient virtual-machine execution. The performance of GUI functions and graphical applications are enhanced through Java's integral multi-threading capability and *Just-in-Time* (JIT) compilation. The applications are also more secure than those that run native code, because the Java run-time system (which is part of the virtual machine) checks all code for viruses and tampering before running the code. Application development is facilitated through code reuse, making it easier to test the code and making it faster to deploy on the Internet or corporate intranet.

Because Java applications originate at the server, clients only receive the code when they need to run the application. If there are changes to the applications, they are made at the server. Programmers and network administrators do not have to worry about distributing all of the changes to every client. The next time the client logs on to the server and accesses the application, it automatically gets the most current code. This method of delivering applications also reduces support costs.

Accessing Legacy Data

Java applets that run on remote or mobile computers can perform emulation, providing easy access to legacy data on mainframe and minicomputer hosts via a remote-access gateway on the corporate intranet or the Internet. Although a locally installed emulator might have the same capabilities as a network-delivered, Java-based emulator, there is more work to be done to install and configure the local emulator than the Java-based emulator that is delivered each time a user needs it. The same local emulator takes up local disk space, whether it is being used or not. The Java-based emulator, in contrast, takes up no local disk space and remains in the browser's cache, which is cleared when the emulator is no longer needed.

The Java emulator supports all standard emulation features, including the following:

- Menu items for common keyboard functions
- Button bar
- Configuration of terminal options, based on user needs
- Font size that changes when a window is resized
- Cut-and-paste options
- Color
- Online help

For added flexibility, the user can open up to multiple, resizable environment windows. The configuration file determines whether the Java emulator initially opens as a separate window. The user can also create separate windows with the Web browser's New button. The Java emulator also supports hotspots for commands and menu selections. By clicking a hot spot, you cause the Java emulator to treat the action as if the user had pressed the equivalent command button or function key (or had entered the menu selection and pressed the Enter key).

Summary

Now, there is widespread recognition that thin clients and PCs are not mutually exclusive and that both are valued for certain tasks. Furthermore, it is fairly easy to integrate the two environments under central management, thereby realizing a significant reduction on TCO for both. All of this information means that there will not be the wholesale replacement of PCs with thin clients, as many vendors originally predicted. But that is not the end of the story. A new model of computing is emerging—namely, Oracle's Internet computing—which seeks to leverage existing investments in IP infrastructure. Although it is still too early to say how well this model will be accepted, Oracle is betting its future on the outcome.

See Also

Client/Server Networks

The Internet

Network-Design Tools

To stay competitive, companies are relying on their networks as never before. Typically, these networks consist of different kinds of transmission facilities, LAN technologies, protocols, and standards—all of which are cobbled together to meet the differing needs of workgroups, departments, branch offices, divisions, subsidiaries, and (increasingly) strategic partners, suppliers, and customers. Building such networks presents special design challenges.

Fortunately, a variety of automated design tools have become available in recent years. With built-in intelligence, these tools take an active part in the design process, from building a computerized model of the network, validating its design and gauging its performance, to quantifying equipment requirements and exploring reliability and security issues before the purchase and installation of any network component. Even faulty equipment configurations, design flaws, and standards violations are identified in the design process.

Data Acquisition

The design process usually starts by opening a blank drawing window from within the design tool, into which various vendor-specific devices—workstations, servers, hubs, and routers—can be dragged from a product library and dropped into place (refer to Figure N-3). The devices are further defined by type of components, software, and protocols (as appropriate). By drawing lines, the devices are linked to form a network, with each link being assigned physical and logical attributes. Rapid prototyping is aided by the capability to copy objects—devices, LAN segments, network nodes, and subnets—from one drawing to the next, editing as necessary until the entire network is built. Along the way, various simulations can be run in order to test virtually any aspect of the design.

The autodiscovery capabilities that are found in management platforms such as Hewlett-Packard's OpenView, IBM's NetView, and Sun's Solstice SunNet Manager—which automatically detect various network elements and represent them with icons on a topology map—are often useful in accumulating the raw data for network design. Some stand-alone design tools

Figure N-3
Typically, a design tool provides workspace into which objects are dragged and dropped from device libraries in order to start the network-design process from scratch. Source: NetFormx (formerly ImageNET, Ltd.)

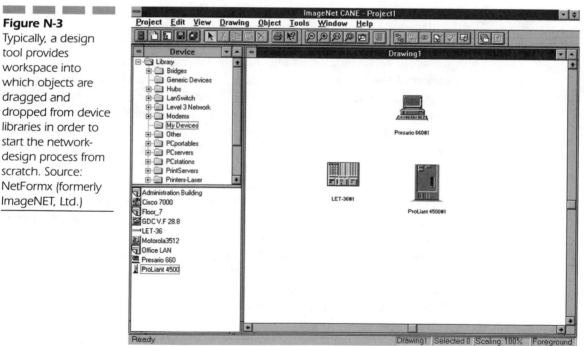

enable designers to import this data from network-management systems, which eases the task of initial data compilation. Although these network-management systems offer some useful design capabilities, they are not as feature rich as high-end, stand-alone tools that also have the capability to incorporate a broader range of network technologies and equipment makes and models.

Designing a large, complex network requires a multi-faceted tool—ideally, one that is graphical, object-oriented, and interactive. This tool should support the entire network life cycle, starting with the definition of end-user requirements and conceptual design to the detailed, vendor-specific configuration of network devices, the protocols that they use, and the various links between them. At each phase in the design process, the tool should have the capability to test different design alternatives in terms of cost, performance, and validity. When the design checks out, the tool should spew out network diagrams and a bill of materials—all before a single equipment vendor or carrier sales representative is contacted (or before an RFP is written).

With the right tools, modules, and device libraries, every conceivable type of network can be designed, including legacy networks such as SNA and DECnet and voice networks including ISDN, as well as T1, X.25, ATM, and TCP/IP networks. Some tools even take into account the use of satellite, microwave, and other wireless technologies.

The designer can take a top-down or bottom-up approach to building the network. In the former, the designer starts by sketching out the overall network; subsequent drawings add increasing levels of detail until every aspect of the network is eventually fleshed out. The bottom-up approach might start with a LAN on a specific floor of a specific building, with subsequent drawings linked to create the overall network structure.

As the drawing window is populated, devices can be further defined by type of component such as chassis, interface cards, and daughter boards. Even the operating system can be specified. Attributes can be added to each device taken from the library—to specify a device's protocol functionality, for example. Once the devices have been configured, a simulation profile is assigned to each device, which specifies its traffic characteristics for purposes of simulating the network's load and capacity.

With each device's configuration defined, lines are drawn between them to form the network. With some design tools, the links can be validated by testing for common protocols and network functions. This prevents Net-Ware clients from being connected to other clients instead of servers, for example. Such on-line analysis can also alert the designer to undefined links, unconnected devices, insufficient available ports in a device, and incorrect addresses in IP networks. Some tools are even able to report violations of network integrity and proper network design practices.

Network Simulation

Once the initial network design is completed, it can be tested for proper operation. The network is simulated by running it against a database that describes how the actual network devices behave under various real-world conditions. The simulator generates network events over time, based on the type of device and traffic pattern recorded in the simulation profile. This enables the designer to test the network's capacity under various what-if scenarios and fine-tune the network for optimal cost and performance. Simulators can be purchased as stand-alone programs, or might be part of the design tool itself.

Some tools are more adept at designing WANs, particularly those that are based on time division multiplexers. With a TDM component taken from the device library, for example, a designer can build an entire T1 network within specified parameters and constraints. The designer can strive for the lowest transmission cost that supports all traffic, for instance, or strive for line redundancy between all TDM nodes. By mixing and matching different operating characteristics of various TDM components, overall design objectives can be addressed, simulated, and fine-tuned. Some tools come with a tariff database to price transmission links and determine the most economical network design.

Such tools might also address clocking in the network design. Clocks are used in TDM networks to regulate the flow of data transmitted between nodes. All clocks on the network must therefore be synchronized to ensure the uninterrupted flow of data from one node to another. The design tool automatically generates a network topology synchronization scheme, taking into account any user-defined criteria, to ensure that there are no embedded clock loops.

Bill of Materials

Once the design is validated, the network design tool generates a bill of materials that includes order codes, prices, and discounts. This report can be exported to any Microsoft Windows application, such as Word or Excel, for inclusion in the proposal for top management review or an RFP issued to vendors and carriers who will build the network. Through the tool's capability to render multiple device views, network planners can choose either a standard schematic or an actual as-built rendering of the cards and the slot assignments of the various devices. Some tools also generate Web-enabled output, which allows far-flung colleagues to discuss and annotate the proposal over the Internet—even allowing each person to drill-down and extract appropriate information from the network device library.

Summary

Today's networks are more complex by orders of magnitude than networks envisioned only a few years ago. New Internet services, new technologies, new trends toward VPNs and voice-data convergence, plus the sheer number of new equipment offerings, have made reliance on traditional manual solutions to network engineering problems simply unworkable. Fortunately, a new generation of intelligent design tools with built-in error-detection, simulation and analysis capabilities, and plug-in modules for ancillary functionality are now available. They do not require managers and planners to be intimately familiar with every aspect of their networks. The essential information can be retrieved on a moment's notice—often with point-and-click ease—analyzed, queried, manipulated, and reanalyzed if necessary, with the results displayed in easy-to-understand graphical form or exported to other applications for further manipulation and study.

See Also

Network Drawing Tools
Network Management

Network Drawing Tools

Network administrators who are faced with managing detailed and often large quantities of information on local and world-wide corporate networks require tools that can accurately depict these complex infrastructures. While the automatic discovery capabilities of high-end network-management systems can help in this regard, they are not too useful for documenting the equipment at the level of detail that is now required by network planners.

A variety of drawing tools have become available that can aid the network design process. Such tools provide the five major features that are considered critical to network planners:

- An easy-to-use drawing engine for general graphics
- An extensive library of predrawn images that represent vendor-specific equipment
- A drill-down capability that enables multiple drawings to be linked, in order to show various views of the network
- A database capability to assign descriptive data to the device images
- A high degree of embedded intelligence that makes images easy to create and update

Most network drawing tools are Windows based and employ the drag-and-drop technique in order to move images of network equipment from a device library to a blank workspace. Many also enable network designs to be published on the corporate intranet or the public Internet, enabling any authorized user to view them with a Web browser. Some drawing tools can automatically discover devices on an existing network in order to ease the task of drawing and documenting the network.

Device Library

A device library holds images of such items as modems, telephones, hubs, PBXs, and CSU/DSUs from different manufacturers. Representations of LANs and WANs, databases, buildings and rooms, satellite dishes, microwave towers, and a variety of line connectors are included. There are also shapes that represent generic accessories such as power supplies, PCs, towers, monitors, keyboards, and switches. There are even shapes for racks, shelves, patch panels, and cable runs (refer to Figure N-4).

Typically, an annual subscription provides unlimited access to the hundreds of new network devices, adapters, and accessories that are added to the device library. Depending on the drawing-tool vendor, new objects might even be downloadable from the company's Web site.

While many drawing tools offer thousands of exact-replica hardware-device images from hundreds of network-equipment manufacturers, some tools have embedded intelligence into the shapes, which enables components such as network cards to snap into equipment racks and to remain in place even when the rack is moved. In addition, each shape can be annotated with product-specific attributes, including vendor, product name, part

Figure N-4
From a library of
network shapes,
items are dragged
and dropped into
place as needed in
order to design a
new node or to build
a whole network.

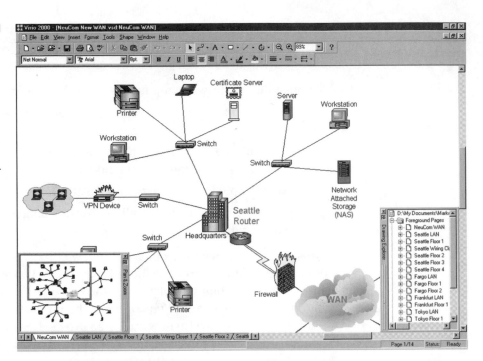

SOURCE: Visio Corporation (As of January 2000, Visio Corporation became the Visio Division operating within Microsoft's Business Productivity Group).

number, and description (refer to Figure N-5). This feature permits users to generate detailed inventory reports for network asset management.

The shapes are even programmable, so they can behave like the objects they represent. This feature reduces the need for manual adjustments while drawing and ensures the accuracy of the final diagram. For example, the shape that represents an equipment rack from a specific vendor can be programmed to know its dimensions. When the user populates the drawing with multiple instances of this shape, it could issue an alert if there is a discrepancy between the space that is available on the floor plan and the space requirements of the equipment racks.

Each shape can also be embedded with detailed information. For example, the user can associate a spreadsheet with any network element—to provide cost information on a new switch node or LAN segment, for example—along with a bar chart to perhaps illustrate the cost data by system component. The spreadsheet data can be manipulated until costs fit within budgetary parameters. The changes will be reflected in the bar chart the next time the chart is opened.

Details about
network equipment
can be stored by
using custom
property fields.
Device-specific data
for each network
shape keeps track of
asset, equipment,
and manufacturer
records that can be
accessed from within
network diagrams.
Source: Visio
Corporation

Network Shape Properties ☒

| Asset | Equipment | Maintenance |

Manufacturer: |Timestep

Product Number: |VPN1400

Part Number: |1251-5135179

Product Description: |Hardware Firewall

[?] [Define...] [OK] [Cancel]

The Drawing Process

To start a network diagram, the user typically opens the template for the manufacturer whose equipment will be placed in the diagram. This action causes a drawing page to appear that contains rules and a grid. The drawing page itself can be sized in order to show the entire network or just a portion of it.

Various other systems and components can be added to the diagram by using the drag-and-drop technique. The user has the option of having or not having the shapes snap into place within the drawing space so that they will be precisely positioned on grid lines. Once placed in the drawing space, the shapes can be moved, resized, flipped, rotated, and glued together. Expansion modules, for example, can be dragged onto the chassis so that the modules' endpoints glue to the connection points on the chassis expansion slots. This functionality enables the chassis and modules to be moved anywhere in the diagram as a single unit. Via the cut-and-paste method, the user can add as many copies of the component as desired in order to quickly populate the network drawing.

To show the connections between various systems and components, the user can choose shapes that represent different types of networks, including LANs, X.25, satellite, microwave, and radio. Alternatively, the user can choose to connect the shapes with simple lines that can have square or curved corners.

Each network-equipment shape has properties that are associated with it. Custom properties can be assigned to shapes that are used in tracking equipment and generating reports, such as inventories. Text can be added to any network system or component, including a Lotus Notes field, specifying the font, size, color, style, spacing, indent, and alignment. Text blocks can be moved and resized. Some tools even include a spell checker and a search-and-replace tool. The user can add words that are not in the standard dictionary that comes with the program. The user can specify a search of the entire drawing, a particular page, or selected text only.

AutoCAD files and clip art can be added to network drawings. The common file formats that are usually supported for importing graphics from other applications include *Encapsulated PostScript* (.EPS), *Joint Photographic Experts Group* (.JPG), *Tag Image File Format* (.TIF), and *ZSoft PC PaintBrush Bitmap* (.PCX).

The various shapes that are used in a network drawing can be kept organized by using layers. A layer is a named category of shapes. For example, the user can assign walls, wiring, and equipment racks to different layers in a space plan, which allows the user to perform the following actions:

- Show, hide, or lock shapes on specific layers so that they can be edited without affecting shapes on other layers
- Select and print shapes based on their layer assignments
- Temporarily change the display color of all shapes on a layer, to make them easier to identify
- Assign a shape to more than one layer, as well as assign the member shapes of a group to different layers

The user can also group shapes into customizable stencils. If the same equipment is used at each node in a network, for example, the user can create a stencil that contains all of the devices. All of the graphics and text that is associated with each device will be preserved in the newly created stencil, which saves time when drawing large-scale networks—especially those that are based on equipment from a variety of manufacturers.

At any step in the design process, the user can share the results with other network planners by sending copies via e-mail. The diagram is converted to an image file, which is displayed as an icon in the message box and is sent as an attachment. When the recipient receives the message, the attached diagram—including all embedded information—can be opened by clicking the icon. The document can then be marked up by creating a separate layer for review comments. Using a separate layer for comments protects the original drawing and makes it easier to view, print, and color separately from the rest of the drawing.

Some network drawing tools provide a utility that converts network designs and device details into a series of hyperlinked HTML documents that can be accessed over the Web. These documents show device configurations, port usage, and even device photographs. Users can activate the links to navigate from device to device in order to trace connectivity and to review device configurations (refer to Figure N-6). In addition to supporting fault identification, the hyperlinked documents aid in planning design changes.

Several ways exist for the network diagrams to be protected against inadvertent changes, especially if they are shared via e-mail or are posted on the Web:

- The shapes can be locked to prevent them from being modified in specific ways.

- The attributes of a drawing file (styles, for example) can be protected against modification.

Figure N-6

This floor plan of a 10BaseT network is a hyperlinked drawing that was rendered by Netscape Navigator. Source: NetSuite Development

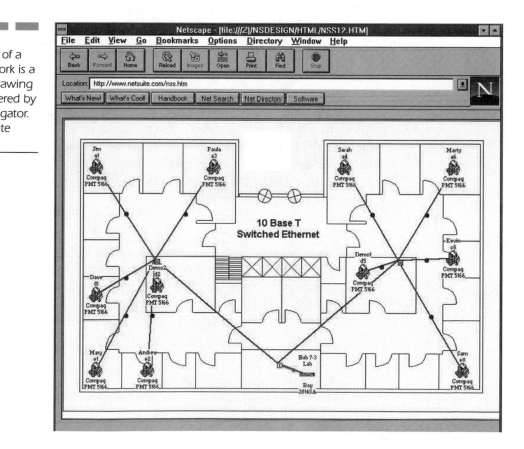

- The file can be saved as read-only, so it cannot be modified in any way.
- The shapes on specific layers can be protected against modification.

Users can password-protect their work in order to prevent attributes of a drawing file from being changed. For example, a background that contains standard shapes or settings can be password-protected. Users can also set a password for a drawing's styles, shapes, backgrounds, or masters. A password-protected item can be edited only if the correct password is entered.

Embedded Intelligence

Some products are so intelligent that they can no longer be considered merely drawing tools. Visio 2000 Enterprise Edition, for example, supports switched WANs through its AutoDiscovery feature. This technology includes support for layer two (switched data link), layer three (IP network), and Frame Relay network environments. AutoLayout technology makes it simple for users to automatically generate network diagrams of the discovered devices, including detailed mappings.

In addition to enabling IT specialists to create conceptual, logical, and physical views of their information systems, Enterprise Edition owners can purchase Visio's add-on solution for monitoring network performance. Working with Enterprise Edition's AutoDiscovery, Real-Time Statistics documents the behavior of a network environment, capturing real-time data from any SNMP-manageable device on LANs and WANs. In being able to monitor the network's performance, managers have the information that they need in order to redistribute network traffic and prevent overloads. Real-Time Statistics then turns this performance data into graphs that can be printed or exported for analysis.

Enterprise Edition enables developers to visualize and to quickly start software-development projects. They can visualize the design architecture of existing systems by reverse-engineering source code from Microsoft Visual Studio. They can also decrease development time by generating fully customizable code skeletons for Visual Basic, C++, and Java from *Unified Modeling Language* (UML)[2] class diagrams.

---

[2]Pioneered by Rational Software Corporation and officially adopted as a standard by the *Object Management Group* (OMG), the *Unified Modeling Language* (UML) is the industry-standard language for specifying, visualizing, constructing, and documenting the elements of software systems. UML simplifies the complex process of software design, making a blueprint for construction.

Summary

Unlike traditional CAD programs, today's drawing tools are specifically designed for network and IT planners. They can improve communications and productivity with their easy-to-use and easy-to-learn graphics capabilities that offer seamless integration with other applications on the Windows desktop. Their graphical representation of complex projects also enables more people to understand and participate in the planning process. Despite their origins as simple drawing tools, this new generation of tools provides a high degree of intelligence, programmability, and Web awareness that makes them well suited for the demanding needs of planners.

See Also

> Asset Management
> Network-Design Tools
> Network Management

Network Integration

Distinct from systems integration, which focuses on getting different computer systems to communicate with each other, network integration is concerned with getting diverse, far-flung LANs and host systems interconnected over the WAN. Typically, network integration requires that attention be given to a plethora of different physical interfaces, protocols, frame sizes, and data-transmission rates. Companies also face a bewildering array of carrier facilities and services to extend the reach of information systems globally over high-speed WANs. Getting everything working properly might require a fair amount of hardware and software customization.

A network integrator brings objectivity to the task of tying together diverse products and systems in order to form a seamless, unified network. To reach this goal, the network integrator draws upon its expertise in *Information Systems* (IS), office automation, LAN administration, telephony, data communications, and network-management systems. Added value is provided through strong business planning, needs analysis, and project-management skills, as well as through accumulated experience in meeting customer requirements in a variety of industry segments and operating environments.

A qualified network integrator will have in place a stable support infrastructure that is capable of handling a high degree of ambiguity and complexity, as well as any technical challenge that might step in the way of the integration effort. In addition to financial stability, this support infrastructure includes staff members that represent a variety of technical and management disciplines and strategic relationships with specialized companies, such as cable installers and software firms.

Integration Services

A number of discrete services are provided by network-integration firms, including the following:

- *Design and development* Involves activities such as network design, facilities engineering, equipment installation and customization, and acceptance testing
- *Consulting* Includes needs analysis, business planning, systems/network architecture, technology assessment, feasibility studies, RFP development, vendor evaluation and product selection, quality assurance, security auditing, disaster-recovery planning, and project management
- *Systems implementation* Entails procurement, documentation, configuration management, contract management, and program management
- *Facilities management* Provides operations, technical support, hotline services, move and change management, and trouble-ticket administration
- *Network management* Includes network optimization, remote monitoring and diagnostics, network restoration, technician dispatch, and carrier and vendor relations

Network integration might be performed by in-house technical staff or through an outsourcing arrangement with a computer company, local or interexchange carrier, management consulting firm, traditional IS-oriented system integrator, or an interconnect vendor. Each type of firm has specific strengths and weaknesses. The wrong selection can delay the implementation of new information systems and LANs, disrupt network expansion plans, and impede applications development—any of which can inflate operating costs over the long term and have adverse competitive impacts.

You should therefore choose an integrator whose products and services are particularly pivotal to the application. For example, if the network-integration application is such that the computer requirements are extremely well defined and no significant computer changes are expected, but a range of new communications services might be involved, a carrier would be a better choice of integrator than a computer vendor. On the other hand, if the project is narrow in scope and the needs are well understood, in-house staff might have the capability to handle the integration project with as-needed assistance from a computer firm or carrier.

Summary

While some companies have the expertise that is required to design and install complex networks, others are turning to network integrators in order to oversee the process. The evaluation of various integration firms should reveal a well-organized and staffed infrastructure that is enthusiastic about helping to reach the customer's networking objectives. This infrastructure includes having the methodologies already in place, the planning tools already available, and the required expertise already on staff. Beyond that, the integrator must be able to show that its resources have been successfully deployed in previous projects of a similar nature and scope.

See Also

> *Business-Process Engineering* (BPE)
> Outsourcing
> Systems Integration

Network-Management Systems (NMSs)

The task of keeping multi-vendor networks operating smoothly with a minimum of down time is an ongoing challenge for most organizations. While many companies prefer to retain total control of their network resources, others rely on computer vendors and carriers to find and correct problems on their networks or depend on third-party service firms. Wherever these responsibilities ultimately reside, the tool set that is used for monitoring the status of the network and initiating corrective action is the *Network-Management System* (NMS).

With an NMS, technicians can remotely diagnose and correct problems that are associated with each type of device on the network. Although today's network manager is primarily concerned with diagnosing failures, the likelihood of problems to occur can also be predicted so that traffic can be diverted from failing lines or equipment (with little or no inconvenience to users).

Network management begins with basic hardware components such as modems, data sets (CSUs/DSUs), multiplexers, and dial backup units (refer to Figure N-7). Each component typically has the capability to monitor, self-test, and diagnose problems regarding its own operation and report problems to a central management station. The management station operator can initiate test procedures on systems at the other end of a point-to-point line. On more complex multi-point and multi-drop configurations, the capability to test and diagnose problems from a central location greatly facilitates problem resolution. This capability also minimizes the need to dispatch technicians to remote locations and reduces maintenance costs.

Figure N-7
Each type of device on the network can have its own Element Management System (EMS), which reports to an Integrated Network-Management System (INMS).

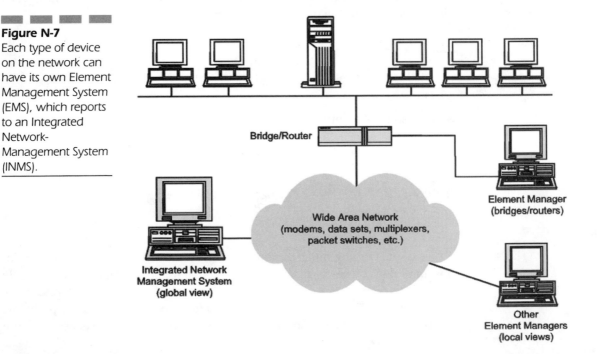

A minimal network-management system consists of a central processing unit, system controller, operating-system software, storage device, and an operator's console. The central processor can consist of a minicomputer or microcomputer. The system controller, the heart of the network-management system, continuously monitors the network and generates status reports from data that is received from various network components. The system controller also isolates network faults and restores segments of the network that have failed or that are in the process of degrading. The controller usually runs on a powerful platform, such as UNIX or Windows NT.

NMS Functions

Although differing by vendor, the basic functions that are common to most network-management systems include the following: topology mapping, administration, performance measurement, control and diagnostics, configuration management, applications management, and security. Some network-management systems include other functions such as network modeling, for example, which would enable the operator to simulate aggregate, node, or circuit failures in order to test various disaster-recovery scenarios.

Topology mapping Many network-management systems have an automatic discovery capability that finds and identifies all devices or nodes that are connected to the network. Based on the discovered information, the NMS automatically draws the required topology maps. Nodes that cannot be discovered automatically can be represented by manually adding custom or standard icons to the appropriate map views, or by using the network-management systems' SNMP-based *Application Programming Interfaces* (APIs) for building map applications without having to manually modify the configuration to accommodate non-SNMP devices.

A network map is useful for ascertaining the relationships of various equipment and connections, for keeping accurate inventory of network components, and for isolating problems on the network. The network map is updated automatically when any device is added to or removed from the network. The device status is displayed via color changes to the map. Any changes to the network map are carried through to the relevant submaps.

Administration The administration element enables the user to take stock of the network in terms of which hardware is deployed and where it is located. This element also tells the user which facilities are serving various locations and which lines and equipment are available for implementing alternate routing. The vehicle for storing and using this information is the *Relational Database Management System* (RDBMS).

For administrative tasks, multiple, specialized databases are used that relate to each other. One of these databases accumulates trouble-ticket information. A trouble ticket contains information such as the date and time the problem occurred, the specific devices and facilities that were involved, the vendor from which the product was purchased or leased, and the service contact. This ticket also contains the name of the operator who initially responded to the alarm, any short-term actions that were taken to resolve the problem, and space for recording follow-up information. This information might include a record of visits from the vendor's service personnel, dates on which parts were returned for repair, serial numbers of spares that were installed, and the date of the problem's final resolution.

A trouble-ticket database can be used for long-term planning. The network manager can call up reports on all outstanding trouble tickets, trouble tickets that involve particular segments of the network, trouble tickets that were recorded or resolved within a given period, trouble tickets that involve a specific type of device or vendor, and even trouble tickets over a given period that were not resolved within a specific time frame. The user can customize report formats in order to meet unique needs.

Such reports provide network managers with insight into the reliability of a given network-management station operator, the performance record of various network components, the timeliness of on-site vendor maintenance and repair services, and the propensity of certain segments of the network to fail. Also, with information about both active and spare parts, network managers can readily support their decisions concerning purchasing and expansion. In some cases, cost and depreciation information on the network's components is also provided.

Performance measurement   Performance measurement refers to network response time and network availability. Many network-management systems measure response time at the local end, from the time the monitoring unit receives a *Start-of-Transmission* (STX) or *End-of-Transmission* (EOT) signal from a given unit. Other systems measure end-to-end response time at the remote unit. In either case, the network-management system displays and records response-time information and generates operator-specified statistics for a particular terminal, line, network segment, or the network as a whole. This information can be reported in real time or stored for a specified time frame for future reference.

Network personnel can use this information to track down the cause of the delay. When an application exceeds its allotted response time, for example, a network administrator can decide whether to reallocate terminals, place more restrictions on access, or install faster communications equipment in order to improve response time.

Availability is a measure of actual network up time, either as a whole or by segments. This information can be reported as total hours that are available over time, average hours that are available within a specified time, and *Mean Time between Failure* (MTBF). With response time and availability statistics, which are calculated and formatted by the NMS, managers can establish current trends in network usage, predict future trends, and plan the assignment of resources for specific present and future locations and applications.

Control and diagnostics With control and diagnostic capabilities, the NMS operator can determine from various alarms (i.e., an audio or visual indication at the operator's terminal) what problems have occurred on the network and pinpoint the sources of those problems so that corrective action can be taken. Alarms can be correlated to certain events and triggered when a particular event occurs. For example, an alarm can be set to go off when a line's *Bit-Error Rate* (BER) approaches a predefined threshold. When that event occurs and the alarm is issued, automated procedures can be launched without operator involvement. In this case, traffic can be diverted from the failing line and routed to an alternate line or service. If the problem is equipment-oriented, another device that is on hot standby can be placed into service until the faulty system can be repaired or replaced.

Configuration management Configuration management gives the NMS operator the capability to add, remove, or rearrange nodes, lines, paths, and access devices as business circumstances change. If a T1 link degrades to the point that it can no longer handle data reliably, for example, the network-management system can automatically reroute traffic to another private line or through the public network. When the quality of the failed line improves, the system reinstates the original configuration. Some integrated network-management systems—those that unify the host (LAN) and carrier (WAN) environments under a single management umbrella—are even capable of rerouting data but leaving voice traffic where it is.

Voice and data traffic can even be prioritized. This NMS capability is important, because failure characteristics for voice and data are different.

Voice is more delay sensitive, and data is more line-error sensitive. On networks that serve multiple business entities and on statewide networks that serve multiple government agencies, the capability to differentiate and prioritize traffic is important.

On a state-wide network, for example, state police have critical requirements 24 hours a day, seven days a week, whereas motor vehicle branch offices use the network to conduct relatively routine administrative business only eight hours a day, five days a week. Consequently, the response-time objectives of each agency are different, as would be their requirements for network restoration in case of an outage. On the high-capacity network, there can be two levels of service for data and another for voice. Critical data will have the highest priority in terms of response time and error thresholds and will take precedence over other classes of traffic when it comes to restoration. Because routine data will have the capability to tolerate a longer response time, the point at which restoration is implemented can be prolonged. Voice is more tolerant than data with regard to error, so restoration might not be necessary at all. The capability to prioritize traffic and reroute only when necessary ensures maximum channel fills, which impacts the efficiency of the entire network (and consequently, the cost of operation).

Configuration management not only applies to the links of a network but to equipment, as well. In the WAN environment, the features and transmission speeds of software-controlled modems can be changed. If a nodal multiplexer fails, the management system can call its redundant components into action or invoke an alternate configuration. And when nodes are added to the network, the management system can devise the best routing plan for the traffic that it will handle.

Applications management   Applications management is the capability to alter circuit routing and bandwidth availability in order to accommodate applications that change by time of day. Voice traffic, for example, tends to diminish after normal business hours, while data traffic might change from transaction-based to wide-band applications that include inventory updates and remote printing tasks.

Applications management includes having the capability to change the interface definition of a circuit so that the same circuit can alternatively support both asynchronous and synchronous data applications. This management also includes having the capability to determine appropriate data rates in accordance with response-time objectives or to conserve bandwidth during periods of high demand.

Security   Network-management systems have evolved to address the security concerns of users. Although voice and data can be encrypted in order to protect information against unauthorized access, the management system represents a single point of vulnerability to security violations. Terminals that are employed for network management might be password protected in order to minimize disruption to the network through database tampering. Various levels of access can be used to prevent accidental damage. A senior technician, for example, might have a password that enables him to make changes to the various databases, whereas a less-experienced technician's password enables him to only review the databases without making any changes. Other possible points of entry, such as gateways, bridges, and routers, can be protected with hardware-defined or software-defined partitions that restrict internal access.

Individual users, too, might be given passwords that permit them to make use of certain network resources but deny them access to others. A variety of methods are even available to protect networks from intruders who might try to access network resources with dialup modems. For instance, the management system can request a password and hang up if it does not obtain one within 15 seconds. Or, it can hang up and call back over an approved number before establishing the connection. To frustrate persistent hackers, the system can limit unsuccessful call attempts before denying further attempts. All successful and unsuccessful attempts at entry are automatically logged to monitor access and to aid in the investigation of possible security violations.

Summary

Today's network-management systems have demonstrated their value in permitting technicians to control individual segments or the entire network remotely. In automating various capabilities, network-management systems can speed up the process of diagnosing and resolving problems with equipment and lines. The capabilities of network-management systems permit maximum network availability and reliability, thus enhancing the management of geographically-dispersed operations while minimizing revenue losses from missed business opportunities that might occur as a result of network down time.

See Also

Outsourcing

Network Integration

Network Restoration

Simple Network-Management Protocol (SNMP)

Network Operations Centers (NOCs)

A *Network Operations Center* (NOC) comprises equipment, management tools, and technical personnel whose function is to keep a private or carrier network continuously available and operating at peak performance. Depending on the size of the network and its coverage, the carrier might have multiple NOCs. A large carrier's NOC might be manned 24 hours a day.

Interexchange Carriers (IXCs), *Incumbent Local-Exchange Carriers* (ILECs), *Competitive Local-Exchange Carriers* (CLECs), and *Internet Service Providers* (ISPs) are among the types of carriers that maintain NOSs. The larger carriers supervise their networks on a 24-hour-a-day, seven-day-a-week basis. Among other things, these operations centers monitor the performance of the network, reroute traffic in order to smooth out peak-hour flows, and isolate problems until on-site repairs can be dispatched.

AT&T has one of the most sophisticated network-management systems of this kind. The NOC in Bedminster, New Jersey has the responsibility for the bulk movement of traffic across the AT&T Worldwide Intelligent Network. The network-management centers in Denver, Colorado and Conyers, Georgia (near Atlanta) are responsible for managing traffic that comes onto or leaves the AT&T network from the LECs.

The network-management centers use real-time information for managing the AT&T network and can implement controls directly from their centers in order to manage traffic flow. Personnel use specific network techniques and have skills for managing network situations during earthquakes, hurricanes, and other natural disasters, so that a high degree of call completion can be maintained during a crisis situation.

Typical Tasks

The typical carrier NOC is dedicated to operations and maintenance of the network, support of customers in the provisioning of new services, the optimization of network performance, and the implementation of system or service upgrades or reconfigurations.

Within the NOC, technicians and network managers monitor and control all facilities and nodes (and, in many cases, customer-premises equipment). The NOC is responsible for performance management, fault management, configuration management, and security management. Performance management includes the following:

- Monitoring and managing the transmission performance of the network-ending components, including cables, cable stations, and back-haul links
- Developing and implementing corrective actions
- Performance reporting
- Managing scheduled maintenance operations

Fault management includes the following:

- Monitoring and responding to network faults by using network-management tools and fault identification and sectionalization procedures
- Administering and activating repair processes and restoration plans

Configuration management includes the following:

- Configuring, upgrading, testing, and commissioning network elements
- Network capacity and customer-service provisioning
- Managing third-party suppliers of equipment and support services

Security management includes the following:

- Monitoring the cable station environment and security/intrusion alarms
- Responding to alarms and initiating corrective action immediately, day or night

Customer Care Centers

In addition to one or more NOCs, some carriers have separate customer care centers that act as the first point of contact with the customer. If problems cannot be resolved by customer-care staff, a trouble ticket is opened that gets passed to the NOC for action. The ongoing responsibilities of the customer care center differ by carrier. Among them are capacity management, service interfacing, billing administration, and customer reporting. Staff members have access to all records and resources that are needed to provide timely and effective responses to customer inquiries and service requests.

Capacity management includes the following:

- End-to-end provisioning and maintenance-order management
- Administration of *Capacity Purchase Agreements* (CPAs)

- Customer interface for provisioning activities
- Centralized interface to back-haul providers
- Monitoring network usage to forecast the need for a system upgrade

Service interface activities include the following:

- Point of contact for service or maintenance requests and inquiries
- Single point of contact with back-haul providers

Billing administration includes the following:

- Billing for capacity purchases and *Operations, Administration, and Maintenance* (OA&M) services
- Billing inquiries and adjustments

Customer reporting functions include the following:

- Reporting network performance
- Reporting to customers on utilization, repair intervals, and restoration operations

At a well-equipped NOC, an operator can view the health and status of the entire network by using a map of the network that is shown on a graphical display. If a network error occurs or if traffic exceeds predefined limits, the operator is notified by an alarm on the display screen. The displayed trouble area can be expanded to see local details and the parameter that triggered the alarm. Based on type of trouble, a network engineer can be notified or a technician can be dispatched. Often, the trouble can be handled right from the NOC. For example, specialized software at the NOC can reroute a customer's Frame Relay *Permanent Virtual Circuits* (PVCs) from a bad switch port to a spare switch port—usually within a few minutes.

While NOC staff can respond to alarms that are coming in from various systems on the network, it is also possible to create a proactive environment in which network problems are anticipated and corrected before customers become aware that anything is wrong. By continuously monitoring the network, the NOC can build performance baselines that define the normal behavior of network nodes and facilities. The NOC can issue an alert when behavior moves out of the normal range, so action can often be taken to head off problems before they impact customers.

Networks have a tendency to grow, and a NOC can provide tools for planning such growth. Historical data collected by the NOC can be used to identify optimum network configurations. Information about top talkers, high usage paths, and local versus remote communications can all be factored into this type of analysis. Then, before any network changes are actually made, a network-simulation software tool can be used to test the proposed configuration in order to discover optimum distribution of routers, switches, hubs, end users, and other resources.

Summary

NOCs are vital to maintaining the health of carrier and enterprise networks. Using various management tools, the NOC presents alarms and other information to network operators as well as to subscribers. The NOC also provides for provisioning, troubleshooting, researching, analyzing, and tracking of physical network problems. The NOC processes historical data for trend analysis and generates reports that can be used for accounting, capacity planning, and future growth planning. Ideally, the management tools used in the NOC comply with the *Telecommunications Management Network* (TMN) model, which provides a cooperative framework for integrated management (both within and across multiple networks).

See Also

Help Desks

Network-Management Systems (NMSs)

Telecommunications Management Network (TMN)

Network Restoration

Businesses are increasingly relying on their communication networks to improve customer service, exploit new market opportunities, and secure strategic competitive advantages. Therefore, when these networks become severely congested or fail, effective restoration solutions must be implemented as soon as possible.

Although local and long-distance carriers build reliability into their networks at the design stage and monitor performance of the network on a continuing basis, there is always the chance that unforeseen problems will occur. When they do, automated processes perform such functions as raise alarms, reroute traffic, activate redundant systems, perform diagnostics, isolate the cause of the problem, generate trouble tickets and work orders, dispatch repair technicians, and return primary facilities and systems to their original service configuration.

Network Redundancy

Most networks are designed with a certain amount of redundancy built in. There usually is a duplicate or backup system that can be called into service immediately upon failure of the primary system.

For example, central office switches are equipped with dual processors, so that if one processor fails, the second one can take over automatically. The switches are designed to run self-diagnostic tests periodically to help ensure proper operation. If a problem occurs, systems can often automatically fix themselves by rebooting software, for example, or switching automatically to a backup system so the primary system can reinitialize itself. These systems also have the capability to alert technicians and network managers if the problem cannot be corrected automatically.

Redundancy also applies to *Signal Transfer Points* (STPs). These are the computers used to route messages over the carrier's packet-based signaling network, which is used to set up calls and implement intelligent services. Each STP has two computers that operate at just under 50 percent capacity. The STP pairs are not collocated, but are usually many miles away from each other. If something happens to one STP, its mate can pick up the full load and operate until the repair or replacement of the damaged STP can be accomplished. Should both halves of a mated pair of STPs fail, the switch that normally relies on them can access the signaling network through helper switches that use a different STP pair.

Network Control Points (NCPs), the customer databases for advanced services such as 800 or VPN services, not only have dual processors—but also, if the second processor should fail, there is a backup NCP for the protection of all customer configuration information.

Digital Interface Frames (DIFs) provide access to and from interoffice switches for processing long-distance calls. The digital interface units that actually handle this work have spares that take over immediately when a problem occurs. Guiding the overall work of the DIF are two controllers running simultaneously, so that if one experiences a problem, the backup controller can take over without customers even noticing that a problem has occurred. As an option, however, a customer's traffic can be sent to another DIF at another interexchange office if the primary switch encounters a problem. This procedure ensures that the customer's calls continue to flow, should the primary DIF experience a prolonged outage.

The power systems used to operate the carrier's network also have backup protection. In normal operation, the carrier's power system provides direct current from redundant rectifiers fed by commercial power. If commercial power fails, batteries, which are charged by the rectifiers, provide backup power. An additional level of redundancy is provided by diesel oil-powered generators, which can replace commercial power during prolonged outages.

Network Diversity

Diversity is the concept of providing as many alternative paths as possible to ensure survival of the network when some kind of natural or man-made disaster strikes. Like redundancy, diversity is built into the network during the design stage.

One way carriers ensure diversity is to arrange transmission lines as a series of circles or loops to form an interconnecting grid. Should any particular loop be cut—such as a backhoe operator hitting a buried cable—traffic can be sent over another facility on one or more adjacent loops. In the case of fiber-optic rings around major metropolitan areas, the use of dual counter-rotating fiber-ring configurations enables carriers to offer disaster recovery services. In the event of a node failure on the fiber-optic network, traffic is automatically routed to the other ring in a matter of milliseconds.

In larger metropolitan areas, carriers often offer their business customers building diversity. In having the capability to reach the carrier's network from two distinct points, businesses can enhance the reliability of their mission-critical applications. To further ensure uninterrupted service, carriers also offer route diversity for their signaling systems. Each pair of STPs is connected to every other pair of STPs by multiple links. To ensure that connectivity will always be available, these links are laid out over multiple geographically separated routes. Should something happen along one route to disrupt service, the other routes remain available to keep the carrier's signaling system operational.

Optional Restoration Services

For most businesses, a temporary interruption of service lasting only a few minutes does not present a problem. For businesses that need a much shorter restoration period, carriers offer optional services that can be tailored to meet specific requirements. These services can range from having the carrier plan and build a complete private network in order to meet certain reliability and performance specifications, to selecting one or more of the following lower-cost alternatives:

- For businesses that have toll-free 800 service and virtual private networks, the capability to receive calls is of primary importance. The carrier provides routes from two separate switches to the corporate location. In the event of a network disruption, calls are automatically directed to the working switch.

- For businesses that have toll-free 800 service, if traffic is blocked at one location for any reason, calls are automatically sent to another corporate location.

- For businesses that use digital services, the carrier can provide a geographically separate backup facility, enabling traffic to be switched to the standby link(s) within milliseconds of a service interruption on the primary link(s).

- For the access portion of a circuit, the carrier can mitigate the effects of certain network failures by automatically transferring service to a dedicated, separately routed access circuit.

Customer-Controlled Reconfiguration (CCR) is a carrier-provided service that gives businesses a way to organize and manage their own circuits from an on-premises terminal that issues instructions to the carrier's *Digital Cross-Connect System* (DCS). If a circuit drops to an unacceptable level of performance or fails entirely, the network manager can issue rerouting instructions to the DCS. If several circuits have failed, the network manager can upload a pre-tested rerouting program to the DCS to restore the affected portion of the network.

Some carriers offer businesses a reservation service in which one or more dedicated digital facilities are brought online after the customer verbally requests it with a phone call. This restoration solution requires that the customer presubscribe to the service and that access facilities already be in place with the local carriers at each end.

Site Recovery Options

Some carriers offer optional site recovery options. This type of service is meant to deal with the loss of a primary data center that runs mission-critical applications. If a customer's data center suffers from a catastrophic fire or natural disaster, for example, traffic will be quickly rerouted to another comparably equipped site. This service is far more economical than having to set up and maintain another data center and links.

To offer site recovery services, the carrier typically partners with an established firm such as Comdisco Disaster Recovery Systems. When disaster strikes, the customer calls the carrier and requests activation of links to the alternate site—a process that might take about two hours to complete and that might entail uploading new routing tables to each router in order to reflect the changes.

Summary

Traditionally, most organizations relied on their long-distance carrier for maintaining acceptable network performance. More often than not, the carriers were not up to the task. This situation led to the emergence of private networks in the 1980s, which enabled companies to exercise close control of leased lines with an in-house staff of network managers and technicians. In their eagerness to recapture lost market share, the carriers have made great strides in improving their response to network congestion and outages. This process has gone a long way toward restoring lost confidence that once prompted companies to set up and maintain their own networks. Today, companies are once again comfortable in relying on the carriers for maintaining acceptable network performance. Many carriers now back their performance claims with *Service-Level Agreements* (SLAs) and credit the customer's next invoice if performance falls below a certain threshold.

See Also

> *Network-Management Systems* (NMSs)
>
> *Service-Level Agreements* (SLAs)

Network Security

Protecting vital information from unauthorized access has always been a high-priority concern among most companies. While access to distributed data networks improves productivity by making applications, processing power and mass storage readily available to a large and growing user population, it also makes those resources more vulnerable to abuse and misuse.

Among the risks are unauthorized access to mission-critical data, information theft, and malicious file tampering that can result in immediate financial loss—and, in the long term, loss of customer confidence and damage to competitive position. Various protective measures can be taken, however, to safeguard information that is in transit as well as information that is stored at various points on the network, including servers and desktop computers.

Physical Security

Protecting data in distributed environments starts with securing the premises. Such precautions as locking office doors and wiring closets, restricting access to the data center, and having employees register when they enter sensitive areas can greatly reduce risk. Issuing badges to visitors, installing electronic locks on doors, providing visitor escorts, and having a security guard station in the lobby can reduce risk even further.

Other measures such as keyboard and disk drive locks are also effective in deterring unauthorized access to unattended workstations. These are important security features, especially because some workstations might provide management access to wiring hubs, LAN servers, bridge-routers, and other network-access points. In addition, locking down workstations to desks can help protect against equipment theft.

Access Controls

Access controls can prevent unauthorized local access to the network and control remote access through dialup ports. Network administrators can assign three levels of access to different users based on need: public, private, and shared access. Public access enables all users to have read-only access to file information. Private access gives specific users read-and-write file access, while shared access enables all users to read and write to files.

When a company offers network access to one or more databases, it should restrict and control all user query operations. Each database should have a protective key (or a series of steps) that is known only to those individuals who are entitled to access the data. To ensure that intruders cannot duplicate the data from the system files, users should first have to sign on with passwords and then prove that they are entitled to access the data by responding to a challenge with a predefined response. This process is the basis of a security procedure that is known as authentication.

Log-on Security

NOSs or add-on software can offer effective log-on security, which requires that the user enter a logon ID and password to access local or remote systems. Passwords not only can identify the user, but they can also associate the user with a specific workstation, as well as a designated shift, workgroup, or department. The effectiveness of this measure hinges on users' abilities maintain password confidentiality.

A user ID should be suspended after a certain number of passwords have been entered, in order to thwart trial-and-error attempts at access. Changing passwords frequently—especially when key personnel leave the company—and using a multilevel password-protection scheme can enhance security. With multi-level passwords, users can gain access to a designated security level as well as to all lower levels. With specific passwords, on the other hand, users can access only the intended level and not the others that are above or below. Finally, users should not be allowed to make up their own passwords; rather, they should be assigned a password by using a random password generator. Although such schemes entail an increased administrative burden, the effort is usually worthwhile.

The effectiveness of passwords can be enhanced by using them in combination with other control measures, such as a keyboard lock or card reader. Biometric devices also can be used that identify an authorized user based on such characteristics as a hand print, voice pattern, or the layout of capillary blood vessels in the retina of the eye. Of course, the choice of control measure will depend on the level of security that is desired.

Data Encryption

Protecting data (and voice) as it traverses the network requires the data to be scrambled with an encryption algorithm. One of the most effective encryption algorithms is offered by *Pretty-Good Privacy* (PGP), a method that uses a public key to protect computer and e-mail data. The program generates two keys that belong uniquely to the user. One PGP key is secret and stays in the user's computer. The other key is public and is given to people with whom the user wants to communicate. The public key can be distributed as part of the message.

PGP does more than encrypt; it also has the capability to produce digital signatures, enabling the user to sign and authenticate messages. A digital signature is a unique mathematical function derived from the message being sent. A message is signed by applying the secret key to it before it is sent. By checking the digital signature of a message, the recipient can make sure that the message has not been altered during transmission. The digital signature can also prove that a particular person originated the message. The signature is so reliable that not even the originator can deny creating it.

Firewalls

A firewall is a method of protecting one network from another untrusted network. The actual mechanism whereby this goal is accomplished varies widely, but in principle, the firewall can be thought of as a pair of mechanisms: one that blocks traffic, and another that permits traffic. Some firewalls place a greater emphasis on blocking traffic, while others emphasize permitting traffic.

One way firewalls protect networks is through packet filtering, which can be used to restrict access from or to certain machines or sites. Packet filtering can also be used to limit access based on time or day or day of week, by the number of simultaneous sessions that are permitted, service host(s), destination host(s), or service type. In addition to dedicated firewall systems, this kind of functionality can be set up on various network routers, communications servers, or front-end processors.

Transparent proxies are also used to provide secure out-bound communication to the Internet from the corporation's internal network. The firewall software achieves this task by appearing to be the default router that provides access to the internal network. When packets hit the firewall, however, the software does not route the packets but immediately starts a dynamic, transparent proxy. The proxy connects to a special intermediate host, which actually connects to the desired service.

Proxies are often used instead of router-based traffic controls, to prevent traffic from passing directly between trusted and untrusted networks. Many proxies contain extra logging or support for user authentication. Because proxies must understand the application protocol that is being used, they can also implement protocol-specific security (e.g., an FTP proxy might be configurable to permit incoming FTPs and block outgoing FTPs).

Remote Access Security

With an increasingly decentralized and mobile workforce, organizations are coming to rely on remote access arrangements that enable telecommuters, traveling executives, salespeople, and branch offices to dial into the corporate network. This situation calls for appropriate security measures in order to prevent unauthorized access to corporate resources. One or more of the following security methods can be employed:

- *Authentication* This method involves verifying the remote caller by user ID and password, thus controlling access to the server. Security is enhanced if the ID and password are encrypted before going out over the communications link.

- *Access restrictions* This method involves assigning each remote user a specific location (i.e., directory or drive) that can be accessed in the server. Access to specific servers also can be controlled.

- *Time restrictions* This involves assigning each remote user a specific amount of connection time, after which the connection is dropped.

- *Connection attempts* This method involves limiting the number of consecutive connection attempts and/or the number of times connections can be established on an hourly or daily basis.

Among the most popular remote access security schemes are *Remote-Access Dial-in User Service* (RADIUS) and *Terminal Access-Controller Access-Control System+* (TACACS+). Of the two, RADIUS is the more popular. Users are authenticated through a series of communications between the client and the server. When the client initiates a connection, the communications server puts the name and password into a data packet called the authentication request, which also includes information identifying the specific server sending the authentication request and the port that is being used for the connection. For added protection, the communications server, acting as a RADIUS client, encrypts the password before passing it on to the authentication server.

When an authentication request is received, the authentication server validates the request and decrypts the data packet to access the username and password information. If the username and password are correct, the authentication server sends back an authentication acknowledgment that includes information about the user's network system and service requirements. The acknowledgment can even contain filtering information in order to limit the user's access to specific network resources.

The older security system is TACACS, which has been updated by Cisco into a version called TACACS+. Although the protocols are different, the proprietary TACACS+ offers many of the same features as RADIUS but is used mainly on networks that consist of Cisco remote-access servers and related products. Companies that have mixed-vendor environments tend to prefer the more open RADIUS.

Callback Systems

Callback security systems are useful in remote access environments. When a user dials into the corporate network, the answering modem requests the caller's identification, disconnects the call, verifies the caller's identification against a directory, and then calls back the authorized modem at the number matching the caller's identification. This scheme is an effective way to ensure that data communication occurs only between authorized devices—more so when used in combination with data encryption.

Security procedures can even be implemented before the modem handshaking sequence, rather than after it, as is usually the case. This situation effectively eliminates the access opportunity from potential intruders. This method uses a precision high-speed analog security sequence that is not even detectable by advanced line-monitoring equipment.

While these callback techniques work well for branch offices, most callback products are not appropriate for mobile users whose locations vary on a daily basis. There are now products on the market that accept roving callback numbers. This feature enables mobile users to call into a remote access server or host computer, type their user ID and password, and then specify a number where the server or host should call them back. The callback number is then logged and can be used to help track down security breaches.

To safeguard sensitive information, you can add third-party authentication systems to the server. These systems require a user password and also a special credit card-sized device that generates a new ID every 60 seconds, which must be matched by a similar ID number-generation process on the remote user's computer.

Link-Level Security

When peers at each end of a serial link support the PPP suite, link level security features can be implemented. PPP can integrally support the *Password Authentication Protocol* (PAP) and *Challenge Handshake Authentication Protocol* (CHAP) in order to enforce link security. PPP is a versatile WAN connection standard that can be used for tying dispersed branch offices to the central backbone via dialup serial links. PPP is actually an enhanced version of the older *Serial-Line Internet Protocol* (SLIP). SLIP is limited to the IP-only environment, while PPP is used in multi-protocol environments. Because PPP is protocol-insensitive, it can be used to access AppleTalk, IPX, and TCP/IP networks, for example.

PAP uses a two-way handshake for the peer to establish its identity. This handshake occurs only upon initial link establishment. An ID/password pair is repeatedly sent by the peer to the authenticator until verification is acknowledged or the connection is terminated. Passwords are sent over the circuit in text format, however, which offers no protection from interception and playback by network intruders.

CHAP periodically verifies the identity of the peer using a three-way handshake. This technique is employed throughout the life of the connection. With CHAP, the server sends a random token to the remote workstation. The token is encrypted with the user's password and sent back to the server. Then, the server performs a lookup to see whether it recognizes the password. If the values match, the authentication is acknowledged; otherwise, the connection is terminated. Every time remote users dial in, they are given a different token—which provides protection against playback, because the challenge value changes in every token. Some vendors of remote-node products support both PAP and CHAP, while low-end products tend to support only PAP (which is the less robust of the two authentication protocols).

Policy-Based Security

With today's LAN administration tools, security goes far beyond mere password protection to include implementation of a policy-based approach characteristic of most mainframe systems. Under the policy-based approach to security, files are protected by their description in a relational database. In other words, newly created files are automatically protected—not at the discretion of each creator, but consistent with the defined security needs of the organization.

Some products use a graphical calendar through which various assets can be made available to select users only during specific hours of specific days. For each asset or group of assets, a different permission type can be applied: Permit, Deny and Log. Permit enables a user or user group to have access to a specified asset. Deny enables an exception to be made to a Permit (not enabling writes to certain files, for example). Log enables an asset to be accessed but stipulates that such access will be logged.

Although the LAN administrator usually has access to a full suite of password controls and tracking features, today's advanced administration tools also provide the capability to determine whether or not a single login

ID can have multiple terminal sessions on the same system. Through the console, the LAN manager can review real-time and historical violation activity online, along with other system activity.

Summary

To protect valuable information, companies must establish a sound security policy before an intruder has an opportunity to violate the network and do serious damage. This process means identifying security risks, implementing effective security measures, and educating users on the importance of following established security procedures.

See Also

Firewalls

Telephone Fraud

Network Support

Today's networks have increased in functionality and complexity, pushing support issues into the forefront of management concerns. Whether problems are revealed through network-management tools—alarms, diagnostics, predictive methods—or through user notification, the need for timely and qualified network-support services is of critical importance, especially in the WAN environment. Recognizing these concerns, the long-distance carriers now offer network support options in conjunction with their services and facilities. In some cases, these support services are specialized, as in wide-area SNA management.

Types of Services

The support concept encompasses dozens of individual activities from which the customer can select. Generally, these activities include (but are not limited to) the following:

- Site engineering, utilities installation, cable laying, and rewiring
- Performance monitoring of the system or network, alarm interpretation, and initiation of diagnostic activities

- Identification and isolation of system faults and degraded facilities on the network
- Notification of the appropriate hardware vendor or carrier for restorative action
- Testing of the restoration action to verify proper operation of the system or network
- The repair or replacement of the faulty system or component
- Monitoring of the repair/replacement process and the escalation of problems
- Trouble ticket and work-order administration, inventory tracking, maintenance histories, and cost control
- Administration of moves, adds, and changes
- Network design, tuning, and optimization
- Systems documentation and training
- Preventive maintenance
- Management reports

A variety of other types of support are also available, such as 24-hour telephone (hotline) assistance, short-term equipment rental, fast equipment exchange, and guaranteed response time to trouble calls. In addition, the carrier or vendor might offer customized cooperative maintenance plans that qualify the organization for premium reductions if an internal help desk is established to weed out routine problems, many of which are applications related and beyond the support purview of the carrier or vendor. An increasingly popular support offering is remote diagnostics and network management from the vendor or carrier's network control center.

Levels of Support

Carriers also offer multiple levels of technical support. The most basic form of technical support is toll-free telephone access to technical specialists during normal business hours. This type of service assists customers in resolving hardware or software problems. Typically, there is no charge for this service, and calls are handled on a first-come-first-served basis. There is usually no expiration date for this service; it is available to customers for as long as they use the carrier's services or facilities.

Extended or priority technical assistance is provided via phone 24 hours a day, seven days a week in order to assist customers in resolving hardware or software problems. As an extra-cost service, the assistance professional ensures that customers are called back within 30 minutes during normal business hours and within one hour after normal business hours.

Some carriers offer subscription services, which provide the most up-to-date technical product information on maintaining network efficiency and reliability. Written by the carrier's own engineers and field service personnel, this kind of service usually emphasizes how to more effectively operate and manage various data communications and network access products. This information can come in a variety of forms, including technical bulletins, product application notes, software release notes, user guides, and field bulletins—in print or on CD-ROM. Increasingly, the Web is being used to distribute such information. Because access is limited to customers, a valid user ID and password are usually required.

Remote dial-in software support addresses the needs of customers operating mission-critical networks. Technical specialists remotely dial-in to the customer's network to resolve software problems via diagnostic testing, or by modifying a copy of the system configuration and then downloading the revised configuration file directly to the affected equipment.

Carriers can also assume single-point responsibility for remote network management, providing customers with a proactive approach to service delivery. Technical staff at a central control facility continuously monitor network performance and immediately respond to and resolve any fault resulting from hardware, software, or circuits. From the control facility, network faults are identified and alarm conditions resolved through continuous end-to-end diagnostics. Once a problem is recognized, the latest diagnostic equipment and isolation techniques are used to identify the source of the problem and provide effective resolution. Often, problems are identified and corrected before they become apparent to network users.

If the problem originates with the carrier, it assumes ownership until it is resolved. If the problem originates from a local telephone company or competitive access provider, the long-distance carrier reports the problem, makes appropriate status inquiries—and, if necessary, escalates the problem within the other company's organization.

Summary

The long-distance carriers are competing with equipment vendors and third-party service firms in the provision of network support services, providing customers with a broad range of plans that encompass just about every aspect of problem identification, diagnostics, and resolution. Sometimes the support is applications related—as in the case of managing SNA over the WAN—the cost of which is bundled with a service such as Frame Relay.

See Also

Managed SNA Services

Network Integration

Outsourcing

CHAPTER O

Object-Oriented Networks

Object-oriented technology has been in practical use on the public telecommunications network in one form or another for several years. The object orientation permits carriers to easily administer and manage various network elements and gives corporate users the means to easily upgrade, change, and customize telecommunications services without carrier involvement.

Object-oriented technology permits applications to be broken into classes of objects that can be reused and easily modified for use elsewhere. This feature greatly reduces application development time, simplifies maintenance, and increases reliability. With each object being viewed as a separate functional entity, reliability is improved because there are fewer chances that a change will produce new bugs in previously stable sections of code.

The use of objects also improves the productivity of programmers. Each instance of an object draws upon the same piece of error-free code, resulting in less application development time. Over time, this approach also makes it easier to maintain program integrity with changes in personnel.

Objects

In a network-management application, the functions of a switch, multiplexer, bridge or router—any device that exists on the network—can be described as an object. Each object swaps messages with the network-management system, triggering events such as status and performance reports. Through messaging, the reports can be sent to another object, such as a printer or disk drive.

In the TCP/IP environment, the collection of object definitions that a given management system can work with is called the *Management Information Base* (MIB). The MIB is a repository of information necessary to manage the various devices on the network. The MIB contains a description of SNMP-compliant objects on the network and the kinds of management information that they provide. The objects can be hardware, software, or a logical association (such as a connection or virtual circuit). The attributes of an object might include such things as the number of packets sent, routing table entries, and protocol-specific variables for IP routing.

The messaging functions between an object and the network-management system are carried out via datagrams that traverse virtual circuits. These datagrams contain commands that request various types of information from the object (such as its status) or that collect performance information such as the number of packets sent. For example, `frCircuitSentFrames` is an object definition for the number of frames that are sent from a specified Frame Relay virtual circuit since it was created. Upon request, the appropriate response is sent back to the network-management station.

Intelligent Networks

Emerging intelligent networks offer another illustration of how object-oriented principles are being applied. Intelligent networks provide the means with which carriers can create and uniformly introduce and support new services and features via a common architectural platform. The creation and support of new services is accomplished through the manipulation of software objects that are accessible at intelligent nodes that are distributed throughout the network.

Instead of investing heavily in premises-based equipment and leasing lines in order to provide a high level of performance, functionality, and control via private networks, corporate users can tap the service logic of intelligent nodes that are embedded in the public network for services, advanced calling features, and bandwidth on demand. Users can build their own networks, create services, and customize features without carrier involvement simply by combining and recombining the available objects at a workstation.

The required resources, in the form of functional components, are then assembled automatically by the intelligent network in accordance with the user's design specifications. You can also test the integrity of network models by simulation, prior to actual implementation. For example, you might need to predict the delay performance of particular links in order to ensure that certain applications will not time out. Even additional bandwidth can be made available through object manipulation. In essence, companies can manage their portion of the public network as though it were a private network.

Carriers benefit from this object orientation, as well. With all services and features defined in software (as objects), and the programs distributed among fewer locations (intelligent nodes) instead of at every switch, carriers can more quickly deploy new services. Once new services are developed, they can be made immediately available to customers from intelligent network nodes. In accessing these nodes, customers can instantly implement a uniform set of services for maximum efficiency and economy, regardless of the location of their business units.

In enabling users to design their own networks at a management workstation and giving them the means to add or change services without their involvement, carriers are relieved of much of the administrative burden associated with customer service. The decrease in demand for customer support reduces the carrier's manpower requirements—and, consequently, the cost of network operation. Cost savings can be passed on to customers in the form of lower service rates.

Summary

The object-oriented paradigm signals a fundamental shift in the way networks, applications, databases, and operating systems are put together—as well as how they are used, upgraded, and managed. Although object-oriented networking is available in the TCP/IP environment, where the concept originated as an intrinsic capability of SNMP, telephone companies and other carriers have been slow in extending the technology to customers, enabling them to manage and control service and feature elements without carrier involvement.

See Also

> *Advanced Intelligent Network* (AIN)
>
> *Open-Network Architecture* (ONA)
>
> *Simple Network-Management Protocol* (SNMP)

Open-Network Architecture (ONA)

Open Network Architecture (ONA) refers to the overall design of an *Incumbent Local Exchange Carrier*'s (ILEC) network; specifically, its capability to provide *Competitive Local-Exchange Carriers* (CLECs) with comparably efficient interconnection. This interconnection is intended to prevent ILECs from discriminating against CLECs by denying them efficient access to network facilities, forcing them to buy unneeded services and features, or overcharging them for the necessary connections—any of which can prevent a competitor from establishing a viable presence in the market.

Implementation of ONA requires existing feature-group access arrangements to be unbundled and new access-charge subelements, known as *Basic Service Elements* (BSEs), and *Basic Serving Arrangements* (BSAs) to be established. The *Federal Communications Commission* (FCC) supervises the efforts of the ILECs to open their networks in this manner—which, in turn, determines whether the ILECs can participate in markets that have previously been closed to them.

Regulatory History

Over the last three decades, the FCC initiated three major inquiries that focused on the same fundamental issue: which regulations, if any, should apply to services that combine computer processing with pure transmission to provide "value-added" applications? The Computer I inquiry opened the question and the FCC addressed it by establishing that computer data processing service providers are not subject to common carrier regulation, whereas common carriers seeking to offer data services must offer them through a separate affiliate.

In Computer II, the FCC focused on the need to develop a workable categorical definition of both regulated telecommunications services and unregulated data services. The result: the creation of the categories of "basic" and "enhanced" services. The FCC defined "basic service" as common carrier telecommunications offerings like telephone service, which entails providing pure transmission capacity for the movement of information. The FCC defined "enhanced services" as a service that employs computer processing applications that act on the format, content, code, protocol, or similar aspects of the subscriber's transmitted information; provide the subscriber with additional, different, or restructured information; or involve subscriber interaction with stored information.

In the first stage of implementing Computer III, the FCC required the BOCs to obtain its approval for service-specific CEI plans prior to offering individual enhanced services on an integrated basis. In these CEI plans, the FCC required the BOCs to demonstrate how they would provide competitors with equal access to all basic, underlying network services that the BOCs used to provide their own enhanced services.

During the second stage of Computer III, the BOCs developed and implemented *Open Network Architecture* (ONA) plans that detailed the unbundling of basic network services. After the FCC approved these ONA plans and the BOCs filed tariffs for ONA services, they were permitted to provide integrated, enhanced services without filing service-specific CEI plans. ONA incorporates CEI equal-access requirements and provides for the further unbundling of network-service elements (not limited to those elements that are associated with specific BOC-enhanced services). After the implementation of ONA, the BOCs were still required to offer network services to competitors on a CEI equal-access basis, although they were no longer required to file a CEI plan for each service that they wished to offer.

Comparably Efficient Interconnection

Comparably efficient interconnection is achieved when the ILEC can demonstrate that it offers the following:

- Standardized interfaces to provide access to the transmission, switching, and signaling resources of the network
- Unbundled basic services
- Common basic service rates
- Common basic service performance characteristics
- Common installation, maintenance, and repair services
- Common end-user access
- Common knowledge of impending availability of new basic-service features
- Comparable interconnection costs for competitors

Since 1996, with passage of the Telecommunications Act, CEI has been expanded in order to include access to the ILECs' *Operations Support Systems* (OSS). These are databases and information that ILECs use to provide services and features to their customers. Among the functions that are supported by an OSS are preordering, ordering, provisioning, maintenance and repair, and billing. The FCC considers access to OSS functions to be necessary for meaningful competition.

ONA Building Blocks

As noted, implementation of ONA requires that feature group access arrangements be unbundled and that access-charge subelements, known as *Basic Service Elements* (BSEs), and *Basic Serving Arrangements* (BSAs) be established.

Basic serving arrangements   The BSA specifies the access links and transport elements that comprise a basic transmission service. For example, circuit-switched trunk-side access is a BSA that provides a trunk-side access connection to the CLEC's premises. This service can be provided directly from an end office or optionally from a tandem switch, in order to deliver one-way originating traffic to the CLEC. This service includes a seven-digit number with which users can access the service.

Another BSA provides dedicated connections between end users and the CLEC, so that a channel of up to 9.6Kbps can be used for applications such as the transmission of alarm signals from subscriber locations to a central alarm-monitoring company.

Other examples of BSAs include X.25 and X.75 interfaces to packet switches, broad-band links for video transmission, in-band signaling, and central office announcements.

Basic service elements Through a series of *Basic Service Elements* (BSEs), a variety of network capabilities can be offered. Under CEI, the BSEs must be offered on an unbundled basis. Originally, the CLECs and ILECs decide which BSEs are appropriate to support a given service, and the CLEC pays for only those BSEs. Four general categories of BSEs existed:

- *Switching* Supports services that require call processing, routing, and management
- *Signaling* Supports monitoring services that require a derived channel over subscriber lines
- *Transmission* Allocates appropriate bandwidth to a customer application
- *Network management* Provides the means to monitor system performance and to reallocate assigned capabilities

The BSEs that are associated with circuit-switched services might include call forwarding, distinctive ringing, three-way calling, and *Automatic Number Identification* (ANI). BSEs that are associated with private lines might include an out-of-band diagnostic channel, line conditioning, customer-controlled reconfiguration, and route diversity.

Room for interpretation existed among ILECs in determining which elements are classified as BSAs or BSEs. For example, while multi-line hunt groups are universally considered BSEs, detection of telecommunications company line breaks within 60 seconds might be considered a BSA by one ILEC and a BSE by another ILEC.

Ancillary services Services that provide utility to the service provider, but are not associated with a specific network feature or function, fall under the category of ancillary services. These services typically include maintenance and diagnostics, billing services, and the collection of traffic statistics.

Because telephone companies differ widely in their interpretation of what constitutes an ancillary service, the FCC directed that all regulated services must be classified as BSAs or BSEs, and that only unregulated services can be classified as ancillary services.

***Complementary Network Services* (CNS)** *Complementary Network Services* (CSNs) are those features that are applied to the end user's local service in order to make it interact more efficiently with the service provider's BSAs or BSEs. Examples of CNS might include the multi-faceted call forwarding feature:

- *Call Forward Busy Line/Don't Answer* Enables user calls to a busy line or to an unanswered line to be forwarded to another number for call completion
- *Call Forward Don't Answer With Variable Ring Count* Enables user calls to be forwarded after a specified number of rings on a Don't Answer condition
- *Customer Control of Call Forward Busy Line/Don't Answer* Enables the service provider's operator to override the Call Forward Busy Line/Don't Answer feature on a demand basis

Summary

ONA provides the framework for competition in the telecommunications market. Under the concept of comparably efficient interconnection, the ILECs must provide CLECs with the same economic and technical efficiencies as they use to provide telecommunications services to their subscribers. Achieving CEI is a prerequisite for ILEC entry into other markets in which they have previously been excluded (i.e., long-distance service). The rules for CEI were codified in the Telecommunications Act of 1996. The list of unbundled network elements were last clarified by the FCC in September 1999 (refer to Unbundled Access).

See Also

> *Operations Support Systems* (OSSs)
>
> Telecommunications Act of 1996
>
> Unbundled Access

Open Systems Interconnection (OSI)

The seven-layer *Open Systems Interconnection* (OSI) Reference Model was first defined in 1978 in ISO standard 7498. The lower layers (1 to 3) represent local communications, while the upper layers (4 to 7) represent end-to-end communications (refer to Figure O-1). Each layer contributes protocol functions that are necessary to establish and maintain the error-free exchange of information between network users.

The model provides a useful framework for visualizing the communications process and comparing products in terms of standards conformance and interoperability potential. This layered structure not only aids users in visualizing the communications process, but it also provides vendors with the means for segmenting and allocating various communications requirements within a workable format. This functionality can reduce much of the confusion that is normally associated with the complex task of supporting successful communications.

Figure O-1
The seven-layer OSI
Reference Model

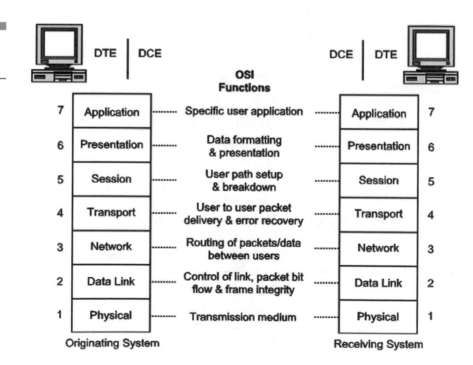

Layers

Each layer of the OSI model exchanges information with a comparable layer at the other side of the connection—a process known as peer-protocol communications. The functionality that is associated with each layer is as follows:

Application layer The highest layer in the OSI reference model is the application layer, which serves as the window for users and application processes to access network services. This level applies to the actual meaning, rather than the format or syntax (as in Layer 6) of applications and permits communication between users. According to the model, each type of application must employ its own Layer 7 protocol, and with the wide variety of available application types, Layer 7 offers definitions for each, including the following:

- Resource sharing and device redirection
- Remote file access
- Remote printer access
- Inter-process communication
- Network management
- Directory services
- Electronic messaging (such as e-mail)
- Network virtual terminals

Presentation layer Layer 6 deals with the format and representation of data that applications use; specifically, it controls the formats of screens and files. Layer 6 defines elements such as syntax, control codes, special graphics, and character sets. Additionally, this level determines how variable alphabetic strings will be transmitted, how binary numbers will be presented, and how data will be formatted.

The presentation layer formats the data to be presented to the application layer. This layer can be viewed as the translator for the network, because it translates data from a format that is used by the application layer into a common format at the sending station, then translates the common format to a format that is known to the application layer at the receiving station. Specifically, the presentation layer provides the following:

- *Character code translation*   For example, ASCII to *Extended Binary Coded Decimal Interexchange Code* (EBCDIC)
- *Data conversion*   Bit order, *Carriage Return* (CR) or *Carriage Return / Line Feed* (CR/LF), integer floating point, and other functions
- *Data compression*   Reduces the number of bits that need to be transmitted on the network
- *Data encryption*   Encrypts data for security purposes

Session layer   The session layer manages communications. For example, it sets up, maintains, and terminates virtual circuits between sending and receiving devices. This layer also sets boundaries for the start and end of messages and establishes how messages will be sent: half-duplex, with each computer taking turns sending and receiving, or full-duplex, with each computer sending and receiving at the same time. These details are negotiated during session initiation.

The session layer enables session establishment between processes that are running on different stations. Specifically, the session layer provides the following:

- *Session establishment, maintenance, and termination*   Enables two application processes on different machines to establish, use, and terminate a connection (called a session)
- *Session support*   Performs the functions that enable these processes to communicate over the network, performing security, name recognition, logging, and other functions

Transport layer   Layer 4 handles end-to-end transport. If there is a need for reliable, end-to-end, sequenced delivery, then the transport layer performs this function. For example, each packet of a message might have followed a different route through the network toward its destination. The transport layer re-establishes packet order through a process called sequencing, so that the entire message is received exactly the way in which it was sent. At this layer, lost data is recovered, and flow control is implemented. With flow control, the rate of data transfer is adjusted in order to prevent excessive amounts of data from overloading network buffers.

Layer 4 might also support datagram transfers; that is, transactions that need not be sequenced. This functionality is required for voice and video, which might tolerate loss of information but need to have low delay and low variance in transmittal time. This flexibility is the result of the protocols that are implemented in this layer, ranging from the five OSI protocols—TP0 to TP4—to TCP and UDP in the TCP/IP suite, and many others in proprietary suites. Some of these protocols do not perform retransmission, sequencing, checksums, and flow control. Specifically, the transport layer provides the following:

- *Message segmentation* Accepts a message from the (session) layer above it, splits the message into smaller units (if they are not already small enough), and passes the smaller units down to the network layer. The transport layer at the destination station reassembles the message.
- *Message acknowledgment* Provides reliable end-to-end message delivery with acknowledgments
- *Message traffic control* Tells the transmitting station to back off when no message buffers are available
- *Session multiplexing* Interleaves several message streams or sessions onto one logical link and keeps track of which messages belong to which sessions (refer to the session layer)

Network layer Layer 3 formats the data into packets, adds a header containing the packet sequence and the address of the receiving device, and specifies the services that are required from the network. The network does the routing to match the service requirement. Sometimes a copy of each packet is saved at the sending node until it receives confirmation that it has arrived undamaged at the next node, as is done in X.25 packet-switched networks. When a node receives the packet, it searches a routing table to determine the best path for that packet's destination—without regard for its order in the message. In a network where not all nodes can communicate directly, this layer takes care of routing packets through the intervening nodes. Intervening nodes may reroute the message to avoid congestion or node failures. Specifically, the network layer provides:

- *Routing* Routes frames among networks
- *Subnet traffic control* Routers (network-layer intermediate systems) can instruct a sending station to scale back its frame transmission when the router's buffer fills up
- *Frame fragmentation* If it determines that a downstream router's *Maximum Transmission Unit* (MTU) size is less than the frame size, a router can fragment a frame for transmission and reassembly at the destination station.

- *Logical-physical address mapping*   Translates logical addresses, or names, into physical addresses
- *Subnet usage accounting*   Has accounting functions to keep track of frames that are forwarded by subnet intermediate systems, in order to produce billing information

Data-link layer   All modern communications protocols use the services that are defined in Layer 2. The data-link layer provides the lowest level of error control, detecting errors and requesting the sending node to retransmit the data. This layer has assumed a greater role as communications lines have become less noisy through the replacement of analog lines with digital lines, while end-stations have become more intelligent through the use of more powerful processors and high-capacity memory. Combined, these factors have lessened the need for high-level information protection mechanisms in the network, moving them to the end systems. Layer 2 does not know what the information or packets that it encapsulates mean or where they are headed. Networks that can tolerate this lack of information are rewarded by low transmission delays. Specifically, the functions that are provided by the data-link layer include the following:

- *Link establishment and termination*   Establishes and terminates the logical link between two nodes
- *Frame traffic control*   Tells the transmitting node to back off when no frame buffers are available
- *Frame sequencing*   Transmits/receives frames sequentially
- *Frame acknowledgment*   Provides/expects frame acknowledgments, in addition to detecting and recovering from errors that occur in the physical layer by retransmitting non-acknowledged frames and handling duplicate frame receipts
- *Frame delimiting*   Creates and recognizes frame boundaries
- *Frame error checking*   Checks received frames for integrity
- *Media-access management*   Determines when the node has the right to use the physical medium

Physical Layer   The lowest OSI layer is the physical layer. This layer represents the actual interface, electrical and mechanical, which connects a device to a transmission medium. Because the physical interface has become so standardized, it is usually taken for granted in discussions of OSI connections. Yet, physical connections—cables and connectors, with their pinouts and transmission characteristics—can still be a problem in designing a reliable network if they do not conform to a common model. Specifically, the functions that are provided at the physical layer include the following:

- *Data encoding*   Modifies the simple digital signal pattern (1s and 0s) that is used by the PC to better accommodate the characteristics of the physical medium, and to aid in bit and frame synchronization
- *Physical medium attachment*   Accommodates various possibilities in the medium, such as the number of pins that a connector has and what each pin is used for
- *Transmission technique*   Determines whether the encoded bits will be transmitted by digital or analog signaling
- *Physical medium transmission*   Transmits bits as electrical or optical signals that are appropriate for the physical medium, and determines how many volts/dB should be used to represent a given signal state, using a given physical medium

Conformance versus Interoperability

Twelve laboratories are accredited by the *National Institute of Standards and Technology* (NIST) to run a suite of tests that certify vendor products for conformance to the OSI reference model. While the products of different vendors might conform to the OSI model, however, they are not necessarily interoperable.

Conformance testing is the process of comparing a vendor's protocol implementation against a model of the protocol. Conformance test results are sent to NIST for approval. Approved results for each product are then entered into NIST's registry of OSI-conforming products. By itself, however, conformance does not guarantee that the product of one vendor will work with the product of another vendor, although both products have passed the same conformance test. OSI product-conformance testing only increases the probability of successful interoperability in a customer's multi-vendor OSI network. To ensure that the products of both vendors do indeed work together on the network, they must be specifically tested for interoperability at the highest level of OSI: the applications layer. This process involves running both vendors' protocol implementations of FTAM or X.400, for example, to see whether they work properly across their respective products.

Summary

Throughout the 1980s, the prediction was often made that OSI would replace TCP/IP as the preferred technique for interconnecting multi-vendor networks. Now, we know that this situation will not happen in the United States. Several reasons for this statement exist, including the slow pace of OSI standards progress in the 1980s and the expense of implementing complex OSI software and having products certified for OSI interoperability. Furthermore, TCP/IP was already widely available, and plug-in protocols continue to be developed to add functionality. The situation is different in Europe, where OSI compliance was mandated early on by the regulatory authorities in many countries.

See Also

International Organization for Standardization (ISO)

Simple Network-Management Protocol (SNMP)

Transmission Control Protocol / Internet Protocol (TCP/IP)

Operations Support Systems (OSSs)

Operations Support Systems (OSSs) consist of databases and information that a *Local Exchange Carrier* (LEC) uses to provide telecommunications services to its customers. Among the functions of OSS are preordering, ordering, provisioning, maintenance and repair, and billing.

The *Federal Communications Commission* (FCC) has determined that OSS functions fall within the definition of an *Unbundled Network Element* (UNE) and that it is technically feasible for incumbent LECs to provide access to OSS functions on an unbundled basis to requesting *Competitive Local Exchange Carriers* (CLECs). The FCC considers access to OSS functions as necessary for meaningful competition, and that failing to provide such access could impair the capability of requesting telecommunications carriers to provide competitive service.

The FCC had set a January 1, 1997 deadline for the incumbent LECs to provide access to OSS functions on a parity basis. No standards exist for OSS, however, and as of year-end 1999, these systems were still not fully automated. Many CLECs complain about faulty OSSs, claiming that the ILECs are not giving priority attention to this matter.

Interface Evaluation

A primary objective of the Telecommunications Act of 1996 was to facilitate the emergence of competition for local communications services, which would have the effect of reducing service costs to consumers and businesses. The Telecommunications Act is designed to facilitate new entrants' uses of different entry strategies, including resale, UNEs, and facilities competition. Each of these strategies depends heavily on the computer systems, databases, and personnel of the LEC; specifically, their operations and support systems.

The ILECs must enable competitors to access their OSSs before they will be allowed into the long-distance market. In denying various petitions from ILECs to enter in-region long-distance services, the FCC established a two-step test for evaluating ILEC OSS interface operations:

- Has the ILEC deployed the systems and personnel that are necessary to provide adequate access to OSS functions?
- Is the OSS interface operationally ready for use?

The nagging problem with OSSs is that the ILECs use a variety of software, hardware systems, and procedures. Even within the same organization, there is not always integration among discrete systems that belong to the OSS, and each system might operate at a different level of performance. Without a unified system, it is not easy to extend access to competitors.

Performance Measurement

In April 1998, the FCC responded to the requests from LCI, Comptel, and the *National Association of Regulatory Utility Commissioners* (NARUC) for rules on OSS performance measures. OSS performance relates directly to the ILECs' capabilities (and obligations under federal law) to provide just, reasonable, and nondiscriminatory interconnection, UNEs, and resale.

Appropriate measurements and reporting requirements can be of considerable value in promoting successful access to OSS. In this area, detail matters; significant disparities in any one of multiple areas of performance can seriously undermine the prospects for competition.

For example, if a CLEC can successfully order unbundled loops, ports, and other network elements—but its new customers are less likely to be identified accurately in E911 databases—it is reasonable to expect that the CLEC might be impeded in its efforts to compete. If dial-tone service is cut over promptly from the ILEC to the CLEC, but interim number portability is commonly cut over at a different point in time, incoming calls will go astray—and again, competition and consumers will suffer.

OSS performance measurements can capture these problems. They can assist the LECs with self-assessments, so that corrective actions can be taken before disputes arise. Alternatively, when disputes do arise, appropriate measurement data might make it easier to distinguish isolated incidents from recurrent problems. In addition, measurement guidelines will enable the state commissions and the FCC to use a common framework to monitor what are, typically, regional rather than single-state systems and databases. Guidelines will also provide state *Public Utility Commissions* (PUCs) with the flexibility to address state-specific circumstances and needs.

Summary

In the more than three years since passage of the Telecommunications Act of 1996, local competition in telephone services is still in its formative stage. One of the last technical obstacles to be overcome is the extension of OSS capabilities to competitors, which will accelerate the pace of competition in the market for local service.

See Also

Telecommunications Act of 1996
Unbundled Access

Operator Language-Translation Services

Language can be a significant problem for those callers who wish to place international calls. Callers might find it hard to use these services because they are confronted with operators who do not speak their language and have trouble processing the call. Because of the limited numbers of international operators, it is often difficult for callers to get through to place a collect call.

To overcome these problems, the large global telecommunications companies such as AT&T, MCI WorldCom, and Sprint offer language-translation services in conjunction with collect or credit card calls. AT&T's Direct In-Language Service offering, for example, provides callers with access to a bilingual AT&T operator who completes the collect call to the United States. The call is processed the same as a regular call, except that the operator completes the call in the preferred language of the caller. AT&T currently provides language-translation service for more than 20 countries.

To place a AT&T Direct In-Language Service call, a caller dials the appropriate access code and is connected with an AT&T operator in the United States who speaks the caller's language. The operator takes the name and number of the person who is being called and completes the call. If the call is completed in a language other than English, the operator remains on the line with the caller through completion of the call. Because calls are placed in a queue while they are ringing, a different operator might complete the call than the one who originally answered the call. When the call is completed, it is billed to the called party.

This type of operator service is also being provided by telecommunications carriers in other countries, making translation services available to their citizens who are traveling in the United States. This feature is available to more than 70 countries, and callers can access direct service numbers from both the U.S. mainland and Hawaii. The *Post Telephone and Telegraph* (PTTs) agencies and AT&T are handling the service together. Calls must be collect or billed to a PTT credit card. For example, a German traveler in the United States who wants to call back to Germany can call the German Direct Service and be connected to a German-speaking operator in Germany who places the collect or PTT credit card call.

MCI WorldCom provides this kind of service through its WorldPhone offering. To use WorldPhone, the caller dials the toll-free access number of the country from which they are calling, and an operator who speaks the person's language will connect the call. Fifteen languages are available for translation.

WorldPhone offers several other travel-related features. For example, callers can speak with an MCI Traveler's Assist Specialist who will give emergency, local, medical, legal, translation, restaurant, and entertainment referrals.

Summary

As more companies in different countries participate in the global economy and more people travel to international locations for both business and vacation, service providers see an increasing need to support their communications offerings with multi-lingual operators who can facilitate call completion. Early in the next century, operators might even be dispensed with entirely for this task as computerized language-conversion systems are added to the network.

See Also

Telecommunications Relay Services

Outsourcing

For many companies, it makes sense to outsource tasks that tend to consume a disproportionate share of corporate resources. Running information systems and communication networks involves a commitment of time and money that is becoming increasingly difficult to justify in the face of other pressing concerns. Consequently, many companies are turning to service firms that specialize in such functions as running data centers, managing networks, integrating diverse computer systems, and developing software. This arrangement is called outsourcing.

Given the increasing complexity of today's communications networks and the need of companies to focus more on core business in order to succeed in the global economy, many companies are seeking ways to offload communications-management responsibilities to those who have more knowledge, experience, and hands-on expertise than they alone can afford.

Outsourcing firms typically provide an analysis of an organization's business objectives, application requirements, and current and future communications needs. The resulting network design may incorporate the services and lines of multiple carriers and include equipment from may vendors. Acting as the client's agent, the outsourcing firm coordinates the activities of equipment vendors and carriers to ensure efficient and timely installation and service turn-up.

Typical Services

In a typical outsourcing arrangement, an integrated control center—located at the outsourcing firm's or client's premises—serves as a single point of service support where technicians are available 24 hours a day, 365 days a year to monitor network performance, contact the appropriate carrier or dispatch field service as needed, perform network reconfigurations, and take care of any necessary administrative chores.

Integration Today's communications networks consist of a number of different elements: legacy hosts, clients and servers, LANs and wiring hubs, bridges and routers, PBXs and key systems, and WAN facilities. The selection, installation, integration, and maintenance of these elements requires a broad range of expertise that is not usually found within a single organization. Consequently, many companies are increasingly turning to the services of outsourcing firms.

Briefly, the integration function is concerned with unifying disparate computer systems and transport facilities into a coherent, manageable enterprise-wide utility. This process typically involves the reconciliation of different physical connections and protocols. The outsourcing firm also ties in additional features and services that are offered through the *Public-Switched Telephone Network* (PSTN). The objective is to provide compatibility and interoperability among different products and services, making access transparent to users.

Project Management Project management entails the coordination of many discrete activities, starting with the development of a customized project plan based on the client's organizational needs. For each ongoing task, critical requirements are identified, lines of responsibility are drawn, and problem-escalation procedures are defined.

Line and equipment ordering can also be included in project management. Acting as the client's agent, the outsourcing firm interfaces with multiple suppliers and carriers to economically upgrade and/or expand the network without sacrificing predefined performance requirements. Before new systems are installed at client locations, the outsourcing firm performs site survey coordination and preparation, ensuring that all power requirements, air conditioning, ventilation, and fire-protection systems are properly installed and in working order.

When an entire node must be added to the network or a new host must be brought into the data center, the outsourcing firm will stage all equipment for acceptance testing before bringing it online, thus minimizing potential disruption to normal business operations. When new lines are ordered from various carriers, the outsourcing firm will conduct the necessary performance testing before making them available to user traffic.

Trouble-ticket administration   In assuming responsibility for daily network operations, a key service performed by the outsourcing firm is trouble-ticket processing, which is typically automated. The sequence of events is as follows:

1. An alarm indication is received at the network control center that is operated by the outsourcing firm.

2. The outsourcing firm uses various diagnostic tools in order to isolate and identify the cause of the problem.

3. Restoral mechanisms are initiated (manually or automatically) in order to bypass the affected equipment, network node, or transmission line until the faulty component can be brought back into service.

4. A trouble ticket is opened. If the problem is with hardware, a technician is dispatched to swap the appropriate board. If the problem is with software, analysis might be performed remotely. If the problem is with a particular line or service, the appropriate carrier is notified.

5. The client's help desk is kept informed of the problem's status, so that the help desk can assist local users.

6. Before closing out the trouble ticket, the repair is verified with an end-to-end test by the outsourcing firm.

7. Upon successful end-to-end testing, the primary CPE or facility is turned back over to user traffic, and the trouble ticket is closed.

Vendor/carrier relations   Another benefit of the outsourcing arrangement comes in the form of improved vendor/carrier relations. Instead of having to manage multiple relationships, the client only needs to manage one: the outsourcing firm. Dealing with only one firm has several advantages:

■ Improves response time to trouble calls/alarms

■ Eliminates delays caused by vendor/carrier finger-pointing

- Expedites order processing
- Reduces time spent in invoice reconciliation
- Frees staff time for applications development and planning
- Reduces the total cost of network ownership

Maintenance/repair/replacement   Some outsourcing arrangements include maintenance, repair, and replacement services. Not only does this arrangement eliminate the need for ongoing technical training, but the company is also buffered from the effects of technical staff turnover. Repair and replacement services can increase the availability of systems and networks while eliminating the cost of maintaining a spare parts inventory, test equipment, and asset-tracking system.

Disaster recovery   Disaster recovery includes numerous services that can be customized in order to ensure maximum network availability and performance:

- Disaster impact assessment
- Network-recovery objectives
- Evaluation of equipment redundancy and dial backup
- Network inventory and design, including circuit allocation
- Vital records recovery
- Procedure for initiating the recovery process
- Location of hot site, if necessary
- Installation responsibilities
- Test run guidelines
- Escalation procedures
- Recommendations to prevent network loss

Long-term planning support   An outsourcing firm can provide numerous services that can assist the client with strategic planning. Specifically, the outsourcing firm can assist the client in determining the impact of the following:

- Emerging services and products
- Industry and technology trends
- International developments in technology and services

With experience drawn from a broad customer base, as well as its daily interactions with hardware vendors and carriers, the right outsourcing firm can have a lot to contribute to clients in the way of strategic planning assistance

Training Outsourcing firms can fulfill the varied training requirements of users, including the following:

- Basic communications concepts
- Product-specific training
- Resource management
- Help-desk operator training

The last type of training is particularly important, because 80 percent of reported problems are applications oriented and can be solved without the outsourcing firm's involvement. This situation can speed up problem resolution and reduce the cost of outsourcing. For this training to be effective, however, the help-desk operator must know how to differentiate between applications problems, system problems, and network problems. Basic knowledge can be gained by training and can be improved with experience.

Equipment leasing An outsourcing arrangement can include equipment leasing. A number of financial reasons exist for including leasing in the outsourcing agreement, depending on the organization's financial situation. Leasing can improve a company's cash position, because costs are spread over a period of years. Leasing can free capital for other uses and even cost-justify technology acquisitions that would normally prove too expensive to purchase. With new technology becoming available every 12 months to 18 months, leasing can prevent the organization from becoming saddled with obsolete equipment.

Summary

While outsourcing promises numerous benefits, determining whether such an arrangement makes sense is a difficult process that requires the organization to consider a range of factors. Besides calculating the baseline cost of managing one's own information systems and communications networks and determining their strategic value, the decision to outsource often hinges on the company's business direction, its present systems and network architecture, the internal political situation, and the company's readiness to deal with the culture shock that inevitably occurs when two firms must work closely on a daily basis.

See Also

Application Service Providers (ASPs)

Help Desks

Leasing

Network Integration

Network Support

Packet-Switched Network—X.25

The X.25 packet-switched network relies on a feature-rich protocol set that enables data to be transported reliably over analog lines. The X.25 protocols were standardized by the CCITT (now the ITU) in the early 1970s. The CCITT specified data communication in a networking environment that was dominated by copper lines and electromechanical switches, which were subject to a variety of impairments that made data transmission difficult. To deal with this environment, packet switches were deployed by using the X.25 protocol. Among the many features of this protocol is error correction, which enables any node on the network to request a retransmission of error-filled data from the node that sent the data, thus overcoming the poor performance of analog lines and equipment. For some applications, this functionality makes X.25-based packet networks of value today, even with the availability of high-speed digital networks such as Frame Relay.

Applications

X.25's error-correction capabilities entail an overhead burden that limits network throughput. This functionality, in turn, limits X.25 to niche applications (such as terminal-to-host interactive services like point-of-sale transaction processing, where the reliable transmission of credit card numbers and other financial information—not speed—is the overriding concern).

Architecture

The X.25 standard defines three protocol layers that are used to interface various *Data-Terminal Equipment* (DTE) at the customer premises with *Data-Communications Equipment* (DCE) on a service provider's network.

Physical layer   Layer 1 defines the physical, electrical, functional, and procedural characteristics that are required to establish the communications link between two devices. X.25 specifies the use of several standards for the physical connection of equipment to an X.25-based network. These standards include X.21, X.21bis, and V.24—with the latter two being virtually identical to the EIA-232 standard. The physical layer operates as a full-duplex, point-to-point, synchronous circuit.

Data-link layer X.25's data-link layer corresponds to the second layer of the OSI model. At this layer, *Link Access Procedure-Balanced* (LAPB) is used to provide efficient and timely data transfer, to synchronize the data-link signals between the transmitter and receiver (flow control), to perform error checking and error recovery, and to identify and report procedural errors to higher levels of the system architecture. LAPB ensures the accurate transmission of packets that are delivered by the network layer and that are contained in HDLC information frames between the DTE and the network.

This layer also defines the unit of data transfer: the frame (refer to Figure P-1). The specific data-link protocol determines the organization and interpretation of each field in the frame. The general definitions of each field are as follows:

- *Opening Flag (8 bits)* Delimits or marks the beginning (opening flag)
- *Address Field (8 bits)* As a portion of the header, this field identifies the destination of the frame.
- *Control Field (8 bits)* Also part of the header, this field specifies the type of message (i.e., command or response), the frame sequence number, and other control information. The frame sequence number prevents a duplicate frame from being received unintentionally.

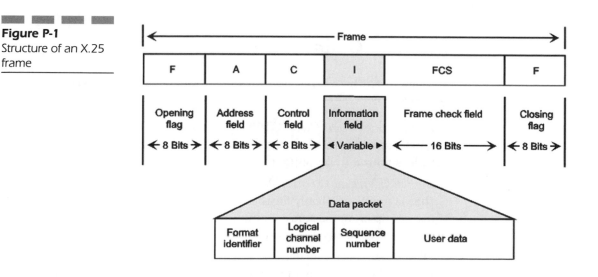

Figure P-1
Structure of an X.25 frame

- *Information Field (variable length)*   Contains the Format Identifier, Logical Channel Number, Sequence Number, and User Data
- *Frame Check Sequence (16 bits)*   Transmitted after the data bits are sent, this field provides error checking via the *Cyclic Redundancy Check* (CRC) method. Frames that are received with errors are retransmitted.
- *Closing Flag (8 bits)*   Delimits or marks end of a frame (closing flag). In some applications, the closing flag also acts as the opening flag for the next frame.

Network layer   The network layer is the highest-level protocol that is stipulated in X.25. This layer provides access to services that are available on a public, packet-switched network. When users subscribe to an X.25 service, the *Packet-Data Network* (PDN) provides configuration parameters. These parameters include the following:

- *Gateway Address*   Can be dedicated or non-dedicated and uses an adapter card that plugs into a communications server on a LAN or a dedicated gateway device
- *Frame Size*   Specifies the maximum number of characters that can be sent on the line at one time
- *Window Size*   The maximum number of packets that can be transmitted without an acknowledgment from the destination host or LAN
- *Logical Channel Number*   Identifies either a switched or permanent virtual circuit

Types of Connections

X.25 specifies three types of connections for routing information over the network:

- *Permanent Virtual Circuit* (PVC)   Resembles a leased line in that it is a permanent path between through a network that never changes, unless manually reconfigured
- *Switched Virtual Circuit* (SVC)   A temporary path through a network that is maintained only for the duration of a data transmission. SVCs are set up on request, are maintained for the duration of the call, and then are released on request.
- *Datagram*   A simple delivery service that operates on a best-efforts basis, depending on bandwidth availability. Each message or packet

contains enough information for the datagram to be routed to the appropriate destination DTE without requiring a call to be established.

Packet Transmission

X.25 specifies the means by which DTEs establish, maintain, and clear virtual circuits. The network layer uses packet-interleaved, statistical multiplexing to enable a DTE to set up concurrent virtual circuits with multiple DTEs. This multiplexing technique makes some basic assumptions: typical virtual circuits do not always carry data; frames of data are interleaved or mixed together in order to form full packets; and packet size varies by network.

X.25 assigns a logical channel number that corresponds to an SVC or PVC to each packet. This channel number applies to both transmission directions. Logical channels are the equivalent of dial-in ports in a conventional time-share network; they are a conceptual (rather than physical) path between the DTE and the network. When idle, logical channels are free to handle new calls that are requested by a local or remote DTE. After a call is established, the logical channel remains busy until the call is released.

A single, logical channel number is used for a DTE that supports only one virtual circuit. If multiple virtual circuits are involved, the service company assigns a range of channel numbers to the DTE user. If a subscriber uses both PVCs and SVCs, the X.25 service company statistically assigns individual PVCs within a range of numbers—beginning with one—while virtual calls are assigned a second range of numbers above the first. SVC, logical channel numbers are dynamically assigned during call establishment and identify all of the packets that are sent while that call is in progress.

Virtual Circuits

X.25 supports both PVCs and SVCs. A PVC, which is generally suited for a fixed, host-to-host connection, is used for point-to-point interfaces that do not require higher-level, dynamic session-control features and the use of a data-switching network. Host-to-host and host-to-terminal communications through a switched network generally use an SVC.

A PVC performs the following functions:

- Assembles data into one logical channel by using asynchronous time-division multiplexing
- Packetizes the data stream into one logical channel
- Performs packet-level error control by using packet sequence numbers
- Performs logical-channel flow control
- Permits end-to-end confirmation of data delivery

An SVC has all of the characteristics of a PVC, plus the following:

- Provides a means to request the dynamic establishment of virtual circuits
- Enables the host to accept or reject virtual call requests from other hosts
- Enables the host to take down a virtual call when a data transfer is complete

Delivery Confirmation

X.25's Delivery Confirmation procedure enables either the network or the DTE to select the maximum number of data packets on a virtual circuit. The network limits the number of packets based on network-performance criteria, including throughput and resource availability. The DTE controls the maximum number of packets on the network with a higher-level DTE-to-DTE error-control protocol. If a DTE wants to receive end-to-end acknowledgment of the data that it sends, the DTE sets the Delivery Confirmation bit in the packet's header to one. The packet-receive sequence numbers, which are embedded in frames that are sent by the receiver, acknowledge data receipt. When a DTE activates delivery confirmation, the DTEs determine the maximum number of packets on the network. Delivery confirmation limits the amount of unconfirmed data on a network and thereby facilitates error recovery. If the Delivery Confirmation bit is set to zero, acknowledgments have only local significance between the DTE and the DCE. In this case, the network determines the maximum number of packets within the throughput limits of the DTEs.

Throughput

A virtual circuit's maximum throughput varies according to allocated switch resources and the statistical multiplexing of data transmission. A virtual circuit's throughput is further limited by access-line characteristics, including line speed, flow-control parameters, and other call traffic at both the local and remote DTE network boundaries. Use of the Delivery Confirmation procedure also affects throughput; the packet transfer rate is affected by the packet-delivery confirmation rate from the receiving DTE. In addition, different types of national and international calls can cause the throughput limit to vary.

On the other hand, the X.25 network maximizes throughput when the DTE access links at both ends of the virtual circuit are properly engineered, when the receiving DTE does not control flow from the DCE, or when the transmitting DTE sends full data packets.

Extended Packet Length

An X.25-based network can accommodate extended-length packets. The X.25 network can logically chain together data packets in order to convey a large block of related information. This feature improves throughput and minimizes delay by requiring fewer acknowledgments. This procedure is implemented when the packets have the Delivery Confirmation (D) bit set to zero and the More-Data-To-Follow indicator bit (M) set to one (active) until the last packet in the chain, where D is set to one and M is set to zero. This mechanism is also significant for flow control.

After the network establishes the virtual circuit, data packets can be sent across the logical channel. X.25 numbers each data packet and limits the maximum number of packets that can be sent without additional authorization from the receiving DTE, DCE, or network to seven (the default value is usually set to two). The actual limit is either set at subscription time or during call setup. (There is an extended mode of operation within LAPB that supports up to 127 packets.)

Flow Control

Data packets carry a packet-receive sequence number that aids flow control. This sequence number authorizes the maximum number of unconfirmed packets that the logical channel can transmit. Either a DTE or the network can authorize transmission of one or more packets by sending a Receive Ready packet to the calling DTE. The packet-receive sequence numbers ensure that no error-free frames are lost or interpreted out of order.

When the Delivery Confirmation (D) bit is set to zero, the packet-receive sequence number provides local flow-control information (i.e., packet acknowledgment has only local significance). When the Delivery Confirmation bit is set to one, the packet-receive sequence number provides delivery-confirmation information between the sending and receiving DTEs. Two communicating DTEs can operate at their locally determined packet size if the user includes the More-Data-To-Follow (M) indicator, either in a full packet or in any packet that has the Delivery Confirmation bit set to one. This indicator then informs the network and receiver that there is a logical continuation of data in the next packet on a particular logical channel.

The DTE can transmit interrupt packets even when the data packets are flow controlled. These packets do not contain either send or receive sequence numbers. Therefore, in order to maintain packet integrity, a network can contain only one unconfirmed interrupt packet at a time between sending and receiving DTEs.

Error Recovery

A typical data-communications network performs error detection and recovery on various levels, some of which overlap:

- X.25 specifies several error-checking levels.
- The network might provide some level of error control.
- The DTE/DCE software might contain error-control mechanisms.

  X.25 provides the following guidelines for handling packet-level errors:

- Procedural errors that occur during call establishment and clearing are reported to the calling DTE with a diagnostic packet that clears the call.
- Procedural errors that occur during the data transfer phase (such as loss of synchronization) are reported to the sending DTE with a diagnostic packet that resets the sequence counters of both the DTE and DCE.
- A diagnostic field, which is included in the packet, provides additional information to the DTE and to the network.
- Time-outs that resolve some deadlock conditions are defined for two major areas: the length of time that the DTE has to respond to an incoming call (the minimum is typically three minutes), and the amount of time that the DCE has to wait for confirmation of a reset,

clear, or restart packet. To avoid looping conditions, the DCE takes an appropriate action for the indication packet and continues operation.

- Misalignments of subscription options between the DTE and the DCE can cause DTE procedural errors.

- Error tables, which define the actions to be taken by the DCE upon receipt of various packet types in various stages of the interface and the state to which the DCE enters, define the diagnostic code that is generated for each error condition.

X.25 also identifies a number of special error cases, such as a packet that is received on an unassigned logical channel, that cause a diagnostic packet to be sent to the DTE (rather than resetting or clearing the logical channel). A diagnostic packet includes the logical channel number on which the error occurred and a diagnostic code. Diagnostic codes exist for reset, clear, and restart packets. Because the diagnostic packet is non-procedural, it does not affect the normal meanings of call-progress signals—nor is a DTE required to take action on receipt of a diagnostic packet. The DTE logs diagnostic packets for troubleshooting information.

The transmitting DTE, the receiving DTE, and the network can detect errors in transferred data packets. If an error is detected by a DTE, it informs the other DTE and requests the affected packets to be resent. If the network detects an error, it informs both DTEs by sending a reset call-progress signal. These signals include remote DTE out-of-order (PVC only), procedural error at the remote DTE/network boundary, network congestion, or the inability of the remote DTE to support a particular function.

Data that is generated before and after an error-caused reset occurs is handled in one of two ways. If a reset occurs before data reaches its destination, that data either continues to its destination—or, more likely, is discarded by the network. Data that is generated after both local and remote ends recover from the reset continues to its destination. Data that is generated by a remote DTE before it receives the error indication from the local DTE either continues to its destination—or, more likely, the network discards it. In this case, the appropriate DTE resends discarded packets. The assigned resources for a given virtual circuit and the network end-to-end transmission delay and throughput characteristics determine the maximum number of packets that can be discarded.

Optional User Facilities

The various optional features that apply to the subscriber's network are determined at the time of subscription or as requested specifically as part of the call establishment procedure. The following X.25 user facilities can be activated within the call request packet:

Closed user-group facility   As an alternative to having a private data network for manageability and security needs, companies can establish closed user groups on the PDN between a group of users and the network administrator for a specified length of time.

Flow-control parameter selection   A network administrator can restrict access at the data level by using specific packets and window sizes in order to prevent unauthorized users from communicating with the X.25 gateway. With this option, any network user who does not have the correct configuration is denied access. Specified either at the time of subscription or during call establishment, flow-control parameters include packet size and window size. (Window size determines the maximum number of packets on a network without additional authorization from the receiving DTE.) X.25 supports the following packet sizes: 16, 32, 64, 128, 256, 512, 1024, 2048, and 4096 bytes. The maximum window size is seven, with two being the most commonly used.

Throughput class negotiation   Throughput class is the measure of the throughput that is not normally exceeded on a virtual circuit. This feature is a characteristic of virtual circuits and is a function of the amount of network resources that are allocated to the circuit. The X.25 network and the user DTE decide default values for the maximum throughput class that is associated with a virtual circuit, but these values might not always be attained because of overall link utilization, network congestion, and host processing.

One-way, outgoing, logical channel   This optional feature restricts the use of a range of logical channels to outgoing calls only. This restraint does not affect the full-duplex data-transfer process.

Incoming or outgoing call barring   X.25 provides two call-barring service options. The first option bars the presentation of incoming calls to the DTE, although the DTE can initiate outgoing calls. The second option, outgoing call barring, prevents the DCE from accepting calls from a DTE; however, the DTE can receive incoming calls.

Fast-select facility   The fast-select facility, a variation on switched virtual-circuit service, is designed to satisfy short, low-volume, transaction-based applications, such as point-of-sale, funds transfer, credit checks, and meter reading. Fast select enables the inclusion of up to 128 bytes of data in the call establishment and clearing procedures for a switched virtual circuit.

Dial X.25 (X.32) Dial X.25, or X.32, enables users to dial synchronously into a PDN over public telephone lines. This service option is designed for companies that use a PDN only occasionally or that are just beginning to use an X.25 service. X.32 saves a company the cost of leasing a dedicated, packet-switched line, and it enables users to access the network from unsupported locations while gaining complete error-detection facilities.

Other Protocols

Other standards also govern various aspects of X.25 packet-switched networks. Some of the most commonly used standards include the following:

- *X.3* Defines the functions of the *Packet Assembler/Disassembler* (PAD), which is used to communicate with a remote X.25 device that is connected to the PDN
- *X.28* Defines the procedures that an asynchronous terminal uses to connect with a PAD
- *X.29* Defines the procedures that enable a packet mode device to control the operation of a PAD
- *X.75* Defines the gateway procedures for interconnecting X.25 PDNs, giving end hosts the appearance of a single X.25 network

The relationships of these packet network standards are shown in Figure P-2.

Summary

X.25 offers error-free communications and guaranteed delivery, making it the best choice for financial transactions and for companies that must establish international networks in countries that still have analog-based communications infrastructures in place. X.25 provides connectivity with legacy mainframes, minicomputers, and LANs. Despite the emergence of

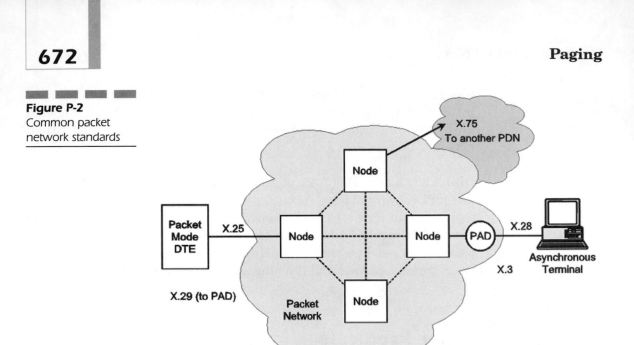

higher-speed cell and frame-switched services that operate over more reliable digital links, such as ATM and Frame Relay, there is still strong demand for X.25 products and services.

See Also

> *Asynchronous Transfer Mode* (ATM)
>
> Frame Relay

Paging

Paging is a wireless service that provides one-way or two-way messaging to give mobile users continuous accessibility to family, friends, and business colleagues while they are away from telephones. Typically, the mobile user carries a palm-sized device (the pager or some other portable device with a paging capability) that has a unique identification number. The calling party inputs this number, usually through the public telephone network, to the paging system—which then signals the pager to alert the called party.

Alternatively, call-back numbers and short-text messages can be sent to pagers via messaging software that is installed on a PC or that is inputted into forms that are accessed on the Web for delivery via an Internet gateway. Regardless of delivery method, the called party receives an audio or visual notification of the call, which includes a display of the phone number to call back. If the pager has alphanumeric capabilities, messages can be displayed on the pager's screen.

Paging Applications

Many applications exist for paging, and among the most popular are the following:

- *Mobile messaging*   Enables messages to be sent to mobile workers. They can respond with confirmation or additional instructions.
- *Data dispatch*   Enables managers to schedule work appointments for mobile workers. Upon activating their pagers each morning, their itinerary will be waiting for them.
- *Single-key call back*   Enables the user to read a message and to respond instantly with a predefined stored message that is selected with a single key.

Some message-paging services work with text-messaging software programs, enabling users to send messages from their desktop or notebook computers to individuals or groups. This kind of software also keeps a log of all messaging activity. This method also offers privacy, because messages do not have to go through an operator before delivery to the recipient.

Types of Paging Services

Several types of paging services are available:

Selective operator-assisted voice paging   Early paging systems were non-selective and operator-assisted. Operators at a central control facility received voice input messages, which were taped as they arrived. After an interval of time—15 minutes or so—these messages were then broadcast and received by all of the paging system subscribers. In other words, subscribers had to tune in at appointed times and listen to all of the messages that were broadcast to see whether there were any messages for them. Not only did this method waste air time, but the system was inconvenient, labor-intensive, and offered no privacy.

These disadvantages were overcome with the introduction of address encoders at the central control facility and associated decoders in the pagers. Each pager was given a unique address code. Messages that were intended for a particular called party were inputted into the system, preceded by this address. In this way, only the person to whom the message was addressed was alerted to switch on his or her pager in order to retrieve messages. With selective paging, tone-only alert paging became possible. The called party was alerted by a beep tone to call the operator (or a prearranged home or office number) to have the message read back.

Automatic paging   Traditionally, an operator was always needed either to send the paging signal or to play back/relay messages for the called party. With automatic paging, a telephone number is assigned to each pager, and the paging terminal can automatically signal for voice input, if any, from the calling party—after which it will automatically page the called party with the address code and relay the input voice message.

Tone and numeric paging   Voice messages take up a lot of air time, and as the paging market expands, frequency overcrowding becomes a potentially serious problem. Tone-only alert paging saves on air time usage but has the disadvantage that the alerted subscriber knows only that he has to call certain prearranged numbers, based on the kind of alert tone that is received.

With the introduction of numeric display pagers in the mid-1980s, the alert tone is followed by a display of a telephone number to call back or a coded message. This method resulted in great savings on air-time usage because it was no longer necessary to add a voice message after the alert tone. This method is still the most popular form of paging.

Alphanumeric paging   Alphanumeric pagers display text or numeric messages that are entered by the calling party or operator by using a modem-equipped computer or a custom page-entry device that is designed to enter short text messages. Although alphanumeric pagers have captured a relatively small market in recent years, the introduction of value-added services that include news, stock quotes, sports scores, traffic bulletins, and other specialized information services has heated up the market for such devices.

Ideographic paging   Pagers that are capable of displaying different ideographic languages—Chinese, Japanese, and others—are also available. The particular language that is supported is determined by the firmware (computer program) that is installed in the pager and in the page-entry device. The pager is similar to a pager that is used in alphanumeric display paging.

Paging System Components

The key components of a paging system include an input source, the existing wireline telephone network, the paging encoding and transmitter control equipment, and the pager itself.

Input source A page can be entered from a phone, from a computer that has a modem or another type of desktop page-entry device, from a PDA, or through an operator who takes a phone-in message and enters the message on behalf of the caller. Various forms posted on the Web can also be used to input messages to pagers.

The Web form of MCI WorldCom, for example (refer to Figure P-3), enables users to send a text message consisting of a maximum of 240 characters to subscribers of its one-way alphanumeric service and 500 characters to subscribers of its enhanced one-way, interactive (two-way), and QuickReply Interactive services. In addition, users can send a text message consisting of a maximum of 200 characters to subscribers of MobileComm. The form even provides a means to check the character count before the message is sent. AT&T Wireless also offers a similar Web form for inputting messages to pagers (refer to Figure P-4).

Telephone network Regardless of exactly how the message is entered, the message eventually passes through the *Public-Switched Telephone Network* (PSTN) to the paging terminal for encoding and transmission through the wireless paging system. Typically, the encoder accepts the incoming page, checks the validity of the pager number and looks up the directory or database for the subscriber's pager address, and converts the address and message into the appropriate paging signaling protocol. The encoded paging signal is then sent to the transmitters (base stations), through the paging transmission control systems, and is broadcast across the coverage area on the specified frequency.

Encoder Encoding devices convert pager numbers into pager codes that can be transmitted. Two ways exist in which encoding devices accept pager numbers: manually or automatically. In manual encoding, a paging system

MCI WorldCom offers a Web pager that enables anyone to send a page to anyone else who has an MCI WorldCom or SkyTel pager.

operator enters pager numbers and messages via a keypad that is connected to the encoder. In automatic encoding, a caller dials an automatic paging terminal and uses the phone keypads to enter pager numbers. Regardless of the method that is used, the encoding device then generates the paging code for the numbers that are entered and sends the code to the paging base station for wireless transmission.

Base-station transmitters The base-station transmitters send page codes on an assigned radio frequency. Most base stations are specifically designed for paging, but those that are designed for two-way voice can be used, as well.

Figure P-4
AT&T Wireless offers a Web pager that enables anyone to send a message to people who carry AT&T PCS phones, alphanumeric pagers, and CDPD-compliant PocketNet phones.

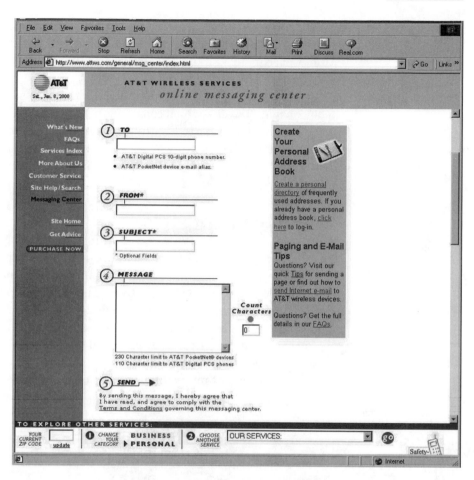

Pagers Pagers are essentially FM receivers that are tuned to the same RF frequency as the paging base station. A decoder unit that is built into each pager recognizes the unique code that is assigned to the pager and rejects all other codes for selective alerting. Pagers can be assigned the same code for group paging, however. Pagers also exist that can be assigned multiple page codes, typically up to a maximum of four, enabling the same pager to be used for a mix of individual and group paging functions.

Despite all of the enhancements that have been built into pagers and paging services in recent years, the market is slowing down. Alphanumeric services—which provide word messages instead of just phone numbers—have failed to attract a wide audience, largely because paging subscribers still need a phone in order to respond to messages. Some providers have tried without success to offer services that would enable callers to leave voice messages on pagers, but this solution has also failed to gain momentum. Consequently, about four out of five of today's paging customers still rely on cheap numeric services.

Two-way paging networks might be the industry's last hope for survival. They enable a pager—which comes equipped with a mini-keyboard—to talk to another pager (refer to Figure P-5) through a telephone, e-mail address, or fax machine. Some two-way services enable consumers to reply to messages with predetermined responses.

Signaling Protocols

In a paging system, the paging terminal—after accepting an incoming page and validating it—will encode the pager address and message into the appropriate paging-signaling protocol. The signaling protocol enables individual pagers to be uniquely identified/alerted and to be provided with the additional voice message or display message (if any).

Various signaling protocols are used for the different paging service types, such as tone-only, tone and voice, etc. Most paging networks have the capability to support many different paging formats over a single frequency.

Figure P-5

With Motorola's PageWriter 2000, users can send text messages to other two-way pagers and alphanumeric pagers, directly or through e-mail.

Many paging formats are manufacturer specific and are often proprietary, but there are public-domain protocols such as the *Post Office Code Standardization Advisory Group* (POCSAG) that enable different manufacturers to produce compatible pagers.

POCSAG is a public-domain digital format that has been adopted by many pager manufacturers around the world. POCSAG can accommodate two million codes (pagers), each of which are capable of supporting up to four addresses for paging functions such as tone-only, tone and voice, and numeric display. POCSAG operates at data rates of up to 2400bps. At this rate, in order to send a single tone-only page, you need only 13 milliseconds. This rate is about 100 times faster than two-tone paging.

With the explosion of wireless technology and dramatic growth in the paging industry in many markets, existing networks are becoming more and more overcrowded. In addition, RF spectrum is not readily available because of demands by other wireless applications. In response to this problem, Motorola has developed a one-way messaging protocol called *Feature-Rich, Long-Life Environment for Executing* (Flex) messaging applications, which is intended to transform and broaden paging from traditional low-end, numeric services into a range of PCS/PCN and other wireless applications.

Relative to POCSAG, Flex can transmit messages at up to 6,400bps and permit up to 600,000 numeric pagers on a single frequency, compared to POCSAG's 2,400bps transmission rate and 300,000 users per frequency. In addition, Flex provides enhanced bit-error correction and much higher protection against signal fades that are common in FM simulcast paging systems. The combination of increased bit-error correction and improved fade protection increases the probability of receiving a message intact—especially longer alphanumeric messages and data files that will be sent over PCS/PCN. Motorola has also developed ReFlex, a two-way protocol that will enable users to reply to messages, and InFlexion, a protocol that will enable high-speed voice messaging and data services at up to 112Kbps.

Summary

The hardware and software that are used in radio paging systems have evolved from simple operator-assisted systems to terminals that are fully computerized, with such features as message handling, scheduled delivery, user-friendly prompts to guide callers to a variety of functions, and automatic reception of messages. After tremendous growth in the last decade, from 10 million subscribers in 1990 to 60 million today, the paging industry has slowed down markedly—in large part because of cut-throat competition and the increasing use of digital mobile phones.

See Also

Electronic Mail

Personal Communication Services (PCS)

Personal Digital Assistants (PDAs)

Passive Optical Networks (PON)

A *Passive Optical Network* (PON) uses high-capacity optical fiber to transmit voice, data, and video. This network is called passive because it eliminates the active electronics in various network elements, which require AC power. A PON reduces the carrier's costs because it eliminates the need for an external power source. Electronics does not come into play until the ends of the network, where the laser transmitters and photodiode detectors reside.

An example of a network device for the PON environment is Lucent's WaveStar LambdaRouter. The device uses a series of microscopic mirrors to instantly direct and route optical signals from fiber to fiber in the network, without first converting them to electrical form as is done today. The router will save service providers up to 25 percent on operational costs and will enable them to direct network traffic 16 times faster than electrical switches.

Eliminating active electronic components also makes the carrier's fiber network more reliable. With the elimination of such components, there are fewer chances of equipment malfunctions and no need for maintenance. Also, because passive components do not need external power sources, there is no chance of service disruptions from power outages.

For all of these reasons, carriers are now beginning to favor PON technology. With an awareness of the long-term payoff of PONs, NTT in Japan and regional service providers in the United States are already deploying the technology or are seriously considering doing so. Because PON bandwidth is so high, operators can carry a variety of services over it, meeting a wide variety of customer needs. One customer might require 100BaseT Ethernet data connections for his or her enterprise WAN, while another person might need an *Asynchronous Transfer Mode* (ATM) link to the network.

Still others might request a more conventional DS3 channel for digital transport. Whatever the need, PONs frame these channels at the access point and carry them as laser pulses into the host terminal at a *Central Office* (CO), where equipment efficiently passes the pulses into network channels for local distribution.

On the distribution side of a PON, signals are routed over the local fiber link with all signals along that link going to all interim transfer points. Optical splitters route signals through the network, much like vehicles that are being routed through a freeway network with on-ramps and off-ramps. Optical receivers at intermediate points and subscriber terminals that are tuned for specific wavelengths of light pick off the signals that are intended for their groups of subscribers. At the final destination, only the signal that is intended for a specific residence or business can be correctly interpreted. In this way, PONs offer more security than current distribution networks that are based on active components.

The subscriber terminal, or *Optical Network Unit* (ONU), is a key component of this network because it has the capability to accommodate multiple services in order for the carrier to save on costs. In addition to POTS, ISDN, and ATM services, the ONU supports two-way video and CATV.

Bell South is among the U.S. carriers that are testing PON technology. The company has installed a new fiber-optic system to connect its switching offices to some 400 homes in Atlanta. Participating residential customers receive a variety of services, including the following:

- Internet access at super high speeds through a 100Mbps interface
- 120 channels of digital video entertainment
- 70 channels of analog video entertainment
- 31 channels of CD-quality digital audio service

Bell South first installed fiber to customer homes more than 10 years ago in select markets. Since that time, and especially during the last several years, the carrier has made *Fiber to the Curb* (FTTC) its preferred solution to providing telecommunications services for new subdivisions. Bell South's latest step, the PON trial in Atlanta, provides the final link for an all-fiber connection from its switch all the way to the home, instead of terminating FTTC. *Fiber to the Home* (FTTH) is Bell South's ultimate platform for satisfying customers' voracious appetites for bandwidth (which are growing at exponential rates). The PON ultimately will be used to deliver voice, video, and high-speed data access via the ATM data-networking protocol—the fastest technology in use today.

Summary

The demands of telecommunications users are driving a number of world-wide advances that are increasing network capacity and bringing more bandwidth to individual users. PON offers an effective and economical solution for local access. In addition, it is easy to maintain and offers privacy and security. PON prepares the entire network, from interoffice links to the subscriber, to take full advantage of all the advances that optical networking will bring in the future: unlimited bandwidth and enormous flexibility in services.

See Also

Fiber in the Loop (FITL)

Fiber-Optic Technology

Wave-Division Multiplexing (WDM)

Pay-Per-Call Services

Pay-per-call is a revenue-generating tool for businesses and is usually associated with 900 numbers that provide live or recorded services. Organizations ranging from Fortune 500 companies to start-up entrepreneurial firms are using 900 pay-per-call service to accomplish the following goals:

- Generate income from services that they previously provided for free
- Pay for operations that used to cost them money
- Profit from information and knowledge
- Create all-new revenue streams where none previously existed

Virtually no limit is placed on the ways in which 900 pay-per-call services can be put to work in almost every industry. The reason is because information sells. Consumers and businesspeople now realize the value of information, and they are willing to pay for it. They prefer the convenience and immediacy of information. Pay-per-call services give businesses an easy way to deliver the information that they already have and to be paid for each inquiry.

Applications

Among the information services that can be provided on a pay-per-call basis are as follows:

- Technical/product support
- Lottery services
- Adult entertainment
- Banking and financial services
- News and information services
- Call-in polls and surveys
- Health care information services
- Fundraising
- Marketing and promotions

Even government agencies can easily and economically recoup some expenses for the taxpayer by charging a modest fee to callers who use their value-added services. The *Federal Communications Commission* (FCC), for example, has used AT&T MultiQuest 900 Service during auctions for wireless *Personal Communications Services* (PCS) licenses. Working from a remote PC, bidders access the FCC's *Wide-Area Network* (WAN) via the 900 service. Then, they participate in activities such as bidding electronically for licenses or querying the FCC licensing database.

Agencies that offer the MultiQuest 900 Service need dedicated access to the AT&T switched network. Each agency decides the amount that callers will pay for obtaining information. The agencies do not need an in-house billing or collection system, because all fees appear on callers' AT&T long-distance telephone bills. AT&T collects the fees and sends a monthly check to the agency.

Implementation

To implement a pay-per-call system, the information provider is assigned one or more 900 numbers for its application. When a caller dials the information provider's 900 number, the call is routed to a specified location, where a live representative or an automated voice-response system takes the call and provides the information or help that the caller needs.

The carrier's billing and collection system then charges each caller through his or her local telephone company bill. The actual charge for the call depends on the fee structure that the information provider has established for the application. When the customer pays the bill, a transport charge and a billing fee is deducted from the total cost of the call. The 900 service provider sends the information provider the accumulated net proceeds every month.

Keys to Success

Although consumers are quite willing to pay for information, entertainment, and services that are offered through 900 numbers, a few key principles can improve the success rate of this business:

- Callers are reluctant to use a 900 number unless it provides something that they want, need, or enjoy. The more interesting and useful the content, the more customers are willing to use a 900 number.

- When callers understand (in advance) the cost of a 900 call, they are more willing to dial. Also, they tend to be more comfortable with flat-fee or capped fee structures. Some carriers, such as AT&T, require information providers to show the cost of the call in all advertisements and promotions for the 900 number and to announce the costs at the beginning of each call.

- Customers prefer to deal with reputable companies. If the information provider is not well known, the chances for success can be greatly improved by taking extra steps to treat callers professionally and to deliver what has been advertised.

Startup Concerns

For companies that want to become information service providers via 900 number pay-per-call, several issues must be considered. The carrier typically works with the information provider to tailor the 900 service to its individual needs, as follows:

- If the information provider plans to have representatives or consultants handle the calls live, consideration must be given to how many agents would be needed, what kind of knowledge or skill they should have, and where they will receive the calls. The 900 calls can be routed to one central location or can be distributed to many different locations.

■ If the service will play back recordings, consideration must be given to how many options will be provided to the caller. If several options will be made available to callers, a menu must be developed from which callers can select the items of interest.

■ Callers can be served anywhere in the United States. If the information provider's service only attracts callers from certain regions, however, it can limit the service to certain areas and can reduce the cost of operation.

■ To attract callers to the 900 application, they have to know about the number. Therefore, consideration must be given to where and how to publicize the service. Newspaper advertisements, magazines, direct mail, radio, TV, billboards, the Internet, and catalogs are some of the advertising possibilities. The choice will depend on who the audience is and where they are most likely to be found. Even the most compelling idea needs strong advertising support, which requires an investment.

■ For a recorded service, there must be some way to answer calls, present the choices, and play back the recordings. Most recorded applications use voice-response equipment that requires a significant investment. For information service providers that do not have the capital available, there are service bureaus available that can handle the calls on a per-call fee basis. This alternative lowers the cost of entry into the 900 pay-per-call service market.

■ For a live service, attendants or representatives will need to handle the calls. If the information provider expects to have several representatives handling calls, systems that distribute and manage the call flow will also be required. The same equipment that handles 800 calls (i.e., Automatic Call Distributors, or ACDs) can probably be used for 900 calls. If the information provider does not have the staff or equipment yet, many service bureaus can provide turnkey services that include them.

Carriers offer two different services, depending on the volume of calls that the information service provider must handle. If the service is for businesses only and fewer than 100 hours of calls per month are expected, the carrier will support the service over ordinary facilities. For larger-scale applications, such as consumer or business applications, the carrier will dedicate special high-capacity facilities in order to support the service.

Research shows that callers have definite ideas about what they are willing to pay for information and services. Charge too much, and the information provider risks missing too many callers; charge too little, and there is the risk of missing out on revenue. With AT&T, for example, information providers can set pay-per-call fees in three different ways:

- Per call, such as $5 per call (regardless of length)
- Per minute, such as $2 for the first minute and $1 for each additional minute
- Pre-selected, such as $5 for the first five-minute period and 95 cents for each additional minute

Information service providers can establish almost any amount in these formats. AT&T even provides a way to adjust the charges for each call (on the spot), depending on what the caller needs or who the caller is. The actual revenue from a pay-per-call depends on the nature of the service, how it is promoted, and how many callers are attracted to the service.

Regulation

In 1991, the FCC adopted regulations governing interstate pay-per-call services to address complaints from consumers of widespread abusive practices involving 900 services. Among other protective measures, the FCC established the following policies:

- Requiring pay-per-call programs to begin with a preamble that discloses the cost of the services and that affords the caller an opportunity to hang up before incurring charges
- Requiring *Local-Exchange Carriers* (LECs), where technically feasible, to offer telephone subscribers the option of blocking access to 900 numbers
- Prohibiting common carriers from disconnecting basic telephone service for failure to pay pay-per-call charges

To expand upon this regulatory framework, Congress enacted legislation in 1992 requiring both the FCC and the *Federal Trade Commission* (FTC) to adopt rules that were intended to increase consumers' protection from fraudulent and deceptive practices and to promote the development of legitimate pay-per-call services. In response to complaints from consumers, businesses, and organizations—alleging that they had been billed for calls that were made from their phones to toll-free numbers—this legislation also mandated explicit restrictions on the use of 800 and other toll-free numbers to provide information services.

In mid-1993, the FCC amended its pay-per-call regulations to be consistent with the Congressional mandate. The new rules required that all interstate pay-per-call services be provided through 900 numbers. In other words, use of 800 numbers or any other number that is advertised or that is widely understood to be toll-free cannot be used to charge callers for information services.

Even with these safeguards, carriers and information providers are still free to use 800 numbers to provide a wide variety of information services. For example, information services charged on a per-call basis can be made by using 800 numbers when they are charged to a credit card or provided under a written presubscription arrangement. The safeguards simply recognize the significant governmental interest in shielding consumers from deceptive practices that are associated with a service that the public widely perceives as free.

Summary

Pay-per-call service has been available to the commercial sector for more than a decade. This form of information distribution has become a key revenue-generating tool for businesses and government agencies. As of year-end 1999, there were more than 10,000 pay-per-call information service providers in the United States.

See Also

Automatic Call Distributors (ACDs)

Integrated Voice Response

 # PCS 1900

PCS 1900 is the *American National Standards Institute* (ANSI) radio standard for 1900MHz *Personal Communications Service* (PCS) in the United States. As such, it is compatible with the *Global System for Mobile Com-*

munications (GSM), an international standard that was adopted by 160 operators supporting 55 million subscribers in 110 countries as of year-end 1999. Network operators who aligned to the GSM standard have 35 percent of the world's wireless market.

PCS 1900 can be implemented with either TDMA or CDMA technology. TDMA-based technology enjoys an initial cost advantage over rival technology CDMA equipment, because suppliers who are making TDMA infrastructure equipment and handsets have already reached economies of scale. In contrast, CDMA equipment is still in its first generation and is therefore generally more expensive.

At present, the CDMA (IS-95) standard has been chosen by about half of all of the PCS licensees in the United States, giving it the lead in the total number of potential subscribers. The first operational PCS networks have been using PCS 1900 as their standard, however, mainly because of the maturity of the GSM-based technology.

Although similar in appearance to analog cellular service, PCS 1900 is based on digital technology. As such, PCS 1900 provides better voice quality, broader coverage, and a richer feature set. In addition to improved voice quality, fax and data transmissions are more reliable. Laptop computer users can connect to the handset with a PCMCIA card and can send fax and data transmissions at higher speeds, with fewer opportunities for error.

Architecture

The PCS 1900 system architecture consists of the following major components:

- *Switching system*   Controls call processing and subscriber-related functions
- *Base station*   Performs radio-related functions
- *Mobile station*   The end-user device that supports voice and data communications, as well as short message services
- Operation and Support System (*OSS*)   Supports the operation and maintenance activities of the network

The Switching System for PCS 1900 service contains the following functional elements (refer to Figure P-6):

- Mobile Switching Center (*MSC*)   Performs the telephony switching functions for the network. This center also controls calls to and from

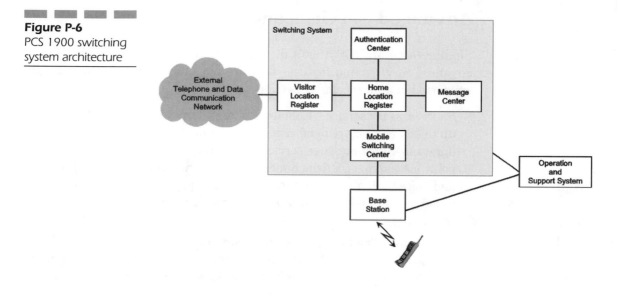

Figure P-6
PCS 1900 switching
system architecture

other telephone and data communications networks, such as *Public-Switched Telephone Networks* (PSTNs), *Integrated Services Digital Networks* (ISDNs), *Public Land Mobile Radio Services* (PLMRS) networks, *Public Data Networks* (PDNs), and various private networks.

■ Visitor Location Register (*VLR*)   A database that contains all temporary subscriber information that the MSC needs in order to serve visiting subscribers

■ Home Location Register (*HLR*)   A database for storing and managing subscriptions. The HLR contains all permanent subscriber information, including the subscriber's service profile, location information, and activity status.

■ Authentication Center (*AC*)   Provides authentication and encryption parameters that verify the user's identity and ensure the confidentiality of each call. This functionality protects network operators from common types of fraud that are found in the cellular industry today.

■ Message Center (*MC*)   Supports numerous types of messaging services, such as voice mail, facsimile, and e-mail, for example.

Advanced Services and Features

Similar to GSM, PCS 1900's digital orientation makes possible several advanced services and features that are not efficiently and economically supported in analog cellular networks:

- *Short message service*   Enables alphanumeric messages that consist of up to 160 characters to be sent to-and-from PCS 1900-compatible handsets. Short message service applications include two-way point-to-point messaging, confirmed message delivery, cell-based messaging, and voice mail alert. These messaging and paging capabilities create a broad array of potential new revenue-generating opportunities for carriers.

- *Voice mail*   The PCS 1900 network provides one central voice mail mailbox for both wired and wireless service. In addition, the voice mail alert feature ensures that subscribers do not miss important messages.

- *Personal call management*   Offers subscribers a single telephone number for all of their physical telecommunication devices. For example, a single number can be assigned for home and mobile use or for office and mobile use, which enables subscribers to receive all calls regardless of their physical location.

- *Data applications*   Wireless data applications that can be supported by PCS 1900 networks include Internet access, e-commerce, and fax transmission.

Smart Cards

The PCS 1900 standard supports the Smart Card, which provides similar features as GSM's *Subscriber Identity Module* (SIM). The size of a credit card, Smart Cards contain embedded computer chips that contain user profile information. By removing the Smart Card from one PCS 1900 phone and inserting it into another PCS 1900 phone, the user can receive calls at that phone, make calls from that phone, or receive other subscribed services such as wireless Internet access. The handsets cannot be used to place calls (except 911 emergency calls) until the subscriber inserts the Smart Card and enters a *Personal Identification Number* (PIN).

The profile information that is stored in the Smart Card also enables international roaming. When traveling in the United States, international GSM customers can rent handsets, insert their SIM, and access their services as if they are home. By the same token, when U.S. subscribers travel internationally to cities that have compatible networks and mutual roaming agreements, they only need to take their Smart Card with them to access the services that they subscribed to back home via the local GSM network.

Similar to SIMs, Smart Cards also provide storage for features such as frequently called numbers and short messages. Smart Cards also include the AT command-set extensions, which integrate computing applications with cellular data communications. In the future, Smart Cards and PCS 1900 technology will also link subscribers to applications in e-commerce, banking, and health care.

Summary

GSM looks to be a perennial third in North America's digital markets, mainly because the technology has not been adopted for use in cellular 800MHz frequencies. GSM has shown impressive gains in the 1900MHz PCS realm. PCS 1900 is a frequency-adapted version of GSM, which operates at 1800MHz in Europe and elsewhere. Otherwise, PCS 1900 and GSM are similar in all other respects, including the network architecture and types of services that are supported. An advantage that U.S. carriers have in supporting the PCS 1900 standard is that it is interoperable with the worldwide GSM standard, which means that users can roam globally.

See Also

Global System for Mobile (GSM) Telecommunications

Personal Communications Services (PCS)

Personal-Access Communications Systems (PACS)

Personal-Access Communications Systems (PACS) is a standard that was adopted by ANSI for *Personal Communications Services* (PCS). Adopted in June 1995, PACS provides an approach for implementing PCS in North America that is fully compatible with the local-exchange telephone network and interoperable with existing cellular systems. Based on the *Personal Handyphone System* (PHS) that was developed in Japan and the *Wireless-*

Access Communications System (WACS) that was developed by Bellcore (now known as Telcordia Technologies), PACS is designed to support mobile and fixed applications in the 1900MHz frequency range. PACS promises low installation and operating costs while providing high-quality voice and data services. In the United States, trials of PACS equipment began in 1995, and equipment rollout began in 1996.

Most of the standards—including upbanded versions of CDMA, TDMA, and GSM—look like cellular systems in that they have high transmit powers and receivers that are designed for the large delay spreads of the macrocell environment, and they typically use low bit-rate voice coders (vocoders). PACS fills the niche between these classes of systems, providing high-quality services, high data capability, and high user density in indoor and outdoor microcellular environments. PACS equipment is simpler and less costly than macrocell systems, yet it is more robust than indoor systems.

PACS capabilities include pedestrian and vehicular-speed mobility, data services, licensed and unlicensed systems, simplified network provisioning, maintenance, and administration. Key features of PACS include the following:

- Voice and data services that are comparable in quality, reliability, and security with wireline alternatives
- Optimized to provide service to the in-building, pedestrians, and city traffic operating environments
- Most cost effective to serve high-density traffic areas
- Small inexpensive, line-powered radio ports for unobtrusive pole or wall mounting
- Low-complexity-per-circuit signal processing
- Low transmit power and efficient sleep mode, requiring only small batteries to power portable subscriber units for hours of talk time and multiple days of standby time

Similar to PHS, PACS uses 32Kbps ADPCM waveform encoding that provides virtual land-line voice quality. ADPCM has demonstrated a high degree of tolerance to the cascading of vocoders, as experienced when a mobile subscriber calls a voice mail system and the mail box owner retrieves the message from a mobile phone. Using other mobile technologies, the message becomes unintelligible. With PACS, the message is clear. Similarly, the compounding of delays in mobile to PCS through satellite calls—a routine situation in Alaska and in many developing countries—can be troublesome. PACS provides extremely low delay.

The low complexity and transmit power of PACS yield limited cell sizes, which makes it well suited for urban and suburban applications where user density is high. Antennas can be installed inconspicuously, piggy-backing on existing structures (which avoids the high costs and delays that are associated with obtaining permits for the construction of high towers).

Applications

Wireless local loop, pedestrian venues, commuting routes, and indoor wireless are typical PACS applications. Additionally, PACS is designed to offer high-capacity, superior voice quality, and ISDN data services. Interoperability with ISDN is provided by aggregating two 32Kbps time slots in order to form a 64Kbps channel. A 64Kbps channel can also support 28.8Kbps voice-band data by using existing modems.

PACS can also be used for providing wireless access to the Internet. The packet-data communications capabilities that are defined in the PACS standards, together with the capability to aggregate multiple 32Kbps channels, makes it possible for users to access the Internet from their PCs that are equipped with suitable wireless modems at speeds of up to 200Kbps. When using the packet mode of PACS for Internet traffic, radio channels are not dedicated to users while they are on active Internet sessions (which can be extremely long). Rather, radio resources are only used when data is actually being sent or received, resulting in efficient operation and minimally impacting the capacity of the PACS network to support voice communications.

PACS was designed to support the full range of *Advanced Intelligent Network* (AIN) services, including custom calling features and personal mobility. As new AIN features are developed, the PACS-compliant technology will evolve in order to facilitate incorporation of the new services.

Summary

The market for PCS is competitive. Already, PCS is exerting downward price pressure on traditional analog cellular services, where the two compete side by side. PACS enables PCS operators to differentiate their offerings through digital voice clarity, high bit-rate data communications, and AIN services—all in a lightweight handset. Moreover, the cost savings and ease-of-use that is associated with PACS makes it economical for residential and business environments, compared to competitive high-powered wide-area systems.

See Also

Adaptive, Differential Pulse-Code Modulation

Advanced Intelligent Network (AIN)

Personal Handyphone System (PHS)

Personal Communications Services (PCS)

Wireless Communications Services

Personal Communications Services (PCS)

PCS is a set of wireless communication services that are personalized to the individual. Subscribers can tailor their service package to include only the services that they want, which can include stock quotes, sports scores, headline news, voice mail, e-mail and fax notification, and caller ID. Telephone numbers that are used in PCS handsets are tied specifically to an individual and belong to that person for as long as he or she wants, regardless of whether they move to another location. These handsets have full roaming capability, enabling anywhere-to-anywhere communication.

Unlike the existing cellular network, PCS is completely digital. The digital nature of PCS enables antennas, receivers, and transmitters to be smaller, in addition to enabling the simultaneous transmission/reception of data and voice with no performance penalty. Eventually, PCS will overtake analog cellular technology as the preferred method of wireless communication.

Several technologies are being used to implement PCS. In Europe, the underlying digital technology for PCS is GSM. *Personal Communications Networks* (PCN) in Europe operate at 1800MHz. GSM has been adapted for operation at 1900MHz for PCS in the United States (i.e., PCS 1900). Other versions of GSM are employed to provide PCS services in other countries, such as the *Personal Handyphone System* (PHS) in Japan. Many service providers in the United States have standardized their PCS networks on CDMA.

The PCS Network

A typical PCS network operates around a system of microcells—smaller versions of a cellular network's cell sites—each of which are equipped with a base-station transceiver. The microcell transceivers require less power to operate but cover a more limited range. The base stations that are used in the microcells can even be placed indoors, enabling seamless coverage as the subscriber walks into and out of buildings.

Similar to packet radio networks today, terminal devices stay connected to the network even when they are not in use, enabling the network to locate an individual within the network via the nearest microcell and routing calls and messages directly to the subscriber's location. For a follow-me service that incorporates more than one device, a subscriber might be required to turn on a pager, for example, in order to receive messages on that device. If the subscriber receives a phone call while the pager is on, the network might store the call, take a message, or send the call to a personal voice mail system and simultaneously page the subscriber. In this way, PCS enables the concept of universal messaging to be fully realized.

Currently, cellular switching systems operate separately from the public-switched network. When cellular subscribers call a land-line phone (and vice versa), the two systems are interconnected in order to complete the communications circuit. In the future, many PCS services will be switched on the same switches that service the wireline network, so that the only distinction between a wireless and wireline call will be the medium that is used to complete the last leg of the communication at each end of the circuit. In some places, wireless PCS and *Cable Television* (CATV) are already integrated in a unified CDMA-based architecture called PCS-over-cable. The use of CATV enables PCS to reach more potential subscribers with a lower start-up cost for service providers.

Broad-Band and Narrow-Band

Two technically distinct types of PCS exist: narrow-band and broad-band, each of which operate on a specific part of the radio spectrum and has unique characteristics.

Narrow-band PCS is intended for two-way paging and for other types of communications that handle small bursts of data. These services have been assigned to the 900MHz frequency range; specifically, 901MHz to 912MHz, 930MHz to 931MHz, and 940MHz to 941MHz. Broad-band PCS is intended for more sophisticated data services. These types of services have been assigned a frequency range of 1850MHz to 1990MHz.

Narrow-band PCS and broad-band PCS license ownership have been determined by public auctions conducted by the FCC. PCS service areas are divided into 51 regional service areas that are subdivided into a total of 492 metropolitan areas. Competition by at least two service providers exists in each service area. There are 10 national service providers, plus six regional providers in each of five multi-state regions called major trading areas.

An unlicensed portion of the PCS spectrum has been allocated from 1890MHz to 1930MHz. This service is designed to enable the unlicensed operation of short-distance—typically indoor or campus-oriented environments—voice and data services that are provided by wireless LANs and wireless PBXs.

One of the largest PCS networks is operated by Sprint PCS. At year-end 1999, the company's CDMA-based wireless network served more than five million customers. In addition to offering voice services from 280 major metropolitan markets, including more than 4,000 cities and communities in the United States, the company offers nation-wide availability of FOXSports.com on the Sprint PCS Wireless Web. Sports fans who have an Internet-ready Sprint PCS phone can access the latest news and up-to-the-minute scores for major in-season sports, virtually anytime and anywhere on the Sprint PCS network.

In addition, users can shop online at Amazon.com from Internet-ready Sprint PCS phones. The service supports two-way transactional e-commerce services in order to provide users with easy and convenient shopping on the Internet via their Sprint PCS phones. Users can also access the Sprint PCS Wireless Web to check e-mail, news, stock portfolios, or flight schedules.

Summary

PCS is bringing about a variety of new mobile and portable devices such as small, light-weight telephone handsets that work at home, in the office, or on the street; advanced, smart paging devices; and wireless e-mail and other Internet-based services. At this writing, PCS services are available in all regions of the United States. Most of the smaller PCS networks are now interconnected to the nation-wide PCS networks, providing users with extensive roaming areas.

See Also

Cellular Voice Communications

Code-Division Multiple Access (CDMA)

Global System for Mobile (GSM) Telecommunications

PCS 1900

Personal Handyphone System (PHS)

Personal Digital Assistants (PDAs)

Personal Digital Assistants (PDAs) are hand-held computers that are equipped with an operating system, applications software, and communications capabilities for short-text messaging, e-mail, fax, news updates, and voice mail. They are intended for mobile users who require instant access to information, regardless of their location at any given time (refer to Figure P-7).

The Newton MessagePad, introduced by Apple Computer in 1993, was the first true PDA. Trumpeted as a major milestone of the information age, the MessagePad was soon joined by similar products from companies such as Hewlett-Packard, Motorola, Sharp, and Sony.

Figure P-7
Palm Computing offers one of the most popular lines of PDAs. This Palm V, shown with cradle charger and HotSync serial cable, weighs in at only 4 ounces.

These early hand-held devices were hampered by poor performance, excessive weight, and unstable software. Without a wireless communications infrastructure, there was no compelling advantage of owning a PDA. With the performance limitations largely corrected and the emergence of new wireless *Personal Communications Services* (PCS)—plus continuing advances in operating systems, connectivity options, and battery technology —PDAs are now well on the way toward fulfilling their potential.

Applications

Real-estate agents, medical professionals, field-service technicians, and delivery people are just a few of the people who use PDAs. Real-estate agents can use PDAs to conveniently browse through property listings at client locations. Health-care professionals can use PDAs to improve their abilities to access, collect, and record patient information. Numerous retailers and distributors can collect inventory data on the store and warehouse floor and later export that information into a spreadsheet on a PC. Insurance agents, auditors, and inspectors can use PDAs to record data in the field and then instantly transfer that data to PCs and databases at the home office.

PDA Components

Aside from the case, PDA components include a screen, keypad, or other type of input device, an operating system, memory, and a battery. Many PDAs can be outfitted with fax/modem cards and a docking station in order to facilitate direct connection to a PC or LAN for data transfers and file synchronization. Some PDAs, such as Palm Computing's Palm VII, have a wireless capability that enables information retrieval from the Internet. Of course, PDAs run numerous applications to help users stay organized and productive. Some PDAs have integral 56Kbps modems and serial ports that enable them to be attached via cable to other devices. Add-on components are even available, such as voice recorders and digital cameras that clip onto the PDA.

A unique PDA is Handspring's Visor, which uses the Palm OS operating system. What makes the Visor unique, is that it is expandable via an external expansion slot called Springboard. In addition to a backup storage and flash storage modules, the slot enables users to add software and hardware modules that completely change the function of the Visor. Springboard modules enable the Visor to become an MP3 player, pager, modem, GPS receiver, or video-game device. The Springboard modules are plug-and-play and require no software drivers or special adapters to install.

Display   The biggest limitation of PDAs is the size of their screens. Visibility is greatly improved through the use of non-glare screens and backlighting, which aid viewing and entering information in any lighting condition. In a dim indoor environment, backlighting is a virtual necessity —but it drains the battery faster. Some PDAs offer user-controllable backlighting, while others enable the user to set a timer that shuts off the screen automatically after the unit has been idle for a specified period of time. Both features greatly extend battery life.

Other PDAs, such as Compaq's Aero Series, feature displays with 16-level gray-scale. The 240-by-320 non-glare display provides more realistic images than competing two-color displays. Compaq's Aero 2100 line offers a reflective TFT color display with 64 colors—viewable even in direct sunlight. Hewlett-Packard's Jornada and Casio's Cassiopeia E100 offer full-color displays.

Keypad   Some PDAs have keypads, but they are too small to permit touch typing. The addition of a stylus speeds up and simplifies navigation, also making drag-and-drop task selection and text editing much easier. Of course, the instrument can be used for handwritten notes, but the recognition technology still needs improvement.

A promising scheme developed by Palm Computing employs a small 2-by-1-inch writing surface and a special character set for data input. The character set uses the same symbol for both upper-case and lower-case letters. Stylus gestures alert the recognizer that the user is about to enter numbers, letters, or symbols. Limiting the character set, as well as limiting the way to draw the characters, greatly improves handwriting recognition. Once the character set is learned, entry of up to 15 words per minute is possible.

Operating systems   A PDA's operating system provides the foundation upon which applications run. The operating system can offer handwriting recognition, for example, and include solutions for organizing and communicating information via fax or e-mail—as well as the capability to integrate with Windows and Mac OS-based computers in enterprise environments. The operating system might also include built-in support for a range of modems and third-party paging and cellular-communication solutions.

Because memory is limited in a PDA, usually not more than 16MB, the operating system and the applications that run on it must be compact. Microsoft's Windows CE, for example, is a greatly scaled-down version of the Windows operating environment that is commonly used on laptop and desktop machines.

Some operating systems come with useful utilities. There are utilities that set up direct connections between the PDA and desktop applications to transfer files between them. A synchronization utility ensures that the user is working from the latest version of a file. Some operating systems offer tools called intelligent agents, which automate routine tasks. An intelligent agent can be programmed to set up a connection to the Internet, for example, and check for e-mail. To activate this process, the user might only have to touch an icon on the PDA's screen with a pen.

Memory   Although PDAs come with a base of applications that are built into *Random Operating Memory* (ROM)—usually a file manager, word processor, and scheduler—users can install other applications, as well. New applications and data are stored in *Random-Access Memory* (RAM). At a minimum, PDAs come with only 2MB, while others offer up to 16MB or 24MB. When equipped with 2MB of memory, the PDA can store approximately 6,000 addresses, five years of appointments, 1,500 to-do items, and 1,500 memos. Some PDAs have a PC card (formerly PCMCIA) slot that can accommodate storage cards that are purchased separately.

Power   Many PDAs use ordinary AAA alkaline batteries. Manufacturers claim a battery life of 45 hours when users search for data five minutes out of every hour that the unit is turned on. Of course, using the backlight display will drain the batteries much faster. Using the backlight will reduce battery life by about 22 percent. Other power sources that are commonly used with PDAs include an AC adapter and a lithium-ion battery.

Fax/modems   Some PDAs come with external fax/modems to support basic message needs when hooked up to a telephone line. Others offer a PC card slot (formerly PCMCIA) that not only can accept fax/modems but can also accept storage cards, as well. With fax/modems, PDA users can receive a fax from their office, annotate it, and fax it back with comments written on it in electronic ink.

Docking station   A noteworthy peripheral is a docking unit that permits the PDA to be used as a companion to a desktop PC via a serial cable. While many PDAs include PC connectivity, others are designed specifically as a

PC peripheral. The user simply drops the PDA into the docking unit and presses a button to automatically synchronize desktop files with those that are held in the PDA.

A standard for wireless data exchange between portable devices and desktop computers and peripherals was finalized in 1993 by the *Infrared Data Association* (IrDA). Prior to that time, companies used proprietary protocols and wireless technologies for exchanging data between devices. Adoption of the first IrDA standard paved the way for interoperability between the wireless devices of different manufacturers, enabling users to send documents to printers or to and from wireless modems.

With an IR-enabled PDA, users can beam business cards, phone lists, memos, and add-on applications to other IR-enabled PDAs. IR-enabled PDAs can also use third-party beaming applications with IR-enabled phones, printers, and other devices.

Hybrid Devices

A potential competitor to the PDA is the hybrid cell phone, which can be used for fax, e-mail, and short-text messaging (as well as for voice conversations). The key advantage of the hybrid cell phone is that it eliminates the need for users to carry around separate devices (one for voice communication and another for data communication).

Perhaps the ultimate hybrid device is QualComm's pdQ smartphone, which combines CDMA digital wireless phone technology with the Palm platform. The pdQ smartphone features a flip-down phone keypad for easy call dialing and a large, 160-by-240 pixel LCD touch-screen for displaying information and data entry. Any number in the pdQ smartphone's address book can be dialed by simply tapping that name with the included stylus or by dialing from the pdQ smartphone's keypad. Users can input data three ways: through Palm Computing's Graffiti Power writing software, the on-screen keyboard, or through a personal computer's keyboard when connected to the pdQ smartphone with the included synchronization utility.

Summary

Improvements in technology and the availability of wireless communications services, including PCS, overcome many of the limitations of early

PDA products—making today's hand-held devices attractive to mobile professionals. In the process, PDAs might find acceptance beyond vertical markets and finally become popular among consumers, particularly those who are looking for an alternative to notebook computers.

See Also

> *Global System for Mobile* (GSM) Telecommunications
>
> *Personal Communications Services* (PCS)
>
> *Personal Handyphone System* (PHS)

Personal Handyphone System (PHS)

The *Personal Handyphone System* (PHS) offers high-quality, low-cost mobile telephone services via a fully digital system that operates in the 1.9GHz spectrum. Originally developed by NTT, the Japanese telecommunications giant, PHS is based on GSM technology. PHS made its debut in Japan in July 1995, where service was initially offered in metropolitan Tokyo and Sapporo. Although PHS was originally developed in Japan, it is now considered a pan-Asian standard.

Advantages over Cellular

PHS phones (Handyphones) operate at 1.9GHz, whereas cellular phones operate at 800MHz. To achieve superior wireline voice quality, PHS uses a portion of its capacity advantage to support a high-performance voice-encoding algorithm called *Adaptive Differential Pulse Code Modulation* (ADPCM). With this algorithm, PHS can support a much higher data throughput (32Kbps) than a cellular based system, enabling PHS to support fax and voice mail services and emerging multimedia applications such as high-speed Internet access and photo and video transmission.

PHS supports the hand-off of calls from one microcell to the next during roaming. PHS goes a step farther than cellular, however, by giving users the flexibility to make calls at home (similar to conventional cordless phones), at school, in the office, while riding the subway, or while roaming the streets. PHS also gives subscribers more security and complete privacy. Also, unlike cellular phones, the identity of PHS phones cannot be "cloned" (i.e., duplicated) for fraudulent use.

Another key advantage of PHS over cellular is cost. PHS can provide mobile communications more economically that cellular. Through its efficient microcell architecture and use of the public network, startup and expansion costs are minimized. As a result, total per-subscriber costs tend to be much lower than traditional cellular networks. Because PHS is a low-tier microcellular wireless network, it offers far greater capacity per dollar of infrastructure than existing cellular networks, which results in lower calling rates. In Japan, the cost of a three-minute call using a PHS handset is only about 10 yen (about 10 cents) more than making that same call on a public phone.

Handsets

Not only are PHS handsets extremely small and lightweight—almost half the size and weight of cellular handsets—but the battery life of PHS handsets is superior to that offered by cellular handsets. PHS phones output 10 milliwatts to 20 milliwatts, whereas most cellular phones output between 1 watt and 5 watts. Whereas the typical cellular handset has a battery life of three hours talk time, the typical PHS handset has a battery life of six hours talk time. Whereas the typical cellular handset has a battery life of 50 hours in standby mode, the typical PHS handset has a battery life of 200 hours in standby mode (more than a week). The low power operation of PHS handsets is achieved through strict, built-in power management and sleep functions in individual circuits.

Applications

A number of applications of PHS technology exist. In the area of mobile telecommunications, users can establish communications by accessing public cell stations that are installed throughout an area. PHS phones can also be used with a home base station as a residential cordless telephone at the *Public-Switched Telephone Network* (PSTN) tariff.

When used in the local loop, PHS provides the means to access the PSTN in areas where conventional local loops—consisting of copper wire, optical fiber, or coaxial cable—are impractical or not available. Wireless technologies are fast being adopted throughout the Asia/Pacific region, because elimination of the costly and slow process of laying copper wire and optical fiber means faster installation for network customers and faster returns on investment for service providers. Also, because companies can charge customers lower rates, there is also increased cash flow (because customers typically pay their bills sooner). In addition, relocation or expansion of the network is as simple as rewriting the database. A PHS *Wireless Local Loop* (WLL) can be easily expanded as its customer base increases or can be quickly and efficiently scaled back if the customer base decreases.

PHS can also be adopted as a digital cordless PBX for office use, providing readily expandable, seamless communications throughout a large office building or campus. Users carry PHS handsets with them and are no longer chained to their desks by their communications systems. As a digital system, PHS provides a level of voice quality not normally associated with a cordless telephone. In addition, the digital signal employed by PHS provides security for corporate communications, and the system's microcell architecture can be easily reconfigured to accommodate increases or decreases in the number of users. Another benefit of PHS is the ease and minimal expense with which the entire network can be dismantled and set up again in another facility (if, for example, a business decides to relocate its offices).

For personal use as a cordless telephone at home, PHS is an low-cost mobile solution that enables the customer to use a single handset at home and outside, with a digital signal that provides improved voice quality for a cordless phone and enough capacity for data and fax transmissions that are increasingly a part of users' home communications.

Network Architecture

The PHS radio interface has a four-channel *Time-Division Multiple Access* (TDMA) capability with *Time-Division Duplexing* (TDD), which provides one control channel and three traffic channels for each cell station.

The base station allocates channels dynamically and is not constrained by a frequency reuse scheme, thus deriving the maximum advantage of carrier-switched TDMA. In other words, PHS handsets that communicate to a base station might all be on different carrier frequencies.

The PHS system uses a microcell configuration that creates a radio zone with a 100-meter to 300-meter diameter. The base stations themselves are spaced at a maximum of 500 meters apart. In urban areas, the microcell configuration is capable of supporting several million subscribers. This configuration also makes possible smaller and lighter handsets and the more efficient reuse of radio spectrum to conserve frequency bands. In turn, this functionality permits low transmitter power consumption—and as a result, much longer handset talk times and standby times than are possible with cellular handsets.

A drawback of the lower operating power level is the smaller radius that a PHS base station can cover: only 100 meters to 300 meters, versus at least 1500 meters for cellular base stations. The extra power of cellular systems improves penetration of the signals into buildings, whereas PHS might require an extra base station inside some buildings. Another drawback of PHS is that the quality of reception can diminish significantly when mobile users are traveling at a rate greater than 15 miles per hour.

Because PHS uses the public network, rather than dedicated facilities between microcells, the only service startup requirements are handsets, cell stations, PHS server and a database of services to support PHS network operation. With no separate transmission network needed for connecting cell stations and for call routing, carriers can introduce PHS service with little initial capital investment.

Service Features

PHS is a feature-rich service, giving Handyphone users access to a variety of call-handling features, including the following:

- *Call Forwarding*   To a fixed line, to another PHS phone, or to a voice mail mailbox
- *Call Waiting*   Alerts the subscriber to an incoming call
- *Call Hold*   Enables the subscriber to alternate between two calls
- *Call Barring*   Restricts any incoming local or international call
- *Calling Line Identification* (CLI)   Displays the number of the incoming call, informing the subscriber of the caller's identity
- *Voice Mail*   The subscriber receives recorded messages, even when the phone is busy or is turned off
- *Text Messaging*   Enables the subscriber to send and receive text messages through the PHS phone
- *International Roaming*   Enables subscribers to use their PHS phones in another country and to be billed by the service provider in their home country

Depending on the implementation progress of the service provider, the following value-added data services might also be available to Handyphone users:

- *Virtual Fax*   Enables subscribers to retrieve fax messages anywhere, to have fax messages sent to a Handyphone, or to have the messages redirected to any fax machine

- *Fax*   By attaching the Handyphone to a laptop or desktop computer, subscribers can send and receive faxes anywhere.

- *E-mail/Internet access*   Enables subscribers to retrieve e-mail from the Internet through the Handyphone

- *Conference Calls*   Enables subscribers to talk to as many as four other parties at the same time

- *News, sports scores, and stock quotes*   Enables subscribers to obtain a variety of information on a real-time basis

Many other types of services can be implemented over PHS. The world's first video phone cellular service already exists in Japan. Kyocera Corporation offers a cellular phone that has the capability to transmit a caller's image and voice simultaneously. Two color images are transmitted per second through a camera that is mounted on the top of the handset. The recipient can view the caller via a two-inch active matrix LCD. Because the transmission technology sends data at only 32Kbps, jerky video images result. By comparison, another wireless standard called wide-band CDMA will have the capacity to send data at 380Kbps, or more than 10 frames per second—creating smooth, seamless video. This service will become available in Japan in 2001.

Summary

While 32Kbps transmission is now available in Japan, research is now underway to achieve a transmission rate of 64Kbps through the use of two channels. With this much capacity, PHS can be extended to a variety of other services in the future, including better-quality video.

In combination with a small, lightweight portable data terminal, PHS might also be used to realize Oracle Corporation's concept of network computing, whereby users would access application software stored on the Internet for use when needed. With the limited memory and disk storage capacity of such network computers, the applications and associated programs would stay on the Internet, preventing the PHS devices from becoming overwhelmed.

See Also

> Cellular Communications
>
> *Global System for Mobile* (GSM) Telecommunications
> *Personal Access Communications Systems* (PACS)
>
> *Personal Communications Services* (PCS)
>
> *Personal Digital Assistants* (PDAs)

Point-to-Point Protocol (PPP)

Point-to-Point Protocol (PPP) provides the means to transfer data across any full-duplex (i.e., two-way) circuit, including dialup links to the Internet via modems, ISDN, and high-speed SONET over fiber-optic lines. PPP is an enhanced version of the older *Serial-Line Internet Protocol* (SLIP). While SLIP is typically used in an IP-only environment, PPP is more versatile in that it can be used in multi-protocol environments. In addition, PPP enables traffic for several protocols to be multiplexed across the link, including IP, IPX, DECnet, ISO and others. PPP also carries bridged data over complex internetworks.

Features

PPP supports authentication, link configuration, and link monitoring capabilities via several subprotocols, including the following:

- Link Control Protocol (*LCP*)   Negotiates details and desired options for establishing and testing the overall serial link
- *Authentication protocols*   The *Password Authentication Protocol* (PAP) and the *Challenge Handshake Authentication Protocol* (CHAP). PAP uses a two-way handshake for the peer to establish its identity at the time of link establishment. CHAP periodically verifies the identity of the peer using a three-way handshake, which is employed throughout the life of the connection. Of the two authentication protocols, CHAP is the more robust.

- *Network control protocols* Used to dynamically configure different network layer protocols, such as IP and IPX. For each type of network-layer protocol, there is a network control protocol that is used to initialize, configure, and terminate its use.

- *Link-quality monitoring* Provides a standardized way of delivering link quality reports on the quality/accuracy of a serial link

During TCP sessions, PPP's compression feature can reduce the typical 40-byte TCP header to only 3 bytes to 5 bytes, providing significant savings in transmission time. This goal is accomplished by sending only the changes in a PPP frame's header values. Because most header information does not change from one frame to the next, the savings can be quite substantial.

Multi-Link PPP (MPPP)

Multi-Link PPP (MPPP) combines multiple B channels of an ISDN link into a single, higher-speed channel. Although extra channels can be added to an established ISDN connection, MPPP does not offer dynamic control. This problem is remedied by the *Bandwidth Allocation Control Protocol* (BACP), which works in conjunction with MPPP. With BACP, ISDN channels can be added as needed and dropped when they are no longer required to support the application.

BACP enables bandwidth to change on demand through a standard set of rules, while minimizing the need for the end user to be involved in complex connection configuration issues. BACP can even interact with the resource *Reservation Protocol* (RSVP) to provide enhanced functionality. For example, if a bandwidth reservation is queued for lack of bandwidth somewhere on the network, this action could trigger the creation of additional channels to support the application. If the application is a video conference, for example, when the router senses that network load has gone down because participants are dropping out of the session, it starts terminating B channels. This process minimizes the usage charges that are associated with ISDN.

Summary

While SLIP is typically used to connect computers to an IP network via a dialup link, it has severe limitations in that it cannot support any other protocol and does not perform error checking. PPP is more versatile in terms of the protocols it can handle and is more functional, particularly with regard to authentication, link configuration, and link monitoring. PPP has displaced SLIP in recent years. Other protocols, such as MPPP, BACP, and RSVP, are used with PPP to support more sophisticated applications (such as video conferencing).

See Also

Integrated Services Digital Network (ISDN)

Pre-Paid Phone Cards

Pre-paid calling cards enable users to conveniently pay for long-distance calls without having to carry coins for use at a pay phone. When using the card, the user is typically greeted with a custom message that is created by the card distributor, which reinforces the message printed on the card. These cards have become popular as marketing tools, enabling companies to provide their customers a valuable and useful premium for the following:

- Promotional giveaways
- Customer reward gifts
- Employee rewards or gifts
- Inclusion in tour/travel packages and a value-added service that the purchasing company can provide to its end users for business and/or leisure travel

The cards can be embossed with the organization's logo to promote its image and to drive brand awareness, or as an advertising vehicle to enhance sales or acquire new customers. Cards can be used continuously to support promotions and marketing efforts simply by changing the artwork on the card itself and by changing the customized promotional announcements in the carrier's pre-paid card network, in order to deliver a series of individual advertisements to each card user.

Card Creation

Virtually all carriers issue calling cards and sell customized versions in bulk to their corporate customers. The carrier and the purchasing company work together to design a promotional card by applying the appropriate packaging design and production coupled with properly scripted network announcements. The carrier will produce the product using either recyclable UV-coated paper stock or recyclable plastic stock. The front of the card might have the logos of the carrier and the purchasing company. The card provides the user with easy-to-follow calling instructions, a unique account number, and customer-service information. A brief message pertaining to the purchasing company—usually fewer than 25 words—might also be included on the card back, but if additional promotional or instruction information is required, card carriers, inserts, or other collateral can be used.

Instructions for Use

From a touch-tone phone, the user enters the 800 number printed on the back of the card. Then, the user enters the carrier's pre-paid card PIN number, which is also printed on the back of the card. Once the PIN number is verified, the user will hear the balance remaining on the card, and the carrier's dial tone.

To place a call within the United States, to Canada, or to the Caribbean, the user presses 1, then dials the area code and the local telephone number. (For calls to any other country, the user enters the country code, the city code and the local telephone number.) Upon completion of the call, the user can wait to hear the balance, stay on the line to place another call, or hang up. When there is one minute remaining on the card, an announcement notifies the user. If the user wants to continue the call, an additional pre-paid card number can be entered.

Summary

The pre-paid calling-card industry has exploded into a billion-dollar-a-year business, but it also has attracted its share of fly-by-night operators who take consumers' money but do not deliver the phone services. Consumers might not be able to contact the carrier when problems arise and advance payments are not adequately protected by current performance bond guidelines. This situation has many state regulators proposing protections for pre-paid calling-card users.

In some states pre-paid calling cards must now contain the following printed information: expiration date, company's registered name and toll-free number and authorization code. At the store where the phone cards are sold, additional consumer information must be displayed, including the maximum charge per minute; approved surcharges; expiration and recharge policies; and the toll-free phone number of the state regulatory agency, in case the consumer is unable to resolve a complaint with the company.

Some states require companies offering pre-paid phone cards to give consumers an idea of how much time is left on the card before each call, to indicate when it is about to be depleted and to honor a customer's request for refunds on a card's unused time. Others require pre-paid phone-card companies to include a 24-hour toll-free customer-service number, provide a call-detail report at a customer's request, and charge customers for conversational time only. Busy signals or unanswered calls would not be counted as completed calls, and therefore, they would not be deducted from the pre-paid telephone card.

In some states, a company must receive regulatory approval to sell pre-paid phone cards. Such companies must meet stringent requirements to prove they can meet their financial obligations: businesses must maintain a corporate debt rating; a performance bond; or an escrow account large enough to make refunds to customers on unused portions of the cards if the company goes out of business. In addition, some states have issued rules that include fines and other penalties for unlicensed businesses caught selling pre-paid phone cards.

See Also

Calling Cards

Price Caps

Price caps are applied to the ILECs—the former RBOCs and GTE—in order to prevent them from using pricing flexibility to deter efficient entry of competitors or engage in exclusionary pricing practices. Price caps are also intended to prevent the ILECs from increasing rates to unreasonable levels for customers who lack competitive alternatives.

The price-cap model has been in effect since 1990, when it replaced traditional rate-of-return regulation. In 1990, the Commission replaced rate-of-return regulation for the BOCs and GTE with an incentives-based system of regulation that encourages companies to do the following:

■ Improve their efficiency by developing profit-making incentives to reduce costs

■ Invest efficiently in new plant and facilities

■ Develop and deploy innovative service offerings

The price-cap plan was designed to replicate some of the efficiency incentives that are found in fully competitive markets and to act as a transitional regulatory scheme until actual competition made price-cap regulation unnecessary.

Original Plan

Under the original price-cap plan, interstate access services were organized into four categories or baskets: common line, traffic-sensitive, special access, and interexchange transport. Later, the FCC combined special access and interexchange transport services into a single, newly created trunking basket.

Each basket was subject to a *Price-Cap Index* (PCI), which limits the total charges an ILEC can impose for interstate access services in that basket. The PCI is adjusted annually by a measure of inflation, minus a productivity factor (also known as an X-Factor). A separate adjustment is made to the PCI for exogenous cost changes, which are changes outside the carrier's control that are not otherwise reflected in the price-cap formula.

Within the traffic-sensitive and trunking baskets, services are grouped into service categories and subcategories. Rate revisions for these services are limited by upper and, in the original price-cap plan, lower pricing bands established for that particular service. Originally, the pricing band limits for most of the service categories and subcategories were set at 5 percent above and below the *Service Band Index* (SBI). In 1995, the FCC increased the lower pricing bands to 10 percent for those service categories in the trunking and traffic-sensitive baskets and 15 percent for those services subject to density zone pricing.

These pricing bands gave ILECs the capability to raise and lower rates for elements or services, as long as the *Actual Price Index* (API) for that basket did not exceed the PCI and the prices for each category of services within the basket were within the established pricing bands. Together, the PCI and pricing bands restrict an ILEC's capability to offset price reductions for services that are subject to competition with price increases for services that are not subject to competition.

Since 1997, the FCC has committed itself to a market-based approach to implementing price caps. Under this approach, the ILECs receive increased pricing flexibility in the provision of interstate access services, but only as competition for those services continues to develop.

Revised Plan

Although the original price-cap regime gave ILECs some pricing flexibility and considerable incentives to operate efficiently, significant regulatory constraints remained. As the market becomes more competitive, such constraints become counter-productive. Accordingly, in August 1999, the FCC revised its rules governing the provision of interexchange access services by ILECs that are subject to price-cap regulation, recognizing that the variety of access services available on a competitive basis has increased significantly in recent years.

In response to changing market conditions, the FCC has granted the ILECs immediate flexibility to deaverage services in the trunking basket and to introduce new services on a streamlined basis. The FCC has also removed certain interstate interexchange services from price-cap regulation upon implementation of intraLATA and interLATA toll dialing parity. In addition, the FCC established a framework for granting ILECs further pricing flexibility upon satisfaction of certain competitive showings.

Immediate Regulatory Relief

The original rate structure for interstate switched transport services required ILECs to charge averaged rates throughout a study area. The FCC subsequently found that this requirement forced the ILECs to price above cost in high-traffic, lower-cost areas where competition is more likely to develop. To overcome this situation, the FCC created a density-zone pricing plan that allowed some degree of deaveraging of rates for switched transport services. The commission concluded that relaxing the pricing rules in this manner would enable the ILECs to respond to increased competition in the interstate switched transport market.

Although the density-zone pricing plan afforded some pricing flexibility to price the ILECs, it contained several constraints, such as the increased scrutiny applicable to plans with more than three zones. The FCC now concludes that market forces, as opposed to regulation, are more likely to compel IECs to establish efficient prices. Accordingly, for purposes of deaveraging rates for services in the trunking basket, the FCC has eliminated the limitations that are inherent in the density-zone pricing plan and enables the ILECs to define the scope and number of zones within a study area. The only caveat is that each zone, except the highest-cost zone, accounts for at least 15 percent of the trunking basket revenues in the study area—and that annual price increases within a zone do not exceed 15 percent. In addition, the FCC eliminated the requirement that ILECs file zone-pricing plans prior to filing their tariffs.

Streamlined Service Introduction

The FCC also permits the ILECs to introduce new services on a stream-lined basis, without prior approval. The commission modified its rules to eliminate the public interest showing and the new services test, except in the case of loop-based, new services. These modifications will eliminate the delays that previously existed for the introduction of new services as well as encourage efficient investment and innovation.

Removal of Price Caps

Certain interstate interexchange services that are provided by ILECs are found in the interexchange basket, including interstate intraLATA services and certain interstate interLATA services called corridor services. The FCC now enables the ILECs to remove from the interexchange basket (and, consequently, from price-cap regulation) their interstate intraLATA toll services and corridor services, provided that the price cap LEC has implemented intraLATA and interLATA toll dialing parity in all of the states in which it provides local exchange service.

The presence of competitive alternatives for these services, coupled with implementation of dialing parity, should prevent the ILECs from exploiting any market power they can possess with respect to these services and thus warrants removal of these services from price-cap regulation.

Framework for Further Regulatory Relief

The FCC has adopted a framework for granting further regulatory relief upon satisfaction of certain competitive showings. Relief generally will be granted in two phases and on a *Metropolitan Statistical Area* (MSA) basis. To obtain Phase I relief, the ILECs must demonstrate that competitors have made irreversible investments in the facilities needed to provide the services at issue. For instance, for dedicated transport and special access services, the ILECs must demonstrate that unaffiliated competitors have collocated in at least 15 percent of the ILEC's wire centers within an MSA, or collocated in wire centers accounting for 30 percent of the ILEC's revenues from these services within an MSA.

Higher thresholds apply, however, for channel terminations between an ILEC end office and an end user customer. In that case, the ILEC must demonstrate that unaffiliated competitors have collocated in 50 percent of the price cap ILEC's wire centers within an MSA or collocated in wire centers accounting for 65 percent of the price cap LEC's revenues from this service within an MSA. For traffic-sensitive, common line, and the traffic-sensitive components of tandem-switched transport services, an ILEC must show that competitors offer service over their own facilities to 15 percent of the ILEC's customer locations within an MSA.

Phase I relief permits ILECs to offer, on one day's notice, volume and term discounts and contract tariffs for these services, as long as the services that are provided pursuant to contract are removed from price caps. To protect those customers that might lack competitive alternatives, however, ILECs receiving Phase I flexibility must maintain their generally available, price-cap-constrained tariffed rates for these services.

To obtain Phase II relief, the ILECs must demonstrate that competitors have established a significant market presence—that competition for a particular service within the MSA is sufficient to preclude the incumbent from exploiting any individual market power over a sustained period —for provision of the services at issue. Phase II relief for dedicated transport and special access services is warranted when an ILEC demonstrates that unaffiliated competitors have collocated in at least 50 percent of the ILEC's wire centers within an MSA or collocated in wire centers, accounting for 65 percent of the ILEC's revenues from these services within an MSA.

Again, a higher threshold applies to channel terminations between an ILEC end office and an end-user customer. In that case, a price cap LEC must show that unaffiliated competitors have collocated in 65 percent of the ILEC's wire centers within an MSA or collocated in wire centers accounting for 85 percent of the ILEC's revenues from this service within an MSA. Phase II relief permits ILECs to file tariffs for these services on one day's notice.

Summary

The goal of the FCC is to continue to foster competition and enable market forces to operate where they have gained a foothold. Accordingly, in 1999, the FCC revised its price-cap rules that govern the provision of interstate access services by ILECs subject to price-cap regulation. The revisions were deemed necessary to advance the pro-competitive, deregulatory national policies embodied in the Telecommunications Act of 1996. With these revisions, the FCC continues the regulatory reform process it began in 1997 in order to accelerate the development of competition in all telecommunications markets and to ensure that its own regulations do not unduly interfere with the operation of these markets as competition develops.

See Also

Access Charges

Federal Communications Commission (FCC)

Regulatory Process

Telecommunications Act of 1996

Unbundled Access

Presubscribed Interexchange Carrier Charge (PICC)

The *Presubscribed Interexchange Carrier Charge* (PICC) is a fee that long-distance companies pay to local telephone companies to help them recover the costs of providing the local loop. Local loop is a term that refers to the outside telephone wires, underground conduit, telephone poles, and other facilities that link each telephone customer to the telephone network.

The monthly service fee that consumers pay for local telephone service is not enough to cover all of the costs of the local loop. Historically, the local telephone companies have recovered the shortfall through per-minute access charges to long-distance companies. Now, however, part of these costs are recovered through flat-rated charges to long-distance companies, who use the local networks to complete their long-distance calls. Because the costs of the local loop do not depend on usage, this flat-rated charge better reflects the local telephone company's costs of providing service.

A long-distance company pays this charge for each residential and business telephone line that is presubscribed to that long-distance company. If a consumer or business has not selected a long-distance company for its telephone lines, the local telephone company might still bill the consumer or business for the PICC.

As of July 1, 1999, the PICC went up, but the per minute charge that long-distance companies pay to local companies for each call made by their customers was reduced by an even greater amount. Consumers should therefore expect to continue to see reductions in the per-minute rates they pay for long-distance calls.

The maximum PICC paid by the long-distance companies for primary residential lines and single-line business lines is $1.04 per line per month. For non-primary residential lines, the maximum PICC paid by the long-distance companies is $2.53 per line per month. (Local telephone companies treat a line as non-primary when it serves the same address as the primary line, even if the bill is in a different name at the same address.)

These amounts represent maximum PICC levels. The actual charge that is paid by the long-distance companies can vary, based on the actual cost of providing local phone service in each area, and can be less than this maximum amount. The maximum PICC paid by the long-distance companies for each multi-line business line is $4.31. Similar to the residential PICC, this amount is a maximum; the actual charge might be less than this maximum amount.

Each year, the maximum multi-line business PICC will increase by $1.50, as adjusted by inflation. As various phases of the FCC's plan are implemented, however, it is estimated that the average PICC for multi-line business lines will dip below $1 in 2001 (and, in most places, will eventually be zero).

The FCC does not require long-distance companies to put the PICC—or any other charges or surcharges—on customer telephone bills. Because the long-distance market is competitive, the FCC does not directly regulate long-distance company charges for service. As a result of this flexibility, long-distance companies are taking different approaches to how they are changing charges to their customers to reflect the PICC that they pay. Some long-distance companies might not charge any separate fees related to the PICC. Others have added charges to their customers' bills—such as a national access fee—in order to recover the charges that they pay to local telephone companies.

Summary

The PICC payments long-distance companies pay to the local telephone companies help ensure that all Americans have affordable access to telephone services. The PICC is largely offset by reductions in the amount of per-minute charges the companies pay for each call made by their customers. Because there is competition for long-distance service, the FCC does not regulate how long-distance companies compute their charges or the amount of those charges. The FCC did not tell the long-distance companies how to adjust their customers' rates in response to changes in access charges, including the companies' new PICC payments. The long-distance companies have decided what to do, and some have implemented charges significantly different from other companies.

See Also

Access Charges

Dialing Parity

Local Loop

Local Number Portability

Universal Service

Private-Branch Exchanges (PBXs)

Over the years, the *Private-Branch Exchange* (PBX) has emerged to become the undisputed cornerstone of today's corporate voice network. In the simplest terms, the PBX is a private telephone switch that, through control signaling, performs three basic functions:

- In response to a call request, establishes end-to-end connectivity among its subscribers (on the network) and from its own subscribers to remote subscribers (off the network) through intermediate nodes, which might consist of other PBXs or central office switches on the *Public-Switched Telephone Network* (PSTN). The connected path is dedicated to the user for the duration of the call.

- Supervises the circuit to detect call request, answer, signaling, busy, and disconnect (hang-up)

- Tears down the path upon call termination (disconnect) so that another user can access the resources that are available over that circuit

These functions closely parallel those of the central office switch. In fact, the PBX evolved from the operator-controlled switchboards that were used on the public telephone network. The first of these simple devices was installed in 1878 by the Bell Telephone Company to serve 21 subscribers in New Haven, Connecticut. The operator had full responsibility for answering call requests, setting up the appropriate connections, supervising for answer and disconnect, and tearing down the path upon call completion. Interconnectivity among subscribers was accomplished via cable connections at a patch panel.

Today's PBXs are much more complicated of course, but they provide the same basic functionality as the first generation of circuit switching devices. The difference is that the process of receiving call requests, setting up the appropriate connections, and tearing down the paths upon call completion is now entirely automated. And because the intelligence necessary to do all this resides on the user's premises, the PBX enables organizations to exercise more control over internal operations and incorporate communications planning into the long-term business strategy.

In a little more than 100 years, PBXs have evolved from simple patch panels to sophisticated systems that are capable of integrating voice and data. The six generations characterizing the development of PBXs might be summarized as follows:

- First-generation PBXs consisted of the operator-controlled patch panels.

- Second-generation PBXs evolved from the electromechanical central office switches: step-by-step (Strowger) and cross-bar. Automatic dialing and space division switching are the capabilities that differentiated second-generation from first-generation switches.

- Third-generation PBXs include the attributes of second-generation PBXs. Instead of electromechanical control, third-generation PBXs use electronic components under stored program control, making possible the addition of many new features as well as distributed architectures.

- Fourth-generation devices are computer-based to permit the integration of capabilities that previously had to be added on with external components, such as automatic call distribution and voice mail. Fourth-generation PBXs also use time-division switching, which permits the integration of voice and data over T1 trunks and ISDN services.

- Fifth-generation PBXs are more data-oriented, adding LAN and Internet connectivity plus support for ATM and management via SNMP.

Basic Features

Many PBX features are under direct user control and can be implemented right from the telephone keypad, including the following:

- *Add-on conference*   Enables the user to establish another connection while having a call already in progress

- *Call forwarding*   Enables a station to forward incoming calls to another station, including forwarding calls when the station is busy or unattended (or as needed)

- *Call hold*   Enables the user to put the first party on hold so that an incoming call can be answered

- *Call waiting*   Enables the user to know that an incoming call is waiting. While a call is in progress, the user will hear a special tone that indicates another call has come through.

- *Camp-on*   Enables the user to wait for a busy line to become idle, at which time a ring signal notifies both parties that the connection has been made

- *Last number redial*   Enables users to press one or two buttons on the keypad to activate dialing of the previously dialed number

- *Message waiting*   Enables the user to signal an unattended station that a call has been placed. Upon returning to the station, an indicator tells the person that a message is waiting.

- *Speed dialing*   Enables the user to implement calls with an abbreviated number. This feature also enables users to enter a specified number of speed dial numbers into the main database. These numbers can be private or shared among all users. Entering and storing additional speed dial numbers is accomplished via the telephone keypad.

A number of PBX capabilities are also available that operate in the background and are transparent to the user, including the following:

- *Automated attendant*   Enables the system to answer incoming calls and to prompt the caller to dial an extension or leave a voice message without going through the operator

- Automatic Call Distribution (*ACD*)   Enables the sharing of incoming calls among a number of stations, so that the calls are served in order of their arrival. This capability is usually optional, but it might be integral to the PBX or purchased separately as a stand-alone device.

- *Automatic least-cost routing* Ensures that calls are completed over the most economic route available. This feature can be programmed so that mail room staff always get the cheapest carrier, while executives get to choose whichever carrier they want.

- Call-Detail Recording (*CDR*) Enables the PBX to record information about selected types of calls for management and cost control

- *Call pickup* Enables incoming calls made to an unattended station to be picked up by any other station in the same trunk group

- *Class-of-service restrictions* Controls access to certain services or shared resources. Access to long-distance services, for example, can be restricted by area code or exchange. Access to the modem pool for transmission over analog lines can be similarly controlled.

- *Database redundancy* Enables the instructions stored on one circuit card to be dumped to another card as a protection against loss

- Direct Inward Dialing (*DID*) Enables incoming calls to bypass the attendant and to ring directly on a specific station

- Direct Outward Dialing (*DOD*) Enables outgoing calls to bypass the attendant for completion anywhere over the PSTN

- *Hunting* A capability that routes calls automatically to an alternate station when the called station is busy

- *Music on hold* Indicates to callers that the connection is active while the call waits in queue for the next-available station operator

- *Power-fail transfer* Permits the continuance of communication paths to the external network during a power failure. This capability works in conjunction with an *Uninterruptible Power Supply* (UPS), which kicks in within a few milliseconds after detecting a power outage.

- *System redundancy* Enables sharing of the switching load so that in the event of failure, another processor can take over all system functions

PBX Components

Aside from the line and trunk interfaces, there are three major elements that comprise the typical PBX: processor, memory, and switching matrix. Together, these elements provide all of the intelligence necessary to place calls anywhere on the public or private network, without the need for human intervention.

Processor   The processor is responsible for controlling the various operations of the PBX, including monitoring all lines and trunks that provide connectivity, establishing line-to-line and line-to-trunk paths through the switching matrix, and tearing down connections upon call completion. The processor even controls optional capabilities such as voice mail and the recording of traffic statistics and billing information. Because the processor is programmable, features and services can be easily added or changed at a management terminal.

Many PBXs can be optioned for two or more processors, which adds to the reliability of the system in that if one fails, another takes over. Programs and configuration information are automatically downloaded from the main processor to the standby processor in order to ensure uninterrupted service.

Memory   The processor uses the memory element to implement the sophisticated functions of the PBX. Those functions are defined and implemented in software, instead of hardware. There are two types of memory: non-volatile and volatile. The former is fixed, whereas the latter can be changed as needed. Non-volatile memory contains the operating instructions and stores system-configuration information. Volatile memory, or *Random-Access Memory* (RAM), is used for temporary storage of frequently used programs or for workspace.

In the event that a system failure destroys the contents of non-volatile memory, a reserve program, which is also stored in non-volatile memory, is put into operation automatically. (Some systems, however, still require a spare program to be loaded manually via disk or tape cartridge after a catastrophic failure.) As the term implies, non-volatile memory can withstand a power outage, retaining its contents in the process. Information that is stored in non-volatile *Electronically Erasable Read-Only Memory* (EEROM) is automatically dumped into RAM for access by the redundant processor. Although EEROM is non-volatile, it is changeable by a programmer.

Switch matrix   The switch matrix, under control of the processor, interconnects lines and trunks.[1] This process can be accomplished through space-division switching or time division switching. Space-division switching originated in the analog environment. As its name implies, a space-

---

[1]Although the terms *line* and *trunk* are often used interchangeably, an important difference exists. A line refers to the link between each station (telephone set) and the switch, whereas a trunk refers to the link between switches. The term *tie line*, then, really refers to a tie trunk.

division switch sets up signal paths that are physically separate from one another, or divided by space. Each connection establishes a physical point-to-point circuit through the switch that is dedicated entirely to the transfer of signals between the two endpoints. The basic building block of the space-division switch is a metallic cross-point (relay contact) or semiconductor gate that can be enabled and disabled by the processor or control unit. Thus, physical interconnection is achieved between any two lines by enabling the appropriate cross-point.

Although the single-stage space-division switch is virtually non-blocking,[2] it has several limitations, the most serious of which is the number of cross-points that are required as the number of input lines and output lines grows. This situation is not only costly but also results in ever greater inefficiencies in the utilization of available cross-points. These limitations are mitigated through the use of multi-stage cross-point matrices. Although requiring a more complex control scheme, the use of multiple stages reduces the number of cross-points while increasing overall reliability.

IP-PBXs

A relatively new development in PBXs is the IP-based PBX. Vendors such as Selsius, Inc. (acquired by Cisco Systems in 1998) offer scaleable IP-based PBXs that can transport intra-office voice over an Ethernet LAN and wide-area voice over the PSTN or a managed IP network. Full-featured digital phone sets link directly to the Ethernet LAN via a 10BaseT interface, without requiring direct connection to a desktop computer. Phone features can be configured using a Web browser. Existing analog devices, such as phones and fax machines, can be linked to the LAN via a gateway. In addition to IP networks, calls can be placed or received by using T1, PRI ISDN, or traditional analog telephone lines.

All of the desktop devices have access to the calling features that are offered through the IP PBX management software running on a LAN server. The call-management software enables hundreds of client devices on the network, such as phones and computers, to perform functions such as call hold, call transfer, call forward, call park, and calling party ID. Other sophisticated PBX functions, such as multiple lines per phone or

---

[2]Blocking refers to the inaccessibility of the switch due to the unavailability of cross-points, which establish the connections between various endpoints. In theory, all switches can experience blocking, no matter how well they are designed. As a practical matter, however, some switch designs are less prone to blocking than others.

multiple phones per line, also are implemented by the call-management software.

The Selsius IP PBX is rated to deliver in excess of 75,000 busy-hour call completions. This equipment provides a truly linear PBX that can keep pace with corporate growth through the deployment of additional phones and gateway ports while accommodating substantial increases in traffic.

Other vendors, such as Lucent Technologies and Nortel, offer IP interfaces to their conventional PBX systems. Lucent, for example, offers an IP trunk interface for certain models of its DEFINITY product line. The IP interface supports 24 ports and enables businesses to integrate least cost routing and class of service features, giving network managers the capability to add the Internet and corporate intranet as alternative routes for voice and fax services. The capability to add the IP trunk interface directly into the DEFINITY also reduces the cost of obtaining Internet telephony capabilities by eliminating the need to buy a separate IP-PSTN gateway. Nortel also offers a 24-port IP interface for its Meridian communication system, enabling the routing of real-time voice and fax calls over IP data networks, rather than over the PSTN.

PBXs can also be interconnected over IP networks. PSINet offers a service called Voice iPEnterprise, a fully managed service that delivers internal voice traffic over the same connection as corporate Internet data (Figure P-8). This service leverages existing investments in PBXs and uses a T1 dedicated connection to the PSINet network, enabling businesses to carry voice services between their offices.

PSINet's managed service is equal in quality to traditional tie-line or dedicated voice services between PBXs. The service is aimed at companies with branch offices, especially those that must scale rapidly due to internal growth or operate in widely dispersed geographic locations. The service could save corporate customers 50 percent over traditional carrier services. IP desktop fax, conference calling, and unified messaging capabilities can be added to the service.

Summary

Despite its humble beginnings more than 100 years ago, PBX systems still have plenty of room for innovation. Today's PBXs emphasize integration into enterprise networking infrastructures, thereby addressing the applications that will be most in demand for the rest of the 1990s and beyond:

■■■ ■■■ ■■■ ■■

Figure P-8

PSINet's Voice iPEnterprise solution for interconnecting corporate PBXs over the carrier's managed IP backbone.

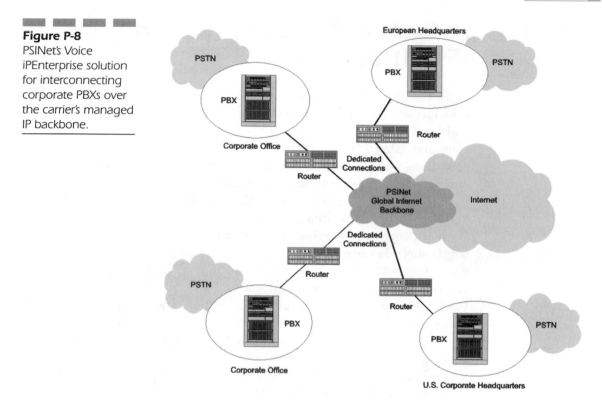

LAN interconnectivity, video conferencing, and multimedia applications. PBX add-ons are available that support in-building mobile communications using wireless technology. PBXs can also be connected to LANs and IP networks, enabling users to send e-mail and run real-time and multimedia applications (such as voice calls and video conferences) more economically.

See Also

> *Automatic Call Distributors* (ACDs)
>
> Central Office Switch
>
> Centrex
>
> *Communications Services Management* (CSM)
>
> Key Systems
>
> *Telecommunications Management Systems* (TMS)

Private Land Mobile Radio Services (PLMRS)

Since the 1920s, the *Private Land Mobile Radio Services* (PLMRS) have been meeting the internal communication needs of private companies, state and local governments, and other organizations. These services provide voice and data communications that enable users to control their business operations and production processes, to protect worker and public safety, and to respond quickly in times of natural disaster or other emergencies.

In 1934, shortly after its establishment, the *Federal Communications Commission* (FCC) identified four private-land mobile services: Emergency Service, Geophysical Service; Mobile Press Service; and Temporary Service, which applied to frequencies used by the motion picture industry. Over the years, the FCC refined these categories. Today, PLMRS consists of 21 services spread among six service categories: Public Safety, Special Emergency, Industrial, Land Transportation, Radiolocation, and Transportation Infrastructure. PLMRS are used by organizations that are engaged in a wide variety of activities. Police, fire, ambulance, and emergency-relief organizations such as the American Red Cross use private wireless systems to dispatch help when emergency calls come in or disaster strikes. Utility companies, railroad and other transportation providers, and other infrastructure-related companies use their systems to provide vital day-to-day control of their systems (including monitoring and control and routine maintenance and repair) and also to respond to emergencies and disasters—often working with public safety agencies. A wide variety of businesses, including package delivery companies, plumbers, airlines, taxis, manufacturers, and even the *American Automobile Association* (AAA) rely on private wireless systems to monitor, control, and coordinate their production processes, personnel, and vehicles.

Although commercial services can serve some of the needs of these organizations, private users generally believe that their own systems provide them with capabilities, features, and efficiencies that commercial services cannot. Some of the requirements and features that PLMRS users believe make their systems unique include the following:

- Immediate access to a radio channel (no dialing required)
- Coverage in areas where commercial systems cannot provide service
- Peak usage patterns that could overwhelm commercial systems
- High reliability

- Priority access, especially in emergencies
- Specialized equipment required by the job

The FCC has continually allocated more frequencies for PLMRS in several different spectrum bands to overcome congestion and keep pace with advances in technology that make possible the use of higher frequencies. Currently, more than one million licensed stations are authorized to operate more than 12 million transmitters in support of PLMRS. This number represents a total investment of more than $25 billion.

Public Safety Radio Services

The first uses of mobile radio date to the 1920s, when police departments, marine fireboats, and industries that worked in remote locations were the primary users. When the FCC was established in 1934, these uses were classified into four emergency services: marine fire, municipal police, state police, and special emergency. Since then, they have evolved into six different public-safety radio services: Local Government, Police, Fire, Highway Maintenance, Forestry-Conservation, and Emergency Medical.

Special Emergency Radio Service

The Special Emergency Radio Service was once categorized under the public-safety services but now is governed separately, although it serves similar needs. Special Emergency originally applied to telegraph stations that were used by power companies when regular communications were disrupted by storms or other emergencies. In 1946, however, most power companies switched to a new service for public utilities, and the term *special emergency* became restricted to matters that were directly related to public safety and the protection of life and property. Special emergency is now classified as a service for the communications needs of hospitals and clinics, ambulance and rescue services, veterinarians, handicapped persons, disaster-relief organizations, school buses, beach patrols, persons or organizations in isolated areas, and emergency standby and repair facilities for telephone and telegraph systems.

Industrial Radio Services The FCC created an industrial classification of radio services in 1949, after most of the industrial uses of radio had already been established under the miscellaneous or experimental categories. By that time, such diverse industries as oil exploration, mining, news reporting, and motion-picture production had been using radio for about 20 years.

Similar to other early users of radio, these industries first relied on radio mainly for the safety of work crews in remote locations, but they quickly learned the value of mobile radio as an economical tool for carrying company instructions to remote operations, for dispatching and diverting work vehicles, and for coordinating the activities of workers and machines on location. Today, there are nine industrial radio services: Power, Petroleum, Forest Products, Film and Video Production, Relay Press, Special Industrial, Business, Manufacturers, and Telephone Maintenance.

Land Transportation Radio Services

Land Transportation Radio Services also became a special classification in 1949. Before that date, most transportation industries shared two frequency bands in the experimental General Mobile Radio Service. One band served vehicles operating over highways, such as inter-city buses and large trucks, and the other band served urban vehicles such as taxicabs, delivery vans, and tow trucks. As the various transportation industries grew, the FCC allocated more frequencies and created exclusive radio services for different types of transportation: railroad, urban transit, taxicab, inter-city bus, highway truck, and automobile emergency. Today, railroad, taxicab, and automobile emergencies remain separate categories, but urban transit, inter-city bus, and highway trucks are now classified together under the Motor Carrier Radio Service. Land transportation radio stations might not be used for passenger communications.

Radiolocation Service

Radiolocation is the use of radio waves to determine an object's distance, direction, speed, or position for any purpose except navigation. The radiolocation service authorizes persons engaged in commercial, industrial, scientific, educational, or government activities to use radiolocation devices in connection with those activities. Various types of radar (such as police radar and weather radar) are examples of radiolocation applications.

Transportation Infrastructure Radio Service

The Transportation Infrastructure Radio Services category was created in 1995 to integrate radio-based technologies into the nation's infrastructure and to develop and implement the nation's intelligent transportation systems. This service category includes the *Location and Monitoring Service* (LMS). LMS systems are used to determine the location and status of vehicles and equipment. The railroad industry, for example, operates an extensive automatic equipment identification system that enables companies to track, identify, and monitor the movement and location of more than 1.3 million railroad cars and equipment throughout the United States.

Other Private Radio Services

Several other private radio services are available, such as Private Land Mobile Paging and Private Operational Fixed Microwave Systems.

A private paging service is not for profit and serves the licensee's internal communications needs. Private paging systems (in general) provide the same applications offered by commercial paging services: tone, tone-voice, numeric or alphanumeric. Shared-use, cost-sharing, or cooperative arrangements, multiple licensed systems that use third-party managers, or users combining resources to meet compatible needs for specialized internal communications facilities are presumed to be private paging services.

In addition to their mobile operations, many private companies, public utilities, and state and local governments, also make extensive use of Private Operational-Fixed Microwave Systems. These systems connect specific locations in either a point-to-point or point-to-multi-point configuration and can carry or relay voice, data (including teletype, telemetry, facsimile, and other digital communications), and video communications. Such fixed links often connect mobile radio base stations or far-flung offices but are also used for a variety of other purposes, including the remote control of unattended equipment, valves, and switches; the recording of data, such as pressure, temperature, or speed of machines; the monitoring of voltage and current in power lines; and for other control or monitoring functions, such as would be necessary for rivers, railroads, and highways.

Summary

Private radio systems serve a great variety of communication needs that common carriers and other commercial service providers traditionally have not had the capability or the will to fulfill. Companies large and small use their private systems to support their business operations, safety, and emergency needs. The one characteristic that all of these uses share—and that differentiates private wireless use from commercial use—is that private wireless licensees use radio as a tool to accomplish their missions in the most effective and efficient ways possible. Private radio users employ wireless communications as they would any other tool or machine. Radio contributes to their production of some other good or service. For commercial wireless service providers, by contrast, the services offered over the radio system are the end products. Cellular, PCS, and SMR providers sell service or capacity on wireless systems, permitting a wide range of mobile and portable communications that extend the national communications infrastructure. This difference in purpose is significant, because it determines how the services are regulated.

See Also

> Cellular Communications
>
> *Federal Communications Commission* (FCC)
>
> *Personal Communications Services* (PCS)
>
> Specialized Mobile Radio
>
> Telemetry

Protocol Analyzers

A category of test equipment known as the protocol analyzer is used to monitor and diagnose performance problems on LANs and WANs by decoding upper-layer protocols. Protocol analyzers exist for all types of communications circuits, including Frame Relay, X.25, T1 and ISDN, and ATM. Protocol analyzers also exist that offer full seven-layer decodes of NetBIOS, SNA, SMB, TCP/IP, DEC LAT, XNS/MS-NET, NetWare, and VINES, as well as the various LAN cabling, signaling, and protocol architectures (including those for AppleTalk, ARCnet, Ethernet, StarLAN, and Token Ring).

In the case of a LAN, the protocol analyzer connects directly to the cable as if it was just another node, or to the test port of *Data-Terminal Equipment* (DTE) or *Data-Communication Equipment* (DCE) where trouble is suspected (refer to Figure P-9).

Troubleshooting Features

Many protocol analyzers have sophisticated features, such as data capture to RAM or disk, automatic configuration, counters, timers, traps, masks, and statistics. These features can dramatically shorten the time that it

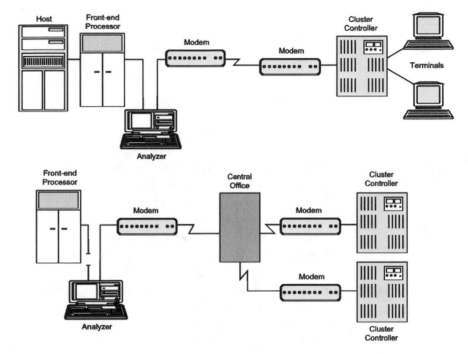

Figure P-9

A protocol analyzer in the monitor mode (top) enables the user to check events taking place between the front-end processor and the local modem over a synchronous link. A protocol analyzer in the simulation mode (bottom) enables the user to run a program that exhibits proper operation of the suspect front-end processor over a synchronous multi-drop link, which includes two cluster controllers.

takes to isolate a problem. Some sets are programmable and offer simulation features.

Monitoring and simulation   Protocol analyzers are generally used in either a passive monitoring application or a simulation application. In the monitoring application, the analyzer sits passively on the network and monitors both the integrity of the cabling and the level of data traffic, logging such things as excessive packet collisions and damaged packets that can tie up an Ethernet LAN, for example. The information on troublesome nodes and cabling are compiled for the network manager. In a monitoring application, the protocol analyzer merely displays the protocol activity and user data (packets) that are passed over the cable, providing a window into the message exchange between network nodes.

In the simulation application, the protocol analyzer is programmed to exhibit the behavior of a network node, such as a gateway, communications controller, or *Front-End Processor* (FEP). This functionality makes it possible to replace a suspect device on the network with a simulator that is running a program, in order to simulate proper operation. This functionality also enhances the capability for fault isolation. For example, a dual-port protocol analyzer can monitor a gateway while running a simulation. More sophisticated protocol analyzers can run simulations that are designed to stress-test individual nodes in order to verify their conformity to standards. Protocol simulation is most often used to verify the integrity of a new installation.

Trapping   The trapping function enables the troubleshooter to command the protocol analyzer to start recording data into its buffer or onto disk when a specific event occurs. For example, the protocol analyzer could be set to trap the first error-filled frame that it receives. This feature permits the capture of only essential information. Some protocol analyzers enable the user to set performance thresholds according to the type of traffic on the network. When these performance thresholds are exceeded, an alarm message is triggered, indicating that a problem exists.

Filtering   With the protocol analyzer's filtering capability, the user can exclude certain types of information from capture or analysis. For example, the technician might suspect that errors are being generated at the data-link layer, so network-layer packets can be excluded until the problem is located. At the data-link layer, the analyzer tracks information such as where the data was generated and whether the data contains errors. If no problems are found, the user can set the filter to include only network-layer packets. At this layer, the protocol analyzer tracks information such as where the data is destined and the type of application under which it was generated. If the troubleshooter has no idea where to start looking for problems, then all of the packets might be captured and written to disk. Various filters can be applied later for selective viewing.

Bit error-rate testing   Bit error-rate testing is used to determine whether data are being passed reliably over a carrier-provided communications link. This goal is accomplished by sending and receiving various bit patterns and data characters to compare what is transmitted with what is received. The bit-error rate is calculated as a ratio of the total number of bit errors divided by the total number of bits received. Any difference between the two is indicated and is displayed as an error. Additional information that might be presented includes sync losses, sync loss seconds, errored seconds, error-free seconds, time unavailable, elapsed time, frame errors, and parity errors.

Packet generation   In being able to generate packets, the protocol analyzer enables the user to test the impact of additional traffic on the network. Using a set of configuration screens, the technician can set the following parameters of the packets:

- The source address from which the packets will be sent
- The destination address to which the packets will be sent
- The maximum and minimum frame size of the packets
- The spacing between the packets, expressed in microseconds
- The number of packets that are sent out with each burst

The technician can also customize the contents of the data field section of the packets to simulate real or potential applications. When the packets are generated, the real time impact of the additional traffic on the network can be observed on the monitor of the protocol analyzer. Packets might also be generated to force a suspected problem to reoccur, thereby expediting the identification of the problem.

Load generation   A related capability is load generation, whereby varying traffic rates on the network can be created. By loading the network from 1 percent to 98 percent, network components such as repeaters, bridges, and transceivers can be stressed for the purpose of identifying any weak links on the network before they become serious problems later.

Timing   Protocol analyzers contain timers that measure the time interval between events. By setting up two traps—one for the transmit path and one for the receive path—the troubleshooter can verify whether a handshake procedure has exceeded its maximum time interval, for example. Although this type of problem is most often better handled from the central network control point or host location, a protocol analyzer that supports simple timing measurements between data events can often solve the problem without invoking these resources.

Terminal emulation   Asynchronous terminal emulation is a feature that is found in some analyzers. If there is a requirement to communicate with intelligent network devices for configuration management, or to access remote databases, this feature can save the cost and inconvenience that is associated with carrying a separate terminal. Implementations of terminal emulation can vary considerably among analyzers, with some supporting complete 24-by-80 characters on a single display and others requiring windowing to access an entire page.

Cable testing   Cable problems are the source of more than 50 percent of LAN network failures. Many protocol analyzers include the capability to test for cable breaks and improperly terminated connections, by using a technique called *Time-Domain Reflectometry* (TDR). This test procedure involves sending a signal down the cable and then receiving and interpreting its echo. The status of the cable and connections can be reported simply as: no fault detected, no carrier sense, open on coax, or short on coax. The distance to the problem is also reported. These devices are capable of pinpointing such problems as shorts, crimps, and water-related faults.

Mapping   The mapping capability of some protocol analyzers can save network managers hours of work. In automatically documenting the physical location of LAN nodes, many hours of work can be eliminated in rearranging the network map when devices are added, deleted or moved. The mapping software enables the network manager to name nodes. Appropriate icons for servers and workstations are included. The icon for each station also provides information about the type of adapter used, as well as the node's location along the cable. When problems arise on the network the network manager can quickly locate the problem by referring to the visual map. Some protocol analyzers can depict network configurations according to the usage of network nodes, arranging them in order of highest to lowest traffic volume.

Programmability   The various tasks of a protocol analyzer can be programmed, enabling performance information to be collected automatically. While some analyzers require the use of programming languages, others employ a setup screen, enabling the operator to define a sequence of tests to be performed. Once preset thresholds are met, an appropriate test or sequence of tests is performed automatically. This capability is especially useful for tracking down intermittent problems. An alternative to programming or defining analyzer operation is to use off-the-shelf software that can be plugged into the data analyzer in support of various test scenarios.

Automatic configuration   Some protocol analyzers have an auto-configure capability that enables the device to automatically configure itself to the protocol characteristics of the line-under test. This functionality eliminates the need to go through several manually established screens for setup, which can save a lot of time and frustration.

English translation   Some protocol analyzers are unique in their capability to decode packets and display their contents in English-like notation, in addition to hexadecimal or binary code. Further details about a specific protocol can be revealed through the analyzer's zoom capability, which enables the troubleshooter to display each bit field along with a brief explanation of its status. This feature can be applied to virtually any protocol, including SNA, X.25, or TCP/IP.

Editing   Some analyzer software include a text editor that can be used in conjunction with captured data, which enables the user to delete unimportant data, enter comments, print reports, and even create files in common database formats.

Switched Environments

Protocol analyzers were originally designed for shared networks. They pick up and examine all traffic as it is broadcast across a shared wire. With LAN switches growing in popularity, diagnosing problems has become a complex undertaking. Switches break shared networks into segments, and traffic is only broadcast over a particular segment. Although this process improves performance by cutting down on contention and devoting more bandwidth to each station, it also makes diagnosing problems more difficult. A protocol analyzer can usually hear traffic only on the segment to which it is connected, which makes it difficult to obtain an overall picture of what is happening through the switch.

One way to deal with this situation is for the technician to set the switch to operate in Promiscuous mode, which sends all Ethernet packets to all ports on the switch, enabling the analyzer to see all traffic. This situation, however, results in a measurement that does not reflect real-world switch conditions. Another technique is port mirroring, which copies traffic going through one port to a port where a protocol analyzer is connected. The problem with this approach is that it limits the analyzer's view to one segment at a time.

The functionality of protocol analyzers recently has been expanded to monitor switches so network administrators can get traffic statistics across a switch's ports and the switch itself, detect configuration problems in virtual LANs, and track problems between switches and desktop computers. Today's protocol analyzers can discover VLAN configurations in a switch, for example, and detect problems in the configuration. Network managers can set thresholds for traffic levels through a switch port. When the threshold is reached, the analyzer takes the traffic going through that port, mirrors it to the port with the analyzer on it, and alerts the help desk.

Another technique for gathering statistics through an entire switch is called port looping. The analyzer uses port mirroring to look at each port on the switch for only a short time. By sampling traffic through each port, one at a time, the analyzer can build statistics about traffic through the switch.

Summary

Protocol analyzers have long been among the key diagnostic tools of network managers, helping them quickly isolate trouble spots. However, the use of protocol analyzers was almost always reactive—they captured trace information and network statistics that were later interpreted by network technicians. This required a skilled analyst to interpret the data. Today's protocol analyzers can detect deteriorating conditions that lead to errors and suggest remedial actions to head off problems before they affect performance. Tools are even available that track the performance of data offerings that are covered by *Service-Level Agreements* (SLA), so that the organization can be assured that it is receiving the performance for which it is paying.

See Also

> Analog-Line Impairment Testing
>
> Help Desks
>
> *Service-Level Agreements* (SLAs)

Public-Switched Telephone Network (PSTN)

Since the invention of the telephone by Alexander Graham Bell in 1876, the *Public-Switched Telephone Network* (PSTN) has evolved to become a highly reliable method of communication. For most of this period, spanning almost 125 years, the PSTN carried voice conversations by telephone. The circuits over which these conversations take place are set up by interconnected switches and remain in place for the duration of the call.

Today, the PSTN increasingly carries data as well, mostly from users with modems who want to access the Internet at speeds of up to 56Kbps. Others might run data over ISDN, whose bearer channels are carried over the PSTN at speeds of 56Kbps/64Kbps, 128Kbps, 384Kbps, or 1.536Mbps. Higher-speed data is diverted to data switches and is passed on to separate data networks, in order to offload the voice-optimized PSTN of this burden.

The PSTN is so complex that in most network diagrams, it is represented merely as a cloud. Inside this cloud are all of the discrete components that play a role in setting up and tearing down circuits, implementing services, providing value-added features, and managing the infrastructure.

The traditional PSTN began as a human-operated analog switching system with operators connecting calls via cords at a patch panel. Over the years, this type of system evolved to electromechanical switching, which eliminated the need for operators to connect every call. Today's switching systems are now almost completely electronic and digital and are under stored program control.

Customer-Premise Equipment (CPE)

CPE includes individual equipment or whole systems that serve a particular customer. The customer can be any subscriber of the telephone service, including an individual, a company, or a government agency. The equipment or system interfaces with the public network.

CPE ranges from simple house wiring and telephone sets to advanced *Private-Branch Exchanges* (PBX) and/or *Key Telephone Systems* (KTS) and all of their associated equipment, including an *Automated Call Distributor* (ACD), voice-mail system, and fax machines. Modems are also considered CPE. These devices connect to computers and transform data into an acoustical signal, which can be carried over a standard voice-grade line.

Local Loop

The local loop is the connection between the customer premises and the local central office switch. Usually, this connection is a twisted-pair, 24 *American Wire Gauge* (AWG) line that provides the transmission path that enables the CPE to be connected with the major carriers, both local and long-distance. Although the local loop is comprised mostly of copper wire, new architectures incorporate optical fiber and coaxial cabling into the mix to support broad-band data applications. In some cases, the local loop is bypassed through the use of wireless technologies.

Local loop includes the distribution plant that radiates from the CO, consisting mostly of *Subscriber-Line Carrier* (SLC) systems that use fiber-optic cables to extend the reach of the CO to remote locations. This structure enables the telephone company to provide service to outlying homes and businesses as if they were directly attached to the central office.

Switching

The switching systems, resident in the carriers' *Central Offices* (COs), set up a dedicated transmission path that connects the calling and called parties for two-way communication. When the conversation is finished and the parties hang up, the path is torn down—freeing the network resources to handle another call.

Path setup and tear down through the long-distance portion of the PSTN is handled by a separate data network known as *Signaling System 7* (SS7). This high-speed data network transfers network control and routing information so that telephone network resources are never committed to handling a call, unless SS7 determines that the call can actually go through to completion.

Trunking

Trunking refers to the high-capacity links between switches in the PSTN. This interoffice trunking is accomplished primarily over fiber-optic links. At a minimum, interoffice trunks operate at the DS3 level, or 44.736Mbps. Many of today's interoffice trunks operate at the SONET OC-3 and OC-12 levels or at 155Mbps and 622Mbps, respectively. Along high-traffic corridors, higher-capacity interoffice trunks in the gigabits-per-second range are used.

Summary

While most carriers view the PSTN as voice-oriented and have preferred to run data over a separate overlay network, the trend is clearly toward merging voice and data over a single network. The two choices for accomplishing this convergence hinge around *Asynchronous Transfer Mode* (ATM) or *Internet Protocol* (IP) platforms, with some carriers committing to one or the other and some committing to multi-service platforms that handle both ATM and IP. Still other carriers have not made any commitment, preferring instead to continue investing in their circuit-switched voice networks. Under this course, however, the carrier risks being saddled with an obsolete voice network that is not capable of supporting future market opportunities, which will be data-oriented.

See Also

> *Central Office* (CO) Switches
>
> *Local Access and Transport Areas* (LATAs)

Local Loop

Network Operations Centers (NOS)

Operations Support Systems (OSSs)

Signaling System 7 (SS7)

Telephone

Public Telephone Service

Public telephone service, also known as pay-phone service, is either outgoing only or two-way service and can be coin or coinless. At least four public telephone service options are available:

- *Coin-operated pay-phone service (classic or dumb pay phone)*   This service relies on an operating company's central office controls to collect and return coins at the pay-phone set. End users have access to local, toll, and operator network services. The payment options that are available to end users are cash or billing as a collect call, a charge to a third-party number, or a calling card.

- *Coin-operated pay-phone service (smart pay phone)*   This service uses pay-phone sets that are capable of rating local calls, collecting and returning coins, rating and routing toll calls, and providing station-based operator services. The payment options that are available to the user are cash or billing as collect, third party, or calling card.

- *Card and coin pay-phone service (advanced pay phone)*   This service is similar to the coin-operated pay phone service but provides the additional payment options of using commercial credit cards or debit cards.

- *Coinless pay-phone service*   This service uses pay-phone sets that rely on central office and operator systems to rate and bill for calls that are placed from the sets. This system has a unique screening feature that prohibits coin-paid, direct-dialed call billing. Users have access to local and toll services. The payment options that are available to the user are limited to billing as collect, third party, or calling card.

Some carriers offer semi-public telephone service and shared pay-phone service. These services utilize a dumb pay-phone set that relies on central office control to collect and return coins. The end users have access to local, toll, and operator network services. The payment options that are available to the user are cash or billing as collect, third party, or calling card. The semi-public telephone service and shared payphone service recover a portion of the carrier's costs through separate monthly service charges that are billed to the location site provider.

Inmate calling service also exists, which utilizes a pay-phone set that relies on central office and operator or premises-based call-management systems for rating and billing. This system has unique screening features that prohibit direct-dialed calls, calling-card calls, and third-number billing. This service option provides users with access to local and toll services on a collect billing arrangement.

Pay-Phone Market

Bell South Public Communications is the nation's largest stand-alone pay-phone services provider. Each day, more than 3.2 million calls are completed on Bell South pay phones. The pay-phone industry is now a level playing field where thousands of pay-phone operators can compete fairly for customers. This competition is accomplished through *Comparably Efficient Interconnection* (CEI) rules that are issued by the FCC, which ensures that the *Incumbent Local Exchange Carriers* (ILECs) provide telephone line services to independent pay-phone service providers on a non-discriminatory basis.

Two types of pay-phone service providers exist. An *Independent Pay-Phone Provider* (IPP) is neither a local-exchange telephone company nor an affiliate of a local-exchange telephone company. In some states, other terms such as *Customer-Owned Coin-Operated Telephone* (COCOT) is used to refer to these providers.

A *Pay-Phone Service Provider* (PSP) can be a local-exchange telephone company, an affiliate of a local-exchange telephone company, an independent pay-phone provider, a long-distance carrier, or a competing local-exchange telephone company. The term PSP was coined by the FCC to emphasize that all pay-phone providers will be treated on an equitable basis from a regulatory standpoint.

Summary

The FCC regulates pay-phone service to the extent that the carrier is not permitted to subsidize its pay-phone service directly or indirectly from its telephone exchange or exchange-access service operations. In addition, the ILEC cannot discriminate in favor of its pay-phone service.

See Also

Calling Cards

Pre-Paid Calling Cards

Public Utility Commissions (PUCs)

In the United States, each state has a commission charged with the responsibility for regulating public utilities and intrastate service rates. Utilities include privately-owned corporations providing electric, natural gas, water, sewer, telephone, and radio service to the public and can also include railroads, motor buses, truck lines, ferries, and other transportation companies. The *Public Utility Commissions* (PUCs) supervise and regulate utilities and public transportation in the state, in order to ensure that service and facilities are made available at rates that are just and reasonable.

PUC policies benefit rate payers through lower rates, new and improved utility products and services, and protect consumers where competition otherwise does not. The PUC uses competitive markets to accomplish these goals where possible and appropriate. Generally, regulated utilities must seek approval from the PUC to introduce new services and change rates or services.

The PUCs are independent quasi-judicial regulatory bodies whose jurisdiction, powers, and duties are delegated by the state legislature. The PUCs have quasi-legislative and quasi-judicial authority in that they establish and enforce administrative regulations—and, like a court, might take testimony, subpoena witnesses and records, and issue decisions and orders. Usually, the PUCs conduct public hearings on applications, petitions, and complaints.

In its decision-making, the PUC seeks to balance the public interest and need for reliable, safe utility services at reasonable rates with the need to assure utilities operate efficiently, remain financially viable, and provide stockholders with an opportunity to earn a fair return on investment. The PUC encourages rate payers, utilities, and consumer and industry organizations to participate in its proceedings and seeks their assistance in resolving complex issues.

The commissioners usually are appointed by the governor, with the approval of the state legislature, for a term of six years. The commissioners make all final policy, procedural, and other decisions. Their terms might be staggered in order to assure that the PUC always has the benefit of experienced members.

Summary

In cases where state and federal jurisdictions might overlap, a federal-state joint board is established to arrive at a mutually agreeable course of action. Each joint board is comprised of three FCC commissioners, four state commissioners, and is moderated by the chairman of the FCC. The issue of universal service is a case where federal and state jurisdictions overlap. A federal-state joint board on universal service was established to make recommendations to the FCC with respect to the implementation of the universal service provisions of the Telecommunications Act of 1996. The public meetings included panel discussions held to address issues raised in the universal service proceeding. The panels discussed competition and universal service, and universal service for consumers in rural, insular, and high-cost areas and for low-income consumers—as well as the provision of advanced telecommunications services to schools, classrooms, libraries, and health-care providers.

See Also

Federal Communications Commission (FCC)

Universal Service

CHAPTER **R**

Radio Communication Interception

When it comes to the interception of radio communications, the *Federal Communications Commission* (FCC) has the authority to interpret Section 705 of the Communications Act, 47 U.S.C. Section 605, which deals with "Unauthorized Publication of Communications."

Although the act of intercepting radio communications might violate federal or state statutes, this provision generally does not prohibit the mere interception of radio communications. For example, if someone happens to overhear a conversation on a neighbor's cordless telephone, this event is not a violation of the Communications Act. Similarly, if someone listens to radio transmissions on a scanner, such as emergency service reports, this action is not a violation of Section 705.

A violation of Section 705 would occur, however, if someone were to divulge or publish what they hear or use this information for their own or someone else's benefit. An example of using an intercepted call for a beneficial use in violation of Section 705 would be someone listening to accident reports on a police channel and then driving or sending one of their own tow trucks to the reported accident scene in order to obtain business.

The Communications Act does allow for the divulgence of certain types of radio transmissions. The statute specifies that there are no restrictions on the divulgence or use of radio communications that have been transmitted for the use of the general public, such as transmissions of a local radio or television broadcast station. Likewise, there are no restrictions on divulging or using radio transmissions that originate from ships, aircraft, vehicles, or persons in distress. Transmissions by amateur radio or *Citizens Band* (CB) radio operators are also exempt from interception restrictions.

In addition, courts have held that the act of viewing a transmission (such as a pay television signal) that the viewer was not authorized to receive is a "publication" that violates Section 705. This section also has special provisions governing the interception of satellite television programming that is transmitted to cable operators. The section prohibits the interception of satellite cable programming for private, home viewing if the programming is either scrambled or not scrambled but is sold through a marketing system. In these circumstances, authorization must be obtained from the programming provider in order to legally intercept the transmission.

The Act also contains provisions that affect the manufacture of equipment used for listening or receiving radio transmissions, such as scanners. Section 302(d) of the Communications Act, 47 U.S.C. Section 302(d), prohibits the FCC from authorizing scanning equipment that is capable of receiving transmissions in the frequencies allocated to domestic cellular services, that is capable of readily being altered by the user to intercept cellular communications, or that might be equipped with decoders that convert digital transmissions to analog voice audio. Also, since April 26, 1994 (47 CFR 15.121), such receivers cannot be manufactured in the United States or imported for use in the United States. FCC regulations also prohibit the sale or lease of such scanning equipment (47 CFR 2.803).

Summary

The FCC receives many inquiries regarding the interception and recording of telephone conversations. To the extent that these conversations are radio transmissions, there would be no violation of Section 705 if no divulgence or beneficial use of the conversation takes place. Again, however, the mere interception of some telephone-related radio transmissions—whether cellular, cordless or land-line conversations—might constitute a criminal violation of other federal or state statutes.

See Also

Telephone Fraud

Redundant Array of Inexpensive Disks (RAID)

An increasingly popular method of protecting data is the *Redundant Array of Inexpensive Disks* (RAID). Instead of risking all data on one high-capacity disk, this solution distributes the data across multiple smaller disks. RAID technology is also being used to deliver *Video-on-Demand* (VOD) programming to cable TV subscribers, enabling multiple viewers to access one copy of the same program on the video server. Rather than storing an entire movie on a single hard drive, segments of a movie or program are spread or striped across multiple disks in the RAID configuration. In addition to eliminating blocking of a single title to all users, this fault-tolerant storage method prevents video stream interruptions. More than 20 percent of storage disks would have to fail before viewing would be affected. The result is a more efficient, cost-effective, and reliable VOD server that is capable of delivering hundreds of movies to thousands of households simultaneously.

RAID Categories

RAID products are usually grouped into the following categories:

- *RAID Level 0*   These products are technically not RAID products at all. Although data is striped block-by-block across all of the drives in the array, such products do not offer parity or error correction of the data.

- *RAID Level 1*   These products duplicate data that is stored on separate disk drives. Also called mirroring, this approach ensures that critical files will be available in case of individual disk drive failures. Each disk in the array has a corresponding mirror disk and the pairs run in parallel. Blocks of data are sent to both disks at the same time. While highly reliable, Level 1 is costly because every drive requires its own mirror drive, which doubles the hardware cost of the system.

- *RAID Level 2*   These products distribute the code used for error detection and correction across additional disk drives. The controller includes an error-correction algorithm, which enables the array to reconstruct lost data if a single disk fails. As a result, no expensive mirroring is required. But the code requires that multiple disks be set aside to do the error-correction function. Data is sent to the array one disk at a time.

- *RAID Level 3*   These products store user data in parallel across multiple disks. The entire array functions as one large logical drive. Its parallel operation is ideally suited to supporting applications that require high data transfer rates when reading and writing large files. RAID Level 3 is configured with one parity (error-correction) drive. The controller determines which disk has failed by using additional check information recorded at the end of each sector. Because the drives do not operate independently, however, every time a file must be retrieved, all of the drives in the array are used to fulfill that request. Other user requests are put into a queue.

- *RAID Level 4*   These products store and retrieve data using independent writes and reads to several drives. Error correction data is stored on a dedicated parity drive. In RAID Level 4, data striping is accomplished in sectors, not bytes (or blocks). Sector-striping offers parallel operation in that reads can be performed simultaneously on independent drives, which enables multiple users to retrieve files at the same time. While multiple reads are possible, multiple writes are not—because the parity drive must be read and written to for each write operation.

- *RAID Level 5*   These products interleave user data and parity data, which are then distributed across several disks. Because data and parity codes are striped across all of the drives, there is no need for a dedicated parity drive. This configuration is suited for applications that require a high number of I/O operations per second, such as transaction processing tasks that involve writing and reading large numbers of small data blocks at random disk locations. Multiple writes to each disk group are possible because write operations do not have to access a single common parity drive.

- *RAID Level 6*   These products improve reliability by implementing drive mirroring at the block level so that data is mirrored on two drives instead of just one. In other words, up to two drives in the five-drive disk array can fail without loss of data. If a drive in the array fails with RAID 5, for instance, data must be rebuilt from the parity information that spans the drives. With RAID 6, however, the data is simply read from the mirrored copy of the blocks found on the various striped drives (no rebuilding is required). Although this situation results in a slight performance advantage, it requires at least 50 percent more disk capacity to implement.

- *RAID Level 10*   Some vendors offer products that combine the performance advantages of RAID 0 with the data availability and consistent high performance of RAID 1 (also referred to as striping over a set of mirrors). This method offers high performance and high availability for mission-critical data.

Summary

Businesses today have multiple data storage requirements. Depending on the application, performance might be valued more than availability; other times, the reverse might be true. Today, it is even common to have different data structures in different parts of the same application. Until recently, the choice among specific RAID solutions involved tradeoffs between cost, performance, and availability: once installed, they cannot be changed to take into account the different storage needs of applications that might arise in the future. Vendors have responded with storage solutions that support a mix of RAID levels (or non-RAID) simultaneously. Individual disk drives or groups of drives can now be configured via a PC-based resource manager for high performance, high availability, or as an optimized combination of both. This situation solves a classic data storage dilemma: meeting the exacting requirements of current applications while staying flexible enough to adapt to long-term needs.

See Also

Hierarchical Storage Management (HSM)

Storage Media

Regulatory Process

The FCC is the primary agency in the United States that regulates the telecommunications industry. The regulatory process consists of discrete steps that ensure that the issues and views of all interested parties are taken into account before a decision is made. This process is open to individuals and consumer groups, not just to lawyers and lobbyists from big telecommunications companies.

Notice of Inquiry

The FCC's decision-making process starts with a *Notice of Inquiry* (NOI). This step is designed primarily for fact gathering—a way to seek information about a broad subject or generate ideas concerning a specific issue, such as streamlining. This document will generally ask questions and provide few conclusions.

The NOI describes where and when comments can be submitted, where and when the comments others have made can be reviewed, and how to respond to those comments. After reviewing comments from the public, the FCC will issue either a Notice of Proposed Rulemaking or a Memorandum Opinion and Order concluding the NOI. For easy reference, both the NOI and the NPRM contain docket numbers that are printed on the document's front page.

Notice of Proposed Rulemaking

An NPRM is issued to detail proposed changes to FCC rules and to seek public comment on either focused, or specific proposals and/or to ask questions on an issue or set of issues. The NPRM describes where and when comments can be submitted, where and when interested parties can review comments others have made, and how to respond to those comments.

After the comments review period, and in conjunction with or before issuing a *Report & Order* (R&O), the FCC might also choose to issue a Further NPRM regarding issues raised in comments to provide an opportunity for the public to comment further on a related alternative proposal. After that, the FCC will issue a R&O, adopting any rule changes.

Report & Order (R&O)

After considering comments and reply comments to a Notice of Proposed Rulemaking, the FCC can issue a R&O that amends the rules or makes a decision not to do so. Summaries of the R&O are published in the Federal Register, letting everyone know when a rule change will become effective.

After an R&O is issued, changes can still be made. If people feel that certain issues were not really defined or resolved, a Petition for Reconsideration can be filed within 30-days from the date the R&O appears in the Federal Register. The petition will be acted upon by the commission, at which time the FCC can issue a *Memorandum Opinion and Order* (MO&O) or an Order of Reconsideration amending the rules or stating that the rules will not be changed.

En Banc Sessions

The FCC receives information in several other ways, such as through en banc sessions. An en banc session is held by the commissioners to hear various presentations on specific topics by diverse parties, usually in panel groups. Specific witnesses are asked to present information at an en banc hearing, typically following the issuance of a public notice announcing the hearing. The commissioners question the presenters; the information given can be used by the FCC when it makes or proposes rules.

Ex Parte Meetings

Another way the commissioners gather information is through ex parte meetings between interested parties and decision-making staff of the FCC, such as the agency's legal or technical personnel who are assigned to an issue. There are ex parte rules that ensure all participants in an agency proceeding are given a fair opportunity to present information and evidence in support of their positions.

An ex parte presentation can be made in written or oral form. If written, the communication is not served on the parties to the proceeding. If oral, the communication is made without advance notice to the parties to the proceeding and without opportunity for them to be present.

Restricted and non-restricted ex parte presentations exist. Generally, ex parte presentations are prohibited in restricted proceedings. The prohibition stays in effect until the proceeding has been decided or until a settlement or agreement has been approved by the commission and the matter is no longer subject to a reconsideration by the commission or review by any court.

Ex parte presentations in a proceeding that could become restricted also are prohibited—even if it is not restricted at the time. For example, if that person intends to file a mutually exclusive application that could cause the proceeding to become restricted or that intends to file an opposition, complaint, or objection that would cause the proceeding to become restricted, any ex parte presentation would be prohibited.

In non-restricted proceedings, ex parte presentations are generally permissible but are subject to disclosure. A person who makes or submits a written ex parte presentation must provide on the same day an original and one copy of the presentation to the FCC's secretary for inclusion in the public record. The ex parte rules apply to anyone who seeks to influence the outcome of a particular proceeding, whether or not that person is a party to the proceeding.

Filing Comments Electronically

As noted, individual consumers can submit comments to the FCC for consideration in the regulatory process. The FCC has made this process easier by setting up a Web page containing a comment form. Sending a comment is a two-step process. The first step is to complete and send the cover sheet. After the cover sheet is sent, one of two transmittal methods must be selected: sending a file, or sending a short message that can be typed directly on the Web page.

The *Electronic Comment Filing System* (ECFS) is designed to give the public access to FCC rulemaking and docket proceedings by accepting comments via the Internet. ECFS also enables users to research any document in the system including non-electronic documents that have been scanned into the system. ECFS includes data and images from 1992 onward. The ECFS can be accessed at `http://www.fcc.gov/e-file/ecfs.html`.

Summary

For the regulatory process to work effectively, the FCC must have information on which to base its decisions. Information is solicited from all interested parties—consumers as well as telecommunications carriers and large businesses that rely on their services. The FCC has set up several pathways for receiving information. These include en banc sessions, ex parte communications, and electronic transmittal of comments from Web-based forms. The FCC's plans for addressing various issues and their outcomes are made public through various means, including presentations at Congressional hearings, the Federal Register, press releases, and its own Web page at `www.fcc.gov`.

See Also

> *Federal Communications Commission* (FCC)
>
> Telecommunications Act of 1996

Remote Control

Remote control is a software-based solution for remotely accessing another computer. In a typical scenario, there are two types of computers: the remote system and the host system. The remote system can be a branch office PC, a PC located at home, or a portable computer whose location varies on a daily basis. The host system can be any computer that a remote user wishes to access, including LAN-attached workstations and stand-alone PCs that are equipped with a modem. Both the remote and host computers must be equipped with the same remote control software, and the user must be authorized to access a particular host.

With remote control, the remote computer user takes full control of a host system sitting on the corporate network. All of the user's keystrokes and mouse movements are sent to the host system, and the image on that screen is forwarded back to the remote PC for display as if the user were sitting in front of the host system.

Applications

Remote-control software has many applications. This software enables a mobile user to check e-mail at the office, to query a corporate contact database from a hotel room, or to access an important file stored on a server. Help-desk operators and technicians can use remote control software to troubleshoot problems with specific systems. Trainers can periodically monitor the performance of users to determine their need for additional training.

With remote control, however, security can become an important concern. The host system's monitor displays all of the information that is being manipulated remotely, enabling any casual observer to view all screen activity (including e-mail, financial data, and confidential documents). A possible solution would be to turn off the monitor and leave the computer running, but this solution is not always possible.

Features

The leading remote control programs are feature-rich, even permitting the remote user to access the host system through the Internet to save on long-distance charges on dialup connections. Such programs provide useful features such as the following:

- *Callback*   This feature enhances security by having a remote host call back to a guest. With fixed callback, the guest is automatically called back at a previously specified number. With a roving callback, the guest has the opportunity to enter the callback number.
- *Chat*   Enables two connected users to type messages to each other. Some products enable messaging during a remote control or file-transfer session.
- *Clipboard*   Enables the user to highlight and copy material between the remote computer and host
- *Compression*   Compresses transferred data by a factor of at least 2-to-1, in order to cut dialup-line costs
- *Directory synchronization*   Synchronizes directories on the host and remote machines, ensuring that they contain the same directory trees
- *Drag-and-drop*   Enables the user to copy or move files from one machine to the other with the drag-and-drop technique

- *Drive mapping*   Makes the disk drives on both the remote and local machines seamlessly accessible to users on both ends. This feature would enable a user on a remote host to open a document stored on a local drive from within a word processor. Without remote drive mapping, the user would have to transfer the document to the remote host before being able to open it.

- *Emulation*   Some vendors offer limited modem support in their remote control software, supporting only TTY terminal emulation and only ASCII and Xmodem file transfer protocols. Others provide extensive emulation support, even enabling users to logon to commercial services such as *America Online* (AOL).

- *Encryption*   Encrypts the data stream during transmission so that remote sessions cannot be monitored

- *File management*   Files can be displayed and sorted by name, size, modification date and time, and attribute flags. The directory can be filtered to display only specified types of files.

- *File synchronization*   Synchronizes files on the remote machine, ensuring that they contain the same files as the host. File transfers can be expedited by having the capability to copy only the parts of files that have changed. This process also saves on long-distance connect charges.

- *Printer redirection*   The capability to reroute printer output from the host to the local printer

- *Screen data caching*   Caches screen data from the host locally so that it all does not have to be retransmitted. The cache can be set to retain data between sessions.

- *Scripting facility*   Scripts enable unattended file transfer operations and enable various tasks such as the selection of connections and file transfers to be automated. A script can even automate logon procedures. Multiple tasks can be automated in a single script.

- *Transfer restart*   If a connection is lost during a file transfer, the remote-control program can restart the transfer where it left off.

- *Virus scanning*   Automatically scans files for viruses as they are being transferred between the remote computer and host

- *Voice-to-data switching*   Establishes a remote-control connection from a voice call without having to hang up and redial. This feature is especially convenient for technical support applications.

One of the most popular remote control solutions is Symantec's pcTelecommute. The product's DayEnd Sync feature enables telecommuters to keep their office PC up-to-date with the most current versions of files when they work from home. The program even reminds the user at the end of the workday which files have changed and prompts them to start automatic file synchronization with their office PC.

Summary

Remote-control software enables users to dial a specific modem-equipped or LAN-attached PC in order to access its files or to troubleshoot a problem, even through the Internet. As long as the software is installed on both ends, the remote computer assumes the capabilities of the host. Everything on the host's screen is mirrored on the remote computer's screen. Today's increasing, distributed workforce makes this kind of connectivity a virtual necessity, especially for computer users who divide their work time between home and office.

See Also

Remote Node
Telecommuting

Remote Monitoring

The common platform from which to monitor multi-vendor networks is SNMP's *Remote Monitoring* (RMON) MIB. Although a variety of SNMP MIBs collect performance statistics to provide a snapshot of events, RMON enhances this monitoring capability by keeping a past record of events that can be used for fault diagnosis, performance tuning, and network planning.

Hardware-based and/or software-based RMON-compliant devices (i.e., probes) that are placed on each network segment monitor all data packets sent and received. The probes view every packet and produce summary information on various types of packets, such as undersized packets, and events, such as packet collisions. The probes can also capture packets according to predefined criteria set by the network manager or test technician. At any time, the RMON probe can be queried for this information by a network management application or an SNMP-based management console, so that detailed analysis can be performed in an effort to pinpoint where and why an error occurred.

The original Remote Network Monitoring MIB defined a framework for the remote monitoring of Ethernet. Subsequent RMON MIBs have extended this framework to Token Ring and to other types of networks, including FDDI. A map of the RMON MIB for Ethernet and Token Ring is shown in Figure R-1.

RMON Applications

A management application that views the internetwork, for example, gathers data from RMON agents that are running on each segment in the network. The data is integrated and correlated to provide various internetwork views that provide end-to-end visibility of network traffic, both LAN and WAN. The operator can switch between a variety of views.

For example, the operator can switch between a media access control (MAC) view (which shows traffic going through routers and gateways), a network view (which shows end-to-end traffic), or apply filters to see only traffic of a given protocol or suite of protocols. These traffic matrices provide

Figure R-1
A map of SNMP's remote monitoring management information base, RMON MIB.

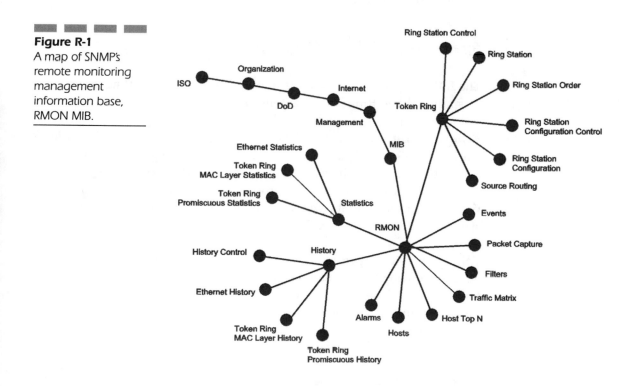

the information necessary to configure or partition the internetwork in order to optimize LAN and WAN utilization.

In selecting the MAC-level view, for example, the network map shows each node of a segment separately, indicating intra-segment node-to-node data traffic. The map also shows total inter-segment data traffic from routers and gateways. This combination enables the operator to see consolidated internetwork traffic and how each end-node contributes to the traffic.

In selecting the network level view, the network map shows end-to-end data traffic between nodes across segments. By connecting source and ultimate destination, without clouding the view with routers and gateways, the operator can immediately identify specific areas that are contributing to an unbalanced traffic load.

Another type of application enables the network manager to consolidate and present multiple segment information, configure RMON alarms, provide complete Token Ring RMON information, as well as perform baseline measurements and long-term reporting. Alarms can be set on any RMON variable. Notification via traps can be sent to multiple management stations. Baseline statistics enable long-term trend analysis of network traffic patterns that can be used to plan for network growth.

Ethernet Object Groups

The RMON specification consists of nine Ethernet/Token Ring groups and 10 specific Token Ring RMON extensions.

Ethernet statistics group   The statistics group provides segment-level statistics (refer to Figure R-2). These statistics show packets, octets (or bytes), broadcasts, multicasts, and collisions on the local segment, as well as the number of occurrences of dropped packets by the agent. Each statistic is maintained in its own 32-bit cumulative counter. Real-time packet size distribution is also provided.

Ethernet history group   With the exception of packet-size distribution, which is provided only on a real-time basis, the history group provides historical views of the statistics that are provided in the statistics group. The history group can respond to user-defined sampling intervals and bucket counters, enabling some customization in trend analysis.

The RMON MIB comes with two defaults for trend analysis. The first provides for 50 buckets (or samples) of 30-second sampling intervals over a period of 25 minutes. The second provides for 50 buckets of 30-minute sam-

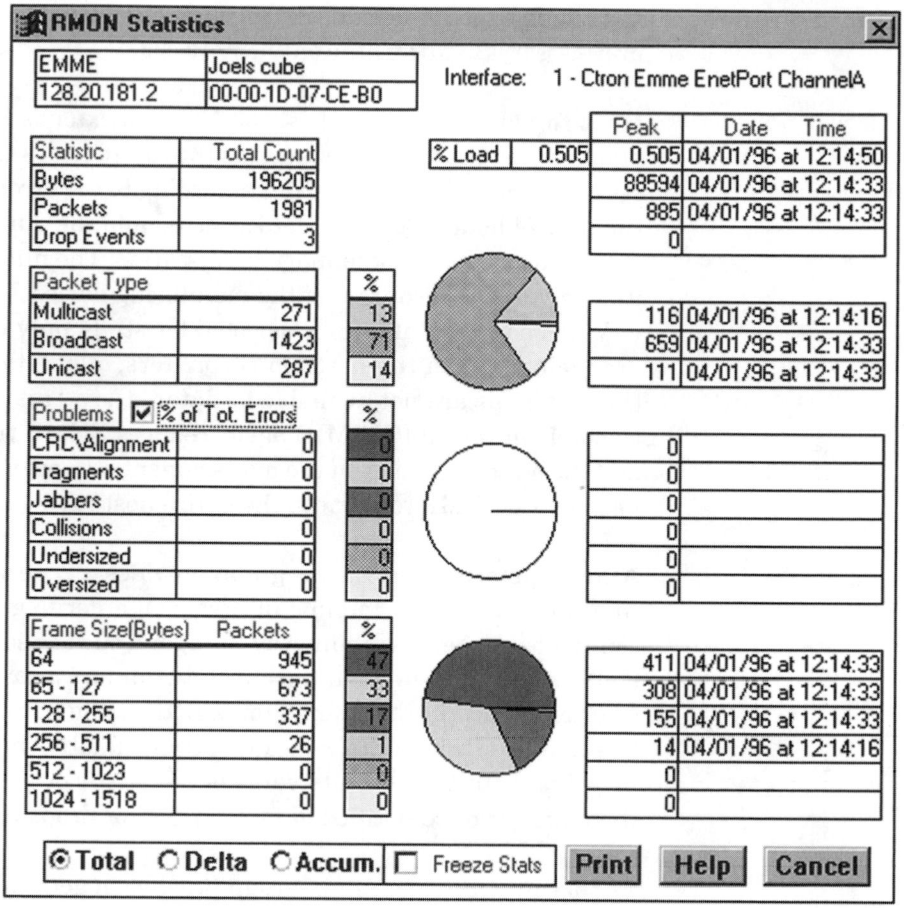

Figure R-2

The Ethernet Statistics
Window, accessed
from Cabletron
Systems' Spectrum
Element Manager for
Windows. The
column on the left
side of the window
displays the statistic
name, total count,
and percentage; the
column on the right
displays the peak
value for each
statistic and the date
and time that value
occurred. Note that
peak values are
always delta values.

pling intervals over a period of 25 hours. Users can modify either of these
or add additional intervals to meet specific requirements for historical
analysis. The sampling interval can range from one second to one hour.

Host table group The RMON MIB specifies a host table which includes
node traffic statistics: packets sent and received, octets sent and received,
and as broadcasts, multicasts, and error-filled packets that are sent. In the
host table, the classification of errors sent is the combination of packet
undersizes, fragments, CRC/alignment errors, collisions, and oversizes that
are sent by each node.

The RMON MIB also includes a host timetable that shows the relative
order in which the agent discovered each host. This feature is not only useful for network management purposes but also assists in uploading those

nodes to the management station of which it is not yet aware. This reduces unnecessary SNMP traffic on the network.

Host top N group   The host top N group extends the host table by providing sorted host statistics, such as the top 10 nodes that are sending packets or an ordered list of all nodes according to the errors that were sent over the past 24 hours. Both the data selected and the duration of the study are defined at the network management station. The number of studies is limited only by the resources of the monitoring device.

When a set of statistics is selected for study, only the selected statistics are maintained in the host top N counters; other statistics over the same time intervals are not available for later study. This processing-performed remotely in the RMON MIB agent reduces SNMP traffic on the network and the processing load on the management station, which would otherwise need to use SNMP to retrieve the entire host table for local processing.

Alarms group   The alarms group provides a general mechanism for setting thresholds and sampling intervals, in order to generate events on any counter or integer maintained by the agent, such as segment statistics, node traffic statistics defined in the host table, or any user-defined packet match counter defined in the filters group. Both rising and falling thresholds can be set, each of which can indicate network faults. Thresholds can be established for both the absolute value of a statistic or its delta value, enabling the manager to be notified of rapid spikes or drops in a monitored value.

Filters group   The filters group provides a generic filtering engine that implements all packet capture functions and events. The packet capture buffer is filled with only those packets that match the user-specified filtering criteria. Filtering conditions can be combined using the Boolean parameters "and" or "not." Multiple filters are combined with the Boolean "or" parameter.

Packet capture group   The type of packets collected is dependent upon the filter group. The packet capture group enables the user to create multiple capture buffers and to control whether the trace buffers will wrap (overwrite) when full or stop capturing. The user can expand or contract the size of the buffer to fit immediate needs for packet capturing, rather than permanently commit memory that will not always be needed.

Notifications (events) group   In a distributed management environment, traps can be delivered by the RMON MIB agent to multiple management stations that share a single community name destination specified for the trap. In addition to the three traps already mentioned—rising threshold and falling threshold (see alarms group) and packet match (see packet capture group)—there are seven additional traps that can be specified:

- *coldStart* This trap indicates that the sending protocol entity is reinitializing itself such that the agent's configuration or the protocol entity implementation can be altered.
- *warmStart* This trap indicates that the sending protocol entity is reinitializing itself such that neither the agent configuration nor the protocol entity implementation is altered.
- *linkDown* This trap indicates that the sending protocol entity recognizes a failure in one of the communication links that are represented in the agent's configuration.
- *linkUp* This trap indicates that the sending protocol entity recognizes that one of the communication links represented in the agent's configuration has come up.
- *authenticationFailure* This trap indicates that the sending protocol entity is the addressee of a protocol message that is not properly authenticated. While implementations of the SNMP must be capable of generating this trap, they must also be capable of suppressing the emission of such traps via an implementation-specific mechanism.
- *egpNeighborLoss* This trap indicates that an *External Gateway Protocol* (EGP) neighbor for whom the sending protocol entity was an EGP peer has been marked down and that the peer relationship is no longer valid.
- *enterpriseSpecific* This trap indicates that the sending protocol entity recognizes that some enterprise-specific event has occurred.

The notifications (events) group enables users to specify the number of events that can be sent to the monitor log. From the log, any specified event can be sent to the management station. The log includes the time of day for each event and a description of the event written by the vendor of the monitor. The log overwrites when full, so events might be lost if they are not uploaded to the management station periodically.

Traffic matrix group The RMON MIB includes a traffic matrix at the *Media Access Control* (MAC) layer. A traffic matrix shows the amount of traffic and number of errors between pairs of nodes-one source and one destination address per pair. For each pair, the RMON MIB maintains counters for the number of packets, number of octets, and error packets between the nodes. Users can sort this information by source or destination address.

Applying remote monitoring and statistics gathering capabilities to the Ethernet environment offers a number of benefits. The availability of critical networks is maximized, because remote capabilities allow for a more timely resolution of the problem. With the capability to resolve problems remotely, operations staff can avoid costly travel to troubleshoot problems on-site. With the capability to analyze data collected at specific intervals over a long period of time, intermittent problems can be tracked down that would normally go undetected and unresolved.

Token Ring Extensions

As noted, the first version of RMON defined media-specific objects for Ethernet only. Later, media-specific objects for Token Ring became available.

Token Ring MAC-Layer Statistics

This extension provides statistics, diagnostics and event notification associated with MAC traffic on the local ring. Statistics include the number of beacons, purges, and 803.5 MAC management packets and events; MAC packets; MAC octets; and ring soft-error totals.

Token Ring promiscuous statistics This extension collects utilization statistics of all user data traffic (non-MAC) on the local ring. Statistics include the number of data packets and octets, broadcast and multicast packets, and data frame size distribution.

Token Ring MAC-Layer History

This extension offers historical views of MAC-layer statistics based on user-defined sample intervals, which can be set from one second to one hour to enable short-term or long-term historical analysis.

Token Ring promiscuous history This extension offers historical views of promiscuous (i.e., unfiltered) statistics based on user-defined sample intervals, which can be set from one second to one hour in order to enable short-term or long-term historical analysis.

Ring Station Control Table

This extension lists status information for each ring being monitored. Statistics include ring state, active monitor, hard-error beacon-fault domain, and the number of active stations.

Ring station table This extension provides diagnostics and status information for each station on the ring. The types of information collected includes station MAC address, status, and isolating and non-isolating soft-error diagnostics.

Source Routing Statistics

The extension for source routing statistics is used for monitoring the efficiency of source-routing processes by keeping track of the number of data packets routed into, out of, and through each ring segment. Traffic distribution by hop count provides an indication of how much bandwidth is being consumed by traffic-routing functions.

Ring Station Configuration Control

The extension for station configuration control provides a description of the network's physical configuration. A media fault is reported as a fault domain—an area that isolates the problem to two adjacent nodes and the wiring between them. The network administrator can discover the exact location of the problem—the fault domain—by referring to the network map. Faults that result from changes to the physical ring, including each time a station inserts or removes itself from the network, are discovered by comparing the start of symptoms with the timing of physical changes.

The RMON MIB not only keeps track of the status of each station, but it also reports the condition of each ring being monitored by a RMON agent. On large Token Ring networks with several rings, the health of each ring segment and the number of active and inactive stations on each ring can be monitored simultaneously. Network administrators can be alerted to the location of the fault domain should any ring go into a beaconing (fault) condition. Network managers can also be alerted to any changes in backbone ring configuration, which could indicate loss of connectivity to an interconnect device such as a bridge or to a shared resource such as a server.

Ring Station Configuration

The ring station group collects Token Ring specific errors. Statistics are kept on all significant MAC-level events to assist in fault isolation, including ring purges, beacons, claim tokens, and error conditions such as burst errors, lost frames, congestion errors, frame-copied errors, and soft errors.

Ring Station Order

Each station can be placed on the network map in a specified order relative to the other stations on the ring. This extension provides a list of stations that are attached to the ring in logical ring order. This extension lists only stations that comply with 802.5 active monitoring ring poll or IBM trace tool present advertisement conventions.

RMON II

The RMON MIB is basically a MAC-level standard. Its visibility does not extend beyond the router port, meaning that it cannot see beyond individual LAN segments. As such, it does not provide visibility into conversations across the network or connectivity between the various network segments. Given the trends toward remote access and distributed workgroups, which generate a lot of intersegment traffic, visibility across the enterprise is an important capability to have.

RMON II extends the packet capture and decoding capabilities of the original RMON MIB to Layers 3 through 7 of the OSI reference model, which enables traffic to be monitored via network-layer addresses. This process enables RMON to see beyond the router to the internetwork and to distinguish between applications.

Analysis tools that support the network-layer can sort traffic by protocol, rather than just report on aggregate traffic. In other words, network managers will be able to determine, for example, the percent of IP versus IPX traffic traversing the network. In addition, these higher-level monitoring tools can map end-to-end traffic, giving network managers the ability to trace communications between two hosts—or nodes—even if the two are located on different LAN segments. RMON II functions that will enable this level of visibility include the following:

- *Protocol directory table* Provides a list of all of the different protocols that a RMON II probe can interpret
- *Protocol distribution table* Permits tracking of the number of bytes and packets on any given segment that have been sent from each of the protocols that are supported. This information is useful for displaying traffic types by percentage in graphical form.
- *Address mapping* Permits identification of traffic-generating nodes or hosts by Ethernet or Token Ring address, in addition to MAC address. Address mapping also discovers switch or hub ports to which the hosts are attached, which is helpful in node discovery and network topology applications for pinpointing the specific paths of network traffic.
- *Network-layer host table* Permits tracking of bytes, packets, and errors by host, according to individual network-layer protocol
- *Network-layer matrix table* Permits tracking by network-layer address of the number of packets that are sent between pairs of hosts
- *Application-layer host table* Permits the tracking of bytes, packets, and errors by host and according to application
- *Application-layer matrix table* Permits tracking of conversations between pairs of hosts by application
- *History group* Permits filtering and storing of statistics according to user-defined parameters and time intervals
- *Configuration group* Defines standard configuration parameters for probes, including parameters such as network address, serial-line information, and SNMP trap destination information

RMON II is focused more on helping network managers understand traffic flow for the purpose of capacity planning rather than for the purpose of physical troubleshooting. The capability to identify traffic levels and statistics by application has the potential to greatly reduce the time it takes to troubleshoot certain problems. Without tools that can pinpoint which software application is responsible for gobbling up a disproportionate share of the available bandwidth, network managers can only guess. Often, it is easier just to upgrade a server or a buy more bandwidth, which inflates operating costs and shrinks budgets.

Summary

Applying remote monitoring and statistics gathering capabilities to the Ethernet and Token Ring environments via the RMON MIB offers a number of benefits. The availability of critical networks is maximized, because remote capabilities allow for more timely problem resolution. With the capability to resolve problems remotely, operations staff can avoid costly travel to troubleshoot problems on-site. With the capability to analyze data collected at specific intervals over a long period of time, intermittent problems can be tracked down that would normally go undetected and unresolved. Also, with RMON II, these capabilities are enhanced and extended up to the applications level across the enterprise.

See Also

Network Agents

Open Systems Interconnection (OSI)

Simple Network Management Protocol (SNMP)

Remote Node

Remote node is a method of remote access that permits users to dial into the corporate network and perform tasks as if they were locally attached to the LAN. With remote node, the remote system performs client functions, while the office-based host systems perform true server functions. This functionality enables remote users to take advantage of a host's processing capabilities. Only the results are transmitted over the connection back to the remote computer.

With remote node, the user's modem-equipped PC dials into the LAN and behaves as if it were a local LAN node. Instead of keystrokes and screen updates, the traffic on the remote node's dialup line is essentially normal network traffic. The remote PC does not control another PC, as in remote control; rather, it runs regular applications as if it were directly attached to the LAN.

Remote node usually relies on a remote access server that is set up and maintained at a central location, which enables many remote users, all at different locations, to share the same resources. The server usually has Ethernet and/or Token Ring ports and built-in modems for dialup access at up to 56Kbps. Instead of modems, some vendors offer high-speed asynchronous ports. Depending on vendor, there might be optional support for higher-speed connections over services such as ISDN or Frame Relay.

Remote node offers many of the same features as remote control—and, in some cases, surpasses them in functionality. This statement is especially true in key areas such as management, event reporting, and security.

Management

Most vendors enable the remote access server's routing software to be configured from a local management console, a remote Telnet session, or an SNMP management station. SNMP support facilitates ongoing management and integrates the remote access server into the management environment that is commonly used by network administrators.

Management utilities enable status monitoring of individual ports, the collection of service statistics, and the viewing of audit trails on port access and usage. Other statistics include port address, traffic type, connect time/break connection, and connect time exceeded. Filters can be applied to customize management reports with only the desired type of information.

The SNMP management station, in conjunction with the vendor-supplied *Management Information Base* (MIB), can display alerts about the operation of the server. The network administrator is notified when an application processor has been automatically reset due to a time out, for example, or when there is a hardware failure on a processor, which triggers an anti-locking mechanism reset. This type of reset ensures that all users are not locked out of the server by the failing processor.

Event Reporting

Remote-node products typically include management software that is installed at the server in order to provide usage statistics, including packets that are sent and received and transmission errors, as well as who is logged on, how long they have been connected, and what types of modems are attached to the device.

Support for SNMP enables the server to trap a variety of meaningful events. Special drivers pass these traps to any SNMP-based network management platform, such as Hewlett-Packard's OpenView, IBM's NetView, or Sun's Solstice SunNet Manager.

Security

Remote-node products generally offer more levels of security than remote control products. Depending on the size of the network and the sensitivity of the information that can be remotely accessed, one or more of the following security methods can be employed:

- *User ID and password*   These items are routinely used to control access to the server. Security is enhanced if the ID and password are encrypted before going out over the communications link.

- *Authentication*   Authentication involves the server verifying the identity of the remote caller by issuing a challenge to which he or she must respond with the correct answer.

- *Access restrictions*   These restrictions involve assigning each remote user a specific location (i.e., a directory or drive) that can be accessed in the server. Access to specific servers also can be controlled.

- *Time restrictions*   These restrictions involve assigning each remote user a specific amount of connection time, after which the connection is dropped.

- *Connection restrictions*   These restrictions involve limiting the number of consecutive connection attempts and/or the number of times connections can be established on an hourly or daily basis.

- *Protocol restrictions*   These restrictions involve limiting users to a specific protocol for remote access.

Other options exist for securing the LAN, such as callback and cryptography. With callback, the remote client's call is accepted, the line is disconnected, and the server calls back after checking that the telephone number is valid. While this function works well for branch offices, most callback products are not appropriate for mobile users whose locations vary on a daily basis. Products that accept roving callback numbers exist on the market, however. This feature enables mobile users to call into a remote access server or host computer, type their user ID and password, and then specify a number where the server or host should call them back. The callback number is then logged and can be used to help track down security breaches.

To safeguard sensitive information, third-party security systems can be added to the server. Some of these systems require a user password and also a special credit card-sized device that generates a new ID every 60 seconds, which must be matched by a similar ID number-generation process on the remote user's computer.

In addition to callback and encryption, security can be enforced via IP filtering and login passwords for the system console and for Telnet and FTP-server programs. Many remote-node products also enforce security at the link level using the *Point-to-Point Protocol* (PPP) with the *Challenge Handshake Authentication Protocol* (CHAP) and *Password Authentication Protocol* (PAP).

Summary

Companies are being driven to provide effective and reliable remote access solutions—remote control, as well as remote node—in order to meet the productivity needs of today's increasingly decentralized workforce. While remote control usually relies on dialup connections that are established by modems, remote-node connections to the remote access server can be established by dialup bridge/routers as well as by modems. Dialup bridge/routers serve branch offices, while modems are typically used by mobile workers and telecommuters.

See Also

Network Security
Remote Control

Repeaters

A repeater is a device that extends the inherent distance limitations of various networks by regenerating the signals. As such, the repeater operates at the lowest level of the *Open Systems Interconnection* (OSI) reference model: the physical layer (refer to Figure R-3).

Signal regeneration is necessary because signal strength weakens with distance. The longer the path a signal must travel, the weaker the signal becomes. This condition is known as signal attenuation. On a telephone call, a weak signal will cause low volume, interfering with the parties' abilities

Figure R-3
Repeaters operate at
the physical layer of
the OSI reference
model.

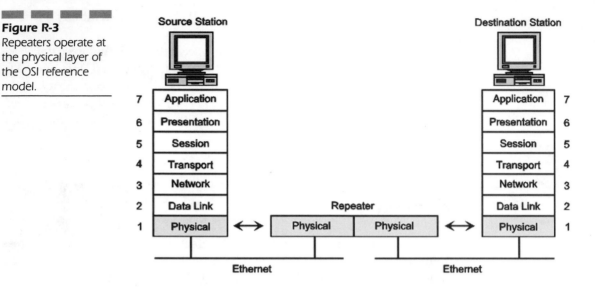

to hear each other. In cellular networks, when a mobile user moves beyond the range of a cell site, the signal fades to the point of disconnecting the call. In the LAN environment, a weak signal can result in corrupt data, which can substantially reduce throughput by forcing retransmissions when errors are detected. When the signal level drops low enough, the chances of interference from external noise increase, rendering the signal unusable.

In the LAN environment, the protocols limit the use of repeaters. With Ethernet, for example, repeaters are usually required every 500 feet. The IEEE 802.3 design guidelines for Ethernet LANs specify that the span between the two farthest users, including the cable connecting each user to the LAN, cannot exceed 500 meters (about 1,500 feet). Even with repeaters, the typical Ethernet LAN application requires that the entire path not exceed 1,500 meters end-to-end. This limit is established for the proper operation of Ethernet's media access control mechanism, which is known as *Carrier-Sense Multiple Access with Collision Detection* (CSMA/CD).

Aside from signal regeneration, repeaters can also be used to link different types of network media—fiber to coaxial cable, for example. Often LANs are interconnected in a campus environment by means of repeaters that form the LANs into connected network segments. The segments can employ different transmission media—thick or thin coaxial cable, twisted-pair wiring, or optical fiber. Often, the hub performs the repeater function.

Regenerators

Often, the terms *repeater* and *regenerator* are used interchangeably, but there is a subtle difference between the two. In an analog system, a repeater boosts the desired signal strength but also boosts the noise level, as well. Consequently, the signal-to-noise ratio on the output side of the repeater remains the same as on the input side. In other words, once noise is introduced into the desired signal, it is impossible to get the signal back into its original form again on the output side of the repeater.

In a digital system, regenerators are used instead of repeaters. The regenerator determines whether the information-carrying bits are 1s or 0s, based on the received signal on the input side. Once the decision of 1 or 0 is made, a fresh signal representing that bit is transmitted on the output side of the regenerator. Because the quality of the output signal is a perfect replication of the input signal, it is possible to maintain a high level of performance over a range of transmission impairments.

Summary

Stand-alone repeaters have transceiver interface modules that provide connections to various media. Fiber-optic transceivers, coaxial transceivers, and twisted-pair transceivers are available. Some repeaters have the intelligence to detect packet collisions and will not repeat collision fragments to other cable segments. They also can disconnect themselves from the hub when there are excessive errors on the cable segment and can submit performance information to a central management station.

See Also

Bridges

Routers

Gateways

Request for Proposal (RFP)

The *Request for Proposal* (RFP) is a formal, technical document that completely describes an organization's requirements for new systems or networks. After the RFP is written, it is distributed to qualified vendors or service providers who must respond by a specified date with a proposal that describes how they plan to meet the organization's requirements and at what cost. Among the goals of the RFP is to provide the organization with the following components:

- A consistent set of vendor responses that are narrow in scope, for easy comparison

- A formal statement of requirements from which a purchase contract can be written and against which vendor performance can be benchmarked

- A mechanism within which vendors, fostered by the implied competition of a general solicitation for bids, assure the terms and conditions

A properly written RFP and carefully managed evaluation process can accomplish these goals. The RFP should not be so rigid, however, that it locks out vendor-recommended alternative solutions that might be more efficient and economical. At the same time, an overly broad RFP that invites vendors to propose the optimum solution every step of the way is essentially no RFP at all. Not only does this open-ended approach produce responses that are difficult to compare, but it also leaves too much room for generalities and obfuscation on the part of vendors.

An overly interpretive RFP can also open the door to challenge from the losing bidders, which can tie up corporate resources and delay installation. To avoid this situation, it is best to know as much as possible about the business objectives and feasible solutions and to describe them clearly and concisely so that all vendor responses will be focused enough for easy comparison and fair evaluation.

RFP Alternatives

The purpose of the RFP is bid solicitation. Other types of documents are used when different forms of assistance are required. For example, the *Request for Quotation* (RFQ) is used is used when planning the bulk purchase of commodity products such as PCs, printers, modems, and applications software. The RFQ is used when the most cost-effective solution is the overriding concern.

The *Request for Information* (RFI) is used when the organization is looking for the latest information on a particular technology, but has no immediate need. The purpose of the RFI is merely to get briefed on new technologies, how vendors plan to employ a particular technology in the future, or to obtain vendors' perspectives on the feasibility of using or integrating a particular technology in a current network. Compared to the RFP and RFQ, the RFI is an informal document. Vendor responses tend to be brief, and they might or might not include information on product pricing and availability. Nevertheless, the RFI responses can be useful for planning purposes and for deciding which vendors might qualify for a future RFP.

Summary

A technically knowledgeable team usually develops the RFP with input from all parts of the organization. The RFP is packaged in a professional manner; it is organized simply and logically, thereby making it easy for vendors to follow and helping them to develop a timely response that addresses all of the important issues. A good RFP is also written in a style that invites input from the vendors or service providers bidding on the project. A team manager plays the leading role in overseeing the development of the RFP, evaluating vendor proposals and conducting meetings with vendors, sometimes with the aid of a consultant.

See Also

Business Process Re-Engineering (BPE)

Routers

A router operates at the network layer of the OSI Reference Model, distinguishing among network layer protocols and making intelligent packet-forwarding decisions. The router can be used to segment a network with the goals of limiting broadcast traffic and providing security, control, and redundant paths. A router also provides firewall service and provides multiple types of interfaces, including those for T1, Frame Relay, ISDN, ATM, cable networks and *Digital Subscriber Line* (DSL) services, among others.

A router is similar to a bridge in that both provide filtering and bridging functions across the network. But while bridges operate at the physical and data-link layers of the OSI reference model, routers join LANs at the network layer (Figure R-4). Routers convert LAN protocols into WAN protocols and perform the process in reverse at the remote location. They can be

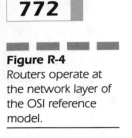

Figure R-4
Routers operate at
the network layer of
the OSI reference
model.

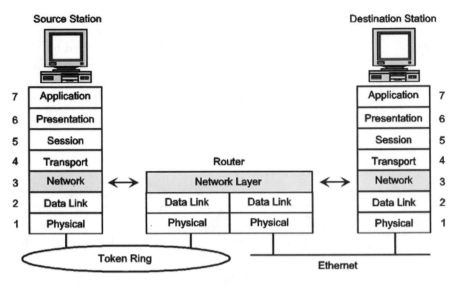

deployed in mesh as well as in point-to-point networks—and, in certain situations, can be used in combination with bridges.

Although routers include the functionality of bridges, they differ from bridges in that they generally offer more embedded intelligence—and, consequently, offer more sophisticated network management and traffic-control capabilities than bridges. Perhaps the most significant distinction between a router and a bridge is that a bridge delivers packets of data on a best-effort basis; specifically, by discarding packets that it does not recognize onto an adjacent network. Through a continual process of discarding unfamiliar packets, data gets to its proper destination: on a network where the bridge recognizes the packets as belonging to a device attached to its network. By contrast, a router takes a more intelligent approach to getting packets to their destination: by selecting the most economical path (i.e., least number of hops) based on its knowledge of the overall network topology, as defined by its internal routing table. Routers also have flow control and error-protection capabilities.

Types of Routing

Two types of routing are available: static and dynamic. In static routing, the network manager configures the routing table to set fixed paths between two routers. Unless reconfigured, the paths on the network never change.

Although a static router will recognize that a link has gone down and will issue an alarm, it will not automatically reroute traffic.

A dynamic router, on the other hand, reconfigures the routing table automatically and recalculates the most efficient path in terms of load, line delay, or bandwidth. Some routers even balance the traffic load across multiple links, which enables the various links to better handle peak traffic conditions.

Routing Protocols

Each router on the network keeps a routing table and moves data along the network from one router to the next by using routing protocols such as *Open Shortest Path First* (OSPF), *Intra-Autonomous System to Intra-Autonomous System* (IS-IS), *External Gateway Protocol* (EGP), *Border Gateway Protocol* (BGP), *Inter-Domain Policy Routing* (IDPR), and *Routing Information Protocol* (RIP).

Although still supported by many vendors, RIP does not perform well in today's increasingly complex networks. As the network expands, routing updates grow larger under RIP and consume more bandwidth to route the information. When a link fails, the RIP update procedure slows route discovery, increases network traffic and bandwidth usage, and might cause temporary looping of data traffic. Also, RIP cannot calculate routes based on such factors as delay and bandwidth, and its line selection facility is capable of choosing only one path to each destination.

The newer routing standard, OSPF, overcomes the limitations of RIP and even provides capabilities that do not exist in RIP. The update procedure of OSPF requires that each router on the network transmit a packet with a description of its local links to all other routers. On receiving each packet, the other routers acknowledge the packet, and in the process, distributed routing tables are built from the collected descriptions. Because these description packets are relatively small, they produce a minimum of overhead. When a link fails, updated information floods the network—enabling all of the routers to simultaneously calculate new tables.

Types of Routers

Multi-protocol nodal (or hub) routers are used for building highly meshed wide-area internets. In addition to enabling several protocols to share the same logical network, these devices pick the shortest path to the end node, balance the load across multiple physical links, reroute traffic around points of failure or congestion, and implement flow control in conjunction with the end nodes. They also provide the means to tie remote branch offices into the corporate backbone, which might use such WAN services as TCP/IP, T1, ISDN, and ATM. Some vendors also provide an interface for SMDS.

Access routers are typically used at branch offices. These are usually fixed-configuration devices available in Ethernet and Token Ring versions, which support a limited number of protocols and physical interfaces. They provide connectivity to high-end multi-protocol routers, enabling large and small nodes to be managed as a single logical enterprise network. Although low-cost, plug-and-play bridges can meet the need for branch office connectivity, low-end routers can offer more intelligence and configuration flexibility at a comparable cost.

The newest access routers are multi-service devices, which are designed to handle a mix of data, voice, and video traffic. They support a variety of WAN connections through built-in interfaces that include dual ISDN BRI interfaces, dual analog ports, T1/Frame Relay port, and an ISDN interface for video conferencing. Such routers can run software that provides IPSec VPN, firewall and encryption services.

Mid-range routers provide network connectivity between corporate locations in support of workgroups or the corporate intranet, for example. These routers can be stand-alone devices or packaged as modules that occupy slots in an intelligent wiring hub or LAN switch. In fact, this type of router is often used to provide connectivity between multiple wiring hubs or LAN switches over high-speed LAN backbones such as ATM, FDDI, and Fast Ethernet.

A new class of consumer routers are capable of providing shared access to the Internet over such broad-band technologies as cable and DSL. The Linksys Instant Broadband EtherFast Cable/DSL Router (refer to Figure R-5), for example, is used to connect a small group of PCs to a high-speed Internet connection or to an Ethernet backbone. Configurable through any networked PC's Web browser, the router can be set up as a firewall and *Dynamic Host Configuration Protocol* (DHCP) server, enabling it to act as an externally recognized Internet device with its own IP address for the home LAN. Unlike a typical router, which can only share 100Mbps over all of its connections, the Linksys device is also equipped with a four-port Ethernet switch that dedicates 100Mbps to every connected PC.

The Linksys router also supports *Network Address Translation* (NAT), a feature that translates one public IP address, given by the cable or DSL Internet provider, and assigns automatically up to 253 private IP addresses to users on the LAN. All of the users who are given an IP address by the

Figure R-5
The Linksys Instant Broadband EtherFast Cable/DSL Router is representative of the new breed of user-friendly, multi-function routers that are aimed at the emerging home-networking market.

router are safe behind the firewall so incoming and outgoing requests are filtered, keeping unwanted requests off the LAN.

At the same time, the Linksys router supports a feature called DMZ /Expose Host, which dissembles one of the 253 private IP addresses to become a public IP address so outside users can access that PC without getting blocked by the firewall. An example would be a gamer playing another gamer via the Internet. They want to access each other's computers so that they can play the game.

Summary

Routers fulfill a vital role in implementing complex mesh networks. They also have become an economical means of tying branch offices into the enterprise network and providing PCs that are tied together on a home network with shared access to broad-band Internet services, such as cable and DSL. Similar to other interconnection devices, routers are manageable via SNMP, as well as the proprietary management systems of vendors. Just as bridging and routing functions made their way into a single device, routing and switching functions are being combined in the same way and even add firewall, DHCP, and NAT capabilities.

See Also

Branch-Office Routing

Bridges

Gateways

Repeaters

Rural Radiotelephone Service

Operating in the paired 152MHz/158MHz and 454MHz/459MHz bands, rural radiotelephone service is a fixed wireless service that enables common carriers to provide telephone service to the homes of subscribers who are living in extremely remote, rural areas where it is not feasible to provide telephone service by wire or other means. Rural radiotelephone stations employ standard duplex analog technology in order to provide telephone service to subscribers' homes.

The quality of conventional rural radiotelephone service is similar to that of pre-cellular mobile telephone service. Several subscribers might have to share a radio channel pair (similar to party-line service), each waiting until the channel pair is not in use by the others before making or receiving a call. Rural radiotelephone service is generally considered by state regulators to be a separate service that is interconnected to the PSTN. This service has been available to rural subscribers for more than 25 years.

Summary

Carriers must apply to the FCC for permission to offer rural radiotelephone service. Among other things, each application for a central office station must contain an exhibit showing that it is impractical to provide the required communication service by means of land-line facilities.

See Also

Air-Ground Radiotelephone Service

Fixed Wireless Access

Satellite Communications

The idea of using satellites as relay stations for an international microwave radiotelephone system goes back to 1945 when Arthur C. Clarke (refer to Figure S-1) proposed the scheme in a British technical journal. Clarke, then a young scientist and officer of the Royal Air Force, later became a leading science fiction writer and co-author of the motion picture "2001: A Space Odyssey." The first satellite, however, was not put into orbit until 1957. Although only a beacon whose primary purpose was to announce its presence in the sky, the Russian satellite Sputnik had a successful orbital deployment that sparked a technological revolution in communications that has continued to this day. Currently, there are 2,559 satellites in orbit, along with over 6,000 pieces of debris that are tracked by the *National Aeronautics and Space Administration* (NASA).

The first communications satellites, Echo 1 and Echo 2, were launched by the United States in the early 1960s. They were little more than metallic balloons that simply reflected microwave signals from point A to point B. Although highly reliable, these passive satellites could not amplify the signals. Reception was often poor, and the range of transmission was limited. Ground stations had to track them across the sky, and communication between two ground stations was only possible for a few hours a day when both had visibility with the satellite at the same time. Later, geosynchronous (or geostationary) satellites overcame this problem. Such systems were high enough in orbit to move with the earth's rotation, in effect, giving them fixed positions so they could provide communications coverage to specific areas.

Satellites are categorized by their type of orbit and area of coverage, as follows:

- Geostationary Earth-Orbit (*GEO*) satellites orbit the equator in a fixed position about 22,000 miles above the Earth. Three GEO satellites can cover most of the planet, with each unit being capable of handling 20,000 voice channels. Because of their large coverage footprint, these satellites are ideal for radio and television broadcasting and long-distance domestic and international communications.

- Middle Earth-Orbit (*MEO*) satellites circle the Earth at about 6,000 miles up. About 12 satellites are necessary to provide global coverage. The lower orbit reduces power requirements and transmission delays that can affect signal quality and service interaction.

Figure S-1
Arthur C. Clarke articulated his vision of satellites in a British technical journal in 1945.

■ Low Earth-Orbit (*LEO*) satellites circle the Earth only 600 miles up (refer to Figure S-2). As many as 200 satellites might be required to provide global coverage. Because their low altitude means that they have non-stationary orbits and that they pass over a stationary caller rather quickly, calls must be handed off from one satellite to the next in order to keep the session alive. The omni-directional antennas of these devices do not have to be pointed at a specific satellite. Little propagation delay also exists. Also, the low altitude of these satellites means that Earth-bound transceivers can be packaged as low-powered, inexpensive, hand-held devices.

The *International Telecommunications Union* (ITU) is responsible for all frequency/orbit assignments. The *Federal Communications Commission* (FCC) regulates all service rates, competition among carriers, and inter-state and international telecommunications traffic in the United States, ensuring that U.S. satellite operators conform to ITU frequency and orbit assignments. The FCC also issues licenses to domestic satellite-service providers.

Figure S-2
Low Earth-Orbiting
(LEO) satellites hold
the promise of
ubiquitous Personal
Communications
Services (PCS),
including telephone,
pager, and two-way
messaging services
worldwide.

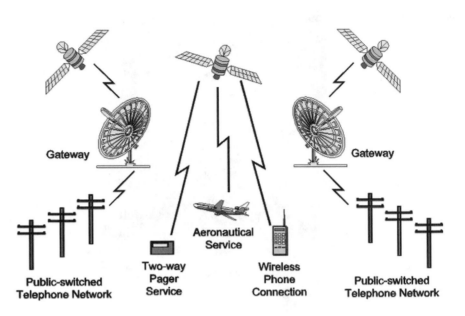

Satellite Technology

Each satellite carries transponders, which are devices that receive radio signals at one frequency and convert them to another for transmission. The uplink and downlink frequencies are separated to minimize interference between transmitted and received signals.

Satellite channels enable one sending station to broadcast transmissions to one or more receiving stations simultaneously. In a typical scenario, the communications channel starts at a host computer that is connected through a traditional telephone company medium to the central office (i.e., master Earth station or hub) of the satellite communications vendor. The data from this loop and from other local loops is multiplexed into a fiber-optic or microwave signal and is sent to the satellite vendor's Earth station. This signal becomes part of a composite transmission that is sent by the Earth station to the satellite (uplink) and then is transmitted by the satellite to the receiving Earth stations (downlink). At the receiving Earth station, the data is transferred by a fiber-optic or microwave link to the satellite carrier's central office. The composite signal is then separated into individual communications channels that are distributed over the PSTN to their destinations.

Satellite communication is reliable for data transmission. The *Bit-Error Rate* (BER) for a typical satellite channel is in the range of one error in one billion bits transmitted. A potential problem with satellite communication, however, is delay. Round-trip satellite transmission takes approximately 500 milliseconds, which can hamper voice communications and create significant problems for real-time, interactive data transmissions. For voice communications, digital echo cancelers can correct voice echo problems that are caused by the transmission delay.

A variety of techniques are employed to nullify the effects of delay during data transmissions via satellite. One technique that is employed by Mentat, Inc. increases the performance of Internet and intranet access over satellites by transparently replacing the *Transmission Control Protocol* (TCP) over the satellite link with a protocol that is optimized for satellite conditions. The company's SkyX Gateway intercepts the TCP connection from the client and converts the data to a proprietary protocol for transmission over the satellite. The gateway on the opposite side of the satellite link translates the data back to TCP for communication with the server (refer to Figure S-3). The result is vastly improved performance, while the

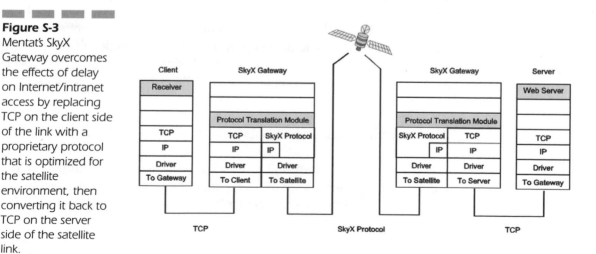

Figure S-3

Mentat's SkyX Gateway overcomes the effects of delay on Internet/intranet access by replacing TCP on the client side of the link with a proprietary protocol that is optimized for the satellite environment, then converting it back to TCP on the server side of the satellite link.

process remains transparent to the end user and is fully compatible with the Internet infrastructure. No changes are required to the client or server, and all applications continue to function without modification.

Very Small Aperture Terminals (VSATs)

VSAT networks have evolved to become mainstream communication-networking solutions that are affordable to both large and small companies. Today's VSAT is a flexible, software-intensive system that is built around standard communications protocols. With a satellite as the serving office and using *Radio Frequency* (RF) electronics instead of copper or fiber cables, these systems can be truly considered packet-switching systems in the sky.

VSATs can be configured for broadcast (one-way) or interactive (two-way) communications. The typical star topology provides a flexible and economical means of communications with multiple remote or mobile sites. Applications include the following: broadcasting database information, insurance agent support, reservations systems, retail point-of-sale credit checking, and interactive inventory data sharing.

Today's VSATs are used for supporting high-speed message broadcasting, image delivery, integrated data and voice, and mobile communications. VSATs are increasingly used for supporting LAN-to-LAN connectivity and LAN-to-WAN bridging, as well as for providing route and media diversity for disaster recovery.

To make VSAT technology more affordable, VSAT providers offer compact hubs and submeter antennas that provide additional functionality at approximately 33 percent less than the cost of full-size systems. Newer submeter antennas are even supporting direct digital TV broadcasts to the home.

Network Management

The performance of the VSAT network is continuously monitored at the hub location by the network control system. A failure anywhere on the network automatically alerts the network control operator, who can reconfigure capacity among individual VSAT systems. In the case of signal fade due to adverse weather conditions, for example, the hub detects the weak signal —or the absence of a signal—and alerts the network operations staff so that corrective action can be taken.

Today's network-management systems indicate whether power failures are local or remote. They can also locate the source of communications problems and determine whether the trouble is with the software or hardware. Such capabilities often eliminate the expense of dispatching technicians to remote locations. Also, when technicians must be dispatched, the diagnostic capabilities of the network-management system can ensure that service personnel have the appropriate replacement parts, test gear, software patch, and documentation with them to solve the problem in a single service call.

Overall link performance is determined by the bit-error rate, network availability, and response time. Because of the huge amount of information transmitted by the hub station, uplink performance requirements are more stringent. A combination of uplink and downlink availability, coupled with BER and response time, provides the network control operator with overall network performance information on a continuous basis.

The VSAT's management system provides an interface to the major enterprise management platforms for single-point monitoring and control and offers a full range of accounting, maintenance, and data-flow statistics, including those for inbound versus outbound data flow, peak periods, and total traffic volume by node. Also provided are capabilities for identifying fault conditions, performing diagnostics, and initiating service restoration procedures.

Communications Protocols

Within the VSAT network, there are three categories of protocols: those that are associated with the backbone network, those of the host computer, and those that are concerned with transponder access. The scope and functionality of protocol handling differ markedly among VSAT network providers.

The backbone network protocol is responsible for flow control, retransmissions of bad packets, and running concurrent multiple sessions. The backbone network protocol could be associated with either the host or the communications link. The host protocol is related to the user application interface, which provides a compatible translation between the backbone protocol and the host communications protocol. Several host protocols are used in VSAT networks, including SNA/SDLC, 3270 BSC, Poll Select, and HASP. Multiple protocols can be used at the same VSAT location.

Transponder access protocols are used to assign transponder resources to various VSATs on the network. The three key transponder access protocols that are used on VSAT networks include *Frequency-Division Multiple Access* (FDMA), *Time-Division Multiple Access* (TDMA), and *Code-Division Multiple Access* (CDMA).

FDMA With FDMA, the radio frequency is partitioned so that bandwidth can be allocated to each VSAT on the network. This functionality permits multiple VSATs to simultaneously use their portion of the frequency spectrum.

TDMA With TDMA, each VSAT accesses the hub via the satellite by the bursting of digital information onto its assigned radio frequency carrier. Each VSAT bursts at its assigned time relative to the other VSATs on the network. Dividing access in this way (by time slots) is inherently wasteful, because bandwidth is available to the VSAT in fixed increments (whether or not the bandwidth is needed). To improve the efficiency of TDMA, other techniques are applied to ensure that all the available bandwidth is used, regardless of whether the application contains bursty or streaming data. A reservation technique can even be applied to ensure that bandwidth is available for priority applications.

CDMA With CDMA, all VSATs share the assigned frequency spectrum and can also transmit simultaneously. This function is possible through the use of spread spectrum technology, which employs a wide-band channel, as opposed to the narrow-band channels employed by other multiple access techniques such as FDMA and TDMA. Over the wide-band channel, each transmission is assigned a unique code—a long row of numbers resembling a combination to a lock. The outbound data streams are coded so that they can be identified and received only by the station(s) having that code. This technique is also used in mobile communications as a means of cutting down interference and increasing available channel capacity by as much as 20 times.

Mobile Satellite Communications

Mobile satellite communications are used by the airline, maritime, and shipping industries. In fact, maritime mobile satellite services are quite advanced, primarily because of the efforts of the *International Maritime Satellite Organization* (INMARSAT).

Formed in the late 1970s, INMARSAT's original, primary goal was to design and construct a world-wide maritime communications system. Today, as a result of its efforts, more than 7,000 ships, off-shore drilling rigs, and other vessels and maritime structures are equipped with satellite communications capabilities—including telex, voice, data, and video transmission. Currently, COMSAT, a private U.S. company that was formed to establish communications via satellites, is the only authorized U.S. organization that can directly access the INMARSAT system.

Two land-based mobile satellite services are currently being implemented in the United States: *Radio Determination Satellite Service* (RDSS) and *Mobile Satellite Service* (MSS).

RDSS is a set of radio telecommunications and computational techniques that helps users determine their precise geographical location so that they can relay their position and other digital information to other subscribers or to a central control center. The system constantly maintains and processes the information flow between a control center, satellites, and the mobile terminals.

MSS technology provides both voice and data communications to mobile users and offers wider geographic coverage than RDSS by utilizing relay satellites. The communications link identifies a transmitting vehicle's position between the mobile units and the satellite by using Loran-C navigational technology. MSS's uplink operates in the L-band, while the downlink operates in the K-band.

Summary

Satellites provide a reliable, economical way of providing communications to remote locations or supporting mobile telecommunications. When taking into account the large number of satellites that can be employed, along with their corresponding radio frequency assignments, you will find that satellite communications systems offer ample room for expansion. Conversion of satellite transmissions from analog to digital, and the use of more sophisticated multiplexing techniques, will further increase satellite transmission capacity. Other technological advances are focusing on the higher frequency bands—applying them in ways that decrease signal degradation.

See Also

Global Positioning System (GPS)

Service-Creation Environment

In today's service-creation environment, new services are hard to deliver. Telephone companies and interexchange carriers and their competitors are dependent on the manufacturers of circuit switches to design and program these new services, which can mean a long development cycle and, ultimately, a longer delivery cycle for new and improved services to customers. Service providers typically must pay exorbitant fees to switch manufacturers for each new service.

With the cost of voice calls falling to the five-cent to seven-cent per minute range, service-provider margins are becoming very thin, leaving less money available for the development and rollout of new services. Nevertheless, in this new market, service providers need the ability to innovate and get ahead of the competition, improving margins and profitability while meeting the emerging demands of their customers. The ability to create and deploy new services in today's market depends on the ability of carriers end their dependence on circuit-switch platform vendors by implementing a service-creation environment for rapid applications development.

The New Model

The new network model, based on the convergence of voice and data over an IP infrastructure, enables service offerings to be delivered faster, at lower cost, and at higher profit margins to service providers. The new service-creation model is one in which services are built on open, standards-based platforms, such as Sun workstations, and are developed on commercially available tools. In this manner, carriers can develop their own applications for new services, dramatically cutting the time to market for revenue generating opportunities while also having the flexibility to provision services anywhere in their network.

While it is generally accepted that the movement of voice traffic onto IP networks will result in substantial operational savings for carriers, the creation of wholly new voice-enabled services and their associated revenue streams is the single most compelling driver for converged voice and data over IP. The key to realizing this potential is the ability to rapidly develop new services through an open service-creation environment.

One of the few companies that are addressing the need for an IP-based service-creation environment is Sonus Networks. Through its *Open-Services Architecture* (OSA), the company provides an open framework upon which new services can be added more quickly and economically than is possible in the traditional circuit-switch model. By tapping OSA's policy-based and script-based service definitions and coupling them with open *Application Programming Interfaces* (APIs) and internetworked links to *Signaling System 7* (SS7), carriers can create enhanced services and make them available transparently to users on their networks. The architecture also enables carriers to offer services that their customers can configure and manage themselves via open interfaces, such as Web browsers.

Open-Services Architecture (OSA)

Sonus Networks' OSA supports any application developed on standards-based IP servers, giving carriers the capability to create and offer new services or applications that seamlessly integrate voice and data or voice and the Web. Support for IP servers is implemented through open APIs, enabling equipment from multiple vendors to work together. Carriers can flexibly combine service developed internally, by network equipment providers, or by third parties. The APIs adhere to all industry standards. The OSA also supports legacy applications delivered from *Service Control Points* (SCPs) on the SS7 network, enabling carriers to continue offering them. Among these applications are 800-number translation and *Local Number Portability* (LNP).

Another benefit of an open architecture is improved performance, which comes from the capability of the service provider to size the application platforms for the *Quality of Service* (QoS) that is required. This situation is analogous to determining the type and size workstation for a Web server. In this way, the carrier will not be burdened with a proprietary platform that is tied to the legacy SCP applications.

A key component of the Sonus OSA is the provision of policy-based networking, which features centralized management of the network's services. With a script orientation, control of the services in the switch is easy and flexible. The Sonus SoftSwitch handles all incoming calls and applies intelligence to each call that is handled. Calls can be provisioned on a subscriber-by-subscriber basis and in conjunction with a *Virtual Private Network* (VPN). The SoftSwitch uses SS7 information together with its databases to decide which services will be applied to the call. The application can also draw upon the service-handling capabilities of legacy SCPs. Alternatively, the SoftSwitch can upload scripts to the switch in order to implement call and mid-call handling or invoke the services of any of the application servers that are attached to the IP network.

Sonus enables third-party application developers to apply their expertise in the critical area of enhanced data and voice services through its *Open-Services Partner Alliance* (OSPA). OSPA enables carriers, *Application Service Providers* (ASPs), third-party *Independent Service Developers* (ISDs), and telephony-system software providers to rapidly develop and deliver competitive services and applications based on the OSA.

Summary

The capability to rapidly create and deploy new value-added services is the most significant benefit of building converged voice-data packet networks. Expecting a wholesale upgrade from circuit-based networks to packet-based networks is not realistic, however. Instead, there will be a migration to *Voice-over-IP* (VoIP) in the immediate future. By slowly introducing voice-over data technologies while maintaining the circuit-switched network in parallel, multiple objectives can be achieved. These objectives include maintaining legacy connectivity to the PSTN by ordinary phones, offloading data traffic from legacy networks, testing the new IP-based networks, gradually deploying new services on packet networks, and protecting investments in legacy equipment and service platforms.

See Also

Multi-Service Networking

Voice-Data Convergence

Service-Level Agreements (SLAs)

Service-Level Agreements (SLAs) are contracts between service providers and users that define mutually agreed upon levels of network performance. SLAs typically cover the data services telecommunications carriers and Internet service providers (ISPs) offer to corporate customers. For example, AT&T, MCI WorldCom, and Sprint build guarantees against network delay into their Frame Relay offerings that vary from 50 milliseconds to 65 milliseconds, depending on the number of PVCs and the applications running over them. SLAs are available for other services as well, from traditional T-carrier services to IP-based VPNs, intranets, and extranets. Telecommunications service providers feel compelled to offer written performance guarantees to differentiate themselves in the increasingly competitive market that is brought about by deregulation.

SLAs are even becoming popular among ISPs as a means to lure business applications out of the corporate headquarters to outsourced Web hosting arrangements. Performing this task successfully requires the ISP to operate a carrier-class data center, to offer reliability guarantees, and to have the technical expertise to fix any problem, day or night. The SLA might also include penalties for poor performance, such as credits against the monthly invoice if network up time falls below a certain threshold. Some ISPs even guarantee levels of accessibility for their dialup remote-access customers.

Intra-Company SLAs

SLAs are not just for carrier-provided data services. They can also be used for the services an IT department provides to other business units within the organization. Here, SLAs are gaining more attention because fundamental changes are underway in the way IT departments are run. Today, many IT departments are being operated as profit centers and there is more awareness of customer satisfaction. Consequently, platform and third-party vendors are providing IT managers with more tools for SLA monitoring and enforcement. For example, IBM's SecureWay Communications Server now monitors and enforces SLAs. If a priority application is being denied the bandwidth called for by the SLA, the performance monitor alerts IT staff that the SLA has been breached. The system administrator can then make adjustments by choking off bandwidth for another application and assigning it to the priority application. In the future, IBM will automate the monitoring and adjustment processes.

In other companies, there are efforts at recentralization, which entails putting more emphasis on data-center resources, rather than equipping every employee with expensive computers and applications. In such environments, there is a lot of pressure on the IT group to document how good of a job it is doing. If IT is going to take control of computing resources away from a department, it must have the tools to prove that it is still meeting the department's needs.

Performance Parameters

The SLAs from service providers define such parameters as the type of service, data rate, and what the expected performance level is to be in terms of delay, error rate, port availability, and network up time. Response time to system repair and/or network restoration also can be incorporated into the SLA, as can penalties for non-compliance. The increasing use of SLAs has resulted in a proliferation of performance measurement and reporting tools to provide users with the documentation they need to confront service providers when network performance falls below mutually agreed-upon thresholds.

Among the issues to consider with regard to SLAs is having appropriate tools to monitor the performance of the service provider. There are a variety of performance metrics that require measurement to ascertain whether the carrier (or IT department) is delivering the grade of service promised in the SLA. These performance metrics differ according to the type of network service. A variety of vendors provide measurement tools, and each uses a different approach to address a narrow range of metrics. In having the capability to independently measure various performance metrics, however, companies have the means to effectively manage the SLA.

Although SLAs exact financial penalties when network performance does not meet expectations, you should remember that they do not indemnify the organization for lost business. Particularly, such guarantees do not help much if the organization's network were to go down for an extended period of time, as was demonstrated in August 1999 when MCI WorldCom's Frame Relay network went out of service for 10 days as the result of a faulty software upgrade. Although MCI WorldCom doubled the credits to 20 days to compensate users for the network outage, its customers were left to fend for themselves in dealing with lost business opportunities.

Summary

In order to meet the challenges of this new competitive era, telecommunications service providers are exploring new approaches to doing business. A major step has been taken with SLAs that offer performance guarantees and credits to customers if various metrics are not sustained. In some cases, the customer need not even report the problem and provide documentation to support its claim. The carrier or ISP will report the problem to the customer and automatically apply appropriate credits to the invoice, as stipulated in the SLA. At the same time, IT departments are implementing SLAs to support the networking and system resource needs of business units. The purpose of the intra-company SLA is to specify, in mutually agreeable metrics, what the various end-user groups can expect from IT in terms of resource availability and system response. SLAs also specify what IT can expect from end users in terms of system usage and cooperation in maintaining and refining the service levels over time. SLAs also provide a useful metric against which the performance of the IT department can be measured. How well the IT department fulfilled its obligations, as spelled out in the SLA, can determine future staffing levels, budgets, raises, and bonuses.

See Also

Network-Management Systems (NMSs)

Signaling System 7 (SS7)

Placing a telephone call successfully requires the use of a signaling protocol to convey information about the call (at a minimum, the call's origin and destination). Early signaling systems used dial pulses or tone pulses to represent the called number. Over the years, these in-band signaling systems were expanded to generate and detect tone pulses that represented the status of the called number (i.e., subscriber free, busy, etc.) or the calling number for *Automatic Number Identification* (ANI). The number of signal types, however, are limited by the number of available tones in the signaling system, and call setup time is extended each time more features are added.

Common Channel Signaling (CCS) overcomes the limitations of in-band signaling systems by using a separate signaling network to convey call information. By separating the signaling and voice networks, not only can signaling occur during a call without affecting voice traffic, but more sophisticated services and features can also be supported. Compared to in-band signaling, common channel signaling enables the following:

- Faster call-setup times
- More efficient use of voice circuits
- Support for services that require signaling during a call
- Better control over fraudulent network usage

With the emergence of ISDN and the increased use of 800 and credit card services in the 1970s, the existing *Common Channel Signaling System 6*

(CCSS6) network architecture had to be revamped to handle not only circuit-related messages, but database-access messages as well. These database-access messages convey information between toll centers and centralized databases in order to permit real-time access to billing-related information and other sophisticated services. Because high-speed digital facilities were being employed to support these new services, a new end-to-end advanced signaling arrangement was required. This new system became known as *Common Channel Signaling System 7* (SS7).

This signaling system is an ITU standard and is in use by all carriers in the United States and in most industrialized countries. Large carriers such as AT&T, Sprint, and MCI WorldCom operate their own SS7 networks, as do many of the large telephone companies. Many smaller telephone companies and CLECs cannot afford SS7 networks of their own. In such cases, they can subscribe to the SS7 networks of third-party service providers. Application developers see the link between data networks and SS7 intelligence as crucial in their efforts to roll out such value-added IP services as voice or fax over IP and *Virtual Private Networking* (VPN).

Network Elements

The SS7 network is comprised of packet-switching elements that are tied together with digital transmission links (refer to Figure S-4):

Figure S-4
Key components of an SS7 network

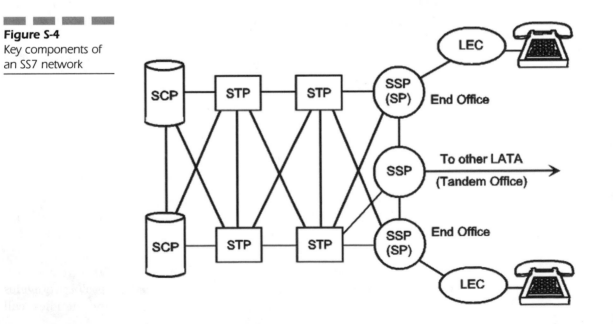

- *Signal Transfer Point* (STP)   A packet switch that routes signaling messages within the network
- *Service Control Point* (SCP)   A network element that interfaces with the STP and contains the network configuration and call-completion database
- *Service Switching Point* (SSP)   Usually an end office (local switch) or access tandem office (long-distance switch) that contains the network signaling protocols and can access the SCP

These service elements are also key building blocks of the *Advanced Intelligent Network* (AIN). In fact, SS7 is what makes the intelligent network intelligent.

Message Types

Two types of signaling messages are conveyed via SS7: circuit-related messages and database-access messages. Circuit-related messages are used to establish and disconnect calls between two signaling points (i.e., SSPs). The STP conveys these messages over the appropriate circuit from the originating signaling point to the terminating signaling point. The information that is contained in these messages includes the identity of the circuit that connects both signaling points, the called number, answer indication, release indication, and release completion indication.

Database-access messages retrieve information that is stored in the *Service Control Point* (SCP). An inquiry message is transmitted from the originating signaling point via the STP to the SCP, requesting the necessary data in order to complete the call. The SCP sends a response message containing the requested data over the SS7 network to the originating signaling point, indicating that the network resources are available to complete the call.

SS7 and OSI

The hardware and software functions of the SS7 protocol are divided into functional abstractions called levels. These levels are somewhat analogous to the seven-layer *Open Systems Interconnect* (OSI) reference model (refer to Figure S-5), which is defined by the *International Standards Organization* (ISO).

***Message Transfer Part* (MTP)**   The MTP of the SS7 protocol provides communication between intelligent network nodes and handles call

Figure S-5
Relationship of SS7 to
the OSI reference
model

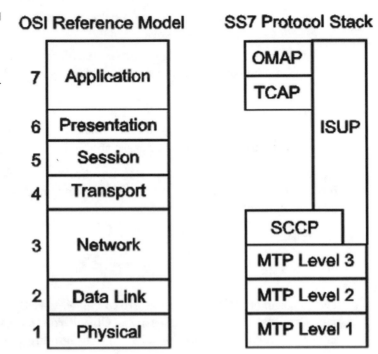

establishment, supervision, disconnection, and billing. MTP is divided into three levels, as follows:

- *MTP Level 1*   The lowest level (Level 1) is equivalent to the OSI physical layer. MTP Level 1 defines the physical, electrical, and functional characteristics of the digital signaling link. Physical interfaces that are defined include E1 (2.048Mbps), DS1 (1.544Mbps), V.35 (64Kbps), DS0 (64Kbps), and DS0-A (56Kbps).

- *MTP Level 2*   MTP Level 2 ensures accurate end-to-end transmission of a message across the signaling link. When an error occurs on the link, the message (or set of messages) is retransmitted. SS7 signal units (messages) are formatted as follows (refer to Figure S-6):

  - Flag   Indicates the beginning and end of a signal unit

  - Backward Sequence Number (*BSN*)   Used to acknowledge the receipt of signal units by the remote signaling point

Figure S-6
SS7 message signal
unit format

Flag	BSN	BIB	FSN	FIB	LI		SIO	SIF	CRC	Flag
Bits 8	7	1	7	1	6	2	8	8	16	8

- Backward Indicator Bit (*BIB*)   Indicates a negative acknowledgment by the remote signaling point
- Forward Sequence Number (*FSN*)   Contains the sequence number of the signal unit
- Forward Indicator Bit (*FIB*)   Used for error recovery
- *Length Indicator* (*LI*)   Indicates the number of octets that follow the LI and precede the CRC
- Service Information Octet (*SIO*)   Contains the network indicator (e.g., national or international) and the message priority
- Signaling Information Field (*SIF*)   Contains the routing label and signaling information
- Cyclic Redundancy Check (*CRC*)   Calculated and appended to the forward message. Upon receiving the message, the remote signaling point checks the CRC and copies the value of the FSN into the BSN of the next available message scheduled for transmission back to the initiating signaling point. If the CRC is correct, the backward message is transmitted. If the CRC is incorrect, the remote signaling point indicates negative acknowledgment by toggling the BIB prior to sending the backward message. When the originating signaling point receives a negative acknowledgment, it retransmits all forward messages, beginning with the corrupted message, with the FIB toggled.
- *MTP Level 3*   MTP Level 3 provides message routing and network-management functions. Messages are routed based on the routing label in the *Signaling Information Field* (SIF) of message signal units. MTP Level 3 also provides network-management and maintenance functions to reconfigure the SS7 network in the event of link failures and to control signaling traffic when congestion occurs.

Signaling Connection Control Part (**SCCP**)   SCCP is used as the transport layer for TCAP-based services, including 800/888, 900, and calling-card services. In support of such services, SCCP provides connectionless and connection-oriented network services and global title translation capabilities. A global title is an address—a dialed 800 or 888 number, for example—which is translated by SCCP into a destination point code and subsystem number. Each service has a unique subsystem number (e.g., 800 service SSN = 254), which identifies the SCCP user at the destination signaling point. SCCP messages are contained within the *Signaling Information Field* (SIF) of a message signal unit. The SIF contains the routing label, followed by the SCCP message contents.

***Transaction Capabilities Applications Part* (TCAP)**   TCAP enables the deployment of AIN services by supporting non-circuit related information exchange between signaling points by using the SCCP connectionless service. An SSP uses TCAP to query an SCP to determine the routing number(s) that is associated with a dialed 800, 888, or 900 number. The SCP uses TCAP to return a response containing the routing number(s)—or an error or reject component—back to the SSP. In the wireless environment, TCAP query and response messages make possible the following functions:

- Roamer registration/validation
- PCS home database
- Roamer service profile delivery
- PCS routing
- Roamer call delivery
- PCS controller access

These transactions are performed in real time. Calling card calls are also validated by using TCAP query and response messages.

***ISDN User Part* (ISUP)**   ISUP messages are used to control the setup and release of trunk circuits that carry voice and data between the calling party and the called party at different terminating line exchanges. (Calls that originate and terminate at the same switch do not use ISUP signaling.) ISUP information is carried in the *Signaling Information Field* (SIF) of a message signal unit.

Various types of ISUP messages are exchanged between intelligent nodes and STPs within the intelligent network, including the following:

- *Forward address message*   Used to set up a circuit between endpoints
- *General setup message*   Conveys additional information that is required during call setup and provides the means for ensuring that a circuit that is crossing multiple ISDNs maintains the desired transmission characteristics across all networks

- *Backward setup message*   Supports call setup and initiates appropriate call accounting and charging procedures

- *Call-supervision message*   Supports call establishment with additional information, including whether the call was answered or not, and the capability for manual operator intervention on ISDN calls between national boundaries

- *Circuit-supervision message*   Supports three functions on a pre-established circuit: release, which terminates the call; suspend and resume; and blocking outgoing calls, which also permits incoming calls on an established (but inactive) circuit

- *Circuit group supervision message*   Performs the same functions as above, but on a group of circuits that is treated as a single unit for purposes of control

- *In-call modification message*   Used to alter the characteristics or associated network facilities of an active call

- *Node-to-node message*   Enables the management and control of closed user groups, so that incoming or outgoing or both types of calls are permitted only between members of the group

Operations, Maintenance, Administration, and Provisioning (OMAP)   OMAP specifies network-management functions and messages that are related to the common operations and maintenance procedures of the carriers.

Summary

SS7 has become the primary mode for signaling and information transfer in today's wireless and wireline networks. Information elements such as calling-party number, routing information that is related to 800 numbers, and current location information for roaming wireless subscribers are all carried over SS7 signaling networks. SS7 is much more powerful and flexible than the earlier signaling systems because it is a message-based system designed to operate on separate digital facilities. This situation led to SS7 growing beyond the initial support of voice services to support for enhanced services for the mass market, such as credit card and debit-card validation, follow-me services, 800 number routing, wireless applications, automatic call-distribution capabilities, and PSTN-to-IP connectivity for voice calls. Work continues within the ITU to evolve and improve SS7, and numerous software companies develop network applications that take advantage of SS7's capabilities.

See Also

>*Advanced Intelligent Network* (AIN)
>
>*Open Systems Interconnection* (OSI)

Simple Network Management Protocol (SNMP)

Since 1988, the *Simple Network Management Protocol* (SNMP) has been the de facto standard for the management of multi-vendor TCP/IP-based networks. SNMP specifies a structure for formatting messages and for transmitting information between reporting devices and data-collection programs on the network. The SNMP-compliant devices on the network are polled for performance-related information, which is passed to a network-management console. Alarms are also passed to the console. There, the gathered information can be viewed in order to pinpoint problems on the network (or can be stored for later analysis).

SNMP runs on top of TCP/IP's datagram protocol—the *User Datagram Protocol* (UDP), which is a transport protocol that offers a connectionless-mode service. In other words, a session does not need to be established before network-management information can be passed to the central control point. Although SNMP messages can be exchanged across any protocol, UDP is well suited to the brief request/response message exchanges characteristic of network-management communications.

SNMP is a flexible network-management protocol that can be used to manage virtually any object, even OSI objects. An object refers to hardware, software, or a logical association, such as a connection or virtual circuit. An object's definition is written by the equipment vendor. The definitions are held in a *Management Information Base* (MIB), which is simply a list of switch settings, hardware counters, in-memory variables, or files that are used by the network-management system to determine the alarm and reporting characteristics of each device on the network, including those that are connected over LANs.

As noted, SNMP is basically a request/response protocol. The management system retrieves information from the agents through SNMP's get and `get-next` commands. The `get` request retrieves the values of specific objects from the MIB, and the MIB lists the network objects for which an agent can return values. These values might include the number of input packets, the number of input errors, and routing information. The `get-next` request permits navigation of the MIB, enabling the next MIB object to be retrieved, relative to its current position. A `set` request is used to request a logically remote agent to alter the values of variables. In addition to these message types, there are trap messages, which are unsolicited messages conveyed from management agent to management stations. Other commands are available that enable the network manager to take specific actions to control the network. Some of these commands look similar to SNMP commands but are really vendor-specific implementations. For example, some vendors use a `stat` command to determine the status of network connections.

All of the major network-management platforms support SNMP, including Hewlett-Packard's OpenView, IBM's NetView, and Sun's Solstice Sun-Net Manager. In addition, many of the third-party systems and network-management applications that plug into these platforms support SNMP. The advantage of using such products is that they take advantage of SNMP's capabilities while providing a *Graphical User Interface* (GUI) to make SNMP easier to use (refer to Figure S-7). Even MIBs can be selected for display and navigation through the GUI.

Another advantage of commercial products is that they can use SNMP to provide additional functionality. For example, OpenView, NetView and Sun-Net Manager are used to manage network devices that are IP addressable and run SNMP. Their automatic discovery capability finds and identifies all IP nodes on the network, including those of other vendors that support SNMP. Based on discovered information, the management system automatically draws the required topology maps. Nodes that cannot be discovered automatically can be represented in either of two ways: first, by manually adding custom or standard icons to the appropriate map views; second, by using SNMP-based APIs for building map applications without having to manually modify the configuration to accommodate non-SNMP devices.

Architectural Components

SNMP is one of three components comprising a total network-management system (refer to Figure S-8). The other two are the *Management Information Base* (MIB) and the *Network Manager* (NM). The MIB defines the controls embedded in network components, while the NM contains the tools

Figure S-7
Castle Rock
Computing, Inc.
offers SNMPc for
Windows NT, which
provides a graphical
display that supports
multi-level
hierarchical mapping
(among other
functions).

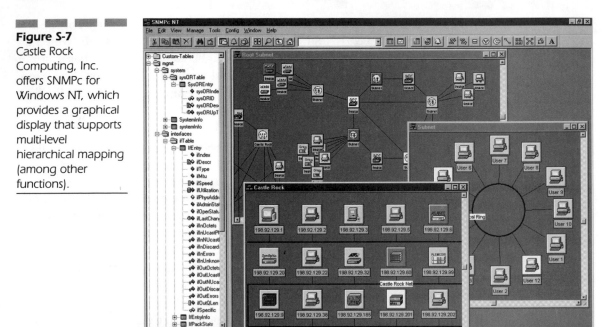

Figure S-8
SNMP architecture

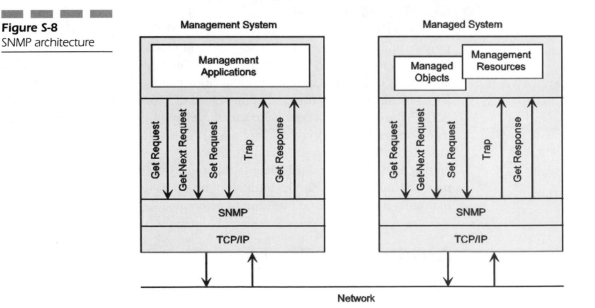

that enable network administrators to comprehend the state of the network
from the gathered information.

Network Manager Network Manager is a program that can run on one host or more than one host, with each host being responsible for a particular subnet. SNMP communicates network-management data to a single site, called a *Network-Management Station* (NMS). Under SNMP, each network segment must have a device, called an agent, which can monitor devices (called objects) on that segment and report the information to the NMS. The agent might be a passive monitoring device whose sole purpose is to read the network, or it might be an active device that performs other functions as well, such as bridging, routing and switching. Devices that are non-SNMP compliant must be linked to the NMS via a proxy agent.

The NMS provides the information display, communication with agents, information filtering, and control capabilities. The agents and their appropriate information are displayed in a graphical format, often against a network map. Network technicians and administrators can query the agents and read the responses on the NMS display. The NMS also periodically polls the agents, searching for anomalies. Detection of an anomaly results in an alarm at the NMS.

MIB The MIB is a list of information that is necessary to manage the various devices on the network. The MIB contains a description of SNMP-compliant objects on the network and the kind of management information that they provide. An object refers to hardware, software, or a logical association, such as a connection or virtual circuit. The attributes of an object might include items such as the number of packets that are sent, the routing table entries, and the protocol-specific variables for IP routing.

The first MIB was primarily concerned with IP routing variables used for interconnecting different networks. 110 objects form the core of the standard SNMP MIB. The latest generation MIB, known as MIB II, defines more than 160 objects and extends SNMP capabilities to a variety of media and network devices, marking a shift from Ethernets and TCP/IP WANs to all media types used on LANs and WANs. Many vendors want to add value to their products by making them more manageable, so they create private extensions to the standard MIB, which can include 200 or more additional objects.

Many vendors of SNMP-compliant products include MIB tool kits that generally include two types of utilities. One, a MIB compiler, acts as a translator that converts ASCII text files of MIBs for use by an SNMP management station. The second type of MIB tool converts the translator's output into a format that can be used by the management station's applications or graphics. These output handlers, also known as MIB editors or MIB walkers, enable users to view the MIB and select the variables to be included in the management system. Some vendors of SNMP management stations do not offer MIB tool kits; rather, they offer an optional service whereby they will integrate into the management system any MIB a user requires for a given network. This service includes debugging and technical support.

Summary

SNMP's popularity stems from the fact that it works, that it is reliable, and that it is widely supported. The protocol itself is in the public domain. SNMP capabilities have been integrated into just about every conceivable device that is used on today's LANs and WANs, including intelligent hubs and carrier services such as Frame Relay.

See Also

Network Management

Open-Systems Interconnection (OSI)

Remote Monitoring (RMON)

Slamming

Since the start of long-distance competition in the telephone industry in the early 1980s, when consumers could switch from AT&T to another long-distance carrier of their choice for the first time, telecommunications companies have been using various marketing techniques to win over new customers. Some of these—media advertising, telemarketing, frequent-flyer bonuses, and lower rate offers—are legitimate. Other techniques, such as slamming, are not.

Slamming is the illegal practice of having a long-distance or other telecommunications service switched without the subscriber's knowledge or permission. According to the FCC, slamming is one of the top three complaints that the agency receives from consumers.

Federal law requires a customer's approval before his or her long-distance carrier can be changed. Some carriers, however, might inform the local telephone company that a customer has given the carrier authorization to carry his or her long-distance call, when in fact no such authorization was provided. Some typical ways that slamming can occur include the following:

- A telemarketer might call with an offer to save the subscriber money if he or she switches long-distance carriers. Even if the subscriber says no, they might wind up being slammed if the telemarketer still reports that the subscriber has agreed to switch.

- The subscriber might receive a check in the mail that states in small print that by signing and cashing the check, he or she agrees to switch to a new carrier.

- The subscriber can fill out what he or she believes to be a contest entry blank that is actually also a consent form to be switched.

In September 1995, the FCC issued rules designed to strengthen protections for consumers while preserving their right to choose or change long-distance carriers. Signatures on contest and sweepstakes forms and other such gimmicks might no longer be used as authorizations to switch companies.

An endorsement on a bonus check will be accepted as an authorization to change a long-distance provider, but only if the authorization form to switch carriers is separate from promotional material and if it is written in plain language and in a print size that is comparable to accompanying advertising copy. In addition, the form must clearly state that its purpose is to change a customer's long-distance carrier.

Case Study

In December 1998, the FCC proposed a forfeiture of $2 million against *Long Distance Direct, Inc.* (LDDI) for apparently engaging in the practice of slamming. In a *Notice of Apparent Liability* (NAL), the FCC determined that LDDI apparently violated the Communications Act and FCC rules by substituting itself as the long-distance carrier for 25 consumers without their consent. In addition, LDDI apparently placed unauthorized product and service charges on the customers' telephone bills, a practice called cramming.

The Commission found these violations particularly egregious because LDDI, a long-distance reseller, appeared to have submitted the long-distance switches and billed consumers for membership fees that were simply based upon the consumers' calls to a psychic hotline service—or, in some cases, without any evidence of contact with the consumer prior to switch or charge.

In many instances, consumers who discovered that they were switched to LDDI were unable to reach the company to complain, even after repeated attempts. Those who were successful in reaching LDDI were often told that the switch had been authorized by a call made from the consumer's telephone to the Psychic Friends hotline. After the consumers contacted LDDI to complain, they discovered on their next telephone bills that LDDI not only had failed to stop providing service, but that they also had imposed additional, unauthorized charges. Based on these findings, the FCC found LDDI liable for a forfeiture in the amount of $40,000 for each of the slamming violations and $40,000 for each of the cramming violations, resulting in a total forfeiture amount of $2,000,000.

Summary

Despite state and federal laws against the practice, slamming is still a problem. In 1998 alone, the FCC proposed $13 million in fines against slammers. To prevent becoming a slamming victim, customers can call their local phone company's business office to request a freeze of their long-distance phone carrier. In that way, the long-distance carrier cannot be switched unless the local company receives the customer's authorization, including a personal ID code, verbal consent, and written consent.

See Also

Telephone Fraud

Truth in Billing

Smart Buildings

So-called smart buildings are multi-tenant office or residential buildings that provide the latest telecommunications technologies—usually at a faster rate and cheaper cost than traditional carriers. The Telecommunications Act of 1996 spurred competition and innovation in the local telecommunications industry, which led to the development of smart buildings. Using cutting-edge technologies such as fixed wireless and fiber networks, *Competitive Local Exchange Carriers* (CLECs) arrange with the building owner to supply the services that tenants want and need, including the following:

- Local and long-distance calling
- High-speed Internet access
- Paging

- Faxing
- Voice messaging
- Data services
- Video conferencing

Tenants get one-stop shopping for all of the telecommunications needs, and their services cost less—up to 50 percent less than traditional phone companies. Previously, customers had one company for local phone service, another for long-distance, and still another company for services such as Internet access. Under the smart-building concept, customers can choose to have all these services bundled together, all under a single monthly bill from one company.

Upgrading the Buildings

Upgrading an existing building to smart building standards can be done in various ways. One upgrade method is to deploy a fixed wireless technology called *Local Multi-Point Distribution Service* (LMDS). With LMDS, antennas the size of a dinner plate are mounted on the rooftop. The installation process is simple, taking one or two days, and requires little wiring because the system depends on over-the-air transmissions. Only rarely can these receivers be viewed from street level.

Another way to upgrade an existing building is through wireline technology. The service provider brings fiber-optic cable to the basement of the building then runs the cabling from its equipment to customer sites throughout the building. The upgrade is simple when the service provider has access to the inside of the building.

While both the telecommunications and real-estate industries are subject to free market forces, however, building facilities are not. People who want access to the new technology and services offered by the CLECs are often denied it because the service providers are often prevented from entering and making upgrades to current building facilities. Without these upgrades, customers cannot have access to the better, faster, and cheaper services that they want.

Legislative Remedies

Currently, there is no consistent, pro-competitive framework in place to foster building technology upgrades in a way that balances the needs of customers with the concerns of property owners. Many in the industry believe that a legislative solution is required to establish the national framework that is necessary to meet the demand for smart buildings.

Such a legislative provision, modeled on legislation at the state level in Texas and Connecticut, would address occupied buildings and access to services, wiring, compensation, regulations and civil penalties. The Connecticut law, for instance, defines the rights and conditions under which an owner of an occupied building must permit wiring to provide communications service by a telecommunications provider in their building. Only state-certified telecommunications providers can upgrade exiting structures to smart-building standards (and then, only with the written request of someone who works or lives in the building).

The smart building industry points out that buildings that offer advanced communications services are higher-valued buildings. These communications services create work and living spaces that are more competitive and sought-after in the marketplace. People who work and live in smart buildings are more satisfied and are less likely to leave. In turn, owners face less exposure to financial loss.

Service Providers

Among the many service providers that are specifically targeting the smart building market are Nextlink, Teligent, and Winstar. All three companies own the largest share of spectrum for LMDS, a broad-band microwave technology that is used for building wireless local loops. Nextlink is the largest holder of broad-band fixed wireless spectrum in North America, with licenses covering 95 percent of the population.

Nextlink was founded in 1994 by telecommunications pioneer Craig O. McCaw to provide high-quality broad-band communications services to businesses over fiber-optic and broad-band wireless facilities. Nextlink had been focused on providing business users with local dial tone, but it is now trying to evolve into a much broader company with multiple voice and data service offerings. The company's wireless capabilities complement and extend the reach of its local fiber-optic networks in the markets where it has spectrum. With its LMDS-extended fiber network, Nextlink intends to provide integrated, end-to-end telecommunications services.

Nextlink uses small antennas on the top of customers' buildings that send signals to a hub site. The hub then sends the signal through Nextlink's fiber network to the company's switching equipment, which routes the data and voice to local and national inter-city networks. The company's broad-band wireless hub sites have capabilities to carry data and voice capacity up to OC-3 rates (155Mbps) in point-to-point configurations or 672 T-1 (1.544Mbps) connections to multiple buildings in a point-to-multi-point configuration.

Summary

Concern exists in the smart-building industry that property owners are acting as gatekeepers, preventing the people who work and live in their buildings from receiving better communications options. Most state laws do not help these consumers who want more telecommunications options. The gas, electric, water, telephone, and other utilities have possessed that right for decades. If more states do not take action to permit competition in this market, federal action might be warranted to remove this last barrier to fulfilling the pro-competitive goals of the Telecommunications Act of 1996.

See Also

Fiber-Optic Technology

Local Multi-Point Distribution Services (LMDS)

Telecommunications Act of 1996

Specialized Mobile Radio (SMR)

Specialized Mobile Radio (SMR) enables carriers to provide two-way radio dispatch service for the public safety, construction and transportation industries. The 800MHz SMR service was established by the FCC in 1979, while the 900MHz SMR service was established in 1986. Although SMR is primarily used for voice communications, such systems also support data and facsimile services. Generally, SMR systems provide dispatch services for companies with multiple vehicles by using the push-to-talk method of communication.

A traditional SMR system consists of one or more base station transmitters, one or more antennas, and end-user equipment that usually consists of a mobile radio unit that is obtained from the SMR operator for a fee (or purchased from a retail source). SMR has limited roaming capabilities, but its range can be extended through interconnection with the public telephone network, as if the user were a cellular subscriber. Both types of services operate over different assigned frequencies within the range of 800Mhz to 900MHz. Cellular services are assigned to bands between 824MHz to 849MHz and 869MHz to 894MHz.

SMR networks traditionally used one large transmitter to cover a wide geographic area, which limited the number of subscribers—because only one subscriber could talk on one frequency at any given moment. The number of frequencies that are allocated to SMR is smaller than for cellular, and there have been several operators in each market. Because dispatch messages are short, SMR services worked reasonably well.

Types of Systems

The 800MHz SMR systems operate on two 25KHz channels paired, while the 900MHz systems operate on two 12.5KHz channels paired. Due to the different sizes of the channel bandwidths that are allocated for 800MHz and 900MHz systems, the radio equipment that is used for 800MHz SMR is not compatible with the equipment that is used for 900MHz SMR systems.

SMR systems consist of two distinct types: conventional and trunked systems. A conventional system enables an end user to use only one channel. If someone else is already using that end user's assigned channel, that user must wait until the channel is available. In contrast, a trunked system combines channels and contains processing capabilities that automatically search for an open channel. This search capability enables more users to be served at any one time. A majority of the current SMR systems are trunked systems.

Summary

In 1993, Congress reclassified most SMR licensees as *Commercial Radio Service* (CMRS) providers and established the authority to use competitive bidding to issue new licenses.

With the development of digital systems, the SMR marketplace now offers new services such as acknowledgment paging and inventory tracking, credit-card authorization, automatic vehicle location, fleet management, inventory tracking, remote database access, and voice mail.

See Also

Cellular Voice Communications

Spectrum Auctions

In recent years, the FCC has assigned licenses for wireless spectrum by putting it up for auction. Since 1994, the FCC has conducted 21 spectrum auctions. The idea behind the auction process is that is encourages companies to roll out new services as soon as possible, in order to recover their investments in the licenses. In placing spectra into the hands of those who initially value it the most, competitive bidding also facilitates efficient spectrum aggregation, rather than fragmented secondary markets.

In the past, the FCC often relied on comparative hearings, in which the qualifications of competing applicants were examined to award licenses in cases where two or more applicants filed applications for the same spectrum in the same market. This process was time-consuming and resource-intensive. The FCC also used lotteries to award licenses, but this procedure created an incentive for companies to acquire licenses on a speculative basis and to resell them.

Of the three methods of assigning spectra, competitive bidding has proven to be the most effective way to ensure that licenses are assigned quickly and to the company that values them the most, while recovering the value of the spectrum resource for the public.[1] In addition, auctions avoid the perception of the government making decisions that are biased towards or against individual industry players. The rules and procedures of the auctions are clearly established, and the outcomes are definitive.

Auction Process

The FCC's auctions of electromagnetic spectrum assign licenses using a unique methodology called electronic, simultaneous, multiple-round auctions. This methodology is similar to a traditional auction, except that rather than selling licenses one at a time, a large set of related licenses are

---

[1] The revenues that are generated from spectrum auctions go to the United States Treasury, not to the FCC.

auctioned simultaneously—and bidders can bid on any license that is offered. The auction closes when all bidding activity has stopped on all licenses.

Another characteristic of the auction process is that it is automated. When the FCC began to design auctions for the airwaves, it became apparent that manual auction methods could not adequately allocate large numbers of licenses when thousands of interdependent licenses were being auctioned to hundreds of bidders at the same time. The FCC's *Automated Auction System* (AAS) provides the necessary tools to conduct large auctions efficiently. The system accommodates the needs of bidders by enabling them to bid from their offices by using a PC and a modem through a private and secure WAN. The system can also accommodate on-site bidders and telephonic bidding.

Bidders and other interested parties are able to track the progress of the auctions through the *Auction Tracking Tool* (ATT), a stand-alone application that enables a user to track detailed information about an auction. During an auction, the FCC releases result files after every round with details on all of the activity that occurred in that round. Users can use the ATT to import these round result files into a master database file and then view a number of different tables that contain a large amount of data in a spreadsheet view. Users can sort, filter, and query the tables to track the activity of an auction in virtually any way they desire. Canned tables also exist that contain simple summary data, in order to enable more casual observers to track the progress of the auction in general.

The FCC also provides the capability to plot maps of auction winners, high bidders by round, and more general auction activity. Through a *Geographical Information System* (GIS), interested parties can use a Web browser-based application to construct queries against the database for a particular auction and have the results displayed in a map format. The GIS presents its query results primarily in maps which the user can export to easily transportable graphical formats. The GIS also enables the user to display data in tabular format. Currently, three queries can be executed against any closed or open auction in the GIS:

- *Market analysis by number of bids* Enables users to see which licenses received a bid in a given round and how many bids each market received

- *Round results summary* Provides a high-level summary of activity for the selected round, depicting markers for which a new high bid was received, markets for which a bid was withdrawn, and markets that had no new activity

- *Bidder activity* Enables users to query the database to generate a map that shows all of the licenses for which a particular area has a high bid in a given round

Companies that are interested in participating in spectrum auctions must submit an electronic application to the FCC, disclosing their owner-ship structure and identifying the markets/licenses on which they intend to bid. Approximately two weeks after the filing deadline and two weeks before the start of the auction, potential bidders must submit a refundable deposit that is used to purchase the bidding units that are required to place bids in the auction. This deposit is not refundable after the auction closes.

At a minimum, an applicant's total up-front payment must be enough to establish eligibility to bid on at least one of the licenses applied for, or else the applicant will not be eligible to participate in the auction. In calculating the up-front payment amount, an applicant should determine the maximum number of bidding units it might wish to bid on in any single round, and submit a payment that covers that number of bidding units. Bidders have to check their calculations carefully because there is no provision for increas-ing a bidder's maximum eligibility after the up-front payment deadline.

About 10 days before the auction, qualified bidders receive their confi-dential bidding access codes, AAs software, telephonic bidding phone num-ber, and other documents necessary to participate in the auction. Five days before the start of the auction, the FCC sponsors a mock auction that enables bidders to work with the software, become comfortable with the rules and the conduct of a simultaneous multiple-round auction, and famil-iarize themselves with the telephonic bidding process.

When the auction starts, it continues until all bidding activity has stopped on all licenses. To ensure the competitiveness and integrity of the auction process, the rules prohibit applicants for the same geographic license area from communicating with each other during the auction about bids, bidding strategies, or settlements.

The winning bidders for spectrum in each market are awarded licenses. Within 10 business days, each winning bidder must submit sufficient funds (in addition to its up-front payment) to bring its total amount of money on deposit to 20 percent of its net winning bids (actual bids less any applicable bidding credits). Up-front payments are applied first in order to satisfy the penalty for any withdrawn bid, before being applied toward down pay-ments. If a company fails to pay on time, the FCC takes back the licenses and holds them for a future auction. The licenses are granted for a 10-year period, after which the FCC can take them back if the holder fails to provide service over that spectrum.

Summary

The FCC's simultaneous multiple-round auction methodology and the AAS design have generated interest worldwide. The FCC has demonstrated the system to representatives of many countries, including Argentina, Brazil, Canada, Hungary, Peru, Russia, South Africa, and Vietnam. Mexico licensed the FCC's copyrighted system and has used it successfully in a spectrum auction. In addition, in 1997 the FCC was awarded a bronze medal from the Smithsonian Institution for recognition of the visionary use of information technology.

See Also

Federal Communications Commission (FCC)

Spread-Spectrum Radio

Spread spectrum is a digital coding technique in which the signal is taken apart, or spread, so that it sounds more like noise to the casual listener. The patent for spread spectrum is held jointly by actress Hedy Lamarr (refer to Figure S-9) and composer George Antheil. Their patent for a Secret Communication System, which was issued in 1942, was based on the frequency-hopping concept—with the keys on a piano representing the different frequencies and frequency shifts that are used in music.[2]

Lamarr had become intrigued with radio-controlled missiles and the problem of how easy it was to jam the guidance signal. She realized that if the signal jumped from frequency to frequency quickly—like changing stations on a radio—and if both sender and receiver changed in the same order at the same time, then the signal could never be blocked without knowing exactly how and when the frequency changed. Although the frequency-hopping idea could not be implemented due to technology limitations at that time, it eventually became the basis for cellular communication based on *Code-Division Multiple Access* (CDMA) and wireless Ethernet LANs based on infrared technology.

---

[2]In 1942, the technology did not exist for a practical implementation of spread spectrum. When the transistor finally became available, the Navy used the idea in secure military communications. When transistors became really cheap, the idea was used in cellular phone technology to keep conversations private. By the time the Navy used the idea, the original patent had expired, and Lamarr and Antheil never received any royalty payments for their idea.

Figure S-9
Actress Hedy Lamarr
(1913-2000), co-
developer of spread
spectrum technology

Spreading

Spread spectrum is a digital coding technique in which the signal is taken apart, or spread, so that it sounds more like noise to the casual listener. The coding operation increases the number of bits transmitted and expands the bandwidth used. Using the same spreading code as the transmitter, the receiver correlates and collapses the spread signal back down to its original form. With the signal's power spread over a larger band of frequencies, the result is a more robust signal that is less susceptible to impairment from electromechanical noise and other sources of interference. Spreading also makes voice and data communications more secure.

Technology

Spread spectrum uses the industrial, scientific, and medical (ISM) bands of the electromagnetic spectrum. The ISM bands include the frequency ranges at 902MHz to 928MHz and 2.4GHz to 2.484GHz, which do not require a site license from the FCC.

Spread spectrum is a highly robust wireless data transmission technology that offers substantial performance advantages over conventional narrowband radio systems. As noted, the digital coding technique used in spread spectrum takes the signal apart and spreads it over the available bandwidth, making it appear as random noise. The coding operation increases the number of bits transmitted and expands the bandwidth that is used. Noise has a flat, uniform spectrum with no coherent peaks and can generally be removed by filtering. The spread signal has a much lower power density but has the same total power.

This low power density, spread over the expanded transmitter bandwidth, provides resistance to a variety of conditions that can plague narrowband radio systems, including the following:

- *Interference*   A condition in which a transmission is being disrupted by external sources, such as the noise that is emitted by various electromechanical devices, or internal sources (such as cross-talk)

- *Jamming*   A condition in which a stronger signal overwhelms a weaker signal, causing a disruption to data communications

- *Multi-path*   A condition in which the original signal is distorted after being reflected off a solid object

- *Interception*   A condition in which unauthorized users capture signals in an attempt to determine the signal content

Non-spread spectrum narrow-band radio systems transmit and receive on a specific frequency that is just wide enough to pass the information, whether the information is in a voice or data format. By assigning users different channel frequencies, confining the signals to specified bandwidth limits, and restricting the power that can be used to modulate the signals, undesirable cross-talk—interference between different users—can be avoided. These rules are necessary, because any increase in the modulation rate widens the radio signal bandwidth (which increases the chance for cross-talk).

The main advantage of spread spectrum radio waves is that the signals can be manipulated to propagate fairly well through the air, despite electromagnetic interference, to virtually eliminate cross-talk. In spread-spectrum modulation, a signal's power is spread over a larger band of frequencies. This situation results in a more robust signal that is less susceptible to interference from similar radio-based systems, because they too are spreading their signals (but with different spreading algorithms).

Two spreading techniques are in common use today: direct sequence and frequency hopping.

Direct sequence In direct-sequence spreading—the most common implementation of spread-spectrum technology—the radio energy is spread across a larger portion of the band than is actually necessary for the data. This process is done by breaking each data bit into multiple subbits called chips, in order to create a higher modulation rate. The higher modulation rate is achieved by multiplying the digital signal with a chip sequence. If the chip sequence is 10, for example, and it is applied to a signal carrying data at 300Kbps, then the resulting bandwidth will be 10 times wider. The amount of spreading is dependent upon the ratio of chips to each bit of information.

Because data modulation widens the radio carrier to increasingly larger bandwidths as the data rate increases, this chip rate of 10 times the data rate spreads the radio carrier to 10 times wider than it would otherwise be for data alone. The rationale behind this technique is that a spread spectrum signal that has a unique spread code cannot create the exact spectral characteristics as another spread-coded signal. Using the same code as the transmitter, the receiver can correlate and collapse the spread signal back down to its original form, while other receivers using different codes cannot perform this function.

This feature of spread spectrum makes it possible to build and operate multiple networks in the same location. By assigning each one its own unique spreading code, all transmissions can use the same frequency band, yet remain independent of each other. The transmissions of one network appear to the other network as random noise and are filtered out because the spreading codes do not match.

This spreading technique would appear to result in a weaker signal-to-noise ratio, because the spreading process lowers the signal power at any one frequency. Normally, a low signal-to-noise ratio would result in damaged data packets that would require retransmission. The processing gain of the despreading correlator, however, recovers the loss in power when the signal is collapsed back down to the original data bandwidth but is not strengthened beyond what would have been received had the signal not been spread.

The FCC has set rules for direct sequence transmitters. Each signal must have 10 or more chips. This rule limits the practical raw data throughput of transmitters to 2Mbps in the 902MHz band and 8Mbps in the 2.4GHz band. The number of chips is directly related to a signal's immunity to interference. In an area with a lot of radio interference, users will have to give up throughput to successfully limit interference.

Frequency hopping   Frequency hopping entails the transmitter jumping from one frequency to the next at a specific hopping rate in accordance with a pseudo-random code sequence. The order of frequencies that is selected by the transmitter is taken from a predetermined set as dictated by the code sequence. For example, the transmitter might have a hopping pattern of going from channel 3 to channel 12 to channel 6 to channel 11 to channel 5, and so on (refer to Figure S-10). The receiver tracks these changes. Because only the intended receiver is aware of the transmitter's hopping pattern, then only that receiver can make sense of the data that is being transmitted.

Other frequency hopping transmitters will be using different hopping patterns that usually will be on non-interfering frequencies. Should different transmitters coincidentally attempt to use the same frequency and the data of one or both become garbled at that point, retransmission of the affected data packets is required. Those data packets will be sent again on the next hopping frequency of each transmitter.

The FCC mandates that frequency-hopped systems must not spend more than 0.4 seconds on any one channel each 20 seconds, or 30 seconds in the 2.4GHz band. Furthermore, they must hop through at least 50 channels in the 900MHz band or 75 channels in the 2.4GHz band. These rules reduce the chance of repeated packet collisions in areas with multiple transmitters.

Summary

Direct-sequence spread spectrum offers better performance, but frequency-hopping spread spectrum is more resistant to interference and is preferable in environments that have electromechanical noise and more stringent security requirements. Direct sequence is more expensive than frequency-hopping and uses more power. Although spread spectrum generally provides more secure data transmission than conventional narrow-band radio systems, the transmissions are not immune from interception and decoding by knowledgeable intruders who have sophisticated tapping equipment. For this reason, many vendors provide optional encryption for added security.

See Also

Code-Division Multiple Access (CDMA)

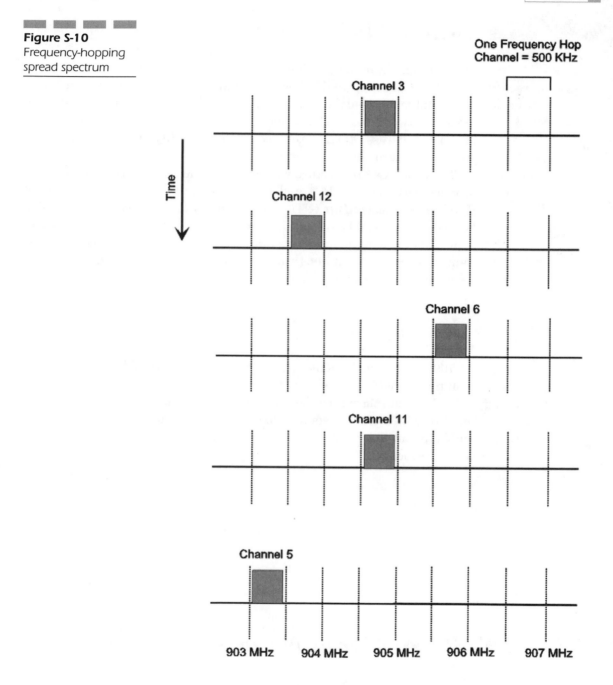

Figure S-10
Frequency-hopping
spread spectrum

StarLAN

AT&T originally developed StarLAN to satisfy the need for a low-cost, easy-to-install LAN that would offer more configuration flexibility than Token Ring and more availability than Ethernet. The hub-based StarLAN was offered in two versions: 1Mbps and 10Mbps. Because it was based on IEEE 802.3 standards, StarLAN offered interoperability with Ethernet and Token Ring through driver software.

The StarLAN architecture was based on the use of one or more hubs. Connectivity between PCs and hubs was achieved through the use of *Unshielded Twisted-Pair* (UTP) wiring. In multi-hub networks, up to five levels of hubs could be cascaded, with one hub designated as the header to which one or more intermediate hubs were connected. The maximum distance between adjacent hubs was 250 meters. The maximum span of a five-level network was 2,500 meters.

Summary

In 1991, AT&T and NCR merged under the name AT&T Global Information Solutions, where responsibility for StarLAN resided until 1996. That year, AT&T Global Information Solutions changed its name back to NCR Corp. in anticipation of being spun off to AT&T shareholders as an independent, publicly-traded company. Around that time, NCR discontinued the StarLAN product line.

See Also

   ARCnet

   Ethernet

   Token Ring

Storage Area Networks (SANs)

The demand for increased performance, availability, and manageability of storage—combined with the emergence of Fibre Channel technology—is driving the convergence of storage and networking architectures. To harness the full capabilities and performance of storage hardware and connectivity, a new network-based storage topology has emerged in recent years: the *Storage Area Network* (SAN). In providing any-to-any connectivity for storage resources on a dedicated high-speed network, the SAN offloads storage traffic from daily network operations while establishing a direct connection between storage elements and servers (refer to Figure S-11).

SAN Concepts

Essentially, a SAN is a specialized network that enables fast, reliable access among servers and external or independent storage resources. In a SAN, a storage device is not the exclusive property of any particular server. Rather, storage devices are shared among all networked servers as peer resources. Just as a LAN can be used to connect clients to servers, a SAN can be used to connect servers to storage, servers to each other, and storage to storage.

Figure S-11
Storage Area
Network (SAN)

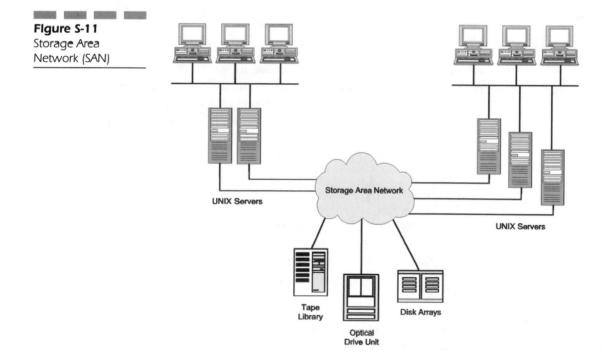

SANs provide an open, extensible platform for storage access in data intensive environments such as those used for video editing, prepress, *Online Transaction Processing* (OLTP), data warehousing, storage management, and server-clustering applications. SANs offer a number of benefits. The redundancy that is an inherent part of SAN architectures makes high availability more cost-effective for a wider variety of application environments. The pluggable nature of SAN resources—storage, nodes, and clients—enables much easier scaleability while preserving ubiquitous data access. Also, with centralized storage, more efficient management of the data for tasks such as optimization, reconfiguration, and backup/restore become much easier.

SANs are particularly useful for backups of mission-critical data. Previously, there were two choices: either a tape drive had to be installed on every server and someone went around changing the tapes, or a backup server was created and the data was moved across the network, which hogged all the bandwidth. Performing data backups over the LAN can be excruciatingly disruptive and slow. A daily backup can suddenly introduce gigabytes of data into the normal LAN traffic. With SANs, organizations can have the best of both worlds: high-speed backups under centralized management.

SAN Origins

SANs have existed for years in various forms. In the IBM mainframe environment, the high-speed data connection took the form of *Enterprise Systems Connection* (ESCON). In mid-range environments, the high-speed data connection was primarily *Small Computer System Interface* (SCSI), a point-to-point connection that is severely limited in terms of the number of connected devices that it can support, as well as the distance between devices.

An alternative to network-attached storage was developed in 1997 by Michael Peterson, president of Strategic Research (Santa Barbara, California). He believed that network-attached storage was too limiting because it relied on network protocols and did not guarantee delivery. He proposed that SANs could be interconnected using network protocols such as Ethernet, and that the storage devices themselves could be linked via non-network protocols.

In a traditional storage environment, a server controls the storage devices and administers requests and backup. With a SAN, instead of being involved in the storage process, the server simply monitors it. By optimizing the box at the head of the SAN to do only file transfers, users are able to get much higher transfer rates, such as 100Mbps via Fibre Channel. Traditional SCSI connections offer transfer rates of only 40 Mbps, while the newer Ultra2 SCSI offers 80Mbps.

Using Fibre Channel as the connection between storage devices also increases distance options. While traditional SCSI enables only a 25-meter (about 82 feet) distance between machines and Ultra2 SCSI enables only a 12-meter distance (about 40 feet), Fibre Channel supports spans of 10 kilometers (about 6.2 miles). SCSI can only connect up to 16 devices, whereas Fibre Channel can link as many as 126. By combining LAN networking models with the core building blocks of server performance and mass storage capacity, SAN eliminates the bandwidth bottlenecks and scaleability limitations imposed by previous SCSI bus-based architectures.

SAN Features

The following items are key features of SANs:

- Storage and archival traffic are routed over a separate network, offloading the majority of data traffic from the enterprise network.
- Data transfers are fast with Fibre Channel—up to 100Mbps with a single loop configuration and up to 200Mbps with a dual-loop configuration.
- A shared data storage pool can be easily accessed by remote workstations and servers.
- The SAN can be easily expanded to a virtually unlimited size with hubs or switches.
- Nodes on the SAN can be easily added or removed with minimal disruption to the active network.
- For totally redundant operation, the SAN can be easily configured to support mission-critical applications.

A key feature of SANs is zoning. This is the division of a SAN into subnets that provide different levels of connectivity between specific hosts and devices on the network. In effect, routing tables are used to control access of hosts to devices. This functionality gives IT managers the flexibility to support the needs of different groups and technologies without compromising data security. Zoning can be performed by cooperative consent of the hosts or can be enforced at the switch level. In the former case, hosts are responsible for communicating with the switch to determine if they have the right to access a device.

You can enforce zoning in several ways. With hard zoning, which delivers the highest level of security, IT managers program zone assignments into the flash memory of the hub. This process ensures that there can be absolutely no data traffic between zones. Virtual zoning provides additional flexibility because it is set at the individual port level. Individual ports can be members of more than one virtual zone, so groups can have access to more than one set of data on the SAN. Broadcast zoning can be used to restrict the scope of broadcasts. For example, IP *Address-Resolution Protocol* (ARP) broadcasts can be kept from SCSI ports on the switch. These IP broadcasts can otherwise cause storage devices to crash.

Technology Mix

As the SAN concept has evolved, it has moved beyond association with any single technology. In fact, just as LANs and WANs use a diverse mix of technologies, so can SANs. This mix can include FDDI, ATM, and IBM's *Serial Storage Architecture* (SSA), as well as Fibre Channel. SAN architectures also allow for the use of a number of underlying protocols, including TCP/IP and all of the variants of SCSI.

Instead of dedicating a specific kind of storage to one or more servers, a SAN enables different kinds of storage—mainframe disk, tape, and RAID—to be shared by different kinds of servers, such as Windows NT, UNIX and OS/390. With this shared capacity, organizations can acquire, deploy and use storage devices more efficiently and cost-effectively.

SANs also enable users who have heterogeneous storage platforms to utilize all of the available storage resources. In other words, within a SAN users can backup or archive data from different servers to the same storage system. They can also enable stored information to be accessed by all servers, create and store a mirror image of data as it is created, and share data between different environments.

With a SAN, there is no need for a physically separate network to handle storage and archival traffic. This is because the SAN can function as a virtual subnet that operates on a shared network infrastructure. For this function to work, however, different priorities or classes of service must be established. Fortunately, both Fibre Channel and ATM provide the means to set different classes of service.

Although early implementations of SANs have been local or campus-based, there is no technological reason why they cannot be extended much farther over the WAN. As WAN technologies such as SONET and ATM mature, and especially as class-of-service capabilities improve, the SAN can be extended over a much wider area—perhaps globally in the future.

SANs also promise easier and less expensive network administration. Today, administrative functions are labor-intensive and time-consuming, and IT organizations typically have to replicate management tools across multiple server environments. With a SAN, only one set of tools is needed, which eliminates the need for replication and associated costs.

SAN Components

Several components are required to implement a SAN. A Fibre Channel adapter is installed in each server, and these are connected via the server's PCI bus to the server's operating system and applications. Because Fibre Channel's transport-level protocol wraps easily around SCSI frames, the adapter appears to be a SCSI device. The adapters are connected to a single Fibre Channel hub, running over fiber-optic cable or copper coaxial cable. Category 5 cable, the high-end twisted pair that is rated for Fast Ethernet and 155Mbps ATM, can also be used.

A LAN-free backup architecture can include some type of automated tape library that attaches to the hub via Fibre Channel. This machine typically includes a mechanism capable of feeding data to multiple tape drives and can be bundled with a front-end Fibre Channel controller. Existing SCSI-based tape drives can be used also through the addition of a Fibre Channel-to-SCSI bridge.

Storage-management software that is running in the servers performs contention management by communicating with other servers via a control protocol to synchronize access to the tape library. The control protocol maintains a master index and uses data maps and time stamps to establish the server-to-hub connections. Currently, control protocols are specific to the software vendors. Eventually, the storage industry will likely standardize on one of the several protocols now in proposal status before the Storage Network Industry Association.

From the hub, a standard Fibre Channel protocol called *Fibre Channel-Arbitrated Loop* (FC-AL) functions similarly to Token Ring in order to ensure collision-free data transfers to the storage devices. The hub also contains an embedded SNMP agent for reporting to network-management software.

Role of Hubs

Similar to Ethernet hubs in LAN environments, Fibre Channel hubs provide fault tolerance in SAN environments. On an FC-AL, each node acts as a repeater for all other nodes on the loop, so if one node goes down, the entire loop goes down. For this reason, hubs are an essential source of fault isolation in Fibre Channel SANs. The hub's port by-pass functionality will automatically bypass a problem port and avoid most faults. Stations can be powered off or added to the loop without serious loop effects. Storage-management software is used to mediate contention and synchronize data—activities that are necessary for moving backup data from multiple servers to multiple storage devices. Hubs also support the popular physical star cabling topology for more convenient wiring and cable management.

To achieve full redundancy in a Fibre Channel SAN, two fully independent, redundant loops must be cabled. This scheme provides two independent paths for data with fully redundant hardware. Most disk drives and disk arrays that are targeted for high-availability environments have dual ports specifically for the purpose. Wiring each loop through a hub provides higher-availability port bypass functionality to each of the loops.

Many organizations will have a need for multiple levels of hubs. Hubs can be cascaded up to the FC-AL limit of 126 nodes (127 nodes with an FL or switch port). Normally, the distance limitation between Fibre Channel hubs is 3 kilometers. Hewlett-Packard, however, offers technology that extends the distance between hubs to 10 kilometers, enabling organizations to link servers situated on either side of a campus (or even spanning a metropolitan area).

Management

Today's storage environment requires proactive management software to ensure that the enterprise gets the most out of its SAN, such as safeguarding mission-critical applications. With the right management tools, users can remotely monitor performance of an entire SAN, or the individual devices throughout the enterprise, and quickly identify device faults to minimize SAN down time. Among the growing number of vendors that are addressing the management needs of the SAN environment is Vixel Corporation, which offers a Java-based tool called SAN InSite.

SAN InSite provides comprehensive status, control, diagnostics and advanced capabilities for switches, hubs and *Gigabit Interface Converters*

(GBICs), all from a single console, to keep an entire SAN operating at its full potential. Built on the Java programming language, SAN InSite coexists with all server platforms that are operating independently on the SAN.

Among other items, the software provides network managers a historical view of bandwidth consumption, from seconds to months, on each port of a hub or switch. Performance monitoring enables administrators to correctly size the SAN for high-availability storage applications such as e-commerce, data warehousing, video, multimedia, and other Internet applications.

The proactive management provided by SAN InSite enables network managers to set policies that specify in advance how the system should react when problems occur. In this way, problems can be detected and resolved before users realize anything is wrong. In addition, the software's event notifications inform network managers of system faults. When a fault condition occurs, SAN InSite's SNMP trap support alerts system managers and other SNMP management software packages produced by Tivoli, CA Unicenter, HP OpenView, Veritas, and Legato.

Summary

The move to SANs in recent years provides a new level of scaleability to system administrators—and, via Fibre Channel, enables a much greater degree of flexibility than the traditional attached-storage paradigm. SCSI, however, has not yet exhausted its potential as a storage connectivity option. SCSI is installed in more than 90 percent of networks, and the latest SCSI variant—Ultra160/m SCSI—might even pose a challenge to Fibre Channel. In addition to supporting data transfers at up to 160Mbps, Ultra160/m SCSI offers intelligent data management. Products that incorporate Ultra160/m intelligently test and manage the storage network so that the maximum reliable data transfer rate is used. If the cabling, backplanes, and terminators can support the desired speed, the data flows at full throttle. If not, the hardware will be smart enough to negotiate a lower transfer rate automatically. The result will be more system autonomy and less IT-manager involvement.

See Also

Fibre Channel

Synchronous Optical Network (SONET)

The *Synchronous Optical Network* (SONET) is an industry standard for high-speed transmission over optical fiber. The SONET standard was developed by the *Alliance for Telecommunications Industry Solutions* (ATIS), formerly known as the *Exchange Carriers Standards Association* (ECSA), with input from Bellcore (now known as Telcordia Technologies), the former research and development arm of the original seven RBOCs. The standard was published and distributed by the *American National Standards Institute* (ANSI).

SONET facilities are used on enterprise backbone networks and by carriers and competitive-access providers for long-haul routes and fault-tolerant rings around major metropolitan areas. SONET-based services have a performance objective of 99.99 percent error-free seconds and an availability rate of at least 99.99 percent. SONET combines bandwidth and multiplexing capabilities, enabling users to fully integrate voice, data, and video over a single fiber-optic facility.

Advantages

SONET is expected to provide the transport infrastructure for the next three to four decades, in much the same way that T1 (1.544Mbps) and its extension, T3 (44.736Mbps), have provided the transmission infrastructure of the past two decades. SONET offers numerous benefits to carriers and users.

Bandwidth   The enormous amounts of bandwidth available with SONET and its integral management capability permit carriers to create global intelligent networks capable of supporting the next generation of services. SONET-based information superhighways support bit-intensive applications such as three-dimensional *Computer-Aided Design* (CAD), medical imaging, collaborative computing, interactive virtual-reality programs, and multi-point video conferences—as well as new consumer services such as video on demand and interactive entertainment. Under SONET, bandwidth is scaleable from 51.84Mbps to 13Gbps (and potentially much higher).

The SONET standard specifies a hierarchy of rates and formats for optical transmission, ranging from 51.84Mbps to 13.271Gbps, as summarized in the following table.

OC Level	Line Rate
OC-1	51.84Mbps
OC-3	155.520Mbps
OC-9	466.560Mbps
OC-12	622.080Mbps
OC-18	933.120Mbps
OC-24	1.244Gbps
OC-36	1.866Gbps
OC-48	2.488Gbps
OC-192	9.95Gbps
OC-256	13.271Gbps

Although the SONET standards definition goes only to OC-256 speeds, OC-768 speeds of 40Gbps are possible. According to Hitachi, which has developed the necessary indium-phosphide transmitter circuits that make this speed possible, OC-768 will initially show up in metropolitan rings and in other applications in which a concatenated version of OC-192, OC-192c, is used.

The base signal rate of SONET is the *Synchronous Transport Signal— Level 1* (STS-1), which provides a transmission rate of 51.84Mbps. The optical equivalent of STS-1 is called *Optical Carrier—Level 1* (OC-1). These signals can be multiplexed in a hierarchical fashion to form higher-rate signals.

The STS-1/OC-1 frame, from which all larger frames are constructed, has a 9-by-90-byte format that permits efficient packing of data rates in a payload of 783 bytes, plus 27 bytes for transport overhead, for a total of 810 bytes. This functionality results in a usable payload of 48.384Mbps (refer to Figure S-12). The overhead bytes are used for real-time error monitoring, self-diagnostics, and fault analysis. The signal is transmitted byte-by-byte beginning with byte one, scanning left to right from row one to row nine. The entire frame is transmitted in 125 microseconds. Higher-level signals (STS-n) are integer multiples of the base electrical rate, which are interleaved and converted to optical signals (OC-n).

A key feature of SONET framing is its capability to accommodate existing synchronous and asynchronous signal formats. The SONET payload

Figure S-12
A SONET STS-1/OC-1
frame

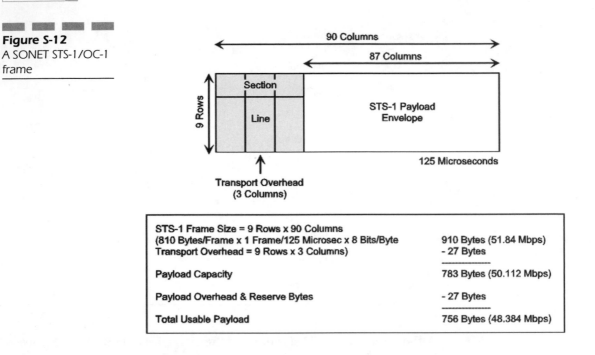

can be subdivided into smaller envelopes called *Virtual Tributaries* (VTs), in order to transport lower-capacity signals. Because VTs can be placed anywhere on higher-speed SONET payloads, they provide effective transport for existing North American and international formats. The following table highlights some of these VTs:

VT Level	Line Rate (bps)	Standard
VT1.5	1.728M	DS1
VT2	2.304M	CEPT1
VT3	3.456M	DS1C
VT6	6.912M	DS2
VT6-N	Nx6.9M	Future
Async DS3	44.736M	DS3

Standardization   Because T3 implementation is unique to each equipment vendor, carriers are severely limited in terms of product selection and configuration flexibility. SONET standards make possible seamless interconnectivity among compliant equipment, eliminating the need to deploy equipment from the same vendor end-to-end and making it much easier to interconnect compliant networks, even among international locations.

In eliminating proprietary optics, carriers are freed from the necessity of dealing with a single vendor for all of their equipment needs. Instead, they can buy equipment based on price and performance and mix and match hardware from multiple vendors. SONET also gives corporate users flexibility in choosing customer-premises equipment, instead of being locked into the carrier's preferred vendor.

Bandwidth management A major advantage of SONET is its ability to easily manage huge amounts of bandwidth. Within the SONET infrastructure, carriers can tailor the width of information highways in a standard way. Carriers can parcel out specific amounts of this bandwidth to meet the needs of a broad and diverse array of user applications. Such parceling can be accomplished without adding equipment to the network or manually reconnecting cables. SONET eliminates central office reliance on metallic DSX technology with its cumbersome manual cabling and jumpers and replaces it with remotely configurable, optical cross-connects. SONET provides more efficient switching and transport by eliminating the need for mid-level network elements such as back-to-back M13 multiplexers, for example, that normally cross-connect T1 facilities.

With remotely configurable SONET equipment, carriers can more expeditiously support the connectivity requirements of their customers. For example, a DS3 signal with a rate of 44.736Mbps can be mapped directly into an STS-1 at 51.84Mbps. The remaining STS-1 bytes are used for overhead and stuffing. Or, two 51.84Mbps channels can be combined to support LAN traffic between *Fiber-Distributed Data Interface* (FDDI) backbones that operate at 100Mbps.

Real-time monitoring SONET permits sophisticated self-diagnostics and fault analysis to be performed in real time, making it possible to identify problems before they disrupt service. Intelligent network elements, specifically the SONET *Add-Drop Multiplexer* (ADM), can automatically restore service in the event of failure via a variety of restoration mechanisms.

SONET's embedded control channels enable the tracking of end-to-end performance and identification of elements that cause errors. With this capability, carriers can guarantee transmission performance, and users can readily verify it without having to go offline to implement various test procedures. For network managers, these capabilities enable a proactive approach to problem identification, which can prevent service disruptions. Along with the self-healing capabilities of ADMs, these diagnostic capabilities ensure that properly configured SONET-compliant networks experience virtually no down time.

Survivable networking   SONET offers multiple ways to recover from network failures, including the following:

- *Automatic protection switching*   The capability of a transmission system to detect a failure on a working facility and to switch to a standby facility in order to recover the traffic. One-to-one protection switching and one-to-n protection switching are supported.

- *Bidirectional line switching*   Requires two fiber pairs between each recoverable node. A given signal is transmitted across one pair of fibers. In response to a fiber facility failure, the node preceding the break loops the signal back toward the originating node, where the data traverses a different fiber pair to its destination.

- *Unidirectional path switching*   Requires one fiber pair between each recoverable node. A given signal is transmitted in two different paths around the ring. At the receiving end, the network determines and uses the best path. In response to a fiber-facility failure, the destination node switches traffic to the alternate receive path.

Universal connectivity   As the foundation for future high-capacity backbone networks, SONET carries a variety of current and emerging traffic types, including *Asynchronous Transfer Mode* (ATM), Frame Relay, *Switched Multi-Megabit Data Services* (SMDS), and *Broad-Band ISDN* (BISDN). To ensure universal connectivity down to the component level equipment vendors support common product specifications, including pin-for-pin compatible transmitters and receivers. Uniformity and compatibility are achieved by using common specifications for module pin-out, footprints, logic interfaces, optical performance parameters, and power supplies.

Network Elements

SONET network infrastructures consist of various types of specialized equipment including ADM, BDCS, WDCS, digital loop carriers, regenerators, and SONET *Customer-Premises Equipment* (CPE).

ADM   The ADM provides an interface between network signals and SONET signals. It is a single-stage multiplexer/demultiplexer that converts

DS-n signals into OC-n signals. The ADM can be used in terminal sites and intermediate (add-drop) sites; at an add-drop site, it can drop lower-rate signals down or pull lower-rate signals up into the higher-rate OC-n signal.

BDCS   The BDCS interfaces various SONET signals and legacy DS3s and is the synchronous equivalent of the DS3 digital cross-connect. The BDCS accepts optical signals and enables overhead to be maintained for integrated *Operations, Administration, Maintenance, and Provisioning* (OAM&P). Most asynchronous systems prevent overhead from being passed from signal to signal, but the BDCS makes two-way cross-connections at the DS3 level. BDCS is typically used as a SONET hub that grooms traffic for broadband restoration purposes or for routing traffic.

WDCS   The WDCS is a digital cross-connect that terminates SONET and DS3 signals, and maintains the basic functionality of a VT and/or DS1-level cross-connects. WDCS is the optical equivalent to the DS3/DS1 digital cross-connect and accepts optical-carrier signals, as well as DS1s and DS3s. In a WDCS, switching is done at the VT, DS1, or DS0 level. Because SONET is synchronous, low-speed tributaries are visible in VT-based systems and are directly accessible within the STS-1 signal, which enables tributaries to be extracted and inserted without demultiplexing. Finally, the WDCS cross-connects constituent DS1s between DS3 terminations and between DS3 and DS1 terminations.

Digital-loop carrier   Similar to the DS1 digital-loop carrier, this network element accepts and distributes SONET optical-level signals. Digital-loop carriers enable the network to transport services that require large amounts of bandwidth. The integrated overhead capability of the digital-loop carrier enables surveillance, control, and provisioning from the central office.

Regenerator   A SONET regenerator drives a transmitter with output from a receiver and stretches transmission distances far beyond what is normally possible over a single length of fiber.

Protocol Stack

The SONET transmission protocol consists of four layers: photonic, section, line, and path:

- *Photonic layer*   The photonic layer is the electrical and optical interface for the transport of information bits across the physical medium. Its primary function is to convert STS-N electrical signals into OC-N optical signals. This layer performs functions associated with the bit rate, optical-pulse shape, power, and wavelength (it uses no overhead).

- *Section layer*   The section layer deals with the transport of the STS-N frame across the optical cable and performs a function that is similar to the Data-Link Layer (DLL), or layer 2, of bit-oriented protocols such as *High-Level Data-Link Control* (HDLC) and *Synchronous Data-Link Control* (SDLC). This layer establishes frame synchronicity and the maintenance signal; functions include framing, scrambling, error monitoring, and orderwire communications.

- *Line Layer*   The line layer provides the synchronization, multiplexing, and *Automatic Protection Switching* (APS) for the path layer. Primarily concerned with the reliable transport of the path-layer payload (voice, data, or video) and overhead, this layer enables automatic switching to another circuit if the quality of the primary circuit drops below a specified threshold. Overhead includes line-error monitoring, maintenance, protection switching, and express order wire.[3]

- *Path Layer*   The path layer maps services such as DS3, FDDI, and ATM into the SONET payload format. This layer provides end-to-end communications, signal labeling, path maintenance, and control and is accessible only through terminating equipment. A SONET ADM accesses the path-layer overhead; a cross-connect system that performs section and line-layer processing does not require access to the path-layer overhead.

Channelization

Channelized interfaces provide network-configuration flexibility and contribute to lower telecommunications costs. For example, channelized T1 delivers bandwidth in economical 56Kbps/64Kbps DS0 units, each of which can be

---

[3]The term *order wire* originated in the early days of telephony when requests for new telephone service were fulfilled by manually configuring a patch panel. Today, order wire is software controlled, enabling services of all types to be installed and terminated via instructions issued at a computer keyboard. Express order wire refers to the installation or termination of service on a priority basis.

used for voice or data and routed to different locations within the network or aggregated as needed to support specific applications. Likewise, channelized DS3 delivers economical 1.544Mbps DS1 units, which can be routed separately or aggregated as needed. Channelized interfaces also apply to the SONET world—channelized OC-48, OC-12, and OC-3 can all be channelized down through DS3 speeds. OC-48, for example, can be channelized as follows:

- Four OC-12 tributaries, all configured for IP (packet over SONET) framing or all configured for ATM framing

- Four OC-12 tributaries, with two configured for IP (packet over SONET) framing and the other two for ATM framing

- Two OC-12, with one configured for IP framing and the other for ATM framing; 8 OC-3, with four configured for IP framing and four configured for ATM framing

- Two OC-12, with one configured for IP framing and the other for ATM framing; 6 OC-3, with three configured for IP framing and three configured for ATM framing; 6 DS3, with three configured for IP framing and three configured for ATM framing

- 48 DS3s, with 24 configured for IP framing and 24 configured for ATM framing

Multi-service, channelized SONET can be implemented on a single OC-48 line card, which provides IP packet and ATM cell encapsulation in order to support business data and Internet services from the same hardware platform. One vendor that offers such products is Argon Networks. The company's *GigaPacket Node* (GPN) is a native IP router as well as a native ATM switch in a single, modular platform, which enables any port to be configured for either *Packet over SONET* (PoS) or ATM service. The configurations are implemented through keyboard commands at service time, rather than permanently assigned at network build-out time.

Summary

The continued deployment of SONET-compliant equipment on public and private networks will have a significant impact on telephone companies, interexchange carriers, and corporate users. SONET offers virtually unlimited bandwidth, integral fault recovery and network management, interoperability between public services and corporate enterprise-wide networks, and multi-vendor equipment interoperability. As SONET becomes more widely available and the promised efficiencies and cost savings materialize, it is changing the way networks are designed, operated and managed. For carriers, timely and effective deployment of SONET determines the types of broad-band services that can be offered to subscribers, because SONET comprises the physical layer upon which broad-band services are built.

See Also

> *Asynchronous Transfer Mode* (ATM)
>
> Fiber-Optic Technology
>
> *Passive Optical Networks* (PONs)
>
> *Wavelength-Division Multiplexing* (WDM)

Streaming Video

Streaming video is a method of delivering content to subscribers over a network. The content can be in the form of a training module, video-on-demand program, news feed, movie preview, DVD segment, or any other type of video application. Usually, the stream is sent over the most economical transmission media possible, which is an IP network such as the public Internet or private intranet. But streaming video can be sent over Frame Relay and ATM networks, as well.

In an IP environment, the video stream can be sent in either of two ways: unicast or multicast. Unicast delivery involves sending a separate data stream to each recipient. Multicast delivery involves sending only one data stream into the network, which is replicated only as many times as necessary to distribute the stream to the nodes (i.e., routers) that have registered subscribers attached.

Applications

Multicast has a number of applications. It is ideal for content providers with real-time applications such as news and entertainment events, and for the distribution of dynamic content, such as financial information and sports scores. The application itself can be audio, video, or text—or any combination of these. For content providers, IP multicast is a low-cost way to supplement current broadcast feeds. In fact, CNN is one of the biggest users of IP multicast.

Corporations can use IP multicast to deliver training to employees and keep them informed of internal news, benefits programs, and employment opportunities within the organization. They can also use IP multicast to broadcast annual meetings to shareholders, or introduce new products to their sales channels. Associations can use IP multicast to broadcast conference sessions and seminars to members who would not otherwise be able to attend in person. Political parties and issue advocacy groups can use IP multicast to keep their members informed of late-breaking developments and call them to action. Entrepreneurs can use IP multicast to offer alternative programming to the growing base of Internet-access subscribers.

Performance

Concern exists among potential users of IP multicast about the effects of delay on performance. After all, they have experienced the long delays accessing multimedia content on the Web. They see video that is slow and jerky and hear audio that pauses periodically until something called a buffer has a chance to fill. When video and audio run together, often the two are out of synchronization. So, they wonder how the Internet can handle a real-time multicast with acceptable quality.

Currently, multicast works best on a managed IP backbone network, where a single company or carrier has control of all the equipment, protocols, and bandwidth end-to-end. This functionality is not possible on the public Internet because there is no central management authority. While simple, real-time applications (such as text tickers) might work well enough, the performance of a sophisticated, graphically enriched, real-time multimedia application suffers.

Not only can the performance of a private IP network be controlled to eliminate potential points of congestion and to minimize delay, but the company or carrier can also place dedicated multicast routers throughout its network. This type of router replicates and distributes the content stream in a highly efficient way that does not require massive amounts of bandwidth.

For example, instead of sending out 100 information streams to 100 subscribers, only one information stream is sent. The multicast routers replicate and distribute the stream within the network to only the nodes that have subscribers. Users no longer need to purchase enormous amounts of bandwidth to accommodate large multimedia applications or buy a high-capacity server to send out all the data streams. Instead, a single data stream is sent, the size of which is based on the type of content. This data size can be as little as 5Kbps for text, 10Kbps for audio, and 35Kbps for video.

Operation

A multicast can reach potentially anyone who specifically subscribes to the session—whether they have a dedicated connection or whether they use a dialup modem connection. Of course, the content originator can put distance limits on the transmission and restrict the number of subscribers that will be accepted for any given program.

A variety of methods can be used to advertise a multicast. A program guide can be sent to employees and other appropriate parties via e-mail or it can be posted on a Web site. If the company already has an information channel on the Web that delivers content to subscribers, the program guide can be one of the items that is pushed to users when they access the channel. When a person wants to receive a program, they enroll through an automated registration procedure. The request is handled by a server running the multicast application, which adds the end-station's address to its subscription list. In this way, only users who want to participate will receive packets from the application.

The user also selects a multicast node from those listed in the program guide. Usually, this will be the router closest to the user's location. The user becomes a member of this particular node. Group membership information is distributed among neighboring routers so that multicast traffic gets routed only along paths that have subscribers at the end nodes. From the end node, the data stream is delivered right to the user's computer.

Once the session is started, users can join and leave the multicast group at any time. The multicast routers adapt to the addition or deletion of network addresses dynamically, so the data stream gets to new destinations when users join, and stops the data stream from going to destinations that no longer want to receive the session.

Infrastructural Requirements

Several basic requirements have to be met in order to implement IP multicast. Most of these requirements are already met by the products used on IP nets today. First, the host computer's operating system and TCP/IP stack must be multicast-enabled. Among other things, it must support the *Internet Group Management Protocol* (IGMP). Fortunately, virtually all of today's operating systems can accommodate IP multicast and IGMP, including Windows NT, Windows 95, Windows 98, and most versions of UNIX.

Second, the adapter and its drivers must support IP multicast, which enables the adapter to filter DLL addresses that are mapped from network-layer IP multicast addresses. Newer adapters and network drivers are already capable of implementing IP multicast. Third, the network infrastructure—the routers, bridges, and switches—must support IP multicast. If they do not, perhaps it is only a matter of turning on this feature. If the feature is unavailable, a simple software upgrade is all that is usually needed. Finally, applications must be written or updated in order to take advantage of IP multicast.

Developers can use application program interfaces to add IP multicast to their applications. Although implementing IP multicast might require changes to end-stations and infrastructure, many of the steps are simple— such as making sure that hardware and software can handle IP multicast and making sure the content server has enough memory. Other steps are more complicated, such as making the network secure.

Anyone who is serious about IP multicast should have a working knowledge of the latest protocols for real-time delivery and be familiar with the management tools for monitoring such things as QoS parameters and usage patterns. This information is often useful in isolating problems and fine-tuning the network. For companies that understand the value of multicast but prefer not to handle it themselves, multicast host services are available from sources such as GlobalCast Communications, PSINet, and UUNET. UUNET, for example, offers a multicast hosting service called UUCast, which provides six data streams of varying size in order to accommodate virtually any real-time data transmission.

Data-Stream Size	Application Example	End-User Access Speed
5Kbps	Ticker banner	Dialup modem up to 56Kbps
10Kbps	Ticker banner or audio message	Dialup modem up to 56Kbps
25Kbps	Audio/video applications	Dialup modem up to 56Kbps
35Kbps	Audio/video applications	Dialup modem up to 56Kbps
64Kbps	High-quality audio/video	ISDN up to 128Kbps
128Kbps	High-quality audio/video	ISDN up to 128Kbps

UUCast requires a dedicated UUNET connection. The subscribing organization supplies its own content and equipment. Upon installation, the organization is given a unique multicast group address for each of the data streams. The organization's router is configured with a virtual point-to-point connection to the multicast router located in a local UUNET point-of-presence, or POP. Because these multicast routers only transmit data streams to the corresponding multicast group address, there is no interference with other traffic sources.

UUNET has equipped all domestic POPs with multicast routers, so any of its dialup customers can also receive the data stream (if it is made available to them). A dial-access router, located within each POP, recognizes the request for a particular multicast data stream and begins transmitting the appropriate content to the end user's desktop.

Costs

Telecommunications costs are always an important consideration. Companies are continually looking for ways to save money in this area and are understandably interested in the cost of multicasting.

The cost to implement multicast on an internal IP network is minimal because an existing infrastructure is simply being leveraged. There might be some upgrades to hardware and software, and possibly the need for management tools to monitor the multicasts. Organizations that are serious about IP multicast often have a dedicated full-time administrator.

An overlooked cost comes in the form of continuously developing the multicast content. This task could take a whole staff of creative people with specialized skills—writers, editors, graphic artists, audio/video production staff, and a multimedia server administrator—not to mention all of the expensive equipment and facilities that they will need.

The cost of production will hinge on the type of content to be developed. Obviously, it will be much cheaper to use text only, but not very many multicast applications will attract viewers if only text is involved. Production costs jump dramatically as audio and video components are added, because special equipment and expertise are required. An alternative to the do-it-yourself approach is to outsource production to specialized firms. This can cut development costs by as much as 60 percent.

For companies that prefer to outsource IP multicast, companies such as UUNET offer a predictable price for a large-scale Internet broadcast and they take care of all the server and router management. UUNET multicast hosting is available in dedicated configurations. The monthly fee is based on a server component, starting at $3,000 per month, and a multicast stream size, starting at $2,200 per month for a 5Kbps stream, $10,000 per month for a 25Kbps stream, and $15,000 for a 35Kbps stream.

The following table compares the cost of a unicast transmission to 1,000 users on a private IP network (coast-to-coast) with a multicast transmission to 1,000 users over the managed Internet backbone of UUNET:

Traditional Unicast Transmission	UUCast Transmission
1,000 dial users x 28.8Kbps per user = 28.8Mbps of bandwidth required	One 35Kbps data stream required
One T-3 connection (45Mbps). This unicast transmission will utilize approximately 50 percent of the T-3.	Tremendous amounts of bandwidth are no longer needed to support a single broad cast.
Monthly cost for full T-3 = $54,000	Monthly cost: $15,000
Total savings = $39,000 per month, or $468,000 per year	

The cost-savings of multicasting over traditional unicasting is quite compelling. Of course, this scenario assumes that the private T3 link is dedicated to unicasting and that the carrier does not yet offer Fractional T3, in which case the monthly cost could be cut in half, to about $27,000. Even in this case, the monthly savings is $12,000 or $144,000 per year when outsourcing.

Summary

As noted, the Internet is not centrally managed, making it difficult to convey sophisticated multimedia content in real-time with any consistently in performance. Congestion and delays are still obstacles that must be overcome for applications such as video streaming. Other real-time applications such as IP telephony experience the same problem when run over the public Internet. The next big step toward full deployment of IP multicast is convincing more Internet service providers to offer the service to customers. The argument is not hard to make, especially because multimedia applications are obviously running on the Internet anyway and multicasting actually conserves bandwidth. With only modest costs, mostly for router upgrades, wider support of multicast would improve the Internet's performance for everyone. Until then, multicasting video streams will be most effective on private intranets and carrier-managed backbones where performance can be controlled end to end.

See Also

Internet Protocol

Voice-Data Convergence

Switched Multi-Megabit Data Services (SMDS)

Initially offered in 1991, SMDS is a carrier-provided, connectionless, cell-switched service developed by Bellcore (now known as Telcordia Technologies) and standardized by the *Institute of Electrical and Electronics Engineers* (IEEE). As such, SMDS provides organizations with the flexibility they need for distributed computing and supporting bandwidth-intensive applications.

As a connectionless service, SMDS eliminates the need for carrier switches to establish a call path between two points before data transmission can begin. Instead, SMDS access devices, or routers, pass customer traffic to the carrier's data network, which is always available. The switch reads addresses and forwards cells one-by-one over any available path. SMDS addresses ensure that the cells arrive at the right destination. The benefit of this connectionless any-to-any service is that it puts an end to the need for precise traffic-flow predictions and for dedicated connections between locations. With no need for a predefined path between devices, data can travel over the least congested route in an SMDS network, providing faster transmission, increased security, and greater flexibility to add or drop network sites.

Applications

SMDS is useful for a number of broad-band applications, including the following:

- Channel-attached IT services, such as dark data centers, data vaults, and print factories
- Disaster recovery using mirroring, shadowing, and off-site magnetic tape storage
- Networked computer centers

- Distributed supercomputer applications, such as crash and process simulations
- Distribution of expensive, bit-hungry resources such as document image and CAD/CAM files
- Bulk file transfer for program testing, verification, and interchange
- Remote access to high-resolution satellite images, X-rays, and CAT scans

Benefits

Although offered by MCI WorldCom as a nation-wide service, SMDS is primarily used for metropolitan or regional LAN internetworking. As a technology-independent service, SMDS offers several advantages:

- *Simplicity* Virtual connections are made as needed; there are no permanent, fixed connections between sites.
- *E.164 addressing* SMDS addresses are similar to standard telephone numbers. If a user knows the SMDS address of another user, he or she can call and begin sending and receiving data.
- *Call control* SMDS supports call blocking, validation, and screening for the secure interconnection of LANs and distributed client/server applications
- *Multicasting* SMDS supports group addressing for the transmission of data to multiple locations simultaneously.
- *Multi-protocol support* SMDS supports the key protocols used in LANs and WANs, including TCP/IP, Novell, DECNet, AppleTalk, SNA, and OSI.
- *Management* SMDS is easier to manage than other services such as Frame Relay, which can have a multitude of virtual circuits that make setup, reconfiguration, and testing much more difficult.
- *Reconfiguration* Sites can be connected and disconnected easily and inexpensively, usually within 30 minutes, without impacting other network equipment. This functionality makes SMDS suitable for external corporate relationships that change frequently.
- *Security* SMDS includes built-in security features that enable the intra-company transmission of confidential data.
- *Scaleability* SMDS is easily and cost-effectively scaled as organizations grow or application requirements change.

- *Migration path*   Customers have a well-defined migration path to *Asynchronous Transfer Mode* (ATM) technology, because both services use the same 53-byte cell structure.

- *Cost*   Because SMDS is a switched service, users pay only for the service when it is used—which can make it less expensive than other services, such as Frame Relay.

Architecture

SMDS defines a three-tiered architecture (refer to Figure S-13):

- A switching infrastructure comprising SMDS-compatible switches that might or might not be cell-based

- A delivery system made up of T1, T3 and lower-speed circuits called *Subscriber Network Interfaces* (SNIs)

- An access-control system for users to connect to the switching infrastructure without having to become a part of the infrastructure

LANs provide the connectivity for end users on the customer premises. The LAN is attached to the SMDS network via a bridge or router, with an SMDS-capable CSU/DSU at the front-end of the connection. T1 SNIs are used to access 1.17Mbps SMDS offerings, while T3 SNIs are used to tap into 4Mbps, 10Mbps, 16Mbps, 25Mbps, or 34Mbps offerings. A fractional T3 circuit can be used to access intermediate-speed SMDS offerings. Some carriers offer low-speed SMDS access at 56Kbps, 64Kbps, and in increments of 56Kbps/64Kbps. This functioanlity enables smaller companies, large com-

Figure S-13
SMDS architecture

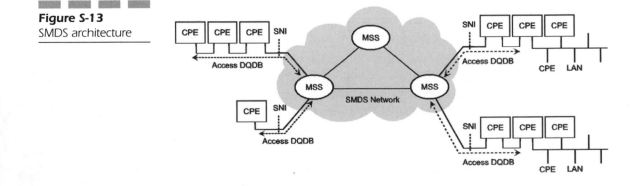

panies that have branch offices, and current users of Frame Relay technology to also take advantage of SMDS.

Each subscriber has a private SNI, and can connect multiple user devices (CPEs) to it. At this interface point, the CPE attaches to a dedicated access facility that connects to an SMDS switch. Security is enforced, because only data originating from or destined for that subscriber will be transported across that SNI.

The *SMDS Interface Protocol* (SIP), operating across the SNI, is based on the IEEE 802.6 *Distributed Queue Dual Bus* (DQDB) *Media-Access Control* (MAC) scheme. The SIP consists of three protocol layers that describe the network services and how these services are accessed by the user. The SIP defines the frame structure, addressing, error control, and data transport across the SNI.

The SMDS network itself is a collection of *SMDS Switching Systems* (SS). The SS is a high-speed packet switch—most likely an ATM switching platform—that provides the SMDS service interface. An SS will typically be located in a service provider's central office. Interconnecting several SS locations forms the foundation for a metropolitan or regional network.

The *Inter-Switching Systems Interface* (ISSI) provides communications between different switching vendors within the same network, while the *Interexchange Carrier Interface* (ICI) enables local telephone companies and interexchange carriers to interconnect SMDS networks.

Summary

SMDS provides users with the cost-effectiveness of a public-switched network; the benefits of fully meshed, wide-area interconnection; and the privacy and control of dedicated, private networks. The key benefits subscribers can realize with SMDS include widespread current availability and increased LAN performance. SMDS provides data management features, flexibility, bandwidth on demand, network security and privacy, multi-protocol support, and technology compatibility.

Although SMDS has fallen out of favor, it is still offered on a regional basis by Bell Atlantic, while MCI WorldCom is the only long-distance carrier that offers SMDS (including the required CPE and performance reports that can be accessed on the Web). Ameritech had long ago migrated its SMDS customers to ATM and Pacific Bell is looking to implement a similar plan. Despite the many advantages of SMDS, the service lags in popularity to Frame Relay and ATM. Many industry watchers believe that the ILECs have simply done a poor job of marketing SMDS.

See Also

Asynchronous Transfer Mode (ATM)

Frame Relay

T-Carrier Facilities

T-carrier is a type of digital transmission system that is employed over copper, optical fiber, or microwave to achieve various channel capacities for the support of voice and data. The most popular T-carrier facility is T1, which is implemented by a system of copper wire cables, signal regenerators, and switches that provides a transmission rate of up to 1.544Mbps by using *Digital Signal Level 1* (DS1). In Europe, the United Kingdom, Mexico, and other countries that abide by ITU standards, the equivalent facility is E1, which provides a transmission rate of 2.048Mbps.

T-carrier had its origins in the 1960s and was first used by telephone companies as the means of aggregating multiple voice channels onto a single, high-speed digital backbone facility between central-office switches. The most widely deployed T-carrier facility is T1, which has been commercially available since 1983.

Digital Signal Hierarchy

To achieve the DS1 transmission rate, selected cable pairs with digital signal regenerators (repeaters) are spaced approximately 6,000 feet apart. This combination yields a transmission rate of 1.544Mbps. By halving the distance between the span-line repeaters, the transmission rate can be doubled to 3.152Mbps, which is called DS1C. Adding more sophisticated electronics and/or multiplexing steps makes higher transmission rates possible, creating a range of digital signal levels as shown in the following table.

For example, a DS3 signal is achieved in a two-step multiplexing process (refer to Figure T-1) whereby DS2 signals are created from multiple DS1 signals in an intermediary step. DS1C is not commonly used, except in highly customized private networks where the distances between repeaters is short, such as between the floors of an office building or between the buildings in a campus environment. Some channel banks and multiplexers support DS2 by performing multiplexing to achieve 96 voice channels over a single T-carrier facility. DS4 is used mostly by carriers for trunking between central offices.

Quality Objectives

The quality of T-carrier facilities is determined by two criteria: performance and service availability.

Digital Signal Hierarchy				
North America				
Signal Level	**Bit Rate**	**Channels**	**Carrier System**	**Typical Medium**
DS0	64Kbps	1	—	Copper wire
DS1	1.544Mbps	24	T1	Copper wire
DS1C	3.152Mbps	48	T1C	Copper wire
DS2	6.312Mbps	96	T2	Copper wire
DS3	44.736Mbps	672	T3	Microwave/fiber
DS4	274.176Mbps	4032	T4	Microwave/fiber
International (ITU)				
Signal Level	**Bit Rate**	**Channels**	**Carrier System**	**Typical Medium**
0	64Kbps	1	—	Copper wire
1	2.048Mbps	30	E1	Copper wire
2	8.448Mbps	120	E2	Copper wire
3	34.368Mbps	480	E3	Microwave/fiber
4	44.736Mbps	672	E4	Microwave/fiber
5	565.148Mbps	7680	E5	Microwave/fiber

The performance objective refers to the percentage of seconds per day when there are no bit errors on a circuit. The service-availability objective refers to the percentage of time that a circuit is functioning at full capability during a three-month period. If these objectives are not met, the carrier issues credits to the user. Each carrier has its own T-carrier quality objectives that are based on circuit length.

The quality objectives for AT&T's Fractional T1 service, for example, is nine errored seconds per day, which translates to 99.99 percent error-free seconds per day, four severely errored seconds per day, and 99.96 percent service availability per year. According to AT&T, severely errored means that 96 percent of all frames transmitted in a second have at least one error.

Figure T-1
DS1 multiplexing to
DS3 requires the
intermediate step to
DS2.

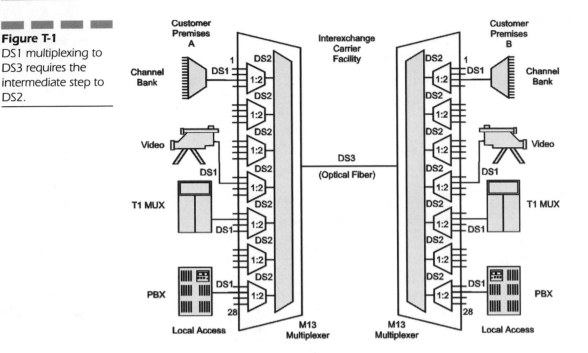

A related measure of performance is failed seconds, which AT&T defines as the time starting after 10 consecutive, severely errored seconds and ending when there have been 10 consecutive seconds that are not severely errored. *Channel Service Units/Data Service Units* (CSUs/DSUs) at each end of the circuits collect and issue reports on this type of information.

Wireless T-Carrier

Wireless T-carrier loops can provide flexibility in areas where the cabling infrastructure to support dedicated lines is lacking, as is often the case in rural communities, for example. Wireless T-carrier uses direct-sequence spread-spectrum technology in the unlicensed 2.4GHz to 5.8GHz *Industrial, Scientific, and Medical* (ISM) band.

Western Multiplex developed the original spread-spectrum T1 product, which it introduced in 1992. Its third-generation offering, the Lynx.sc, is delivered from the factory preconfigured for specific radio channel-pairs, making it easy for customers to install. Unlike other vendors who offer modular units that separate the modem from the *Radio Frequency* (RF) unit, Western Multiplex's Lynx.sc is built as a single unit. The product offers a wide selection of capacities, from 56Kbps to 8 x 1.544Mbps. The maximum range of the system is 50 miles, line of sight.

At a hub site, multiple Lynx radios can be interconnected and monitored by daisy-chaining the diagnostic and data ports. Western Multiplex's management strategy is based on the *Telemetry Byte-Oriented Serial* (TBOS) standard that was developed by Bellcore (now known as Telcordia Technologies) and AT&T. Western Multiplex's TBOS-based OpenLynx is a monitoring tool that enables a technician to view configuration, performance, and alarm parameters from up to eight radios. Lynx.sc also includes a service channel to carry alarm and management traffic without affecting T1 performance. An order-wire interface enables intercom capabilities between units. Remote monitoring is accomplished via front-panel controls.

These products and the comparable products of other vendors enable *Competitive Local-Exchange Carriers* (CLECs) and other network operators to migrate from legacy T1/ E1 connections (for example, using PBX and voice circuits) to IP-based WAN/MAN facilities, greatly facilitating the transition. Operating in the 5.8GHz ISM band, these radios provide wireless interconnect up to 50 miles, with no throughput degradation over any distance. Also, unlike microwave technologies such as *Local Multi-Point Distribution Service* (LMDS), transmissions are unaffected by rainfall and dense fog.

Summary

T-carrier underlies just about every type of carrier facility that is available today, including T1 and T3 and their fractional derivatives—dedicated or switched. These facilities, in turn, support services such as Frame Relay and ISDN and provide access to SMDS, ATM, and VPNs. Through multiplexing techniques, companies can subdivide T-carrier facilities in order to achieve greater bandwidth efficiency and cost savings. With wireless technology, the benefits of T-carrier can be extended well beyond the local loop, bypassing the *Incumbent Local-Exchange Carrier* (ILEC) and eliminating access charges.

See Also

Channel Banks

Microwave

Fiber Optics

Local Multi-Point Distribution Service (LMDS)

Spread-Spectrum Radio
T1 Lines

T1 Lines

T1 lines are digital facilities that provide a transmission rate of up to 1.544Mbps using *Digital Signal Level 1* (DS1). The available bandwidth is divided into 24 channels that operate at 64Kbps each, plus an 8Kbps channel for basic supervision and control. Voice is sampled and digitized via *Pulse-Code Modulation* (PCM). T1 lines are used for more economical and efficient transport over the *Wide-Area Network* (WAN).

Economy is achieved by consolidating multiple lower-speed voice and data channels via a multiplexer or channel bank and sending the traffic over the higher-speed T1 line. This method is more cost-effective than dedicating a separate lower-speed line to each terminal device. The economics are such that only five to eight analog lines are needed to cost-justify the move to T1.

Efficiency is obtained by compressing voice and data to make room for even more channels over the available bandwidth. Individual channels also can be dropped or inserted at various destinations along the line's route. Network-management information can be embedded into each channel for enhanced levels of supervision and control.

Usually, a T1 multiplexer provides the means for companies to realize the full benefits of T1 lines, but channel banks offer a low-cost alternative. The difference between the two devices is that T1 multiplexers offer higher line capacity, support more types of interfaces, and provide more management features than channel banks.

D4 Framing

T1 multiplexers and channel banks transmit voice and data in frames that are called D4 frames. Frames are bounded by framing bits that perform two functions: they identify the beginning of each frame and help locate the channel carrying signaling information. For voice, this bit is carried in the eighth bit position of frames 6 and 12 (refer to Figure T-2).

D4 frames consist of 193 bits, which equates to 24 channels of 8 bits each (plus a single framing bit). Each frame contains a framing bit or signaling bit in the 193rd position, which permits the management of the DS1 facil-

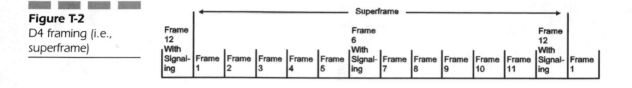

ity itself. This process is completed by robbing the least-significant bit from the data stream, which alternatively carries information or signaling data. Another bit is used to mark the start of a frame. Twelve D4 frames comprise a superframe.

Extended Superframe Format (ESF)

Extended Superframe Format (ESF) is an enhancement to T-carrier, which specifies methods for error monitoring, reporting, and diagnostics. Use of ESF enables technicians to maintain and test the T1 line while it is in service and to often fix minor troubles before they become service-affecting. ESF extends the normal 12-frame superframe structure of the D4 format to 24 frames. By doubling the number of bits that are available, more diagnostic functionality also results (refer to Figure T-3).

Of the 8Kbps bandwidth (repetition rate of 193rd bit or framing bit) that is allocated for basic supervision and control, 2Kbps are used for framing, 2Kbps are used for *Cyclic Redundancy Checking* (CRC-6), and 4Kbps are used for the *Facilities Data Link* (FDL). With CRC, the entire circuit can be segmented so that it can be monitored for errors, without disrupting normal data traffic. In this manner, performance statistics can be generated to monitor T1 circuit quality. Via FDL, performance-report messages are relayed to the customer's equipment—usually a *Channel Service Unit* (CSU)—at one-second intervals. Alarms can also use the FDL, but performance-report messages always have priority.

ESF diagnostic information is collected by the CSU at each end of the T1 line for both carrier and user access. CSUs gather statistics concerning items such as clock-synchronization errors and framing errors, errored seconds, severely errored seconds, failed seconds, and bipolar violations. A supervisory terminal that is connected to the CSU displays this information, furnishing a record of circuit performance.

Originally, the CSU compiled performance statistics every 15 minutes. This information would be kept updated for a full 24 hours so that a complete one-day history could be accessed by the carrier. The carrier would have to poll each CSU to retrieve the collected data and clear its storage

Figure T-3
A comparison of the
193rd bit in D4 and
ESF formats

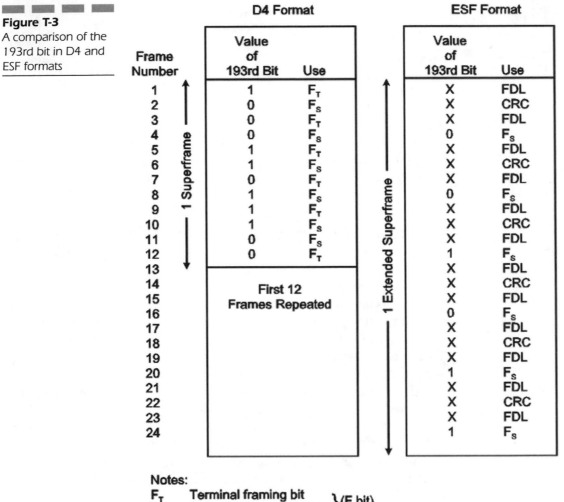

Notes:
F_T	Terminal framing bit	} (F bit)
F_S	Multiframe alignment bit	
FDL	4 Kbps data link bit (M bit)	
CRC	Cyclic redundancy check bit (C bit)	
X	Data dependent	

register. By equipping the CSU with dual registers, one for the carrier and one for the user, both carrier and user alike have full access to the performance history. Today, the CSU is not required to store performance data for 24 hours. Also, the CSU no longer responds to polled requests from carriers; rather, it simply transmits ESF performance messages every second.

ESF also enables end-to-end performance data and sectionalized alarms to be collected in real time. This feature enables the customer to narrow down problems between carrier access points and on interoffice channels—and to find out the direction in which the error is occurring.

E1 Frame Format

In today's increasingly global economy, more and more companies are expanding their private networks beyond the United States and Canada to European locations. In doing so, the first item that they notice is that the primary bit-rate service is not T1, but rather E1. Whereas T1 has a maximum bit rate of 1.544Mbps, E1 has a maximum bit rate of 2.048Mbps. Between the two, there is only one common characteristic: the 64Kbps channel or DS0. A T1 line carries 24 DS0s, while an E1 carries 32 DS0s. Despite this commonality, the DS0s of a T1 line and the DS0s of an E1 line are not compatible.

Although each uses PCM to derive a 64Kbps voice channel, the form of PCM encoding differs. T1 is based on mu-law, while E1 is based on A-law companding. This difference is not as great as it might seem. Most multiplexers and carrier switches have the integral capability to convert between them. Conversion includes both the signaling format and companding method.

In E1, as in T1, there is the need to identify the DS0s to the receiver. The E1 format uses framing for this function, as does T1; that is, there are 8,000 frames per second, with each frame containing one sample from each time slot, numbered 0 to 31.

The frame synchronization in E1 uses half of time slot 0. Signaling occupies time slot 16 in the 0 to 31 sequence, resulting in 30 channels left for user information. All 30 channels cannot signal within the 8 bits that are available in time slot 16, however. The channels must therefore take turns using time slot 16. Two channels send their signaling bits in each frame. The 30 user channels then take 15 frames to cycle through all of the signaling bits. One additional frame is used to synchronize the receiver to the signaling channel, so the full multi-frame ends up having 16 frames. This multi-frame corresponds to the T1 superframe.

Summary

In recent years, T1 lines and services have become the basic building blocks of digital networks. They can support voice, data, and video channels. The individual channels can be added to or dropped from the aggregate bandwidth in order to improve the efficiency and economy of a private network.

In addition, these channels can traverse the *Public-Switched Telephone Network* (PSTN) to bring off-net locations into the private network. Alternatively, the channels can go through a carrier's Frame Relay network or *Asynchronous Transfer Mode* (ATM) network, enabling even greater efficiencies and economies for certain applications.

See Also

> CSU/DSUs
>
> Channel Banks
>
> Multiplexers
>
> T-Carrier Facilities
>
> *Time-Division Multiplexing* (TDM)
>
> Voice Compression

Tariffs

Tariffs are documents that are issued by various types of carriers that describe their voice and data services and the rates that are charged for each. These documents are filed with state and federal regulatory agencies and are accessible by the public, often over the Internet.

Tariffs that are filed at the FCC fall into three broad categories: dominant carriers, non-dominant carriers, and operator service providers. Dominant carriers include AT&T and the ILECs. Non-dominant carriers include MCI WorldCom and Sprint and various resale carriers who offer interstate long-distance services. CLECs also are considered by the FCC to be non-dominant carriers. Operator service providers offer long distance—and, in some cases, local telephone services from public telephones, such as pay telephones or public telephones that are located in hotels/motels, hospitals, airports, restaurants, gas stations, convenience stores, and other public locations. These operators are not necessarily the same telephone companies that offer local service.

All carriers must still file tariffs for new services. Until recently, these tariffs had to be in effect for at least 30 days before the issuing carrier was permitted to revise the tariffs. This rule was originally intended to limit rate churn—rapid rate increases and decreases in a short period. Rate churn can be disruptive or confusing for customers, making it difficult to determine what rates are applicable at any given time. Rate churn also makes it difficult for customers, especially businesses, to accurately budget telecommunications costs.

Tariff Streamlining

In January 1997, the FCC adopted the policy of tariff streamlining, as required by the Telecommunications Act of 1996. The 1996 Act provides that, absent FCC action, a tariff reflecting a reduction in rates would become effective seven days from the date of filing, and a tariff reflecting an increase in rates would become effective in 15 days from the date of filing. Section 204 of the 1996 Act also provides that tariffs that are filed on a streamlined basis are deemed lawful.

Consistent with judicial precedent, the FCC interpreted "deemed lawful" to mean that a pending tariff is conclusively presumed reasonable and lawful upon its effective date, unless the Commission suspends the tariff prior to that date. LEC tariffs will still be subject to post-effective review by the Commission, either through an investigation or a compliance proceeding. The Commission also adopted rules establishing new filing procedures for the streamlined tariffs.

The Commission determined that all ILEC tariffs, whether involving rate increases, decreases, and/or changes to the rates, terms, and conditions of existing services or introducing new services, are eligible for streamlined filing. The Commission said that this approach would give greatest effect to Congressional intent to streamline the tariff review process, noting that making all tariffs eligible will simplify the process by eliminating the need to identify different kinds of filings. The Commission also determined that only tariffs exclusively containing rate decreases will become effective on seven-days' notice. All other tariffs will become effective on 15-days' notice. In August 1999, however, the FCC agreed with Bell Atlantic's argument that price cap ILECs should be permitted to file tariffs for new services on only one-day's notice. The FCC concluded that Bell Atlantic's request is in the public interest and that the 15-day notice period is no longer warranted.

Several factors underscored the FCC's decision. The primary focus of the FCC's review of new service tariffs had been to determine whether the ILEC complied with the new service test. By eliminating the new service test, the commission greatly reduces the need for reviewing ILEC new service tariff filings. In addition, no customer is required to purchase the new service. Furthermore, a longer notice period would delay the introduction of new services, which was the primary reason for revising the price cap rules that had been put into effect years earlier.

Electronic Filing

Along with streamlining the tariff filing process, the FCC found that establishment of an electronic tariff filing program would afford a quick and economical means to file tariffs while giving interested parties prompt access to the tariffs. The Commission concluded that in order for the full benefit of electronic filing to be realized, all affected carriers are required to file tariffs electronically. Other participants in the LEC tariff review process can (but are not required to) file electronically. The Commission delegated the authority to the Common Carrier Bureau to establish the electronic filing program.

The FCC's Common Carrier Bureau was established July 1, 1998, which is the date upon which ILECs must use ETFS to file tariffs and associated documents with the commission. The *Electronic Tariff Filing System* (ETFS) is an Internet-based system through which ILECs submit official tariffs and associated documents to the FCC, in lieu of filing paper copies with the secretary's office. In addition, the public can also use ETFS via the Internet to view the electronically filed tariffs and associated documents.

Summary

A tariff is a document that carriers file with the FCC or a state PUC, detailing the services, equipment, prices, and terms that the carrier offers. Even the unique services of a carrier that might apply to only one customer must be described in a tariff. Consequently, an estimated 20,000 tariffs are on file, more than half of which are still active at this writing. A custom tariff is designed to give a particular customer a low rate. Any other company that uses the same telecommunications service and has similar traffic volumes has 30 days to request the same deal. The custom tariff, however, is usually worded in such a way that few other customers qualify for the same treatment.

See Also

> Access Charges
>
> *Federal Communications Commission* (FCC)
>
> Price Caps
>
> *Public Utility Commissions* (PUCs)
>
> Telecommunications Act of 1996

Transmission Control Protocol/Internet Protocol (TCP/IP)

Transmission Control Protocol / Internet Protocol (TCP/IP) is a suite of networking protocols that is valued for its capability to interconnect diverse computing platforms—from PCs, Macintoshes, and UNIX systems to mainframes and supercomputers. The protocol suite originated from the work that was done by four key individuals more than 30 years ago: Leonard Kleinrock, Vinton Cerf, Robert Kahn, and Lawrence Roberts. Each disagrees on who deserves the lion's share of credit in the development of the Internet. Although the early experiments of Kleinrock made his computer the first node on the early *Advanced Research Projects Agency Network* (ARPANET), Cerf and Roberts generally get the credit for designing the network architecture that eventually became known as the Internet.

The further development of TCP/IP was funded by the U.S. government's *Advanced Research Projects Agency* (ARPA) in the 1970s. As noted, the protocol suite was developed to enable different networks to be joined, in order to form a virtual network known as an internetwork. The original Internet was formed by converting an existing conglomeration of networks that belonged to ARPANET over to TCP/IP, which evolved to become the backbone of today's Internet. The *Internet Engineering Task Force* (IETF) oversees the development of the TCP/IP protocol suite.

Several factors have driven the acceptance of TCP/IP for mainstream business and consumer use over the years. These include the technology's capability to support local and wide-area connections, its open architecture, and a set of specifications that are freely available in the public domain. Although not the most functional or robust transport available, TCP/IP offers a mature, dependable environment for corporate users who need a common denominator for their diverse and sprawling networks.

Key Protocols

The key protocols in the suite include the *Transmission Control Protocol* (TCP), the *Internet Protocol* (IP), and the *User Datagram Protocol* (UDP). Also, there are application services that include the Telnet protocol, providing virtual terminal service; the *File Transfer Protocol* (FTP); and the *Simple Mail-Transfer Protocol* (SMTP). Management is provided by the *Simple Network-Management Protocol* (SNMP).

TCP   TCP forwards data that is delivered by IP to the appropriate process at the receiving host. Among other functions, TCP defines the procedures for breaking up the data stream into packets and reassembling the packets in the proper order to reconstruct the original data stream at the receiving end. Because the packets typically take different routes to their destination, they arrive at different times and arrive out of sequence. All packets are temporarily stored until the missing packets arrive, so that they can be placed in the correct order. If a packet arrives damaged, it is simply discarded and another one is resent.

To accomplish these tasks and others, TCP breaks the messages or data stream into a manageable size and adds a header to form a packet. The packet's header (refer to Figure T-4) consists of the following components:

- *Source Port (16 bits)/Destination Port (16 bits) Address*   The source and destination ports correspond to the calling and called TCP applications. The port number is usually assigned by TCP whenever an application makes a connection. Well-known ports are associated with standard services such as Telnet, FTP, and SMTP.

- *Sequence Number (32 bits)*   Each packet is assigned a unique sequence number that enables the receiving device to reassemble the packets in sequence, in order to form the original data stream.

- *Acknowledgment Number (32 bits)*   The acknowledgment number indicates the identifier or sequence number of the next expected packet. Its value is used to acknowledge all packets that are transmitted in the data stream up to that point. If a packet is lost or

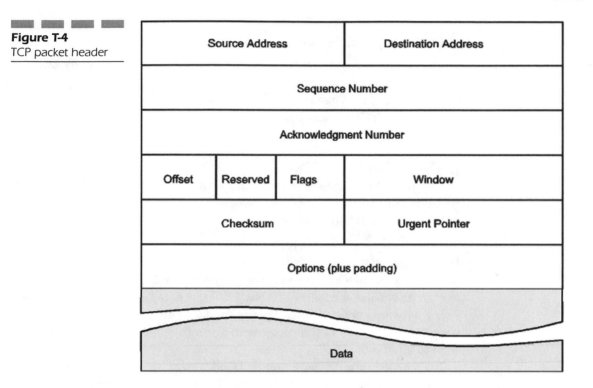

corrupted, the receiver will not acknowledge that particular packet. This negative acknowledgment triggers a retransmission of the missing or corrupted packet.

- *Offset (4 bits)* The offset field indicates the number of 32-bit words in the TCP header. This field is required because the TCP header can vary in length, according to the options that are selected.

- *Reserved (6 bits)* This field is not currently used but can accommodate some future enhancement of TCP.

- *Flags (6 bits)* The flags field serves to indicate the initiation or termination of a TCP session, to reset a TCP connection, or to indicate the desired type of service.

- *Window (16 bits)* The window field, also called the receive window size, indicates the number of 8-bit bytes that the host is prepared to receive on a TCP connection. This field provides precise flow control.

- *Checksum (16 bits)* The checksum is used to determine whether the received packet has been corrupted in any way during transmission.

- *Urgent Pointer (16 bits)*   The urgent pointer indicates the location in the TCP byte stream where urgent data ends.

- *Options (0 or more 32-bit words)*   The options field is typically used by TCP software at one host to communicate with TCP software at the other end of the connection. This field passes such information as the maximum TCP segment size that the remote machine is willing to receive.

The bandwidth and delay of the underlying network impose limits on throughput. Poor transmission quality causes packets to be discarded, which in turn results in retransmissions. Too many retransmissions result in effective bandwidth being cut.

IP   An internet is comprised of a series of autonomous systems or subnetworks, each of which is locally administered and managed. The subnetworks can consist of Ethernet LANs, X.25 packet networks, ISDN, or Frame Relay networks, for example. IP delivers data between these different networks through routers that process packets from one *Autonomous System* (AS) to another.

Each node in the AS has a unique IP address. IP adds its own header and check sum to make sure that the data is properly routed (refer to Figure T-5). This process is aided by the presence of routing update messages that keep the address tables in each router current. Several different types of update messages are used, depending on the collection of subnets that are involved in a management domain. The routing tables list the various nodes on the subnets as well as the paths between the nodes. If the data packet is too large for the destination node to accept, it will be segmented into smaller packets.

The IP header consists of the following fields:

- *IP version (4 bits)*   The current version of IP is 4; the next generation of IP is 6.

- *IP header length (4 bits)*   Indicates header length; if options are included, the header might have to be padded with extra zeros so that it can end at a 32-bit word boundary (which is necessary because header length is measured in 32-bit words).

- *Precedence and type of service (8 bits)*   Precedence indicates the priority of data-packet delivery, which can range from 0 (lowest priority) for normal data to 7 (highest priority) for time-critical data (i.e., multimedia applications). Type of service contains *Quality of*

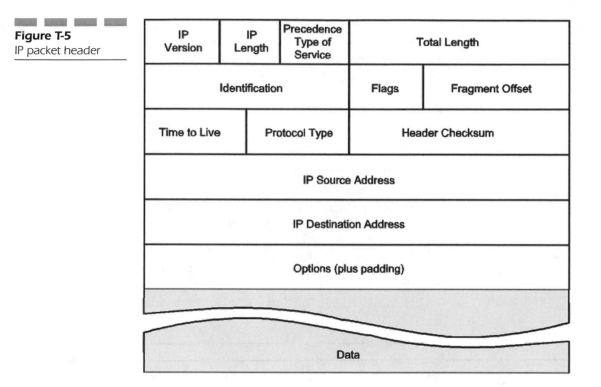

Figure T-5
IP packet header

Service (QoS) information that determines how the packet is handled over the network. Packets can be assigned values that maximize throughput, reliability, or security and that minimize monetary cost or delay. This field will play a larger role in the future as internets evolve to handle more multimedia applications.

■ *Total packet length (16 bits)* Total length of the header plus the total length of the data portions of the packet

■ *Identification (16 bits)* A unique ID for a message that is used by the destination host to recognize packet fragments that belong together

■ *Flags (3 bits)* Indicates whether or not the packets can be fragmented for delivery. If a packet cannot be delivered without being fragmented, it will be discarded—and an error message will be returned to the sender.

■ *Fragmentation offset (13 bits)* If fragmentation is allowed, this field indicates how IP packets are to be fragmented. Each fragment has the same ID. Flags are used to indicate that more fragments are to follow, as well as to indicate the last fragment in the series.

- *Time to live (8 bits)*   This field indicates how long the packet can exist on the network in its undelivered state. The hop counter in each host or gateway that receives this packet decrements the value of the time-to-deliver field by one. If a gateway receives a packet with the hop count decremented to zero, the packet will be discarded. This process prevents the network from becoming congested by undeliverable packets.

- *Protocol type (8 bits)*   Specifies the appropriate service to which IP delivers the packets, such as TCP or UDP

- *Header checksum (16 bits)*   This field is used to determine whether the received packet has been corrupted in any way during transmission. The checksum is updated as the packet is forwarded, because the time-to-live field changes at each router.

- *IP source address (32 bits)*   The address of the source host (e.g., 130.132.9.55)

- *IP destination address (32 bits)*   The address of the destination host (e.g., 128.34.6.87)

- *Options (up to 40 bytes)*   Although seldom used for routine data, this field enables one or more options to be specified. Option four, for example, time stamps all stops that the packet made on the way to its destination. This action enables measurement of overall network performance in terms of average delay and nodal processing time.

Internet performance is dependent on the resources that are available at the various hosts and routers—transmission bandwidth, buffer memory, and processor speed—and how efficiently these resources are used. Although each type of resource is manageable, there are always tradeoffs between cost and performance.

UDP   The protocols themselves offer yet another example of the cost-performance tradeoff. While TCP offers assured delivery, it does so at the price of more overhead. UDP, on the other hand, functions with minimum overhead (refer to Figure T-6); it merely passes individual messages to IP for transmission. Because IP is not reliable, there is no guarantee of delivery.

Nevertheless, UDP is useful for certain types of communications, such as quick database lookups. For example, the *Domain Name System* (DNS) consists of a set of distributed databases that provide a service that translates between system names and their IP addresses. For simple messaging between applications and these network resources, UDP does the job. The UDP header consists of the following fields:

Source Port	Destination Port
Length	Checksum
Data	

- *Source port (16 bits)* This field identifies the source port number.
- *Destination port (16 bits)* This field identifies the destination port number.
- *Length (16 bits)* Indicates the total length of the UDP header and data portion of the message
- *Checksum (16 bits)* Validates the contents of a UDP message. Use of this field is optional. If this field is not computed for the request, it can still be included in the response.

Applications using UDP communicate through a specified numbered port that can support multiple virtual connections, which are called sockets. A socket is an IP address and port, and a pair of sockets (source and destination) forms a TCP connection. One socket can be involved in multiple connections (refer to Figure T-7).

Some ports are registered (well-known) and can be found on many TCP/IP implementations. Well-known ports are numbered from 0 to 1023.

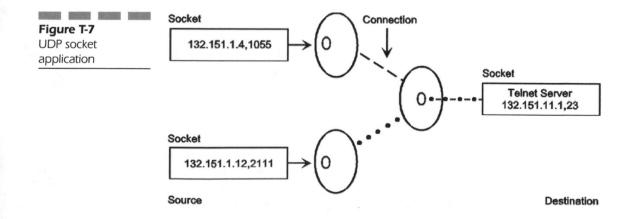

Socket
132.151.1.4,1055

Connection

Socket
Telnet Server
132.151.11.1,23

Socket
132.151.1.12,2111

Source

Destination

Telnet, for example, always uses port 23 for communications, while FTP uses port 21. The well-known ports are assigned by the *Internet Assigned Numbers Authority* (IANA) and on most systems can only be used by system (or root) processes or by programs that are executed by privileged users. Other examples of UDP well-known ports are included in the following table:

Service	Port	Description
Users	11	Shows all users on a remote system
Quote	17	Returns a quote of the day
Mail	25	Used for e-mail via SMTP
Domain Name Server	53	Translates system names and their IP addresses
BOOTpc	68	Client port that is used to receive configuration information
TFTP	69	Trivial FTP that is used for initializing diskless workstations
World Wide Web	80	Provides access to the Web via the *Hypertext Transfer Protocol* (HTTP)
snagas	108	Provides access to an SNA Gateway-Access Server
nntp	119	Provides access to a newsgroup via the *Network News Transfer Protocol* (NNTP)
SNMP	161	Used to receive network-management queries via SNMP

In addition to the well-known ports, there are also registered ports that are numbered from 1024 to 49151 and private ports that are numbered from 49152 to 65535.

High-Level TCP/IP Services

The TCP/IP model includes three simple types of services for file transfers, e-mail, and virtual terminal sessions:

■ File Transfer Protocol (*FTP*)   A protocol that is used for the bulk transfer of data from one remote device to another. Usually implemented as an application-level program, FTP uses the Telnet and TCP protocols. Most FTP offerings have the options to support the unique aspects of each vendor's file structures. Data in the FTP environment consists of a stream of data that is followed by an end-of-file marker, enabling only entire files to be transferred (not selected records within a file). Sending a file via FTP to a user who is on another TCP/IP network requires a valid user ID and password for a host on that network.

■ Simple Mail Transfer Protocol (*SMTP*)   A protocol for exchanging mail messages between systems, without regard for the type of user interface or the functionality that is available locally. SMTP sessions consist of a series of commands, starting with both ends exchanging handshake messages to identify themselves. This handshake is followed by a series of commands that indicate that a message is to be sent and that receipts are needed, and by commands that actually transfer the data. Separating the data message from the address field enables a single message to be delivered to multiple users and to verify that there is at least one deliverable addressee before sending the contents. SMTP modifies every message that it receives by adding a time stamp and a reverse path indicator into each message. In other words, a mail message in the SMTP environment usually consists of a fairly long header with information from each node that handled the message. Many user interfaces can automatically filter out this kind of information, however.

■ *Telnet Virtual Terminal Service*   The Telnet protocol defines a network-independent virtual terminal through which a user can log in to remote TCP/IP hosts. The user goes through the standard login procedure on the remote TCP/IP host and must know the characteristics of the remote operating system in order to execute host-resident commands. Telnet enables remote terminals to access different hosts by fooling an operating system into thinking that a remote terminal is locally connected. Most telnets operate in full-duplex mode, meaning that they are capable of sending and receiving at the same time. A half-duplex mode exists in order to accommodate IBM hosts, however. In this case, a turnaround signal switches the sending of data to the other side of the connection.

Summary

In its early years of development and implementation, TCP/IP was considered of interest only to research institutions, academia, and defense contractors. Today, corporations have embraced TCP/IP as a platform that can meet their needs for multi-vendor, multi-network connectivity. Because it was developed in large part with government funding, TCP/IP code is in the public domain; this availability has encouraged its use by hundreds of vendors who apply it to support nearly all types of computers. Because of its flexibility, comprehensiveness, and non-proprietary nature, TCP/IP has captured a considerable and growing share of the commercial internetworking market.

See Also

Internet

Open Systems Interconnection (OSI)

Simple Network-Management Protocol (SNMP)

Telecommunications Act of 1996

The Telecommunications Act of 1996, which became law in the United States on February 8, 1996, establishes a pro-competitive, deregulatory framework for telecommunications. This law makes sweeping changes that affect all consumers and telecommunications service providers. The intent of this law is to rapidly accelerate private-sector deployment of advanced telecommunications and information technologies and services to all Americans by opening all telecommunications markets to competition.

For years, competitive service providers have wanted to offer local telecommunications services, a market dominated by the traditional monopolistic telephone companies. A major goal of the Telecommunications Act is the availability of competitive local telecommunications services, including advanced and innovative services, to all Americans. The Act contemplates that competitors will offer local telecommunications services by reselling the incumbent LECs' services, using unbundled elements of the incumbent LECs' networks or building their own network facilities. Only facilities-based competition, however, can bring the full benefits of competition to consumers and break down the dominant position of the incumbent LECs. At the same time, the incumbent LECs have wanted to expand into long-distance services. Among the key provisions of the Telecommunications Act of 1996 is that competition in both areas is allowed for the first time. In addition, electric-utility companies and cable TV companies can (for the first time) offer telecommunications services.

State regulators and the FCC, however, must review *Bell Operating Company* (BOC) compliance with a comprehensive checklist before the

BOCs can provide long-distance telephone service. This measure is meant to ensure that the BOCs do not impose burdensome restrictions on would-be competitors, even as they themselves branch out into profitable new markets. Section 271 of the Telecommunications Act provides a 14-point checklist that is meant to ensure a truly competitive market for local and long-distance services.

Checklist Item 1

The Act requires "interconnection in accordance with the requirements of sections 251(c)(2) and 252(d)(1)." In other words, all ILECs must enable interconnection to their networks (1) for exchange service and exchange access; (2) at any technically feasible point; (3) that is at least equal in quality to what the local exchange carrier gives itself, its affiliates, or anyone else; and (4) on rates terms and conditions that are just, reasonable, and nondiscriminatory (251(c)(2)).

Any interconnection, service, or network element that is provided under an approved agreement shall be made available to any other requesting telecommunications carrier upon the same terms and conditions as those provided in the agreement (252(i)). Prices for interconnection shall be based on cost (without reference to any rate-based proceeding) and shall be nondiscriminatory, and they can include a reasonable profit (252(d)(1)).

Checklist Item 2

The Act requires "nondiscriminatory access to network elements in accordance with the requirements of sections 251(c)(3) and 252(d)(1)." In other words, all ILECs must provide (to any requesting telecommunications carrier) a telecommunications service and nondiscriminatory access to network elements on an unbundled basis at any technically feasible point on rates, terms, and conditions that are just, reasonable, and nondiscriminatory. These unbundled network elements will be provided in a manner that enables carriers to combine the elements in order to provide the telecommunications service (251(c)(3)). A network element is a facility or equipment that is used in the provision of a telecommunication service, including features, functions, and capabilities such as subscriber numbers, databases, signaling systems, and information that is sufficient for billing and collection or that is used in transmission, routing, or provision of a telecommunications service (3(a)(45)).

In determining which network elements will be made available, the FCC shall consider, at a minimum, whether (A) access to network elements that are proprietary is necessary and (B) whether failure to provide access to these network elements would impair the capability of a carrier to provide the services that it wishes (251(d)(2)). Prices shall be based on cost (without reference to any rate-based proceeding) and shall be nondiscriminatory, and they can include a reasonable profit (252(d)(1)). As part of their competitive checklist, BOCs are required to unbundle loop transmission, trunk-side local transport, and local switching (271(c)(2)(B)(iv)-(vi)).

Checklist Item 3

The Act requires "nondiscriminatory access to the poles, ducts, conduits, and rights-of-way owned or controlled by the Bell operating company at just and reasonable rates in accordance with the requirements of section 224." In other words, each LEC is to afford nondiscriminatory access to the poles, ducts, conduits, and rights-of-way to competing providers of telecommunications services, but they can deny access for reasons of safety, reliability, and generally applicable engineering purposes (251(b)(4), 224(f)). Within two years, the FCC must prescribe regulations for charges for pole attachments that are used by telecommunications carriers (not ILECs) to provide telecommunications services when the parties fail to agree. Charges must be just, reasonable, and nondiscriminatory (224(a)(5), (e)(1)). Pole-attachment charges shall include costs of usable space and other space (224(d)(1)-(3), (e)(2)). Duct and conduit charges shall be no greater than the average cost of duct or conduit space (224(d)(1)). A utility must impute and charge affiliates its pole-attachment rates (224(g)).

Checklist Item 4

The Act requires "local loop transmission from the central office to the customer's premises, unbundled from local switching or other services." In other words, BOCs must unbundle loop transmission (271(c)(2)(B)(iv)). This service is to be provided at any technically feasible point and in a way that is nondiscriminatory, including rates, terms, and conditions that are just, reasonable, and nondiscriminatory. Unbundled network elements will be provided in a manner that enables carriers to combine the elements in order to provide the telecommunications service (251(c)(3)). In determining which network elements will be made available, the FCC shall consider, at a minimum, whether (A) access to network elements that are proprietary is necessary and (B) whether failure to provide access to these network elements would impair the capability of a carrier to provide the services that it wishes (251(d)(2)). Prices shall be based on cost (without reference to any rate-based proceeding) and shall be nondiscriminatory, and they can include a reasonable profit (252(d)).

Checklist Item 5

The Act requires "local transport from the trunk side of a wireline local exchange carrier switch unbundled from switching or other services." In other words, BOCs must unbundle trunk-side local transport (271(c)(2)(B)(v)). This service is to be provided at any technically feasible point and in a way that is nondiscriminatory, including rates, terms, and conditions that are just, reasonable, and nondiscriminatory. Unbundled network elements will be provided in a manner that enables carriers to combine the elements in order to provide the telecommunications service (251(c)(3)). In determining which network elements will be made available, the FCC shall consider, at a minimum, whether (A) access to network elements that are proprietary is necessary and (B) whether failure to provide access to these network elements would impair the capability of a carrier to provide the services it wishes (251(d)(2)). Prices shall be based on cost (without reference to any rate-based proceeding) and shall be nondiscriminatory, and they can include a reasonable profit (252(d)).

Checklist Item 6

The Act requires "local switching unbundled from transport, local loop transmission, or other services." In other words, BOCs must unbundle local switching (271(c)(2)(B)(vi)). This service is to be provided at any technically feasible point and in a way that is nondiscriminatory, including rates, terms, and conditions that are just, reasonable, and nondiscriminatory. Unbundled network elements will be provided in a manner that enables carriers to combine the elements in order to provide the telecommunications service (251(c)(3)). In determining which network elements will be made available, the FCC shall consider, at a minimum, whether (A) access to network elements that are proprietary is necessary and (B) whether failure to provide access to these network elements would impair the capability of a carrier to provide the services it wishes (251(d)(2)). Prices shall be based on cost (without reference to any rate-based proceeding) and shall be nondiscriminatory, and they can include a reasonable profit (252(d)).

Checklist Item 7

The Act requires "nondiscriminatory access to: (I) 911 and E911 services; (II) directory assistance services to allow the other carrier's customers to obtain telephone numbers; and (III) operator call completion services."

Checklist Item 8

The Act requires "White pages directory listings for customers of the other carrier's telephone exchange service." In other words, access or interconnection that is provided or generally offered by a BOC to other telecommunication carriers must include White Pages directory listings for customers of the other carrier's telephone-exchange service (271(c)(2)(B)(viii)).

Checklist Item 9

The Act states, "Until the date by which telecommunications numbering administration guidelines, plan, or rules are established, nondiscriminatory access to telephone numbers for assignment to the other carrier's telephone exchange service customers. After that date, compliance with such guidelines, plan, or rules." In other words, the FCC must create or designate one or more impartial entities to administer telecommunications numbering and to make numbers available on an equitable basis. The FCC has exclusive jurisdiction over the U.S. portion of the North American Number Plan but can delegate any or all jurisdiction to State commissions or other entities (251(e)(1)). BOCs are required to provide nondiscriminatory access to telephone numbers for assignment by other carriers until telecommunications numbering administration guidelines, plan, or rules are established. Once these guidelines, plans, or rules are established, BOCs must comply with them (271(c)(2)(B)(ix)).

Checklist Item 10

The Act requires "nondiscriminatory access to databases and associated signaling necessary for call routing and completion." In other words, access or interconnection that is provided or generally offered by a BOC to other telecommunication carriers shall include nondiscriminatory access to databases and associated signaling that is necessary for call routing and completion (271(c)(2)(B)(x)). In determining which of these network elements will be made available, the FCC shall consider, at a minimum, whether (A) access to network elements that are proprietary is necessary and (B) whether failure to provide access to these network elements would impair the capability of a carrier to provide the services it wishes (251(d)(2)). Prices of network elements shall be based on cost (without reference to any rate-based proceeding) and shall be nondiscriminatory, and they can include a reasonable profit (252(d)(1)).

Checklist Item 11

The Act states, "Until the date by which the Commission issues regulations pursuant to section 251 to require number portability, interim telecommunications number portability through remote call forwarding, direct inward dialing trunks, or other comparable arrangements, with as little impairment of functioning, quality, reliability, and convenience as possible. After that date, full compliance with such regulations."

In other words, all LECs must provide number portability, to the extent feasible, and in accordance with the FCC's requirements (251(b)(2)). Number portability enables customers to retain, at the same location, their existing telecommunications numbers without impairment of quality, reliability, or convenience when switching from one telecommunications carrier to another (3(a)(46)). Until the date that the FCC establishes number portability, BOCs are required to provide interim number portability through remote-call forwarding, direct inward-dialing trunks, or other comparable arrangements, with as little impairment of functioning, quality, reliability, and convenience as possible. BOCs must fully comply with all FCC number-portability regulations (271(c)(2)(B)(xi)).

Checklist Item 12

The Act requires "nondiscriminatory access to such services or information as are necessary to allow the requesting carrier to implement local dialing parity in accordance with the requirements of section 251(b)(3)." In other words, access or interconnection that is provided or generally offered by a BOC to other telecommunication carriers shall include nondiscriminatory access to such services or information as necessary, to enable the requesting carrier to implement local dialing parity in accordance with the 251 (b)(3) policy (271(c)(2)(B)(xii)). All LECs have the duty to provide dialing parity to competing providers of telephone-exchange service and telephone toll service and have the responsibility to permit all such providers to have nondiscriminatory access to telephone numbers, operator services, directory assistance, and directory listing (with no unreasonable dialing delays) (251(b)(3)).

Checklist Item 13

The Act requires "reciprocal compensation arrangements in accordance with the requirements of section 252(d)(2)." In other words, all LECs must establish reciprocal compensation arrangements for transport and termination of telecommunications traffic (251(b)(5)). The terms and conditions shall enable each carrier to cover its additional costs of terminating the traffic, including offsetting of reciprocal obligations such as bill-and-keep. Commissions cannot engage in any rate proceedings nor require record keeping to determine the additional costs of the calls (252(d)(2)).

Checklist Item 14

The Act states, "Telecommunications services are available for resale in accordance with the requirements of sections 251(c)(4) and 252(d)(3)." In other words, all LECs must not prohibit and not impose unreasonable or discriminatory restrictions on resale (251(b)(1)). ILECs must offer wholesale rates for any telecommunications service that is provided at retail to customers who are not telecommunications carriers (251(c)(4)(A)). Wholesale prices shall be based on retail prices less the marketing, billing, collection, and other costs that will be avoided by selling the service at wholesale (252(d)(3)). State commissions can, to the extent that is permitted by the FCC, prohibit a reseller from buying a service that is available only to one category of customers and reselling it to a different category of customers (251(c)(4)(B)).

Initial applications to state and federal regulators from 1996 to 1999 for permission to offer interLATA long-distance telephone service have been denied. Although the BOCs believed that they met the requirements of the 14-point checklist, upon further examination the regulators found that the BOCs only articulated a plan to meet the checklist requirements and had not actually implemented many of them. In other cases, the FCC has turned down requests to enter the long-distance market because the BOCs could not demonstrate that viable local competition existed in their LATAs.

Summary

Under the Telecommunications Act of 1996, a local Bell company wanting to offer long-distance service is required to show that it has opened up its local monopoly to give rivals a chance to compete before it could offer long-distance. Others had tried, but Bell Atlantic became the first to convince the FCC that it met the requirements. In December 1999, Bell Atlantic received approval to offer its own long-distance service to local customers and in January 2000 began offering long-distance service in New York State, where it is the local phone company for 6.6 million households. This situation marks the first time since the court-ordered breakup of AT&T in 1984 that consumers can obtain both local and long-distance service from one of the former AT&T companies.

See Also

　　Colocation Agreements
　　Federal Communications Commission (FCC)
　　Interconnection Agreements
　　Unbundled Access

Telecommunications-Industry Mergers (TIMs)

In the deregulated, competitive telecommunications industry, *Mergers and Acquisitions* (M&As) are a fact of life. Several hundreds of billions of dollars have changed hands over the six-year period from 1993 to 1999—ostensibly to improve telecommunications service, pricing, and coverage, but also to improve the market share and competitive position of the companies that are involved in these transactions.

Among the hundreds of mergers and acquisitions that occurred during this period, Bell Atlantic merged with NYNEX in 1994, and SBC had acquired Pacific Bell in 1997 and Ameritech in 1999. MCI and WorldCom merged in 1998, and AT&T acquired IBM's global network in mid-1999.

Qwest and US WEST merged in 1999, as did Global Crossing and Frontier Corporation.

Mergers and acquisitions are not solely an American way of doing business. A report that was issued in June 1999 by the investment bank Broadview International notes that European telecommunications, media, and information-technology companies have launched an aggressive and unprecedented blitz of mergers and acquisitions in the United States. According to the report, European firms spent $72.8 billion on 94 acquisitions in the United States during the first half of 1999 alone, more than tripling the value of deals from the same period last year.

According to Broadview, for the first time in at least a decade, more than half of the 10 largest technology deals were carried out by European buyers —among them was Germany's Siemens, France's Alcatel, and Sweden's Ericsson. So-called Internet fever is driving merger and acquisition activity in the United States and Europe, with 447 digital media deals worth $37.5 billion announced in the first six months of 1999—more than six times the number during that same period in 1998.

In the second half of 1999, one deal dwarfed the total value of all previous deals. The telecommunications industry's largest merger came in mid-1999, when the United Kingdom's Vodafone completed its acquisition of AirTouch in the United States for $62 billion. At this writing, Vodafone and Bell Atlantic are in merger discussions. If a deal is struck and approved by the FCC, the combined companies would constitute the premiere wireless-service provider in the United States.

Later in 1999, other mergers followed that were even bigger than the Vodafone-AirTouch merger. SBC merged with Ameritech in a transaction worth $74 billion. MCI WorldCom and Sprint agreed to merge in a transaction worth $129 billion, pending regulatory approval. In January 2000, *America Online* (AOL), the leading *Internet Service Provider* (ISP), announced a merger with Time Warner, the world's largest media and entertainment company—creating a new company called AOL Time Warner in a transaction valued at $166 billion. The following month, Vodafone AirTouch plc agreed to buy Mannesmann AG, Germany's largest cellular-service company, in a $198.9 billion stock swap. Together, the companies start off with 42 million customers worldwide.

These monster mergers reflect how the worlds of telecommunications and information distribution are changing. Demand is shifting from voice to data, from narrow-band to broad-band, and from wireline to wireless. The combination of other companies such as CBS and Viacom, as well as AOL and Time Warner, promise to fundamentally change the way people obtain information, communicate with others, buy products, and are entertained.

More combinations are possible as smaller companies begin to sense that they are being out-muscled in this new era of industry consolidation and regulatory acquiescence.

Role of the FCC

In addition to all of its other activities, the FCC also has authority over mergers and acquisitions in the telecommunications industry. The FCC has the power to determine whether the merger or acquisition would serve the public's interest. While the *Department of Justice* (DoJ) shares concurrent responsibility with the Commission with respect to the antitrust issues arising from M&A activity, the Commission's responsibility is both broader and more distinct from the DoJ's role.

First, the Commission must consider the effect of the merger or acquisition on competition when weighing the public's interest. Second, the Commission's inquiry can extend beyond antitrust principles and judicial standards of evidence. Third, the Commission must implement and enforce the national policies that are articulated in the 1996 Telecommunications Act.

Specifically, the merger or acquisition should provide short-term and long-term competitive benefits to state and local communities. Furthermore, the merger should not adversely affect competition in state and local communities. In evaluating the merger, the Commission incorporates the interests and experiences of state and local communities that would be affected by the merger. To accomplish this goal, the FCC might conduct local forums, such as town meetings, in a representative set of affected communities.

In late 1999, the FCC approved the $74 billion merger of SBC Communications and Ameritech. SBC had previously bought Pacific Bell and Southern New England Telephone. With its merger with Ameritech, SBC now controls nearly a third of the nation's local telephone lines. In approving the merger, the FCC gave the combined company 30 months to expand into 30 major markets and to offer competitive local telephone service. If the company fails to meet this schedule, the company will have to pay $40 million in penalties for each city that it misses. An additional $1.1 billion in penalties could apply if regulators find that the SBC is discriminating against competitors in its local markets. The FCC believes these conditions will be a boon to consumers, bringing more choices into homes and businesses.

The FCC has noted the increase in mergers and acquisitions and expects to see many more in the future. Throughout the telecommunications industry, companies that previously focused on a single-market sector now want to expand their horizons by reshaping themselves into vertically integrated companies that provide content, distribution, software, and equipment. Some might seek to provide one-stop shopping by bundling local and long-distance service, video, and Internet access. Some companies might even throw in local service for free in an effort to capture market share.

While some mergers or joint ventures might work, others might suffer from the inevitable clash of corporate cultures. Also, foreign interests, which are attracted by the lucrative (and now open) U.S. telecommunications markets will seek ways to enter new global alliances with U.S. companies, just as U.S. companies will continue to expand their holdings abroad. In encouraging competition, the Telecommunications Act of 1996 enables enormous change and opportunities for diversification in the industry. The FCC is charged with carrying out the Act's provisions in this regard—and, in the process, must ensure that the public's interest is served.

In response to the unprecedented consolidation in the telecommunications industry, the FCC has revamped its merger review process. The goal is to review even the most complex transactions within 180 days while ensuring that the public's interest is met in this era of consolidation and convergence. The FCC now has procedures in place that will ensure that applicants know what is expected of them, what will happen when, and the current status of their application.

Summary

According to a 1999 study funded by Bell Atlantic and released by the New Millennium Research Council, a Washington, D.C. public-policy research group, consumers stand to gain about $93 billion in savings by 2005 as a result of gains from the economies of scale that result from telecommunications mergers. At the same time, about 650,000 new jobs will be created in voice, data, and video. According to the study, "Deregulation and Consolidation of the Information Transport Sector: A Quantification of Economic Benefits to Consumers," telecommunications mergers are leading to higher efficiency and lower consumer prices on more services. But federal regulators might be slowing the progress of those benefits to consumers because of regulations that are based on older, separate industry classifications, such as telephone and cable television, instead of a converged voice, data, and video transport market. The report advises regulators to adopt a broader definition of the industry and market that reflects convergence and consumer options.

See Also

Federal Communications Commission (FCC)

Telecommunications Management Network (TMN)

Telecommunications carriers today are faced with the task of rapidly introducing and managing new competitive services. To reach this goal, they must quickly integrate communications equipment from multiple vendors. In response to the challenge, the carriers and international standards bodies have defined a solution called the *Telecommunications Management Network* (TMN), which consists of a series of interrelated national and international standards and agreements that provide for the surveillance and control of telecommunications service-provider networks on a worldwide scale. The result is the capability to achieve higher service quality, reduced costs, and faster product integration. TMN is also applicable in wireless communications, cable television networks, private overlay networks, and other large-scale, high-bandwidth communications networks.

The *International Telecommunications Union-Telecommunications* (ITU-T) began work on TMN in 1988. The ITU-T (formerly known as CCITT) extended the ISO/OSI standards for systems management by adding its own recommendations for architecture (M.3000 series), modeling (G series), and interfaces (Q series). The resulting standard M.3010 is the high-level specification of the framework, which provides for the observation, control, coordination, and maintenance of telecommunications networks. Because the TMN model is based on open interfaces, it is attractive to telecommunications operators and service providers as a way to solve common management problems without being limited by proprietary solutions.

By making all internal network-management functions available through standardized interfaces, service providers can achieve more rapid deployment of new services and maximum use of automated functions. Vendors of network elements can offer specialized management systems that are known as element managers, which can integrate readily into a service provider's larger management hierarchy. Groups of service providers can enter business-level agreements and deploy resource-sharing arrangements that can be administered automatically through interoperable interfaces.

Under TMN, management tasks are arranged into Network Element, Element Management, Network Management, Service Management, and Business Management layers. OSI's *Common Management Information Protocol* (CMIP) is used to communicate between the adjacent layers.

Object-oriented techniques are used for creating data structures and their access methods for the *Management Information Base* (MIB). Traditional manager/agent concepts are used for the hierarchical exchange of management information between systems. A MIB is located at an agent and provides an abstraction of network resources for the purpose of facilitating management. The agent provides service elements, which are a standard set of access methods to the MIB.

Functional Architecture

TMN breaks the tasks that are performed in a management network into functional blocks, as follows:

- Network Element Function (*NEF*)   Contains the telecommunications functions that are the actual subject of management
- Operations Systems Function (*OSF*)   Processes information that is related to management
- Workstation Function (*WSF*)   Provides the means to interpret TMN information for the management-system operator
- Q Adaptor Function (*QAF*)   Connects a non-TMN NEF or OSF to a TMN via protocol conversion to a standard TMN interface
- Mediation Function (*MF*)   Mediates between a non-OSF and an NEF or a QAF and presents the information in a different form for the OSF
- Data-Communication Function (*DCF*)   Transfers telecommunications network-management information

Physical Architecture

TMN standards define two types of telecommunications resources: managed systems, which are generally known as *Network Elements* (NE), and managing systems, of which the *Operation System* (OS) is the most prominent. TMN also defines reference points (f, g, m, q, and x) that can exist between these function blocks. Reference points become interfaces when they occur at locations that require data communications between elements. The key interface in the TMN model is the OS-NE (or Q3) interface, which uses a CMIP manager/agent pair to provide access to a standardized MIB.

For each functional block, a physical block can be implemented—thus leading to a physical architecture. Reference points are a significant part of the TMN functional architecture and are realized within the physical architecture by physical interfaces within systems or equipment. The implementation of the reference points are represented by capital letters (Q, F, X) and form the common boundary between associated building TMN blocks. The F interface is found between workstations and the TMN; the Q (Q3) interface is found between TMN devices; and the X interface is found between the devices of one TMN and the devices of another TMN via a *Data-Communication Network* (DCN). Figure T-8 describes the physical architecture, showing the implementation of reference points (i.e., interfaces) within TMN.

A telecommunications network provides voice and data services to customers (this function is outside the jurisdiction of the TMN). The network

Figure T-8
TMN physical
architecture

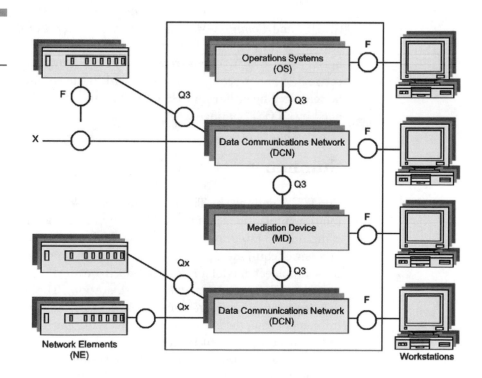

consists of various network elements that can provide TMN functions and services, however. Central to the TMN is the DCN, which typically uses portions of the carrier's telecommunication network for transmission of management information. Telecommunication-network NEs, which provide transmission to the DCN, are part of the TMN's DCN. Most NEs provide services for a facility or a telecommunication network, such as environmental alarm notification or telecommunication transmission. NEs that provide a TMN management interface as well as non-TMN functions straddle the TMN boundary.

The operation system provides the mechanism for interaction, observation, and execution of many of the functions within the TMN. All higher-level functionality, however, does not have to be physically located within the OS, NE, or any other single point in the TMN.

TMN elements use object-oriented techniques to represent resources under their control as managed objects. A common understanding of these objects is shared between the element controlling the resources, a *Mediation Device* (MD) or NE, and the manager of the resource, an OS or MD. TMN uses a manager/agent paradigm where the manager issues instructions and the agent carries out the instructions. These operations are accomplished by using the *Common Management Information Services* (CMIS) and *Common Management Information Services Protocol* (CMIP) standards. In response to instructions to the agent, methods in the managed object are invoked—resulting in a variety of information storing, retrieving, and processing activities within the NE or MD. A NEF agent can be managed by either an OSF or MD manager, while an MF agent is managed by an OSF manager.

Summary

The ITU-T defined the TMN standards to specify the management of global telecommunications networks. The advantage of TMN is that it enables companies to integrate diverse network elements, including legacy systems and newer equipment, into the same network-management structure. TMN uses an object model to describe each network element being managed (such as ATM switch or a GSM base station). The object model eases the development of management applications that are independent of the implementation details of network elements, such that a single application can manage equipment from many suppliers. Among the popular management platforms that support TMN are Hewlett Packard's OpenView, IBM's NetView, Sun's Solstice Enterprise Manager, and Tivoli's Management

Framework (an IBM company). The companies also offer TMN development tools that enable the rapid integration of applications into TMN networks.

See Also

  Open Systems Interconnection (OSI)

Telecommunications Management Systems

Telecommunications management systems capture call records from the PBX or key system for the purpose of identifying and managing costs, keeping the corporate telephone network optimized for maximum savings and availability, and providing decision-making information on the need for more lines and equipment.

The captured call records are stored on a PC or other collection device. The collected data is then processed into a variety of cost and usage reports. With billing reports, for example, call charges can be allocated across the organization. Traffic statistics reports provide a detailed look at trunk usage, call volumes, and calling patterns.

In addition to traditional call accounting functions, which provide cost allocation and other reports on employee and department telephone usage, numerous other related applications and options are available with telecommunications management systems. For instance, there are software modules that can be added to the system to track equipment and cable assets within the organization.

Equipment Inventory

Equipment inventory applications track information on the quantity, type, and location of installed and spare voice equipment (PBX and station), as well as circuits/trunks (e.g., DID, WATS, and private lines). This application usually tracks inventory and indicates if it is committed or available for use. Some systems even provide key sheets or face layouts that include pictures (graphical representations) of each component (e.g., telephone, turret, market data terminal, etc.) within the system, specifying the features and types of circuits that are associated with each key or button. The emphasis of many equipment inventory systems is to work with call accounting in order to provide internal billing for equipment usage.

Invoice Management

Invoice management systems are designed to compare vendor invoices to equipment inventory databases, and create reports that flag erroneous charges or duplicate billings. Invoice management software also makes billing comparisons using previous period and/or previous year invoices in order to automate approval of vendor invoices, allow telecommunications managers to analyze usage trends, and facilitate budgeting activities.

Alarm Monitoring and Reporting

Alarm monitoring and reporting applications monitor voice and data equipment alarm ports, organize alarms by priority, and generate plain-English alarm reports. Some systems automatically dial out to a central reporting center and generate trouble tickets for alarms. The alarm reports that are generated by alarms monitoring and reporting software provide performance statistics concerning PBX hardware and software components.

In addition to monitoring and reporting alarms for the telephone system, some telecommunications management systems can act on alarms from other types of equipment, including voice-mail systems, environmental control systems, and security systems.

Trouble Reporting and Tracking

Trouble reporting and tracking applications maintain a log of all system problems, track repair progress, and provide reports on vendor response times and system performance. These applications often provide an interface to help desks or service desks, providing historical problem-resolution information to the operators.

Work-Order Processing and Tracking

When a problem occurs, a telecommunications manager will generate either a paper or electronic work order and give it to a technician. The work order contains information about the work to be performed, the machine or item on which work is to be performed, the required completion date, the cost of work, and billing information. Work-order processing and tracking applications can automate the activities that are involved with processing move, add, and change requests. The orders can be categorized as pending, overdue, or completed to help manage vendor and staff resources.

In some work-order processing and tracking applications, the software even schedules work orders and automatically updates the equipment and cable inventory databases upon completion of the job. Some applications also commit inventory and schedule vendors and produce reports that flag overdue orders or scheduling problems associated with large projects.

Network Design

Network-design applications were originally developed to optimize the use of WATS lines—and, for a time, private-line services. Now, however, they are in greater demand for intra-state trunk and toll analysis. A tariff database is used that is available on a subscription basis to perform what-if analyses. When given data concerning existing calling patterns, the network-design application applies various rate plans from different carriers to find the most economical blend of services.

Summary

Organizations of all types and sizes are turning to telecommunications management systems to optimize their voice networks for efficiency and cost containment. This type of system includes call-detail collection and reporting capabilities but also goes beyond these basic features to include advanced capabilities such as inventory tracking, invoice management, trouble reporting and tracking, work-order processing and tracking, and network design. These capabilities can be included in a single, integrated product or can be purchased separately as complementary modules that plug into the vendor's telecommunications-management system.

See Also

Asset Management

Call-Detail Reporting

Network Design

Telecommunications Relay Services (TRS)

Title IV of the *Americans with Disabilities Act* (ADA) of 1990 requires the FCC to ensure that *Telecommunications Relay Services* (TRS) are available, to the extent possible and in the most efficient manner, to the more than 30 million Americans who have hearing or speech disabilities. TRS is a telephone transmission service that is designed to give people who have hearing or speech disabilities "functionally equivalent" access to the telephone network. TRS has been available on a uniform, nation-wide basis since July 26, 1993. The FCC sets minimum operational, functional, and technical standards for TRS, certifies state TRS programs, and oversees the administration of the interstate TRS cost-recovery fund. Currently, TRS provides access to the voice telephone network for more than 30 million Americans who have hearing and speech disabilities. The service is offered seven days a week, 24 hours a day in all 50 states, the District of Columbia, Puerto Rico, and the United States Virgin Islands.

Operation

TRS relies on *Communications Assistants* (CAs) to relay the content of calls between users of *Text Telephones* (TTYs) and users of traditional handsets (voice users). For example, a TTY user can telephone a voice user by calling a TRS provider (or relay center), where a CA will place the call to the voice user and relay the conversation by transcribing spoken content for the TTY user and reading text aloud for the voice user. To access TRS, the user dials the local service provider, either through TTY or through voice telephone. The call is answered by a CA, who takes down the number of the party to contact and completes the call. The CA relays the conversation to the called party, verbatim and in real time.

The CA cannot intentionally alter a relayed conversation and cannot limit the length of calls. The CA also cannot disrupt the continuity of a TRS call, except when necessary for facilitation of the call (for example, to ask for clarification of unintelligible messages).

The TRS provider number is listed in the local telephone directory, usually in the information section, or is available from directory assistance. Some states have separate TRS provider numbers for TTY and voice callers. The FCC also publishes a TRS Directory.

Funding

The cost of interstate TRS is recovered from all providers of interstate telecommunications services as a percentage of their gross revenues and as a contribution factor that the FCC determines annually. Contributions are administered in a TRS fund by the *National Exchange Carrier Association* (NECA), an association of local telephone companies. TRS providers are compensated for interstate TRS minutes of use, based on a payment rate that the FCC also determines annually.

The FCC has established an interstate TRS Fund Advisory Council to advise the TRS fund administrator regarding funding issues. The Council is comprised of consumer representatives, TRS users, state regulatory officials, TRS providers, and state relay administrators. The Advisory Council's meetings are open to the public.

Telecommunications Act of 1996

Section 255 of the Telecommunications Act requires all manufacturers of telecommunications equipment and providers of telecommunications services to ensure that such equipment and services are designed and developed to be accessible to and usable by individuals with disabilities, if readily achievable. The FCC enforces implementation of this provision through its rule-making authority.

In 1999, the FCC noted that TRS users do not always have the same access to their carrier of choice or special pricing plans as non-TRS users. The Commission had been informed that some TRS users had been unable to place TRS calls through their chosen carrier or had been unable to make dial-around calls by using a carrier-specific access code. If TRS users are not able to use their carrier of choice and are forced to select an alternative provider, they might pay rates that are higher than those that are charged by their preferred carrier, or they might not have access to particular services.

The FCC noted that this result is inconsistent with the ADA and the Commission's own rules. The FCC should take appropriate measures to ensure that callers in the areas that they serve have access to their services through TRS. Because the Commission's rules require each common carrier to enable access via TRS to their services throughout the area(s) in which they offer service, the Commission is reviewing the extent to which TRS users are unable to access their chosen carrier. The Commission intends to work with TRS users, the industry, TRS centers, state commissions, and other interested parties to ensure that TRS users have equal access to their carrier of choice. The Commission reminded carriers, however, that it might consider enforcement action (including forfeitures) should this obligation not be met.

Summary

TRS users pay rates that are no greater than those rates that are paid by other users of functionally equivalent voice communication services with respect to such factors as the duration of the call, the time of day, and the distance from the point of origin to the point of termination. As long as individuals who have hearing and speech disabilities are required to purchase specialized *Customer-Premises Equipment* (CPE), however, including TTYs, they might not enjoy the same access to the telephone network as voice telephone users. Such specialized equipment can be much more expensive than a regular telephone. The FCC currently permits common carriers to provide, under tariff, appropriate CPE. Some state legislatures have also passed legislation enacting such equipment-distribution programs.

See Also

Telecommunications Act of 1996

Telephone Access for the Disabled

Telecommuting

Telecommuting is a cooperative arrangement between companies and employees that enables employees to work outside the office—usually in the home or at a satellite office that is set up for that purpose—on a part-time or full-time basis. Some industry estimates have pegged the number of telecommuters in the United States at 30 million.

Advances in computer and communications technologies, including wireless products and services, have made telecommuting a viable option for many types of jobs. Telecommuting provides participants with a greater sense of job autonomy, which can increase productivity and satisfaction, while providing employees with a means to better manage their work and family lives. A telecommuting program can also help trim corporate overhead expenses by providing savings on office space and other facilities.

The jobs that are best suited for telecommuting include those of white-collar workers who are engaged in research, consulting, auditing, illustration, writing, computer programming, and Web-page design. Telecommuting has also proved effective for call-center agents, telemarketers, and customer-service representatives whose jobs involve lots of keyboard and/or telephone interaction. Telecommuting can be useful for people who are highly mobile throughout most of the workday: salespeople, real estate agents, field service personnel, and insurance claims adjusters, to name a few.

The telecommuting concept reached national prominence with amendments to the Clean Air Act of 1990, which took effect in November 1992. The amendments require companies with 100 or more employees in certain high-pollution metropolitan areas to develop plans to reduce their number of commuters by 25 percent by 1996. In addition, several states, notably California, have instituted their own laws to reduce commuter traffic. In other states, subsidized public transportation, ride sharing, or van pooling are typical approaches to reducing traffic. These programs have failed to meet expectations; however, work-at-home programs and satellite offices offer practical alternatives.

Another piece of legislation that encourages companies to set up telecommuting programs is the Family and Medical Leave Act, one of the first laws that was signed by President Clinton. This law entitles employees to 12 weeks of unpaid, job-protected leave each year for specified family and medical reasons. The act leaves room for companies to develop telecommuting plans that enable employees to perform some work at home. In this situation, telecommuting can help employees earn some or all of their pay, keep their careers on track, and preserve their seniority—all while attending to family medical problems.

Management Issues

Although telecommuting offers the promise of greater employee productivity and lower overhead costs for the company, it will likely be limited to a few individuals unless companies reorient first-line supervisors to accept it. Many senior first-line supervisors continue to believe that productivity will decrease if workers are not closely watched. Numerous studies have confirmed, however, that employees who work at home are generally 20 percent more productive than their counterparts at the office. Some companies and government agencies report productivity gains as high as 40 percent.

Supervising remote employees requires a different set of skills. To ensure that their supervisors have the necessary skills, some companies have issued detailed guidelines followed by specific supervisory training that emphasizes interpersonal communications and sensitivity to the problems that are experienced by remote workers and their office-bound colleagues.

Telecommuting does entail some risk, primarily organizational and social. For example, people who spend too much time away from the office are concerned about damaging their relationships, particularly with management. They believe that out-of-sight and out-of-mind perceptions can stifle or eliminate their advancement opportunities or cause them to miss opportunities to cultivate new relationships with face-to-face networking. Over time, telecommuters can feel isolated from the rest of the company and out of step with organizational changes.

In addition, there is the potential for telecommuting to cause resentment among office workers who feel that telecommuters are receiving special treatment. Only trained managers and supervisors can keep the negative feelings of employees from getting out of hand if they should arise. If left unaddressed, these negative feelings can dampen morale and impede productivity—situations that telecommuting programs are intended to improve. Not everyone will qualify for the program, so there will be the need to explain to others that the arrangement is work-related and job-related, not simply a company perk.

One way to expose potential problems with telecommuting is to implement it as a pilot program. Department managers and first-line supervisors find it much easier to agree to a three-month project than to an indefinite change, and they might agree to a pilot if they can help define the metrics (such as the time frame and what measures constitute a successful test).

Summary

The promises of telecommuting have been touted by advocates as long as there have been PCs. Due to improvements in computing and communications technologies, plus lower costs for hardware and services, telecommuting is now widely viewed as a practical alternative to working in the office. This situation is occurring just in time, too, with traffic congestion becoming worse in many metropolitan areas and pollution posing health hazards. New work arrangements must be explored to preserve the quality of life.

A noteworthy aspect of telecommuting is that the objectives of reducing travel and associated vehicle pollution need not be achieved at the expense of employee productivity or at great cost to participating companies. With proper planning and sensitivity training for managers and supervisors, telecommuting typically results in lower overhead costs for companies with the added benefit of increasing the productivity and loyalty of employees. The critical factor in the success of any telecommuting program, however, is selecting the right jobs and people for the program.

See Also

Distance Learning

Telemedicine

Teleconferencing

The term teleconferencing refers to the broad category of communications by which three or more people communicate in audio-only mode via telephone lines. The key benefit of teleconferencing is that it eliminates travel while enhancing communications by enabling many people to share information directly and simultaneously. Businesses are implementing teleconferencing strategies to handle large volumes of calls on a daily basis, thereby reducing non-telecommunications costs while improving productivity. Educational institutions are using teleconferencing for cost-effective distance-learning programs, while government agencies are using teleconferencing for crisis management as well as for daily information exchange.

Teleconferencing can be implemented in a variety of ways. For a basic telephone conference involving a limited number of participants, a telephone set with either a three-way calling feature on the line or a conferencing feature that is supported by the PBX or key system is required. Alternatively, a teleconference can be set up by a system attendant. For a teleconference with more than three parties, the attendant-console operator can establish the connections through the PBX and add more participants to the call than can be accommodated from a single set.

As another option, a company can initiate a conference through the telephone company's conference operator. A conference can be established at a prearranged time, or it can be organized so that the participants can phone in to a preassigned toll-free number at a designated time. After dialing the number, users must enter the assigned conference code to join the conference. This code is for security purposes (to ensure the privacy of the conference).

While individuals use their telephone to participate in a conference, several people at a given location can also participate in the conference as a group. In such cases, specialized equipment is required. The main component is the audio system, which consists of a control unit with an integral keypad, an omni-directional microphone, and a speaker that is typically positioned in the center of a table.

The Audio System

At the heart of any teleconferencing system is the audio-control unit, because high-quality audio is the key component of teleconferencing. Yet, audio quality can be substantially diminished by a condition known as echo.

Two sources of echo exist: multi-path and direct. Multi-path echo results when sound being emitted from a loudspeaker reflects off surfaces and objects within the room and is passed into the microphone, creating multiple echoes of the original speech. The microphone transmits the sound to the remote site(s), where it is heard as an echo. Direct echo results when sound from the far end is broadcast by the loudspeaker and is passed directly into the microphone without reflecting off surfaces in the room.

Most audio systems employ some form of echo control. Nevertheless, audio quality can suffer as a result of an ineffective echo control system. In such cases, the following conditions might become apparent to the conference participants:

- *Clipping* The speaker's voice is broken up.
- *Drop out* The voice is suddenly cut off when any noise is introduced at the connecting site.
- *Attenuation* A momentary loss of volume during the conference
- *Artifacts* Unintelligible voice remnants that are heard as stuttering during pauses

Two methods exist for enhancing audio quality by controlling echo: echo suppression and echo cancellation. Echo suppression employs a gating process in which the loudspeaker and microphones are alternately turned on and off to avoid transmitting the multi-path and direct echo. Alternately, if the received level for the microphones is higher than some predefined threshold, the loudspeaker signal is muted. In this way, the system attenuates the loudspeaker or the microphones to suppress echo from being picked up and sent.

Echo cancellation employs a high-speed *Digital-Signal Processor* (DSP) or custom *Application-Specific Integrated Circuit* (ASIC) to electronically compare the microphone signal to the transmitted loudspeaker signal. Any similarities in these two signals is recognized as echo. These similarities are then electronically subtracted from the microphone input, enabling only the original speech to be transmitted. This process eliminates the effects of both direct and multi-path echo. Of the two methods for controlling echo, echo cancellation is the best because it not only permits full-duplex operation, but it is also more dependable.

The Role of Bridges

The most economical way to teleconference is to call the various participants on separate telephone lines, then join the phone lines via a bridge. Bridges can accommodate a number of different types of LAN and WAN interfaces and are capable of linking several hundred participants in a single call. The problem with this method is that the participants cannot always hear each other well. Teleconferencing devices are available, however, that amplify and balance the conversation, enabling everyone to hear everyone else as though the participants were talking one-on-one. The bridge can operate as a stand-alone device or can connect through ports of a PBX or Centrex switch on either the line-side ports or trunk-side ports.

A bridged conference can be implemented in several ways. In one method, each participant can be called and then transferred to one of the PBX ports assigned to support the conference. Another method entails participants dialing a designated phone number at a preset time and automatically being placed on the bridge. Still another method of access involves participants dialing a main number and being transferred onto the bridge manually by a live operator. In any case, whenever a new person joins the party, a tone indicates his or her presence. The more conversations that are brought into the conference, the more free extensions are required.

An optional moderator phone can be used to set up bridged conferences. The moderator phone enables the operator to initiate conferences, actively participate in conversations, and terminate connections—as well as conduct isolated conversations with one party and then either admit or readmit the call to the conference or disconnect the call.

Conference Modes

Depending on the vendor, there are a number of conference modes to choose from that are implemented by the bridge:

- *Operator Dial-Out*   Participants are brought into the conference by an operator, using such methods as manual dialing and abbreviated dialing.

- *Originator Dial-Out*   Additional participants are added to the conference by the conference moderator, who accesses available lines by using a touch-tone telephone.

- *Prearranged*   Conferences are dialed automatically or by a user who dials a predefined code from a touch-tone phone. In either case, the information that is needed to set up the conference is stored in a scheduler.

- *Meet-Me*   Participants call into a bridge at a specified time to begin the conference. If teleconferencing is used frequently, a dedicated 800 number can be justified for this purpose.

- *Security-Code Access*   Participants enter a conference code and are automatically routed to the appropriate conference. If an invalid password is entered, the call is routed to an attendant station (where an operator screens the caller and offers assistance).

- *Automatic Number Identification* (ANI)   ANI automatically processes incoming calls based on the phone number of the caller, including call branding, call routing, and conference identification. Custom greetings can be designed for each incoming call.

System Features

The various features of a teleconferencing system can be grouped into five categories: participant, moderator, conference, administration, and maintenance.

Participant features   Individual participants in any teleconference need only one item of equipment: a touch-tone telephone. Some bridges facilitate touch-tone interaction, dramatically increasing the level of participation and end-user control. The conference administrator can configure *Dual-Tone Multi-Frequency* (DTMF) detection for one or two digits, which enables conference participants to implement various features by pressing one or two buttons on their touch-tone phone. Among the features that a participant can implement are the following:

■ *Help*   If the participant needs help, pressing 0 or *0 signals the conference operator for assistance.

■ *Mute*   If a participant wants to mute the line, possibly to talk without having remarks conveyed to the other conference participants, pressing 6 or *6 (M for mute) will place the line in listen-only mode.

■ *Polling*   Conference participants can participate in voting sessions. By pressing one or two digits to indicate yes or no, for example, each participant's preference can be recorded by the bridge. The results can be read immediately, stored on disk, or printed via the parallel printer port.

■ *Question and answer*   Conference participants can enter a question queue to signal the moderator that a question is waiting. At the appropriate time, the moderator can address each participant's question.

Moderator features   Moderators are given special privileges that provide a higher level of conference control than is afforded to participants. The following privileges can be activated by using their touch-tone phone:

■ *Security*   A moderator can secure a conference by pressing one or two buttons on the touch-tone keypad.

■ *Conference gain*   To level all signals in the conference, the moderator can implement the gain-control feature of the bridge (if it is not already set for automatic gain control).

■ *Lecture*   The moderator can initiate a lecture in which all lines, except those that are designated as moderators, are muted. This action enables the presenter to convey information without interruption from conference participants.

Conference features   Some bridges provide an extensive array of features, all of which contribute to a productive and successful teleconference. Depending on the vendor, these features can include the following:

■ *Polling*   With this feature, the moderator can poll participants by using one of the following methods: yes/no, true/false, multiple choice, or assigned ranges. Participants use their touch-tone phones to make their selections. Results are compiled immediately and can be printed or saved to disk.

■ *Question and answer*   During a lecture, for example, participants can indicate that they have questions by pressing one or two buttons on their touch-tone phones. When the lecture is over, the moderator can take each question randomly or in the order that they are received. Participants also can remove their own line from the queue.

- *Fast dial*   Often called speed dial, this feature enables an operator to quickly initiate outbound calls from an attendant console. The operator accesses a stored list of phone numbers and highlights the individual to be called into the conference. By merely pressing the Enter key, the number is dialed.

- *Auto dial*   This feature is similar to fast dial, except that it enables the operator to highlight all of the phone numbers of individuals who must join a conference. By pressing the Enter key, all of the calls are dialed simultaneously. Each conference participant is greeted by a recorded announcement that provides further instructions, such as prompting for a security code.

- *Lecture*   This mode of operation automatically mutes all lines in the conference (except that of the moderator) for uninterrupted sessions.

- *Security*   This feature provides confidential conferencing, prohibiting unauthorized individuals from entering. Secured conferences lock out the operator and cannot be recorded. By pressing the same buttons on the touch-tone keypad, the moderator can remove security from the conference.

- *Mute*   This feature places a specific line into listen-only mode. Also, lines are automatically muted by the system when certain features are implemented, such as lecture, polling, and Q&A.

- *Music*   Via an external device that is attached to the bridge, music on hold is provided to entertain participants until the conference begins. Music also provides assurance to waiting participants that the connection is still alive and that they should continue to wait.

- *Record/playback*   With this feature, conferences can be recorded and played back. Also, recorded material can be played into a conference via an external system. All lines are muted during playback. Some systems can be configured to enable the moderator to be heard during the playback.

- *Help*   This feature enables conference participants to signal an operator for assistance. Help can be set on a per-line basis, whereby the operator removes the conference participant from the conference in order to provide help. Help can also be set on a conference-wide basis, whereby the operator responds by entering the conference to address the entire group.

- *Conference ID*   This feature provides for the identification of conferences for report generation. The operator can assign the conference ID, or the system can be configured to assign IDs automatically.

- *Conference note*   The operator can record notations during a conference. Notations will appear on the conference report.

- *Conference scan*   An automatic audio scan of the entire conference can be performed at assigned intervals. Secured conferences are not scanned.

- *Operator chat*   Two or more operators can send e-mail to each other without disrupting active conferences.

- *Listen mode*   For quality-control purposes, operators can listen to individual lines or to a range of lines without affecting conference activity.

- *Disconnect notification*   The system can be configured to notify the operator of a disconnect during the conference.

- *Operator alarms*   The system can be configured to provide the operator with audible and visual signals upon disconnect, help request, or queue activity.

Administration features   From a teleconferencing product's system-administration menu, an administrator can perform the following tasks:

- Modify system configurations
- Perform supervision of the system during operation
- Configure the system for auto dial
- Configure operator functions
- Configure channels
- Perform file management
- Implement disk utilities
- Configure the conference scheduler

Maintenance features   From a system's maintenance menu, a maintenance person can access all conference and administrative functions, plus special maintenance features such as the following:

- Power-up diagnostics
- Online diagnostics

- Remote diagnostics
- Warm boot
- Maintenance reports
- Alarms

Internet Teleconferencing

A relatively new development is the use of IP networks for teleconferencing. Internet-telephony software makes it possible for users to engage in long-distance conversations between virtually any location in the world, without worrying about per-minute usage charges. In most cases, all that is needed is an Internet-access connection and a computer that is equipped with tele-phony software, a sound card, a microphone, and speakers or a headset.

A teleconference is established by entering one or more e-mail or IP addresses, which alerts the called parties to join at a particular server on the Internet or corporate intranet. The voice of each party is compressed and packetized, sometimes with encryption, for transmission over the IP network. Because the packets can travel a different route to their destination, there might be significant delay as the packets arrive and are put into the proper order. If some packets do not arrive in time, they are discarded. If enough packets are discarded, there can be noticeable gaps in a person's speech.

Companies can fashion their own internal telephone networks by lever-aging their existing intranets and conduct teleconferences (and video conferences) at virtually no extra charge. Alternatively, subscription services are emerging that offer teleconferencing services over private IP networks that can be accessed from the corporate intranet or from the Internet at greatly reduced rates.

Summary

For a variety of reasons, many people have lost the daily face-to-face contact that they once had with fellow employees. In addition, the rate of change in the business environment requires information to be distributed more rapidly. These factors and others are making teleconferencing an essential tool for all people in business today, enabling them to stay in contact and to be informed without resorting to expensive and time-consuming travel. While the same can be said about video conferencing, teleconferencing continues to be the more available, more economical, and a simpler solution for the majority of business needs.

See Also

Internet Telephony

Video Conferencing

Telegraphy

Telegraphy is a form of telecommunication that is based on the use of a signal code. The word *telegraphy* comes from the Greek *tele*, meaning distant, and *graphein*, to write (therefore, *writing at a distance*).

The inventor of the first electric telegraph was Samuel Finley Breese Morse, an American inventor and painter (refer to Figure T-9). On a trip home from Italy, Morse became acquainted with the many attempts to create usable telegraphs for long-distance telecommunication. He was fascinated by this problem and began to study books about physics for two years, in order to acquire the necessary scientific knowledge.

Early Attempts

Morse focused his research on the characteristics of electromagnets, examining how they became magnets only while the current flowed. The intermittence of the current produced two states—magnet and no magnet—from which he developed a code for representing characters, which eventually became known as Morse Code. (International Morse Code is a system of dots and dashes that can be used to send messages by a flash lamp, telegraph key, or other rhythmic devices, such as a tapping finger.)

His first attempts at building a telegraph failed, but he eventually succeeded with the help of some friends who were more technically knowledgeable. The signaling device was simple and consisted of a transmitter containing a battery and a key, a small buzzer as a receiver, and a pair of wires connecting the two. Later, Morse improved the device by adding a second switch and a second buzzer to enable transmission in the opposite direction, as well.

In 1837, Morse succeeded in a public demonstration of his first telegraph. Although he received a patent for the device in 1838, he worked for

Figure T-9
Samuel Finley Breese Morse (1791-1872), c. 1865

six more years in his studio at New York University to perfect his invention. Finally, on May 24, 1844, with a $30,000 grant from Congress, Morse unveiled the results of his work. Over a line strung from Washington, D.C. to Baltimore, Maryland, Morse tapped out the message, "What hath God wrought?" The message reached Morse's collaborator, Alfred Lewis Vail, in Baltimore, who immediately sent the message back to Morse.

With the success of the telegraph assured, the line was expanded to Philadelphia, New York, Boston, and to other major cities and towns. The telegraph lines tended to follow the rights-of-way of railroads—and, as the railroads expanded westward, the nation's communications network expanded, as well.

Morse Code

International Morse Code uses a system of dots and dashes that are tapped out by an operator via a telegraph key. (Morse Code can also be used to communicate via radio and flash lamp.) Various combinations of dots and dashes represent characters, numbers, and symbols that are separated by spaces. The Morse Code for letters, numbers, and symbols that are used in the United States is shown in the following table.

Morse Code is the basis of today's digital communication. Although it has disappeared in the world of professional communication, it is still used in the world of amateur radio (HAM radio) and is kept alive by history buffs. There are even pages on the Web that teach telegraphy and perform translations of text into Morse Code.

Wireless Telegraphy

An Italian inventor and electrical engineer, Guglielmo Marconi (1874-1937) pioneered the use of wireless telegraphy. Telegraph signals had previously been sent through electrical wires. In experiments that he conducted in 1894, Marconi (refer to Figure T-10) demonstrated that telegraph signals could also be sent through the air.

A few years earlier, Heinrich Hertz had produced and detected the waves across his laboratory. Marconi's achievement was producing and detecting the waves over long distances, laying the groundwork for what we today know as radio. So-called Hertzian waves were produced by sparks in one circuit and were detected in another circuit a few meters away. By continuously refining his techniques, Marconi soon detected signals over several kilometers, demonstrating that Hertzian waves could be used as a medium for communication.

The results of these experiments led Marconi to approach the Italian Ministry of Posts and Telegraphs for permission to set up the first wireless telegraph service. He was unsuccessful, but in 1896 his cousin, Henry Jameson-Davis, arranged an introduction to Nyilliam Preece, engineer-in-chief of the British Post Office. Encouraging demonstrations in London and on Salisbury Plain followed, and in 1897 Marconi obtained a patent and established the Wireless Telegraph and Signal Company Limited, which opened the world's first radio factory at Chelmsford, England in 1898.

Experiments and demonstrations continued. Queen Victoria at Osborne House received bulletins by radio about the health of the Prince of Wales while he was convalescing on the royal yacht off Cowes. Radio transmission was pushed to greater and greater lengths, and by 1899, Marconi had sent a signal nine miles across the Bristol Channel and then 31 miles across the English Channel to France. Most people believed that the curvature of the earth would prevent sending a signal much farther than 200 miles, so when

Letters	Numbers	Symbols	
A .-	1 .—	period (.)	.-.-.-
B -...	2 ..—	question mark (?)	..--..
C -.-.	3 ...-	slash (/)	-..-.
D -..	4 ....-	equals sign (=)	-...-
E .	5		
F ..-.	6 -....		
G --.	7 --...		
H	8 ---..		
I ..	9 ----.		
J .---	0 -----		
K -.-			
L .-..			
M --			
N -.			
O ---			
P .--.			
Q --.-			
R .-.			
S ...			
T -			
U ..-			
V ...-			
W .--			
X -..-			
Y -.--			
Z --..			

Marconi was able to transmit across the Atlantic in 1901, it opened the door to a rapidly developing wireless industry. Commercial broadcasting was still in the future—the *British Broadcasting Company* (BBC) was established in 1922—but Marconi had achieved his aim of turning Hertz's laboratory demonstration into a practical means of communication.

▬▬ ▬▬ ▬▬ ▬▬

Figure T-10
Guglielmo Marconi at
age 22, behind his
first patented wireless
receiver (1896)

Summary

By Morse's death in 1872, the telegraph was being used worldwide and would pave the way for the telephone. Western Union had the monopoly on commercial telegraph service but spurned Alexander Graham Bell, who approached them with an improvement that would convey voice over the same wires. Bell had to form his own company, the American Bell Telephone Company, to offer a commercial voice-communication service. Since then, voice and data technologies have progressed through separate evolutionary paths. Only in recent years have companies pursued voice-data integration as a means of saving on telecommunications costs.

See Also

> *Hertz* (Hz)
> Telephone

Telemedicine

Telemedicine refers to the delivery of health-care services to remote locations by using a combination of high-speed digital networks, video conferencing, and medical technologies. In its capability to exchange patient records and diagnostic images while permitting interactive consultations among medical specialists at different locations, telemedicine promises not only to save lives and extend sophisticated medical services to rural areas but also to cut the overhead costs of hospitals and clinics, saving an estimated $250 billion of the $1 trillion that Americans now spend each year on health care.

Applications

Telemedicine has numerous applications. Hospitals, walk-in clinics, mobile medical units, and health-care management organizations now routinely exchange information concerning patient eligibility, claims processing, scheduling, and medical records. What is relatively new is the use of networks for the transmission of diagnostic images and multi-site doctor-patient consultations.

For example, radiologists at major medical centers can read *Magnetic Resonance Imaging* (MRI) and other types of image scans that are transmitted from other hospitals over wireline or wireless networks. Receiving such information in a timely fashion greatly enhances the ability of doctors to offer a quick second opinion or to tap into specialized expertise at major medical centers.

Big hospitals that can afford special equipment for nuclear medicine, for example, can now transmit diagnostic tests and images to satellite locations that are not so elaborately equipped. This arrangement brings economies of scale to otherwise prohibitively expensive medical diagnostic procedures. The central facility can recover capital investments by billing smaller hospitals and clinics for its services. This functionality would not be possible if several hospitals and clinics in the same area competed for patients with their own nuclear-medicine facilities.

Firms now exist that specialize in providing diagnostic and consulting services to patients and physicians. Online interpretation of X-rays, *Electrocardiograms* (EKGs), pathology slides, and other medical images can be performed over networks that use compressed, interactive video links.

The telemedicine concept also has been applied to the delivery of mental-health services. While physicians are in short supply in rural communities, there is an even greater shortage of skilled mental-health professionals. Consequently, rural residents must travel even longer distances for psychiatric care.

To improve accessibility to mental-health professionals in rural areas, video conferencing-based training arrangements have been used with great success. Site coordinators at local facilities schedule a room and equipment for training that is provided by hospital staff via a video conference. The local coordinators handle equipment setup and operation, as well as registrations for the medical-education courses that are delivered over the network. A project coordinator at the hospital maintains a master schedule to facilitate connection of the local sites and to prevent conflicts in network usage.

System Requirements

Equipment at the rural sites typically includes a video codec, monitor, two video cameras, an audio system, control devices, graphics tablets, and an interface. At the main location, from which the video conference is moderated, a control console is used to set up the connections to each remote site. If a *Multi-Point Control Unit* (MCU) is used, the participants at each remote site can see each other. The site in the picture is determined by voice-activated switching; that is, when a participant is speaking, that site is displayed on the monitor.

Data from digital medical instruments can be networked for collaborative patient diagnosis. For example, a nurse in the field can use an *Ear, Nose, and Throat* (ENT) videoscope to relay video images of the patient's eardrum to doctors in the nearest city. There, physicians can determine whether the problem is a perforated eardrum. Real-time ultrasound examinations also can be conducted from the field. Portable ultrasound equipment that is connected to a computer and phone line relays the image to a hospital's ultrasound department, where physicians can interpret the images.

Another device called a video dermascope can be used to examine various skin conditions. A nurse in the field can use the video dermascope to examine the layers of skin in a lesion, for example, while hospital physicians watch on a video screen to determine whether the condition is malignant. Likewise, an instrument called an endoscope enables images from inside a human digestive tract to be transmitted over a phone-modem or ISDN link for viewing by doctors at major medical centers. Similar to other such instruments, the endoscope system enables staff at hospitals to spend less time traveling to patient locations and more time actually diagnosing problems and treating them.

Such instruments, when coupled to computers and relayed over networks for display on video-conferencing systems, can greatly improve health care in rural areas. In fact, a new medical field is emerging called telediagnosis. According to its advocates, telediagnosis is a cost-effective solution to managed care that will not only result in greater access to care but will also improve the quality of care that is provided in a high-touch, face-to-face application.

Not content with connecting health-care facilities to one another, many telemedicine advocates insist that the technology must eventually reach into the home if it is to fulfill its promise of optimal health-care access. Already, there is a portable defibrillator to restart a heart by remote control over a standard or cellular telephone line. Other devices enable doctors to remotely monitor heart rate, blood pressure, and temperature.

Eventually, so-called telehealth networks will be set up that would not only connect health-care providers to patients but also tie in any organization that is connected to the health-care process within a given community. Community-wide telehealth networks could combine the use of LANs, MANs, and WANs to provide networking within health-care facilities and to link those facilities to each other, to insurance carriers, to pharmacies, and to academic research centers.

Summary

Telemedicine has the potential to revolutionize health-care delivery and access. The integration of various telecommunications technologies, video conferencing, and sophisticated medical instrumentation brings interactive visual communications into operating rooms and rural clinics alike, providing access to the best doctors while reducing costs and increasing the availability and quality of health care.

See Also

Distance Learning

Video Conferencing

Telemetry

Telemetry is the monitoring and control of remote devices from a central location via wireline or wireless links. Applications include utility-meter reading, load management, environmental monitoring, vending machines, and security alarms. Companies are deploying telemetry systems to reduce the cost of manually reading and checking remote devices. For example, vending machines do not need to be visited daily to check for proper operation or out-of-stock conditions. Instead, this information can be reported via modem to a central control station, so that a repair technician or supply person can be dispatched as necessary. Such telemetry systems greatly reduce service costs.

When telemetry applications use wireless technology, additional benefits accrue. The use of wireless technology enables systems to be located virtually anywhere, without depending on the telephone company for line installation. For instance, a kiosk that is equipped with a wireless modem can be located anywhere in a shopping mall, without incurring line installation costs. Via wireless modems, data is collected from all of the kiosks at the end of the day for batch processing by a mainframe. The kiosk also runs continuous diagnostics to ensure proper operation. If a malfunction occurs, problems are reported via the wireless link to a central control facility, which can diagnose and fix the problem remotely or dispatch a technician if necessary.

Security Application

A variety of wireless security systems are available for commercial and residential use (refer to Figure T-11). The use of wireless technology provides installation flexibility, because they can be placed anywhere without regard to phone jacks. A variety of sensors are available to detect such elements as temperature, frequency, or motion. The system can be programmed to automatically call-monitor station personnel, police, or designated friends and

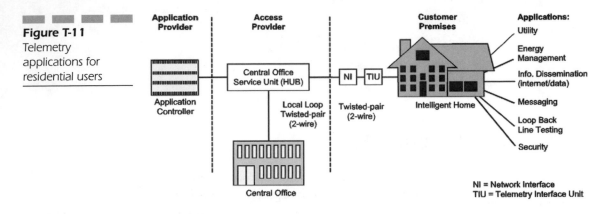

Figure T-11
Telemetry applications for residential users

neighbors when an alert is triggered. Such systems can be set to randomly turn lights on and off at designated times to give the appearance of someone being home. Depending on the vendor, the system can even perform continuous diagnostics to report low battery power or tampering.

In a typical implementation, the security-system console monitors the sensors that are placed at various potential points of entry. The console expects the sensor to send a confirmation signal at preset intervals (say, every 90 seconds). If the console does not receive the signal, it knows that something is wrong. For example, a sensor that is attached to a corner of a window or to a glass panel is specially tuned to vibrations that are caused by breaking glass. When it detects the glass breaking, the sensor opens its contact and sends a wireless signal to an audio alarm that is located on the premises, at the police station, or at a private security firm.

The use of wireless technology for security applications can actually improve service reliability. A security service that does not require a dedicated phone line is not susceptible to intentional or accidental outages when phone lines are down or when there is bad weather. Wireless links offer more immunity to such problems.

Traffic Monitoring

An increasingly popular application of wireless technology is its use in traffic monitoring. For example, throughout the traffic signal-control industry there has been a serious effort to find a substitute for the underground, hard-wired, inductive loop that is in common use today to detect the presence of vehicles at stoplights. Although the vehicle-detection loop is inherently simple, it has many disadvantages:

- Slot cutting for loop and lead-in wire is time consuming and expensive.
- Traffic is disrupted during installation.
- Reliability depends on geographical conditions.
- Maintenance costs are high, especially in cold climates.

These problems can be overcome by a wireless proximity detector. Its signals activate traffic lights in the prescribed sequence. The detector, usually mounted on a nearby pole, focuses the wireless signal narrowly on the road to represent a standard loop. The microprocessor-based detector provides real-time information while screening out environmental variations such as temperature, humidity, and barometric pressure. By tuning out environmental variations, the detectors provide consistent output, which increases the reliability of traffic-control systems. Using a laptop computer with a Windows-compatible setup package, a user can exchange information with the detector via an infrared link. From the laptop, the pole-mounted detector can be set up remotely, calibrated, and put through various diagnostic routines in order to verify proper operation.

Role of Cellular Carriers

Cellular carriers are well positioned to offer wireless telemetry services. The cellular telephone system has a total of 832 channels, half of which are assigned to each of the two competing cellular carriers in each market. Each cellular carrier uses 21 of its 416 channels as control channels. Each control channel set consists of a *Forward Control Channel* (FOCC) and a *Reverse Control Channel* (RECC).

The FOCC is used to send general information from the cellular base station to the cellular telephone. The RECC is used to send information from the cellular telephone to the base station. The control channels are used to initiate a cellular telephone call. Once the call is initiated, the cellular system directs the cellular telephone to a voice channel. After the cellular telephone has established service on a voice channel, it never goes back to a control channel. All information concerning handoff to other voice channels and termination of the telephone call are handled via communication over the voice channels.

This situation leaves the control channels free to provide other services such as telemetry, which is achieved by connecting a gateway to a port at the local *Mobile Switching Center* (MSC) or regional facility. The gateway can process the telemetry messages according to the specific needs of the applications.

For instance, if telemetry is used to convey a message from an alarm panel, the gateway will process the message on a real-time, immediate basis and will pass the message to the central alarm-monitoring service. On the other hand, if a soft-drink vending company uses telemetry to poll its machines at night for their stock status, the gateway will accumulate all of the data from the individual vending machines and process it in batch mode, so the management reports can be ready for review the next morning.

Individual applications can have different responses from the same telemetry radio. While the vending machine uses batch processing for its stock status, it could have an alarm message conveyed to the vending company on an immediate basis, indicating a malfunction. A similar scenario is applicable for utility-meter reading. Normal meter readings can be obtained on a batch basis during the night and delivered to the utility company the following morning. Real-time meter readings, however, can be made at any time during the day for customers who desire to close or open service and who require an immediate, current meter reading. Telemetry can even be used to turn utility service on or off remotely by the customer-service representative.

Summary

Telemetry services, once implemented by large companies over private networks, are becoming more widely available for a variety of mainstream business and consumer applications. Wireless technology permits more flexibility in the implementation of telemetry systems and can save on line-installation costs. Telemetry systems are inherently more reliable when wireless links are used to convey status and control information, because they are less susceptible to outages due to tampering and severe weather.

See Also

Cellular Communications

Telephone

The telephone was invented by Alexander Graham Bell (refer to Figure T-12). Born in 1847, Bell moved to Boston from Edinburgh, Scotland. His invention of the telephone actually arose from his work on trying to

Figure T-12
On March 7, 1876, the United States Patent Office granted Alexander Graham Bell a patent covering "the method of, and apparatus for, transmitting vocal or other sounds telegraphically . . . by causing electrical undulations, similar in form to the vibrations of the air accompanying the said vocal or other sounds." Source: The New York Herald Tribune

improve the telegraph. In fact, over the years, Bell accumulated 18 patents in his name alone (and 12 that he shared with his collaborators). These patents included 14 for the telephone and telegraph, four for the photophone, one for the phonograph, five for aerial vehicles, four for hydroairplanes, and two for a selenium power cell.

Largely self-educated, Bell had little skill in building working models of his telephones. For this task, he hired Thomas Watson, a repair mechanic and model maker, who built the equipment to Bell's specifications. After inventing the telephone, Bell continued his experiments in communication, which culminated in the invention of the photophone for the transmission of sound on a beam of light—a precursor of today's fiber-optic systems. He also worked in medical research and invented techniques for teaching speech to the deaf. In 1888, he founded the National Geographic Society.

In 1878, Bell set up the first telephone exchange in New Haven, Connecticut. In essence, this manually operated patch panel was the first central office switch; it also provided the basis for the *Private-Branch Exchange* (PBX). By 1884, lines were strung up on poles between Boston and New

York City, marking the start of the long-distance telecommunications industry. Later, in 1892, the first telephone connection between Chicago and New York was put into service (refer to Figure T-13).

AT&T was incorporated on March 3, 1885 in New York as a wholly owned subsidiary of the American Bell Telephone Company. Its original purpose was to manage and expand the burgeoning toll (or long-distance) business of American Bell and its licensees. AT&T continued as the long-distance company until December 30, 1899, when in a corporate reorganization it assumed the business and property of American Bell and became the parent company of the Bell System.

Telephone Components

Regardless of the model or manufacturer, the telephone includes three key components: a transmitter, a receiver, and a dial mechanism. The transmitter converts sound waves that are emitted by human vocal cords into a fluctuating electric current, while the receiver converts the electric current back into sound waves that can be picked up by the human ear. The dial mechanism, whether pulse or tone, provides the means to reach a telephone with an assigned number at the other end of the circuit. The call itself is switched through a series of telephone company central offices until a dedicated path is created between the calling and called parties. This path, or circuit, stays in place for the duration of the conversation. Only when the parties hang up are the lines and central-office equipment free to handle other calls.

Call Processing

Calls are circuit-switched to their destinations via a system of dialed digits. All telephones are connected to a local central office, each of which is assigned a three-digit exchange number. Each subscriber's telephone is assigned a four-digit identification number. Together, the exchange number and subscriber number are what we commonly refer to as a telephone number, which takes the following format:

XXX-XXXX

When trying to reach a telephone that is located outside the local service region, customers might have to dial a three-digit area code, as well. The use of an area code tells the local-exchange switch to contact an interexchange switch, which can be operated by a different carrier. Eventually, a

Figure T-13

The first telephone connection between Chicago and New York was put into service on October 18, 1892. Source: Mansell Collection

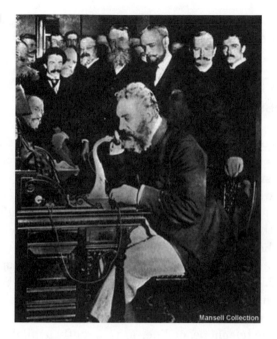

path is established through multiple local and interexchange switches—and, when the circuit is complete, the parties at each end can communicate. The area code is entered before the seven-digit phone number. The entire number takes the following format:

XXX-XXX-XXXX

If the call is to a location outside the country, two additional numbers come into play: a three-digit international access code and a two-digit or three-digit country code. The three-digit international-access code sends the call from the local-exchange switch to one or more interexchange switches and then to a special gateway switch that just handles calls between other gateway switches at international locations. For example, an international call from the United States to Sydney, Australia takes the following format:

011 61 2 local number

where 011 is the international-access code, 61 is the country code for Australia, and 2 is the city code for Sydney. Following these codes is the called party's local number.

The dedicated pair of wires running from the subscriber's telephone to the local central-office switch is called a line circuit or loop. The central office places a voltage of about 48 volts across the local loop to power the telephones and to monitor call activity for billing purposes.

Address Signaling

Two types of address signaling exist: pulse and tone. Pulse dialing is used when the telephone is connected to an older, analog, electromechanical central-office switch, whereas tone dialing is used when the telephone is connected to a newer, digital central-office switch. Most telephones that are sold today for residential use support both types of address signaling.

With either method of address signaling, placing a call starts when the user lifts the handset off the cradle. This action closes a contact relay in the telephone and permits current to flow through the loop, which signals the central office that the user would like to place a call. When the user hears a continuous tone from the central office switch (i.e., dial tone), an idle line has been secured to handle the call. On the other hand, if a fast busy tone is sent to the telephone, no lines are currently available to handle the call. A recorded announcement is activated, indicating that "all circuits are busy, please try again . . . ".

Pulse dialing Upon detecting loop current, the central office searches for an unused dial-pulse register to store the dialed digits as they are received. The register is connected, and a dial tone is sent down the line. Upon hearing the dial tone, the user can proceed to dial the desired telephone number.

With rotary dial or pulse dial, the telephone set rapidly opens and closes the loop at a rate of about 10 pulses per second. The number of pulses corresponds to the digit that is dialed (refer to Figure T-14). This process continues until all of the digits of the telephone number have been dialed. If the number is a local subscriber's, the connection is made immediately. If the number belongs to a subscriber who is outside the local calling area, the central office looks for a trunk (interexchange or inter-office line) that will connect it with the appropriate central office.

A trunk circuit provides a signal path between two central offices. Unlike a line circuit, or loop, a trunk circuit is shared by many users—although only one uses a trunk circuit at any given time. There might be 100 or more trunk circuits between these central offices, and as one tele-

Figure T-14
In pulse dialing, the number of pulses equates to the number that is dialed.

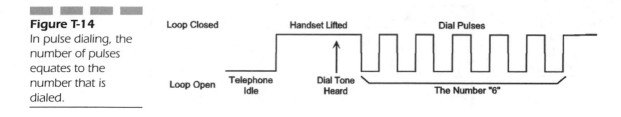

phone call ends, a trunk circuit is released and is made available to handle another call.

When a path has been established, the called party is alerted to the presence of the incoming call. This action is performed by the local central office, which sends an 88-volt, 20-hertz signal down the loop to drive a bell inside the telephone. This signal is repeated—two seconds on, four seconds off—to create the familiar telephone ring. When the phone's handset is lifted, current flows in the telephone loop, telling the central office to disconnect the ringing and establish the two-way voice connection.

Tone dialing   While pulse dialing is still used in many areas, a newer method of address signaling called tone dialing is replacing pulse and has become more common. With tone dialing, each digit is formed by selecting two out of seven possible frequencies. These frequencies are selected to minimize the possibility of accidental duplication by voice.

For example, pressing the 7 digit on the phone's keypad sends an 852-hertz tone and a 1209-hertz tone simultaneously (refer to Figure T-15). This process is sometimes referred to as *Dual-Tone Multi-Frequency* (DTMF) signaling. AT&T has branded the scheme as TouchTone. The central-office switch recognizes these different frequencies and associates them with the numbers that are dialed. The advantage of tone dialing over pulse dialing is faster call processing.

Summary

Bell was not the only person who was involved with the invention of the telephone. Bell's rival was Elisha Grey, who had filed his intention to invent the telephone with the United States Patent Office (now the Patent and Trademark Office). Bell gets credit for the invention, however, because he

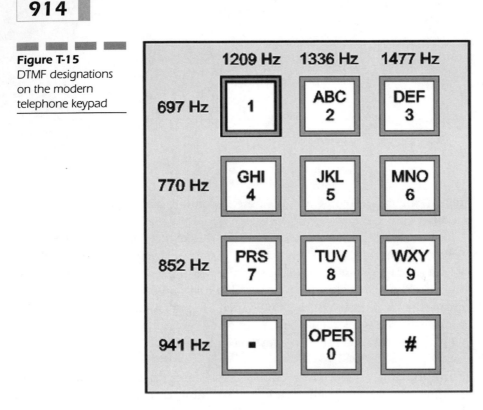

filed a claim for a workable device only two hours earlier (supposedly after bribing a patent clerk). With support from Western Union, Gray disputed Bell's claim in a lawsuit, but Bell prevailed. Gray made a fortune with other inventions and helped found the Western Electric Company in Chicago. Bell's patent was also contested by Thomas Edison, who claimed that he had invented the telephone years earlier. Edison subsequently made a great improvement in the telephone receiver, which was still in use decades later. Alexander Graham Bell helped form the Bell Telephone Company in 1877 but did not take an active part in the emerging new industry.

This role was left to Theodore N. Vail, the first president of AT&T, who believed that the chaotic telephone industry, by the nature of its technology and the need for interoperability, would operate most efficiently and economically as a monopoly. The U.S. government accepted Vail's argument, first informally and then legislatively. As political philosophy evolved, however, federal administrations investigated the telephone monopoly in light of general antitrust law and alleged company abuses. One notable result was an antitrust suit filed in 1949, which led in 1956 to a consent decree signed by AT&T and the Department of Justice, whereby AT&T agreed to restrict its activities to the regulated business of the national telephone system and government work.

Changes in the telecommunications industry during these years eventually led to an antitrust suit by the U.S. government against AT&T. The suit began in 1974 and was finally settled in January 1982, when AT&T agreed to divest itself of its wholly owned *Bell Operating Companies* (BOCs) that provided local-exchange service. In return, the U.S. Department of Justice agreed to lift the constraints of the 1956 decree. Divestiture took place on January 1, 1984, marking the end of the Bell System. In its place emerged a new AT&T, plus seven regional telephone holding companies—some of which have merged in recent years.

See Also

Central-Office Switch

Key Telephone Systems

Private-Branch Exchange (PBX)

Telephone Access for the Disabled

In mid-1999, the FCC adopted rules and policies to implement provisions of the Telecommunications Act of 1996 that require equipment manufacturers and carriers to ensure (if readily achievable) that their products and services are accessible to and usable by people who have disabilities. These rules are intended to give people with disabilities access to a broad range of products and services, including telephones, cell phones, pagers, call waiting, and operator services, that they cannot use today.

In the United States, an estimated 54 million Americans have disabilities. People who have disabilities are the largest minority group in the United States, yet despite their numbers, they do not experience equal participation in society. Access to telecommunications can bring independence.

The FCC's action on this matter represented the most significant opportunity for people who have disabilities since the passage of the *Americans with Disabilities Act* (ADA) in 1990. The rules require manufacturers and service providers to design telecommunications equipment and services with the needs of people with disabilities in mind. In developing these rules, the FCC relied heavily on the Access Board guidelines for equipment, extensive discussions with interested parties from the disability community and industry, and an analysis of the appropriate precedent under the ADA and other statutes that are designed to remove access barriers.

During the FCC's policy-making proceedings, the disability community made the Commission aware of the frustration of not being able to check the balance of a checking account by using Telecommunications Relay Service, not being able to tell whether a wireless phone was turned on, or not being able to use a calling card because of inadequate time to enter the appropriate numbers. The FCC also received numerous reports from relatives of senior citizens, saying that their elderly parents could live on their own if only they had telecommunications equipment that they could use.

The benefits of increased accessibility to telecommunications are not limited to people with disabilities. Just as people without disabilities benefit from the universal design principles in the ADA and the Architectural Barriers Act (for example, a parent pushing a stroller over a curb cut), many people without disabilities are expected to also benefit from accessible telecommunications equipment and services.

Many people already benefit from accessibility features in telecommunications today: vibrating pagers do not disrupt meetings; speakerphones enable us to use our hands for other activities; and increased volume control on public pay phones enables users to talk in noisy environments. The FCC expects many similar results from the rules that are adopted for the disabled. More importantly, everyone benefits when people with disabilities become more active in the community and in society as a whole.

Statistically, most Americans will have a disability or experience a limitation at some point in their lives. While 5.3 percent of those people in the 15-24 age bracket have some kind of functional limitation, 23 percent in the 45-54 age range experience functional limitation. The percentage of those who are affected by functional limitations increases with age: 34.2 percent between ages 55-64; 45.4 percent between ages 65-69; 55.3 percent between ages 70-74; and 72.5 percent who are 75 and older. The number of people with functional limitations will also increase with time. Today, only about 20 percent of Americans are older than 55, but by the year 2050, 35 percent of our population will be older than 55.

Today, most Americans rely on telecommunications for routine daily activities (for example, to make doctors' appointments, to call home when they are late for dinner, to participate in conference calls at work, and to make an airline reservation). Moreover, diverse telecommunications tools such as distance learning, telemedicine, telecommuting, and video conferencing enable Americans to interface anytime from virtually any location. Understanding that communications is now an essential component of American life, Congress intended the 1996 Telecommunications Act to provide people with disabilities access to employment, independence, emergency services, education, and other opportunities.

More specifically, telecommunications is a critical tool for employment. If this technology is not accessible to and usable by people who have disabilities, many qualified individuals will not be able to work or achieve their full potential in the workplace. Congress recognized the importance of creating employment opportunities for people who have disabilities when it crafted Title I of the ADA, which addresses the employer's responsibilities in making the workplace accessible to employees who have disabilities.

At a time when Americans are experiencing the lowest unemployment rate in years, unemployment among people who have severe disabilities is roughly 73 percent, and when employed, they earn only one-third of what people without disabilities earn. The rules that the FCC adopted in mid-1999 gave employers expanded tools with which to employ and to accommodate people with disabilities.

Summary

Under FCC-approved rules, telephones and communications services will become increasingly accessible to people who have disabilities. Manufacturers and service providers had resisted making products more accessible. Prior to these rules, they feared losing a competitive edge. The disabled, as a group, have been regarded as a small niche market. Now that all equipment makers and service providers must comply with the same rules, they do not have to be afraid that they are going to take a huge business hit.

See Also

Telecommunications Relay Services (TRS)

Telephone Fraud

Like many businesses, the telecommunications industry is a target of fraud and abuse. This statement is especially true of mobile-phone services. The underlying wireless technologies make it fairly easy for criminals to intercept signals and to obtain *Electronic Identification Numbers* (EINs) that can be used to clone phones and to have their usage billed to legitimate subscribers. In the United States alone, the annual losses from this kind of fraud amount to about $600 million.

Cellular fraud is an extension of phone phreaking, which is the term that is given to a method of pay-phone fraud that originated in the 1960s that employed an electronic box held over the speaker. When a user is asked to insert money, the electronic box plays a sequence of tones that fool the billing computer into thinking that money has been inserted. Since then, criminals and hackers have applied their knowledge and expertise to cellular phones.

Fraud Techniques

There are many ways to implement mobile telephone fraud. Some methods of fraud, such as cloning and rechipping, require technical expertise. The method of least risk entails opening a subscriber account and then running up the phone bill as high as possible until the service provider catches on to the scam and terminates the service.

Cloning   Dishonest dealers clone large numbers of handsets and sell them to targeted classes of users. Immigrants—legal and illegal—have long been a favorite target of cloners, because they know such people want to stay in touch with friends and relatives back home but usually have no money to call frequently. Migrant workers, youth gangs, and those who are engaged in the drug trade represent other potentially lucrative markets for cloned phones. Sometimes the phones are not sold at all. Instead, illegal phone cells are set up in apartments or abandoned buildings, and people off the street pay a per-minute charge just to use a clone.

For analog systems, each mobile phone carries handshake information that consists of the *Electronic Serial Number* (ESN) and the *Mobile Identification Number* (MIN). These numbers can fall into the wrong hands in a variety of ways. For instance, lax internal security can enable disgruntled employees to obtain this information from internal computer systems or customer files and sell it to cloners. When a vehicle is left unattended at a public parking lot or repair facility or is driven off by a valet, dishonest individuals can copy the ESN from the car phone and locate the MIN details from phone documents in the glove compartment. Special counterfeiting software is then used to recode the chips of other handsets with the stolen ESN.

Rechipping This technique is frequently used to recycle stolen mobile phones that have been reported to network operators and barred from further services. The software gives a stolen mobile phone a new electronic identity that enables it to be reconnected to the network until it is discovered and barred again.

Scanning A criminal does not have to steal a mobile phone to clone the ESN. With the latest scanning equipment and a laptop computer, a criminal can tune in to a call and steal the ESN from the air. Typically, the method used is to sit along a busy highway, intersection, or rest area and wait for unsuspecting users to make calls from their vehicles. The radio scanner can then pick up a phone's ESN, which is broadcast at the beginning of each call. In only a few hours of scanning, a criminal can walk away with hundreds of legitimate ESNs.

While mobile phones that are used on analog networks are easy to intercept, phones that are used on digital networks such as the *North American Personal Communication Services* (PCS), the *European Global System for Mobile* (GSM) communications, and the pan-Asian *Personal Handyphone System* (PHS) are not. The signals can be encrypted, making them impossible to interpret.

GSM signal encryption is performed via a programmable smart card, the *Subscriber Identification Module* (SIM), which slips into a slot that is built into the handset. Each customer has a personal smart card that holds personal details (short codes, frequently called numbers, etc.), an *International Mobile-Subscriber Identity* (IMSI) (which is equivalent to an MIN for analog systems), and an authentication key on the microprocessor. Plugging the smart card into another phone will enable that phone to be used as if it were the customer's own phone.

False accounts By far, the easiest and most risk-free way to engage in mobile-phone fraud is to open an account and pose as a legitimate subscriber. In this case, false identification, false Social Security cards, and false addresses are used to open these accounts. Because of competitive pressures, carriers are signing up new customers quickly and are not always able to screen all of this information in order to weed out unqualified individuals. Often, new subscribers can begin using the service within hours of opening an account. As the carrier closes in on bad accounts when false information does not check out, criminals are already a step ahead and are opening new accounts.

Fraud-Control Technologies

A number of technologies have been developed to discourage attempts at mobile-phone fraud, including the following:

■ *Personal Identification Numbers*   Similar to an ATM bank card, a PIN is a private number that only the subscriber knows. The PIN is required in order to place a call. Even if the phone's signal is captured, thieves would not be able to use the phone without the PIN code.

■ *Calling-Pattern Analysis*   When a subscriber's phone deviates from its normal call activity, it trips an alarm at the service provider's fraud-management system. There, the case is put into a queue where a fraud analyst ascertains whether the customer has been victimized and then remedies the situation. Newer systems establish and maintain the calling profiles of subscribers. Out-of-profile call requests are identified, and a PIN can be requested in order to ensure that only legitimate calls can be placed.

■ *Authentication*   Uses advanced encryption technology that involves the exchange of a secret code based on an intricate algorithm between the phone and the switch. The cellular network and the authentication-ready phones operating on it carry matching information. When a user initiates a call, the network challenges the phone to verify itself by performing a mathematical equation only that specific phone can solve. A valid phone will match the challenge, confirming that it is being used by a legitimate subscriber. If it does not match, the network determines that the phone number is being used illegally, and service to that phone is terminated. This process takes place in a fraction of a second.

■ *Radio-Frequency Fingerprinting*   Through digital analysis technology that recognizes the unique characteristics of radio signals that are emitted by cellular phones, a fingerprint can be made that can distinguish individual phones within a fraction of a second after a call is made. Once the fraudulent call is detected, it is immediately disconnected. The technology works so well that it has cut down on fraudulent calls by as much as 85 percent in certain high-crime markets, including Los Angeles and New York.

- *Voice Verification* These systems are based on the uniqueness of each person's voice and the reliability of the technology that can distinguish one voice from another by comparing a digitized sample of a person's voice with a stored voice print. The front-end analysis recognizes and normalizes conditions such as background noise, channel differences, and microphone variances. The voice-verification system can reside on a public or private network as an intelligent peripheral or can be placed as an adjunct system serving a PBX or *Automatic Call Distributor* (ACD). In a cellular environment, the system can be an adjunct system to a *Mobile Switching Center* (MSC).

Summary

While there is great progress in cracking down on mobile phone fraud in the United States, other countries are experiencing an increase in this kind of criminal activity. According to some experts, the international arena looms as the next frontier for mobile phone fraud, particularly in locations where U.S.-based multinationals are setting up shop and buying this kind of service. Foreign governments have just not been aggressive in finding and prosecuting this kind of criminal, they note. In some countries such as China, there are even operations that are dedicated to building cell phones that get illegally programmed and then sold on the black market.

Scanning the airwaves for cell phone identification numbers and programming them into clones will be made more difficult with emerging authentication services and digital networks that support encryption. GSM will change the nature of fraud in the future, because authentication facilities are built into the SIM card—forcing many criminals to turn their attention to subscription fraud (which is still time-consuming to track down, even for the largest service providers).

See Also

Network Security

Radio Communication Interception

Telephone Subscription

According to a survey by the FCC that was released in October 1999, an estimated 94.4 percent of all households in the United States have telephone service. This subscription rate is the highest ever that has been reported in the United States. The survey breaks down different subscription levels by state, income level, race, age, household size, and employment status, as follows:

- The telephone subscription penetration rate stands at 76.7 percent for households with annual incomes below $5,000, while the rate for households with incomes greater than $75,000 is 98.9 percent.

- By state, the penetration rates range from 89.1 percent in Mississippi to 98.9 percent in North Dakota.

- Households headed by whites have a penetration rate of 95.4 percent, while those headed by blacks have a rate of 88.2 percent and those headed by Hispanics have a rate of 90.7 percent.

- By age, penetration rates range from 87.5 percent for households that are headed by a person under age 25 to 96.3 percent for households that are headed by a person between 60 and 69.

- Households with one person have a penetration rate of 91.2 percent, compared to a rate of 96.2 percent for households that have four or five people.

- The penetration rate for unemployed adults is 89.6 percent, while the rate for employed adults is 96.0 percent.

This report presents comprehensive data concerning telephone-penetration statistics that were collected by the Bureau of the Census, under contract with the FCC. Recognizing the need for precise periodic measurements of subscription, the FCC requested that the Bureau of the Census include questions on telephones as part of its *Current Population Survey* (CPS), which monitors demographic trends between the decennial censuses. Use of the CPS has several advantages: it is conducted every month by an independent and expert agency; the sample is large; and the questions are consistent. Thus, changes in the results can be compared over time with more confidence than other methods that have been previously used. The Census-Bureau data is based on a nation-wide sample of about 48,000 households in the 50 states and the District of Columbia.

Summary

The number and percentage of households that have telephone service represent the most basic measures of the extent of universal service. Continuing analysis of telephone-penetration statistics enables the FCC to examine the aggregate effects of its actions on households' decisions to maintain, acquire, or drop telephone service.

See Also

Universal Service

Telex

Telex is the international teleprinter-exchange service through which a subscriber can be connected to another Telex subscriber, either locally or internationally, for the transmission of text messages. During Telex transmission, a copy of the message is produced simultaneously on both the sending and receiving teleprinters. Telex service is also known as *Teletypewriter Exchange* (TWX).

The transmission standard, character code, and terminal requirements for Telex are internationally standardized, which enables a teleprinter of any origin or design to directly exchange messages with any other teleprinter that is connected to the Telex network (refer to Figure T-16). The transmission speed is globally standardized at 400 characters per minute (50 baud). Telex transmission can occur over wireless networks as well as over wireline networks, enabling messages to be sent and received by ships at sea—even when the teleprinter is unattended.

Telex service operates in three modes: conversational, store-and-forward, and operator-assisted. With conversational Telex, both the sender and receiver can carry on an interactive, written Telex conversation. With store-and-forward Telex, complete text messages are sent over the Telex network to one or more addressees. Operator-assisted Telex transmission is used when there is difficulty reaching the destination number.

You can access the Telex network in several ways: dial-up, dedicated line, or packet network. Dial-up is used by subscribers who do not have enough

message traffic to justify the cost of a dedicated connection. Dedicated lines are used by businesses that use the Telex service on a continuous basis. Whether dial-up or dedicated, the connection is made to a carrier's switch that provides access to the Telex network. Companies that already have packet networks can also access the Telex network.

Telex service entails a minimum charge of one minute and subsequent increments of one minute. Telex subscribers receive a copy of the *International Telecommunications Rate Table*, a guidebook that contains a complete list of country codes and pricing.

Applications

Telex has several applications. For example, a company can send out updated product pricing to hundreds of international locations at once.

High-volume store-and-forward Telex enables the subscriber to store all Telex messages on the network so that they can be sent to multiple locations simultaneously.

The Telex network is linked to many types of databases, enabling subscribers to receive news, commodity prices, and exchange rates. The network can also be used to make travel reservations. Depending on the service provider that is selected, several delivery options are available. Telex messages can be sent to a facsimile machine or to an e-mail account, or they can even be delivered in hard-copy form (like a telegram).

Features

Telex subscribers have access to a full range of convenient, time-saving features, including single-digit dialing, automatic redialing of busy numbers, and multiple calls in a single connection. Other features include the following:

- The International Telex Exchange will automatically print the chargeable duration if it encounters HHHH (the character H four times) at the end of the document.

- Mistakes in typing the destination Telex number can be corrected immediately by keying EEE (the character E three times), followed by the correct number.

- When a sender encounters difficulties making a connection (busy line, power failure, etc.), the message can be addressed to the Exchange and forwarded later to the intended number. This feature incurs an extra charge.

- The International Telex Exchange can transmit a single message to multiple destination numbers, thereby saving the sender time. This feature incurs an extra charge.

- A unique Telex answer-back ensures clear identification of the sender and receiver before transmission begins. The recipient can confirm receipt of the message before signing off.

An added advantage of Telex communication is its recognized legality for business transactions, such as tender offers, counter-offers, and acceptances. This widely recognized legality distinguishes Telex from other forms of communications, such as fax and e-mail.

Summary

Despite the availability of more advanced, higher-speed telecommunications services, Telex remains a valuable source of written communication all over the world. There are approximately 1.8 million Telex users in more than 200 countries and locations worldwide.

See Also

Electronic Mail

Facsimile

Time-Division Multiple Access (TDMA)

In older FM radio systems, the radio spectrum is divided into 30KHz channels, with one user assigned to each channel. *Frequency-Division Multiple Access* (FDMA) improves bandwidth utilization and overall system capacity by dividing the 30KHz channel into three narrower channels of 10KHz each. Newer digital technologies, such as *Time-Division Multiple Access* (TDMA), enable even more users to be supported by the same channel.

TDMA systems have been providing commercial, digital cellular service since mid-1992. TDMA increases the capacity of existing analog cellular systems based on *Advanced Mobile Phone Service* (AMPS). With TDMA, each 30KHz channel is divided into three time slots. Users are assigned their own time slot into which voice or data is inserted for transmission via synchronized timed bursts. The bursts are reassembled at the receiving end and appear to provide continuous smooth communication, because the process is fast. The digital bit streams that correspond to the three distinct voice conversations are encoded, interleaved, and transmitted using a digital modulation scheme called *Differential Quadrature Phase-Shift Keying* (DQPSK). Together, these manipulations reduce the effects of most common radio-transmission impairments.

If one side of the conversation is silent, however, the time slot goes unused. Enhancements to TDMA use dynamic time-slot allocation to avoid the wasted bandwidth when one side of the conversation is silent. This technique almost doubles the bandwidth efficiency of TDMA to about 10:1 over analog. TDMA systems also now use half-rate voice coders, which enable up to six users to share one 30KHz channel. Hierarchical cell topologies further enhance network capacity.

TDMA is a *Telecommunications Industry Association* (TIA) standard (IS-54) for digital cellular in the United States. IS-136 is the next generation of TDMA, which makes use of the control channel to provide advanced call features and messaging services.

Framing

In a TDMA system, the digitized voice conversations are separated in time, with the bit stream organized into frames, typically on the order of several milliseconds. A 6-millisecond frame, for example, is divided into six 1-millisecond time slots, with each time slot assigned to a specific user. Each time slot consists of a header and a packet of user data for the call that is assigned to it (refer to Figure T-17). The header generally contains synchronization and addressing information for the user data.

If the data in the header becomes corrupted as a result of a transmission problem—signal fade, for example—the entire slot can be wasted, in which case, no more data will be transmitted for that call until the next frame. The loss of an entire data packet is called frame erasure. If the transmission problem is prolonged (i.e., deep fade), several frames in sequence can be lost, causing clipped speech or forcing the retransmission of data. Most transmission problems, however, will not be severe enough to cause frame erasure. Instead, only a few bits in the header and user data will become corrupted—a condition that is referred to as single-bit errors.

Network Functions

The IS-54 standard defines the TDMA radio interface between the mobile station and the cell site radio. The radio downlink—from the cell site to the mobile phone—and the radio uplink—from the mobile phone to the cell site—are functionally similar. The TDMA cell site radio is responsible for speech coding, channel coding, signaling, modulation/demodulation, channel equalization, signal strength measurement, and communication with the cell-site controller.

The TDMA system's speech encoder uses a linear predictive coding technique and transfers pulse-code-modulated (PCM) speech at 64 Kbps to and

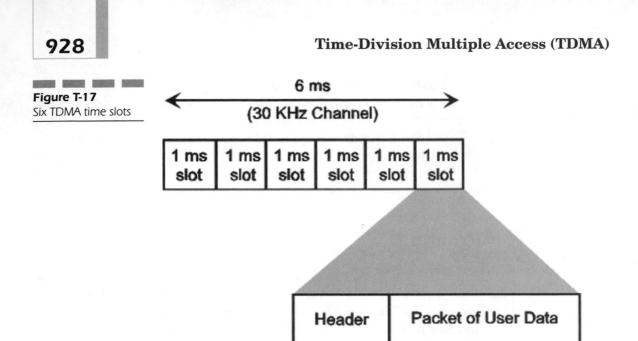

from the network. The channel coder performs channel encoding and decoding, error correction, and bit interleaving and de-interleaving. The coder also processes speech and signaling information, builds the time slots for the channels, and communicates with the main controller, modulator, and demodulator. The modulator receives the coded information and signaling bits for each time slot from the channel coders and performs DQPSK modulation to produce the necessary digital components of the transmitter waveform. These waveform samples are converted from digital to analog signals. The analog signals are then sent to the transceiver, which transmits and receives digitally modulated RF control and information signals to and from the cellular phones.

The modulator/equalizer receives signals from the transceiver and performs filtering, automatic gain control, receive-signal strength estimation, adaptive equalization, and demodulation. The demodulated data for each of the time slots is then sent to the channel coder for decoding. New voice coding technology is available that produces near-land-line speech quality in wireless networks based on the IS-136 TDMA standard. One technology introduced by Lucent Technologies uses an *Algebraic Code Excited Linear Predictive* (ACELP) algorithm, an enhanced internationally-established code for dividing waves of sound into binary bits of data. The ACELP coders can easily be integrated into current wireless base-station radios, as well as new telephones. As ACELP-capable phones become widely available, they will enable users to take advantage of the improved digital clarity over both North American frequency bands—850MHz cellular and 1.9GHz *Personal Communications Services* (PCS).

Call Handoff

Receive signal-strength estimation is used in TDMA's mobile-assisted call handoff process. The traditional handoff process involves the cell site currently serving the call, the switch, and the neighboring cells that can potentially continue the call. The neighboring cells measure the signal strength of the potential call to be handed off and report that data to the serving cell, which uses it to determine which neighboring cell can best handle the call. TDMA systems, on the other hand, reduce the time needed and the overhead required to complete the handoff by assigning some of this signal-strength data gathering to the cellular phone, relieving the neighboring cells of this task and reducing the handoff interval.

Digital Control Channel

The capabilities of TDMA are undergoing continual enhancement. For example, the *Digital Control Channel* (DCCH) described in IS-136 gives TDMA new features that can be added to the existing platform through software updates. Among these new features are the following:

- *Over-the-air activation* Enables new subscribers to activate cellular or PCS service with just a phone call to the service provider's customer-service center
- *Messaging* Enables users to receive visual messages up to the maximum length that is allowed by industry standards (200 alphanumeric characters). Transmission of messages permits a mobile unit to function as a pager.
- *Sleep mode* Extends the battery life of mobile phones and enables subscribers to leave their portables powered on throughout the day, ready to receive calls

■ *Fraud prevention*   The cellular system is capable of identifying legitimate mobile phones and blocking access to invalid ones.

A number of other advanced features can also be supported over the DCCH, such as voice encryption and secure data transmission, caller ID, and voice-mail notification.

Summary

Many vendors and service providers have committed to supporting either TDMA or CDMA. Those who have committed to CDMA claim they did so because they consider TDMA to be too limited in meeting the requirements of the next generation of cellular systems. Although TDMA provides service providers with a significant increase in capacity over AMPS, the standard was written to fit into the existing AMPS channel structure for easy migration. Proponents of TDMA, however, note that the inherent compatibility between AMPS and TDMA, coupled with the deployment of dual-mode/dual-band terminals, offers full mobility to subscribers with seamless hand-off between PCS and cellular networks. They also note that the technology is field-proven in many of the world's largest wireless networks and is providing reliable, high-quality service without additional development or redesign. These TDMA systems can be easily and cost-effectively integrated with existing wireless and land-line systems, and the technology itself can evolve economically to meet new quality and service requirements.

See Also

    Code-Division Multiple Access (CDMA)

    Frequency-Division Multiple Access (FDMA)

Token Ring

The 4Mbps Token Ring architecture for LANs was introduced by IBM in 1985 in response to the commercial availability of Ethernet, which was developed jointly by Digital Equipment, Intel, and Xerox. When Ethernet was introduced, IBM did not endorse it, mainly because its equipment would not work in that environment. Later in 1989, 16Mbps Token Ring became available.

The ring is essentially a closed loop, although various wiring configurations that employ a *Multi-Station Access Unit* (MAU)[1] and patch panel might cause it to resemble a star topology (refer to Figure T-18). In addition, today's intelligent wiring hubs and Token Ring switches can be used to cre-

Figure T-18
Token Ring topologies: closed ring and star-wired

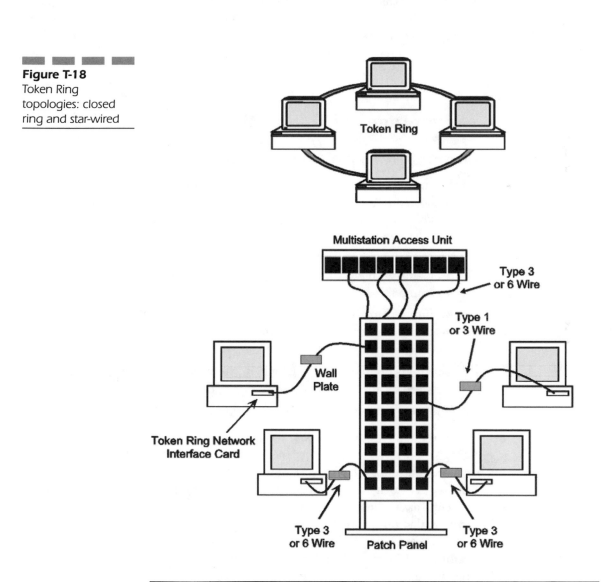

---

[1]A non-intelligent concentrator that can be used as the basis for implementing Token Ring LANs

ate dedicated pipes between rings and provide switched connectivity between users on different rings.

The cable distance of a 4Mbps Token Ring is limited to 1600 feet between stations, while the cable distance of a 16Mbps Token Ring is 800 feet between stations. Because each node acts as a repeater in that data packets and the token are regenerated at their original signal strength, Token Ring networks are not as limited by distance as is bus-type networks. Similar to its nearest rival, Ethernet, Token Ring networks normally use twisted-pair wiring (shielded or unshielded).

Advantages of Token Ring

The ring topology offers several advantages:

■ Because access to the network is not determined by a contention scheme, a higher throughput rate is possible in heavily loaded situations and is limited only by the slowest element: sender, receiver, or link speed.

■ With all messages following the same path, there are no routing problems with which to contend. Logical addressing might be accommodated to permit message broadcasting to selected nodes.

■ Adding terminals is easily accomplished by unplugging one connector, inserting the new node, and plugging it back into the network. Other nodes are updated with the new address automatically.

■ Control is simple, requiring little in the way of additional hardware or software to implement.

■ The cost of network expansion is proportional to the number of nodes.

Another advantage of Token Ring is that the network can be configured to give high-priority traffic precedence over lower-priority traffic. The Token Ring can put data onto the network only if a station has traffic that is equal to or higher in priority than the priority indicator that is embedded in the token.

The Token Ring, in its pure configuration, is not without liabilities, however. Failed nodes and links can break the ring, preventing all the other terminals from using the network. At extra cost, a dual ring configuration with redundant hardware and bypass circuitry is effective in isolating faulty nodes from the rest of the network, thereby increasing reliability. Through the use of bypass circuitry, physically adding or deleting terminals to the Token Ring network is accomplished without breaking the ring. Specific procedures must be used to ensure that the new station is recognized by the others and is granted a proportionate share of network time. The process for obtaining this identity is referred to as neighbor notification. This situation is handled quite efficiently, because each station becomes acquainted with the address of its predecessor and successor on the network upon initialization (power-up) or at periodic intervals thereafter.

Frame Format

The frame size that is used on 4Mbps Token Rings is 4048 bytes, while the frame size that is used on 16Mbps Token Rings is 16192 bytes. The IEEE 802.5 standard defines two data formats: tokens and frames (refer to Figure T-19). The token, three octets in length, is the means by which the right to access the medium is passed from one station to another. The frame format of Token Ring differs only slightly from that of Ethernet. The following fields are specified for IEEE 802.5 Token Ring frames:

- Start Delimiter (*SD*)   Indicates the start of the frame
- Access Control (*AC*)   Contains information about the priority of the frame and a need to reserve future tokens, which other stations will grant if they have a lower priority
- Frame Control (*FC*)   Defines the type of frame, either *Media-Access Control* (MAC) information or information for an end station. If the frame is a MAC frame, all stations on the ring read the information. If the frame contains information (i.e., user data), it is only read by the destination station.

Figure T-19
Format of IEEE 802.5
token and frame

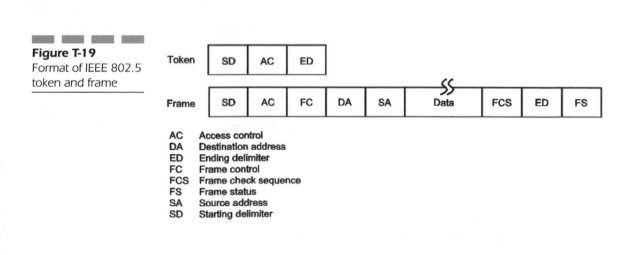

AC	Access control
DA	Destination address
ED	Ending delimiter
FC	Frame control
FCS	Frame check sequence
FS	Frame status
SA	Source address
SD	Starting delimiter

- Destination Address (*DA*)   Contains the address of the station that is to receive the frame. The frame can be addressed to all stations on the ring.

- Source Address (*SA*)   Contains the address of the station that sent the frame

- *Data*   Contains the data payload. If the frame is a MAC frame, this field can contain additional control information.

- Frame-Check Sequence (*FCS*)   Contains error-checking information in order to ensure the integrity of the frame to the recipient

- End Delimiter (*ED*)   Indicates the end of the frame

- Frame Status (*FS*)   Provides indications of whether one or more stations on the ring recognized the frame, whether the frame was copied, or whether or not the destination station was available

Operation

A token is circulated around the ring, giving each station in sequence a chance to put information on the network. The station seizes the token, replacing it with an information frame. Only the addressee can claim the message. At the completion of the information transfer the station reinserts the token on the ring. A token-holding timer controls the maximum amount of time that a station can occupy the network before passing the token to the next station.

A variation of this token-passing scheme enables devices to send data only during specified time intervals. The capability to determine the time interval between messages is a major advantage over the contention-based access method that is used by Ethernet. This time-slot approach can support voice transmission and video conferencing, because latency is controllable.

To protect the Token Ring from potential disaster, one terminal is typically designated as the control station. This terminal supervises network operations and does important housecleaning chores, such as reinserting lost tokens, taking extra tokens off the network, and disposing of lost packets. To guard against the failure of the control station, every station is equipped with control circuitry so that the first station detecting the failure of the control station assumes responsibility for network supervision.

Dedicated Token Ring (DTR)

Dedicated Token Ring (DTR), which is also known as full-duplex Token Ring, enables devices that are directly connected to a Token Ring switch to transmit and receive data simultaneously at 16Mbps, effectively providing each station with 32Mbps of throughput.

Under the IEEE 802.5r standard for DTR, which defines the requirements for end stations and concentrators that operate in full-duplex mode, all new devices will coexist with existing Token Ring equipment and will adhere to the token-passing access protocol. The DTR concentrator consists of C-Ports and a *Data-Transfer Unit* (DTU). The C-Ports provide basic connectivity from the device to Token Ring stations, traditional concentrators, or other DTR concentrators. The DTU is the switching fabric that connects the C-Ports within a DTR concentrator. In addition, DTR concentrators can be linked to each other over a LAN or WAN via data transfer services such as ATM.

High-Speed Token Ring

With 16Mbps, Token Ring connections between switches easily become congested at busy times, and high-performance servers become less capable to deliver their full bandwidth potential over a 16Mbps Token Ring connection. The need for a high-speed solution for Token Ring has become readily apparent in recent years. Other high-speed technologies were already available for inter-switch links and server connections—FDDI, Fast Ethernet, and ATM—but they are inadequate for the Token Ring environment.

In 1997, several Token Ring vendors teamed up to address this situation by forming the *High-Speed Token Ring Alliance* (HSTRA). A year later, the alliance issued a specification for *High-Speed Token Ring* (HSTR), which offers 100Mbps and preserves the native Token Ring architecture. To keep costs to a minimum and to shorten its development time, however, HSTR is based on the IEEE 802.5r standard for DTR, adapted to run over the same 100Mbps physical transmission scheme used by dedicated Fast Ethernet. HSTR links can be run in either half-duplex or full duplex mode, just like DTR.

HSTR uses existing switches, hubs, bridges, routers, *Network-Interface Cards* (NIC), and cabling. This introduces greater throughput where the enterprise needs it most—at the server and backbone. Upgrading these connections with HSTR only requires that an HSTR uplink be plugged into a Token Ring switch and that the existing 16Mbps server network NIC be replaced with a 100Mbps HSTR NIC. To complete the upgrade, the two devices are connected with appropriate cabling. The 100Mbps HSTR operates over both Category 5 UTP and IBM Type 1 STP cable, as well as multimode fiber-optic cabling.

You can also connect desktop systems to Token Ring switches on dedicated 100Mbps HSTR connections. Token Ring vendors offer 4Mbps/16Mbps/100Mbps adapter cards that enable companies to standardize on a single network adapter and prepare their infrastructure for the eventual move to HSTR. While the HSTR standard does not define an auto-negotiation algorithm, individual vendors have a number of ways to implement the feature while adhering to the standard. With this feature, HSTR products operate at the maximum connection speed, automatically determining whether to transmit at 4Mbps, 16Mbps, or 100Mbps. Many corporations install auto-negotiating 4Mbps/16Mbps/100Mbps NICs in today's desktops, although the need for 100Mbps throughput to the desktop might be years away. When the hub or switch at the other end of the connection is later upgraded to 100Mbps HSTR, the Token Ring desktop will automatically adjust transmission to 100Mbps.

Because Ethernet packets can be carried over Token Ring links, HSTR makes an ideal choice of backbone medium for the mixed-technology LAN. With support for the maximum Token Ring frame size, an HSTR backbone segment will have the capacity to handle Ethernet and Token Ring frames on the same VLAN connection, which Fast Ethernet would not have the capability to do without a lot of processing to break down the larger Token Ring frames.

Summary

Token Ring is a stable technology with a proven capacity for handling today's applications. At the same time, network managers can protect their current investments in Token Ring by understanding application performance and the capacity of the network, and tuning it accordingly. The DTR standard prolongs the useful life of Token Ring networks, while meeting the increased bandwidth requirements of emerging applications such as document imaging, desktop video conferencing, and multimedia.

See Also

ARCnet

Ethernet

Fiber-Distributed Data Interface (FDDI)

StarLAN

Figure T-20
A simple Bell telephone bill from February 1887. In those days, the bill simply came on a post card, charging the customer for the number of messages. Each call was forwarded through the main switchboard (in this case Buffalo, New York).

Truth in Billing

Telephone bills were once easy to understand (refer to Figure T-20). One reason why phone bills have become more complex in recent years is because there is a growing number of companies offering a wider variety of services, including Internet access, second lines, and voice mail. Many of these charges—including long-distance—usually end up in one consolidated bill that customers receive from their local phone company. Another reason for more complex telephone bills is that the phone companies, fearing customer backlash, are breaking out the government-ordered subsidies that once were included in rates. Some service providers take advantage of this confusion to engage in fraudulent practices that fleece customers.

In April 1999, the FCC adopted guidelines to make it easier for consumers to read and understand their telephone bills. In 1998 alone, more than 60,000 consumers contacted the Commission expressing confusion, anxiety and concern about their telephone bills. The truth-in-billing guidelines make telephone bills more consumer-friendly by providing information needed to make more informed choices in a competitive telecommunications marketplace and to protect against unscrupulous practices, such as slamming and cramming.

The Commission enacted broad guidelines that implement three basic principles. Consumers should know (1) who is asking them to pay for service, (2) what services they are being asked to pay for, and (3) where they can call to get more information about the charges appearing on their bill. For example, under the guidelines, new service providers must be highlighted on the bill. Thus, as a result of the Commission's action, consumers will be better able to detect when their carrier of choice has been changed without their authorization. This illegal practice, known as slamming, is the number one complaint of consumers filed at the Commission. The guidelines make it more difficult for the purveyors of fraud to deceive customers and get away with it by taking advantage of confusing telephone bills.

The Commission's guidelines also combat the illegal practice known as cramming, the appearance of unwanted charges on customers' bills. Cramming has emerged as an equally troubling fraud against consumers, and is now the second largest category of written complaints received by the Commission. Vague or inaccurate descriptions of charges make it difficult for consumers to know exactly what they are paying for and whether they received the services that they ordered. The Commission's guidelines eliminate this confusion by requiring that bills contain full and non-misleading descriptions and a clear identification of the service provider that is responsible for each charge on their bill.

The Commission's guidelines will also reduce customer confusion about the nature of the services being billed. The guidelines require that carriers clarify when they can withhold payment for service, for example, to dispute a charge, without risking the loss of their basic local service. Consumers will not, out of confusion, pay charges that they do not understand for fear that their basic telephone service would be terminated if they challenged the charge.

Finally, the Commission's rules permit customers to understand and compare the charges appearing on their bills that are related to federal regulatory action. The Commission has received thousands of calls from consumers confused about these charges, which is exacerbated by the myriad of names carriers have used to describe the various charges. The billing guidelines address this problem directly by requiring that carriers which choose to place line items on bills use standard labels to identify these charges. The Commission is seeking further comment, particularly from consumer groups and the industry, as to the specific appropriate labels.

The broad principles that underlie the truth-in-billing rules apply to all telecommunications carriers, both wireline and wireless. Similar to wireline carriers, wireless carriers also must be fair, clear and truthful in their billing practices. The FCC noted, however, that the record does not reflect the same high volume of customer complaints in the wireless context, nor does the record indicate that wireless billing practices fail to provide consumers with the clear and non-misleading information they need to make informed choices.

Summary

At this writing, the Commission only imposes certain requirements on wireless carriers: that the name of the service provider and a contact number be included on the bill, and uniform labeling of federal charges, if and when such requirements are adopted. The FCC is seeking comment on whether it should apply the remaining truth-in-billing rules adopted in the wireline context to wireless carriers, and whether forbearing from these rules would be appropriate.

See Also

> *Federal Communications Commission* (FCC)
>
> Slamming
>
> Telephone Fraud

Twisted-Pair Wiring

Twisted-pair wiring is the most common transmission medium; it is currently installed in most office buildings and residences. Twisted-pair wiring consists of one or more pairs of copper wires. To reduce cross-talk or electromagnetic induction between pairs of wires, two insulated copper wires are twisted around each other. For some business locations, twisted-pair wiring is enclosed in a shield that functions as a ground, which is known as *Shielded Twisted-Pair* (STP) wiring. Ordinary wire to the home is *Unshielded Twisted-Pair* (UTP) wiring.

In the local loop, hundreds of insulated wires are bundled into larger cables, with each pair color-coded for easy identification. Most telephone lines between central offices and local subscribers consist of this type of cabling, which is mounted on poles or laid underground. Bundling facilitates installation and reduces costs. Special sheathing offers protection from natural elements. In the business environment, Category 5 unshielded

Figure T-21
Typical Category 5 UTP cables contain four pairs that consist of a solid color and the same solid color striped onto a white background.

twisted pair cable is usually delivered as a bundle of 25 pairs. Each color-coded wire is solid conductor, 24 *American Wire Gauge* (AWG).

The same unshielded twisted-pair wiring has become the most popular transmission medium for local-area networking (refer to Figure T-21). The pairs of wires in UTP cable are colored-coded so that they can be easily identified at each end. The most common color scheme is the one that corresponds to the Electronic Industry Association/Telecommunications Industry Association's Standard 568B. The following table summarizes the proper color scheme.

The cable connectors and jacks that are most commonly used with category 5 UTP cables are RJ45. The RJ simply means Registered Jack, and the 45 designation specifies the pin-numbering scheme. The connector is attached to the cable, and the jack is the device that the connector plugs into, whether it is in the wall, in the NIC in the computer, or in the hub.

Wire Pair	Color Code
#1	White/blue
	blue
#2	White/orange
	orange
#3	White/green
	green
#4	White/brown
	brown

In response to the growing demand for data applications, cable performance has been categorized into various levels of transmission performance, as summarized in the following table. The levels are hierarchical in that a higher category can be substituted for any lower category.

The use of UTP wiring has several advantages. The technology and standards are mature and stable for voice and data communications. Telephone systems, which use twisted-pair wiring, are present in most buildings, and unused pairs usually are available for LAN connections. When required, additional twisted-pair wiring can be installed relatively easily, and the cost of Category 5 cabling is relatively inexpensive.

Of course, unshielded twisted pair wiring has a few disadvantages as well. It is sensitive to electromechanical interference, so new installations must be planned to route around sources of EMI. Unshielded twisted pair is also more susceptible to eavesdropping, which makes encryption and other security precautions necessary to safeguard sensitive information. An additional requirement is a wiring hub. Although this involves another expense, the hub actually facilitates the installation of new wiring, keeps all wires organized, and makes it easier to implement moves, adds and changes to the network. Over the long term, a hub saves much more than it costs.

Summary

UTP cable has evolved over the years, and different varieties are available for different needs. Improvements over the years, such as variations in the

Category	Maximum Bandwidth	Application	Standards
7	600MHz	1000BaseT and faster	Under development
6	250MHz	1000BaseT	Under development
5e	100MHz	Same as CAT 5 plus 1000BaseT	Under development
5	100MHz	1000BaseT	TIA/EIA 568-A (Category 5)
		100Mbps TPDDI (ANSI X 319.5)	NEMA (Extended Frequency)
		155Mbps ATM	ANSI/ICEA S-91-661
4	20MHz	10Mbps Ethernet (IEEE 802.3)	TIA/EIA 568-A (Category 4)
		16Mbps Token Ring (IEEE 802.5)	NEMA (Extended Distance)
			ANSI/ICEA S-91-661
3	16MHz	10Mbps Ethernet (IEEE 802.3)	TIA/EIA 568-A (Category 3)
			NEMA (Standard Loss)
			ANSI/ICEA S-91-661
2	4MHz	IBM Type 3	IBM Type 3
		1.544Mbps T1	ANSI/ICEA S-91-661
		1 Base 5 (IEEE 802.3)	ANSI/ICEA S-80-576
		4Mbps Token Ring (IEEE 802.5)	
1	less than 1MHz	*Plain Old Telephone Service* (POTS)	ANSI/ICEA S-80-576
		RS 232 & RS 422	ANSI/ICEA S-91-661
		ISDN Basic Rate	

twists or in individual wire sheaths or overall cable jackets, have led to the development of EIA/TIA-568 standard-compliant categories of cable that have different specifications on signal bandwidth. Because UTP cable is lightweight, thin, and flexible, as well as versatile, reliable, and inexpensive, millions of nodes have been and continue to be wired with UTP cable, even for high data-rate applications. For the best performance, UTP cable should be used as part of a well engineered structured cabling system. Businesses that require gigabit-per-second data-transmission speeds, however, should give serious consideration to moving to optical fiber (rather than to Category 6 or 7 UTP).

See Also

Fiber-Optic Technology

Hybrid Fiber / Coax (HFC)

Microwave Communications

Satellite Communications

T-Carrier Facilities

Unbundled Access

As part of the rules to ensure competition in the provision of local telecommunications services, the *Incumbent Local Exchange Carriers* (ILECs) must provide access to network elements on an unbundled basis to any requesting competitor. These unbundled network elements must be offered at any technically feasible point and at rates, terms, and conditions that are just, reasonable, and non-discriminatory. These rules are described in the Telecommunications Act of 1996, which is enforced by the *Federal Communications Commission* (FCC).

Network elements encompass everything that is required to implement telephone service, including loops, transport, signaling and call-related databases, and *Operations Support Systems* (OSSs). Unbundled access is important because it gives competitive local service providers economical access to the network elements that they need in order to provision and support a competitive service, without being forced to buy network elements that they do not need.

According to the competitive guidelines that are established in the Telecommunications Act of 1996, the network elements should be available on a selective basis and not as part of a package deal. Such bundling inflates the operating costs of competitors (and ultimately, the price that customers must pay for services). The incumbent telephone companies challenged the FCC on this issue and on other issues, which culminated in the United States Supreme Court decision affirming the authority of the FCC to carry out the pro-competition provisions of the Telecommunications Act. The Supreme Court, however, also advised the Commission to re-evaluate the standard that it uses to determine which network elements the incumbent local phone companies must offer on an unbundled basis.

In September 1999, the FCC responded to a United States Supreme Court decision by adopting rules specifying the portions of the nation's local telephone networks that the ILECs must make available to competitors. Applying the revised standard, the Commission reaffirmed that incumbents must provide unbundled access to six of the original seven network elements that it required in the original order, which was issued in 1996:

- Loops, including loops that are used to provide high-capacity and advanced telecommunications services
- Network-interface devices
- Local circuit switching (except for larger customers in major urban markets)

- Dedicated and shared transport
- Signaling and call-related databases
- OSSs

The Commission determined that it is generally no longer necessary for the ILECs to provide competitive carriers with the seventh element of the original list—access to their operator and directory-assistance services. The Commission concluded that the market has developed since 1996 to the point where competitors can (and do) self-provision these services or acquire them from alternative sources.

The Commission also concluded, in light of competitive deployment of switches in the major urban areas, that subject to certain conditions, the ILECs do not need to provide access to unbundled, local circuit switching for customers who have four or more lines that are located in the densest parts of the top 50 *Metropolitan Statistical Areas* (MSAs).

The Commission also addressed the unbundling obligations for network elements that were not on the original list in 1996. The Commission required incumbents to provide unbundled access to subloops, or portions of loops, and dark fiber loops and transport.

Summary

The FCC's revised rules concerning unbundled network elements seek to strike an appropriate balance. In densely populated places where there is competition, and in markets where new services such as broad-band are being rolled out in competitive fashion, the FCC will ease up on regulation. At the same time, the FCC intends to remain aggressive in ensuring that any competitor who wants to enter the market and compete on the telephone side has access to the basic network elements that are necessary on an unbundled basis.

See Also

Alternative Exchange Points

Collocation Arrangements

Interconnection Agreements

Open-Network Architecture

Unified Messaging

While the ways in which we can communicate have grown and diversified, so too has the number of devices that people must use to receive all of their messages: desk phones, cellular phones, fax machines, alphanumeric pagers, and e-mail systems (to name a few). Unified messaging brings order to this communications chaos by consolidating the reception, notification, presentation, and management of what have been (until now) stand-alone messaging systems. For this reason, unified messaging is emerging as one of the key strategic weapons in today's corporate world.

The unified messaging capability can be provided through message servers that are connected to a corporate PBX (refer to Figure U-1) or to a carrier switch. These servers can have a distributed architecture, enabling unified messaging services to be added incrementally throughout the carrier or corporate network as demand warrants; or, they can have a hub architecture with enough capacity to serve a large number of subscribers. More recently, unified messaging services have become available over the Internet, enabling users to view communication activity through their Web browser.

A unified messaging service deposits each subscriber's e-mail, fax, and voice messages into a universal messaging inbox, so that the subscriber can find all messages in a single place and through a single interface, such as a telephone handset, Web browser, or desktop application (such as Lotus Notes). Refer to Figure U-2.

Figure U-1
A unified messaging application that is implemented through the integration of a corporate LAN and PBX

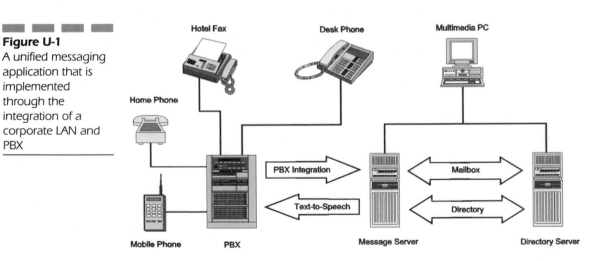

PhoneSoft's Unified MailCall is an example of a complete voice mail, unified messaging, and auto-attendant system for Lotus Notes. Voice messages are simply Notes mail messages that contain compressed .WAV files.

Notification Options

The unified messaging systems notify subscribers whenever a new message arrives. Notification methods include an e-mail notification message, a message-waiting indicator light, a stutter dial tone, a pager, and out-dial.

While it has long been common for voice-messaging subscribers to hear a stutter dial tone upon the arrival of a new voice message, it has not been as common to hear a stutter dial tone when an e-mail or fax message has arrived. With the advent of unified messaging, however, this situation changes, and the system notifies the subscriber of an incoming message whenever a message arrives—regardless of whether the message is a voice, fax, or e-mail message. A stutter dial tone is only one of several potential notification options. The network service provider might choose which notification methods will be offered to subscribers.

If the specified notification method is e-mail, for example, the unified messaging system sends an e-mail-notification message to a subscriber-specified e-mail address. In the body of the notification message, the system embeds a hypertext link that points to the voice or fax message in the system's message store. When the subscriber clicks the hypertext link, the mail application passes the *Uniform Resource Locator* (URL) to the subscriber's Web browser, which opens a window to the subscriber's universal inbox (where the voice or fax message appears).

If the message is a voice message and the subscriber's workstation supports multimedia applications, the subscriber can listen to the voice message over the workstation's audio system. If the message is a fax, the system presents the fax as a graphic image on the workstation's screen. If the message is itself an e-mail message, the system simply passes the message along to the subscriber.

Let's say the subscriber wants to be notified of incoming messages via a telephone's message-waiting indicator. When a new voice, fax, or e-mail message arrives and the subscriber checks for messages, the system indicates how many of each type of message have arrived. The system's synthesized voice might tell the subscriber, "You have two new voice messages, three new fax messages, and five new e-mail messages." Using buttons on the handset, the subscriber can choose to listen to the voice messages, output the fax messages to the nearest fax machine, and save the e-mail messages for viewing at a more convenient time.

The Role of Browsers

Because the Web can present text, graphics, and audio efficiently, it provides an effective medium for the consolidation and presentation of e-mail, fax, and voice messages. Some unified messaging systems can take advantage of Web technology in order to securely handle these messages as binary objects for large numbers of subscribers.

With a Web browser such as Netscape Navigator or Microsoft Internet Explorer, unified messaging-system subscribers can view fax and e-mail messages on their computer screens and listen to voice-mail messages over headphones or speakers that are attached to their workstation. They can also use the Message Compose and Reply features of the browser interface to respond to incoming messages.

For example, a subscriber can send a voice-mail message to another recognized unified-messaging subscriber in response to a voice, fax, or e-mail message. By clicking the Compose Voice button in the browser window and picking the recipient of the message from a directory window, the subscriber records a response. When finished, the user clicks the Send button to deliver the voice mail.

A subscriber can also send an e-mail message in response to an incoming voice, fax, or e-mail message. In response to an incoming voice or fax message, the subscriber clicks the Compose E-mail button in the browser window. If the recipient is a recognized unified-messaging subscriber, his or her address can be selected from the directory. If the recipient is not a recognized subscriber, the e-mail address must be manually entered. After composing the e-mail address, the subscriber clicks the Send button to deliver the e-mail message. If the original message was an e-mail message instead of a voice or fax message, the return address is automatically entered into the reply (whether or not the addressee is a unified-messaging subscriber).

Messages can also be forwarded from the browser interface. A Forward button enables subscribers to send an incoming e-mail message to another subscriber. If the subscriber wants to forward a voice or fax message, the subscriber saves the voice message in an audio-file format (or the fax in a graphic-file format) and sends it as an attachment to an e-mail message.

Subscribers can use the print capabilities of their Web browsers to output hard copies of received fax and e-mail messages. Currently, there is no option for printing voice messages or for listening to text messages, because Internet-based unified messaging systems do not yet support speech-to-text or text-to-speech capabilities. These capabilities are becoming available, however, through dedicated systems that are used in the corporate environment. Lucent Technologies, for example, offers speech recognition for its Intuity Audix unified messaging system. The system is primarily used for voice messaging but is designed to handle and combine various communication forms. This application enables users to attach e-mail or faxes to voice messages and to listen to speech translations of e-mail messages, and the speech-recognition capability can be used to forward attachments to other users.

While certain levels of access are not yet available, such as a viewing a fax or e-mail message over a standard handset, subscribers can use the handset to manipulate even these types of messages. If the fax or e-mail message has been sent from another unified-messaging subscriber, the system can report who sent the fax or e-mail message and what time it arrived. If the fax or e-mail message has been sent by a non-subscriber, the system only reports the type of message and the time that it arrived.

Via the handset, subscribers can also redirect fax messages for output on whatever fax machine they designate. Redirection is accomplished simply by dialing the desired fax phone number and pressing Enter. The fax machine could be in a hotel lobby, at a remote office, at a customer site, or in the subscriber's own home. Incoming e-mail messages can be outputted in a similar manner, except that an image of the e-mail message arrives at the specified fax machine.

By pressing appropriate buttons in response to system prompts, a subscriber can record and send a voice-mail response to a voice or e-mail message that is sent from another subscriber within the unified-messaging environment. The unified-messaging system addresses the reply automatically so that the subscriber does not have to remember (or even know) the address of the person to whom they are sending the reply. To reply to an externally originated message, the subscriber must dial the external number directly and rely on the recording capabilities that are attached to the phone service at the other end of the line.

Although there is usually no support for sending a voice message from the handset in direct response to an incoming fax message, a subscriber can forward voice and fax messages to another recognized unified-messaging subscriber by entering that subscriber's mailbox number at the prompt. Subscribers can also attach voice annotations to messages that they forward from the handset.

Subscribers can also forward fax and e-mail messages to systems outside the unified-messaging environment by using the print capabilities of the unified-messaging system. The subscriber forwards the fax by designating a remote recipient's fax machine or fax mailbox as the target output device. This feature also enables handset users to forward e-mail messages to recipients who are outside the system. As noted previously, the unified messaging system actually faxes an image of the e-mail message to a remote recipient's fax machine. Beyond providing a message to be faxed, the e-mail component of the unified messaging system does not play a role in forwarding messages outside the environment from the handset.

The unified messaging system ensures that actions that are initiated via the browser and handset interfaces are closely synchronized. If a subscriber listens to a new voice message through the browser interface, the message is flagged as read and is not announced as a new message when the subscriber later accesses the inbox via the handset interface. Also, as noted earlier, if a subscriber deletes a message via one interface, the message is deleted from the list of messages that are accessible via any other interface.

Creating Messages

The unified messaging system provides a suite of e-mail and voice-messaging capabilities that are aimed at meeting a wide range of message-creation requirements.

To create and send voice messages from their browser windows, subscribers typically start by clicking the Compose button, which opens an addressing window that enables the message recipient to be specified. Then, the subscriber can speak into the multimedia PC's microphone to record the message. When the message is finished, the system offers to play the message back and enables the subscriber to record the message again if it is not satisfactory. When the message is ready, the subscriber confirms the instruction to send the message by clicking a button in the browser window.

Subscribers can also click a button in the browser window to create and send e-mail messages. No restrictions exist on the destination of an e-mail message, except that it must be valid. The subscriber can choose the e-mail address of another unified-messaging subscriber from the directory or enter the e-mail address manually if the recipient is not a recognized subscriber.

Once the message has been addressed, the subscriber can type the body of the e-mail message. If the subscriber's browser supports attachments (for example, a formatted file such as a Microsoft Word document, a graphic, an audio file, or even a executable binary file), the subscriber can attach a file by clicking a button and selecting the appropriate file.

Subscribers can create a voice message by dialing into the unified messaging system and pressing the phone buttons that are associated with message creation. The system prompts the subscriber to enter an address for the recipient of the message (a mailbox extension, such as 2468). Upon receiving the addressing information, the system prompts the subscriber to record a message. When the subscriber has finished recording the message, the unified messaging system plays back the recorded message, offers the subscriber an opportunity to re-record the message, then sends the message to the designated recipient upon confirmation by the subscriber.

Voice messages can be sent to single or multiple recipients or to a list of recipients in a named group. Subscribers can create group lists (if the network service provider allows that privilege), or they can work with group lists that others in their subscriber community have created for public use.

Summary

Unified messaging solutions are available over the Internet. Numerous companies offer services that consolidate voice mail, e-mail, and faxes in one mailbox. Ranging in price from $10 to $20 a month, these services provide individuals and companies with a telephone number for incoming voice messages and faxes and an e-mail address. The market for unified messaging promises to grow by leaps and bounds as competition among wireless carriers, *Internet Service Providers* (ISPs), and local-exchange carriers heats up. Growth of e-mail, fax, and voice mail is driving the market for unified messaging. More importantly, the introduction of free services on the Internet is driving awareness and development of this market.

See Also

Call Centers

Electronic Mail

Facsimile

Paging

Universal Service

The universal-service system was originally designed to make local telephone service available to all consumers (including low-income consumers) in all regions of the nation. In many cases, universal service policies have required that rates for certain telecommunications services should be set above the cost of providing those services, in order to generate a subsidy that will reduce the rates for local service that is provided to residential customers.

The Telecommunications Act of 1996 updates the traditional universal service system, expanding both the base of companies that contribute to offset communications service rates and the category of customers who benefit from discounts. The Telecommunications Act directs the FCC to accomplish the following tasks that are related to improving universal service:

- Promote the availability of quality services at just, reasonable, and affordable rates
- Increase access to advanced telecommunications services throughout the nation
- Advance the availability of such services to all consumers, including those who are in low-income, rural, insular, and high-cost areas, at rates that are reasonably comparable to those that are charged in urban areas

In addition, the Act states that all providers of telecommunications services should contribute to federal universal service in some equitable and nondiscriminatory manner. The Act also mandates that there should be specific, predictable, and sufficient federal and state mechanisms to preserve and advance universal service. In addition, the Act says that all schools, classrooms, health-care providers, and libraries should generally have access to advanced telecommunications services. The rules for implementing these goals were formalized in mid-1997, and the support mechanisms for universal service (to ensure that customers in high-cost and rural areas will receive telephone service at affordable and reasonably comparable rates) were issued in October 1999.

Supported Services

The specific services that must be given universal service support include the following:

- Voice-grade access to the public switched network, including some usage (at a minimum)
- *Dual-Tone Multi-Frequency* (DTMF) signaling or its equivalent
- Single-party service
- Access to emergency services, including access to 911 (where available)
- Access to operator services
- Access to interexchange services
- Access to directory assistance

Whether local telephone service is affordable depends upon several factors, apart from local rates. Local calling-area size, income levels, cost of living, and other socio-economic indicators help to assess affordability. The states, in their rate-setting roles, make the primary determination of whether rates are affordable and take any necessary actions if they determine that the rates are not affordable. The FCC assesses affordability by monitoring subscription levels. The statutory criteria for receiving universal service support are as follows: The recipient must be a common carrier, and the carrier must offer (throughout a designated service area) all of the services listed previously.

Programs for Low-Income Consumers

In 1985, the FCC established two programs that are still available to assist low-income consumers. The Lifeline program reduces qualified low-income consumers' monthly phone charges with matching federal and state funds.

A state can choose not to participate in the Lifeline program. Currently, 41 states, the District of Columbia, and the U.S. Virgin Islands participate in Lifeline.

The Link Up program provides federal support that reduces qualified low-income consumers' initial local telephone connection charges by up to one half. Link Up is currently funded by contributions from interexchange carriers. Support is currently available only to incumbent wireline local-exchange carriers.

Under the implementation rules for the Telecommunications Act of 1996, the Lifeline and Link Up programs were revised in the following manner:

- Both the Lifeline and Link Up programs have been expanded so that eligible low-income consumers in every state and territory can receive support. Every carrier that is deemed eligible for universal service support is required to participate.

- For each eligible consumer, federal support amounts to $5.25. The federal fund also contributes an additional $1 for every $2 that a state contributes to Lifeline support. The maximum amount of these federal matching funds is $1.75, for a total federal support cap of $7.

- Eligible low-income consumers receive access to the same designated services that are identified for support in rural, insular, and high-cost areas. In addition, these consumers pay no charge for access to toll blocking and toll limitation, but only to the extent that the carrier has the technical capability to provide these services.

- Reduced service deposits if a low-income consumer accepts toll blocking

- Carriers can not disconnect a Lifeline customer's local service for non-payment of toll charges. A limited waiver of this requirement, however, is available for some carriers.

Schools and Libraries

Under the Telecommunications Act of 1996, schools and libraries are eligible for the first time to purchase (at a discount) any telecommunications services, internal connections among classrooms, and Internet access. Higher discounts are possible for economically disadvantaged schools and libraries and those entities that are located in high-cost areas. Discounts are a minimum of 20 percent and range from 40–90 percent for all but the least-disadvantaged schools and libraries. Total expenditures for universal service support for schools and libraries is capped at $2.25 billion per year, although any funds that are not disbursed in a given year can be carried forward and also disbursed.

Health-Care Providers

The approximately 9,600 health-care providers in rural areas in the United States are also eligible to receive telecommunications services that are supported by the universal service mechanism. Health-care providers include teaching hospitals, medical schools, community health centers, migrant health centers, mental health centers, not-for-profit hospitals, local health departments, rural health clinics and consortia, or associations of any of these providers.

Summary

In October 1999, the FCC refined its rules governing the support mechanisms for universal service. The revised rules, based on recommendations that were made by the Federal-State Joint Board on Universal Service, acknowledged that the individual states and the FCC share the responsibility for ensuring that telecommunications services are available to consumers in high-cost areas at affordable and reasonably comparable rates. Specifically, the reforms address how efficient competition in high-cost areas that are served by large, non-rural carriers will be provided. With the new high-cost support mechanism, the FCC rules ensure that rates are reasonably comparable, on average, among states, while the states will continue their historical role of ensuring that rates are reasonably comparable within their borders.

See Also

  Federal Communications Commission (FCC)
  Telecommunications Act of 1996

V-Chip

In Section 551 of the Telecommunications Act of 1996, Congress gave the broadcasting industry the first opportunity to establish voluntary ratings for television programs. The industry established a system for rating programs that contain sexual, violent, or other material that parents might deem inappropriate and committed to voluntarily broadcast signals containing these ratings.

In 1998, the FCC adopted rules requiring all television sets with picture screens 13 inches or larger to be equipped with features to block the display of television programming based upon its content rating. This technology is known as the V-chip. The V-chip reads information that is encoded in the rated program and blocks programs from the set, based upon the rating that is selected by the parent.

The FCC required half of all new television models 13 inches or larger that were manufactured after July 1, 1999 and all sets 13 inches or larger that were manufactured after January 1, 2000 to have V-chip technology. In addition, set-top boxes are available that enable consumers to use V-chip technology on their existing sets.

Rating System

The rating system, also known as TV Parental Guidelines, was established by the *National Association of Broadcasters* (NAB), the *National Cable Television Association* (NCTA), and the *Motion Picture Association of America* (MPAA). These ratings are displayed on the television screen for the first 15 seconds of rated programming—and, in conjunction with the V-Chip, permit parents to block programming with a certain rating from coming into their home.

The following content indicators and respective meanings constitute the rating system as described by the FCC:

- TV-Y   This program is designed to be appropriate for all children. Whether animated or live-action, the themes and elements in this program are specifically designed for a young audience, including children from ages 2-6. This program is not expected to frighten younger children.

- TV-Y7   This program is designed for children ages 7 and older. This program might be more appropriate for children who have acquired the developmental skills that are needed to distinguish between make-believe and reality. Themes and elements in this program might include mild fantasy or comedic violence or might frighten children under the age of 7.

- TV-G   Most parents would find this program suitable for all ages. Although this rating does not signify a program that is designed specifically for children, most parents might allow younger children to watch this program unattended. This program contains little or no violence, no strong language, and little or no sexual dialogue or situations.

- TV-PG   This program contains material that parents might find unsuitable for younger children. Many parents might want to watch this program with their younger children. The theme itself might call for parental guidance, and/or the program might contain one or more of the following items: moderate *Violence* (V), some *Sexual Situations* (S), infrequent, coarse *Language* (L), or some suggestive *Dialogue* (D).

- TV-14   This program contains some material that many parents would find unsuitable for children under 14 years of age. Parents are strongly urged to exercise greater care in monitoring this program and are cautioned against letting children under the age of 14 watch this program unattended. This program contains one or more of the following items: intense *Violence* (V), intense *Sexual Situations* (S), strong, coarse *Language* (L), or intensely suggestive *Dialogue* (D).

- TV-MA   This program is specifically designed to be viewed by adults and therefore might be unsuitable for children under 17. This program contains one or more of the following items: graphic *Violence* (V), explicit *Sexual Activity* (S), or crude, indecent *Language* (L).

Summary

Section 551 of the Telecommunications Act of 1996 encouraged the broadcast and cable industry to "establish voluntary rules for rating programming that contains sexual, violent or other indecent material about which parents should be informed before it is displayed to children" and to voluntarily broadcast signals containing these ratings. The V-chip reads information that is encoded in the rated program and blocks programs based upon the rating that is selected by the parent.

See Also

High-Definition Television (HDTV)

Value-Added Networks (VANs)

Value-Added Network (VAN) service providers offer voice and data-communications services for large companies, particularly those that have international locations. The value-added factor comes in the form of in-country local support, network design, integration, and management assistance, as well as economical access to a variety of feature-rich services and customized billing —tasks that can be resource-intensive if they are performed by the subscriber.

VAN providers offer a variety of value-added services across their networks. These services include network management, e-mail, EDI, *Electronic Funds Transfer* (EFT), X.400 global messaging, LAN interconnection, *Virtual-Private Data Networks* (VPDN), and integrated voice and data. The data services typically include protocol-conversion capabilities between dissimilar network devices, temporary and archival data storage, broadcast services to pre-established distribution lists, timed message delivery, message logging and acknowledgment, usage reports, security, and terminal handling (e.g., polled and scheduled calling). Clients also can set up closed user groups and customize their network requirements.

In addition to providing economical access over a variety of transmission services, such as X.25, Frame Relay, and TCP/IP, the principal advantages of VANs include the following:

- Error control that ensures a high degree of accuracy for critical data during transmission
- Program-conversion utilities that enable a variety of X.25-equipped terminals and computers to intercommunicate without requiring significant software changes
- Protocol conversion that eliminates the need for dedicated protocol-conversion devices and frees front-end processors and mainframes from this processing burden
- Wide-Area Network (*WAN*) support that enables organizations to configure VPNs without incurring the associated setup costs and tie small remote locations or branches to the main network in a cost-effective manner
- *Data security*   Including network authorization, message-recipient authentication, packet filtering, firewall protection, and encryption
- Bandwidth-on-demand that enables users to call up additional bandwidth on an as-needed basis in order to support specific applications, handle peak-period traffic requirements, or provide convenient backup for private-network facilities

■ *Service management*   Including equipment, software, and lines. In-country help desks and a global customer-assistance center are available 24 hours a day, seven days a week—which aids in problem resolution.

X.400 Global Messaging

Among the specific value-added services that VAN offers is X.400 global messaging. For numerous reasons, large organizations often find it impractical to support only one messaging system. VANs offer store-and-forward mail services that enable users to send and receive messages between diverse e-mail systems. This task is accomplished by utilizing X.400 gateways. X.400 is the global messaging standard that is recommended by the *International Telecommunication Union* (ITU). Specifically, this standard is an envelope, routing, and data-format standard for electronic messaging that is especially useful for connecting dissimilar e-mail systems. This standard spares organizations the difficulty of installing and maintaining their own X.400 gateways.

Other Services

Another value-added service is SNA interconnectivity, which provides a cost-effective alternative to companies that currently use multi-drop, private lines to connect remote sites to IBM host systems or to interconnect IBM peer nodes. The VAN service provider supplies the connections between nodes and proactively manages the subscriber's portion of the VAN from end to end. Users do not need to purchase additional equipment to use the service, because the VAN supports native IBM protocols such as *Synchronous Data-Link Control* (SDLC) and *Qualified Logical-Link Control* (QLLC).

Some service providers offer secure Internet access through their VANs. The aim is to provide fluency of access without exposing businesses to unnecessary security risks. Security is provided with firewalls that act as both a buffer and an access filter between the corporate portion of the VAN and the Internet. Services that are offered include *Domain Name Services* (DNSs), the *World Wide Web* (WWW), *Wide-Area Information Services* (WAIS), Gopher, Telnet, and FTP.

With the number of mobile users growing by leaps and bounds, the need for wireless access to VAN services is becoming a virtual necessity. Typically, a user connects to a modem pool that the VAN service provider offers.

From there, users can gain remote access to corporate LANs or to host-based systems as if they were locally attached. The VAN service provider might offer different access plans based on the number of usage hours.

Some VANs offer multimedia services to provide enterprises with the capability to integrate multiple voice, fax, and data networks into a single network. By combining packetized voice and data traffic, an enterprise can maximize its *Return on Investment* (ROI) and significantly reduce the cost of sending on-net voice traffic over a public-switched network. To maintain transmission quality, voice and fax traffic is prioritized—making on-net to on-net voice and fax communication faster, reliable, and more efficient.

Where independence from the public-switched telephone network is desirable, the VAN might offer VSAT satellite technology for network connectivity, which is supported by dedicated satellite or shared satellite connections. In addition, the VSAT service can be bundled with a failure-recovery service for an alternative backup solution to terrestrial line failures. Or, it can be bundled with a managed router service to extend the reach of a company's backbone network.

Summary

As VANs become more economical and feature rich, there will be fewer incentives for organizations to set up and maintain their own networks—particularly when a good portion of these networks spans international locations. Via on-premises management terminals, the VAN service providers are even giving subscribers the capabilities to configure network-routing paths and to schedule their availability based on time of day. Users can also fine-tune the network to increase data rates and choose alternate routing based on trunk speed. Additionally, VANs offer users increasingly sophisticated diagnostic capabilities that are comparable to those that are available in the private-network environment. This functionality gives organizations the control that they need, plus all of the value-added features (but without the expense of operating their own networks).

See Also

Managed SNA Service

Virtual Private Networks (VPNs)

Video Conferencing

Video conferencing is the process whereby individuals or groups at different locations meet online to share information through audio and visual communications. The video conferences can be augmented with text and graphics for display on separate screens. In addition, images, text, and graphics from a variety of sources can be multiplexed over the same video circuit in order to permit interactivity among conference participants (without the need to establish additional communications links).

In recent years, video conferencing has become an increasingly accepted form of communication among businesses, government agencies, and other organizations that want to save on travel costs, encourage collaborative efforts among staff at far-flung locations, and enhance overall productivity among employees. Originally, video conferencing was seen as a method of linking people at remote locations over WANs. More recently, video conferencing is being used to link desktop computer users over LANs in an effort to obtain the same benefits within a building or campus environment. Today, the technology has progressed to the point that virtually anyone can participate in video conferences over the Internet with relatively inexpensive hardware and software.

Applications

Organizations benefit from video conferencing in a variety of ways. Although important, containing travel costs is not the only motivation for implementing video conferencing systems. Instead of consuming valuable time with the logistics of travel for face-to-face meetings at individual locations, executives can take advantage of video-conferencing systems to conduct general meetings and individual sessions with appropriate personnel. In the process, the quality and the timeliness of decision making can be greatly improved.

Because video conferencing adds the capability to exchange information in a visually compelling way, it can be applied to almost any situation in order to enhance the quality and effectiveness of the communications process. In the medical profession, video conferencing is often referred to as telemedicine. In academia, video conferencing is a key component of distance learning. For corporate employees who work full-time or part-time from their homes, video conferencing makes telecommuting easier and more effective.

Whether used for product introductions, sales promotions, employee training, management messages, or collaborative projects between widely dispersed corporate locations, video conferencing is increasingly viewed not as a prestige technology, but as a practical and immediately useful tool that can yield competitive advantages.

Types of Systems

Video-conferencing systems fall into four categories: room-based systems, mid-range or roll-about systems, desktop systems, and videophones.

Room-based systems Room-based systems usually entail the use of one or more large screens in a dedicated meeting room that is equipped with environmental controls. The system components consist of screens, cameras, microphones, and auxiliary equipment, which can be permanently installed because they will not be moved to another room or building. These systems provide high-quality video and synchronized audio. Prices for room-based systems start at around $100,000 (U.S. dollars) but can go much higher as more sophisticated equipment and features are added.

Mid-range systems If an organization does not rely extensively on video conferencing but considers it an important capability when the need arises, a mid-range or roll-about video-conferencing system is a viable solution. Typically, these portable systems use one screen and no more than two cameras and three microphones. Prices range from $20,000 to $50,000, depending on the options.

Desktop systems Desktop video conferencing is becoming popular because it enables organizations to leverage existing computer and peripheral assets. Video conferencing becomes just another application that runs on the desktop. When equipped for video conferencing, the desktop machine can be used for video mail over LANs as well as for video conferencing over WANs. The capability to use widely available Ethernet networks makes video-conferencing technology more accessible, less costly, and easy to deploy. Data sharing can be accomplished through an optional whiteboard capability.

Some desktop video-conferencing systems support TCP/IP, which is the most commonly used WAN protocol. This support makes the system particularly useful in campus-style settings and for corporations that have remote offices. TCP/IP support ensures that the system can be used in conjunction with standard bridges, routers, and dialup lines, giving users convenient, cost-effective access to video-conferencing capabilities (particularly in areas where ISDN service is not available). Support for TCP/IP also eliminates the need for expensive on-premises switching systems and potentially costly telephone company usage charges.

Fully equipped desktop video-conferencing systems are available in the $1,200 to $5,000 per unit range, depending on the options. At the low end of the price range, the video-conferencing capability is added to an existing PC that has the appropriate hardware and software; at the high end, the vendor provides a PC that is already configured for video conferencing. The difficulty in installing and configuring add-in components makes the latter solution appealing to many organizations.

Videophones   Videophones are used for one-on-one communication. This type of equipment satisfies the desire for impulse video conferencing. The videophone unit includes a small screen, a built-in camera, a video coder/decoder (codec), an audio system, and a keypad. The handset enables the unit to work as an ordinary phone as well as a video-conferencing system. Prices start at about $1,000 for models that work over ordinary phone lines.

Vendor compliance with the emerging world-wide standard for *Third-Generation* (3G) wireless networks enables a number of sophisticated applications to be supported, including video conferencing between cellular phone users. For example, a mobile phone that has a color display screen and an integrated 3G communications module becomes a general-purpose communications and computing device for broad-band Internet access, voice, video telephony, and conferencing (refer to Figure V-1).

Multi-Point Control Units (MCUs)

Video conferencing between more than two locations requires a *Multi-Point Control Unit* (MCU), which is a switch that connects video signals among all locations—enabling participants to see each other, converse, and work simultaneously on the same document or view the same graphic. The multi-point conference is set up and controlled from a management console that is connected to the MCU.

The MCU makes it relatively easy to set up and manage conference calls between multiple sites. Among the features that facilitate multi-point conferencing are the following:

- *Meet-Me*   Enables participants to enter a conference by dialing an assigned number at a prearranged time
- *Dial-Out*   Used to automatically dial out to other locations and add them to the conference at prearranged times and dates

Figure V-1
This prototype of a 3G mobile phone from Nokia supports digital, mobile, multimedia communications (including video telephony). Using the camera eye in the top-right corner of the phone, along with the thumbnail screen below it, the local user can line up his or her image so that it appears properly centered on the remote user's phone.

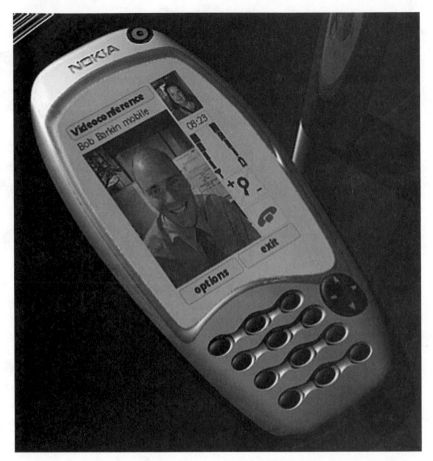

■ *Audio Add-On*　Enables participants to hear or speak to others who do not have video equipment (or compatible video equipment) at their location

■ *Tone Notification*　Provides special tones to alert participants when a person is joining or leaving the conference and when the conference is about to end

■ *Dynamic Resizing and Tone Extension*　Enables locations to be added or deleted (and the duration of a bandwidth reservation to be extended) during a conference, without the session having to be restarted

■ *Integrated Scheduling*　Permits video conferences to be set up and scheduled days, months, or a year in advance by using an integral calendar or scheduler application. The MCU automatically reserves the required bandwidth, configures itself at the designated time, and dials out to participating sites in order to establish the conference.

The MCU also provides the means to precisely control the video conference in terms of who is seeing what at any given time. Some of the advanced conference-control features of MCUs include the following:

- *Voice-Activated Switching* Enables all participants to see the person who is speaking, while the speaker sees the last person who spoke
- *Contributor Mode* Works with voice-activated switching to enable a single presenter to be shown exclusively on a conference
- *Chair Control* Enables a person to request or relinquish control, choose the broadcaster, disconnect a remote site, or discontinue the conference
- *Presentation or Lecture Mode* Enables a speaker to make a presentation and to question participants in several locations. Participants can see the presenter at all times, but the presenter sees whoever is speaking.
- *Moderator Control* Enables a moderator to select which person or site appears on-screen at any given time
- *Roll Call* Enables a conference moderator to switch each participant for the purpose of introducing them to others or to screen the conference for security purposes
- *Subconferencing* Enables the conference operator to transfer participants into and out of separate, private conferences that are associated with the meeting, without having them disconnect and reconnect
- *Broadcast with Automatic Scan* Enables participants to see the presenter at all times. To gauge audience reaction, the speaker sees participants in each location on a timed, predetermined basis.

Depending on the choice of MCU, a number of network-connectivity options are available. Generally, the MCU can be connected to the network either directly via T1 leased lines or ISDN PRI trunks or indirectly through a digital PBX. Some MCUs can be connected to the network by using dual 56Kbps lines or ISDN BRI. Others connect video-conferencing systems over the WAN by using any mix of private and carrier-provided facilities or any mix of switched services, regardless of the carrier (refer to Figure V-2). This capability offers the most flexibility in setting up multi-point conferences.

For users who do not have high-speed links, some MCUs include an inverse-multiplexing capability that combines multiple 56Kbps/64Kbps channels into a single, higher-speed 384Kbps channel on a demand basis, thus improving video quality. Most MCUs use the inverse-multiplexing

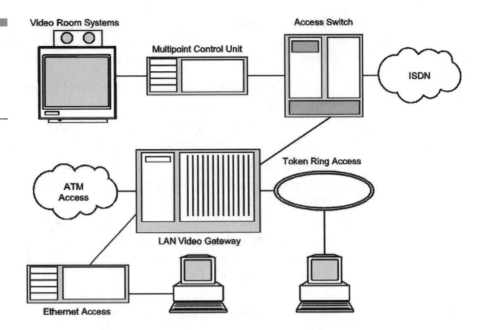

Figure V-2
A typical video-
conferencing
arrangement that is
implemented by an
MCU

method, which has been standardized by the *Bandwidth-on-Demand Inter-operability Group* (BONDING).

While most MCUs are designed for use on the WAN, some MCUs are available for the LAN. These devices are useful for providing multi-point desktop video conferencing over local networks within a campus environment or among many floors in a high-rise building.

Standards

Standards for video conferencing and other transmission technologies are established by the *International Telecommunication Union* (ITU), formerly the CCITT. By establishing world-wide video-conferencing standards, the ITU helps ensure that video-conferencing systems from diverse manufacturers can communicate with one another.

Video-transmission standards Recommendation H.320 is a set of video-conferencing standards that define the operating modes and transmission speeds for video-conferencing system codecs, including the procedures for call setup, call tear-down, and conference control. The codecs that comply with H.320 are interoperable with those of different manufacturers, delivering a common level of performance.

The H.320 video-conferencing standard includes associated specifications that define how the video-conferencing products of different vendors interoperate. Among the key H.320 standards are the following:

- *H.322* A standard for LAN-based video conferencing with guaranteed bandwidth

- *H.323* A standard for LAN-based video conferencing with non-guaranteed bandwidth (non-isochronous), such as Ethernet or Token Ring. H.323 specifies G.711 as the mandatory speech-compression standard.

- *H.324* A standard for video conferencing over high-speed modem connections, using standard telephone lines. H.324 specifies G.723 as the mandatory speech-compression standard.

While the H.324 specification facilitates video conferencing over ordinary phone lines, support for the ITU's V.80 standard is also needed to attain the maximum performance that H.324 promises. V.80 is implemented in V.34 modems to enhance the H.324 software applications that are running on hosts. This protocol is a relatively simple computer-to-modem controller protocol that enables an asynchronous PC interface to talk to a synchronous V.34 modem. The V.80 interface enables the host to define the frame boundaries for the multiplexed video and audio signal and enables the modem to complete the bit-oriented framing. Without V.80, video-conference participants would have to either use a special synchronous modem (not supported by most modem software), put up with start-stop-start video conferencing (losing about 20 percent of the available bandwidth in the process), or use a proprietary, synchronous controller protocol.

Another important component of H.320 is the H.261 video-compression specification, which defines how digital information is coded and decoded. H.261 also permits the signals to be transmitted at a variety of data rates, from 64Kbps to 2.048Mbps in increments of 64Kbps. H.261 also defines two resolutions. One is the *Common Intermediate Format* (CIF), which is a format that is usually used in high-end room systems to provide the highest resolution at 352-by-288 pixels. The other is the *Quarter Common Intermediate Format* (QCIF), a format that is used by most desktop video-conferencing systems and videophones that provides lower resolution at 176-by-144 pixels.

Another standard, H.230, describes the signals that are used by conferencing systems and MCUs to communicate during a conference. These signals enable conferencing systems and the MCU(s) to exchange instructions and status information during the initiation of a conference and while the conference is in progress. A related standard, Recommendation H.243, defines the basic MCU procedures for establishing and controlling communication between three or more video-conferencing systems by using digital channels up to 2Mbps.

Audio-compression standards   A set of ITU recommendations standardizes audio compression for video-conferencing equipment. Three key standards exist in this area:

- *G.711*   Defines the requirements for 64Kbps audio, which is the least compressed and offers the highest-quality audio
- *G.722*   Defines 2:1 audio compression at 32Kbps
- *G.728*   Defines 4:1 audio compression at 16Kbps

Compression is important because it squeezes the audio component of the video conference into a smaller increment of bandwidth, freeing more of the available bandwidth for the video component. This action results in higher-quality video without appreciably diminishing the audio.

Input/output standards   Many video-conferencing systems have ports for auxiliary devices such as televisions, cameras, and VCRs. Regarding the quality of video input/output for these items, the two most pervasive standards are the *National Television Standards Committee* (NTSC), a North American standard; and *Phase Alternating by Line* (PAL), which is used in Europe. NTSC specifies 320-by-240 resolution at 27 to 30 frames per second. PAL specifies 384-by-288 resolution at 22 to 25 frames per second. Most video equipment on the market supports both NTSC and PAL.

Internet Video Conferencing

Vendors and service providers are relying on the Web to simplify the scheduling of video conferences, even to the point of eliminating the need for human conference operators. Such offerings are also designed to shield customers from equipment-incompatibility issues, making it easier for participants from different organizations to hold a conference without worrying about the compatibility of their video gear. In addition, Web-based offerings can add streaming, data collaboration, and other services that might be difficult for enterprises to set up and manage in-house.

Vialog is among the companies that are offering Web-based video conferencing-reservation systems. Customers can log on to Vialog's Web site

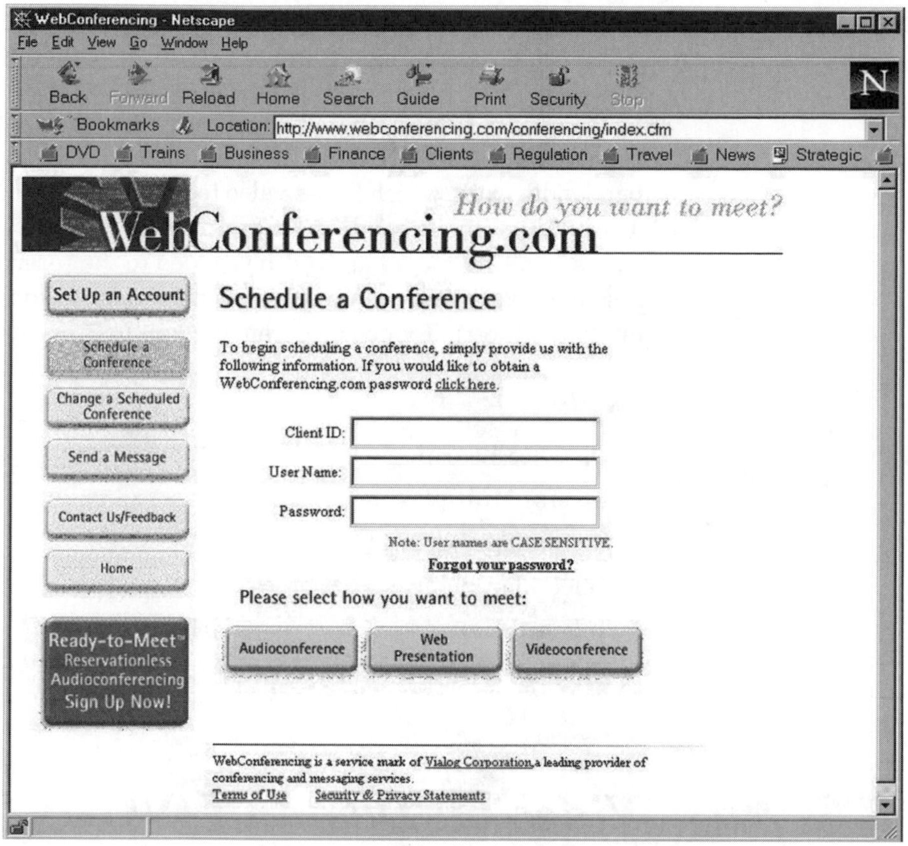

Figure V-3
Vialog's WebConferencing. com service provides instant access to group communications services. Users go to www.webconferencing.com to open an account. Once an account is established, video conferences can be scheduled easily and quickly at the user's convenience.

to list members of a video conference and to schedule a conference call (refer to Figure V-3). The customer can choose to have all parties dial in to the conference, as is typical of most conferences today. Or, perhaps more conveniently, Vialog can dial out to connect all parties to the conference.

Other companies offer similar Web-based video conference-reservation services, including V-Span. As an added benefit, V-Span can record conferences for future playback and offers data-collaboration services for those companies who want participants to edit documents and files during the conference.

Summary

Video conferencing is finally taking its place as a strategically significant corporate-communications tool. The ultimate low-cost and ubiquitous method of video conferencing might be the Internet. Several vendors offer video-conferencing software that works on desktop computers that are equipped with a camera, a sound card, a modem, and an Internet connection. Currently, image quality over the Internet is not the same as the image quality that can be achieved over high-speed LANs and digital WAN services such as ISDN. On the Internet, the problems of variable delay and bandwidth exist, which limits video to only two to five frames per second. As a result, the image tends to be grainy, movement is jerky, and the picture size is small. As more bandwidth is added to the Internet and the use of new resource-reservation protocols becomes more widespread, however, the quality of Internet video conferencing will improve dramatically.

See Also

> Distance Learning
>
> Inverse Multiplexing
>
> Streaming Video
>
> Teleconferencing
>
> Telemedicine
>
> Video on Demand

Video on Demand (VOD)

Video on Demand (VOD) is a service that enables subscribers to select movies for viewing whenever they want, instead of having to rent videotapes from a local store or wait for a scheduled movie to appear on television or on a pay-per-view service.

Programming is delivered in a variety of ways, mostly through the high-capacity *Hybrid/Fiber Coax* (HFC) networks of cable TV companies but also over *Direct-Broadcast Satellite* (DBS), *Multi-Point Multi-Channel Distribution Services* (MMDS), and *Local Multi-Point Distribution Services* (LMDS). Variable bit-rate and constant bit-rate MPEG-2 video compression is applied in order to reduce storage and bandwidth requirements. The encoded programs and navigation screens are stored in online transaction computers and video servers that are located at the cable provider's head-end facility, where the VOD service is provisioned.

Also at the head-end facility is a multiplexer that not only combines and grooms broadcast audio, video, and associated data programming but also supports *Digital Video Broadcast* (DVB) and *Data Encryption Standard* (DES) scrambling to prevent signal piracy. A management system provides a comprehensive suite of software components for billing and content management, as well as for video-asset tracking and device management. A digital link provides the critical bridge between the video server and the cable company's HFC network.

The service enables subscribers to order programs from an electronic program guide that is presented as an on-screen menu consisting of hundreds of program choices each month. The VOD offerings can be grouped into categories such as entertainment (movies and TV shows), children's programming (movies, TV shows, and educational programs), learning and lifestyles programming (documentaries and how-to programs), and home shopping. The electronic program guide is updated and refreshed each month.

Once a selection is made and the program starts, the subscriber has access to all of the features of a VCR (including pause, rewind, and fast forward), which are controlled with a hand-held remote-control device. Depending on the program that is selected, prices range from 49 cents to $4.49, making the service competitive with pay-per-view and video rental.

Several techniques enable multiple viewers to access one copy of the same program on the video server. One company, Diva, uses a system-striping architecture that makes it possible for large numbers of viewers to access and control video programming from a single copy of content that is stored on the server (refer to Figure V-4). Rather than storing an entire movie on a single drive or set of drives, as in a typical *Redundant Array of Inexpensive Disks* (RAID) architecture, segments of a movie or program are spread or striped across RAID-like arrays of processors and hot-swappable hard drives. In addition to eliminating blocking of a single title to all users, this fault-tolerant storage method prevents video-stream interruptions. More than 20 percent of storage disks would have to fail before viewing would be affected. The result is a more efficient, cost-effective, and reliable VOD server that is capable of delivering hundreds of movies to thousands of households simultaneously.

Cable companies and other carriers that do not want to put together their own VOD systems can rely on other companies to perform this job for them on a turnkey basis. In addition to integrating the service into the cable operator's current programming, some of these companies specialize in providing a complete studio-content acquisition and management service to ensure optimum movie-rental selection and availability. They also monitor the performance of the video servers on a continuous basis and monitor the quality of the transmissions over the broad-band network.

Figure V-4
With Diva Systems' OnSet On Demand TV, customers access OnSet Point as the first screen they see when they want to order a program. As a home base, it resembles a compass, pointing customers to several viewing options. Source: Diva Systems Corporation

Summary

About 25 percent of households are expected to purchase VOD service in the first year of availability in their area. Higher penetrations can be expected when service providers offer aggressive pricing to subscribers. These percentages will increase significantly as networks migrate to digital platforms. Cable providers are experimenting with promotional vehicles to encourage viewership, including a barker channel, post cards, and monthly magazines.

See Also

  Hybrid / Fiber Coax (HFC)

  Local Multi-Point Distribution Services (LMDS)

  Multi-Point Multi-Channel Distribution Services (MMDS)

Virtual-Private Networks (VPNs)

Carrier-provided networks that function like private networks are referred to as Virtual-Private Networks (VPNs). With a VPN, corporations can minimize the operating costs and staffing requirements that are associated with private networks. Additionally, corporations have the advantages of dealing with a single carrier, rather than multiple carriers and vendors that are normally involved with setting up and maintaining a private network.

AT&T introduced the first VPN service in 1985. Its *Software-Defined Network* (SDN) was a voice-only service that was offered as an inexpensive alternative to private lines. Since then, VPNs have added more functionalities, including support for data, and have expanded globally. Today, the big-three carriers—AT&T (SDN), MCI WorldCom (Vnet), and Sprint (VPN Service)—each offer VPNs. These networks can include high-speed data and cellular calls that can be combined under a single-service umbrella, expanding opportunities for cost savings within a single discount plan.

Advantages of VPNs

An increasing number of companies are finding VPNs to be a viable alternative method for obtaining private-network functionality without the overhead that is associated with acquiring and managing dedicated private lines. Additionally, there are several other advantages to opting for a VPN, including the following:

- The capability to assign access codes and corresponding class-of-service restrictions to users. These codes can be used for internal billing, in order to limit the potential for misuse of the telecommunications system and to facilitate overall communications management.

- The capability to consolidate billing, resulting in only one bill for the entire network

- The capability to tie small, remote locations to the corporate network economically, instead of using expensive dialup facilities

- The capability to meet a variety of needs (e.g., switched voice and data, travel cards, toll-free service, international and cellular calls, etc.) by using a single carrier

- The availability of a variety of access methods, including switched and dedicated access, 700 and 800 dial access, and remote calling-card access

- The availability of digit-translation capabilities that permit corporations to build global networks by using a single carrier. Digit-translation services can perform seven-to-10-digit, 10-to-seven-digit, and seven-to-seven-digit translations and can convert domestic telephone numbers to *International Direct Distance Designator* (IDDD) numbers via 10-to-IDDD and seven-to-IDDD translation.

- The capability to have the carrier monitor network performance and reroute around failures and points of congestion

- The capability to have the carrier control network maintenance and management, reducing the requirement for high-priced in-house technical personnel, diagnostic tools, and spare inventory

- The capability to configure the network flexibly via on-site management terminals that enable users to meet bandwidth-application needs and control costs

- The capability to access enhanced transmission facilities with speeds ranging from 56Kbps to 384Kbps and 1.536Mbps and to plan for the migration to broad-band services

- The capability to combine network-services pricing, typically based on distance and usage with pricing for other services, to qualify for further volume discounts

- The capability to customize dialing plans in order to streamline corporate operations. A dealership network, for example, can assign a unique four-digit code to the parts department. Then, to call any dealership across the country to find a part, a user would simply dial the telephone-number prefix of that location.

VPN Architecture

The architecture of the VPN makes use of software-defined intelligence that resides in strategic points of the network. AT&T's SDN, for example, consists of *Access-Control Points* (ACPs) that are connected to the PBX via dedicated or switched lines. The ACPs connect with the carrier's *Network-Control Points* (NCPs), where the customer's seven-digit on-net number is converted to the appropriate code for routing through the virtual network (refer to Figure V-5).

Instead of charging for multiple local-access lines to support different usage-based services, the carriers enable users to consolidate multiple ser-

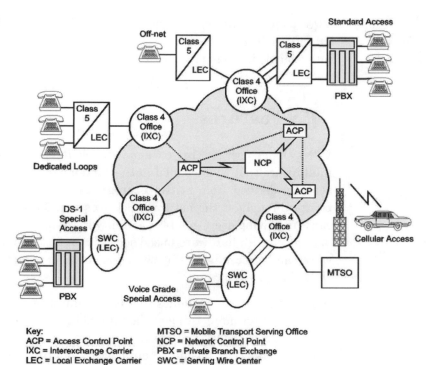

Figure V-5
The architecture of AT&T's Software-Defined Network (SDN)

Key:
ACP = Access Control Point
IXC = Interexchange Carrier
LEC = Local Exchange Carrier

MTSO = Mobile Transport Serving Office
NCP = Network Control Point
PBX = Private Branch Exchange
SWC = Serving Wire Center

vices over a single T1-access line. A user who needs only 384Kbps for a data application can, for example, fill the unused portion of the access pipe with 18 channels of voice traffic in order to justify the cost of the access line. At the carrier's cross-connect system, the dedicated 384Kbps channel and 18 switched channels are split from the incoming DS1 signal. The 384Kbps DS0 bundle is then routed to its destination, while the voice channels are handed to the carrier's Class 4 switch (which distributes the voice channels to the appropriate service).

A variety of access arrangements are available from the VPN service providers, which are targeted for specific levels of traffic, including a single voice-frequency channel, 24 voice channels through a DS1 link, and 44 voice channels through a T1 link that is equipped with bit-compression multiplexers, in addition to a capability that splits a DS1 link into its component 64Kbps DS0s at the VPN-serving office for connection to off-net services.

The same DS1 link can be used for a variety of applications, from 800 service to video conferencing—thereby reducing access costs. Depending on the carrier, there might be optional cellular and messaging links to the

VPN, as well. Even phone-card users can dial into the VPN, with specific calling privileges defined for each card. All of a company's usage can be tied into a single invoicing structure, regardless of the access method.

VPN Features

VPNs enable users to create their own private networks by drawing on the intelligence that is embedded in the carrier's network. This intelligence is actually derived from software programs that reside in various switch points throughout the network. Services and features are defined in the software, giving users greater flexibility in configuring their networks than is possible with hardware-based services. In fact, a subscriber's entire network can be reconfigured by changing the operating parameters in a network database:

- *Flexible Routing*   Enables the telecommunications manager to reroute calls to alternate locations when a node experiences an outage or peak-hour traffic congestion. This feature can also be used to extend customer-service business hours across multiple time zones.

- *Location Screening*   Enables the telecommunications manager to define telephone numbers that cannot be called from a given VPN location, which helps contain call costs by prohibiting certain types of outbound calls

- *Originating Call Screening*   Gives the telecommunications manager the means to create caller groups and screening groups. Caller groups identify individuals who have similar call restrictions while screening groups to identify particular telephone numbers that are prohibited or blocked for each caller. Time intervals can also be used as a call-screening mechanism, prohibiting or blocking calls according to time-of-day and day-of-week parameters.

- *NNX Sharing*   Enables VPN customers to reuse NNXs (i.e., exchange numbers) at different network locations to set up seven-digit on-net numbering plans. This functionality provides dialing consistency across multiple corporate locations.

- *Partitioned Database Management*   Enables corporations to add subsidiaries to the VPN network while providing for flexible, autonomous management when required by the subsidiaries to address local needs. The VPN can even transparently interface with the company's private network or with the private network of a strategic partner. In this case, the VPN caller is not aware that the dialed number is a VPN or private-network location, because the numbering plan is uniform across both networks.

■ Automatic Number Identification (*ANI*)   With ANI, the telephone numbers of incoming calls can be matched to information in a database (for example, the computer and telecommunications assets that are assigned to each employee). When the call comes through to the corporate help desk, the ANI data is sent to a host where it is matched with the employee's file. The help-desk operator can immediately have all relevant data available in order to assist the caller with resolving the problem.

Billing Options

One of the most attractive aspects of VPN services is customized billing. Typically, users can select from the following billing options:

■ The main account can obtain all discounts under the program. In some cases, even the use of wireless voice and data-messaging services can qualify for the volume discount.

■ Discounts can be assigned to each location according to its prorated share of traffic.

■ A portion of the discount can be assigned to each location based on its prorated share of traffic, with a specified percentage assigned to the headquarters location.

■ Usage and access rates can be billed to each location, or subsidiaries can be billed separately from main accounts.

■ Billing information and customized reports can be accessed at customer-premises terminals or can be provided by the carrier on diskette, microfiche, magnetic tape, tape cassette, or CD-ROM (as well as in paper form).

■ A name-substitution feature enables authorization codes, billing groups, telephone numbers, master-account numbers, dialed numbers, originating numbers, and credit-card numbers to be substituted with the names of individuals, resulting in a virtually numberless bill for internal distribution. This functionality prevents sensitive information from falling into the wrong hands.

AT&T, MCI WorldCom, and Sprint all offer rebilling capabilities that can use a percentage or flat-rate formula to mark up or discount internal telephone bills. Billing information can even be summarized in a number of graphical reports, such as bar and pie charts. Carrier-provided software is available that enables users to work with call detail and billing information to generate reports in a variety of formats. Some software even illustrates calling patterns with maps.

Network Management

Management and reporting capabilities are available through a network-management database that enables telecommunications managers to perform tasks without carrier involvement. The network-management database contains information about the network configuration, usage, equipment inventory, and call restrictions. Upon gaining access to the database, the telecommunications manager can set up, change, and delete authorization codes and approve the use of capabilities such as international dialing by caller, workgroup, or department. The telecommunications manager can also redirect calls from one VPN site to another, in order to allow (for example) calls to an East Coast sales office to be answered by the West Coast sales office after the East Coast office closes for the day. Once the telecommunications manager is satisfied with the changes, they can be uploaded to the carrier's network database and will take effect within minutes.

Telecommunications managers can access call detail and network-usage summaries, which can be used to identify network-traffic trends and assess network performance. In addition to being able to download traffic statistics about dedicated VPN trunk groups, users can receive five-minute, 10-minute, and 15-minute trunk group-usage statistics an hour after they occur. These statistics can then be used to monitor network performance and to carry out traffic-engineering tasks. Usage can be broken down and summarized in a variety of ways, such as by location, type of service, and time of day. This information can be used to spot exceptional traffic patterns that might indicate either abuse or the need for service reconfiguration.

Via a network-management station, the carrier provides network alarms and traffic-status alerts for VPN locations by using dedicated access facilities. These alarms indicate potential service outages (e.g., conditions that could impair traffic and lead to service disruption). Alert messages are routed to customers in accordance with preprogrammed priority levels, ensuring that critical faults are reviewed first. The system furnishes the customer with data concerning the specific type of alarm, direction, location, and priority level, along with details about the cause of the alarm (e.g., signal loss, upstream failed signal, or frame slippage). The availability of such detail permits telecommunications managers to isolate faults immediately.

Additionally, telecommunications managers can request access-line status information and schedule transmission tests with the carrier. The network-management database describes common network problems in detail and offers specific advice regarding how to resolve them. The telecommunications manager can submit service orders and trouble reports to the carrier electronically via the management station. Also, telecommunications managers can test network designs and add new corporate locations to the VPN.

Local VPN Service

A new development in the VPN market is the emergence of local service, whereby some *Regional Bell Operating Companies* (RBOCs) enable corporate customers to manage their in-region calls by using the public network as if it were their own private network. This functionality enables customers to perform actions such as accessing their voice network remotely, making business calls from the road or home at business rates, originating calls from remote locations while billing them to the office, and blocking calls to certain telephone numbers or regions. Uniform pricing and billing plans can also be arranged for all of the customer's locations, in order to reduce the administrative costs that are involved with reviewing billing statements (even if each location uses a different carrier).

The service enables large business customers to configure components of the public network like a customized, private network—without the expense of dedicated lines or equipment. The service also is compatible with Centrex services, PBX systems, or other customer-premises equipment.

Summary

VPNs permit the creation of voice and data networks that combine the advantages of both private facilities and public services, drawing on the intelligence that is embedded in the carrier's network. With services and features defined in software and implemented via out-of-band signaling methods, users have greater flexibility in configuring their networks from on-premises terminals and management systems than from services that are implemented with manual patch panels and hard-wired equipment.

See Also

> *Advanced Intelligent Network* (AIN)
> *Signaling System 7* (SS7)
> *Value Added Networks* (VANs)

Voice-Activated Dialing (VAD)

Voice-Activated Dialing (VAD) enables users to place calls simply by saying the name or telephone number of the person whom they wish to call. For example, when a user says the word *home*, the voice-recognition system that is attached to the carrier's switch matches the word to the caller's home telephone number and initiates the call. The user can also simply speak each digit of the called number in order to place a call. VAD is available from cellular carriers as well as from local exchange carriers and ranks among the top two most-desired features among new cellular customers.

Each word that the customer speaks can be stored in *Pulse-Code Modulation* (PCM) format or *Adaptive Delta Pulse-Code Modulation* (ADPCM) format at sampling rates of 8KHz (8 bits per sample) or 4KHz (4 bits per sample), respectively. Usually, the higher sampling rate is used for carrier-provided announcement and instruction messages in order to ensure high fidelity. These messages are locally stored at the carrier's voice-recognition system. The lower sampling rate is used for customer-programmed words and messages for high-volume storage efficiency. These words and messages are stored on the carrier's network file system.

VAD systems employ special error-detection and correction algorithms that use adaptive intelligence to analyze and automatically correct spoken numbers and names. These algorithms look for probable speech-recognition errors by analyzing individual dialing patterns, user statistics, and telephone-number structures. In this way, a voice-activated dialing system can deliver throughput accuracy as high as 98 percent.

System Components

VAD is implemented from an adjunct system that interconnects with the LEC's central-office switch or from a cellular service provider's *Mobile Transport Serving Office* (MTSO) switch. At an MTSO, for example, a voice-recognition system processes each customer call using modular internal cards and software to identify the called number and to manage the call on the wireless network. A file system stores and manages all customer and system information, including words that are programmed by the customer to initiate call dialing. The file system resides in the voice-recognition system in stand-alone configurations or on a separate stand-alone file server in a networked configuration, which enables multiple MTSOs to share a single user database.

In a networked configuration, the voice-recognition system's file server contains application software, including voice-recognition vocabularies, system voice prompts, and local administrative applications. The file server's databases contain customer data, including personal directories, user profiles, and speech samples. The administrative component enables centralized monitoring and control of the file servers and voice-recognition systems. The file server also provides management reports.

The voice-recognition system is installed at the MTSO, where it processes calls from the wireless switch. Its primary function is to receive incoming calls and connect them to their intended destinations. Each system holds one or more 24-channel resource boards that use *Digital-Signal Processing* (DSP) to implement features such as standard prompts and announcements, recording customer voice samples, and enabling customer DTMF interaction. Each channel can independently perform the following tasks:

- Record, digitize, and compress files
- Play back previously recorded files in real time
- Detect and generate *Dual-Tone Multi-Frequency* (DTMF) signals
- Automatically perform DTMF dialing
- DTMF override (i.e., barge in over prompts)

The voice-recognition system also includes boards that are used to listen to and recognize customer speech. Each board enables simultaneous recognition of the following elements:

- Discrete-word speech recognition
- Discrete yes/no speech recognition

- Discrete-digit speech recognition
- Talk-over prompts

The accuracy of voice-recognition systems for English-speaking callers is 98 percent over land-line telephone systems and 96 percent over cellular telephone systems.

User Implementation

The typical implementation scenario for a first-time user of VAD includes the following:

- *Training command words* During the initial VAD training session, customers only need to use two spoken commands: *dial* and *program*.
- *Learning about programmed words* During the initial training session, VAD describes the types of words that customers can choose to associate verbally with telephone numbers. For example, customers can decide to program the words *home* or *office* into the system, which enables VAD to call the telephone number automatically when a single programmed word is spoken.
- *Adding programmed words* Customers can add programmed words following the initial training session.
- *Greeting* Customers are greeted with "Good morning," "Good afternoon," or "Good evening" (based on the time of day) and are asked whether they want to speak a command word or a programmed word.
- *Prompting* Novice VAD users receive a training prompt after a one-second pause on their first call. This prompt interval increases by a half-second on successive calls, enabling experienced users to choose their next action without interruption by a prompt.
- *Redial* Users can simply say *redial*, and the VAD system automatically dials the last number that the person used in dial mode.
- *Hang-up* When using hands-free service (in a car, for example), the user can say *hang up*, and the VAD system will disconnect the call.
- *Fast-path option* Experienced customers can quickly bypass routine questions that the VAD system asks. Customers can use the fast-path option in three different ways:

- Speaking programmed words
- Using number keys
- Using the pound (#) key

To boost productivity, VAD systems can be implemented on private networks through the automated attendant of a PBX or key telephone system. An average of two minutes is necessary to look up a number and to complete a call when using a printed or online directory. In contrast, a VAD system cuts the amount of time to complete a call to only 15 seconds. With hundreds of calls placed by thousands of employees in a single day, large companies and federal agencies can expect to see a substantial increase in employee productivity over the course of a year.

Summary

VAD is a popular add-on feature among cellular telephone customers. They especially value the increased safety that VAD affords them while driving, because it enables them to keep their eyes on the road rather than on the phone. This feature is important, because more than 80 percent of cellular calls take place in the car. Accuracy, convenience, and simplicity are among the other advantages of VAD. For wireline and wireless carriers, VAD represents an additional revenue stream. Customers who have VAD generate 10 percent to 40 percent more minutes of usage per month than non-VAD users. Finally, VAD can reduce costly customer churn. According to some industry estimates, reducing the churn rate by only one percentage point can boost company valuation by approximately $150 million.

See Also

Cellular Voice Communications
Voice Compression

Voice Compression

Numerous compression methods for voice have emerged over the years that employ sophisticated encoding schemes to increase the number of channels on the available bandwidth, enabling businesses to achieve substantial savings on leased lines with only a modest cost for additional hardware. Among the most popular compression methods is *Adaptive Differential Pulse-Code Modulation* (ADPCM), which has been a world-wide standard (G.721) since 1984. ADPCM is used primarily on private T-carrier networks in order to increase the channel capacity of the available bandwidth. Although other compression techniques are available for use on wireline networks[1], ADPCM offers several advantages.

Thousands of subjective tests conducted by Bell Labs and independent-research firms have confirmed that the compressed voice signals that are used in ADPCM are virtually indistinguishable in quality from the original, uncompressed voice signals. ADPCM holds up well in the multi-node environment, where it might undergo compression and decompression several times before arriving at its final destination. Also, unlike many other compression methods, ADPCM does not distort the distinguishing characteristics of a person's voice during transmission.

ADPCM fits well within the ISDN framework. In fact, a T1 multiplexer that is equipped with the *Primary-Rate Interface* (PRI) can enhance ISDN by enabling the 64Kbps B channels to carry ADPCM-compressed voice at 32Kbps or 16Kbps. With speed-selection software that is controlled through the multiplexer's transport-management system, this capability gives users added configuration flexibility that does not come with ISDN alone. Although the ISDN standards specify the entire B channel as the fundamental unit of circuit switching, logical subchannels might be carried on a single circuit between the same pair of subscribers. Network-management information can even be carried within a subchannel to provide users with the kind of end-to-end management and control that is usually available only on a private network.

Pulse-Code Modulation (PCM)

A voice signal takes the shape of a wave, with the top and the bottom of the wave comprising the signal's frequency level, or amplitude. The voice is converted into digital form by an encoding technique called *Pulse-Code Modulation* (PCM). Under PCM, voice signals are sampled at the minimum rate

---

[1]Numerous compression algorithms exist for voice and data. This book discusses only the most widely implemented algorithms. For example, there is a world-wide standard for compressing voice over wireless networks, which is discussed under the *Global System for Mobile* (GSM) communications section in this book.

of two times the highest voice-frequency level of 4000Hz, which equates to 8,000 times per second. The amplitudes of the samples are encoded into binary form by using enough bits per sample to maintain a high signal-to-noise ratio. For quality reproduction, the required digital transmission speed for 4KHz voice signals works out to 8,000 samples per second × 8 bits per sample = 64,000 bps (64Kbps).

The conversion of a voice signal from analog to digital is performed by a coder-decoder, or codec, which is a key component of D4 channel banks and multiplexers. The codec translates amplitudes into binary values and performs mu-law quantizing. The mu-law process (North America only) is an encoding-decoding scheme for improving the signal-to-noise ratio. This concept is similar to Dolby noise reduction, which ensures quality sound reproduction.

Other components of the channel bank or multiplexer function together to interleave the digital signals, representing as many as 24 channels to form a 1.544Mbps bit stream (including 8Kbps for control) that is suitable for transmission over a T1 line. PCM exhibits a high-quality product and is robust enough for switching through the public network without suffering noticeable degradation (and is simple to implement). PCM only allows for 24 voice channels over a T1 line, however.

Compression Basics

The ADPCM device accepts PCM's 8,000-sample-per-second rate and uses a special algorithm to reduce the 8-bit samples to 4-bit words. These 4-bit words, however, no longer represent sample amplitudes; rather, they only represent the difference between successive samples. These items are all that is necessary for a similar device at the other end of the line to reconstruct the amplitudes.

Integral to the ADPCM device is circuitry called the adaptive predictor, which predicts the value of the next signal based only on the level of the previously sampled signal. Because the human voice does not usually change significantly from one sampling interval to the next, prediction accuracy can be high. A feedback loop that is used by the predictor ensures that voice variations are followed with minimal deviation. Consequently, the high accuracy of the prediction means that the difference in the predicted and actual signal is small and can be encoded with only four bits, rather than the eight bits that PCM uses. In the event that successive samples vary widely, the algorithm adapts by increasing the range that is represented by the 4 bits. This adaptation, however, will decrease

Figure V-6
Typical network
configuration that
employs ADPCM

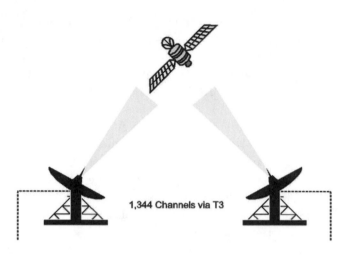

1,344 Channels via T3

the signal-to-noise ratio and reduce the accuracy of voice-frequency reproduction.

At the other end of the T1 line is another compression device (refer to
Figure V-6), where an identical predictor performs the process in reverse in
order to reinsert the predicted signal and restore the original 8-bit code.

By halving the number of bits to accurately encode a voice signal, T1
transmission capacity is doubled from the original 24 channels to 48 channels, providing the user with a 2-for-1 cost savings on monthly charges for
leased T1 lines. ADPCM can also compress voice to 16Kbps by encoding
voice signals with only two bits instead of four bits (as discussed previously). This level of compression provides 96 channels on a T1 line without
significantly reducing signal quality.

Variable-Rate ADPCM

Some vendors have designed ADPCM processors that not only compress
voice but accommodate 64Kbps pass-through, as well. The use of compact
codes enables several different algorithms to be handled by the same
ADPCM processor. The algorithm selection is made by the network manager
through software control. Variable-rate ADPCM offers several advantages.

Compressed voice is more susceptible to distortion than uncompressed
voice—16Kbps more so than 32Kbps. When line conditions deteriorate to
the point where voice compression is not possible without seriously disrupting communication, a lesser compression ratio can be invoked to compensate for the distortion. If line conditions do not permit compression even

at 32Kbps, 64Kbps pass-through might be invoked to maintain quality voice communication. Of course, channel availability is greatly reduced, but the capability to communicate with the outside world becomes the overriding concern at this point (rather than the number of channels).

Variable-rate ADPCM provides opportunities to allocate channel quality based on the needs of different user classes. For example, all intra-company voice links can operate at 16Kbps, while those that are used to communicate externally can be configured to operate at 32Kbps. The number of channels can be increased temporarily by compressing voice to 16Kbps instead of 32Kbps until new facilities can be ordered, installed, and put into service. As new links are added to keep up with the demand for more channels, others can be returned to 32Kbps operation. Variable-rate ADPCM, then, offers much more channel configuration flexibility than products that offer voice compression at only 32Kbps.

Other Compression Techniques

Other compression schemes can be used over T-carrier facilities, such as *Continuously Variable Slope Delta* (CVSD) modulation and *Time-Assigned Speech Interpolation* (TASI).

CVSD   The higher the sampling rate, the smaller the average difference between amplitudes. Also, at a high enough sampling rate—32,000 times a second in the case of 32Kbps voice—the average difference is small enough to be represented by only one bit. This concept drives CVSD modulation, where the one bit represents the change in the slope of the analog curve. Successive ones or zeros indicate that the slope should become steeper and steeper. This technique can result in good voice quality if the sampling rate is fast enough.

Similar to ADPCM, CVSD will yield 48 voice channels at 32Kbps on a T1 line. But CVSD is more flexible than ADPCM in that it can provide 64 voice channels at 24Kbps or 96 voice channels at 16Kbps. The single-bit words are sampled at the signaling rate. Thus, to achieve 64 voice channels, the sampling rate is 24,000 times a second—while 96 voice channels takes only 16,000 samples per second. In reducing the sampling rate to obtain more channels, however, the average difference between amplitudes becomes greater. Also, because the greater difference between amplitudes is still represented by only one bit, a noticeable drop in voice quality results. Thus, the flexibility of CVSD comes at the expense of quality. CVSD can even provide

192 voice channels at 8Kbps, but the quality of voice is so poor that this level of compression is rarely used.

TASI   Because people are not normally able to talk and listen simultaneously, network efficiency at best is only 50 percent. Also, because all human speech contains pauses (which constitute wasted time), network efficiency is further reduced by as much as 10 percent—placing maximum network efficiency at only 40 percent.

Statistical voice-compression techniques such as TASI take advantage of this quiet time by interleaving various other conversation segments together over the same channel. TASI-based systems actually seek and detect the active speech on any line and assign only active talkers to the T1 facility. Thus, TASI makes more efficient utilization of time to double the T1 capacity. At the distant end, the TASI system sorts and reassembles the interwoven conversations on the line to which they were originally intended.

The drawback to statistical compression methods is that they have trouble maintaining consistent quality. Such techniques require a high number of channels (at least 100) from which a good statistical probability of usable quiet periods might be gleaned. With as few as 72 channels, however, a channel gain ratio of 1.5-to-1 might be achieved. If the number of input channels is too few, a condition known as clipping can occur—in which speech signals are deformed by the cutting off of initial or final syllables.

A related problem with statistical-compression techniques is freeze out, which usually occurs when all trunks are in use during periods of heavy traffic. In such cases, a sudden burst in speech can completely overwhelm the total available bandwidth, resulting in the loss of entire strings of syllables. Another liability that is inherent in statistical-compression techniques, even for large T1 users, is that they are not suitable for transmissions that have too few quiet periods (as when facsimile and music on hold is used). Statistical-compression techniques, then, work better in large configurations than in small ones.

Summary

Adding lines and equipment is one way in which organizations can keep pace with increases in traffic. But even when funds are immediately available for such network upgrades, communications managers must contend with the delays that are inherent in ordering, installing, and putting new facilities into service. To accommodate the demand for bandwidth in a timely manner, communications managers can apply an appropriate level of voice compression to obtain more channels from the available bandwidth. Depending on the compression technique that is selected, there should not be a noticeable decrease in voice quality.

See Also

Data Compression

Hertz (Hz)

Voice-Data Convergence

The combination of voice, video, and data signals over the same equipment, line, and protocol is often referred to as voice-data convergence. Examples of voice-data convergence include the following:

- *Ethernet* Combines voice, broad-band data, and video for transmission over Category 5 wiring in the form of 1500-byte packets
- Digital Subscriber Line (*DSL*) Combines voice and broad-band data over the same twisted-pair wires that are used for *Plain Old Telephone Service* (POTS)
- *Cable* Combines voice, broad-band data, and television programming over the same coaxial cable that is connected to homes and businesses
- Local Multi-Point Distribution Service (*LMDS*) Combines voice, broad-band data, and television programming over the same short-haul wireless link operating in the 28GHz to 31GHz range
- *Optical fiber* Carries voice, broad-band data, and video over separate wavelengths within a single fiber strand
- Integrated Services Digital Network (*ISDN*) Carries voice, data, and video within separate bearer channels that are provisioned over a digital line, which are switched through the public telephone network
- Asynchronous Transfer Mode (*ATM*) Interleaves voice, data, and video for transmission through a fiber-based broad-band network as fixed-length 53-byte cells
- Internet Protocol (*IP*) Interleaves voice, data, and video for transmission through the Internet or private intranet as 1500-byte packets

IP for Convergence

A common thread in all of the convergence examples listed previously is IP. Regardless of the specific transmission medium or service, voice-data convergence can be conducted more efficiently and ubiquitously on networks that support IP. With IP, the distinct traffic types are reduced to a single stream of binary ones and zeros that are organized into packets, which can be carried by any delivery platform and can traverse any environment from LAN to WAN. Although IP can be carried over ISDN and ATM, voice-data convergence over IP can also be conducted at much less cost than either ISDN or ATM (which were specifically designed to handle multimedia traffic but are too costly to extend to every desktop).

As part of the TCP/IP protocol suite, IP can be used to bridge the gap between dissimilar computers, operating systems, and networks. IP is supported on nearly every computing platform—from PCs, Macintoshes, and UNIX systems to thin clients and servers, legacy mainframes, and the newest supercomputers. In supporting both local and wide-area connections, IP provides seamless interconnectivity between these traditionally separate environments.

On the WAN, many carriers will use ATM on their fiber-optic backbones. Some carriers, such as AT&T, are using ATM for their on-ramps, as well. Instead of forcing customers to use separate access lines for different types of traffic, this approach entails the use of a customer-premises ATM switch that is also known as an *Integrated-Access Device* (IAD), which consolidates voice, frame, and IP traffic on the same access line. At the IAD, the various traffic types are assigned an appropriate class of service by special prioritization algorithms. The traffic is then fed into the nearest network switch in the ATM cell format.

Umbrella Standard

In 1996, the *International Telecommunication Union* (ITU) issued the H.323 specification for video conferencing over packets, which is part of the H.32X standards family. Increasingly, H.323 is seen as a key element in implementing converged voice-data networks. As an umbrella standard, H.323 incorporates several specifications that govern audio coder/decoder devices, video codecs, data multiplexing, control signaling, and voice and video packet sequencing.

The H.323 specification facilitates the merger of formerly disparate network infrastructures. As carriers and companies migrate to a single voice-capable data network, they will realize substantial benefits in the form of reduced complexity and lower cost of ownership. H.323 also facilitates the development of convergence applications, such as workflow collaboration and unified messaging, in which a single mailbox accepts e-mail, fax, and voice-mail messages.

Toll bypass and LAN telephony are among the newest H.323 applications that are being deployed in the business environment. With toll bypass, an H.323-equipped gateway takes outgoing calls from the company's voice network, compresses voice and places it into data packets, then sends the packets over an IP-enabled WAN link to a second location. At the second location, another H.323 gateway converts the packets back into analog signals for delivery to the company's voice network and for distribution to the called parties.

With LAN telephony, PBX-like voice services and call-handling features can be activated through a Windows NT-based call management server. For internal calls, the voice server creates, manages, and tears down voice connections between user desktops—all without leaving the IP domain. Calls to users outside the company are delivered through H.323 gateways.

Of course, voice-data convergence over IP will greatly benefit by enhancements to H.323, which will provide *Class-of-Service* (CoS) and *Quality-of-Service* (QoS) capabilities—key capabilities that IP currently lacks but that ATM provides. The definitions for these capabilities are being developed by the IEEE's 802.1p and by the IETF's Diff-Serv working groups.

Summary

Similar to other protocols, IP is flexible enough to overcome the traditional boundaries between voice and data services. Unlike other protocols, application developers can come up with innovative new services that combine different content formats and immediately load them onto the existing IP infrastructure for global access. Because particular services are no longer locked into specific forms of infrastructure, voice-data convergence over IP creates new markets and new efficiencies as no other protocol can. The competition from these new markets and efficiencies lowers the cost of communications and fuels the continuous cycle of innovation.

See Also

Asynchronous Transfer Mode (ATM)

Cable Television Networks

Digital Subscriber-Line Technologies (DSLT)

Ethernet

Fiber-Optic Technology

Integrated Access Devices (IADs)

LAN Telephony

Multimedia Networking

Transmission Control Protocol / Internet Protocol (TCP/IP)

Unified Messaging

Wave-Division Multiplexing (WDM)

Voice Mail

Over the years, voice mail has become an effective communications tool that can enhance productivity and permit personal mobility without the risk of being out of touch with friends, family, or colleagues. Businesses especially need to ensure that calls are correctly routed and that messages are reliably delivered—even during busy times and off hours. Voice mail provides the convenience of enabling callers to leave messages for playback by the called party at a more convenient time.

Voice mail can be implemented in a variety of ways: a machine that is connected to a telephone line by a PBX or add-on messaging system, or by a service that is provided by a telephone company or firm that specializes in voice messaging. Voice mail offers many advantages to callers, including the following:

- Ensures that the message is accurate

- Provides the opportunity to leave detailed messages and explanations

- Enables information to be delivered in the caller's own speaking style

- Provides more privacy in contrast to other message-delivery methods, such as operators, receptionists, and secretaries

A business-class voice-mail system for the small to mid-size organization typically comes with multiple ports (expandable from two to eight ports) for interfacing with the telephone system, storage for 35 to 50 hours of messages, and an unlimited number of password-protected voice mailboxes. Custom call routing enables the system administrator to tailor options and menus to the specific needs of the organization.

Users can access the system to customize personal greetings and use dial codes to record, check, and send outgoing messages—with or without a confirmation receipt. Other dial codes enable users to play, skip, save, or delete incoming messages locally or remotely. Users can also recover deleted messages. A copy of received messages can be forwarded to another voice mail mailbox or can be broadcast to a list of mailboxes—with or without annotation. Other features that can be implemented by the user include the following:

- Multiple messaging options, including private, urgent, future delivery, and confirmation receipt
- Playing messages faster, slower, louder, or softer
- Attaching a date/time stamp and sending a mailbox ID with every outgoing message
- Creating multiple greetings per mailbox, including standard, temporary, busy, and time-sensitive greetings
- Pager notification, enabling the voice mail mailbox to contact the user's pager when a message has been left
- Call screening, conveying the calling party's name to the mailbox owner (who can then accept, reject, or forward the call to another extension)

Automated Attendant

Many business-class voice mail systems can manage communications without a live operator. The system will greet callers with an introduction, ask them to enter an extension or select from options, and even ask for the caller's name before transferring so that the recipient can screen the call.

The system can park a call in a hold location and announce the call over the in-house paging system. It can also inform a caller to a busy station where they are located in the station's queue (third in line). Plus, the system can provide frequently requested information, such as directions to the company's headquarters.

Administration

Today's voice-mail systems typically employ Windows-based setup screens to facilitate installation and administration from a laptop or desktop computer. In some cases, programming can even be performed from a telephone set from an off-site location. With the installation wizards, integration with popular telephone systems is easily implemented. In such cases, mailboxes, messages-waiting strings, transfer strings, and time-of-day greetings are automatically implemented.

Voice Mail Service

Instead of purchasing and managing their own voice mail equipment, self-employed individuals and small businesses prefer to subscribe to the voice mail services of a carrier. An added advantage of carrier-provided service is that it works during a local power outage. Users access all the features available through the service right from the telephone's keypad (refer to Figure V-7).

Figure V-7
Ameritech offers several voice-mail services, the features of which can be implemented through this menu system.

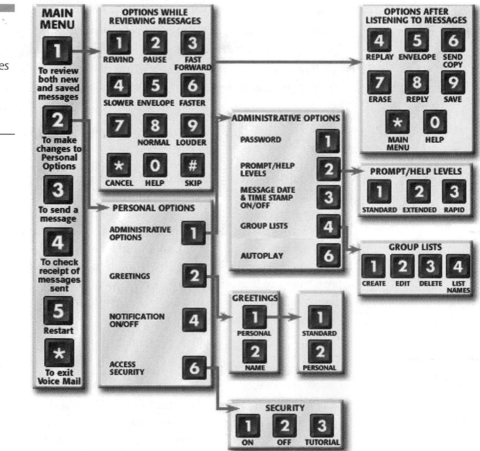

Usually, there is a choice of plans based on the number of voice mail mailboxes and control options that are required. For individuals, basic service usually starts at $5 per month, plus a one-time mailbox setup fee. For businesses that have multiple users who need extra recording time, pager notification, added security, and more message-handling features, the service is priced based on the number of mailboxes and the specific features that are associated with each mailbox. Many carriers have Web pages that offer customers a convenient way to order voice-mail services.

Internet Voice Mail

The most economical way to send voice messages is over the Internet as an attachment to an e-mail message. You can record and send messages in several ways. With encoder software that is loaded on a multimedia-equipped notebook or desktop computer, the user can record a voice message through the microphone and store it on the hard disk as an audio file. The file is then sent as an attachment by using any e-mail software. Among the most popular voice encoders are Progressive Networks' RealAudio, DSP Group's TrueSpeech, and Voxware's ToolVox. These applications can be purchased and downloaded from the companies' Web sites.

In addition, users can access the Sound Recorder (refer to Figure V-8) and Media Player (refer to Figure V-9) applications that come with Windows, which are based on technology that is supplied by TrueSpeech.

Voice-messaging capabilities are also offered in some Internet phone products, such as NetSpeak's WebPhone and VocalTec's Internet Phone. VocalTec even offers a free Voice Mail Player that enables recipients to play back voice-mail messages if they do not happen to have a copy of Internet Phone. Another solution is to use the voice-mail capabilities that are offered

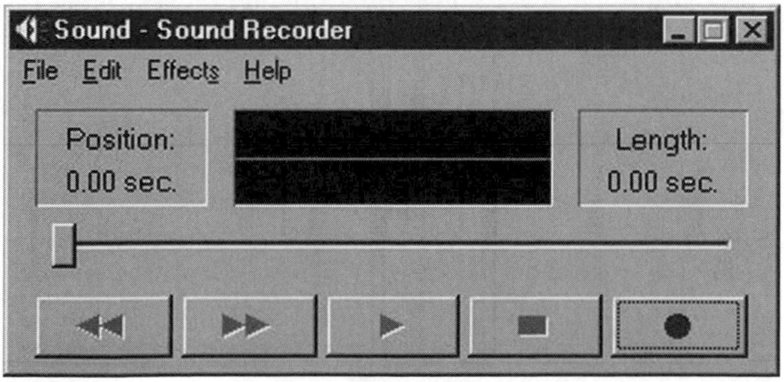

Figure V-8
Sound Recorder comes with all versions of Microsoft Windows and enables users to record and edit voice messages in the .WAV format.

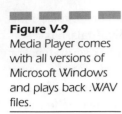

Figure V-9
Media Player comes with all versions of Microsoft Windows and plays back .WAV files.

by some e-mail programs such as Pronto or Bonzi. This approach enables users to open the voice encoder from within the e-mail program, record a message, and send the message just like any other e-mail message. In addition to an Internet phone that has voice-mail capabilities, VocalTec offers a stand-alone voice-mail program and player called Internet Voice Mail.

Summary

Voice mail has been in use by major corporations for more than 15 years and is implemented in the business environment through a PBX or add-on messaging system. Voice mail is also a service that is offered by telephone companies and numerous third-party messaging firms. Similar technology is available for use over the Internet, enabling anyone who has a multi-media PC to record, send, and play back voice messages. Because the Internet spans the globe, voice mail offers an economical way to send personalized messages to friends, family, or colleagues. Regardless of the method that is used, voice mail provides the opportunity to leave detailed messages and explanations in the caller's own speaking style and provides privacy over other message-delivery methods, such as operators, receptionists, and secretaries.

See Also

Electronic Mail

· Unified Messaging

Voice-over-IP (VoIP)

IP telephony has been around since the 1980s, when it was referred to as a voice funnel. In 1983, both the ARPANET and the Internet were being run from the *Network Operations Center* (NOC) facility at the offices of *Bolt, Beranek, and Newman* (BBN) in Cambridge, Massachusetts. There, among the workstations that were dedicated to special projects, was one labeled "Voice Funnel" that digitized voice, arranged the resulting bits into packets, and sent them through the Internet between sites on the east and west coasts. The voice funnel was part of an ARPA research project involving packetized audio. ARPA and its contractors used the voice funnel and related video facilities to conduct three-way and four-way conferencing, saving travel time and money.

The technology was rediscovered in the 1990s but did not become popular until 1995, when improvements in microprocessors, *Digital-Signal Processing* (DSP) technology, and routing protocols all came together to make feasible products for mainstream use. Since then, IP telephony has been adapted for commercial telecommunications service. Some service providers compete directly with established long-distance carriers and *Regional Bell Operating Companies* (RBOCs), enabling consumers to use IP telephony services with a look and feel that is identical to today's phone service. Corporations are adding IP telephony service in order to leverage investments in private intranets and to save on long-distance call charges. They also see the potential of IP telephony for adding value to existing network applications such as call centers, customer support, help desks, and e-commerce.

First-Generation Technology

First-generation IP telephony focused on establishing calls over the public Internet between similarly equipped multimedia PCs. Placing calls involved logging onto the Internet and starting up the phoneware, which provided several ways of establishing a voice link. Users could connect to the vendor's directory server to check the White Pages for other phoneware users who also were logged on to the Internet. The public directories were organized by user name and topic of interest to make it easy for everyone to strike up a conversation with like-minded and willing participants. The directory was periodically updated, reflecting changes as people entered and left the network.

Alternatively, users could click a name in a locally stored, private phone book. Of course, users could simply enter the IP address or e-mail address in order to establish a direct user-to-user connection from the start, without having to first go through a directory server (refer to Figure V-10). Whether it was a public or private listing, the names corresponded with the static IP

Figure V-10

Early phoneware products logged users on to a directory server, which enabled them to receive a list of other registered phoneware users. The connection is then user-to-user, bypassing the vendor's directory server.

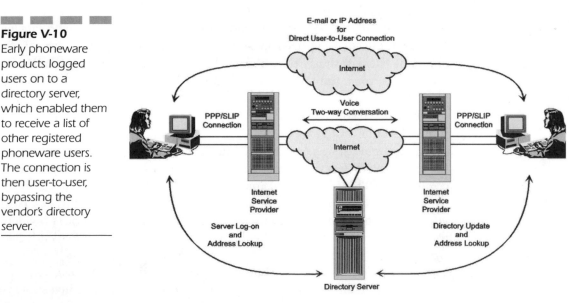

addresses of other users. As *Internet Service Providers* (ISPs) increasingly adopted dynamic IP addresses[2], the names in public directories and private address books corresponded with e-mail addresses instead.

System Requirements

To make calls over the Internet, users required a computer that was equipped with a modem, a sound card, speakers (or a headset), and a microphone. Sound cards came in two types: half-duplex and full-duplex. Half-duplex worked similarly to a *Citizens Band* (CB) radio, where one person could talk at a time and can say "over," indicating that he or she was finished talking. With full-duplex audio cards, both parties could talk at once, just like an ordinary telephone call. If a full-duplex user was connected to a half-duplex user, the conversation defaulted to the half-duplex mode.

In addition to the hardware, three software components were typically required: a TCP/IP dialer program (because most users dialed into the Internet with a modem), a Web browser, and the phone software itself. The critical component is the phoneware, which provides algorithms that compress the recorded speech that is obtained from the sound card and applies optimization techniques to ensure its efficient delivery over the Internet in the form of data packets.

Phoneware vendors use a variety of compression algorithms to minimize bandwidth consumption over the Internet. A software coder/decoder (codec) uses mathematical strategies to reduce the bit-rate requirements as much as possible and yet still provide acceptable reproduction of the original content. Numerous codecs have appeared on the scene in recent years, with more coming all of the time.

---

[2]With dynamic IP addressing, an address is assigned by the Internet service provider each time the user dials into the server. The Internet service provider has a pool of IP addresses for this purpose. With static IP addressing, the same address is used each time the user connects to the Internet. The proliferation of TCP/IP-based networks, coupled with the growing demand for Internet addresses, makes it necessary to conserve IP addresses. Issuing IP addresses on a dynamic basis provides a way to recycle this finite resource. Even companies that have private intranets are increasingly using dynamic IP addresses, instead of issuing unique IP addresses to every machine. The standard that addresses this issue is the *Dynamic Host-Configuration Protocol* (DHCP), developed by the *Internet Engineering Task Force* (IETF). From a pool of IP addresses, a DHCP server doles them out to users as they establish Internet connections. When they log off the network, the IP addresses become available to other users.

For example, NetSpeak's WebPhone started out in 1995 using two audio compression algorithms: GSM and TrueSpeech. GSM is the Global System for Mobile Communications and is a world-wide standard for digital cellular communications. GSM provides close to a 5:1 compression of raw audio with an acceptable loss of audio quality on decompression. TrueSpeech, a product of the DSP Group, provides compression ratios as high as 18:1 with an imperceptible loss of audio quality on decompression. NetSpeaks's WebPhone used GSM compression when it was installed on a 486-based computer and used TrueSpeech when it was installed on a Pentium-based computer. Offering a high compression ratio, TrueSpeech is more CPU-intensive than GSM, so it requires a faster processor to compress the same audio signal in real time.

The strategy of NetSpeak has been to deal with codecs in a plug-and-play fashion to enable new codecs to be incorporated into the client application with little or no effort on the part of the user. As NetSpeak adds new codecs to its repertoire, they are easily incorporated into the clients. The clients can then negotiate between themselves the best codec to be used for an individual session based on available session bandwidth limitations, network delay characteristics, or individual PC resource limitations.

Other phoneware vendors offer proprietary compression algorithms and support one or more accepted industry-standard algorithms. Most now most support the G.7xx international standards (discussed later) as well, which guarantee various levels of speech quality and facilitate interoperability between various H.323-compliant products. H.323 is the umbrella standard which includes the G.7xx audio standards. Recommended by the *International Telecommunication Union* (ITU), H.323 defines how audio and visual conferencing data is transmitted across networks.

In addition to the algorithms that compress/decompress sampled voice, some phoneware products include optimization techniques to deal with the inherent delay of the Internet. The packets can take different paths to their destination and might not all arrive in time to be reassembled in the proper sequence. If this was ordinary data, late or bad packets would simply be dropped, and the host's error-checking protocols would request a retransmission of those packets. This concept cannot be applied to packets that contain compressed audio, however, without causing major disruption to real-time voice conversations.

If only a small percentage of the packets are dropped, say 2 percent to 5 percent, the users at each end might not notice the gaps in their conversation. When packet loss approaches 20 percent, however, the quality of the conversation begins to deteriorate. Some products, such as VocalTec's Internet Phone, employ predictive analysis techniques to reconstruct lost packets—thereby minimizing this problem.

Occasionally, the Internet can become overloaded or congested, resulting in lost packets and choppy sound quality. This problem can be overcome by introducing artificial delay into the signal while regulating the flow of voice to the receiver to smooth out any gaps. This scheme affords extra time to retransmit lost packets using a proprietary algorithm. The end result is better quality, at the expense of increased but predictable delay.

Operation

Once the call is placed, either by IP address or e-mail address, the users at each end speak into the microphones connected to the sound cards in their respective computers. The phone software samples the speech, digitizes and compresses the audio signal, and transmits the packets via TCP/IP over the Internet to the remote party. At the other end, the packets are received and pieced together in the right order. The audio is then decompressed and sent to the sound card's speaker for the other party to hear. The compression algorithm compensates for much of the Internet's inherent delay. As the packets are decompressed and the audio signals are being played, more compressed packets are arriving. This process approximates real-time conversation.

To improve overall sound quality, early users of IP telephony software often found it necessary to fine-tune the sampling rate and compression level in order to suit their modem's speed. For example, to overcome the annoying problem of clipped speech, the user could reduce the sampling rate until smooth speech resumed.

Some products enable users to adjust recording and playback quality in response to the speed of the modem connection. The user can start this manual tuning process by connecting at the default sampling rate, then incrementally increasing the sampling rate. With the new setting, the connection is renegotiated at the higher sampling rate. The user can continue to increment the sampling rate until the other party's speech begins to break up, then back it down until the speech is clear again.

The sampling rate can be set from 4000 bytes to 44000 bytes per second, depending on the capabilities of the sound card and the speed of the Internet connection. In general, the higher the speed of the modem connection, the higher the sampling rate can be set. In conjunction with the sampling rate, users can set the compression level. With a lower-speed modem, a higher compression level can be selected to improve performance (but with some loss in sound quality). With a higher-speed modem, users can select a lower compression level for better sound quality.

Some products, such as VocalTec's Internet Phone, provide real-time statistics that can help users determine the quality of the Internet connection at any given time (refer to Figure V-11). The network statistics window provides a count of incoming and outgoing packets, the average round-trip delay of packets, and the number of lost packets in both directions (incoming and outgoing). This information helps the user pinpoint the source of the problem as originating from the network or locally (e.g., sound card, modem, or software) so that corrective action can be taken.

Features

Internet phone products offer many features and new ones are being added all the time. The benefit of using a computer for telephone calls—rather than an ordinary phone—is that the user can take advantage of integrated voice-data features. The following list provides the most common features that are found across a broad range of products:

- *Adjustable volume control* Enables the volume of the microphone and speakers to be adjusted during the conversation

- *Advanced caller ID* Not only is the calling party identified by name, but some phoneware products also offer a brief introduction message concerning what callers want to talk about, which is displayed as the call comes through. This information can help users decide whether or not to answer the call.

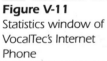

Figure V-11
Statistics window of
VocalTec's Internet
Phone

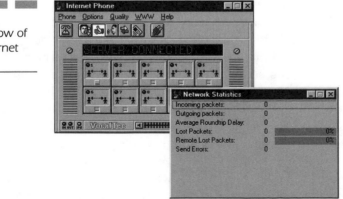

- *Advanced phone book*   Not only holds contact information but also offers a search capability by name, e-mail, country, company, or any other parameter that can identify a particular person
- *Audio date / time stamp*   Notifies unavailable users of call attempts by date and time
- *Automatic notification*   With this feature, the phoneware automatically looks for and provides notification of when specific other users come online so that they can be called.
- *Busy notification*   If a call is placed to someone who is busy with another call, an appropriate busy message is returned. Some products enable callers to send an e-mail message, voice mail, or other notification to the busy party, indicating that they have tried to call.
- *Call blocking*   Annoying or unwanted incoming calls can be blocked by fixed IP or e-mail address.
- *Call conferencing*   The capability to converse with three or more people at the same time
- *Call-duration timer*   Provides an indication of the amount of time that is spent on each call
- *Call hold*   Enables an initial call to be put on hold while the user answers another incoming call. The user can continue the first conversation after holding or hanging up the second call.
- *Call log*   Records information about incoming and outgoing calls, enabling the user to keep track of incoming and outgoing calls
- *Call queue*   A place where incoming calls are held until they can be answered
- *Caller ID*   Identifies the caller by name, nickname, e-mail address, or phone number so that a user can see who is calling before deciding whether to take the call
- *Configuration utility*   Scans the computer system to determine whether the proper hardware is installed to use the phoneware and offers the user advice about configuring various operating parameters, such as IRQs, DMAs, and I/O base-address settings to prevent conflicts with other communications applications
- *Database-repair utility*   The phoneware maintains one or more databases to hold items such as private phone books and configuration data. If a database becomes corrupted or destroyed, the user will be notified and will have the option of running the database repair utility in order to restore the database.

- *Dedicated server*   For those who receive a large volume of incoming calls, some phoneware vendors offer special servers to facilitate call handling.

- *Directory assistance*   A searchable directory of users who are currently online is automatically maintained. Users can initiate a phone call simply by clicking a person's name or by typing in the first few characters of a name.

- *Dynamic, on-screen directory*   Provides the latest information about users who have registered with the server, indicating that they are online and are ready to take or initiate calls. This display is periodically refreshed with new information.

- *Encryption*   To ensure secure voice communication over the Internet, a public-key encryption technology such as *Pretty-Good Privacy* (PGP) can be applied. Depending on the vendor, PGP is integrated into the phoneware or can be licensed separately for use with the phoneware. Other products might not accommodate encryption at all.

- *Event message system*   Enables users to view the ongoing status of the phoneware to determine which features and functions are active at any given moment

- *File transfer*   Enables the user to transmit a file to the other party during a conversation. The file-transfer process takes place in the background and does not interfere with the conversation.

- *Greeting message*   When a user is not available or is too busy to take a call, a recorded message can be played to callers.

- *H.323 compliance*   A world-wide standard for audiovisual communication over packet data networks, such as the Internet. Users of different H.323-compliant products are capable of conversing with each other over the Internet.

- *Last-party redial*   Enables the user to redial the last party that was called without having to look up the address in a directory

- *Map*   Displays the connection of the call against the background of a U.S. or world map, showing the points of origination and destination

- *Multiple calling mechanisms*   Some phoneware products offer multiple methods of initiating calls, including fixed IP address, domain name, e-mail address, saved addresses, and an online directory of registered users.

- *Multiple lines*   Some phoneware products enable users to carry on a conversation on one line and take an incoming call on another line, or to put one call on hold while another call is initiated.

- *Multiple user configurations*   If several people share the same computer, some phoneware products enable each of them to have their own private configuration, including caller ID information and address books.

- *Music on hold*   Plays music to a caller who is on hold until the call can be answered

- *Mute*   A mute button enables private, offline conversations.

- *Online help*   Offers help on the proper use of various phoneware features without having to resort to a manual or opening a separate read-me file

- *Picture compression*   Some phoneware enables the user to call up a photo of the person to whom they are talking (if the remote user supplies a photo). Compression enables fast photo loading over the Internet of the commonly supported file formats.

- *Programmable buttons*   Enables users to configure quick-dial buttons for the people whom they call most frequently. In some cases, buttons are added automatically and are written over, based on the most recent calls.

- *Redial*   If a person is not reachable on the first call attempt, the phoneware can be configured to automatically redial at designated time intervals until the connection is established.

- *Remote information display*   Displays operating system and sound card information for the remote user

- *Remote time display*   Displays the remote time of the called party

- *Selectable codec*   Provides a choice of codecs, depending on the processing power of the computer. A high-compression codec can be used for Pentium-class machines, and a lower-compression codec can be used for 486-based machines. The choice is made during phoneware installation.

- *Silence detection*   Detects periods of silence during the conversation, in order to avoid unnecessary transmission

- *Statistics window*   Enables the user to monitor system performance and the quality of the Internet link

- *Text chat*   Some phoneware products offer an interactive text or chat capability to augment voice conversations. The chat feature can be used before, during, or apart from voice mode.

- *Toolbar*   Icons provide quick access to frequently performed tasks such as hang up, mute, chat, view settings, and help.

- *Toolbox mode*   The interface can be collapsed into a compact toolbox in order to save desktop space, which makes it easier to work in other applications until the phoneware is used for calls.

- User Location Service (*ULS*) Compliance   ULS technology enables Internet phone users to find each other through existing Internet directory services such as Four11, Banyan's Switchboard, WhoWhere, DoubleClick, and BigFoot.

- *User-defined groups*   Enables users to set up private calling circles for calls among members only or to establish new topic groups for public access

- *Video*   Some phoneware products enable the calling and called parties to see each other as they converse, which requires a video-camera connection to the computer.

- *Voice mail*   Enables users to record and play back greeting messages as well as send voice-mail messages for playback by recipients. Depending on the vendor, this functionality might include the capability to give specific messages to callers when they enter a personal code.

- *Voice-mail screening*   Enables users to delete voice messages before they can be downloaded to their computer

- *Web links*   Enables users to put links into their Web pages that, when activated, establish a call with the visitor

- *Whiteboard*   Some multi-function products enable participants to draw or annotate shared text and images while conversing.

While early phoneware products focused on PC-to-PC communication, they had the advantage of combining audio, video, and text capabilities. Such products are still available and are continually being enhanced with new capabilities and features. The business version of NetSpeak's Web-Phone, for example, offers four lines with call holding, muting, do not disturb, and call blocking options. This version also offers a large video-display area with self views and remote views (refer to Figure V-12).

Internet Calls to Conventional Phones

In a demonstration of the feasibility of originating calls on the Internet and receiving calls at conventional phones, the *Free World Dialup* (FWD) *Global Server Network* (GSN) project went online in March 1996. Orga-

Figure V-12
NetSpeaks's
WebPhone has a
video capability that
provides a remote
view (shown) and a
self view.

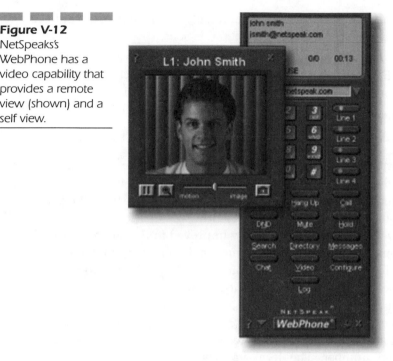

nized by volunteers around the world, the non-commercial project was entirely coordinated in cyberspace via Internet telephony, e-mail, and chat software.

Using popular Internet-telephony software, users could contact a remote server in the destination city of the call. This server patches the Internet phone call to any phone number in the local-exchange area. In other words, a user in Hong Kong can use an Internet-based server in Paris to effectively dial any local phone number and talk with a friend or family member. A global server keeps a list of all servers and the real-time status of each.

The specific steps that are required to place a call through the FWS GSN and to answer the call with a conventional phone are as follows:

1. Connect to the Internet as usual with a PPP or SLIP connection.
2. Start the phoneware and register with the vendor's server, if necessary.
3. Start the FWD client software and connect to a FWD server in a select city.
4. Click the Connect button, and enter the domain name of the server.

5. Once you are connected to the server, a message indicates that the connection is made, and a ring signal is sent.

6. Upon receiving the ring signal, the user enters the telephone number in the FWD client of the person in the local calling area, leaving out the area code.

7. The phoneware dials the number that is entered.

8. When the called person answers, both parties can start talking.

This procedure is certainly more complicated than just picking up the phone and dialing the long-distance number, and its first implementation was limited to processing calls in only one direction—from the Internet to conventional phones. The technology that was demonstrated by the FWD GSN, however, has been continually improved to the point where it is now used in commercial products. Among the dozen or so companies that offer such products is VocalTec. The company's Telephony Gateway not only streamlines the calling process, but it also provides many advanced features that make IP telephony services commercially viable.

To place a call, the user dials an 800 number to access the nearest gateway from any phone or PC. After receiving a dial tone, the user enters the destination number. The local gateway digitizes and compresses the incoming voice signals into packets, which travel over the Internet or intranet to a remote gateway that is near the dialed location. At that point, the packets are decompressed and reconstructed into their original form, making them suitable for transmission over the PSTN. From there, the call is then routed over the local-loop connection to the destination telephone or PBX.

The IP/PSTN gateways offer many features that expedite administration. The VocalTec gateway, for example, provides features for call monitoring, security, and billing. This gateway is even equipped with an *Interactive Voice Response* (IVR) application, which acts as the interface between the PSTN/PBX and the IP network. The application includes an auto attendant that guides users through the calling process.

Service-level guarantees are becoming available with some gateways. Quintum Technologies, for example, offers a voice-over-IP gateway that comes with a written 100 percent guarantee that an enterprise's IP calls will go through to completion. The firm's Tenor Multipath Gateway continually monitors IP network quality, and if it detects congestion, it automatically switches a call from the data network to the PSTN—thus ensuring that the call will be completed.

Summary

Many of the issues that plagued users of first-generation Internet telephony products have been addressed by hardware and software vendors with the goal of facilitating the growth of commercial, carrier-grade VoIP services. There are now scaleable IP switches that compete with traditional central office switches in terms of call-processing capacity and call-handling features. There are roaming agreements between service providers that help ensure the broadest possible coverage for commercial IP telephony service. Mechanisms have been developed to accurately meter usage and charge for voice calls over IP networks. Administrative tools enable individual users or groups of users to be assigned a class of service. Tools even exist that enable network managers to monitor the performance of the IP network in real-time and to check on the quality of service that is being delivered to each user at any given time. These and other developments have prompted carriers, ISPs, and corporations to take this once-spurned technology seriously.

See Also

Cable Telephony

Local-Area Network (LAN) Telephony

Transmission-Control Protocol / Internet Protocol (TCP/IP)

Voice-Data Convergence

Wavelength-Division Multiplexing (WDM)

Wavelength-Division Multiplexing (WDM) technology has been in use by long-distance carriers in recent years to expand the capacity of their trunks by enabling a greater number of signals to be carried on a single fiber. Although the technology has been in existence since the late 1980s, the need among carriers to obtain more performance and flexibility from their fiber-optic networks only arose in the mid-1990s. AT&T, Sprint, and MCI WorldCom (among others) have made long-term commitments to WDM technology and will be using it to ramp up their trunk speeds from 2.5Gbps to more than 40Gbps (without having to install additional fiber).

Applications

WDM will help eliminate capacity constraints in carrier networks that are brought on by the ever-increasing processing power of computers and the need to link multiple users at multiple locations. WDM supports applications such as the simultaneous distribution of full-motion video and medical images, without forcing carriers to install new fiber backbones.

WDM works with a variety of existing protocols and technologies, such as *Synchronous Optical Network* (SONET) services ranging from OC-1 (51.8Mbps) to OC-256 (13.271Gbps) and broad-band *Asynchronous Transfer Mode* (ATM) cell switching.

A relatively new trend in fiber networks is the adaptation and deployment of WDM systems in enterprise networks. Enterprise data communication requirements for greater network bandwidth, lower cost, and absolute reliability are being driven by applications such as data mirroring and vaulting, server clustering, and LAN extension. WDM technology delivers the kind of performance that is needed for these applications.

Among the growing number of vendors addressing the enterprise market is ADVA Optical. The company's SNMP-manageable *Optical Channel Multiplexer* (OCM-16) is a 16-channel *Dense Wavelength-Division Multiplexing* (DWDM) system that is capable of providing an aggregate bandwidth of 1.25Gbps over a distance of up to 37.5 miles (60 km). This amount of bandwidth is ideal for large data centers that require state-of-the-art networks with the highest *Quality of Service* (QoS) and high-capacity management capabilities in terms of bits per wavelength. The system supports native wavelength services in point-to-point, point-to-multi-point, and distributed ring architectures. In addition to legacy protocols such as FDDI and ESCON, the OCM system supports Fast Ethernet, ATM at OC-3/OC-12, Fibre Channel, and Gigabit Ethernet. Thus, an enterprise can take full advantage of WDM's speed potential while fully leveraging its existing investments in fiber and high-capacity systems.

Operation

Unlike older multiplexing techniques, which use separate frequencies or specific time slots, WDM uses light waves of different lengths—each of which constitutes a high-speed data channel. Because the combined wavelengths do not interfere with each other and the systems at each end of the link never process the data, multiple signals that have mixed protocols and speeds travel at their full rated speed through a single optical fiber. A duplicate device at the opposite end receives the combined wavelengths and splits them into their respective end systems (refer to Figure W-1).

Dense WDM

Some optical-transmission systems use a derivative of WDM called *Dense Wavelength-Division Multiplexing* (DWDM), which provides many more wavelengths on a single optical fiber (with each wavelength constituting a channel that can move data at multi-gigabit-per-second speeds).

Figure W-1
With WDM, the different colors of light constitute channels for the transmission of data at high speeds, regardless of the traffic type of protocol.

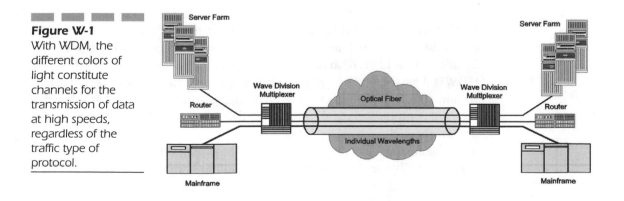

IBM is among the vendors that are offering a DWDM device for customers who are looking to build high-speed interconnections among multiple data centers for distributed applications, disaster recovery, or CPU redundancy. The so-called Fiber Saver for the mainframe S/390 environment enables users to set up high-speed machine mirroring and remote disk-copying functions. Fiber Saver can transmit up to 32 channels over a single pair of fibers or up to 64 channels over two pairs, at distances of up to 31 miles (50 km). The maximum transmission rate for each channel is 1.25Gbps, with an overall maximum capacity of 80Gbps for the system. In contrast, IBM's WDM-based Muxmaster provided only 10 channels and is being phased out.

The Fiber Saver system has been optimized to run IBM mainframe I/O protocols, such as *Fibre Connection* (FICON) and *Enterprise-Systems Connection* (ESCON). FICON handles Fibre Channel traffic to high-volume storage devices, while ESCON is IBM's 17Mbps optical-channel connection. The system also supports Fast Ethernet, Gigabit Ethernet, and ATM data transports at OC-3 to OC-12 speeds and can be managed by any SNMP-management platform.

While DWDM offers unprecedented transmission capacity for next-generation data networks, continued developments in fiber-optic technology promise even more capacity under a new technology called *Ultra-Dense WDM* (UDWDM). With today's experimental techniques developed by Bell Labs, a single strand of fiber can have more than 1,000 channels, with each channel operating at up to 160Gbps. In being able to multiply 160Gbps over additional wavelengths, Bell Labs expects to scale the capacity of optical fiber to trillions of bits per second in the near future.

Summary

WDM technology provides carriers and enterprise users with an economical solution for increasing network capacity without the expense of installing new fiber. With continual improvements, commercial DWDM and UDWDM systems will be deployed in the near future to offer terabits-per-second speeds. Adoption of these technologies is an important step toward the goal of all-optical networking, in which optical-to-electrical conversions are minimized by moving more transport and switching duties into the optical domain. This type of network will be cheaper to deploy and maintain than other fiber architectures that are used in LANs and WANs.

Wide-Area Telecommunications Service (WATS)

Wide-Area Telecommunications Service (WATS) was introduced by AT&T in 1961 to enable customers to receive substantial discounts on telephone calls, provided that they stayed with the service for a specified time and adhered to a minimum monthly revenue commitment. Although the term WATS referred specifically to the discounted toll service of AT&T, other carriers used the term to describe their discounted toll offerings, as well. The term WATS is rarely used anymore, having been replaced by other brand names for discounted toll services.

This type of service was basically a bulk-rate toll service priced according to call distance, or rate band. Customers commit to the service for a specified duration—usually from 18 to 48 months. Discounts expand with the length of the plan and the customer's monthly revenue commitment. If a customer does not meet the monthly minimum revenue level, usage charges are adjusted up to that minimum level. Also, if a customer cancels the plan before it expires, the company is billed for the discount accumulated up to the time of cancellation.

The traditional banded WATS facilities introduced 39 years ago have been replaced by more flexible and manageable WATS-like services, in which billing is based on time of day and call duration, as well as distance. Customers not only qualify for additional savings based on call volume but also on the number of corporate locations that are enrolled in the plan, as well.

An inbound version of the WATS-type service exists, which enables businesses to offer toll-free calls to their customers and other constituents via 800 and 888 services. Carriers offer either switched-access or dedicated-access 800 and 888 service. Switched service provides businesses with the capability to receive 800 and 888 calls over regular telephone lines, while dedicated service provides a private connection from the carrier's network to the business' network.

Each service provides several options. For example, businesses can geographically screen their calls, or block calls from certain parts of the country; other services automatically route calls to specific locations based on customer-specified requirements. The larger IXCs now also offer 800 service for international calls. Callers outside the United States use country-specific numbers to route calls to a company's access line in the United States. Personal 800 service even exists for individuals who work out of the home and have low call volumes.

Detailed billing reports for 800 services, available on customer request, provide call detail and exception reporting. The carriers also offer call detail information in real time, or on a monthly or daily basis. Such services can be used to measure marketing responses, track lost calls, and gauge the effectiveness of call-center operations.

Summary

Over the years, carriers have continually revamped their WATS offerings to take into account changing market conditions. WATS started as an offering for only the largest companies; today, businesses with only one telephone line can realize significant cost savings with WATS-like calling plans tailored to their specific needs.

See Also

Call Centers

Call-Detail Reporting (CDR)

Wireless Communications Services (WCS)

Wireless Communications Service (WCS) is a relatively new category of service that operates in the 2.3GHz band of the electromagnetic spectrum from 2305MHz to 2320MHz and 2345MHz to 2360MHz. Licenses for WCS were issued by the FCC in April 1997 as the result of a spectrum auction. The FCC granted licensees wide latitude in how WCS spectrum is used. Licensees are permitted—within their assigned spectrum and geographic areas—to provide any fixed, mobile, radio location or broadcast-satellite service consistent with the allocation table and associated international agreements concerning spectrum allocations. One use for the WCS band of spectrum is for services that adhere to the *Personal Access Communications System* (PACS) standard. This standard is applied to consumer oriented products, such as personal cordless devices.

WCS is implemented through small relay stations and is designed to interface with the existing telephone network. Where WCS poses interference problems with existing *Multi-Point Distribution Service* (MDS) or *Instructional Television Fixed Service* (ITFS) operations, the WCS licensees must bear the full financial obligation for the remedy. WCS licensees must notify potentially affected MDS/ITFS licensees, at least 30 days before commencing operations from any new WCS transmission site or increasing power from an existing site, of the technical parameters of the WCS transmission facility. The FCC expects WCS and MDS/ITFS licensees to coordinate voluntarily and in good faith to avoid interference problems, which will result in the greatest operational flexibility in each of these types of operations.

Summary

The FCC completed the auction process for WCS spectrum in April 1997. There were only 17 winning bidders for 128 licenses, eight of which qualified as small or very small businesses under the FCC's rules. The auction was expected to bring in about $2 billion to the United States Treasury, but raised only $13.6 million. The FCC was criticized for not providing adequate notice of the auction. Faced with an apparent failure, the FCC claimed that the real issue was not the money raised, but getting spectrum into the marketplace as quickly as possible, and meeting tight Congress-mandated time frames for spectrum distribution. Because of the short notice of the auction, however, many prospective bidders did not have enough time to determine how best to use the WCS spectrum and consequently did not bid at all.

See Also

Personal-Access Communications Systems (PACS)

Spectrum Auctions

Workflow Automation

Workflow is the application of computers and networks to automate previously manual business processes. This kind of automation is especially beneficial to organizations that rely on standardized forms and multiple steps to complete a transaction, such as fulfilling orders, processing claim forms, and reconciling customer accounts. Instead of just dumping all information into a database, for example, logical queues of documents are established, enabling networked workstation operators to obtain the next available document for processing. This functionality makes for more efficient business operations, increased productivity from available staff, faster customer response, and greater accuracy of information.

The Workflow Process

In an insurance claim-processing application (refer to Figure W-2), for example, incoming paper documents are scanned into the system when received. Then, they are indexed by the claim number, name, form type and scan date. All documents for the same case can be grouped in an electronic

Figure W-2
The workflow of a typical insurance-claims process

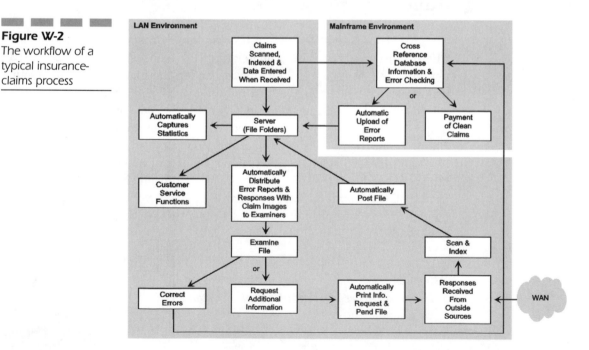

file folder. When a file folder is opened, the whole folder—or individual documents—is routed immediately to the appropriate queue. Queues reside in the image server and are dedicated to particular workstation operators. Other needed information—such as customer account histories—can reside in a mainframe database and can be integrated into the document via terminal emulation. Multiple mainframe sessions can be opened; each of which is displayed in a separate window at the image workstation.

When a processing step is completed, the file is forwarded to the next step by placing it into an appropriate queue where it can be accessed by the next available operator who is assigned to that task. If further work is needed on the document, it can be sent back to the person who previously handled it, or it can be held in suspense until additional information is obtained. Supervisors can easily monitor the movement and status of documents and files to ensure that high-priority cases are handled expeditiously, exception cases get the attention they deserve, the workload is distributed fairly, and operators meet productivity goals.

A workflow script development kit enables workgroup administrators to automate, manage and control the queuing and flow of images, data and text. Basically, a script is a set of instructions, or calls, that can be activated upon command. The instructions consist of functional statements that automate various tasks such as storing documents on disk, retrieving them for processing, moving data from one application to another, and routing information from workstation to workstation. A set of scripts completes the tasks that are assigned to the system, with each script performing one or more steps that relate to the task. The advantage of scripts is that the various steps of a task can be put into the order that best fits the application, providing administrators with greater control over the operation.

Role of Networks

LANs are particularly well suited for workflow applications, because they offer more flexibility and scaleability than either minicomputer or mainframe solutions. LANs have the capability to support the transfer of documents between all nodes—workstations, scanners, storage devices, printers, and facsimile machines—thereby mimicking the movement of documents in the paper-based work environment. LANs also enable greater flexibility in host access, since users typically have a variety of gateways and terminal emulators from which to choose.

LANs also provide more opportunities for performance tuning. Using LAN management systems, analysis tools and report generators, the administrator can accurately measure performance and take immediate steps to make improvements, such as segmenting the network into subnets, upgrading cache memory, using compression, or adding higher-performance peripherals. Because workflow demands can vary on a daily basis, the capacity to respond quickly to changing loads constitutes a key benefit of LAN-based workflow systems over mainframe-based systems.

Network performance can be maintained by putting resource-intensive services, such as scanning, on subnetworks. These subnets can be selectively isolated from the rest of the network using such devices as bridges or routers. This function would enable a large accounts payable department, for example, to scan 10,000 invoices a day without bogging down the main network.

Workflow applications are not limited to running over LANs. With the trend toward distributed operations, often LANs must be linked over the WAN via digital facilities. In such cases, the workflow application supports data compression to minimize activity on the network. Compression is especially important when imaged documents must traverse lower-speed WAN links. The following table provides some comparisons of the number of images per hour that can be sent over links of various speeds in compressed form.

Transmission Times for Compressed Images over the WAN			
Transmission Rate	**Image Size**	**Seconds per Image**	**Images per Hour**
56Kbps	50KB	7.94	168
56Kbps	75KB	11.90	112
56Kbps	100KB	15.87	84
384Kbps	50KB	1.16	1,152
384Kbps	75KB	1.74	768
384Kbps	100KB	2.31	576
1.536Mbps	50KB	.29	4,608
1.536Mbps	75KB	.43	3,072
1.536Mbps	100KB	.58	2,304

Summary

Workflow automation relies on LANs and WANs to move documents within a structured transaction-processing environment. This kind of automation streamlines the processing of documents, speeds the distribution of information to the right people, and enhances the productivity of corporate staff who use the information on a daily basis. In turn, corporate responsiveness to customers, suppliers, and other constituencies can be greatly improved.

See Also

Business Process Re-Engineering

World Trade Organization (WTO)

Established in January 1995, the *World Trade Organization* (WTO) resulted from the Uruguay Round Trade negotiations and is the successor to the *General Agreement on Tariffs and Trade* (GATT). As such, the WTO is the legal and institutional foundation of the multilateral trading system. The WTO provides the principal framework within which governments develop and implement domestic trade legislation and regulations. The WTO is also the platform on which trade relations among countries evolve through collective debate, negotiation and adjudication. One of the principle objectives of the WTO is the reduction of tariffs and other trade barriers, as well as the elimination of discriminatory treatment in international trade relations.

The WTO recognizes that the telecommunications sector has a dual role: it is a distinct sector of economic activity; and it is an underlying means of supplying other economic activities (for example, electronic money transfers). Therefore, a key tenant of the WTO is that governments must ensure that foreign-service suppliers are given access to the public telecommunications networks without discrimination.

Structure

The WTO Secretariat is located in Geneva, Switzerland and has about 450 staff members (headed by a director-general). The highest WTO authority is the Ministerial Conference, which meets every two years. The daily work of the WTO, however, falls to a number of subsidiary bodies, principally the General Council, which also convenes as the Dispute Settlement Body and as the Trade Policy Review Body. The General Council delegates responsibility to three other major bodies: the Council for Trade in Goods, the Council for Trade in Services, and the Council for Trade-Related Aspects of Intellectual Property Rights.

Several other bodies have been established by the Ministerial Conference and report to the General Council: the Committee on Trade and Development; the Committee on Balance of Payments; the Committee on Budget, Finance and Administration; and the Committee on Trade and Environment.

Impact on Telecommunications

The WTO successfully concluded nearly three years of extended negotiations on market access for basic telecommunications services in February 1997. A total of 71 governments—accounting for more than 91 percent of global telecommunications revenues in 1995—agreed to set aside national differences in how basic telecommunications might be defined domestically and to negotiate on all telecommunications services, both public and private, that involve end-to-end transmission of customer supplied information (e.g. simply the transmission of voice or data from sender to receiver).

They also agreed that basic telecommunications services that were provided over network infrastructures (as well as those that were provided through resale over private, leased circuits) would both fall within the scope of market access commitments. As a result, market access commitments will cover not only cross-border supply of telecommunications but also services provided through the establishment of foreign firms, or commercial presence, including the capability to own and operate independent telecommunications-network infrastructures.

Examples of the services under negotiation were voice telephony, data transmission, telex, telegraph, facsimile, private leased circuit services (i.e., the sale or lease of transmission capacity), fixed and mobile satellite systems and services, cellular telephony, mobile data services, paging, and *Personal Communications Services* (PCS).

Value-added services, or telecommunications for which suppliers add value to the customer's information by enhancing its form or content, or by providing for its storage and retrieval, were not formally part of the extended negotiations. Nevertheless, a few participants chose to include them in their offers. Examples include online data processing, online database storage and retrieval, *Electronic Data Interchange* (EDI), e-mail, and voice mail.

Summary

Under the WTO agreement, which became effective January 1, 1997, each of the signatories agreed to permit the resale of current monopoly carrier services, interconnect competitive public networks with existing networks, and enable foreign carriers to buy or build their own networks. The principal result of the agreement will be competitive pressure to eventually knock down the high cost of international calling, which now averages 99 cents a minute. A number of years will be necessary, however, before this situation happens on a wide scale (because of the complex system of international pricing and the extra time that is needed by many countries to implement various provisions of the agreement).

See Also

Telecommunications Act of 1996

World Wide Web

Since its development by Tim Berners-Lee in 1990 at the European Particle Physics Laboratory in Switzerland (CERN), the *World Wide Web* (WWW) has grown to become one of the most sophisticated and popular services on the Internet. The two main mechanisms that make the Web work are the *Hypertext Transfer Protocol* (HTTP) and *Hypertext Markup Language* (HTML).

HTTP is used to transfer hypertext documents among the Web servers on the Internet—and, ultimately, to a client (the end-user's browser-equipped computer). Refer to Figure W-3. Collectively, the hundreds of thousands of servers that are distributed worldwide that support HTTP are known as the World Wide Web. HTML is used to structure information that resides on the servers in a way that can be readily rendered by browser software, such as Microsoft's Internet Explorer and Netscape Navigator, which

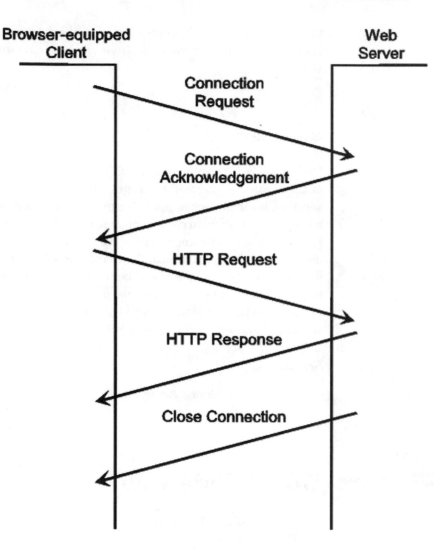

Figure W-3
HTTP delivers
documents from Web
servers to browser-
equipped clients in
response to specific
requests, then closes
the connection until
the client makes a
new request.

is installed on the clients. HTML makes documents portable from one computer platform to another and is intended as a common medium for tying together information from widely different sources.

History

The Web was invented by Tim Berners-Lee, who was named by *Time* magazine as one of the 100 greatest minds of the 20th century (refer to Figure

Tim Berners-Lee,
inventor of the World
Wide Web

W-4). The original vision of Berners-Lee has inspired the Web's further development as a powerful technological force for social change, commerce, and individual creativity. Berners-Lee is now the director of the *World Wide Web Consortium* (W3C), which is based at the *Massachusetts Institute of Technology* (MIT).

Although hypertext systems had been around since the 1980s, they were limited to working across a single database. The contribution of Berners-Lee was to apply hypertext links to multiple databases that were distributed across a network, enabling the links to point to anything (and anywhere).

The idea for the Web occurred to Berners-Lee while working at CERN. He noticed that many people came in and out of the facility with great ideas. They did some work, and when they left, there was no trace of what they had accomplished. He decided the organization needed a method of sharing all this information so that others could benefit from it, rather than merely grabbing somebody at coffee hour for a one-time conversation that would soon be forgotten. The intent of the Web was to enable people to work together as a self-managing team in an ongoing collaborative way, regardless of each participant's location, or what computer platform they have.

Web Characteristics

The Web itself can best be described as a dynamic, interactive, graphically-oriented, distributed, platform-independent, hypertext information system for the following reasons:

- The Web is dynamic because it changes daily. Web servers are continually being added to the Web. New information also is continually being added, as are new hypertext links and innovative services.

- The Web is interactive in that specific information can be requested through various search engines and returned moments later in the form of lists, with each item weighted according to how well it matched the search parameters. Another example of interactivity is text chat, whereby users communicate online in near real-time via their keyboards. Even voice conversations and video conferences can take place over the Web.

- The Web is graphics-oriented. The use of graphics not only makes the Web visually appealing but also easy to navigate. Graphical signposts direct users to other sources of information that are accessed via hypertext links. Sound, animation, and video capabilities may be added to Web pages, as well.

- The Web is distributed, meaning that information resides on hundreds of thousands of individual Web servers around the world. If one server goes down, there is no significant impact on the Web as a whole, except that access to the failed server will be denied until it can be brought back into service. Some servers are mirrored—duplicated at other locations—to keep information available if the primary server crashes.

- The Web is platform-independent, which means that virtually any client can access the Web, whether it uses Windows, OS/2, Macintosh, or a UNIX operating environment. This platform-independence even applies to the Web servers. Although most Web servers are based on UNIX, Windows NT is growing in popularity.

- The Web makes extensive use of hypertext links. A hypertext link, usually identified by an underlined word or phrase or a graphical symbol, points the way to other information. That information may be found virtually anywhere: in the same document, a different document on the same server, or another document on a different server that may be located anywhere in the world. A hypertext link does not necessarily point to text documents; it can point to maps, forms, images, sound and video clips, or applications. The links can even point to other Internet resources such as FTP and Gopher sites and Usenet newsgroups (refer to Figure W-5).

Summary

The Web has become an ideal medium for information distribution, collaborative projects, electronic commerce, and the delivery of support services of all kinds. The capabilities of the Web are continually being expanded. In addition to text and images, the Web is being used for telephony, video con-

Figure W-5
Hypertext documents on the Web can be multimedia in nature, providing links to audio content, image files, video clips, and other Internet services.

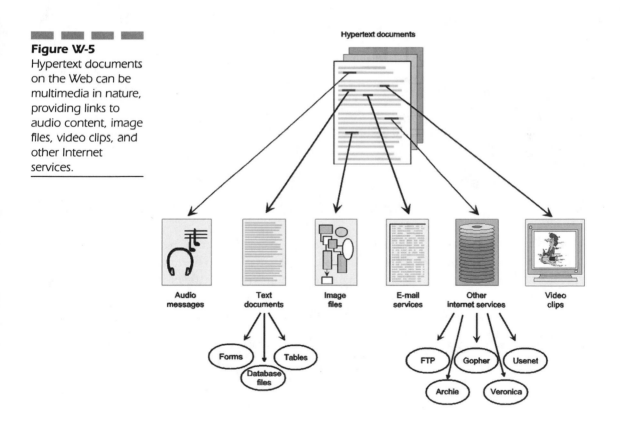

ferencing, faxing, and paging. With the emergence of Java as a popular network-programming language, applets can be embedded into Web pages—which enables users to take advantage of the larger applications that can reside on a corporate Web site.

See Also

E-Commerce

Internet

Intranet

Java

ACRONYMS

A

AAL	ATM Adaptation Layer
AAS	Automated Auction System (FCC)
ABATS	Automated Bit-Access Test System
ABM	Accunet Bandwidth Manager (AT&T)
ABR	Available Bit Rate
AC	Access Control, Address Copied, Alternating Current, Authentication Center
ACD	Automatic Call Distributor
ACELP	Algebraic Code-Excited Linear Predictive
ACP	Access-Control Point
ADA	Americans with Disabilities Act
ADCR	Alternate-Destination Call Routing (AT&T)
ADM	Add-Drop Multiplexer
ADN	Advanced Digital Network (Pacific Bell)
ADPCM	Adaptive Differential Pulse Code Modulation
ADSL	Asymmetrical Digital Subscriber Line
AFP	Apple File Protocol
AGRAS	Air-Ground Radiotelephone Automated Service
AIODC	Automatic Identification of Outward Dialed Calls
AIN	Advanced Intelligent Network
ALI	Automatic Location Information
AM	Amplitude Modulation
AMPS	Advanced Mobile-Phone Service
ANI	Automatic Number Identification
ANR	Automatic Network Routing (IBM Corporation)
ANSI	American National Standards Institute
ANT	ADSL Network Terminator
AOL	America Online

APC	Access-Protection Capability (AT&T)
API	Application Programming Interface
APPC	Advanced Program-to-Program Communications (IBM Corporation)
APPN	Advanced Peer-to-Peer Network (IBM Corporation)
APC	Automatic Protection Switching
ARCnet	Attached Resource Computer Network (Datapoint Corporation)
ARB	Adaptive Rate Based (IBM Corporation)
ARP	Address-Resolution Protocol
ARPA	Advanced Research Projects Agency
ARQ	Automatic Repeat Request
ARS	Action-Request System (Remedy Systems, Inc.)
AS	Autonomous System
ASCII	American Standard Code for Information Interchange
ASIC	Application-Specific Integrated Circuit
ASN.1	Abstract Syntax Notation 1
ASP	Application Service Provider
ASTN	Alternate Signaling Transport Network (AT&T)
AT&T	American Telephone and Telegraph
ATIS	Alliance for Telecommunications Industry Solutions (formerly, ECSA)
ATM	Asynchronous Transfer Mode
ATSC	Advanced Television Systems Committee
ATT	Auction-Tracking Tool (FCC)
AUI	Attachment Unit Interface
AWG	American Wire Gauge

B

B8ZS	Binary Eight Zero Substitution
BACP	Bandwidth Allocation-Control Protocol
BBS	Bulletin-Board System
BCCH	Broadcast Control Channel

BDCS	Broad-Band Digital Cross-Connect System
BECN	Backward Explicit Congestion Notification
Bellcore	Bell Communications Research, Inc.
BER	Bit-Error Rate
BERT	Bit Error-Rate Tester
BGP	Border Gateway Protocol
BHCA	Busy Hour Call Attempts
BIB	Backward Indicator Bit
BIOS	Basic Input-Output System
BMC	Block Multiplexer Channel (IBM Corporation)
BMS-E	Bandwidth Management Service-Extended (AT&T)
BOC	Bell Operating Company
BONDING	Bandwidth-on-Demand Interoperability Group
BootP	Boot Protocol
BPDU	Bridge Protocol Data Unit
BPS	Bits Per Second
BPV	Bipolar Violation
BRI	Basic-Rate Interface (ISDN)
BSA	Basis Serving Arrangement
BSC	Base Station Controller
BSC	Binary Synchronous Communications
BSE	Basic Service Element
BSN	Backward Sequence Number
BTS	Base Transceiver Station

C

CA	Communications Assistant
CAD	Computer-Aided Design
CAM	Computer-Aided Manufacturing
CAN	Campus-Area Network
CAP	Carrierless Amplitude/Phase (modulation)
CAP	Competitive Access Provider

CARS	Cable Antenna Relay Services
CASE	Computer-Aided Software Engineering
CATV	Cable Television
CB	Citizens Band
CBR	Constant Bit Rate
CBRS	Citizens Band Radio Service
CCC	Clear Channel Capability
CCCH	Common Control Channel
CCITT	Consultative Committee for International Telegraphy and Telephony
CCR	Customer Controlled Reconfiguration
CCS	Common Channel Signaling
CCSNC	Common Channel Signaling Network Controller
CCSS 6	Common Channel Signaling System 6
CD	Compact Disc
CDCS	Continuous Dynamic Channel Selection
CD-R	Compact Disc-Recordable
CD-ROM	Compact Disc-Read-Only Memory
CDMA	Code-Division Multiple Access
CDO	Community Dial Office
CDPD	Cellular Digital Packet Data
CDR	Call-Detail Recording
CEI	Comparably Efficient Interconnection
CENTREX	Central-Office Exchange
CGSA	Cellular Geographic Servicing Areas
CHAP	Challenge Handshake Authentication Protocol
CIF	Common Intermediate Format
CIR	Committed Information Rate
CLASS	Custom Local-Area Signaling Services
CLEC	Competitive Local-Exchange Carrier
CLI	Calling-Line Identification
CLP	Cell Loss Priority
CMI	Cable Microcell Integrator

CMIS	Common Management Information Services
CMRS	Commercial Mobile Radio Service
CNR	Customer Network Reconfiguration
CNS	Complementary Network Service
CO	Central Office
COCOT	Customer-Owned Coin-Operated Telephone
CON	Concentrator
COT	Central Office Terminal
CP	Coordination Processor
CPE	Customer-Premises Equipment
CPP	Calling Party Pays
CPS	Cycles per Second (Hertz)
CPU	Central Processing Unit
CRC	Cyclic Redundancy Check
CSA	Carrier Serving Area
CSM	Communications Services Management
CSMA/CD	Carrier-Sense Multiple Access with Collision Detection
CSU	Channel Service Unit
CT	Cordless Telecommunications
CTI	Computer-Telephony Integration
CTIA	Cellular Telecommunications Industry Association
CVSD	Continuously Variable Slope Delta (modulation)

D

D-AMPS	Digital Advanced Mobile-Phone Service
DA	Destination Address
DACS	Digital Access and Cross-Connect System (AT&T)
DAP	Demand Access Protocol
DAS	Dual Attached Station
DASD	Direct Access Storage Device (IBM Corporation)
DAT	Digital Audio Tape
dB	Decibel

DBMS	Database Management System
DBS	Direct-Broadcast Satellite
DBU	Dial Backup Unit
DCCH	Digital Control Channel
DCE	Data Communications Equipment
DCE	Distributed Computing Environment
DCF	Data Communication Function
DCS	Digital Cross-Connect System
DDS	Digital Data Services
DDS/SC	Digital Data Service with Secondary Channel
D/E	Debt/Equity Ratio
DECT	Digital Enhanced (formerly, European) Cordless Telecommunication
DES	Data Encryption Standard
DFSMS	Data Facility Storage Management Subsystem (IBM Corporation)
DID	Direct Inward Dialing
DIF	Digital Interface Frame
DLL	Data-Link Layer
DLCS	Digital Loop Carrier System
DLL	Dynamic-Link Library
DLSw	Data-Link Switching (IBM Corporation)
DLU	Digital Line Unit
DM	Distributed Management
DME	Distributed Management Environment
DMI	Desktop Management Interface
DMT	Discrete Multi-Tone
DMTF	Desktop Management Task Force
DNS	Domain Name Service
DoD	Department of Defense
DOD	Direct Outward Dialing
DOS	Disk Operating System
DOV	Data over Voice

DQDB	Distributed Queue Dual Bus
DQPSK	Differential Quadrature Phase-Shift Keying
DS0	Digital Signal-Level 0 (64Kbps)
DS1	Digital Signal-Level 1 (1.544Mbps)
DS1C	Digital Signal-Level 1C (3.152Mbps)
DS2	Digital Signal-Level 2 (6.312Mbps)
DS3	Digital Signal-Level 3 (44.736Mbps)
DS4	Digital Signal-Level 4 (274.176Mbps)
DSI	Digital Speech Interpolation
DSL	Digital Subscriber Line
DSN	Defense-Switched Network
DSP	Digital Signal Processor
DSS	Decision Support System
DSU	Data Service Unit
DSX1	Digital Systems Cross-Connect 1
DTE	Data Terminal Equipment
DTMF	Dual-Tone Multi-Frequency
DTR	Dedicated Token Ring
DTU	Data Transfer Unit
DTV	Digital Television
DVB	Digital Video Broadcast
DWMT	Discrete Wavelet Multi-Tone
DXI	Data-Exchange Interface

E

E-Mail	Electronic Mail
E-TDMA	Expanded Time-Division Multiple Access
EBCDIC	Extended Binary Coded Decimal Interchange Code (IBM Corporation)
EBS	Emergency Broadcast System
ECFS	Electronic Comment Filing System
ECSA	Exchange Carriers Standards Association

ED	Ending Delimiter
EDI	Electronic Data Interchange
EDRO	Enhanced Diversity Routing Option (AT&T)
EEROM	Electronically Erasable Read-Only Memory
EFRC	Enhanced Full Rate Codec
EFT	Electronic Funds Transfer
EGP	External Gateway Protocol
EHF	Extremely High Frequency (more than 30GHz)
EIA	Electronic Industries Association
EIR	Equipment Identity Register
EISA	Extended Industry Standard Architecture
ELCS	Expanded Local Calling Service
EMI	Electromechanical Inteference
EMS	Element Management System
EOC	Embedded Overhead Channel
EOT	End of Transmission
ESCON	Enterprise System Connection (IBM Corporation)
ESD	Electronic Software Distribution
ESF	Extended Super Frame
ESMR	Enhanced Specialized Mobile Radio
ESN	Electronic Serial Number
ETSI	European Telecommunication Standards Institute

F

4GL	Fourth-Generation Language
FACCH	Fast Associated Control Channel
FASB	Financial Accounting Standards Board
FASC	Fraud Analysis and Surveillance Center (AT&T)
FASTAR	Fast Automatic Restoral (AT&T)
FAT	File-Allocation Table
FC	Frame Control
FC	Fibre Channel

FC-0	Fibre Channel—Layer 0
FC-1	Fibre Channel—Layer 1
FC-2	Fibre Channel—Layer 2
FC-3	Fibre Channel—Layer 3
FC-4	Fibre Channel—Layer 4
FCC	Federal Communications Commission
FCS	Frame-Check Sequence
FDDI	Fiber Distributed Data Interface
FDIC	Federal Deposit Insurance Corporation
FDL	Facilities Data Link
FECN	Forward Explicit Congestion Notification
FEP	Front-End Processor
FIB	Forward Indicator Bit
FIB	Forwarding Information Base
FIFO	First in, First Out
FITL	Fiber in the Loop
FM	Frequency Modulation
FOCC	Forward Control Channel
FOD	Fax on Demand
FRAD	Frame Relay Access Device
FRS	Family Radio Service
FS	Frame Status
FSN	Forward Sequence Number
FTAM	File Transfer, Access, and Management
FT1	Fractional T1
FTP	File Transfer Protocol
FTS	Federal Telecommunications System
FTTB	Fiber to the Building
FTTC	Fiber to the Curb
FTTH	Fiber to the Home
FX	Foreign Exchange (line)
FXO	Foreign Exchange Office

G

GATT	General Agreement on Tariffs and Trade
GBIC	Gigabit Interface Converter
GDS	Generic Digital Services
GEO	Geostationary Earth Orbit
GFC	Generic Flow Control
GHz	gigahertz (billions of cycles per second)
GIS	Geographical Information System
GloBanD	Global Bandwidth on Demand
GMRS	General Mobile Radio Service
GPRS	General Packet Radio Services
GPS	Global Positioning System
GSA	General Services Administration
GSM	Global System for Mobile Telecommunications (formerly Groupe Spéciale Mobile)
GUI	Graphical User Interface

H

H0	High-capacity ISDN channel operating at 384Kbps
H11	High-capacity ISDN channel operating at 1.536Mbps
HDSL	High Bit Rate Digital Subscriber Line
HDTV	High-Definition Television
HEC	Header Error Check
HF	High Frequency (3MHz to 30MHz)
HFC	Hybrid Fiber/Coax
HIC	Head-End Interface Converter
HLR	Home Location Register
HPPI	High-Performance Parallel Interface
HPR	High-Performance Routing (IBM Corporation)
HSCSD	High-Speed Circuit-Switched Data
HSM	Hierarchical Storage Management
HST	Helical Scan Tape

HTML	Hypertext Markup Language
HTTP	Hypertext Transfer Protocol
HVAC	Heating, Ventilation, and Air Conditioning
Hz	Hertz (cycles per second)

I

I/O	Input/Output
IAB	Internet Architecture Board
IANA	Internet Assigned Numbers Authority
ICI	Interexchange Carrier Interface
ICMP	Internet Control Message Protocol
ICR	Intelligent Call Routing
ICS	Intelligent Calling System
ID	Identification
IDDD	International Direct Dialing Designator
IDLC	Integrated Digital Loop Carrier
IDPR	Inter-Domain Policy Routing
IEC	International Electrotechnical Commission
IEEE	Institute of Electrical and Electronic Engineers
IESG	Internet Engineering Steering Group
IETF	Internet Engineering Task Force
IGP	Interior Gateway Protocol
ILEC	Incumbent Local-Exchange Carrier
IMAP	Internet Mail Access Protocol
IMEI	International Mobile-Equipment Identity
IMS/VS	Information Management System/Virtual Storage (IBM Corporation)
IMSI	International Mobile-Subscriber Identity
IN	Intelligent Network
INMARSAT	International Maritime Satellite Organization
INMS	Integrated Network Management System
IOC	Inter-Office Channel

IP	Internet Protocol
IPH	Integrated Packet Handler
IPI	Intelligent Peripheral Interface
IPN	Intelligent Peripheral Node
IPP	Independent Pay Phone Provider
IPX	Internet Packet Exchange
IrDA	Infrared Data Association
IrLAN	Infrared LAN
IrLAP	Infrared Link-Access Protocol
IrLMP	Infrared Link-Management Protocol
IrPL	Infrared Physical Layer
IRQ	Interrupt Request
IrTTP	Infrared Transport Protocol
IS	Information System
IS	Industry Standard
IS-IS	Intra-Autonomous System to Intra-Autonomous System
ISA	Industry-Standard Architecture
ISD	Independent Service Developer
ISDL	ISDN Subscriber Digital Line
ISDN	Integrated Services Digital Network
ISM	Industrial, Scientific, and Medical (frequency bands)
ISO	International Organization for Standardization
ISOC	Internet Society
ISP	Internet Service Provider
ISSI	Inter-Switching Systems Interface
IT	Information Technology
ITFS	Instructional Television Fixed Service
ITR	Intelligent Text Retrieval
ITU-TSS	International Telecommunications Union-Telecommunications Standardization Sector (formerly, CCITT)
IVR	Interactive Voice Response
IXC	Interexchange Carrier

J

JIT	Just in Time
JEPI	Joint Electronic Payments Initiative
JPEG	Joint Photographic Experts Group
JTAPI	Java Telephony Application Programming Interface
JTC	Joint Technical Committee

K

K	*kilo*; one thousand (e.g., Kbps)
KB	Kilobyte
KSU	Key Service Unit
KTS	Key Telephone System
KHz	kilohertz (thousands of cycles per second)

L

LAN	Local-Area Network
LANCES	LAN Resource Extension and Services (IBM Corporation)
LAPB	Link Access Procedure-Balanced
LAT	Local-Area Transport (Digital Equipment Corporation)
LATA	Local Access and Transport Area
LBO	Line Build Out
LCD	Liquid Crystal Display
LCN	Local Channel Number
LCP	Link-Control Protocol
LD	Laser Diode
LEA	Law Enforcement Agency
LEC	Local-Exchange Carrier
LED	Light-Emitting Diode
LEO	Low Earth Orbit
LF	Low Frequency (30KHz to 300KHz)
LI	Length Indicator

LIFO	Last In, First Out
LLC	Logical Link Control
LMDS	Local Multi-Point Distribution System
LMS	Location and Monitoring Service
LPRS	Low-Power Radio Service
LSI	Large Scale Integration
LTG	Line Trunk Group
LU	Logical Unit (IBM Corporation)

M

M	*mega*; one million (e.g., Mbps)
M&A	Mergers and Acquisitions
MAC	Media-Access Control
MAC	Moves, Adds, Changes
MAN	Metropolitan-Area Network
MAPI	Messaging Applications Programming Interface (Microsoft Corporation)
MAU	Multi-Station Access Unit
MB	Megabyte
MCA	Micro-Channel Architecture (IBM Corporation)
MCU	Multi-Point Control Unit
MD	Mediation Device
MDF	Main Distribution Frame
MDS	Multi-Point Distribution Service
MEO	Middle Earth Orbit
MES	Master Earth Station
MF	Mediation Function
MF	Medium Frequency (300KHz to 3MHz)
MFJ	Modified Final Judgement
MHz	megahertz (millions of cycles per second)
MIB	Management Information Base
MIC	Management Integration Consortium

MIF	Management Information Format
MII	Media-Independent Interface
MIME	Multi-Purpose Internet Mail Extensions
MIN	Mobile Identification Number
MIPS	Millions of Instructions per Second
MIS	Management Information Services
MISR	Multi-Protocol Integrated Switch Routing
MJU	Multi-Point Junction Unit
MMDS	Multi-Channel, Multi-Point Distribution Service
MO	Magneto-Optical
MO&O	Memorandum Opinion and Order (FCC)
Modem	Modulation/Demodulation
MPEG	Moving Pictures Experts Group
MPPP	Multi-Link Point-to-Point Protocol
MRI	Magnetic Resonance Imaging
ms	millisecond (thousandths of a second)
MS	Mobile Station
MSC	Mobile Switching Center
MSN	Microsoft Network
MSRN	Mobile Station Roaming Number
MSS	Mobile Satellite Service
MTBF	Mean Time between Failure
MTP	Message Transfer Part
MTSO	Mobile Transport Serving Office
MVC	Multicast Virtual Circuit
MVDS	Microwave Video Distribution System
MVPD	Multi-Channel Video Program Distribution
MVPRP	Multi-Vendor Problem Resolution Process

N

NAL	Notice of Apparent Liability (FCC)
N-AMPS	Narrow-Band Advanced Mobile Phone Service

NNP	National Numbering Plan
NAM	Numeric Assignment Module
NAP	Network Access Point
NASA	National Aeronautics and Space Administration (U.S.)
NAT	Network Address Translation
NAU	Network Addressable Unit (IBM Corporation)
NAUN	Nearest Active Upstream Neighbor
NC	Network Computer
NCP	Network Control Program (IBM Corporation)
NCP	Network Control Point
NE	Network Element
NEBS	New Equipment Building Specifications
NECA	National Exchange Carrier Association
NetBIOS	Network Basic Input/Output System
NEF	Network Element Function
NFS	Network File System (or Server)
NIC	Network Interface Card
NID	Network Interface Device
NiCd	Nickel Cadmium
NiMH	Nickel-Metal Hydride
NIST	National Institute of Standards and Technology
NLM	NetWare Loadable Module (Novell, Inc.)
nm	nanometer
NM	Network Manager
NMS	NetWare Management System (Novell, Inc.)
NMS	Network Management System
NNM	Network Node Manager (Hewlett-Packard Co.)
NNTP	Network News Transfer Protocol
NOI	Notice of Inquiry (FCC)
NOS	Network Operating System
NPA	Numbering Plan Area
NPC	Network Protection Capability (AT&T)
NPRM	Notice of Proposed Rulemaking (FCC)

NPV	Net Present Value
NSA	National Security Agency
NSF	National Science Foundation
NTSA	Networking Technical Support Alliance
NTSC	National Television Standards Committee

O

OSN	Official Services Network
OAM	Operations, Administration, Management
OAM&P	Operations, Administration, Maintenance and Provisioning
OC	Optical Carrier
OC-1	Optical Carrier Signal-Level 1 (51.84Mbps)
OC-3	Optical Carrier Signal-Level 3 (155.52Mbps)
OC-9	Optical Carrier Signal-Level 9 (466.56Mbps)
OC-12	Optical Carrier Signal-Level 12 (622.08Mbps)
OC-18	Optical Carrier Signal-Level 18 (933.12Mbps)
OC-24	Optical Carrier Signal-Level 24 (1.244Gbps)
OC-36	Optical Carrier Signal-Level 36 (1.866Gbps)
OC-48	Optical Carrier Signal-Level 48 (2.488Gbps)
OC-96	Optical Carrier Signal-Level 96 (4.976Gbps)
OC-192	Optical Carrier Signal-Level 192 (9.952Gbps)
OC-256	Optical Carrier Signal-Level 256 (13.271Gbps)
OCR	Optical Character Recognition
OCUDP	Office-Channel Unit Data Port
ODBC	Open Database Connectivity (Microsoft Corporation)
ODS	Operational Data Store
OEM	Original Equipment Manufacturer
OFX	Open Financial Exchange
OLAP	Online Analytical Processing
OLE	Object Linking and Embedding
OMA	Object Management Architecture
OMAP	Operations, Maintenance, Administration & Provisioning

OMF	Object Management Framework
OMG	Object-Management Group
OOP	Object-Oriented Programming
OPX	Off-Premises Extension
ORB	Object Request Broker
OS	Operating System
OS/2	Operating System/2 (IBM Corporation)
OSF	Open Software Foundation
OSF	Operations Systems Function
OSI	Open Systems Interconnection
OSN	Official Services Network
OSS	Operations Support Systems
OTDR	Optical Time Domain Reflectometry

P

PA	Preamble
PACS	Personal-Access Communications System
PAD	Packet Assembler-Disassembler
PAL	Phase Alternating by Line
PAP	Password-Authentication Protocol
PBX	Private-Branch Exchange
PC	Personal Computer
PCB	Printed Circuit Board
PCH	Paging Channel
PCM	Pulse-Code Modulation
PCN	Personal Communications Networks
PCS	Personal Communications Services
PCT	Private Communication Technology
PDA	Personal Digital Assistant
PDN	Packet Data Network
PDU	Payload Data Unit
PEM	Privacy-Enhanced Mail
PGP	Pretty-Good Privacy

PHS	Personal Handyphone System
PHY	Physical Layer
PIM	Personal Information Manager
PIN	Personal Identification Number
PIN	Positive-Intrinsic-Negative
PLMRS	Private Land Mobile Radio Services
PMD	Physical Media Dependent
PnP	Plug and Play
PON	Passive Optical Network
POP	Point of Presence
POP	Post Office Protocol
POS	Point of Sale
POTS	Plain Old Telephone Service
PPP	Point-to-Point Protocol
PRI	Primary Rate Interface (ISDN)
PSAP	Public Safety Answering Point
PSN	Packet-Switched Network
PSP	Pay Phone Service Provider
PSTN	Public-Switched Telephone Network
PT	Payload Type
PTT	Post Telephone and Telegraph
PU	Physical Unit (IBM Corporation)
PUC	Public-Utility Commission
PVC	Permanent Virtual Circuit

Q

QA	Quality Assurance
QAM	Quadrature Amplitude Modulation
QCIF	Quarter Common Intermediate Format
QIC	Quarter-Inch Cartridge
QoS	Quality of Service
QPSK	Quadrature Phase Shift Keying

R

R&O	Report and Order (FCC)
RACH	Random-Access Channel
RAD	Remote Antenna Driver
RAID	Redundant Array of Inexpensive Disks
RAM	Random-Access Memory
RASDL	Rate-Adaptive Digital Subscriber Line
RASP	Remote Antenna Signal Processor
RBES	Rule-Based Expert Systems
RCU	Remote-Control Unit
RDBMS	Relational Database Management System
RDSS	Radio Determination Satellite Service
RECC	Reverse Control Channel
RFC	Request for Comment
RF	Radio Frequency
RF	Routing Field
RFI	Radio Frequency Interference
RFI	Request for Information
RFP	Request for Proposal
RFQ	Request for Quotation
RIP	Routing Information Protocol
RISC	Reduced Instruction-Set Computing
RJE	Remote Job Entry
RMON	Remote Monitoring
ROI	Return on Investment
ROM	Read Only Memory
RPC	Remote Procedure Call
RSVP	Resource Reservation Protocol
RT	Remote Terminal
RTNR	Real-Time Network Routing (AT&T)
RTP	Rapid Transfer Protocol (IBM Corporation)
RX	Receive

S

SA	Source Address
SAFER	Split-Access Flexible Egress Routing (AT&T)
SAP	Second Audio Program
SAR	Segmentation and Reassembly
SAS	Single Attached Station
SBCCS	Single-Byte Command-Code Set (IBM Corporation)
SCC	Standards Coordinating Committees (IEEE)
SCP	Service Control Point
SCSI	Small Computer Systems Interface
SD	Starting Delimiter
SDCCH	Stand-Alone Dedicated Control Channel
SDH	Synchronous Digital Hierarchy
SDLC	Synchronous Data-Link Control (IBM Corporation)
SDM	Subrate Data Multiplexing
SDN	Software-Defined Network (AT&T)
SDP	Service Delivery Point
SDSL	Symmetric Digital Subscriber Line
SET	Secure Electronic Transaction
SFD	Start Frame Delimiter
SHF	Super-High Frequency (3 GHz to 30 GHz)
SHTTP	Secure Hypertext Transfer Protocol
SIF	Signaling Information Field
SIM	Subscriber Identity Module
SIP	SMDS Interface Protocol
SLIC	Serial-Line Interface Coupler (IBM Corporation)
SLIP	Serial Line Internet Protocol
SMDI	Station Message Desk Interface
SMDR	Station Message-Detail Recording
SMDS	Switched Multi-Megabit Data Services
SMR	Specialized Mobile Radio
SMS	Service Management System

SMS	Short Message Service
SMT	Station Management
SMTP	Simple Mail Transfer Protocol
SN	Switching Network
SNA	Systems Network Architecture (IBM Corporation)
snagas	SNA Gateway-Access Server
SNI	Subscriber Network Interface
SNMP	Simple Network-Management Protocol
SONET	Synchronous Optical Network
SPA	Software Publishers Association
SPC	Stored Program Control
SPI	Service Provider Interface
SPX	Synchronous Packet Exchange (Novell, Inc.)
SQL	Structured Query Language
SS	Switching System
SS7	Signaling System 7
SSCP	System Services Control Point (IBM Corporation)
SSCP/PU	System Services Control Point/Physical Unit (IBM Corporation)
SSL	Secure Sockets Layer
SSP	Service Switching Point
STDM	Statistical Time Division Multiplexing
STP	Shielded Twisted Pair
STP	Signal Transfer Point
STP	Spanning Tree Protocol
STS	Shared Telecommunications Services
STS	Synchronous Transport Signal
STX	Start of Transmission
SUBT	Subscriber Terminal
SVC	Switched Virtual Circuit
SWC	Serving Wire Center
SYNTRAN	Synchronous Transmission

T

T1	Transmission service at the DS1 rate of 1.544Mbps
T3	Transmission service at the DS3 rate of 44.736Mbps
TA	Technical Advisor
TA	Technical Advisory
TAG	Technical Advisory Group
TAPI	Telephony Application Programming Interface (Microsoft Corporation)
TASI	Time-Assigned Speech Interpolation
TB	terabyte (trillion bytes)
TBOS	Telemetry Byte-Oriented Serial
Tbps	Terabit-per-Second
TCAP	Transaction Capabilities Applications Part
TCP	Transmission-Control Protocol
TDD	Time-Division Duplexing
TDM	Time-Division Multiplexer
TDMA	Time-Division Multiple Access
TDMA/TDD	Time-Division Multiple Access with Time-Division Duplexing
TDR	Time Domain Reflectometry
TFTP	Trivial File Transfer Protocol
TIA	Telecommunications Industry Association
TIB	Tag Information Base
TIMS	Transmission Impairment Measurement Set
TL1	Transaction Language 1
TMN	Telecommunications Management Network
TRS	Telecommunications Relay Services
TSAPI	Telephony Services Application Programming Interface (Novell, Inc.)

TSI	Time-Slot Interchange
TSR	Terminal Stay Resident
TTRT	Target Token Rotation Time
TTY	Text Telephone
TV	Television
TWX	Teletypewriter Exchange (also known as Telex)
TX	Transmit

U

UART	Universal Asynchronous Receiver/Transmitter
UBR	Unspecified Bit Rate
UDP	User Datagram Protocol
UDWDM	Ultra-Dense Wavelength-Division Multiplexing
UHF	Ultra-High Frequency (300MHz to 3GHz)
UI	Unit Intervals
UMS	Universal Messaging System
UN	United Nations
UNI	User-Network Interface
UPS	Uninterruptible Power Supply
USDLA	United States Distance Learning Association
USNC	United States National Committee
UTP	Unshielded Twisted-Pair

V

VAD	Voice-Activated Dialing
VAR	Value-Added Reseller

VBNS	Very High-Speed Backbone Network Service
VBR	Variable Bit Rate
VC	Virtual Circuit
VCI	Virtual Channel Identifier
VCR	Video Cassette Recorder
VDSL	Very High-Speed Digital Subscriber Line
VF	Voice Frequency
VFN	Vendor Feature Node
VG	Voice Grade
VHF	Very High-Frequency (30MHz to 300MHz)
VLF	Very Low-Frequency (less than 30KHz)
VLR	Visitor Location Register
VLSI	Very Large-Scale Integration
VM	Virtual Machine
VMS	Virtual Machine System (Digital Equipment Corporation)
VOD	Video on Demand
VoFR	Voice over Frame Relay
VPA	Virtual Personal Assistant
VPI	Virtual Path Identifier
VP	Virtual Path
VPN	Virtual Private Network
VSAT	Very Small Aperture Terminal
VT	Virtual Terminal
VT	Virtual Tributary
VTAM	Virtual Telecommunications-Access Method (IBM Corporation)

W–X

WACS	Wireless Access Communications System
WAN	Wide-Area Network
WATS	Wide-Area Telecommunications Service
WCS	Wireless Communications Service
WDCS	Wideband Digital Cross-Connect System
WDM	Wave-Division Multiplexing
WGS	World-Wide Geodetic System
WLAN	Wireless Local-Area Network
WLL	Wireless Local Loop
WORM	Write Once, Read Many
WTO	World Trade Organization
WWW	World Wide Web
W3C	World Wide Web Consortium
XNS	Xerox Network System (Xerox Corporation)

INDEX

Note: Page numbers in boldface type indicate illustrations

T

ABOUT THE AUTHOR

Nathan Muller is an independent consultant in Sterling, Virginia, specializing in advanced technology marketing, research, and education. In his 30 years of experience in the industry, he has written extensively on many aspects of computers and communications, having published 18 books—including three encyclopedias—and more than 1,500 articles about computers and communications in 50 publications worldwide.

In addition, he is a regular contributor to the Gartner Group's Datapro Research Reports. He has participated in market research projects for Dataquest, Northern Business Information, and Faulkner Technical Reports. He also has consulted on custom projects for technology-oriented clients in the computer, telecommunications, and health care industries.

Muller has held numerous technical and marketing positions with companies such as Control Data Corporation, Planning Research Corporation, Cable & Wireless Communications, ITT Telecom, and General DataComm Inc. He has earned an M.A. in social and organizational behavior from George Washington University. His e-mail address is `nmuller@ddx.com`.